8th International Symposium on Quantum Theory and Symmetries 2013

(QTS8)

Journal of Physics: Conference Series Volume 512

Mexico City, Mexico
5 – 9 August 2013

ISBN: 978-1-63439-292-1
ISSN: 1742-6588

Printed from e-media with permission by:

Curran Associates, Inc.
57 Morehouse Lane
Red Hook, NY 12571

Some format issues inherent in the e-media version may also appear in this print version.

Printed by Curran Associates, Inc. (2014)

For permission requests, please contact the Institute of Physics
at the address below.

Institute of Physics
Dirac House, Temple Back
Bristol BS1 6BE UK

Phone: 44 1 17 929 7481
Fax: 44 1 17 920 0979

techtracking@iop.org

Additional copies of this publication are available from:

Curran Associates, Inc.
57 Morehouse Lane
Red Hook, NY 12571 USA
Phone: 845-758-0400
Fax: 845-758-2634
Email: curran@proceedings.com
Web: www.proceedings.com

TABLE OF CONTENTS

FOUNDATIONS OF QUANTUM MECHANICS

Journal of Physics: Conference Series **512** (2014) 011001 doi:10.1088/1742-6596/512/1/011001

Foreword to the Proceedings of the VIII International Symposium on Quantum Theory and Symmetries El Colegio Nacional, Mexico City, 5–9 August 2013

The *Quantum Theory and Symmetries* (QTS) international symposia are periodic biannual meetings of the mathematical physics community with special interest in the methods of group theory in their many incarnations, particularly in the symmetries that arise in quantum systems. The QTSs alternate with the International Colloquia on Group Theoretical Methods in Physics since 1999, when Professor Heinz-Dietrich Doebner organized the first one in Goslar, Germany. Subsequent symposia were held in Kraków, Poland (2001), Cincinnati, USA (2003), Varna, Bulgaria (2005), Valladolid, Spain (2007), Lexington, USA (2009), and Praha, Czech Republic (2011); the eighth QTS was awarded to Mexico (2013), and the next one (2015) will take place in Yerevan, Armenia.

The members of the International Conference Board of the QTS Symposia are:

Laurence Boyle	University of Kent, UK
Vladimir Dobrev	Bulgarian Academy of Sciences
Vladimir Man'ko	Lebedev Institute and University of Naples
Mariano del Olmo	Universidad de Valladolid
George S Pogosyan	Universities of Yerevan and Guadalajara
Alfred Shapere	University of Kentucky
Peter Suranyi	University of Cincinnati
Syed Twareque Ali	Concordia University, Montréal
Heinz-Dietrich Doebner (Chairman)	Universität Clausthal

Periodic meetings of researchers in the widening domains of symmetry methods help to keep their various mathematical dialects and physical application fields in contact and synergy. The scientific community also appreciates venues that provide a window to the cultures of the world. It is indeed a privilege that few professions have. For those of us aware of the course of 21st century civilization, this is a source of worldly knowledge and inspiration. While many previous meetings in Mexico have taken place in attractive tourist locations, we decided to accept the invitation of El Colegio Nacional to hold QTS8 in the heart of the Mexico City megalopolis, sheltered in their excellently refurbished late-17th century installations originally built as a convent and novitiate. Walking to and from the daily sessions, the participants were immersed in the restless multitude of visitors, merchants and shoppers that pursue their trades in the central quarter of ancient Tenochtitlan, which for most of its history has been the largest conurbation of America. An evening at the Palacio de Bellas Artes for a performance of the Ballet Folklórico left a lasting impression on our history and art.

The members of the Local Organizing Committee were:

Roelof Bijker	Instituto de Ciencias Nucleares, UNAM
Octavio Castaños	Instituto de Ciencias Nucleares, UNAM
Rocío Jáuregui	Instituto de Física, UNAM
Renato Lemus	Instituto de Ciencias Nucleares, UNAM
Óscar Rosas-Ortiz	Centro de Investigación y Estudios Avanzados, IPN
Kurt Bernardo Wolf	*Chair*, Instituto de Ciencias Físicas, UNAM-Cuernavaca
Octavio Novaro	*Coordinator*, Instituto de Física, UNAM, and El Colegio Nacional

The International Advisory Committee that supervised the QTS8 program was:

Čestmir Burdik (Czech Technical University)
Alejandro Frank (Instituto de Ciencias Nucleares, UNAM)
Jean Pierre Gazeau (Université de Paris VII)
Alberto Guijosa (Instituto de Ciencias Nucleares, UNAM)
Julio César Gutiérrez Vega (Tecnológico de Monterrey)
Francesco Iachello (Yale University)
Piet van Isacker (GANIL, France)
Tobias Kramer (Universität Regensburg)
Juan Maldacena (Institute for Advanced Study, Princeton)
Bogdan Mielnik (CINVESTAV, Instituto Politécnico Nacional)
Willard Miller Jr (University of Minnesota)
Miriam Mondragón (Instituto de Física, UNAM)
Wolfgang Schleich (Universität Ulm)
Thomas H Seligman (Instituto de Ciencias Físicas, UNAM)
Luis F Urrutia (Instituto de Ciencias Nucleares, UNAM)
Pavel Winternitz (Université de Montréal)

The topics addressed at the parallel sessions of QTS8 were Quantum gravity and string theory, Quantum optics and field theory, Symmetries in many-body systems, Super-integrability and exact solvability, Mathematical physics and symmetries, and Foundations of quantum mechanics.

The Plenary Speakers of QTS8 were:

Mark J Ablowitz	University of Colorado, Boulder CO, USA
Alan Aspuru-Guzik	Harvard University, Cambridge MA, USA
Natig M Atakishiyev	Instituto de Matemáticas, UNAM, Cuernavaca, México
Niklas Beisert	Institut für Theoretische Physik, ETH, Zürich, Switzerland
Giulio Casati	Center for Nonlinear and Complex Systems, Como, Italy
Ernst M Rasel	Institute of Quantum Optics, G W Leibnitz Universität, Germany
Rajiah Simon	The Institute of Mathematical Sciences, Chennai, India
Alexander Turbiner	Instituto de Ciencias Nucleares, UNAM, México
Alfred U'Ren	Instituto de Ciencias Nucleares, UNAM, México
Luc Vinet	Université de Montréal, Montréal, Canada

Both the *VIII International Symposium on Quantum Theory and Symmetries*, and the previous two weeks of the *XLIII Latin American School of Physics*, were hosted as part of the activities of El Colegio Nacional in 2013, coordinated by Member Professor Octavio Novaro Peñalosa. This institution provided the conference halls of its venue in Mexico City, as well as the support staff for projection facilities, coffee breaks and lunches during the week. Support for administration was provided by the Sociedad Mexicana de Física AC and by the Instituto de Ciencias Físicas, UNAM. The Symposium gathered 98 participants, including 50 researchers, 9 postdoctoral associates, 23 graduate students and 16 student auditors. There were 10 plenary presentations, 50 contributed talks and 20 posters. The symposium was well attended by Mexican participants, several of them students who were also present in the Latin American School of Physics.

We express our thanks to the Sponsor institutions of QTS8:

El Colegio Nacional

Universidad Nacional Autónoma de México, through:

Dirección General de Asuntos del Personal Académico
Instituto de Ciencias Físicas
Instituto de Ciencias Nucleares
Instituto de Física

International Association for Mathematical Physics (IAMP)

International Centre for Theoretical Physics (ICTP)

Academia Mexicana de Ciencias AC – FUMEC

The QTS8 website was designed by Ulises Amaya and Alejandro Ricardo Urzúa at ICF-UNAM. We thank their indispensable help in advertising the symposium. All papers published in this volume of *Journal of Physics: Conference Series* have been peer reviewed through processes administered by the Editors. Reviews were conducted by expert referees to the professional and scientific standards expected of a proceedings journal published by IOP Publishing. This Proceedings volume contains 36 reviewed papers, organized by topics and in alphabetical order.

Roelof Bijker (Instituto de Ciencias Nucleares, UNAM)
Guillermo Krötzsch (Instituto de Ciencias Físicas, UNAM)
Óscar Rosas-Ortiz (Centro de Investigación y Estudios Avanzados, IPN)
Kurt Bernardo Wolf (Instituto de Ciencias Físicas, UNAM)

Peer review statement

All papers published in this volume of *Journal of Physics: Conference Series* have been peer reviewed through processes administered by the proceedings Editors. Reviews were conducted by expert referees to the professional and scientific standards expected of a proceedings journal published by IOP Publishing.

Electric-Magnetic duality in (linearized) Hořava-Lifshitz gravity

Ignacio Cortese[a] **and J. Antonio García**[b]

a) Physique Théorique et Mathématique & International Solvay Institutes
Université Libre de Bruxelles, Campus Plaine C.P. 231, B-1050 Bruxelles, Belgium
b) Departamento de Física de Altas Energías, Instituto de Ciencias Nucleares,
Universidad Nacional Autónoma de México,
Apartado Postal 70-543, México D.F. 04510, México

E-mail: icortese@ulb.ac.be, garcia@nucleares.unam.mx

Abstract. We present an implementation of a duality symmetry in the minimal linearized Hořava-Lifshitz gravity. Some difficulties arise when an attempt is made to implement off shell duality symmetries in theories that are not invariant under full spacetime diffeomorphisms. To avoid some of them our approach will consist of two steps: a) Proposing an explicit formulation of a covariant model and applying standard rules to relate a theory with its dual, and b) Implementing a projection of this covariant model to Hořava-Lifshitz gravity using a gauge fixing condition. In this way we are able to present a dual action of the standard formulation of minimal linearized Hořava-Lifshitz gravity.

1. Introduction

Electric-magnetic duality has emerged as a very interesting symmetry of diverse physical theories of increasing complexity, from zero to higher spin theories. In all cases two key properties keep reappearing: (i) The symmetry is an invariance of the action, and not just of the equations of motion, and (ii) The symmetry can be made manifest, while keeping the formulation simple, only at the price of giving up manifest spacetime covariance. The relation between duality and space time covariance exhibits a fascinating complementarity that can be confirmed by the fact that duality imply Lorentz invariance, at least in the simple case of an abelian gauge field. At a more technical level two general features that also appear are: (iii) The need to reformulate the theory in terms of new variables (prepotentials), and (iv) A corresponding doubling of the gauge symmetry group.

In the case of gravitational theories, it is generally expected that gravitational duality could be fundamental in the understanding of hidden symmetries of supergravities and M-theory. While there is overwhelming evidence in favor of the presence of these symmetries, the results obtained so far remain incomplete because of the still somehow mysterious role played by the *dual graviton.* The clue for unlocking the present difficulties might lie in a better grasp of the relationship between the graviton and its dual.

The linearized Einstein equations in D spacetime dimensions can be written as twisted self-duality relations revealing that the linearized curvature tensor of the graviton (rank-two symmetric tensor) is dual to the linearized curvature tensor of the dual graviton described by a tensor of $(D-3,1)$ Young symmetry type. In the case of 4 dimensions, both the graviton and its dual are rank-two symmetric tensors (Young symmetry type (1,1)), while in the case of 11 space-time dimensions relevant to M-theory, the dual graviton is described by a tensor of (8, 1) Young symmetry type.

It is known that KK reduction of 11-dimensional supergravity gives rise to exceptional hidden symmetries. In this context a conjecture has emerged: infinite dimensional Kac-Moody algebra E11 is a symmetry of supergravity (M-theory)[1]. Level decomposition of E11 with respect to SL(D) subgroups reproduce the field content of maximal supergravity in D dimensions. This identification of the level decomposition of E11 and supergravity field content requires in the supergravity side that one add to each field its dual. E11 predicts not only the 3-form of 11 dimensional supergravity but also a 6-form and higher level fields in mixed Young Tableaux representations. The lowest one is the graviton and its dual, the dual graviton.

At linearized level we have two equivalent formulation of Einstein Gravity in D spacetime dimensions: based on the metric $h_{\mu\nu}$ or in terms of its dual mixed Young Tableaux field $C_{\mu_1\cdots\mu_{D-3}|\nu}$ whose covariant action was given by Curtright [2]. At nonlinear level there are obstructions to the implementation of duality in the Einstein Hilbert action. The Curtright action does not admit non-abelian deformations under the standard assumptions of spacetime covariance and locality.

Actually a main trend of current research is the study of deformations of gravity. The basic idea consist in the construction of a new gravity that bypass the basic problems associated with the standard formulation of gravity as a field theory. As is well know the theory is not renormalizable and the corresponding quantum theory is not known. Among the many directions in this trend (the consistent deformations and quantization of gravity) we can mention non commutative gravity, stringy corrections to the action of GR, new massive gravities in many dimensions, double field theory and Hořava-Lifshitz gravity [3, 4, 5, 6, 7]. This last particular deformation pretends to be a UV completion of gravity using an "heretic" proposal [8](breaking the spacetime paradigm), breaking full diff invariance of GR and leaving an invariance subgroup that is still manifest in ADM formalism. As a consequence different treatment of the kinetic and potential terms are necessary. The kinetic term is still quadratic in time derivatives but is deformed by a new parameter λ, and the potential has high order space derivatives of the basic field, the metric g_{ij}. The deformed theory is power counting renormalizable, at large distances higher derivative terms are suppressed and the theory runs to standard GR. The new parameter λ could be considered as the coupling parameter which controls the contribution of the trace of extrinsic curvature and of some higher derivative terms as for example the term proportional to $(R^{(3)})^2$ (see [6]). For $\lambda = 1$ we have a fixed point in the IR, recovering GR. For generic values of λ the theory breaks full diffeomorphism invariance (the theory becomes a non-relativistic theory).

The aim of this note is to implement (in the same way as in the parental action approach) the duality symmetry [9, 10]) in Hořava-Lifshitz theory. The interesting point about this duality is that Hořava-Lifshitz gravity is not invariant under the complete spacetime diffeomorphisms but only under a subgroup of them. However this non full covariance is not an impediment to implement duality and is an example of the role played by the complementarity between covariance and duality. Another very interesting way to construct duality in linearized GR is the pre potentials approach of [11].

We start from the action

$$S_{\rm HL} = 2\int {\rm d}t{\rm d}^3x\sqrt{g}N\Big(K_{ij}K^{ij} - \lambda K^2 + R^{(3)}\Big)$$

in $d = 4$ dimensions and ADM separation, representing the minimal HL deviation from GR, we take the linearized part. Perturbation around flat spacetime $g_{ij} = \eta_{ij} + \varepsilon h_{ij}$, $N = 1 + \varepsilon n$ and

$N_i = \varepsilon n_i$ into $S_{\rm HL}$ gives

$$L^{(2)}_{\rm HL} = \varepsilon^2 \int \mathrm{d}^3x \Big[\frac{1}{2}\dot h_{ij}\dot h^{ij} - \frac{\lambda}{2}\dot h^2 + \partial_i n_j \partial^i n^j + (1-2\lambda)(\partial_i n^i)^2 - 2\partial_i n_j(\dot h^{ij} - \lambda\eta^{ij}\dot h) \\ - \frac{1}{2}\partial_k h_{ij}\partial^k h^{ij} + \frac{1}{2}\partial_i h \partial^i h + \partial^i h_{ij}\partial_k h^{kj} - \partial_i h^{ij}\partial_j h + 2n(\partial_i\partial_j h^{ij} - \Delta h)\Big] \quad (1)$$

($h \equiv \eta^{ij}h_{ij}$). From now on our focus will be the theory defined by this Lagrangian, so we drop the superscript $^{(2)}$ of $L_{\rm HL}$. This Lagrangian deviates from FP theory in those terms proportional to λ, to which this theory reduce in the IR limit given by $\lambda \to 1$. Even thought the modifications due to a value of $\lambda \neq 1$ are subtle, the structural consequences are considerable and have attracted much attention in recent research [1].

In contrast to FP theory, the gauge transformations of this theory are given by the linearized foliation preserving diffeomorphisms $\delta x^i = \xi^i(x,t)$ and $\delta t = \xi^0(t)$, acting on the variables of the theory as

$$\delta h_{ij} = \partial_i\xi_j + \partial_j\xi_i, \quad \delta n_i = \dot\xi_i, \quad \delta n = -\dot\xi_0. \quad (2)$$

This is a subset of the full group of diffeomorphisms $\delta_\xi h_{\mu\nu} = \partial_\mu\xi_\nu + \partial_\nu\xi_\mu$. The latter do not leave the action defined by (1) invariant but rather its variation gives

$$\delta_\xi S_{\rm HL} = \int \mathrm{d}t\mathrm{d}^3x 2(\lambda - 1)(\dot h - 2\partial^i n_i)\Delta\xi_0. \quad (3)$$

Restricting the diffeomorphisms to those that satisfy $\Delta\xi_0 = 0$ we get the isommetries given by eq. (2). The solution of this equation is ξ_0 a constant in space[2], *i.e.*, $\xi_0 = \xi_0(t)$.

2. Dual of linearized minimal HL gravity

The scheme depicted in fig. 1

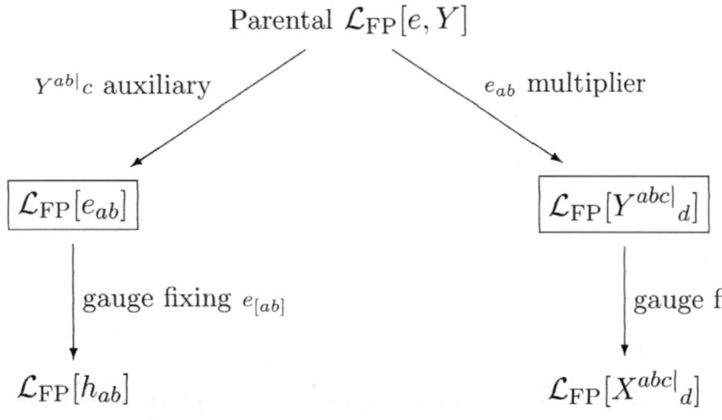

Figure 1. Role of the parental Lagrangian in d dimensions for FP theory. While h_{ab} are the first order perturbations of the metric, interpreted as the graviton field, the variables $X^{abc|}{}_d$ are the components of the dual graviton. Both child theories enjoy arbitrary shifts of some variables as gauge symmetries; those pure gauge fields are marginalized through gauge fixing conditions.

gives the general picture of the role of a parental action for the implementation of the duality symmetry. It is a first order action that delivers both the second order FP action in *vielbein*

[1] An incomplete list of references to this last topic is[12, 13, 14, 15, 16, 17, 18, 19, 20, 21, 22, 23, 24, 25, 8].

[2] Equation $\Delta\xi_0 = 0 \Rightarrow \partial_i\xi_0$ should be the transverse part of a vector, but 0 is the only vector whose transverse part is a gradient, so $\partial_i\xi_0 = 0$ and this imply that ξ_0 does not depend on the space coordinates.

Journal of Physics: Conference Series **512** (2014) 012001 doi:10.1088/1742-6596/512/1/012001

formalism and the one of its dual theory as a subsequent action (see [10]). In order to apply this scheme for the case of linearized HL gravity we will construct an action in the *vielbein* language as our starting point, in the same way as the one that should be obtained by a parental action (see left branch of fig. 2).

In the following subsection we will first construct an action of this sort, which results in an improved action in the sense that it is diffeomorphism invariant (but not local Lorentz invariant). This action will be the linearized HL theory through a gauge fixing procedure. From this point of view we can parallel the case of FP and construct the parental action from which we obtain a dual theory in section 2.2. We wonder if we can also impose on the latter a gauge fixing condition to define the dual of linearized HL theory, as suggested in the right branch of fig. 2.

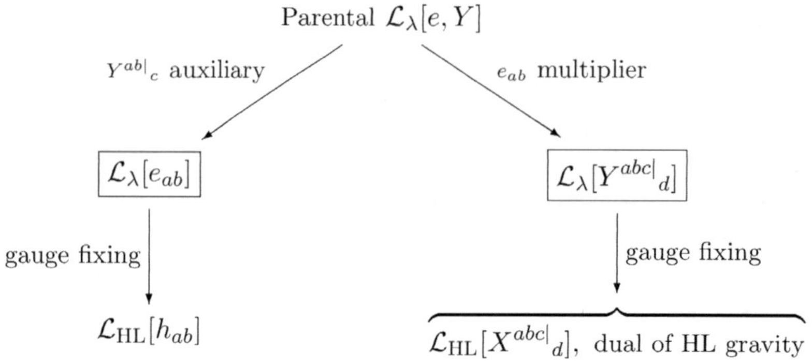

Figure 2. The Lagrangian density $\mathcal{L}_\lambda[e_{ab}]$ depends on the *vielbeins* and it is parametrized by λ, the same parameter appearing in linearized HL gravity defined by Lagrangian (1). This theory is diffeomorphism invariant, so it is not the linearized HL gravity; the latter is obtained from $\mathcal{L}_\lambda[e_{ab}]$ by gauge fixing conditions that in fact eliminate antisymmetric components of e_{ab}. Analogously on the dual side $\mathcal{L}_\lambda[Y^{abc|}{}_d]$ has some variables that can be eliminated through gauge fixing conditions in order to get a Lagrangian depending only on the dual graviton components.

2.1. *Linearized HL gravity in vierbein language*

Linearization of the Einstein-Hilbert theory, written in *vielbein* variables, corresponds to the FP theory

$$\mathcal{L}_{\mathrm{FP}}[e_{ab}] = -\frac{1}{2}\Big(\Omega^{abc}\Omega_{abc} + 2\Omega^{abc}\Omega_{acb} - 4\Omega^a\Omega_a\Big), \tag{4}$$

where $\Omega_{abc} = 2\partial_{[a}e_{b]c}$ are the linearized anholonomy coefficients, and $\Omega_a \equiv \Omega_{ab}{}^b$. Two points are important here: tangent and world space indices are identified, and the coefficients e_{ab} do not have a definite symmetry. This Lagrangian depends on d^2 fields in d space-time dimensions, certainly more than the independent components of the metric. However local Lorentz invariance takes care of this excess of degrees of freedom.

This can be seen by separating e_{ab} in symmetric $h_{ab} \equiv 2e_{(ab)}$ and antisymmetric $f_{ab} \equiv 2e_{[ab]}$ parts. Direct computation shows that the Lagrangian $\mathcal{L}_{\mathrm{FP}}[e_{ab}]$ depends on h_{ab} and f_{ab} in very different ways: while h_{ab} is a dynamical field[3], f_{ab} enters only through a total derivative term,

$$\mathcal{M} \equiv \partial_a\Big(-\frac{1}{2}h^c{}_c\partial_b f^{ab}\Big). \tag{5}$$

[3] In fact, they are the components in tangent space of the perturbation to the metric $h_{\mu\nu} = g_{\mu\nu} - \eta_{\mu\nu}$ around flat spacetime.

The Lagrangian density (4) is invariant under local shifts of f_{ab} with parameters $\omega_{ab}(x) = -\omega_{ba}(x)$. The full gauge invariance is given by $\delta e_{ab} = \partial_a \xi_b + \partial_b \xi_a + \omega_{ab}$, where ξ_a parametrize general coordinate transformations: $\delta x^\mu = \xi^\mu(x)$. In other words, the FP theory is invariant under

$$\delta_\xi h_{ab} = \partial_{(a}\xi_{b)} \quad \text{and} \quad \delta_\omega f_{ab} = \frac{1}{2}\omega_{ab}. \tag{6}$$

Using the gauge freedom for the f_{ab} they can be fixed away and then we can set them to zero. So they can be eliminated from the theory. As a consequence the Lagrangian (4) reduces to a theory for the symmetric fields h_{ab} only.

Linearized HL gravity is a deviation of FP theory parametrized by λ, as in (1). To fix some ideas, we will consider the case of $d = 4$ and as our starting point we propose

$$\mathcal{L}_\lambda[e_{ab}] = \mathcal{L}_{\rm FP}[e_{ab}] + 2(1-\lambda)(\Omega_0)^2, \tag{7}$$

where Ω_0 is the "0" component of the trace of Ω_{abc}. As in the case of FP Lagrangian (4) the above Lagrangian density depends on both the symmetric h_{ab} and antisymmetric f_{ab} components of the *vierbein*. As one can expect from general grounds the presence of the last term for $\lambda \neq 1$ breaks Lorentz invariance, so the dependence of the Lagrangian upon the fields f_{ab} are not of the form of a total derivative term. The antisymmetric part of the *veirbein* turns out to be a set of dynamical fields contributing to the total degrees of freedom of the theory. However, and this is the crux of our argument, we will be able to get rid of them by imposing a set of gauge fixing conditions. At the end we will have a theory with just the symmetric components h_{ab}.

Let's inspect the symmetries of the Lagrangian (7), where

$$\Omega_0 = \frac{1}{2}[\dot{h} - \partial^i(h_{0i} + f_{0i})], \qquad h \equiv \eta^{ij} h_{ij}. \tag{8}$$

For the following it is convenient to make use of the ADM variables, in which $n_i = h_{0i}$. We also define the notation $e_i \equiv f_{0i}$. The Lagrangian (7) does not share all the symmetries of $\mathcal{L}_{\rm FP}$ given in eq. (6).

The symmetry transformations acts on the term $(\Omega_0)^2$ as

$$\delta\Omega_0^2 = \frac{1}{2}\Omega_0(\partial^i\dot{\xi}_i - \Delta\xi_0 - \partial^i\omega_{0i}).$$

Then, this piece is invariant under global Lorentz transformations where ω_{ab} is constant[4]. It is not invariant under local Lorentz transformations though. Instead, in order to have a the gauge symmetry that we need $\partial^i(\dot{\xi}_i - \partial_i\xi_0 - \omega_{0i}) = 0$. This is an equation for ω_{0i} so they can not be arbitrary but are determined by the equation of motion. The general solution for it is given by the sum of the solution to the homogeneous equation (any transverse –divergenceless– 3-vector ζ^i) and a particular solution to the inhomogeneous equation; that is

$$\omega_{0i} = \zeta_i + \dot{\xi}_i - \partial_i\xi_0, \quad \text{such that} \quad \partial^i\zeta_i(x,t) = 0. \tag{9}$$

The parameters ξ_μ, however, remain arbitrary, so the Lagrangian (7) is invariant under general coordinate transformations.

In conclusion, the deviation of FP theory given by eq. (7) have arbitrary shifts of f_{ij} with parameters ω_{ij} among its gauge symmetries, as well as arbitrary shifts of the transverse part of e_i with parameters ζ_i[5]. The transformations of the longitudinal part of e_i is linked with the

[4] The anholonomy coefficients transform as $\delta\Omega_{abc} = 2\partial_c\partial_{[a}\xi_{b]} + 2\partial_{[a}\omega_{b]c}$, and so $\delta\Omega_{abc} = 0$ for global Lorentz transformations.

[5] Actually, the parameters of the gauge transformations of that part would be ζ_i plus the transverse part of $(\dot{\xi}_i - \partial_i\xi_0)$, still arbitrary.

transformations of the longitudinal part of n_i and hence their shifts are not arbitrary. To sum up the gauge symmetries of Lagrangian (7) are

$$\delta_\xi h_{ij} = \partial_{(i}\xi_{j)}, \quad \delta_\xi n_i = \frac{1}{2}(\dot{\xi}_i + \partial_i \xi_0), \quad \delta_\xi n = -\frac{1}{2}\dot{\xi}_0, \tag{10a}$$

$$\delta_\omega f_{ij} = \frac{1}{2}\omega_{ij}, \quad \delta_\omega e_i = \frac{1}{2}(\zeta_i + \dot{\xi}_i - \partial_i \xi_0). \tag{10b}$$

Let's take now a closer look into $\mathcal{L}_\lambda[e_{ab}]$. Using the definition $m_i \equiv n_i + e_i$ a direct computation gives

$$\begin{aligned}
\mathcal{L}_\lambda &= \mathcal{L}_{\rm FP} + \frac{(1-\lambda)}{2}(\dot{h} - \partial^i m_i)^2 \\
&= \frac{1}{2}\Big(\dot{h}_{ij}\dot{h}^{ij} - \lambda \dot{h}^2 - 4\partial_j n_i \dot{h}^{ij} + (4\partial^i n_i - 2\partial^i m_i + 2\lambda \partial^i m_i)\dot{h} \\
&\quad - 2\partial^i n_i \partial^j n_j + (1-\lambda)\partial^i m_i \partial^j m_j + 2\partial_i n_j \partial^i n^j \\
&\quad - \partial_k h_{ij}\partial^k h^{ij} + \partial_i h \partial^i h + 2\partial^i h_{ij}\partial_k h^{kj} - 2\partial_i h^{ij}\partial_j h + 4n(\partial_i\partial_j h^{ij} - \Delta h)\Big) \\
&\quad + \mathcal{M}[h_{ij}, n_i, n; f_{ij}, e_i],
\end{aligned}$$

where we kept the full total derivative $\mathcal{M}$ for clarity of the argument. It is straightforward to see that if we impose the condition

$$\rho \equiv \partial^i n_i - \partial^i e_i = 0 \tag{11}$$

and plug it into the above expression, we have $\partial^i m_i = 2\partial^i n_i$ and then

$$\begin{aligned}
\mathcal{L}_\lambda \mid_{\rho=0} &= \frac{1}{2}\Big(\dot{h}_{ij}\dot{h}^{ij} - \lambda \dot{h}^2 - 4\partial_i n_j(\dot{h}^{ij} - \lambda\eta^{ij}\dot{h}) + 2(1-2\lambda)\partial^i n_i \partial^j n_j + 2\partial_i n_j \partial^i n^j \\
&\quad - \partial_k h_{ij}\partial^k h^{ij} + \partial_i h \partial^i h + 2\partial^i h_{ij}\partial_k h^{kj} - 2\partial_i h^{ij}\partial_j h + 4n(\partial_i\partial_j h^{ij} - \Delta h)\Big) \\
&\quad + \mathcal{M}\mid_{\rho=0} \\
&= \mathcal{L}_{\rm HL} + \mathcal{M}\mid_{\rho=0} .
\end{aligned} \tag{12}$$

So we recover the Lagrangian of linearized HL theory from (7) by imposing eq. (11) and inside $\mathcal{M}$ we have the longitudinal part of e_i in terms of n_i. (This is the full content of $\rho = 0$, since this condition does not have any information about the transverse part.) Now we can go further and get rid of f_{ij} and the transverse part of e_i inside $\mathcal{M}\mid_{\rho=0}$ by setting them to zero using the gauge symmetries (10b). In this way we get a Lagrangian for the symmetric h_{ab}, which is the one of linearized HL gravity.

It can be seen that eq. (11) is actually a gauge fixing condition. The way to see this is by going into the Hamiltonian formalism corresponding to $\mathcal{L}_\lambda$ and performing the Dirac consistency analysis. Among the resulting constraints we find first class constraints generating the gauge transformations

$$\delta_\xi h_{ij} = 0, \quad \delta_\xi n_i = \frac{1}{2}\partial_i \xi_0 = -\delta_\omega e_i, \quad \delta_\xi n = -\frac{1}{2}\dot{\xi}_0,$$

i.e., those of eq. (10) for $\xi^\mu = (\xi^0, 0, 0, 0)$ with $\xi^0 = \xi^0(x,t)$ arbitrary. The condition (11) can be taken as a gauge fixing condition of one of those first class constraints.

Since we are dealing with a singular theory, it is pertinent to ask if plugging the projection $\rho = 0$ into the Lagrangian is consistent, *i.e.* if the dynamics of the linearized HL gravity as the above computation suggests is compatible with the corresponding variational principle. Comparison of the EoM for h_{ij}, n_i, n, e_i coming from $\mathcal{L}_\lambda$ with the EoM for h_{ij}, n_i, n for linearized HL gravity represented by Lagrangian (1) requires indeed the projection $\rho = 0$ to match both dynamics, but there is a subtlety that worths to be mentioned. The dynamical equations for h_{ij} coming from $\mathcal{L}_\lambda$ are projected onto the corresponding equations of linearized HL gravity, and the projected constraint equations arising from variations of n_i, e_i combine to recover the constraint equation arising from variations of n_i in $\mathcal{L}_{\rm HL}$[6]. There remains, however, some constraint equations of the theory defined with that $\mathcal{L}_\lambda$ that are seen only at the stages of secondary constraints found by the consistency analysis of linearized HL gravity.

In conclusion, by considering $\mathcal{L}_\lambda$ simultaneously with $\rho = 0$ we have able to find a theory for the symmetric components h_{ab} of the *vierbeins*, whose dynamics is given by the EoM of the Lagrangian density $\mathcal{L}_{\rm HL}$. Moreover, since the gauge fixing condition given by eq. (11) transforms as

$$\delta\rho = \partial^i(\delta_\xi n_i) - \partial^i(\delta_\omega e_i) = \frac{1}{2}\partial^i(\dot{\xi}_i + \partial_i\xi_0) - \frac{1}{2}\partial^i(\dot{\xi}_i - \partial_i\xi_0) = \Delta\xi_0$$

under (10), then keeping it invariant restricts the gauge symmetries to $\xi^\mu = (\xi_0, \vec{\xi})$ with $\xi^i = \xi^i(x,t)$ completely arbitrary but ξ^0 subject to equation $\Delta\xi_0 = 0$. As a consequence we have $\xi^0 = \xi^0(t)$. These are precisely the foliation preserving diffeomorphisms of linearized HL gravity.

2.2. Parental action and dual theory

We will follow an approach that consist in taking the theory defined by $\mathcal{L}_\lambda[e_{ab}]$ as a first order parental action from which we extract the second order dual theory for the corresponding dual graviton (see fig. 2 –in the following we use the same notation as in [10]–). The first order parental action from which both the "electric" and its "magnetic" dual can be obtained (in d dimensions) depends on the *vielbein* and some auxiliary fields. Neither of them belong to irreducible representations of the Lorentz group, but contain Young *tableaux* that does. In addition to the procedure to obtain the dual of $\mathcal{L}_\lambda[e_{ab}]$, to get the dual of linearized HL gravity for the symmetric components of the *vierbeins* in $d = 4$ we have to impose a "magnetic" analogue of the "electric" condition given by eq. (11).

We start by considering the Lagrangian $\mathcal{L}_\lambda[e_{ab}]$ given in eq. (7). It is not hard to see that the corresponding parental action can be defined with the Lagrangian density

$$\mathcal{L}_\lambda[e,Y] = -2\Big(Y^{ab|c}\Omega_{abc} - Y^{ab|c}Y_{ac|b} + \frac{1}{(d-2)}Y^aY_a\Big) + 2\frac{(\theta-1)}{(d-2)}(Y_0)^2, \tag{13}$$

with θ the function of λ given by

$$\theta = \frac{(d-2)\lambda}{(d-1)\lambda - 1}. \tag{14}$$

Here, besides the *vielbein*, we have some tensor variables $Y^{ab|}{}_c$ antisymmetric in ab indices but with no definite symmetry for c (we use the notation $Y^a \equiv Y^{ab|}{}_b$ for the trace). The above Lagrangian is assumed to be written in the same reference frame as (7).

We see from the EoM of $Y^{ab|}{}_c$ that these are auxiliary fields, solved in terms of e_{ab} and their derivatives as

$$Y_{ab|c} = \frac{1}{2}\Omega_{abc} - \Omega_{c[ab]} + 2\eta_{c[a}\delta_{b]}{}^d\Big(\Omega_d - (1-\lambda)\eta_{d0}\Omega^0\Big). \tag{15}$$

Pluging eq. (15) into the first order Lagrangian (13) we get the second order "electric" Lagrangian (7).

Before going onto the dual theory, notice that the gauge symmetry transformations (10) of $\mathcal{L}_\lambda[e_{ab}]$ are captured by gauge invariances of the parental Lagrangian (13). This is achieved

[6] The constraints arising from variations of n do not even need the projection $\rho = 0$ to match.

by extending the former with proper transformations of the auxiliary $Y^{ab|}{}_c$. In fact, the restriction given in eq. (9) arise also here from general assumptions: using that $\delta e_{ab} = \partial_a\xi_b + \partial_b\xi_a + \omega_{ab} \Rightarrow \delta\Omega_{abc} = 2(\partial_c\partial_{[a}\xi_{b]} + \partial_{[a}\omega_{b]c})$, we get from eq. (15)

$$\delta Y_{ab|c} = 2\partial_c\partial_{[a}\xi_{b]} - \partial_c\omega_{ab} + 2\eta_{c[a}(\eta_{b]d} - (1-\lambda)\eta_{b]0}\delta^0{}_d)(\partial^d\partial^e\xi_e - \partial^e\partial_e\xi^d + \partial_e\omega^{ed}). \tag{16}$$

For the variation of the Lagrangian we get

$$\delta\mathcal{L}_\lambda = -2\Big(\delta Y^{ab|c}\Omega_{abc} + Y^{ab|c}(\delta\Omega_{abc} - 2\delta Y_{ac|b} + \frac{2}{(d-2)}\eta_{bc}\delta Y_a)\Big) + 4\frac{(\theta-1)}{(d-2)}Y_0\delta Y_0 \simeq 0.$$

The first term in this equation involves the e_{ab} variables. Requiring its invariance implies eq. (9), a restriction that eventually gives $\delta Y_0 = 0$ making the last term invariant on its own. The rest of the terms cancel out in account of the antisymmetry of $Y^{ab|}{}_c$ in ab.

Now we compute the dual theory by taking in eq. (13) the e_{ab} variables as Lagrange multipliers of the constraints $\partial_a Y^{ab|}{}_c = 0$. These are solved by some potentials $Y^{abe|}{}_c$ completely antisymmetric in its indices abe and such that $Y^{ab|}{}_c = \partial_e Y^{abe|}{}_c$. Reduction of Lagrangian (13) give us the dual theory:

$$\mathcal{L}_\lambda[Y^{abe|}{}_c] = 2\Big(Y^{ab|c}Y_{ac|b} - \frac{1}{(d-2)}Y^aY_a\Big) + 2\frac{(\theta-1)}{(d-2)}(Y_0)^2 \tag{17}$$

(where the field strengths $Y^{ab|}{}_c$ are functions of the potentials $Y^{abe|}{}_c$). It is not difficult to see that the action defined by this Lagrangian is invariant under transformations (16) for the field strength, the restriction (9) coming here from the last term[7]. However, these do not exhaust the gauge symmetries of the dual theory. There are more gauge symmetries coming from an ambiguity in the potentials solving $\partial_a Y^{ab|}{}_c = 0$. The redundancy is given by the gauge transformations

$$\delta_\phi Y^{abc|}{}_d = \partial_e\phi^{abce|}{}_d \tag{18}$$

with $\phi^{abce|}{}_d$ arbitrary and completely antisymmetric in indices $abce$. These transformations leave the field strengths $Y^{ab|}{}_d$ invariant and so the Lagrangian (17), since the latter is quadratic in those field strengths.

We can get a closer look to the symmetries of the dual "magnetic" theory if we perform the irreducible decomposition (see [10])

$$Y^{abc|}{}_d = X^{abc|}{}_d + \delta_d^{[a}Z^{bc]},$$

which consists of extracting from $Y^{abc|}{}_d$ its trace (the fields $X^{abc|}{}_d$ are completely antisymmetric in abc indices and traceless). Plugging the above decomposition into Lagrangian (17) we end up with the dual theory written as

$$\mathcal{L}_\lambda[X,Z] = \mathcal{L}_{\text{FP}}[X,Z] + \frac{2(d-2)}{9}(\theta-1)(\partial^i Z_{i0})^2. \tag{19}$$

The $\mathcal{L}_{\text{FP}}[X,Z]$ part is the dual of the FP Lagrangian in d dimensions, where

$$\mathcal{L}_{\text{FP}}[X,Z] = 2\,\partial_d X^{abd|c}\partial^e X_{ace|b} + \mathcal{M}[X,Z],$$

[7] With this restriction, using eq. (16) it can be seen that $\delta(\partial_a Y^{ab|}{}_c) = 0$, meaning that the constraint defining the dual potentials $Y^{abc|}{}_d$ is gauge invariant.

the last term being the total derivative

$$\mathcal{M} \equiv \frac{2}{3}\partial_c\Big(2\partial_d X^{cda|b} Z_{ab} + Z_a{}^{[b}\partial_b Z^{ac]}\Big) \tag{20}$$

(and the only place where $\mathcal{L}_{\rm FP}[X,Z]$ depends on Z^{ab}).

At this point in the "magnetic" theory we are in a stage corresponding to the "electric" eq. (7). The FP Lagrangian $\mathcal{L}_{\rm FP}$ is invariant under the arbitrary shifts given by $\delta_{\tilde\omega} Z^{ab} = \tilde\omega^{ab} = -\tilde\omega^{ba}$. These include the projection of the gauge symmetries given in eq. (18) for those variables[8]. It can be seen that $\delta_\phi(\partial_a Z^{ab}) = 0$, and so it is straightforward that these transformations leave also invariant the last term in Lagrangian (19). This term is not invariant, however, under arbitrary shifts with parameters $\tilde\omega_{ab}$. Instead, demanding invariance of this term will restrict them.

Identifying $Z^{0i} \equiv Z^i$ with the components of a vector in $d-1$ dimensions, we split them into transverse and longitudinal parts: $Z^i = Z^i_\perp + Z^i_\parallel$, and doing the same with the parameters $\tilde\omega^{0i} = \tilde\zeta^i + \omega^i$, where the transverse (or divergenceless) $\tilde\zeta^i$ are involved in the local shifts of the transverse part $Z^i_\perp$, the invariance of the last term in Lagrangian (19) requires

$$\delta_{\tilde\omega}(\partial_i Z^i) = \delta_{\tilde\omega}(\partial_i Z^i_\parallel) = \partial_i\tilde\omega^{0i} = \partial_i\omega^i = 0.$$

This relation implies that $Z^i_\perp$ are pure gauge ($\delta_{\tilde\omega} Z^i_\perp = \tilde\zeta^i$ with $\tilde\zeta^i$ arbitrary). On the other hand the parallel components ω^i are restricted by the above equation to be constant along the spatial hypersurface, that is $\delta_{\tilde\omega} Z^i_\parallel = \omega^i(t)$.

We have here a very similar breaking of local Lorentz invariance that we had in the "electric" theory, restricting the gauge invariances to those of eq. (10). In summary, the gauge symmetries for the dual theory are given by

$$\delta_\phi X^{abc|}{}_d = \partial_e\phi^{abce|}{}_d + \frac{3}{(d-2)}\delta^{[a}_d\partial_e\phi^{bc]e}, \tag{21a}$$

$$\delta_{\tilde\omega} Z^{ij} = \tilde\omega^{ij}, \quad \delta_{\tilde\omega} Z^i = \tilde\zeta^i + \omega^i(t), \text{ such that } \partial_i\tilde\zeta^i = 0. \tag{21b}$$

We have constructed the dual of the theory with Lagrangian (7) in d dimensions, and identified its gauge symmetries. We did it by providing the parental action defined with the Lagrangian (13), from which both "electric" and "magnetic" theories are obtained. In the case of the"electric" theory we managed to identify and characterize the way in which linearized HL gravity is obtained from Lagrangian (7), imposing the gauge fixing condition (11). We wonder now if there is an analogous "magnetic" projection for Lagrangian (19) in $d=4$ dimensions that turns the reduced dual theory into the linearized HL gravity again.

To explore this idea we use the field redefinition

$$T_{a_1\ldots a_{d-3}|c} \equiv \frac{1}{3!}\epsilon_{a_1\ldots a_{d-3}efg}X^{efg|}{}_c, \quad \text{such that } T_{[a_1\ldots a_{d-3}|c]} = 0. \tag{22}$$

In $d=4$ dimensions T_{ab} is a symmetric tensor. In these variables the action defined with the dual Lagrangian $\mathcal{L}_{\rm FP}[T,Z]$ is again the FP theory for the dual graviton. We also have that the symmetries (21a) can be written as $\phi^{abcd|}{}_e = \epsilon^{abcd}\chi_e$, with χ_a arbitrary, giving $\delta_\phi T_{ab} = -\partial_{(a}\chi_{b)}$.

[8] They separate as

$$\delta_\phi Y^{abc|}{}_d = \partial_e\phi^{abce|}{}_d \to \begin{cases} \delta_\phi Z^{ab} = -\dfrac{3}{(d-2)}\partial_c\phi^{abc}, \ \phi^{abc} \equiv \phi^{abcd|}{}_d \\ \delta_\phi X^{abc|}{}_d = \partial_e\phi^{abce|}{}_d + \dfrac{3}{(d-2)}\delta^{[a}_d\partial_e\phi^{bc]e} \end{cases}.$$

The $\delta_\phi X^{abc|}{}_d$ leave $\mathcal{L}_{\rm FP}$ invariant modulo a total derivative.

Journal of Physics: Conference Series **512** (2014) 012001 doi:10.1088/1742-6596/512/1/012001

In these variables the action defined with the Lagrangian (19) for the dual theory is written as

$$S_\lambda[T,Z] = \int \mathrm{d}^4x\Big(\mathcal{L}_{\mathrm{FP}}[T,Z] + \frac{4}{9}(\theta-1)(\partial_i Z^i)^2\Big)$$
$$= 2\int \mathrm{d}^4x\Big(\partial^a T^{bc}\partial_a T_{bc} - 2\partial_a T^{ac}\partial^b T_{bc} + 2\partial^b T^a{}_a\partial^c T_{bc} - \partial_a T^b{}_b\partial^a T^c{}_c$$
$$+ \mathcal{M}[T,Z] + \frac{2}{9}(\theta-1)(\partial_i Z^i)^2\Big).$$

We can fix part of the second term (to zero) using the gauge invariance under arbitrary shifts of Z^{ij} and $Z^i_\perp$, the variables that are pure gauge. But the $Z^i_\parallel$ survive on the boundary because the action does not exhibit arbitrary shifts of these variables as a gauge invariance. To describe the theory in terms of the fields T_{ab} we need to impose a "magnetic" anaolgue of the "electric" gauge fixing condition given in eq. (11). With such condition we would like to write $Z^i_\parallel$ (or more precisely, its divergence) in terms of some components of T_{ab}.

For that end we split the Lagrangian (19) in space and time. We have

$$\mathcal{L}_\lambda[T,Z] = -2\Big(\dot{T}_{ij}\dot{T}^{ij} - \dot{T}^2 - 4\partial_i T_{0j}(\dot{T}^{ij} - \eta^{ij}\dot{T}) - 2\partial^i T_{0i}\partial^j T_{0j} - 2\partial_i T_{0j}\partial^i T^{0j} + 2R_T\Big)$$
$$+ \mathcal{M}[T,Z] + \frac{4}{9}(\theta-1)(\partial_i Z^i)^2, \quad (23)$$

where

$$R_T \equiv -\frac{1}{2}\partial_k T_{ij}\partial^k T^{ij} + \frac{1}{2}\partial_i T\partial^i T + \partial_i T^{ij}\partial^k T_{kj} - \partial_i T^{ij}\partial_j T - T_{00}(\partial_i\partial_j T^{ij} - \Delta T)$$

is nothing more but the potential in the Lagrangian (1) as a function of T_{ij} instead of h_{ij}, and $T \equiv \eta^{ij}T_{ij}$.

Taking the following *ansatz*[9] for the "magnetic" projection condition

$$\rho^* \equiv \partial_i Z^i - \varepsilon(\dot{T} - 2\partial^i T_{0i}) = 0, \quad \varepsilon \in \mathbb{R} - \{0\}. \quad (24)$$

Some arithmetics give

$$\mathcal{L}_\lambda[T,Z]\,|_{\rho^*=0} = -2\Big(\dot{T}_{ij}\dot{T}^{ij} - \gamma\dot{T}^2 - 4\partial_i T_{0j}(\dot{T}^{ij} - \gamma\eta^{ij}\dot{T})$$
$$+ 2(1-2\gamma)\partial^i T_{0i}\partial^j T_{0j} - 2\partial_i T_{0j}\partial^i T^{0j} + 2R_T\Big) \quad (25)$$

up to a total derivative, with the parameter γ given by

$$\gamma \equiv 1 + \frac{2}{9}\Big(\frac{1-\lambda}{3\lambda-1}\Big)\varepsilon^2.$$

It is possible to see from the Hamiltonian Dirac analysis of the constraints that condition (24) is also a gauge fixing condition.

From eq. (23) to eq. (25), the latter a theory for T_{ab} only, we were able to get rid of $\mathcal{M}$ because now this term is a total derivative that can be eliminated from the picture with boundary

[9] The *ansatz* could be a general combination of $\dot{T}$ and $\partial^i T_{0i}$, which are the objects involved in the terms proportional to λ in the Lagrangian for linearized HL gravity. It can be seen that those terms should be in a ratio of $\frac{\dot{T}}{\partial^i T_{0i}} \sim -\frac{1}{2}$.

conditions on T_{ab}. The resulting functional is the Lagrangian of linearized HL gravity but parametrized by γ instead of λ. This parameter measures the deviation of the dual theory from the dual of FP action. Hence, the electric and magnetic theories are the same deviation of FP by terms measured with parameters belonging to "weak" and "strong" regimes respectively ($\lambda < 1 \Rightarrow \gamma > 1$, and *vice versa* –the case $\lambda = 1 = \gamma$ being FP theory–).

As in the "electric" theory, preservation of the condition given by eq. (24) restricts again the symmetries (21a), which are $\delta_\phi T_{ab} = -\partial_{(a}\chi_{b)}$. If we want to take the dual Lagrangian (19) together with such condition as the dual theory of linearized HL gravity, we would like to preserve the latter. Then, we have that

$$
\begin{aligned}
0 = \delta\rho^* &= \overset{0}{\cancel{\delta_\omega(\partial_i Z^i)}} - \varepsilon\Big(\delta_\phi(\dot{T}) - 2\delta_\phi(\partial^i T_{0i})\Big) \\
&= -\varepsilon\Big(\eta^{ij}\partial_0\delta_\phi T_{ij} - 2\partial^i\delta_\phi T_{0i}\Big) \\
&= -\varepsilon\Big(\eta^{ij}\partial_0(-\partial_{(i}\chi_{j)}) - 2\partial^i(-\partial_{(0}\chi_{i)})\Big) \\
&= -\varepsilon\Delta\chi_0 .
\end{aligned}
$$

This means that gauge symmetries are restricted to those with $\chi_0 = \chi_0(t)$, the foliation preserving diffeomorphisms of linearized HL gravity.

3. Conclusions

The construction of duality transformations in classical field theories using the powerful tools of covariant parent actions has been implemented in Hořava-Lifshitz gravity. For that end we proceeded in two steps: a) the construction of a covariant model that in an appropriate gauge reduce to Hořava-Lifshitz gravity. We have used a gauge is fixing procedure that it breaks the full diffeomosphism grip of General Realtivity to a subgroup that coincides with the symmetry group of Hořava-Lifshitz gravity., b) The implementation of parent action to construct the duality symmetry in a covariant model. In this way we were able to construct a dual action associated with the standard formulation of Hořava-Lifshitz gravity.

Starting from Lagrangian (7) we have constructed a dual action in D dimensions (eq. (19)) and in particular for $D = 4$ our result is given in eq. (25). The details of our calculations will be published elsewhere. We have also work in progress along a generalisation of this duality implementation when the dual graviton is also present. This example shows that the Lorentz global invariance is not strictly necessary to implement duality symmetry at the variational level.

Acknowledgments

The present work was partially supported by National Council of Science and Technology (CONACyT) grant 104649, as well as DGAPA-UNAM grant IN110312.

IC acknowledge partial support by CONACyTs program Estancias Sabáticas y Posdoctorales al Extranjero (México) 2011-2012, grant number 172676. Also the work of IC is partially supported by IISN-Belgium (conventions 4.4511.06 and 4.4514.08) and by the Communauté Francaise de Belgique through the ARC program and by the ERC through the SyDuGraM Advanced Grant.

References

[1] West P C 2001 *Classical Quantum Gravity* **18** 4443 2
[2] Curtright T 1985 *Phys. Lett. B* **165** 304 2
[3] P Hořava 2011 *Class. Quant. Grav.* **28** 114012 (*Preprint* 1101.1081) 2

[4] P Hořava and C M Melby-Thompson 2010 *Physical Review D* **82** 064027 URL http://link.aps.org/doi/10.1103/PhysRevD.82.064027 2
[5] P Hořava 2009 *JHEP* **03** 020 (*Preprint* 0812.4287) 2
[6] P Hořava 2009 *Phys. Rev. D* **79** 084008 (*Preprint* 0901.3775) 2
[7] P Hořava 2009 *Phys. Rev. Lett.* **102** 161301 (*Preprint* 0902.3657) 2
[8] Kehagias A and Sfetsos K 2009 *Phys. Lett. B* **678** 123–126 (*Preprint* 0905.0477) 2, 3
[9] Casini H, Montemayor R and Urrutia L F 2003 *Phys. Rev. D* **68** 065011 Ver referencias [7] para introducción a dualidad. (*Preprint* hep-th/0304228) 2
[10] Boulanger N, Cnockaert S and Henneaux M 2003 *JHEP* **06** 060 (*Preprint* hep-th/0306023) 2, 4, 7, 8
[11] Henneaux M and Teitelboim C 2005 *Phys. Rev. D* **71** 024018 (*Preprint* gr-qc/0408101) URL http://prd.aps.org/abstract/PRD/v71/i2/e024018 2
[12] Bellorin J, Restuccia A and Sotomayor A 2013 A consistent Hořava gravity without extra modes and equivalent to general relativity at the linearized level arXiv:1302.1357 [hep-th] (*Preprint* 1302.1357) 3
[13] Henneaux M, Kleinschmidt A and Gomez G L 2010 *Phys. Rev. D* **81** 064002 (*Preprint* 0912.0399) 3
[14] Henneaux M, Kleinschmidt A and Lucena Gómez G 2011 Remarks on gauge invariance and first-class constraints vol 272 pp 141–150 ISSN 0081-5438 URL http://www.mathnet.ru/php/archive.phtml?wshow=paper&jrnid=tm&paperid=3257&option_lang=eng 3
[15] Barausse E and Sotiriou T P 2012 *Physical Review Letters* **109** 181101 3
[16] Bellorin J and Restuccia A 2011 *Phys. Rev. D* **84** 104037 (*Preprint* 1106.5766) 3
[17] Blas D, Pujolás O and Sibiryakov S 2010 *Phys. Rev. Lett.* **104** 181302 (*Preprint* 0909.3525) URL http://prl.aps.org/abstract/PRL/v104/i18/e181302 3
[18] Donnelly W and Jacobson T 2011 *Phys. Rev. D* **84** 104019 (*Preprint* 1106.2131) 3
[19] Sotiriou T P 2011 *J. Phys. Conf. Ser.* **283** 012034 (*Preprint* 1010.3218) 3
[20] Griffin T, Hořava P and Melby-Thompson C M 2013 *Phys. Rev. Lett.* **110**(8) 081602 URL http://link.aps.org/doi/10.1103/PhysRevLett.110.081602 3
[21] Blas D, Pujolás O and Sibiryakov S 2009 *Journal of High Energy Physics* **2009** 029 ISSN 1029-8479 URL http://iopscience.iop.org/1126-6708/2009/10/029 3
[22] Bellorin J and Restuccia A 2012 *Int. J. Mod. Phys. D* **21** 1250029 (*Preprint* 1004.0055) 3
[23] Barausse E and Sotiriou T P 2012 Slowly rotating black holes in hořava-lifshitz gravity arXiv:1212.1334 [gr-qc] (*Preprint* 1212.1334) 3
[24] Visser M 2011 *J. Phys. Conf. Ser.* **314** 012002 (*Preprint* 1103.5587) 3
[25] Charmousis C, Niz G, Padilla A and Saffin P M 2009 *Journal of High Energy Physics* **2009** 070 ISSN 1029-8479 URL http://iopscience.iop.org/1126-6708/2009/08/070 3

Corrigendum: Electric-Magnetic duality in (linearized) Hořava-Lifshitz gravity

Ignacio Cortese[a] **and J. Antonio García**[b]

a) Physique Théorique et Mathématique & International Solvay Institutes
Université Libre de Bruxelles, Campus Plaine C.P. 231, B-1050 Bruxelles, Belgium
b) Departamento de Física de Altas Energías, Instituto de Ciencias Nucleares,
Universidad Nacional Autónoma de México,
Apartado Postal 70-543, México D.F. 04510, México

CORRIGENDUM TO: Ignacio Cortese and J Antonio García 2014 *J. Phys.: Conf. Ser.* **512** 012001

We report a problem with the last version of our paper and the one that we sent for the final published version. The published version has an overlap with an already published work. The changes are only in the introduction. The correct introduction section appear below. The scientific content of our article and conclusions remains the same.

1. Introduction

Symmetries and variational principles play an essential role in the construction of interactions. The Noetherian symmetries are a central paradigm in this realm. In recent years the generalization of the concept of symmetry has been a subject of great interest.

The fact that the equations of motion can have more symmetries than the corresponding variational principle ruled out some important symmetries that can be very useful in the analysis and understanding of physical theories. Symmetries like non-Noetherian symmetries (or s-equivalent symmetries) and Dualities are powerful tools in diverse areas like condensed matter and string theory In particular Dualities are very interesting symmetries that relate different physical theories and/or the same theory at different couplings.

Among duality symmetries the electric-magnetic duality is perhaps the best well known of them. Recently the authors of [1, 2] proposed a variational formulation of the this duality, that works not only at the level of the equations of motion. Even though this formulation is constructed giving up manifest spacetime covariance of the original Maxwell action the authors of [3] showed that in the case of linearized gravity it is possible to implement the electric-magnetic duality analogue at the level of an action. This implementation also breaks manifest spacetime covariance in order to have duality invariance manifest.

Motivated by these works we are wondering if Hořava-Lifshitz gravity, a completion of general relativity in the high energy regime that actually breaks spacetime covariance, admits also a duality implementation. The aim of this work is to present in detail such implementation, following the example of the Fierz-Pauli case. We will proceed, however, from another point of view, namely from a parent-action for the covariant Fierz-Pauli action and its dual theory.

Actually a main trend of current research is the study of deformations of gravity. The basic idea consist in the construction of a new gravity that bypass the basic problems associated with the standard formulation of gravity as a field theory. As is well know the theory is not renormalizable and the corresponding quantum theory is not known. Among the many directions in this trend (the consistent deformations and quantization of gravity) we can mention non commutative gravity, stringy corrections to the action of GR, new massive gravities in many dimensions, double field theory and Hořava-Lifshitz gravity [4, 5, 6, 7, 8]. This last particular deformation pretends to be a UV completion of gravity using an "heretic" proposal [9](breaking the spacetime paradigm), breaking full diff invariance of GR and leaving an invariance subgroup that is still manifest in ADM formalism. As a consequence different treatment of the kinetic and potential terms are necessary. The kinetic term is still quadratic in time derivatives but is deformed by a new parameter λ, and the potential has high order space derivatives of the basic field, the metric g_{ij}. The deformed theory is power counting renormalizable, at large distances higher derivative terms are suppressed and the theory runs to standard GR. The new parameter λ could be considered as the coupling parameter which controls the contribution of the trace of extrinsic curvature and of some higher derivative terms as for example the term proportional to $(R^{(3)})^2$ (see [7]). For $\lambda = 1$ we have a fixed point in the IR, recovering GR. For generic values of λ the theory breaks full diffeomorphism invariance (the theory becomes a non-relativistic theory).

The aim of this note is to implement (in the same way as in the parental action approach) the duality symmetry [10, 11]) in Hořava-Lifshitz theory. The interesting point about this duality is that Hořava-Lifshitz gravity is not invariant under the complete spacetime diffeomorphisms but only under a subgroup of them. However this non full covariance is not an impediment to implement duality and is an example of the role played by the complementarity between covariance and duality. Another very interesting way to construct duality in linearized GR is the pre potentials approach of [3].

We start from the action

$$S_{\mathrm{HL}} = 2\int \mathrm{d}t\mathrm{d}^3x\sqrt{g}N\Big(K_{ij}K^{ij} - \lambda K^2 + R^{(3)}\Big)$$

in $d = 4$ dimensions and ADM separation, representing the minimal HL deviation from GR, we take the linearized part. Perturbation around flat spacetime $g_{ij} = \eta_{ij} + \varepsilon h_{ij}$, $N = 1 + \varepsilon n$ and $N_i = \varepsilon n_i$ into S_{HL} gives

$$\begin{aligned} L^{(2)}_{\mathrm{HL}} = \varepsilon^2 \int \mathrm{d}^3x\Big[\frac{1}{2}\dot{h}_{ij}\dot{h}^{ij} - \frac{\lambda}{2}\dot{h}^2 + \partial_i n_j\partial^i n^j + (1-2\lambda)(\partial_i n^i)^2 - 2\partial_i n_j(\dot{h}^{ij} - \lambda\eta^{ij}\dot{h}) \\ - \frac{1}{2}\partial_k h_{ij}\partial^k h^{ij} + \frac{1}{2}\partial_i h\partial^i h + \partial^i h_{ij}\partial_k h^{kj} - \partial_i h^{ij}\partial_j h + 2n(\partial_i\partial_j h^{ij} - \Delta h)\Big] \end{aligned} \quad (1)$$

($h \equiv \eta^{ij}h_{ij}$). From now on our focus will be the theory defined by this Lagrangian, so we drop the superscript $^{(2)}$ of L_{HL}. This Lagrangian deviates from FP theory in those terms proportional to λ, to which this theory reduce in the IR limit given by $\lambda \longrightarrow 1$. Even thought the modifications due to a value of $\lambda \neq 1$ are subtle, the structural consequences are considerable and have attracted much attention in recent research [1].

In contrast to FP theory, the gauge transformations of this theory are given by the linearized foliation preserving diffeomorphisms $\delta x^i = \xi^i(x,t)$ and $\delta t = \xi^0(t)$, acting on the variables of the theory as

$$\delta h_{ij} = \partial_i\xi_j + \partial_j\xi_i, \quad \delta n_i = \dot{\xi}_i, \quad \delta n = -\dot{\xi}_0. \quad (2)$$

[1] An incomplete list of references to this last topic is[12, 13, 14, 15, 16, 17, 18, 19, 20, 21, 22, 23, 24, 25, 9].

This is a subset of the full group of diffeomorphisms $\delta_\xi h_{\mu\nu} = \partial_\mu \xi_\nu + \partial_\nu \xi_\mu$. The latter do not leave the action defined by (1) invariant but rather its variation gives

$$\delta_\xi S_{\rm HL} = \int {\rm d}t{\rm d}^3x 2(\lambda - 1)(\dot{h} - 2\partial^i n_i)\Delta\xi_0. \quad (3)$$

Restricting the diffeomorphisms to those that satisfy $\Delta\xi_0 = 0$ we get the isommetries given by eq. (2). The solution of this equation is ξ_0 a constant in space[2], *i.e.*, $\xi_0 = \xi_0(t)$.

section*References

[1] Deser S and Teitelboim C 1976 *Physical Review D* **13** 1592 URL http://link.aps.org/doi/10.1103/PhysRevD.13.1592 1

[2] Deser S and Seminara D 2005 *Phys. Lett. B* **607** 317–319 (*Preprint* hep-th/0411169) URL http://www.sciencedirect.com/science/article/pii/S0370269304017459 1

[3] Henneaux M and Teitelboim C 2005 *Phys. Rev. D* **71** 024018 (*Preprint* gr-qc/0408101) URL http://prd.aps.org/abstract/PRD/v71/i2/e024018 1, 2

[4] P Hořava 2011 *Class. Quant. Grav.* **28** 114012 (*Preprint* 1101.1081) 2

[5] P Hořava and C M Melby-Thompson 2010 *Physical Review D* **82** 064027 URL http://link.aps.org/doi/10.1103/PhysRevD.82.064027 2

[6] P Hořava 2009 *JHEP* **03** 020 (*Preprint* 0812.4287) 2

[7] P Hořava 2009 *Phys. Rev. D* **79** 084008 (*Preprint* 0901.3775) 2

[8] P Hořava 2009 *Phys. Rev. Lett.* **102** 161301 (*Preprint* 0902.3657) 2

[9] Kehagias A and Sfetsos K 2009 *Phys. Lett. B* **678** 123–126 (*Preprint* 0905.0477) 2

[10] Casini H, Montemayor R and Urrutia L F 2003 *Phys. Rev. D* **68** 065011 Ver referencias [7] para introducción a dualidad. (*Preprint* hep-th/0304228) 2

[11] Boulanger N, Cnockaert S and Henneaux M 2003 *JHEP* **06** 060 (*Preprint* hep-th/0306023) 2

[12] Bellorin J, Restuccia A and Sotomayor A 2013 A consistent Hořava gravity without extra modes and equivalent to general relativity at the linearized level arXiv:1302.1357 [hep-th] (*Preprint* 1302.1357) 2

[13] Henneaux M, Kleinschmidt A and Gomez G L 2010 *Phys. Rev. D* **81** 064002 (*Preprint* 0912.0399) 2

[14] Henneaux M, Kleinschmidt A and Lucena Gómez G 2011 vol 272 pp 141–150 ISSN 0081-5438 URL http://www.mathnet.ru/php/archive.phtml?wshow=paper&jrnid=tm&paperid=3257&option_lang=eng 2

[15] Barausse E and Sotiriou T P 2012 *Physical Review Letters* **109** 181101 2

[16] Bellorin J and Restuccia A 2011 *Phys. Rev. D* **84** 104037 (*Preprint* 1106.5766) 2

[17] Blas D, Pujolás O and Sibiryakov S 2010 *Phys. Rev. Lett.* **104** 181302 (*Preprint* 0909.3525) URL http://prl.aps.org/abstract/PRL/v104/i18/e181302 2

[18] Donnelly W and Jacobson T 2011 *Phys. Rev. D* **84** 104019 (*Preprint* 1106.2131) 2

[19] Sotiriou T P 2011 *J. Phys. Conf. Ser.* **283** 012034 (*Preprint* 1010.3218) 2

[20] Griffin T, Hořava P and Melby-Thompson C M 2013 *Phys. Rev. Lett.* **110**(8) 081602 URL http://link.aps.org/doi/10.1103/PhysRevLett.110.081602 2

[21] Blas D, Pujolás O and Sibiryakov S 2009 *Journal of High Energy Physics* **2009** 029 ISSN 1029-8479 URL http://iopscience.iop.org/1126-6708/2009/10/029 2

[22] Bellorin J and Restuccia A 2012 *Int. J. Mod. Phys. D* **21** 1250029 (*Preprint* 1004.0055) 2

[23] Barausse E and Sotiriou T P 2012 Slowly rotating black holes in hořava-lifshitz gravity arXiv:1212.1334 [gr-qc] (*Preprint* 1212.1334) 2

[24] Visser M 2011 *J. Phys. Conf. Ser.* **314** 012002 (*Preprint* 1103.5587) 2

[25] Charmousis C, Niz G, Padilla A and Saffin P M 2009 *Journal of High Energy Physics* **2009** 070 ISSN 1029-8479 URL http://iopscience.iop.org/1126-6708/2009/08/070 2

[2] Equation $\Delta\xi_0 = 0 \Rightarrow \partial_i\xi_0$ should be the transverse part of a vector, but 0 is the only vector whose transverse part is a gradient, so $\partial_i\xi_0 = 0$ and this imply that ξ_0 does not depend on the space coordinates.

Black strings induced by the Casadio-Fabbri-Mazzacurati Braneworld black holes solutions

Roldão da Rocha[1] **and André M. Kuerten**[2]

[1]Centro de Matemática, Computação e Cognição, Universidade Federal do ABC 09210-170, Santo André, SP, Brazil.

[2]Centro de Ciências Naturais e Humanas, Universidade Federal do ABC 09210-170, Santo André, SP, Brazil.

E-mail: [1]roldao.rocha@ufabc.edu.br, [2]andre.kuerten@ufabc.edu.br

Abstract. This paper aims to evince the bulk metric near the brane and in particular the black string warped horizon associated to the Casadio-Fabbri-Mazzacurati braneworld black holes, and their physical consequences. The black string can be shown to be more unstable to large-scale perturbations than the Schwarzschild standard black strings, namely the Gregory-Laflamme instability. The two solutions of Einstein equations proposed by Casadio, Fabbri, and Mazzacurati, regarding black hole metrics presenting a post-Newtonian parameter measured on the brane are analyzed, in particular when suitable values for the post-Newtonian parameter precluding Hawking radiation on the brane are chosen. The associated black string profile is studied. Further details can be found in [1].

1. Introduction

Extra-dimensional spacetimes are scenarios for extensions of the general relativity, wherein Einstein's field equations can present solutions as black holes in higher dimensions. Furthermore, the hierarchy solution is obtained by leading gravity to leak into extra dimensions [2]. A particularly useful approach to deal with the hierarchy is provided an effective 5D reduction of the Hořava-Witten theory [3, 4, 5, 6].

Motivated by the study of gravity on 5D braneworld scenarios and some applications, in particular the black holes/black strings horizons variations induced by braneworld effects, [7, 8, 9], further aspects concerning generalized black strings and their warped horizons are here reviewed. The Casadio-Fabbri-Mazzacurati metrics on the brane, namely the type I and type II black hole solutions [10, 11] are studied as generators the bulk metric, inducing an associated black string warped horizon [1].

This article is organized as follows: in Sec. 2 the Einstein field equations on the brane are revisited. For a static spherical metric on the brane, the propagating effect of 5D gravity is evinced from the Taylor expansion along the extra dimension of the brane metric, providing the black string warped horizon profile. Such expansion can provide the bulk metric near the brane merely from the metric on the brane. In Sec. 3, the type I and type II Casadio-Fabbri-Mazzacurati black string solutions and their respective warped horizons are obtained, analyzed and depicted. We analyze such solutions in the particular case where the associated post-Newtonian parameter makes the black hole Hawking radiation to be suppressed, for the two models provided by Casadio, Fabbri, and Mazzacurati.

2. Black string warped horizon and the bulk metric near the brane

Hereupon the notation in [12, 13] is adopted, where $\{\theta_\mu\}$ typify a tetrad on a brane embedded in a bulk. A frame $\theta^A = dx^A$ $(A = 0, 1, 2, 3, 4)$ in the bulk is represented in local coordinates. In the brane defined by $y = 0$, [hereon y denotes the associated Gaussian coordinate] the 1-form dy is orthogonal to the brane. The metric

$$g_{\mu\nu}(x^\alpha, y)\, dx^\mu dx^\nu + dy^2$$

endows the bulk. The fine tuning is provided by $\Lambda_4 = \frac{\kappa_5^2}{2}\left(\frac{1}{6}\kappa_5^2\lambda^2 + \Lambda_5\right)$ and the 5D coupling constants are related by $\kappa_4^2 = \frac{1}{6}\lambda\kappa_5^4$. Here Λ_4 denotes the effective brane cosmological constant, and λ is the brane tension. The constant $\kappa_5 = 8\pi G_5$, where G_5 is the 5D Newton gravitational constant, denotes the 5D gravitational coupling, related to the 4D gravitational constant G by $G_5 = G\ell_{\rm Planck}$. The junction condition provides the extrinsic curvature tensor by [14, 15, 12]

$$K_{\mu\nu} = -\frac{1}{2}\kappa_5^2\left(T_{\mu\nu} + \frac{1}{3}(\lambda - T)\, g_{\mu\nu}\right), \tag{1}$$

where $T = T^\mu{}_\mu$ is the trace of the energy-momentum tensor. The trace-free and symmetric components of the 5D Weyl tensor are respectively denoted by $\mathcal{B}_{\mu\nu\alpha} = g_\mu^\rho g_\nu^\sigma C_{\rho\sigma\alpha\beta}n^\beta$ and $\mathcal{E}_{\mu\nu} = C_{\mu\nu\sigma\rho}n^\sigma n^\rho$.

The Einstein brane field equations can be expressed as

$$G_{\mu\nu} = -\mathcal{E}_{\mu\nu} - \frac{1}{2}\Lambda_5 g_{\mu\nu} + \frac{1}{4}\kappa_5^4\left(\frac{1}{2}g_{\mu\nu}\left(T^2 - T_{\alpha\beta}T^{\alpha\beta}\right) + TT_{\mu\nu} - T_{\mu\alpha}T^\alpha_{\ \nu}\right).$$

The Weyl tensor electric term $\mathcal{E}_{\mu\nu}$ carries an imprint of high-energy effects sourcing Kaluza-Klein (KK) modes.

Considering vacuum on the brane, where $T_{\mu\nu} = 0$ outside a black hole, the field equations $G_{\mu\nu} = -\mathcal{E}_{\mu\nu} - \frac{1}{2}\Lambda_5 g_{\mu\nu}$ and $R = 0 = \mathcal{E}^\mu{}_\mu$ hold for braneworlds with $\mathbb{Z}_2$-symmetry. The vacuum field equations in the brane are $\mathcal{E}_{\mu\nu} = -R_{\mu\nu}$, where the bulk cosmological constant is comprised into the warp factor. Although a Taylor expansion of the metric was used to probe properties of a black hole on the brane in, e. g., [16, 12], in order to enhance the range of our analysis throughout this paper a more complete approach to analyze braneworld corrections in the black string profile can be accomplished, based on [9].

A Taylor expansion of the metric along the extra dimension allows us to analyze the black string more deeply. The effective field equations are complemented by other ones, obtained from the 5D Einstein and Bianchi equations in Refs. [13, 12]. Hereupon, since we are concerned with the Taylor expansion of the metric along the extra dimension up to the fourth order, besides the effective field equation $\pounds K_{\mu\nu} = K_{\mu\alpha}K^\alpha{}_\nu - \mathcal{E}_{\mu\nu} - \frac{1}{6}\Lambda_5 g_{\mu\nu}$, the effective equations are considered:

$$\begin{aligned}\pounds\mathcal{E}_{\mu\nu} &= \nabla^\alpha\mathcal{B}_{\alpha(\mu\nu)} + \left(K_{\mu\alpha}K_{\nu\beta} - K_{\alpha\beta}K_{\mu\nu}\right)K^{\alpha\beta} + K^{\alpha\beta}R_{\mu\alpha\nu\beta} - K\mathcal{E}_{\mu\nu} + \frac{1}{6}\Lambda_5\left(K_{\mu\nu} - g_{\mu\nu}K\right)\\ &\quad + 3K^\alpha{}_{(\mu}\mathcal{E}_{\nu)\alpha},\\ \pounds_{\mathbf{n}}\mathcal{B}_{\mu\nu\alpha} &= K_\alpha{}^\beta\mathcal{B}_{\mu\nu\beta} - 2\nabla_{[\mu}\mathcal{E}_{\nu]\alpha} - 2\mathcal{B}_{\alpha\beta[\mu}K_{\nu]}{}^\beta.\end{aligned}$$

These expressions are used to compute the terms in the Taylor expansion of the metric, along the extra dimension, providing the black string profile and further physical consequences as well.

By denoting $K = K_\mu{}^\mu$ and $g_{\mu\nu}(x,0) = g_{\mu\nu}$, the Taylor expansion is given [9, 1], for a vacuum in the brane, by:

$$\begin{aligned} g_{\mu\nu}(x,y) &= g_{\mu\nu} - \frac{1}{3}\kappa_5^2\lambda g_{\mu\nu}\,|y| + \left[\frac{1}{6}\left(\frac{1}{6}\kappa_5^4\lambda^2 - \Lambda_5\right)g_{\mu\nu} - \mathcal{E}_{\mu\nu}\right]y^2 \\ &\quad - \frac{1}{6}\left(\left(\frac{193}{36}\lambda^3\kappa_5^6 + \frac{5}{3}\Lambda_5\kappa_5^2\lambda\right)g_{\mu\nu} + \kappa_5^2 R_{\mu\nu}\right)\frac{|y|^3}{3!} + \\ &\quad + \left[\frac{1}{6}\Lambda_5\left(R - \frac{1}{3}\Lambda_5 - \frac{1}{18}\lambda^2\kappa_5^4 + \frac{7}{324}\lambda^4\kappa_5^8\right)g_{\mu\nu} + \left(R - \Lambda_5 + \frac{19}{36}\lambda^2\kappa_5^4\right)\mathcal{E}_{\mu\nu}\right. \\ &\quad \left. + \frac{1}{6}\left(\frac{37}{36}\lambda^2\kappa_5^4 - \Lambda_5\right)R_{\mu\nu} + \mathcal{E}^{\alpha\beta}\,R_{\mu\alpha\nu\beta}\right]\frac{y^4}{4!} + \cdots \end{aligned} \tag{2}$$

This expression is shown to be prominently relevant for our subsequent analysis.

Hereon, the black hole horizon evolution along the extra dimension — the warped horizon [17] — shall be investigated, exploring the component $g_{\theta\theta}(x,y)$ in (2). Indeed, let us consider any spherically symmetric metric associated to a black hole — in particular the Schwarzschild and the Casadio-Fabbri-Mazzacurati ones here investigated. The black hole solution, namely, the black string solution *on the brane*, is regarded when $\sqrt{g_{\theta\theta}(x,0)} = R$, where R denotes the coordinate singularity, usually calculated by the component $g_{rr}^{-1} = 0$ in the metric in the braneworld context. In the Schwarzschild metric $R = R_S = \frac{2GM}{c^2r}$. The coordinate singularities for the Casadio-Fabbri-Mazzacurati metrics are going to be analyzed in what follows, in the black string context as well. Such singularities shall be shown to be also physical singularities, by analyzing their respective 4D and 5D Kretschmann scalars.

3. Casadio-Fabbri-Mazzacurati braneworld solutions

The analysis of the gravitational field equations on the brane is not straightforward, due to the fact that the propagation of gravity into the bulk does not allow a complete presentation of the brane gravitational field equations as a closed form system [13]. The solutions provided by Casadio, Fabbri, and Mazzacurati for the brane black holes metrics [16, 18, 10] take into account the post-Newtonian parameter β, measured on the brane. The case $\beta = 1$ generates forthwith an exact Schwarzschild solution on the brane, and elicits a black string prototype. Furthermore, it was observed in [10, 11] that $\beta \approx 1$ holds in solar system scales measurements, and β gives information about the vacuum energy of the braneworld [10, 11].

One of the main motivation regarding the Casadio-Fabbri-Mazzacurati setup is that black holes solutions of the Einstein equations on the brane must depart from the Schwarzschild solution. Indeed, it is important to emphasize that, as we shall see for the Casadio-Fabbri-Mazzacurati black string, it might be possible to find out points along the extra dimension for which the Kretschmann scalar ${}^{(5)}K = {}^{(5)}R_{\mu\nu\rho\sigma}{}^{(5)}R^{\mu\nu\rho\sigma}$ diverges, indeed naked singularities along the extra dimension. Since the pure black string configuration is unstable [19], this structure is not physical ab initio. Anyway for $y = 0$ one reproduces the Kretschmann scalars standard 4D behavior. The Casadio-Fabbri-Mazzacurati black string solutions and their respective braneworld corrections are going to be presented and their stability analyzed as well.

Given a general static spherically symmetric

$$g_{\mu\nu}dx^\mu dx^\nu = -B(r)dt^2 + A(r)dr^2 + r^2 d\Omega^2, \tag{3}$$

the Casadio-Fabbri-Mazzacurati 4D black hole solution was obtained in [10, 11]. The Schwarzschild 4D metric is obtained when $B(r) = (A(r))^{-1}$ and $B(r) = 1 - \frac{2GM}{c^2r}$. Its unique extension into the bulk is a black string warped horizon, with the central singularity extending all along the extra dimension, and the bulk horizon singular [10, 11, 20]. If the Schwarzschild metric on the brane is demanded with a regular AdS horizon, there is no matter confinement on the brane: in this case matter percolates into the bulk [22]. The condition $B(r) = (A(r))^{-1}$ holds in the 4D case, although the most general solution is the Reissner-Nordström one [10, 11, 13], related to the case II analyzed in which follows.

3.1. Casadio-Fabbri-Mazzacurati black string: Case I

This case was analyzed by Casadio, Fabbri, and Mazzacurati in [10, 11], regarding the 4D black hole solution (3). They obtained a solution of the Einstein's equations provided by the metric coefficients

$$B(r) \;=\; 1-\frac{2GM}{c^2r} \qquad \text{and} \qquad A(r)=\frac{1-\frac{3GM}{2c^2r}}{\left(1-\frac{2GM}{c^2r}\right)\left(1-\frac{GM}{2c^2r}(4\beta-1)\right)} \tag{4}$$

with respect to (3). The Casadio-Fabbri-Mazzacurati black string classical horizon, in the brane, is the solution of the algebraic equation $(A(r))^{-1} = 0$. In order to extract phenomenological information of numerical calculations, we first consider in this Subsection the case where $\beta = 5/4$. Note that the metric above was also derived as a possible geometry outside a star on the brane [18]. The corresponding Hawking temperature is calculated in [10].

Taking into account the metric in (3), the classical standard black hole radius is given by two solutions of Schwarzschild type $R_S = \frac{2GM}{c^2}$, for the choice $\beta = 5/4$, providing zero Hawking black hole temperature. The 4D Kretschmann scalar $K^{(I)} = R_{\mu\nu\rho\sigma}R^{\mu\nu\rho\sigma}$ diverges for $r = 0$ and $r = \frac{3GM}{2c^2}$. Now, the Gauss equation is well known to relate the 5D and the 4D Riemann curvature tensor as ${}^{(5)}R^{\mu}{}_{\nu\rho\sigma} = R^{\mu}{}_{\nu\rho\sigma} - K^{\mu}{}_{\rho}K_{\nu\sigma} + K^{\mu}{}_{\sigma}K_{\nu\rho}$. By taking the junction conditions into account, is follows that $K_{\mu\nu} = -\frac{1}{6}\kappa_5^2\lambda g_{\mu\nu}$. By inserting it in the Gauss equation, the 5D Kretschmann scalar ${}^{(5)}K = {}^{(5)}R_{\mu\nu\rho\sigma}{}^{(5)}R^{\mu\nu\rho\sigma}$ for the Casadio-Fabbri-Mazzacurati type I black string also diverges for $r = 0$ and $r = \frac{3GM}{2c^2}$: the terms involving the extrinsic curvature in Gauss equation do not cancel the divergence provided by the 4D Kretschmann scalar, in the computation for ${}^{(5)}K$.

Using the same procedure as [12], one can use the metric coefficients (4) in Eq.(2) and calculate the black string warped horizon. In the graphics below, we explicit the value for the black string warped horizon, provided by $\sqrt{g_{\theta\theta}(R_S, y)}$, where R_S is the Schwarzschild radius. Further, $\lambda = \Lambda_5 = 1 = \kappa_5$ hereupon [$M_\odot$ denotes the sun mass]:

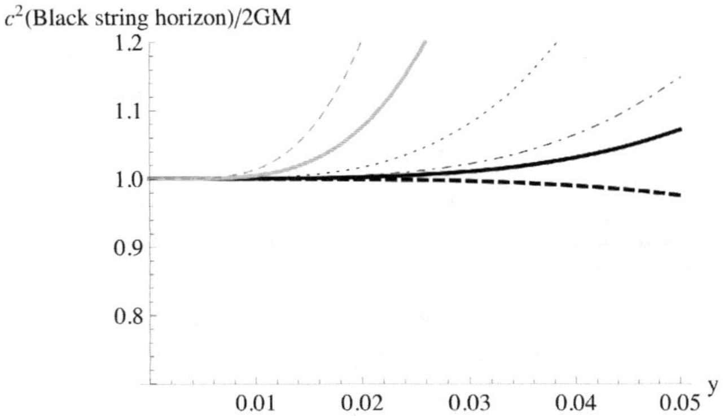

Figure 1. Graphic of the brane effect-corrected black string horizon $\sqrt{g_{\theta\theta}(R_S, y)}$ in the Casadio-Fabbri-Mazzacurati first solution, along the extra dimension y, for different values of the black hole mass M. For the dash-dotted line $M = M_\odot$; for the black dashed line: $M = 10\, M_\odot$; for the thick black line: $M = 10^2\, M_\odot$; for the black dotted line: $M = 10^3\, M_\odot$; for the thick gray line $M = 10^4\, M_\odot$; for the gray dotted line $M = 10^5\, M_\odot$.

Fig. 1 evinces a very interesting profile for the black string horizon behavior along the extra dimension y in Gaussian coordinates. It indicates a critical mass M (indeed our simulations provide $M \sim 73M_\odot$) above which the associated black string warped horizon monotonically increases along the extra dimension. The black string is known to be placed in the bulk, in a tubular neighborhood along the axis of symmetry. We show here that at the coordinate singularities $r = 0$ and $r = \frac{3GM}{2c^2}$ there is a physical singularity for the black string at such values, irrespective of the value for y. In fact, the Kretschmann scalar $K = {}^{(5)}R_{\mu\nu\rho\sigma}{}^{(5)}R^{\mu\nu\rho\sigma}$ diverges for such values (see [1]). Notwithstanding, the black string warped horizon $\sqrt{g_{\theta\theta}(R_S, y)}$ does not equal to zero, as illustrated at the Fig. 1.

3.2. Casadio-Fabbri-Mazzacurati black string: Case II

An alternative solution of (3) is obtained in [10, 11] where the metric coefficients

$$B(r) = 1 - \frac{2GM}{c^2 r} + \frac{2G^2M^2}{c^4 r^2}(\beta - 1), \qquad A(r) = \frac{1 - \frac{3GM}{2c^2 r}}{\left(1 - \frac{2GM}{c^2 r}\right)\left(1 - \frac{GM}{2c^2 r}(4\beta - 1)\right)} \tag{5}$$

are considered in (3). In order that the Hawking temperature be zero on the brane, the choice $\beta = 3/2$ is demanded [10]. The classical solution R for the black hole horizon is given by $R = R_S$ and $R = 5R_S/2$, where R_S denotes the Schwarzschild radius. It implies that the black string horizon now corrected by braneworld effects when (5) is substituted in (2), providing the graphic below.

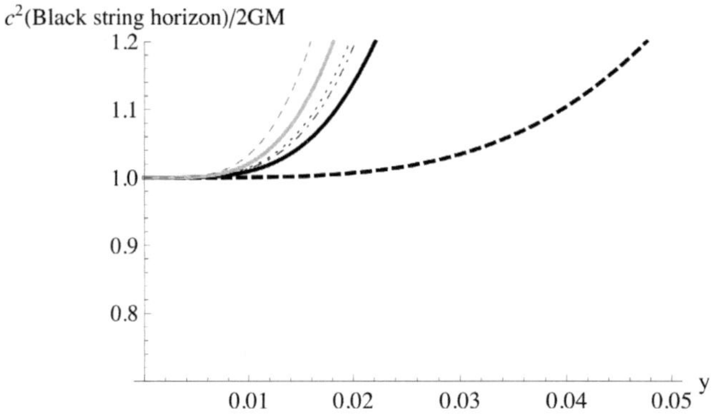

Figure 2. Graphic of the brane effect-corrected Casadio-Fabbri-Mazzacurati type II black string horizon $\sqrt{g_{\theta\theta}(R_S, y)}$, along the extra dimension y, for different values of the black hole mass GM/c^2 in the brane. For the black dashed line $M = M_\odot$; for the gray line $M = 10\,M_\odot$; for the black line: $M = 10^2\,M_\odot$; the dash-dotted line: $M = 10^3\,M_\odot$; for the dotted line: $M = 10^4\,M_\odot$; for the gray dashed line $M = 10^5\,M_\odot$; for the thick gray line $M = 10^6\,M_\odot$.

The black string horizon profile along the extra dimension is qualitatively similar for all values of M depicted here: the warped horizon always increases monotonically. Furthermore, under a similar analysis accomplished this time for the case II Casadio-Fabbri-Mazzacurati, and by taking into account the Kretschmann scalar [1], we conclude that such expression diverges for $r = \frac{3GM}{2c^2}$, for $r = \frac{5GM}{2c^2}$ and $r = \frac{2GM}{c^2}$. Contrary to the Schwarzschild metric, which presents the black hole horizon as a coordinate singularity — which can circumvented by, e. g., the Kruskal-Szekeres coordinates — and not as a physical singularity, the Kretschmann scalar for the Casadio-Fabbri-Mazzacurati case II metric indicates that each black hole horizon on the brane is a physical singularity, since it diverges for such values. Again, the terms involving the extrinsic curvature in the Gauss equation above are not able to cancel the divergence induced by the 4D Kretschmann scalar, when one calculates ${}^{(5)}K$. Hence, the black string also diverges for such values.

Additionally we depict below the CFM type I black string profile along the extra dimension:

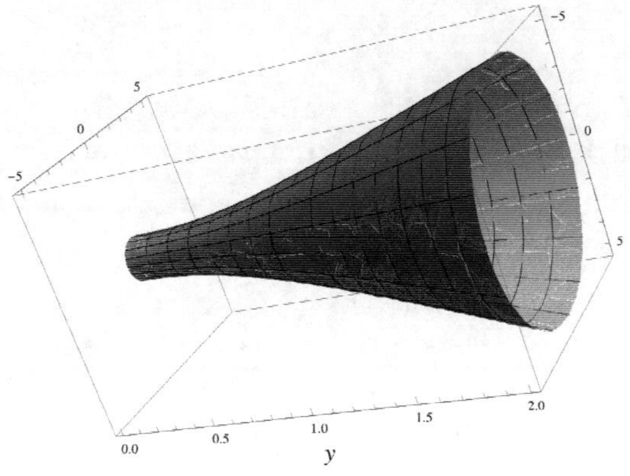

Figure 3. Casadio-Fabbri-Mazzacurati type I black string profile along the extra dimension, for $M = 2.3 \times 10^3 M_\odot$.

Fig. 2 indicates that the Casadio-Fabbri-Mazzacurati (case II) black string horizon always increases. Since the bulk has no fixed metric a priori, but it can be calculated from (2) by taking into account the metric on the brane, we can calculate the bulk curvature using the metric coefficients in (2).

Compact sources on the brane, such as stars and black holes have a description complicated and analytic solutions lacks. Solutions on the brane are provided by [16, 18, 10].

Delving into the analysis concerning the figures above, the general different profile between the Casadio-Fabbri-Mazzacurati black string warped horizon and the Schwarzschild horizon is expected. The warped horizon is provided by $\sqrt{g_{\theta\theta}(R_S, y)}$ in (2). All information about the bulk near the brane can be extracted from the Casadio-Fabbri-Mazzacurati metrics on the brane. We considered terms up to y^4, since irrespective of the black hole horizon radius, the effective distance y along the extra dimension equals the compactification radius [12], as well as the effective size of the extra dimension probed by a graviton. Besides, in particular our procedure considers suitable values for the post-Newtonian parameter in order that the Hawking radiation on the brane is zero. The physical black hole radii $R_S = \sqrt{g_{\theta\theta}(R_S, 0)}$ are now *effectively* dislocated into the bulk, and given by $R_{\rm brane} = \sqrt{g_{\theta\theta}(R_S, y)}$.

4. Conclusion

Any phenomenologically successful theory in which our Universe is viewed as a brane must reproduce the large-scale predictions of general relativity on the brane. It implies that gravitational collapse of matter trapped on the brane provides the Casadio-Fabbri-Mazzacurati solutions on the brane: either a localized black hole or an extended black string solution, possessing a warped horizon. It is possible to intersect this solution with a vacuum domain wall and the induced metric is the ones presented in the analysis in Subsections 3.1 and 3.2. In the case I, our analysis is restricted to the case where $\beta = 5/4$ since for this values there is a zero (Hawking) temperature black hole associated [10, 11]. Since we want to extract physical information on the braneworld effects on the variation of luminosity exclusively, we opted for this value for the parameter β, in such a way that the graphics, concerning this variation on

Journal of Physics: Conference Series **512** (2014) 012002 doi:10.1088/1742-6596/512/1/012002

the luminosity, take into account exclusively the braneworld effects, since Hawking radiation is shown to be suppressed with $\beta = 5/4$ for this metric. Analogously, for the case II the analysis is accomplished taking into account the value $\beta = 3/2$ in Eq.(5), as already discussed. More details can be found in [1]. It can be also shown that the black strings here analyzed are more unstable than the standard ones, in the Gregory-Laflamme context.

Acknowledgments

R. da Rocha is grateful to Conselho Nacional de Desenvolvimento Científico e Tecnológico (CNPq) 303027/2012-6, and 304862/2009-6 for financial support. A. M. Kuerten thanks to CAPES for financial support.

References

[1] Rocha R, Piloyan A, Kuerten A M and Coimbra-Araújo C H 2013 Casadio-Fabbri-Mazzacurati Black Strings and Braneworld-induced Quasars Luminosity Corrections *Class. Quant. Grav.* **30** 045014 (*Preprint* gr-qc/1301.4483)

[2] Arkani-Hamed N, Dimopoulos S and Dvali G R 1998 The Hierarchy Problem and New Dimensions at a Millimeter *Phys. Lett. B* **429** 263-272 (*Preprint* hep-ph/9803315)

[3] Randall L and Sundrum R 1999 A large mass hierarchy from a small extra dimension *Phys. Rev. Lett.* **83** 3370-3373 (*Preprint* hep-ph/9905221)

[4] Randall L and Sundrum R 1999 An alternative to compactification *Phys. Rev. Lett.* **83** 4690-4693 (*Preprint* hep-th/9906064)

[5] Gogberashvili M 2000 Our world as an expanding shell *Europhys. Lett.* **49** 396 (*Preprint* hep-ph/9812365)

[6] Lukas A, Ovrut B A, Stelle K S and Waldram D 1999 The Universe as a Domain Wall *Phys. Rev. D* **59** 086001 (*Preprint* hep-th/9803235)

[7] Abdalla M C B, Hoff da Silva J M and Rocha R 2009 Notes on the Two-brane Model with Variable Tension *Phys. Rev. D* **80** 046003 (*Preprint* gr-qc/1101.4214)

[8] Gregory R, Whisker R, Beckwith K and Done C 2004 Observing braneworld black holes *JCAP* **1004** 013 (*Preprint* hep-th/0406252)

[9] Rocha R and Hoff da Silva J M 2012 Black string corrections in variable tension braneworld scenarios *Phys. Rev. D* **85** 046009 (*Preprint* gr-qc/1202.1256)

[10] Casadio R, Fabbri A and Mazzacurati L 2002 New black holes in the brane-world? *Phys. Rev. D* **65** 084040 (*Preprint* gr-qc/0111072)

[11] Casadio R and Mazzacurati L 2003 Bulk shape of brane-world black holes *Mod. Phys. Lett. A* **18** 651 (*Preprint* gr-qc/0205129)

[12] Maartens R and Koyama K 2010 Brane world gravity *Living Rev. Relativity* **13** 5 (*Preprint* 1004.3962)

[13] Shiromizu T, Maeda K and Sasaki M 2000 The Einstein Equations on the 3-Brane World *Phys. Rev. D* **62** 043523 (*Preprint* gr-qc/9910076)

[14] Israel W 1967 Singular hypersurfaces and thin shells in general relativity *Nuovo Cim.* **44 B** 1-14, Errata-ibid **48 B** [Series 10] (1967) 463

[15] Darmois G 1927 *Memorial des sciences mathematiques* Fasticule XXV ch V (Paris: Gauthier-Villars)

[16] Dadhich N, Maartens R, Papadopoulos P and Rezania V 2000 Black holes on the brane *Phys. Lett. B* **487** 1 (*Preprint* hep-th/0003061)

[17] Seahra S S, Clarkson C and Maartens R 2005 Detecting extra dimensions with gravity wave spectroscopy: the black string brane-world *Phys. Rev.Lett.* **94** 121302 (*Preprint* gr-qc/0408032)

[18] Germani C and Maartens R 2001 Stars in the braneworld *Phys. Rev. D* **64** 124010 (*Preprint* hep-th/0107011)

[19] Gregory R 2000 Black string instabilities in anti-de Sitter space *Class. Quant. Grav.* **17** L125 (*Preprint* hep-th/0004101)

[20] Chamblin I, Hawking S and Reall H S 2000 Brane-World Black Holes *Phys. Rev. D* **61** 065007 (*Preprint* hep-th/9909205)

[21] Giannakis I and Ren H 2001 Possible extensions of the 4D Schwarzschild horizon in the 5D brane world *Phys. Rev. D* **63** 125017 (*Preprint* hep-th/0010183)

[22] Kanti P and Tamvakis K 2002 Quest for Localized 4-D Black Holes in Brane Worlds *Phys. Rev. D* **65** 084010 (*Preprint* hep-th/0110298v2)

[23] Kapner D J 2007 Tests of the Gravitational Inverse-Square Law below the Dark-Energy Length Scale *Phys. Rev. Lett.* **98** 021101

Journal of Physics: Conference Series **512** (2014) 012003 doi:10.1088/1742-6596/512/1/012003

SUSY double-well matrix model as 2D IIA superstring on RR background

F Sugino

Okayama Institute for Quantum Physics, Kyoyama 1-9-1, Kita-ku, Okayama 700-0015, Japan

E-mail: fumihiko_sugino@pref.okayama.lg.jp

Abstract. We discuss an interpretation of a simple supersymmetric matrix model with a double-well potential as two-dimensional type IIA superstrings on a nontrivial Ramond-Ramond background. In particular, we can see direct correspondence between single trace operators in the matrix model and vertex operators in the type IIA theory by computing scattering amplitudes and comparing the results in both sides. Next, we explicitly compute nonperturbative instanton contributions in the matrix model and find that they survive in a double scaling limit realizing the type IIA superstring theory. This suggests that the supersymmetry is spontaneously broken by nonperturbative effect in the target space of the type IIA superstring theory.

1. Introduction

Solvable matrix models for two-dimensional quantum gravity or noncritical string theory were vigorously investigated around 1990, focusing on nonperturbative effects in string theory [1]. While this approach has been successful for bosonic string theory, little has been known for superstring theory, in particular which possesses target-space supersymmetry (SUSY). We would like to consider (solvable) matrix models describing superstring theory with target-space SUSY. In this paper, we discuss correspondence between a simple zero-dimensional SUSY double-well matrix model and two-dimensional type IIA superstring theory on a nontrivial Ramond-Ramond (RR) background. Then, nonperturbative effect of the matrix model is computed in its double scaling limit. As a result, we find that SUSY is spontaneously broken due to instantons in the matrix model. According to the correspondence, this suggests spontaneous SUSY breaking at the nonperturbative level in the type IIA superstring theory. We hope our analysis is helpful to understand nonperturbative dynamics of matrix models of super Yang-Mills type for critical superstring theory [2, 3, 4].

This article is mainly based on the work [5, 6, 7].

2. Double-well SUSY matrix model

Ref. [8] discussed a following simple matrix model:

$$S = N\,\mathrm{tr}\left[\frac{1}{2}B^2 + iB(\phi^2 - \mu^2) + \bar{\psi}(\phi\psi + \psi\phi)\right], \tag{1}$$

Journal of Physics: Conference Series **512** (2014) 012003 doi:10.1088/1742-6596/512/1/012003

where B and ϕ are $N \times N$ hermitian matrices, and ψ and $\bar{\psi}$ are $N \times N$ Grassmann-odd matrices. The action S is invariant under SUSY transformations generated by Q and $\bar{Q}$:

$$Q\phi = \psi, \quad Q\psi = 0, \quad Q\bar{\psi} = -iB, \quad QB = 0, \tag{2}$$

$$\bar{Q}\phi = -\bar{\psi}, \quad \bar{Q}\bar{\psi} = 0, \quad \bar{Q}\psi = -iB, \quad \bar{Q}B = 0, \tag{3}$$

from which one can see that they are nilpotent: $Q^2 = \bar{Q}^2 = \{Q, \bar{Q}\} = 0$. After integrating out B, we have a scalar potential of a double-well shape: $\frac{1}{2}(\phi^2 - \mu^2)^2$. A large-$N$ saddle point equation for the eigenvalue distribution of the matrix ϕ: $\rho(x) \equiv \frac{1}{N}\,\mathrm{tr}\,\delta(x - \phi)$ reads

$$\int dy\, \rho(y)\, \mathrm{P}\frac{1}{x-y} + \int dy\, \rho(y)\, \mathrm{P}\frac{1}{x+y} = x^3 - \mu^2 x. \tag{4}$$

Its solution with filling fraction (ν_+, ν_-) is given by

$$\rho(x) = \begin{cases} \frac{\nu_+}{\pi}\, x\, \sqrt{(x^2 - a^2)(b^2 - x^2)} & (a < x < b) \\ \frac{\nu_-}{\pi}\, |x|\, \sqrt{(x^2 - a^2)(b^2 - x^2)} & (-b < x < -a), \end{cases} \tag{5}$$

where $a = \sqrt{\mu^2 - 2}$ and $b = \sqrt{\mu^2 + 2}$. The filling fractions satisfying $\nu_+ + \nu_- = 1$ indicate that $\nu_+ N$ ($\nu_- N$) eigenvalues are around the right (left) minimum of the double-well. The solution exists for $\mu^2 > 2$. The large-N free energy and the expectation values $\left\langle \frac{1}{N}\,\mathrm{tr}\, B^n \right\rangle$ ($n = 1, 2, \cdots$) evaluated at the solution turn out to all vanish [8]. This strongly suggests that the solution preserves SUSY. Thus, we conclude that the SUSY minima are infinitely degenerate and parametrized by (ν_+, ν_-) at large N. Note that the edges of the support a and b are independent of $\nu_\pm$. It is considered to be a characteristic feature of SUSY matrix models, not observed in bosonic double-well matrix models [9, 10].

There exists a solution having support of a single interval $x \in [-c, c]$ for $\mu^2 < 2$ [11]:

$$\rho(x) = \frac{1}{2\pi}\left(x^2 - \mu^2 + \frac{c^2}{2}\right)\sqrt{c^2 - x^2} \tag{6}$$

with $c = \sqrt{\frac{2}{3}}\left(\mu^2 + \sqrt{\mu^4 + 12}\right)^{1/2}$. Positivity of $\rho(x)$ yields the condition $\mu^2 < 2$. This solution gives nonzero values of $\left\langle \frac{1}{N}\,\mathrm{tr}\, B \right\rangle$ and of the large-N free energy, showing that SUSY is broken. We observed that the third derivative of the free energy with respect to μ^2 is not continuous at $\mu^2 = 2$. The transition between the SUSY phase ($\mu^2 > 2$) and the SUSY broken phase ($\mu^2 < 2$) is of the third order.

In the next section, we will compute various correlation functions at the saddle point (5) and find new logarithmic critical behavior as $\mu^2 \to 2 + 0$. Based on the result, we will discuss correspondence between the matrix model and two-dimensional type IIA superstring theory on a nontrivial Ramond-Ramond (RR) background in sections 4 and 5. The logarithmic critical behavior is somewhat reminiscent of the $c = 1$ matrix model which is a matrix quantum mechanics of a single matrix variable [12]. The Penner model is known as a zero-dimensional matrix model exhibiting the same critical behavior as the matrix quantum mechanics [13]. [1] It describes noncritical string theory propagating on a two-dimensional target space: (Liouville direction) $\times$ (S^1 with self-dual radius). So, it is expected that our matrix model can be regarded as a SUSY version of the Penner model and describes

[1] Also is the normal matrix model [14], which corresponds to $c = 1$ noncritical strings on S^1 with a general radius.

two-dimensional superstring theory with SUSY on the target space (Liouville direction) × (S^1 with self-dual radius). Indeed, two-dimensional type II superstring theory with the identical target space is constructed [15, 16, 17, 18], where target space SUSY exists only at the self-dual radius of the circle.

Our matrix model is interpreted as the $O(n)$ model on a random surface [19] with $n = -2$, whose critical behavior is described by the $c = -2$ topological gravity [20]. The partition function after B, ψ and $\bar{\psi}$ are integrated out is expressed as a Gaussian one-matrix model by the Nicolai mapping $H = \phi^2$, where the H-integration is over the *positive definite* hermitian matrices, not over all the hermitian matrices. Ref. [21] discusses that the difference of the integration region has only effects which are nonperturbative in $1/N$, and the model can be regarded as the standard Gaussian matrix model at each order of genus expansion.

The Nicolai mapping changes the operators $\frac{1}{N}\,\mathrm{tr}\,\phi^{2n}$ $(n = 1, 2, \cdots)$ to regular operators $\frac{1}{N}\,\mathrm{tr}\,H^n$. Hence, the behavior of their correlators is expected to be described by the Gaussian one-matrix (the $c = -2$ topological gravity) at least perturbatively in $1/N$. However, the operators $\frac{1}{N}\,\mathrm{tr}\,\phi^{2n+1}$ $(n = 0, 1, 2, \cdots)$ are mapped to $\pm\frac{1}{N}\,\mathrm{tr}\,H^{n+1/2}$ that are singular at the origin. They are not observables in the $c = -2$ topological gravity, while they are natural observables as well as $\frac{1}{N}\,\mathrm{tr}\,\phi^{2n}$ in the original setting (1). In the next section, we will see that correlation functions among operators

$$\frac{1}{N}\,\mathrm{tr}\,\phi^{2n+1}, \qquad \frac{1}{N}\,\mathrm{tr}\,\psi^{2n+1}, \qquad \frac{1}{N}\,\mathrm{tr}\,\bar{\psi}^{2n+1} \qquad (n = 0, 1, 2, \cdots) \tag{7}$$

exhibit logarithmic singular behavior of powers of $\ln(\mu^2 - 2)$ at the planar topology.

In considering correspondence of the matrix model to superstring theory, the following observation will be helpful. Suppose ψ and $\bar{\psi}$ are regarded as target-space fermions in the corresponding superstring theory. Namely, ψ is interpreted as an operator in the (NS, R) sector and $\bar{\psi}$ in the (R, NS) sector in the RNS formalism. Then, under the so-called $(-1)^{\mathbf{F}_L}$ and $(-1)^{\mathbf{F}_R}$ transformations changing the signs of operators in the left-moving Ramond sector and those in the right-moving Ramond sector respectively, they transform as

$$(-1)^{\mathbf{F}_L}: \quad \psi \to \psi, \qquad \bar{\psi} \to -\bar{\psi}, \tag{8}$$
$$(-1)^{\mathbf{F}_R}: \quad \psi \to -\psi, \qquad \bar{\psi} \to \bar{\psi}. \tag{9}$$

In order for the matrix model action (1) to be invariant under the transformations, B and ϕ should transform as

$$(-1)^{\mathbf{F}_L}: \quad B \to B, \qquad \phi \to -\phi, \tag{10}$$
$$(-1)^{\mathbf{F}_R}: \quad B \to B, \qquad \phi \to -\phi. \tag{11}$$

This indicates that B corresponds to an operator in the (NS, NS) sector, and ϕ in the (R, R) sector.

3. Correlation functions

3.1. Planar one-point functions

The planar one-point function $\left\langle \frac{1}{N}\,\mathrm{tr}\,\phi^n \right\rangle_0$ $(n = 1, 2, \cdots)$ are computed as

$$\begin{aligned}\left\langle \frac{1}{N}\,\mathrm{tr}\,\phi^n \right\rangle_0 &= \int dx\, x^n \rho(x) \\ &= (\nu_+ + (-1)^n \nu_-)(2+\mu^2)^{n/2} F\left(-\frac{n}{2}, \frac{3}{2}, 3; \frac{4}{2+\mu^2}\right),\end{aligned} \tag{12}$$

where the suffix "0" in the left hand side indicates the planar contribution. For n even, the expression reduces to a polynomial of μ^2 giving nonsingular behavior as expected from the $c = -2$ topological gravity. On the other hand, when μ^2 is odd, it exhibits logarithmic singular behavior as $\mu^2 \to 2 + 0$:

$$\left\langle \frac{1}{N} \mathrm{tr}\, \phi^{2k+1} \right\rangle_0 \sim (\nu_+ - \nu_-) \frac{2^{k+2}}{\pi} \frac{(2k+1)!!}{(k+2)!} \omega^{k+2} \ln \omega \tag{13}$$

with $\omega \equiv \frac{1}{4}(\mu^2 - 2)$. The symbol "$\sim$" denotes equality up to additive less singular terms. Explicit form for a first few expectation values reads

$$\begin{aligned}
\left\langle \frac{1}{N} \mathrm{tr}\, \phi \right\rangle_0 &= (\nu_+ - \nu_-) \left[\frac{64}{15\pi} + \frac{16}{3\pi}\omega + \frac{2}{\pi}\omega^2 \ln \omega + \mathcal{O}(\omega^2) \right], \\
\left\langle \frac{1}{N} \mathrm{tr}\, \phi^3 \right\rangle_0 &= (\nu_+ - \nu_-) \left[\frac{1024}{105\pi} + \frac{128}{5\pi}\omega + \frac{16}{\pi}\omega^2 + \frac{4}{\pi}\omega^3 \ln \omega + \mathcal{O}(\omega^3) \right], \\
\left\langle \frac{1}{N} \mathrm{tr}\, \phi^5 \right\rangle_0 &= (\nu_+ - \nu_-) \left[\frac{8192}{315\pi} + \frac{2048}{21\pi}\omega + \frac{128}{\pi}\omega^2 + \frac{160}{3\pi}\omega^3 + \frac{10}{\pi}\omega^4 \ln \omega + \mathcal{O}(\omega^4) \right], \\
&\cdots .
\end{aligned} \tag{14}$$

Matrix models can be seen as a sort of "lattice models" for string theory. In the hypergeometric function $F\left(-\frac{n}{2}, \frac{3}{2}, 3; \frac{1}{1+\omega}\right)$ for n being odd, the logarithmic singular terms can be regarded as universal parts relevant to "continuum physics", whereas polynomials of ω as nonuniversal "lattice artifacts".

3.2. Eigenvalue distribution with source

In computing higher-point correlators $\left\langle \prod_{i=1}^K \frac{1}{N} \mathrm{tr}\, \phi^{n_i} \right\rangle_{C,0}$ at the vacuum with general filling fraction (ν_+, ν_-), it is useful to reduce them to those at the vacuum with $(\nu_+, \nu_-) = (1, 0)$. We can show

$$\left\langle \prod_{i=1}^{K} \frac{1}{N} \mathrm{tr}\, \phi^{n_i} \right\rangle_{C,0}^{(\nu_+,\nu_-)} = (\nu_+ - \nu_-)^{\sharp} \left\langle \prod_{i=1}^{K} \frac{1}{N} \mathrm{tr}\, \phi^{n_i} \right\rangle_{C,0}^{(1,0)} \tag{15}$$

up to $K = 3$, by explicit calculations. Here, the suffix "C" means taking a connected part. The superscripts (ν_+, ν_-) and $(1, 0)$ are put to clarify the filling fractions of the vacua at which the expectation values are evaluated, and $\sharp$ counts the number of odd integers in $\{n_1, \cdots, n_K\}$.

In order to obtain higher-point correlators of $\frac{1}{N} \mathrm{tr}\, \phi^p$ $(p = 1, 2, \cdots)$, we introduce source terms $\sum_{p=1}^{\infty} j_p \,\mathrm{tr}\, \phi^p$ to the partition function:

$$Z_{j_k} = \int d^{N^2}\phi\, e^{-N\,\mathrm{tr}\left[\frac{1}{2}(\phi^2-\mu^2)^2 - \sum_{p=1}^{\infty} j_p \phi^p\right]} \det\left(\phi \otimes \mathbf{1} + \mathbf{1} \otimes \phi\right). \tag{16}$$

In the large-N limit, the eigenvalue distribution $\rho_j(x)$ satisfies the saddle point equation

$$\int dy\, \rho_j(y) \left(\mathrm{P}\frac{1}{x-y} + \mathrm{P}\frac{1}{x+y} \right) = x^3 - \mu^2 x - \sum_{p=1}^{\infty} \frac{p j_p}{2} x^{p-1}. \tag{17}$$

Let us consider the case of the filling fractions $(1, 0)$ with the support of $\rho_j(x)$ $[a_j, b_j]$ $(0 < a_j < b_j)$. We change variables as

$$x^2 = A + B\xi, \quad y^2 = A + B\eta \qquad \text{with} \qquad A \equiv \frac{a_j^2 + b_j^2}{2}, \quad B \equiv \frac{b_j^2 - a_j^2}{2}, \tag{18}$$

and put $\tilde{\rho}(\eta) \equiv \frac{B}{2y}\rho_j(y)$, to simplify (17) as

$$\frac{1}{B}\int_{-1}^{1} d\eta\, \tilde{\rho}(\eta)\, \mathrm{P}\frac{1}{\xi-\eta} = \frac{1}{2}(A-\mu^2+B\xi) - \sum_{p=1}^{\infty}\frac{pj_p}{4}(A+B\xi)^{\frac{p}{2}-1} \tag{19}$$

for $\xi \in [-1,1]$, where $\tilde{\rho}$ is normalized by $\int_{-1}^{1} d\eta\tilde{\rho}(\eta) = 1$.

We act $\int_{-1}^{1} d\xi\, \sqrt{1-\xi^2}\, \mathrm{P}\frac{1}{\zeta-\xi}$ to both sides of (19), and apply the formula

$$\int_{-1}^{1} dy\, \sqrt{1-y^2}\, \mathrm{P}\frac{1}{x-y}\, \mathrm{P}\frac{1}{u-y} = -\pi + \pi^2\sqrt{1-u^2}\,\delta(u-x) \tag{20}$$

for $x, u \in [-1,1]$. Consequently,

$$\begin{aligned}\tilde{\rho}(\zeta) &= \frac{1}{2\pi}\frac{1}{\sqrt{1-\zeta^2}}\left[2-(A-\mu^2)B\zeta - B^2\left(\zeta^2-\frac{1}{2}\right)\right.\\ &\qquad \left. +\sum_{p=1}^{\infty} pj_p\frac{B}{2\pi}\int_{-1}^{1} d\xi\,\sqrt{1-\xi^2}\,\mathrm{P}\frac{1}{\zeta-\xi}(A+B\xi)^{\frac{p}{2}-1}\right].\end{aligned} \tag{21}$$

The condition $\tilde{\rho}(\zeta=\pm 1) = 0$ determines A and B as

$$A = \mu^2 + \sum_{p=1}^{\infty} j_p\frac{p}{2}(A+B)^{\frac{p}{2}-1}F\left(-\frac{p}{2}+1, \frac{1}{2}, 1; \frac{2B}{A+B}\right), \tag{22}$$

$$B = 2\left[1+\sum_{p=1}^{\infty}\frac{j_p}{4}\frac{p}{2}\left(\frac{p}{2}-1\right)B^2(A+B)^{\frac{p}{2}-2}F\left(-\frac{p}{2}+2, \frac{3}{2}, 3; \frac{2B}{A+B}\right)\right]^{1/2}, \tag{23}$$

from which A and B are obtained iteratively with respect to $\{j_p\}$. Up to the first order of $\{j_p\}$,

$$A = \mu^2 + \sum_{p=1}^{\infty} j_p\frac{p}{2}(2+\mu^2)^{\frac{p}{2}-1}F\left(-\frac{p}{2}+1, \frac{1}{2}, 1; \frac{4}{2+\mu^2}\right) + \mathcal{O}(j^2), \tag{24}$$

$$B = 2 + \sum_{p=1}^{\infty} j_p\frac{p}{2}\left(\frac{p}{2}-1\right)(2+\mu^2)^{\frac{p}{2}-2}F\left(-\frac{p}{2}+2, \frac{3}{2}, 3; \frac{4}{2+\mu^2}\right) + \mathcal{O}(j^2), \tag{25}$$

where $\mathcal{O}(j^2)$ means a quantity of the quadratic order of $\{j_p\}$.

3.3. Planar two-point functions (bosons)

Let us express the planar expectation value of $\mathcal{O}$ under the partition function with the source terms (16) as $\langle\mathcal{O}\rangle_0^{(j)}$. The cylinder amplitude at the vacuum with the filling fraction $(1,0)$ is given as

$$\left\langle \frac{1}{N}\,\mathrm{tr}\,\phi^p\,\frac{1}{N}\,\mathrm{tr}\,\phi^q\right\rangle_{C,0}^{(1,0)} = \frac{\partial}{\partial j_p}\left\langle\frac{1}{N}\,\mathrm{tr}\,\phi^q\right\rangle_0^{(j)}\Bigg|_{\{j_p\}=0} = \frac{\partial}{\partial j_p}\int_{-1}^{1} d\zeta\,(A+B\zeta)^{\frac{q}{2}}\tilde{\rho}(\zeta)\Bigg|_{\{j_p\}=0}. \tag{26}$$

Combining (26) and (15) leads to the result for general filling fraction. In what follows, we omit the superscript (ν_+, ν_-) of the correlators when there is no possible confusion. It turns out that the amplitudes are quadratic forms of the hypergeometric functions. For both of p and q even, they are polynomials of ω independent of $(\nu_+ - \nu_-)$, which is expected from the $c = -2$ topological gravity. When p and q are odd and even respectively,

$$\left\langle \Phi_{2k+1} \frac{1}{N} \operatorname{tr} \phi^{2\ell} \right\rangle_{C,0} \sim (\nu_+ - \nu_-)(\text{const.})\omega^{k+1} \ln \omega. \tag{27}$$

When both of p and q are odd,

$$\langle \Phi_{2k+1}\Phi_{2\ell+1}\rangle_{C,0} \sim -(\nu_+ - \nu_-)^2 \frac{1}{2\pi^2} \frac{1}{k+\ell+1} \frac{(2k+1)!}{(k!)^2} \frac{(2\ell+1)!}{(\ell!)^2} \omega^{k+\ell+1} (\ln \omega)^2. \tag{28}$$

Here, in order to subtract nonuniversal contributions appearing in the form (nonuniversal part)$\times$ (universal part), we took a basis of the odd-power operators mixed with lower even-power operators:

$$\Phi_{2k+1} = \frac{1}{N} \operatorname{tr} \phi^{2k+1} + (\nu_+ - \nu_-) \sum_{i=1}^{k} \alpha_{2k+1,2i}(\omega) \frac{1}{N} \operatorname{tr} \phi^{2i} \tag{29}$$

with $\alpha_{2k+1,2i}(\omega)$ being a regular function at $\omega = 0$. For example, we can explicitly construct the basis for the first three operators by considering $\langle \Phi_1\Phi_1\rangle_{C,0}$, $\langle \Phi_1\Phi_3\rangle_{C,0}$, $\cdots$, $\langle \Phi_5\Phi_5\rangle$:

$$\begin{aligned}
\Phi_1 &= \frac{1}{N} \operatorname{tr} \phi, \\
\Phi_3 &= \frac{1}{N} \operatorname{tr} \phi^3 - (\nu_+ - \nu_-)\frac{4}{\pi} \left(1 + \bar{\alpha}^{(1)}_{3,2}\omega + \mathcal{O}(\omega^2)\right) \frac{1}{N} \operatorname{tr} \phi^2, \\
\Phi_5 &= \frac{1}{N} \operatorname{tr} \phi^5 - (\nu_+ - \nu_-)\frac{4}{\pi} \left(1 + \bar{\alpha}^{(1)}_{5,4}\omega + \mathcal{O}(\omega^2)\right) \frac{1}{N} \operatorname{tr} \phi^4, \\
&\quad -(\nu_+ - \nu_-)\frac{8}{3\pi} \left(1 + 3(1 - \bar{\alpha}^{(1)}_{5,4})\omega + \mathcal{O}(\omega^2)\right) \frac{1}{N} \operatorname{tr} \phi^2,
\end{aligned} \tag{30}$$

where $\bar{\alpha}^{(1)}_{3,2}$ and $\bar{\alpha}^{(1)}_{5,4}$ are undertermined constants. They would be determined by considering higher operators.

Note that $(\nu_+ - \nu_-)$ corresponds to an RR charge from the observation at the end of section 2: Φ_{2k+1} measuring an RR charge.

3.4. Planar three-point functions (bosons)

Similar procedure to the case of the two-point functions can be used in computing three-point correlation functions. It turns out that the result is expressed as cubic forms of the hypergeometric functions. The first two amplitudes become

$$\begin{aligned}
\left\langle (\Phi_1)^3 \right\rangle_{C,0} &= (\nu_+ - \nu_-)^3 \left[\frac{1}{16\pi^3} (\ln \omega)^3 + \mathcal{O}((\ln \omega)^2) \right], \\
\left\langle (\Phi_1)^2 \Phi_3 \right\rangle_{C,0} &= (\nu_+ - \nu_-)^3 \left[\frac{2}{\pi^3} + \frac{3}{8\pi^3} \omega (\ln \omega)^3 + \mathcal{O}(\omega (\ln \omega)^2) \right].
\end{aligned} \tag{31}$$

3.5. Planar higher-point functions (bosons)

The results obtained for the one-, two- and three-point functions of operators Φ_{2k+1} $(k = 0, 1, 2, \cdots)$ naturally suggest the form of higher-point functions as

$$\left\langle \prod_{i=1}^{n} \Phi_{2k_i+1} \right\rangle_{C,0} \sim (\nu_+ - \nu_-)^n \, (\text{const.}) \, \omega^{2-\gamma+\sum_{i=1}^{n}(k_i-1)} \, (\ln \omega)^n \tag{32}$$

Journal of Physics: Conference Series **512** (2014) 012003 doi:10.1088/1742-6596/512/1/012003

with $\gamma = -1$. Besides the power of logarithm $(\ln \omega)^n$, it has the standard scaling behavior with the string susceptibility $\gamma = -1$ (the same as in the $c = -2$ topological gravity) and the gravitational scaling dimension k of Φ_{2k+1}, if we identify ω with "the cosmological constant" coupled to the lowest dimensional operator on a random surface [22, 23, 24].

3.6. Planar two-point functions (fermions)

The simplest two-point correlator of fermions is computed as

$$\begin{aligned}\left\langle \frac{1}{N} \operatorname{tr} \psi \, \frac{1}{N} \operatorname{tr} \bar{\psi} \right\rangle_{C,0} &= \frac{1}{2} \int_\Omega dx \, \frac{1}{x} \, \rho(x) = (\nu_+ - \nu_-) \, \frac{1}{2} \, (4(1+\omega))^{-1/2} \, F\left(\frac{1}{2}, \frac{3}{2}, 3; \frac{1}{1+\omega}\right) \\ &= (\nu_+ - \nu_-) \left[\frac{4}{3\pi} + \frac{1}{\pi} \, \omega \ln \omega + \mathcal{O}(\omega)\right] \qquad (\omega \to +0), \qquad (33)\end{aligned}$$

exhibiting singular behavior of $\ln \omega$. SUSY invariance implies that this is equal to $\left\langle \frac{1}{N} \operatorname{tr}(iB) \, \frac{1}{N} \operatorname{tr} \phi \right\rangle_{C,0} = \frac{1}{4} \frac{\partial}{\partial \omega} \left\langle \frac{1}{N} \operatorname{tr} \phi \right\rangle_0$, interestingly which can be seen from (14).

Next, for $\left\langle \frac{1}{N} \operatorname{tr} \psi^3 \, \frac{1}{N} \operatorname{tr} \bar{\psi}^3 \right\rangle_{C,0}$, we should consider an operator mixing similar to the bosonic case (29). Let us take a new basis as

$$\begin{aligned}\Psi_1 &\equiv \tfrac{1}{N} \operatorname{tr} \psi, & \bar{\Psi}_1 &\equiv \frac{1}{N} \operatorname{tr} \bar{\psi}, \\ \Psi_3 &\equiv \tfrac{1}{N} \operatorname{tr} \psi^3 + (\text{mixing}), & \bar{\Psi}_3 &\equiv \frac{1}{N} \operatorname{tr} \bar{\psi}^3 + (\text{mixing}), \\ \Psi_5 &\equiv \tfrac{1}{N} \operatorname{tr} \psi^5 + (\text{mixing}), & \bar{\Psi}_5 &\equiv \frac{1}{N} \operatorname{tr} \bar{\psi}^5 + (\text{mixing}), \\ &\cdots, & &\cdots, \qquad (34)\end{aligned}$$

where "mixing" means operators to be added so that

$$\langle \Psi_{2k+1} \bar{\Psi}_{2\ell+1} \rangle_{C,0} \sim \delta_{k,\ell} \, v_k \, (\nu_+ - \nu_-)^{2k+1} \omega^{2k+1} \ln \omega \qquad (35)$$

with v_k constants holds for $k, \ell = 0, 1$. It turns out that the choice

$$\begin{aligned}\Psi_3 &= \frac{1}{N} \operatorname{tr} \psi^3 + \frac{3}{\sqrt{2}} (1 + \omega + \mathcal{O}(\omega^2)) \frac{1}{N} \operatorname{tr} \{(iB - \phi^2 + \mu^2)\psi\}, \\ \bar{\Psi}_3 &= \frac{1}{N} \operatorname{tr} \bar{\psi}^3 + \frac{3}{\sqrt{2}} (1 + \omega + \mathcal{O}(\omega^2)) \frac{1}{N} \operatorname{tr} \{(iB - \phi^2 + \mu^2)\bar{\psi}\} \qquad (36)\end{aligned}$$

or

$$\begin{aligned}\Psi_3 &= \frac{1}{N} \operatorname{tr} \psi^3 - \frac{3}{\sqrt{2}} (1 + \omega + \mathcal{O}(\omega^2)) \frac{1}{N} \operatorname{tr} \{(iB - \phi^2 + \mu^2)\psi\}, \\ \bar{\Psi}_3 &= \frac{1}{N} \operatorname{tr} \bar{\psi}^3 - \frac{3}{\sqrt{2}} (1 + \omega + \mathcal{O}(\omega^2)) \frac{1}{N} \operatorname{tr} \{(iB - \phi^2 + \mu^2)\bar{\psi}\} \qquad (37)\end{aligned}$$

does the job (35) with $v_0 = \frac{1}{\pi}$ and $v_1 = \frac{6}{\pi}$.

The result (35) tells us that Ψ_{2k+1} and $\bar{\Psi}_{2k+1}$ have the gravitational scaling dimension k same as Φ_{2k+1} besides the logarithmic factor.

4. 2D type IIA superstring

The type II superstring theory discussed in Refs. [15, 16, 17] has the target space $(\varphi, x) \in$ (Liouville direction) $\times$ (S^1 with self-dual radius). The holomorphic energy-momentum tensor on the string world-sheet is

$$T = -\frac{1}{2}(\partial x)^2 - \frac{1}{2}\psi_x\partial\psi_x - \frac{1}{2}(\partial\varphi)^2 + \partial^2\varphi - \frac{1}{2}\psi_\ell\partial\psi_\ell \tag{38}$$

excluding ghosts' part. ψ_x and ψ_ℓ are superpartners of x and φ, respectively. Target-space supercurrents in the type IIA theory

$$q_+(z) = e^{-\frac{1}{2}\phi(z)-\frac{i}{2}H(z)-ix(z)}, \qquad \bar{q}_-(\bar{z}) = e^{-\frac{1}{2}\bar{\phi}(\bar{z})+\frac{i}{2}\bar{H}(\bar{z})+i\bar{x}(\bar{z})} \tag{39}$$

exist only for the S^1 target space of the self-dual radius. ϕ ($\bar{\phi}$) is the holomorphic (anti-holomorphic) bosonized superconformal ghost, and the fermions are bosonized as $\psi_\ell \pm i\psi_x = \sqrt{2}\, e^{\mp iH}$, $\bar{\psi}_\ell \pm i\bar{\psi}_x = \sqrt{2}\, e^{\mp i\bar{H}}$. In addition, we should care about cocycle factors in order to realize the anticommuting nature between q_+ and $\bar{q}_-$. Supercurrents with the cocycle factors are

$$\hat{q}_+(z) = e^{\pi\beta(\frac{1}{2}p_{\bar{\phi}}-i\frac{1}{2}p_{\bar{h}}-ip_{\bar{x}})}\, q_+(z), \qquad \hat{\bar{q}}_-(\bar{w}) = e^{-\pi\beta(\frac{1}{2}p_\phi+i\frac{1}{2}p_h+ip_x)}\, \bar{q}_-(\bar{w}), \tag{40}$$

where $\beta \in \mathbf{Z} + \frac{1}{2}$, and p_ϕ, p_h and p_x ($p_{\bar{\phi}}$, $p_{\bar{h}}$ and $p_{\bar{x}}$) are momentum modes of holomorphic part (anti-holomorphic part) of free bosons [6]. Then the supercharges

$$\hat{Q}_+ = \oint \frac{dz}{2\pi i}\, \hat{q}_+(z), \qquad \hat{\bar{Q}}_- = \oint \frac{d\bar{z}}{2\pi i}\, \hat{\bar{q}}_-(\bar{z}) \tag{41}$$

are nilpotent $\hat{Q}_+^2 = \hat{\bar{Q}}_-^2 = \{\hat{Q}_+, \hat{\bar{Q}}_-\} = 0$, which indeed matches the property of the supercharges Q and $\bar{Q}$ in the matrix model.

The spectrum except special massive states is represented by the NS "tachyon" [2] vertex operator (in (-1) picture):

$$T_k = e^{-\phi+ikx+p_\ell\varphi}, \qquad \bar{T}_{\bar{k}} = e^{-\bar{\phi}+i\bar{k}\bar{x}+p_\ell\bar{\varphi}}, \tag{42}$$

and by the R vertex operator (in $(-\frac{1}{2})$ picture):

$$V_{k,\epsilon} = e^{-\frac{1}{2}\phi+\frac{i}{2}\epsilon H+ikx+p_\ell\varphi}, \qquad \bar{V}_{\bar{k},\bar{\epsilon}} = e^{-\frac{1}{2}\bar{\phi}+\frac{i}{2}\bar{\epsilon}\bar{H}+i\bar{k}\bar{x}+p_\ell\bar{\varphi}} \tag{43}$$

with $\epsilon, \bar{\epsilon} = \pm 1$. Cocycle factors for vertex operators are introduced as

$$\begin{aligned} \hat{T}_k(z) &= e^{\pi\beta(p_{\bar{\phi}}+ikp_{\bar{x}})}\, T_k(z), & \hat{\bar{T}}_{\bar{k}}(\bar{z}) &= e^{-\pi\beta(p_\phi+i\bar{k}p_x)}\, \bar{T}_{\bar{k}}(\bar{z}), \\ \hat{V}_{k,\epsilon}(z) &= e^{\pi\beta(\frac{1}{2}p_{\bar{\phi}}+i\frac{\epsilon}{2}p_{\bar{h}}+ikp_{\bar{x}})}\, V_{k,\epsilon}(z), & \hat{\bar{V}}_{\bar{k},\bar{\epsilon}}(\bar{z}) &= e^{-\pi\beta(\frac{1}{2}p_\phi+i\frac{\bar{\epsilon}}{2}p_h+i\bar{k}p_x)}\, \bar{V}_{\bar{k},\bar{\epsilon}}(\bar{z}). \end{aligned} \tag{44}$$

Locality with the supercurrents, mutual locality, superconformal invariance (including the Dirac equation constraint) and the level matching condition determine physical vertex operators. As discussed in [17], there are two consistent sets of physical vertex operators - "momentum background" and "winding background". Let us consider the "winding background". [3] The

[2] In two dimensions, "tachyon" turns out to be not truely tachyonic but massless.

[3] We can repeat the parallel argument for "momentum background" in the type IIB theory, which is equivalent to the "winding background" in the type IIA theory through T-duality with respect to the S^1 direction.

physical spectrum in the "winding background" is given by

$$\begin{aligned}
&(\mathrm{NS},\,\mathrm{NS}): && \hat{T}_k\,\hat{\bar{T}}_{-k} && (k\in\mathbf{Z}+\tfrac{1}{2}),\\
&(\mathrm{R}+,\,\mathrm{R}-): && \hat{V}_{k,\,+1}\,\hat{\bar{V}}_{-k,\,-1} && (k=\tfrac{1}{2},\tfrac{3}{2},\cdots),\\
&(\mathrm{R}-,\,\mathrm{R}+): && \hat{V}_{-k,\,-1}\,\hat{\bar{V}}_{k,\,+1} && (k=0,1,2,\cdots),\\
&(\mathrm{NS},\,\mathrm{R}-): && \hat{T}_{-k}\,\hat{\bar{V}}_{-k,\,-1} && (k=\tfrac{1}{2},\tfrac{3}{2},\cdots),\\
&(\mathrm{R}+,\,\mathrm{NS}): && \hat{V}_{k,\,+1}\,\hat{\bar{T}}_k && (k=\tfrac{1}{2},\tfrac{3}{2},\cdots),
\end{aligned} \tag{45}$$

where we take a branch of $p_\ell = 1-|k|$ satisfying the locality bound $p_\ell \le Q/2 = 1$ [25]. We can see that the vertex operators

$$\hat{V}_{\frac{1}{2},+1}\,\hat{\bar{V}}_{-\frac{1}{2},-1},\qquad \hat{T}_{-\frac{1}{2}}\,\hat{\bar{V}}_{-\frac{1}{2},-1},\qquad \hat{V}_{\frac{1}{2},+1}\,\hat{\bar{T}}_{\frac{1}{2}},\qquad \hat{T}_{-\frac{1}{2}}\,\hat{\bar{T}}_{\frac{1}{2}} \tag{46}$$

form a quartet under $\hat{Q}_+$ and $\hat{\bar{Q}}_-$:

$$\begin{aligned}
[\hat{Q}_+,\hat{V}_{\frac{1}{2},+1}\,\hat{\bar{V}}_{-\frac{1}{2},-1}] &= \hat{T}_{-\frac{1}{2}}\,\hat{\bar{V}}_{-\frac{1}{2},-1}, & \{\hat{Q}_+,\hat{T}_{-\frac{1}{2}}\,\hat{\bar{V}}_{-\frac{1}{2},-1}\} &= 0,\\
\{\hat{Q}_+,\hat{V}_{\frac{1}{2},+1}\,\hat{\bar{T}}_{\frac{1}{2}}\} &= \hat{T}_{-\frac{1}{2}}\,\hat{\bar{T}}_{\frac{1}{2}}, & [\hat{Q}_+,\hat{T}_{-\frac{1}{2}}\,\hat{\bar{T}}_{\frac{1}{2}}] &= 0,
\end{aligned} \tag{47}$$

$$\begin{aligned}
[\hat{\bar{Q}}_-,\hat{V}_{\frac{1}{2},+1}\,\hat{\bar{V}}_{-\frac{1}{2},-1}] &= -\hat{V}_{\frac{1}{2},+1}\,\hat{\bar{T}}_{\frac{1}{2}}, & \{\hat{\bar{Q}}_-,\hat{V}_{\frac{1}{2},+1}\,\bar{T}_{\frac{1}{2}}\} &= 0,\\
\{\hat{\bar{Q}}_-,\hat{T}_{-\frac{1}{2}}\,\hat{\bar{V}}_{-\frac{1}{2},-1}\} &= \hat{T}_{-\frac{1}{2}}\,\hat{\bar{T}}_{\frac{1}{2}}, & [\hat{\bar{Q}}_-,\hat{T}_{-\frac{1}{2}}\,\bar{T}_{\frac{1}{2}}] &= 0.
\end{aligned} \tag{48}$$

Notice that (47) and (48) are isomorphic to (2) and (3), respectively. It leads to correspondence of single-trace operators in the matrix model to integrated vertex operators in the type IIA theory:

$$\begin{aligned}
\Phi_1 = \frac{1}{N}\,\mathrm{tr}\,\phi \quad &\Leftrightarrow\quad \mathcal{V}_\phi(0) \equiv g_s^2\int d^2z\,\hat{V}_{\frac{1}{2},+1}(z)\,\hat{\bar{V}}_{-\frac{1}{2},-1}(\bar{z}),\\
\Psi_1 = \frac{1}{N}\,\mathrm{tr}\,\psi \quad &\Leftrightarrow\quad \mathcal{V}_\psi(0) \equiv g_s^2\int d^2z\,\hat{T}_{-\frac{1}{2}}(z)\,\hat{\bar{V}}_{-\frac{1}{2},-1}(\bar{z}),\\
\bar{\Psi}_1 = \frac{1}{N}\,\mathrm{tr}\,\bar{\psi} \quad &\Leftrightarrow\quad \mathcal{V}_{\bar{\psi}}(0) \equiv g_s^2\int d^2z\,\hat{V}_{\frac{1}{2},+1}(z)\,\hat{\bar{T}}_{\frac{1}{2}}(\bar{z}),\\
\frac{1}{N}\,\mathrm{tr}\,(-iB) \quad &\Leftrightarrow\quad \mathcal{V}_B(0) \equiv g_s^2\int d^2z\,\hat{T}_{-\frac{1}{2}}(z)\,\hat{\bar{T}}_{\frac{1}{2}}(\bar{z}),
\end{aligned} \tag{49}$$

where the bare string coupling g_s put in the right hand sides is to count the number of external lines of amplitudes in the IIA theory. Note (49) is consistent with the identification in (8)–(11). Furthermore, it can be naturally extended as

$$\begin{aligned}
\Phi_{2k+1} = \frac{1}{N}\,\mathrm{tr}\,\phi^{2k+1} + (\text{mixing}) \quad &\Leftrightarrow\quad \mathcal{V}_\phi(k) \equiv g_s^2\int d^2z\,\hat{V}_{k+\frac{1}{2},+1}(z)\,\hat{\bar{V}}_{-k-\frac{1}{2},-1}(\bar{z}),\\
\Psi_{2k+1} = \frac{1}{N}\,\mathrm{tr}\,\psi^{2k+1} + (\text{mixing}) \quad &\Leftrightarrow\quad \mathcal{V}_\psi(k) \equiv g_s^2\int d^2z\,\hat{T}_{-k-\frac{1}{2}}(z)\,\hat{\bar{V}}_{-k-\frac{1}{2},-1}(\bar{z}),\\
\bar{\Psi}_{2k+1} = \frac{1}{N}\,\mathrm{tr}\,\bar{\psi}^{2k+1} + (\text{mixing}) \quad &\Leftrightarrow\quad \mathcal{V}_{\bar{\psi}}(k) \equiv g_s^2\int d^2z\,\hat{V}_{k+\frac{1}{2},+1}(z)\,\hat{\bar{T}}_{k+\frac{1}{2}}(\bar{z})
\end{aligned} \tag{50}$$

for higher $k(=1,2,\cdots)$. Since the "tachyons" of the negative winding $\int d^2z\,\hat{T}_{-k-\frac{1}{2}}(z)\,\hat{\bar{T}}_{k+\frac{1}{2}}(\bar{z})$ $(k = 0,1,2,\cdots)$ are invariant under $\hat{Q}_+$ and $\hat{\bar{Q}}_-$, they are expected to be mapped to

$\{\frac{1}{N}\,\mathrm{tr}\,(-iB)^{k+1}\}$ $(k=0,1,2,\cdots)$ perhaps with some mixing terms. We see in (50) that the powers of matrices are interpreted as windings or momenta in the S^1 direction of the type IIA theory. This interpretation is similar to what is discussed in refs. [26, 27]: a positive power k of a matrix variable in the Penner model correctly describe the "tachyons" with negative momentum $-k$ in the $c=1$ string on S^1. Furthermore, in the literature the positive momentum "tachyons" are represented by introducing source terms of an external matrix via the Kontsevich-Miwa transformation in the Penner model. In turn, it is natural to expect in our case that the positive winding "tachyons" $\int d^2z\,\hat{T}_{-k-\frac{1}{2}}(z)\,\hat{\tilde{T}}_{k+\frac{1}{2}}(\bar{z})$ $(k=-1,-2,\cdots)$ in the type IIA theory are expressed in a similar manner in the SUSY double-well matrix model.

Note that (R$-$, R$+$) operators are singlets under the target-space SUSYs $\hat{Q}_+,\hat{\tilde{Q}}_-$, and appear to have no counterpart in the matrix model side. Since the expectation value of operators measuring an RR charge $\langle\Phi_{2k+1}\rangle_0$ does not vanish as seen in (13), the matrix model is considered to correspond to the type IIA theory on a nontrivial background of the (R$-$, R$+$) fields. We may introduce the (R$-$, R$+$) background in the form of vertex operators, when the strength of the background $(\nu_+-\nu_-)$ is small.

5. Correspondence between the matrix model and the type IIA theory

Correlation functions among integrated vertex operators in the type IIA theory on the trivial background are given by

$$\left\langle\prod_i \mathcal{V}_i\right\rangle = \frac{1}{\mathrm{Vol.(CKV)}}\int \mathcal{D}(x,\varphi,H,\mathrm{ghosts})\,e^{-S_{\mathrm{CFT}}}e^{-S_{\mathrm{int}}}\prod_i \mathcal{V}_i, \tag{51}$$

where

$$\begin{aligned} S_{\mathrm{CFT}} &= \frac{1}{2\pi}\int d^2z\left[\partial x\bar{\partial}x+\partial\varphi\bar{\partial}\varphi+\frac{1}{2}\sqrt{\hat{g}}\hat{R}\varphi+\partial H\bar{\partial}H+(\mathrm{ghosts})\right],\\ S_{\mathrm{int}} &= \mu_1\mathcal{V}_B^{(0,0)}(0)\equiv\mu_1\int d^2z\,\hat{T}^{(0)}_{-\frac{1}{2}}(z)\hat{\tilde{T}}^{(0)}_{\frac{1}{2}}(\bar{z}). \end{aligned} \tag{52}$$

The 0-picture (NS, NS) "tachyon" is given by

$$\begin{aligned} \hat{T}^{(0)}_{-\frac{1}{2}}(z) &= e^{\pi\beta(ip_{\tilde{h}}-i\frac{1}{2}p_{\tilde{x}})}\,\frac{i}{\sqrt{2}}\,e^{iH-i\frac{1}{2}x+\frac{1}{2}\varphi}(z),\\ \hat{\tilde{T}}^{(0)}_{\frac{1}{2}}(\bar{z}) &= e^{-\pi\beta(-ip_h+i\frac{1}{2}p_x)}\,\frac{i}{\sqrt{2}}\,e^{-i\bar{H}+i\frac{1}{2}\bar{x}+\frac{1}{2}\bar{\varphi}}(\bar{z}). \end{aligned} \tag{53}$$

We consider correlation functions in the IIA theory on a nontrivial (R$-$, R$+$) background as a form

$$\left\langle\!\!\left\langle\prod_i \mathcal{V}_i\right\rangle\!\!\right\rangle \equiv \left\langle\left(\prod_i \mathcal{V}_i\right)e^{W_{\mathrm{RR}}}\right\rangle. \tag{54}$$

The background W_{RR} is invariant under the target-space SUSYs:

$$\begin{aligned} W_{\mathrm{RR}} &= (\nu_+-\nu_-)\sum_{k\in\mathbf{Z}} a_k\,\mu_1^{k+1}\mathcal{V}_k^{\mathrm{RR}},\\ \mathcal{V}_k^{\mathrm{RR}} &\equiv \begin{cases} \int d^2z\,\hat{V}_{k,\,-1}(z)\hat{\tilde{V}}_{-k,\,+1}(\bar{z}) & (p_\ell=1-|k|,\ k\le 0)\\ \int d^2z\,\hat{V}^{(\mathrm{nonlocal})}_{-k,\,-1}(z)\hat{\tilde{V}}^{(\mathrm{nonlocal})}_{k,\,+1}(\bar{z}) & (p_\ell=1+|k|,\ k\ge 1)\end{cases} \end{aligned} \tag{55}$$

with a_k being numerical constants. Although the nonlocal operators in (55) with $p_\ell>1$ do not satisfy the Dirac equation constraint on the trivial background, these operators are necessary to make correspondence with the matrix model as we see later. Since the RR background possibly change the on-shell condition, it would be acceptable. We treat the RR background for $(\nu_+ - \nu_-)$ small as

$$\left\langle\!\!\left\langle \prod_i \mathcal{V}_i \right\rangle\!\!\right\rangle \equiv \left\langle \left(\prod_i \mathcal{V}_i\right) e^{W_{\rm RR}} \right\rangle = \sum_{n=0}^{\infty} \frac{1}{n!} \left\langle \left(\prod_i \mathcal{V}_i\right) (W_{\rm RR})^n \right\rangle, \tag{56}$$

and the picture is adjusted by hand so that the total picture is equal to -2.

In computation of amplitudes in the type IIA theory, we consider the so-called $s = 0$ amplitude in the Liouville theory, which is interpreted as a bulk amplitude insensitive to details of the Liouville wall [28]. It is somewhat similar to considering the leading nontrivial contribution for small $(\nu_+ - \nu_-)$, because higher orders of $(\nu_+ - \nu_-)$ seems to detect a cigar geometry deformed from the two-dimensional target space (Liouville direction)$\times$(S^1 with self-dual radius) [16]. The direction to the Liouville wall corresponds to the direction to the tip of the cigar. Computation in the Liouville theory [6] yields

$$\left\langle \mathcal{V}_B(0)\, \mathcal{V}_\phi(k)\, \mathcal{V}_\ell^{\rm RR} \right\rangle = -g_s^4\, \delta_{k,\ell}\, (2\ln \mu_1)\, e^{i2\pi\beta(-k^2 - \frac{1}{2}k + \frac{1}{4})}, \tag{57}$$

$$\begin{aligned} \left\langle \mathcal{V}_\phi(k_1), \mathcal{V}_\phi(k_2)\, \mathcal{V}_{\ell_1}^{\rm RR}\, \mathcal{V}_{\ell_2}^{\rm RR} \right\rangle &= g_s^4\, \left(\delta_{\ell_1, k_1+k_2}\delta_{\ell_2,-1} + (\ell_1 \leftrightarrow \ell_2)\right) c_L (2\ln \mu_1)^2 \\ &\quad \times \frac{\pi}{2} \left(\frac{(k_1+k_2)!}{k_1! k_2!}\right)^2 e^{-i\pi\beta\{\sum_{i=1}^2 (k_i + \frac{1}{2})^2 + \sum_{i=1}^2 \ell_i^2\}}. \end{aligned} \tag{58}$$

In the derivation of (58), we encountered the integral

$$\int d^2z\, z^\alpha \bar{z}^{\bar{\alpha}} (1-z)^\beta (1-\bar{z})^{\bar{\beta}} = \pi \frac{\Gamma(\bar{\alpha}+1)\Gamma(\bar{\beta}+1)}{\Gamma(\bar{\alpha}+\bar{\beta}+2)} \frac{\Gamma(-\alpha-\beta-1)}{\Gamma(-\alpha)\Gamma(-\beta)} \tag{59}$$

with

$$\alpha = \bar{\alpha} = k_1 + k_2, \qquad \beta = \bar{\beta} = -k_1 - 1, \qquad (k_1, k_2 = 0, 1, 2, \cdots). \tag{60}$$

This expression is indefinite. We computed it by regularizing as

$$\alpha \to \alpha + \epsilon, \qquad \bar{\alpha} \to \bar{\alpha} + \epsilon, \qquad \beta \to \beta + \epsilon, \qquad \bar{\beta} \to \bar{\beta} + \epsilon, \tag{61}$$

where $\epsilon = \frac{1}{c_L V_L}$. $V_L \equiv 2 \ln \frac{1}{\mu_1}$ is the Liouville volume, and c_L is a numerical constant. The point of the regularization is preserving mutual locality of vertex operators due to the homogeneous shifts.

Let us identify the coupling μ_1 of the Liouville interaction $S_{\rm int}$ in (52) with the "cosmological constant" ω by an appropriate shift of the Liouville coordinate. Then, it leads to the identification

$$N\, {\rm tr}(-iB) \cong \frac{1}{4} \mathcal{V}_B^{(0,0)}(0), \tag{62}$$

which is consistent to the last line in (49) (up to the choice of the picture) with

$$\frac{1}{N} \cong g_s. \tag{63}$$

Also, introducing coefficients c_k, d_k, $\bar{d}_k$, we precisely express the correspondence in (49) and (50) as

$$\Phi_{2k+1} \cong c_k \mathcal{V}_\phi(k), \qquad \Psi_{2k+1} \cong d_k \mathcal{V}_\psi(k), \qquad \bar{\Psi}_{2k+1} \cong \bar{d}_k \mathcal{V}_{\bar{\psi}}(k). \tag{64}$$

We put the overall normalization factor $\mathcal{N}$ in identifying the amplitudes in the matrix-model side and those in the IIA theory side:

$$\left\langle N\, {\rm tr}(-iB)\, \Phi_{2k+1} \right\rangle_{C,0} \cong \mathcal{N} g_s^{-2} \left\langle\!\!\left\langle \frac{1}{4} \mathcal{V}_B^{(0,0)}(0)\, c_k \mathcal{V}_\phi(k) \right\rangle\!\!\right\rangle. \tag{65}$$

The left hand side is calculated by using (13):

$$(\text{LHS}) = -\frac{1}{4}\partial_\omega \left\langle \Phi_{2k+1} \right\rangle_0 \sim -(\nu_+ - \nu_-)\frac{2^k}{\pi}\frac{(2k+1)!!}{(k+1)!}\,\omega^{k+1}\ln\omega. \tag{66}$$

On the other hand, under a suitable choice of the picture, leading nontrivial contribution for $(\nu_+ - \nu_-)$ small to the right hand side is

$$\begin{aligned} \frac{1}{4}\mathcal{N}g_s^{-2}c_k \left\langle \mathcal{V}_B(0)\,\mathcal{V}_\phi(k)\,W_{\text{RR}}\right\rangle &= \frac{1}{4}\mathcal{N}g_s^{-4}c_k(\nu_+ - \nu_-)\sum_{\ell\in\mathbf{Z}} a_\ell\,\omega^{\ell+1}\left\langle \mathcal{V}_B(0)\,\mathcal{V}_\phi(k)\,\mathcal{V}_\ell^{\text{RR}}\right\rangle \\ &= -\frac{1}{2}(\nu_+ - \nu_-)\mathcal{N}c_k\,a_k\,\omega^{k+1}(\ln\omega)\,e^{i2\pi\beta(-k^2-\frac{1}{2}k+\frac{1}{4})}, \end{aligned} \tag{67}$$

where (57) was used. The identification (65) leads to

$$\mathcal{N}\,\hat{c}_k\hat{a}_k\,e^{i\pi\beta\frac{3}{4}} = \frac{2}{\pi}\frac{(2k+1)!}{k!(k+1)!} \tag{68}$$

with

$$\hat{c}_k \equiv c_k\,e^{-i\pi\beta(k+\frac{1}{2})^2}, \qquad \hat{a}_k \equiv a_k\,e^{-i\pi\beta k^2}. \tag{69}$$

Next, let us consider the correspondence

$$\left\langle \Phi_{2k_1+1}\Phi_{2k_2+1}\right\rangle_{C,0} \cong \mathcal{N}g_s^{-2}\left\langle\!\left\langle\, c_{k_1}\mathcal{V}_\phi(k_1)\,c_{k_2}\mathcal{V}_\phi(k_2)\,\right\rangle\!\right\rangle. \tag{70}$$

Leading nontrivial contribution to the right hand side is obtained from (58) as

$$\begin{aligned} &\mathcal{N}g_s^{-2}c_{k_1}\,c_{k_2}\left\langle \mathcal{V}_\phi(k_1)\,\mathcal{V}_\phi(k_2)\,\frac{1}{2!}(W_{\text{RR}})^2\right\rangle \\ &= \frac{1}{2}\mathcal{N}g_s^{-2}c_{k_1}\,c_{k_2}\,(\nu_+ - \nu_-)^2\sum_{\ell_1,\ell_2\in\mathbf{Z}} a_{\ell_1}\,a_{\ell_2}\,\omega^{\ell_1+\ell_2+2}\left\langle \mathcal{V}_\phi(k_1)\,\mathcal{V}_\phi(k_2)\,\mathcal{V}_{\ell_1}^{\text{RR}}\,\mathcal{V}_{\ell_2}^{\text{RR}}\right\rangle \\ &= (\nu_+ - \nu_-)^2\mathcal{N}g_s^2 c_L\,\hat{c}_{k_1}\,\hat{c}_{k_2}\,\hat{a}_{k_1+k_2}\,\hat{a}_{-1}\,2\pi\left(\frac{(k_1+k_2)!}{k_1!k_2!}\right)^2\omega^{k_1+k_2+1}\,(\ln\omega)^2, \end{aligned} \tag{71}$$

while the result of the left hand side is given by (28). Comparing these, we find the same dependence on $\nu_\pm$ and ω for any k_1 and k_2. In addition, we have an equation:

$$\left(\frac{\hat{c}_{k_1}}{(2k_1+1)!}\right)\left(\frac{\hat{c}_{k_2}}{(2k_2+1)!}\right)\left(\hat{a}_{k_1+k_2}(k_1+k_2)!(k_1+k_2+1)!\right) = -\frac{1}{4\pi^3}\frac{1}{\mathcal{N}c_L\hat{a}_{-1}}, \tag{72}$$

which is solved as

$$\hat{c}_k = \hat{c}_0\,e^{\gamma k}\,(2k+1)!, \qquad \hat{a}_k = \frac{\hat{a}_0\,e^{-\gamma k}}{k!(k+1)!} \qquad (k = 0,1,2,\cdots) \tag{73}$$

with γ being a numerical constant and

$$\hat{c}_0^2\hat{a}_0 = -\frac{1}{4\pi^3}\frac{1}{\mathcal{N}c_L\,\hat{a}_{-1}}. \tag{74}$$

Remarkably, (68) is consistent to (73). It serves a quite nontrivial check of the correspondence.

Journal of Physics: Conference Series **512** (2014) 012003 doi:10.1088/1742-6596/512/1/012003

Furthermore, the correspondence of the amplitudes containing fermions

$$\langle \Psi_1 \bar{\Psi}_1 \rangle_{C,0} \cong \mathcal{N} g_s^{-2} \left\langle\!\left\langle\, d_0 \mathcal{V}_\psi(0)\, \bar{d}_0 \mathcal{V}_{\bar{\psi}}(0) \,\right\rangle\!\right\rangle, \qquad \langle \Psi_3 \bar{\Psi}_3 \rangle_{C,0} \cong \mathcal{N} g_s^{-2} \left\langle\!\left\langle\, d_1 \mathcal{V}_\psi(1)\, \bar{d}_1 \mathcal{V}_{\bar{\psi}}(1) \,\right\rangle\!\right\rangle \tag{75}$$

yields

$$d_0 \bar{d}_0 = \frac{1}{4} c_0, \qquad d_1 \bar{d}_1 = -\frac{3}{\pi^2} \frac{c_0}{a_0^2}. \tag{76}$$

It leads to the precise correspondence between the supercharges:

$$Q \cong \frac{d_0}{c_0} \hat{Q}_+, \qquad \bar{Q} \cong \frac{\bar{d}_0}{c_0} \hat{\bar{Q}}_-. \tag{77}$$

So far, the correspondence seems consistent at the level of planar or tree amplitudes. Furthermore, the consistency is checked in the torus partition function [6].

6. Nonperturbative SUSY breaking in the matrix model

In this section, we compute instanton effects in the matrix model which are nonperturbative in $1/N$. Although such effects are of the order e^{-N} and vanish in the simple large N limit, we will see that they are nonvanishing in a double scaling limit

$$N \to \infty, \quad \omega \to 0 \quad \text{with } t \equiv N^{2/3}\omega \text{ fixed}. \tag{78}$$

The partition function of the matrix model given by the action (1) is expressed as

$$\begin{aligned} Z &= \int d^{N^2}\phi\, e^{-N\frac{1}{2}\operatorname{tr}(\phi^2-\mu^2)^2} \det(\phi \otimes \mathbf{1} + \mathbf{1} \otimes \phi) \\ &= \tilde{C}_N \int \Big(\prod_{i=1}^N d\lambda_i\Big) \triangle(\lambda)^2 \prod_{i,j=1}^N (\lambda_i + \lambda_j)\, e^{-N\sum_{i=1}^N \frac{1}{2}(\lambda_i^2-\mu^2)^2}, \end{aligned} \tag{79}$$

after integrating out matrices other than ϕ. Here, $\mathbf{1}$ is an $N \times N$ unit matrix, λ_i $(i = 1, \cdots, N)$ are eigenvalues of ϕ, and $\triangle(\lambda)$ denotes the Vandermonde determinant $\triangle(\lambda) = \prod_{i>j}(\lambda_i - \lambda_j)$. $\tilde{C}_N$ is an numerical factor depending only on N given by

$$\frac{1}{\tilde{C}_N} = \int \Big(\prod_{i=1}^N d\lambda_i\Big) \triangle(\lambda)^2 e^{-N\sum_{i=1}^N \frac{1}{2}\lambda_i^2} = (2\pi)^{\frac{N}{2}} \frac{\prod_{k=0}^N k!}{N^{\frac{N^2}{2}}}. \tag{80}$$

Contributions to the partition function are divided by sectors labeled by the filling fraction (ν_+, ν_-) as

$$Z = \sum_{\nu_- N=0}^{N} \frac{N!}{(\nu_+ N)!(\nu_- N)!} Z_{(\nu_+,\nu_-)} \tag{81}$$

with

$$Z_{(\nu_+,\nu_-)} \equiv \tilde{C}_N \int_0^\infty \Big(\prod_{i=1}^{\nu_+ N} d\lambda_i\Big) \int_{-\infty}^0 \Big(\prod_{j=\nu_+ N+1}^{N} d\lambda_j\Big) \Big(\prod_{n=1}^N 2\lambda_n\Big) \Big\{\prod_{n>m} (\lambda_n^2 - \lambda_m^2)^2\Big\} \times e^{-N\sum_{i=1}^N \frac{1}{2}(\lambda_i^2-\mu^2)^2}. \tag{82}$$

Here, it is easy to see

$$Z_{(\nu_+,\nu_-)} = (-1)^{\nu_- N} Z_{(1,0)}, \tag{83}$$

which leads to the vanishing partition function:

$$Z = (1 + (-1))^N Z_{(1,0)} = 0. \tag{84}$$

In order for expectation values normalized the partition function to be well-defined, we regularize the partition function by introducing a factor $e^{-i\alpha\nu_- N}$ with small α in front of $Z_{(\nu_+,\nu_-)}$. The regularized partition function becomes

$$Z_\alpha \equiv \sum_{\nu_- N=0}^{N} \frac{N!}{(\nu_+ N)!(\nu_- N)!} e^{-i\alpha\nu_- N} Z_{(\nu_+,\nu_-)} = \left(1 - e^{-i\alpha}\right)^N Z_{(1,0)}. \tag{85}$$

Notice that calculations in perturbation theory of $1/N$ in sections 2 and 3 concern the partition function in a single sector ($Z_{(\nu_+,\nu_-)}$), in which such a regularization was not needed. On the other hand, since nonperturbative contributions to be computed here possibly communicate among various sectors of filling fractions, we should consider the total partition function (81) and its vanishing value requires the regularization.

The expectation value of $\frac{1}{N}\operatorname{tr}(iB)$ under the regularization (85) is expressed as

$$\left\langle \frac{1}{N}\operatorname{tr}(iB)\right\rangle_\alpha = \left\langle \frac{1}{N}\operatorname{tr}(\phi^2-\mu^2)\right\rangle_\alpha = \frac{1}{N^2}\frac{1}{Z_\alpha}\frac{\partial}{\partial(\mu^2)}Z_\alpha = \left\langle \frac{1}{N}\operatorname{tr}(\phi^2-\mu^2)\right\rangle^{(1,0)} \tag{86}$$

due to a cancellation of the factor $(1-e^{-i\alpha})^N$ in (85) between the numerator and the denominator. The regularized expectation value $\left\langle \frac{1}{N}\operatorname{tr}(iB)\right\rangle_\alpha$ is independent of α and well-defined in the limit $\alpha \to 0$, and thus serves as an order parameter for spontaneous SUSY breaking.

6.1. Orthogonal polynomials

Under the change of variables $x_i = \lambda_i^2 - \mu^2$, the partition function $Z_{(1,0)}$ defined in (82) reduces to Gaussian matrix integrals

$$Z_{(1,0)} = \tilde{C}_N \int_{-\mu^2}^{\infty} \left(\prod_{i=1}^{N} dx_i\right) \triangle(x)^2 \, e^{-N\sum_{i=1}^{N}\frac{1}{2}x_i^2}. \tag{87}$$

It seems almost trivial, but a nontrivial effect possibly arises from the boundary of the integration region. Ref. [21] mentions that the boundary effect is nonperturbative in $1/N$.

Let us consider polynomials

$$P_n(x) = x^n + \sum_{i=0}^{n-1} p_n^{(i)} x^i \qquad (n = 0, 1, 2, \cdots) \tag{88}$$

with the coefficient of the top degree (x^n) fixed to 1. The coefficients $p_n^{(i)}$ are uniquely determined so that the orthogonality relation

$$(P_n, P_m) \equiv \int_{-\mu^2}^{\infty} dx\, e^{-\frac{N}{2}x^2} P_n(x)\, P_m(x) = h_n \delta_{n,m} \tag{89}$$

is satisfied. Similar to the case without a boundary [29], we have recursion relations of the form

$$xP_m(x) = P_{m+1}(x) + S_m P_m(x) + R_m P_{m-1}(x), \tag{90}$$

$$h_m = R_m h_{m-1}. \tag{91}$$

The one-point function is expressed as a sum of the coefficients S_n:

$$\left\langle \frac{1}{N}\,\mathrm{tr}\,(\phi^2-\mu^2)\right\rangle^{(1,0)} = \frac{1}{N}\sum_{n=0}^{N-1} S_n = \frac{1}{N^2}e^{-\frac{N}{2}\mu^4}\sum_{n=0}^{N-1}\frac{1}{h_n}P_n(-\mu^2)^2, \tag{92}$$

where the last equality follows from the identity

$$\int_{-\mu^2}^{\infty} dx\,\frac{d}{dx}\left(e^{-\frac{N}{2}x^2}\,P_n(x)^2\right) = -e^{-\frac{N}{2}\mu^4}\,P_n(-\mu^2)^2. \tag{93}$$

6.2. *One-instanton contribution*

In what follows, we take into account the boundary effects in an iterative manner. First, we simply neglect the boundary effect by changing the lower bound of the integral (89) to $-\infty$. The orthogonal polynomials in this case, denoted by $P_n^{(H)}(x)$, are given by the Hermite polynomials:

$$P_n^{(H)}(x) = \frac{1}{(2N)^{n/2}}\,H_n\left(\sqrt{\frac{N}{2}}\,x\right) \tag{94}$$

with

$$H_n(x) \equiv (-1)^n\,e^{x^2}\,\frac{d^n}{dx^n}\,e^{-x^2} \tag{95}$$

and coefficients

$$S_n^{(H)} = 0, \qquad R_n^{(H)} = \frac{n}{N}, \qquad h_n^{(H)} = \sqrt{2\pi}\,\frac{n!}{N^{n+\frac{1}{2}}}. \tag{96}$$

Note that the boundary $x = -\mu^2$ corresponds to the local maximum of the original double-well potential $\frac{1}{2}(\lambda^2-\mu^2)^2$, and k eigenvalues sitting at the local maximum give k-instanton contribution as discussed in [30]. Computing the one-point function (92) by using the orthogonal polynomials (94) gives rise to one-instanton contribution:

$$\left\langle \frac{1}{N}\,\mathrm{tr}\,(\phi^2-\mu^2)\right\rangle^{(1,0)}\Bigg|_{1-\text{inst.}} = \frac{e^{-z^2}}{\sqrt{2\pi}\,N^{3/2}}\,\frac{1}{2^N\,(N-1)!}\left[H_N(z)^2 - H_{N-1}(z)\,H_{N+1}(z)\right], \tag{97}$$

where

$$z \equiv \sqrt{\frac{N}{2}}\,\mu^2 = \sqrt{2N}\,(1+2\omega), \tag{98}$$

and the relation

$$\sum_{k=0}^{n-1}\frac{1}{2^k k!}\,H_k(x)^2 = \frac{1}{2^n\,(n-1)!}\left[H_n(x)^2 - H_{n-1}(x)\,H_{n+1}(x)\right] \tag{99}$$

was used. Upon taking the double scaling limit in (97), the following asymptotic formula plays a relevant role [7]:

$$e^{-x^2/2}H_n(x) = \pi^{\frac{1}{4}}2^{\frac{n}{2}+\frac{1}{4}}n^{-\frac{1}{12}}\sqrt{n!}\left[\mathrm{Ai}(s) + \mathcal{O}(n^{-2/3})\right] \tag{100}$$

which is valid for large n with

$$x = \sqrt{2n+1} + \frac{s}{\sqrt{2}\, n^{1/6}}, \tag{101}$$

and the Airy function is defined by

$$\mathrm{Ai}(s) \equiv \frac{1}{2\pi}\int_{-\infty}^{\infty} dz\, e^{-isz-\frac{i}{3}z^3}. \tag{102}$$

Applying (100) to (97) yields the result in the double scaling limit (78):

$$\left.\left\langle \frac{1}{N}\,\mathrm{tr}\,(\phi^2-\mu^2)\right\rangle^{(1,0)}\right|_{1-\text{inst.}} = N^{-4/3}\left[\mathrm{Ai}'(4t)^2 - 4t\mathrm{Ai}(4t)^2 + \mathcal{O}(N^{-2/3})\right] \tag{103}$$

$$= N^{-4/3}\,\frac{1}{32\pi t}\,e^{-\frac{32}{3}t^{3/2}}\left[1+\sum_{n=1}^{\infty} a_n^{(1)}\, t^{-\frac{3}{2}n}\right] \tag{104}$$

with $a_1^{(1)} = -\frac{17}{192}$, $a_2^{(1)} = \frac{1225}{73728}$, $a_3^{(1)} = -\frac{199115}{42467328}, \cdots$. The double scaling limit (78) is expected from the $c=-2$ topological gravity with the string susceptibility $\gamma=-1$ because the free energy at the spherical topology behaves as $N^2\omega^{2-\gamma} = t^{2-\gamma}$. However, use of this expectation is established in perturbative computations as pointed out in section 2, while it is nontrivial that the nonperturbative contribution (103) obeys this scaling. The exponential factor $e^{-\frac{32}{3}t^{3/2}}$ in (104) indicates the weight of one-instanton configurations which remains finite in the double scaling limit. Also, the power series with respect to $t^{-3/2}$ can be regarded as perturbative contributions to all orders around the one-instanton background.

6.3. Leading order two-instanton contribution

In order to compute effects from higher instantons, we consider corrections to replacing P_n by $P_n^{(H)}$ in the previous subsection:

$$P_n(x) = P_n^{(H)}(x) + \tilde{P}_n(x),$$
$$S_n = S_n^{(H)} + \tilde{S}_n, \qquad R_n = R_n^{(H)} + \tilde{R}_n, \qquad h_n = h_n^{(H)} + \tilde{h}_n. \tag{105}$$

Treating quantities with tildes at the linearized level gives rise to two-instanton effects. As a result of the computation in [7], we have the two-instanton contribution:

$$\left.\left\langle \frac{1}{N}\,\mathrm{tr}\,(\phi^2-\mu^2)\right\rangle^{(1,0)}\right|_{2-\text{inst.}} = N^{-4/3}\,\frac{1}{(64\pi)^2\, t^{5/2}}\, e^{-\frac{64}{3}t^{3/2}}\left[1+\mathcal{O}(t^{-3/2})\right]. \tag{106}$$

Both effects from one instanton (104) and from two instantons (106) are of the same order in N and equally contribute in the double scaling limit to the quantity

$$N^{4/3}\left\langle \frac{1}{N}\,\mathrm{tr}\,(\phi^2-\mu^2)\right\rangle^{(1,0)} = N^{4/3}\,\frac{1}{N}\sum_{n=0}^{N-1}\left\{\tilde{S}_n^{(1)} + \tilde{S}_n^{(2)} + \cdots\right\}. \tag{107}$$

Thus, we can conclude that the nonperturbative effect dynamically breaks the supersymmetry under wave function renormalization absorbing the factor $N^{-4/3}$. As (86) suggests, the renormalization factor can be understood from the fact that the renormalized one-point function is obtained by the t-derivative of the free energy $F_{(1,0)} = -\ln Z_{(1,0)}$ with the factor $(-\frac{1}{4})$ multiplied. The weight of the exponential in (106) is twice that of (104), as it should be from the interpretation of a two-instanton contribution.

7. Summary and Discussion

We computed planar correlation functions in the double-well SUSY matrix model, and discussed its correspondence to two-dimensional type IIA superstring theory on (R−,R+) background by comparing amplitudes in both sides. This is an interesting example of matrix models for superstrings with target-space SUSY, in which various amplitudes are explicitly calculable.

It is interesting to examine the correspondence at deeper level in higher genus or higher point amplitudes and in amplitudes containing special massive operators. Also, it is important to discuss the correspondence in the off-shell formulation such as the hybrid formalism [18].

Next, we explicitly calculated nonperturbative instanton effects in the matrix model. In particular, a closed form expression was obtained for full one-instanton contribution to the one-point function $\left\langle \frac{1}{N}\,\mathrm{tr}(\phi^2-\mu^2)\right\rangle^{(1,0)}$ including all perturbative fluctuations around the one-instanton background. Also, presented was its analytic expression for the leading two-instanton effect with respect to finite but large t. The result shows that the supersymmetry is spontaneously broken by nonperturbative effects due to instantons. In particular, the instanton effects survive in the double scaling limit, which implies that supersymmetry breaking takes place by nonperturbative dynamics in the target space of the type IIA superstring theory. Corresponding Nambu-Goldstone fermions are identified with $\frac{1}{N}\,\mathrm{tr}\,\bar{\psi}$ and $\frac{1}{N}\,\mathrm{tr}\,\psi$ associated with the breaking of Q and $\bar{Q}$, respectively. It is interesting to investigate dynamics of D-branes in the type IIA theory and to reproduce the instanton contributions from the type IIA theory side.

Moreover, numerical results for full nonperturbative contribution to the one-point function and the free energy $F_{(1,0)}$ are obtained in [7], where the numerical results up to $N = 1,000,000$ are extrapolated to $N = \infty$ in the double scaling limit. The result for the free energy seems to be smooth even at $t = 0$ which corresponds to the strongly coupled limit of the type IIA superstring theory. It might suggest the existence of an S-dual theory. It would be intriguing to obtain an analytic expression for the full nonperturbative contribution and to identify the S-dual theory.

Acknowledgments

The author would like to thank Michael G. Endres, Tsunehide Kuroki and Hiroshi Suzuki for collaboration. He is grateful to the organizers of QTS8, especially Professor Kurt Bernardo Wolf, for the invitation to the wonderful meeting and for warm hospitality.

References

[1] For a review, see Di Francesco P, Ginsparg P H and Zinn-Justin J 1995 2-D Gravity and random matrices *Phys. Rept.* **254** 1 (*Preprint* hep-th/9306153)
[2] Banks T, Fischler W, Shenker S H and Susskind L 1997 M theory as a matrix model: A conjecture *Phys. Rev.* **D 55** 5112 (*Preprint* arXiv:hep-th/9610043)
[3] Ishibashi N, Kawai H, Kitazawa Y and Tsuchiya A 1997 A large-N reduced model as superstring *Nucl. Phys.* **B 498** 467 (*Preprint* arXiv:hep-th/9612115)
[4] Dijkgraaf R, Verlinde E P and Verlinde H L 1997 Matrix string theory *Nucl. Phys.* **B 500** 43 (*Preprint* arXiv:hep-th/9703030)
[5] Kuroki T and Sugino F 2013 New critical behavior in a supersymmetric double-well matrix model *Nucl. Phys.* **B 867** 448 (*Preprint* arXiv:1208.3263 [hep-th])
[6] Kuroki T and Sugino F 2013 Supersymmetric double-well matrix model as two-dimensional type IIA superstring on RR background *Preprint* arXiv:1306.3561 [hep-th]
[7] Endres M G, Kuroki T, Sugino F and Suzuki H 2013 SUSY breaking by nonperturbative dynamics in a matrix model for 2D type IIA superstrings Nucl. Phys. B **876** 758 (*Preprint* arXiv:1308.3306 [hep-th])
[8] Kuroki T and Sugino F 2010 Spontaneous supersymmetry breaking in large-N matrix models with slowly varying potential *Nucl. Phys.* **B 830** 434 (*Preprint* arXiv:0909.3952 [hep-th])
[9] Cicuta G M, Molinari L and Montaldi E 1986 Large N Phase Transitions In Low Dimensions *Mod. Phys. Lett.* **A 1** 125

[10] Nishimura J, Okubo T and Sugino F 2003 Testing the Gaussian expansion method in exactly solvable matrix models *J. High Energy Phys.* JHEP **0310** (2003) 057. (*Preprint* hep-th/0309262)

[11] Kuroki T and Sugino F 2011 Spontaneous supersymmetry breaking in matrix models from the viewpoints of localization and Nicolai mapping *Nucl. Phys.* **B 844** 409 (*Preprint* arXiv:1009.6097 [hep-th])

[12] For a review, see Klebanov I R 1991 String theory in two-dimensions *In *Trieste 1991, Proceedings, String theory and quantum gravity '91* 30-101* (*Preprint* arXiv:hep-th/9108019)

[13] Distler J and Vafa C 1991 A Critical matrix model at $c = 1$ *Mod. Phys. Lett.* **A 6** 259

[14] Alexandrov S Y, Kazakov V A and Kostov I K 2003 2-D string theory as normal matrix model *Nucl. Phys.* **B 667** 90 (*Preprint* hep-th/0302106)

[15] Kutasov D and Seiberg N 1990 Noncritical superstrings *Phys. Lett.* **B 251** 67

[16] Murthy S 2003 Notes on noncritical superstrings in various dimensions *J. High Energy Phys.* JHEP **0311** (2003) 056 (*Preprint* arXiv:hep-th/0305197)

[17] Ita H, Nieder H and Oz Y 2005 On type II strings in two dimensions *J. High Energy Phys.* JHEP **0506** (2005) 055 (*Preprint* arXiv:hep-th/0502187)

[18] Grassi P A and Oz Y 2005 Non-critical covariant superstrings *Preprint* arXiv:hep-th/0507168

[19] Kostov I K 1988 O(n) Vector Model on a Planar Random Lattice: Spectrum of Anomalous Dimensions *Mod. Phys. Lett.* A **4** 217

[20] Kostov I K and Staudacher M 1992 Multicritical phases of the $O(n)$ model on a random lattice *Nucl. Phys.* **B 384** 459 (*Preprint* arXiv:hep-th/9203030)

[21] Gaiotto D, Rastelli L and Takayanagi T 2005 Minimal superstrings and loop gas models *J. High Energy Phys.* JHEP **0505** (2005) 029. (*Preprint* arXiv:hep-th/0410121)

[22] Knizhnik V G, Polyakov A M and Zamolodchikov A B 1988 Fractal Structure of 2D Quantum Gravity *Mod. Phys. Lett.* **A 3** 819

[23] David F, Conformal Field Theories Coupled to 2D Gravity in the Conformal Gauge 1988 *Mod. Phys. Lett.* **A 3** 1651

[24] Distler J and Kawai H 1989 Conformal Field Theory and 2D Quantum Gravity Or Who's Afraid of Joseph Liouville? *Nucl. Phys.* **B 321** 509

[25] Seiberg N 1990 Notes on quantum Liouville theory and quantum gravity *Prog. Theor. Phys. Suppl.* **102** 319

[26] Imbimbo C and Mukhi S 1995 The Topological matrix model of $c = 1$ string *Nucl. Phys.* **B 449** 553 (*Preprint* hep-th/9505127)

[27] Mukhi S 2003 Topological matrix models, Liouville matrix model and $c = 1$ string theory *Preprint* arXiv:hep-th/0310287

[28] Di Francesco P and Kutasov D 1992 World sheet and space-time physics in two-dimensional (Super)string theory *Nucl. Phys.* **B 375** 119 (*Preprint* hep-th/9109005)

[29] Itzykson C and Zuber J B 1980 The Planar Approximation. 2. *J. Math. Phys.* **21** 411

[30] Hanada M, Hayakawa M, Ishibashi N, Kawai H, Kuroki T, Matsuo Y and Tada T 2004 Loops versus matrices: The Nonperturbative aspects of noncritical string *Prog. Theor. Phys.* **112** 131 (*Preprint* hep-th/0405076)

Journal of Physics: Conference Series **512** (2014) 012004 doi:10.1088/1742-6596/512/1/012004

Peres lattices and chaos in the Dicke model

Miguel Angel Bastarrachea-Magnani, Jorge G. Hirsch
Instituto de Ciencias Nucleares, Universidad Nacional Autónoma de México
Apdo. Postal 70-543, México D. F., C.P. 04510

E-mail: miguel.bastarrachea@nucleares.unam.mx

Abstract. Peres lattices are employed as a visual method to identify the presence of chaos in different regions of the energy spectra in the Dicke model. The coexistence of regular and chaotic regions can be clearly observed for certain energy regions, even if the coupling constant is smaller than the critical value to reach superradiance. It also exhibits the presence of two Excited-State Quantum Phase Transitions, a *static* and a *dynamic* one. The diagonalization is performed in a extended bosonic coherent basis which enable us to reach a large number of excited states with good numerical convergence.

PACS numbers: 3.65.Fd, 42.50.Ct, 64.70.Tg

1. Introduction

For decades atom-field systems have attracted attention from researchers. The Dicke model describes the interaction between $\mathcal{N}$ two-level atoms and a single mode of electromagnetic field [1] within a cavity. The model exhibits interesting features like a Quantum Phase Transition (QPT) at zero temperature, form a normal to a superradiant phase [2, 3], quantum chaos [4], Excited-State Quantum Phase Transitions [5], to name a few. The experimental realization of a Dicke-like Hamiltonian where the superradiant phase was observed for the first time [6] renewed the interest in the model. The manipulation of systems with just a few interacting qbits connects it with the field of quantum computing and many other interesting potential applications [7].

The Dicke Hamiltonian can be written (with $\hbar = 1$) as,

$$H = \omega a^\dagger a + \omega_0 J_z + \frac{2\gamma}{\sqrt{\mathcal{N}}}\left(a + a^\dagger\right) J_x, \tag{1}$$

where ω is the field frequency, ω_0 the atomic energy level difference, γ the coupling, and J_x, J_y, and J_z are collective pseudo-spin atomic operators which obey the SU(2) algebra. In what follows we restrict the analysis to the atomic symmetric subspace with $j = \mathcal{N}/2$ (the pseudo-spin lenght), being $j(j+1)$ the eigenvalue of $\mathbf{J}^2 = J_x^2 + J_y^2 + J_z^2$. It has a ground state QPT whose critical coupling is (in the thermodynamical limit $\mathcal{N} \to \infty$) $\gamma_c = \sqrt{\omega_0\omega}/2$.

The Dicke model has an integrable approximation (employing the rotating wave approximation) called the Tavis-Cummings (TC) model [8]. The integrability of the TC model, comes from the conserved quantity Λ, the total number of excitations [9]:

$$\Lambda = a^\dagger a + J_z + j, \tag{2}$$

with eigenvalues $\lambda = n + j + m$, where n is the number of photons, m the eigenvalue of J_z and $j + m$ the number of excited atoms. The presence of the counter-rotating terms in the Dicke

Journal of Physics: Conference Series **512** (2014) 012004 doi:10.1088/1742-6596/512/1/012004

Hamiltonian prevents it to commute with Λ. It does commute, however, with a parity operator Π,

$$\Pi = exp(i\pi\Lambda). \tag{3}$$

It allows to separate the eigenstates into those with positive and negative parity. However, it is not enough to classify all states, implying that the Dicke Hamiltonian is not integrable. The Dicke model becomes integrable in two limits of the Hamiltonian parameters, the zero-coupling limit $\gamma \to 0$, and the zero-gap limit $\omega_0 \to 0$ [10]. As a non-integrable Hamiltonian, there are not analytical solutions, except for asymptotic solutions which are exact for the ground state in the thermodynamical limit [4, 11].

Cavity QED applications [12] are described by the Dicke Hamiltonian with a limited number of qubits, whose eigenstates can only be obtained employing numerical methods. As the bosonic sector is infinite dimensional, a truncation N_{max} in the number of bosons is necessary. To perform the numerical diagonalization of the Hamiltonian we use an extended bosonic coherent basis [10, 13]. This basis enables us to obtain a large fraction of converged states employing a single value of the truncation N_{max}. An upper bound for the numerical precision in the wave function for every state can be assigned [14], as briefly reviewed in Appendix 1. With this tool it is possible to explore a relevant sector of the energy spectrum for relatively large values of the number of atoms $\mathcal{N}$, in the superradiant phase.

A very interesting feature present in some quantum spectra, the Excited-State Quantum Phase Transition (ESQPT) has been found in the Dicke and TC models [5]. An ESQPT takes place along the energy spectrum, for fixed values of the Hamiltonian parameters. It is manifested by singularities in the level density, order parameters, and wave function properties [15]. An ESQPT can have important effects in decoherence [16] and the temporal evolution of quantum quenches [17]. Some average properties of the observables, including the ESQPT, have been analyzed recently in the thermodynamic limit [18].

The purpose of this work is to explore the presence of regular and chaotic regions along the energy spectra employing Peres lattices in the Dicke model. In section 2 we describe Peres method and the Peres lattices idea. In section 3 we show and analyze the Peres lattices in the Dicke model. We employ them to characterize the non-integrability of the Dicke model and to study qualitatively its spectrum. Finally, we show the conclusions.

2. Peres lattices and integrability

The notion of integrability comes from classical physics. A classical integrable system has as many conserved quantities as degrees of freedom, so it can be solved analytically. Its orbits in phase space lie on tori, and if we employ a Poincaré section, the phase space trajectories can be seen as regular curves. The non-integrability in a classical system is strongly bonded to chaos. If we add a perturbation term to the classical Hamiltonian, some quantities are not conserved anymore and the system becomes non-integrable. Also, as the strength of perturbation increases, the tori get destroyed and the regular orbits become chaotic ones. Again, this can be seen easily using a Poincaré section, irregular patterns appear where regular orbits were before. Then, with non-integrability chaos arises [19, 20]. For a quantum system, however, the definition of integrability cannot be the same as in classical physics, mainly because there are no trajectories in phase space where chaotic and regular orbits can be distinguished, and there is not a direct analog of the Poincaré sections that help us to identify the integrability of a Hamiltonian and the presence of chaos.

A quantum system whose classical analog is integrable must have a "regular" spectrum, i. e. its energy levels can be labeled in a natural way by quantum numbers related to the constants of motions. Following this idea, A. Peres proposed a visual method that plays a role similar of Poincaré sections in classical mechanics [21], and helps to identify the integrability of a quantum

Journal of Physics: Conference Series **512** (2014) 012004 doi:10.1088/1742-6596/512/1/012004

system by the regularity in its spectrum. On the other hand, for a non-integrable system the absence of regularity can be observed.

When a quantum system is integrable, associated with an unperturbed Hamiltonian H_0, a plot of the energies of individual states versus the respective eigenvalues of a constant of motion P, which satisfies $[H_0, P] = 0$, form a lattice of regularly distributed points. When the system is perturbed, $H = H_0 + \gamma H_1$, with $[H_1, P] \neq 0$, and becomes non-integrable, P is not anymore a conserved quantity. However, as Peres pointed out, we can use the expectation values of a P (now called Peres operator) in the energy representation and plot them against the Hamiltonian eigenvalues. Peres noted that, in classical physics, the time-average of any bounded quantity depending on canonical variables is trivially a constant of motion. For quantum systems this would correspond to the expectation values of the Peres operator.

The choice of P connects the unperturbed and perturbed cases and such plots are called Peres lattices. If the system is integrable the Peres lattice is regular, as each energy level has a natural way to be labeled by the quantum number. A smooth perturbation doesn't destroy the lattice, instead a set of localized distortions is created, and the rest of the lattice remains regular. This can be seen as the coexistence of chaotic an regular behaviors in a classical system with mixed dynamics, i. e. the remnants of the tori. Finally, as the perturbation increases the system becomes non-integrable and is characterized by an irregular lattice. In this way the Peres method represents a qualitatively sensitive probe of the competition between regular and chaotic behaviors in the quantum spectrum of a system [22]. Moreover, the freedom in choosing the Peres operator makes it possible to focus on various properties of individual states and to closely follow the way how chaos sets in and proliferates in the system.

Therefore, Peres lattices allow us to observe changes in the spectrum qualitatively. We employ the Peres method to analyze the features of the Dicke Hamiltonians spectrum, the onset of chaos and its ESQPT.

3. Peres lattices in Dicke model

In order to build the Dicke model Peres lattices we only need a single Peres operator, as the Dicke Hamiltonian has two degrees of freedom. Thanks to the freedom in the choice of it we can explore various features with different Peres operators. It is worth to be mentioned that, as the Dicke Hamiltonian commutes with the parity operator, it is important to separate states with different parity in order to build the Peres lattices. In appendix 2 we show the extended coherent basis with well defined parity, which we use to diagonalize the Hamiltonian taking advantage of this symmetry. Keeping in mind this considerations we choose as Peres operators J_z, J_x^2 (as J_x is not a good operator because connects states of different parity) and $n = a^\dagger a$. In what follows we show several Peres lattices for different values of the coupling constant γ, in resonance for $\omega = \omega_0 = 1.0$. The critical coupling takes the value $\gamma_c = 0.5$. The number of atoms is fixed at $\mathcal{N} = 40$.

For values of the coupling close to zero, $\gamma = 0.01\gamma_c$, we present in figure 1 the Peres lattices. As it can be seen from the lattices, they exhibit regular patterns. This is just what we expected as the coupling is near the integrable limit $\gamma \to 0$. In this case the energies are $E(n, j, m) \approx n + m$. The Hamiltonian is nearly integrable, we can label each state with their quantum numbers, the number of atoms and the number of photons and thus the Peres lattices are regular. Degenerate states are visualized as vertical lines of dots with the same energy and different expectation values of the Peres operators. The ground state has $n = 0$, $m = -j$ and $\frac{E}{j} = -1$. The degeneracy grows linearly with the energy up to $2j+1$, and remains constant beyond this value, which corresponds to $\frac{E}{j} = 1$, which is clearly seen as a well defined peak in the Peres lattices, (figures 1 a, b and c). Fig. 1 d shows the upper bound of the numerical error in each of the wave functions due to the truncation. All of them are smaller that 10^{-30}.

In figure 2 we present the Peres lattices for a small but non negligible coupling $\gamma = 0.6\gamma_c$.

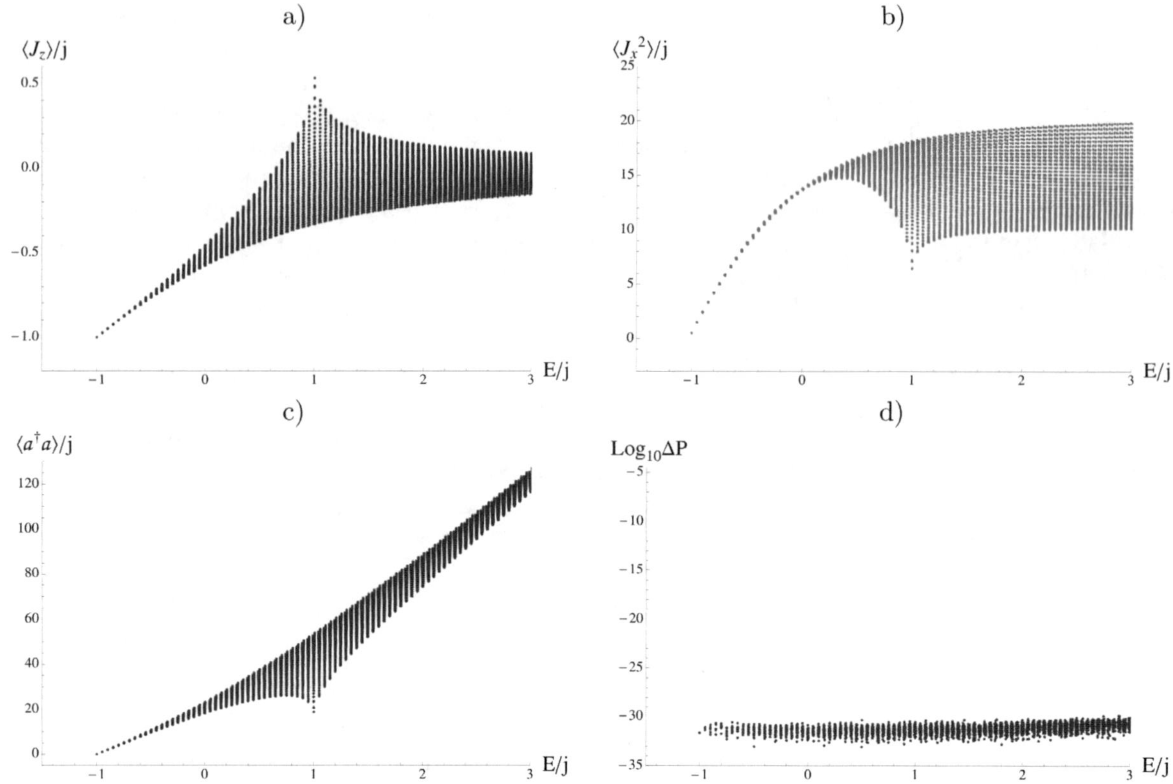

Figure 1. Peres lattices for the Dicke model. Normal phase near zero-coupling $\gamma = 0.01\gamma_c$, with $N_{mnax} = 250$. Peres operators: J_z (a), J_x^2 (b) and n (c). The numerical precision in the wave function for each individual state is shown in (d).

Although the system is in the regular phase, irregular regions in the energy spectrum can be observed for $E > 0$, reflecting the fact that the Hamiltonian is no longer integrable. We can guarantee that the observed effects are robust, because the numerical noise remains around 10^{-30}.

Next, we observe several representative couplings in the superradiant phase where the critical coupling is large than the critical one, $\gamma > \gamma_c$: $\gamma = 1.5\gamma_c$ (figure 3), $\gamma = 2.0\gamma_c$(figure 4), and, for the ultra strong coupling in the superradiant phase $\gamma = 3\gamma_c$ (figure 5). Thanks to the lattices we can observe a pair of important things.

First, the J_z Peres lattice exhibit two ESQPT. There are two changes in the slope of the lattice. One in $E/j = 1$ and another in $E/j = -1$. In fact, the ESQPT can be seen too in figure 1. The $E/j = -1$ ESQPT is the one pointed by Perez-Fernández et. al. [5]. From the J_z lattice and following the results in [18], it can be recognized that the two ESQPT are different in nature. The one in $E/j = 1$ is *static*, because as it can be observed from figure 1 it corresponds to a saturation of the atomic space. For a given $\mathcal{N}$, there is a maximum value of excited atoms, the same $\mathcal{N}$. The *static* ESQPT is independent of the coupling. On the other hand, the second one in $E/j = -1$ is a *dynamical* ESQPT because it depends on the coupling. It is always associated with the state with $n = 0$, $m = -j$, which in the superradiant phase is no longer the ground state. The strong atom-photon interaction builds up many states which combine a large number of photons and excited atoms with energies $E/j < -1$.

The second interesting feature is the break in the regularity. This is what one would expect

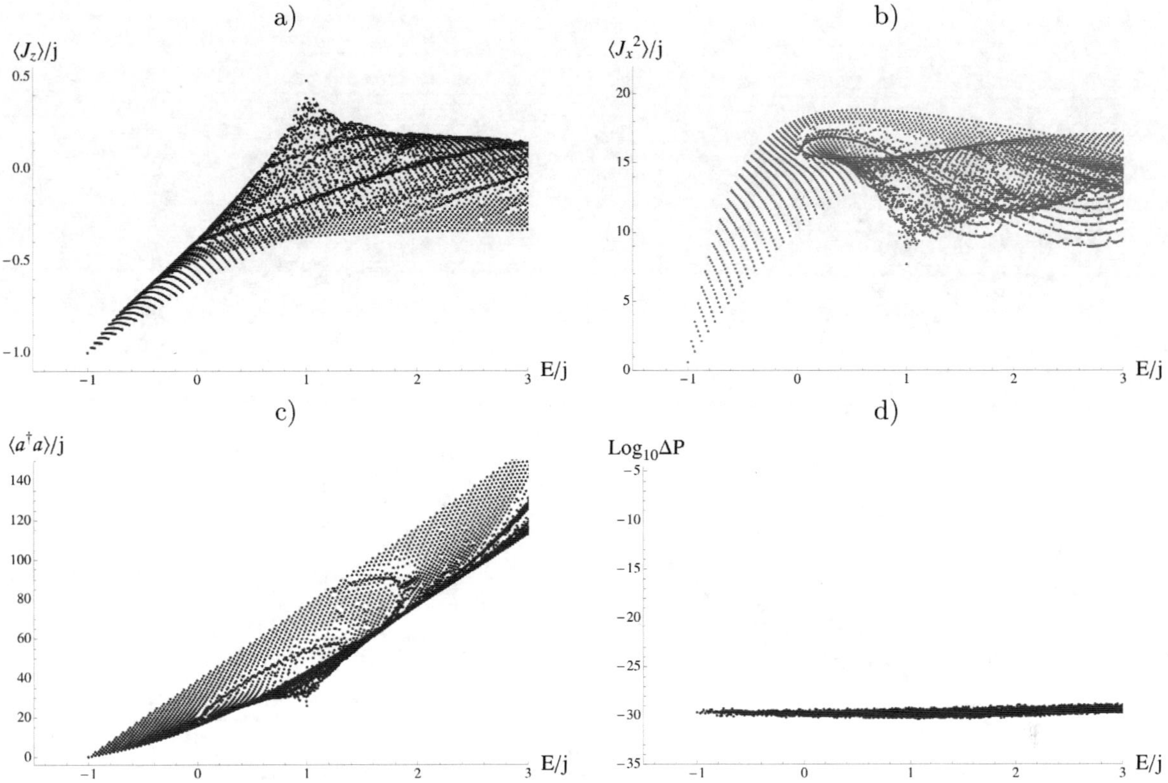

Figure 2. Peres lattices for the Dicke model. Normal phase $\gamma = 0.6\gamma_c$, with $N_{max} = 250$. Peres operators: J_z (a), J_x^2 (b) and n (c). The numerical precision in the wave function for each individual state is shown in (d).

from a non-integrable system. This is clear with every Peres operator. But also, there is a regular part for energies below $E/j = -1$. This regular region is small for weak couplings, but it grows with the coupling. Deep in the superradiant phase (figures 4 and 5) it is clear that there is a transition between regularity and chaos. Besides, it seems the *dynamical* ESQPT could be related with this transition. It would be necessary more work to determine if the onset of chaos is related to the ESQPT.

4. Conclusions

We have employed the Peres lattices as a powerful tool to analyze the spectrum features of quantum systems. They helped us to determine the presence of two Excited-State Quantum Phase Transitions and to distinguish between integrable and non-integrable regions. It gives us a way to define (qualitatively) integrability. The Dicke model is a good choice to study several interesting topics, with the great advantage to be a quantum non-integrable model which has well studied integrable limits. The lattices confirmed this but also reveal us the richness of the spectrum. A non-integrable system does not limit itself to a irregular behavior in the spectrum. In particular, for the Dicke model, Peres lattices help us to observe clearly the presence of two ESQPT, their nature, and a sort of transition between chaos and regularity. In a forthcoming work we present a join study involving the Peres method and quantitative analysis of quantum spectra, both numerical and analytical, which go beyond the scope of the present article [23].

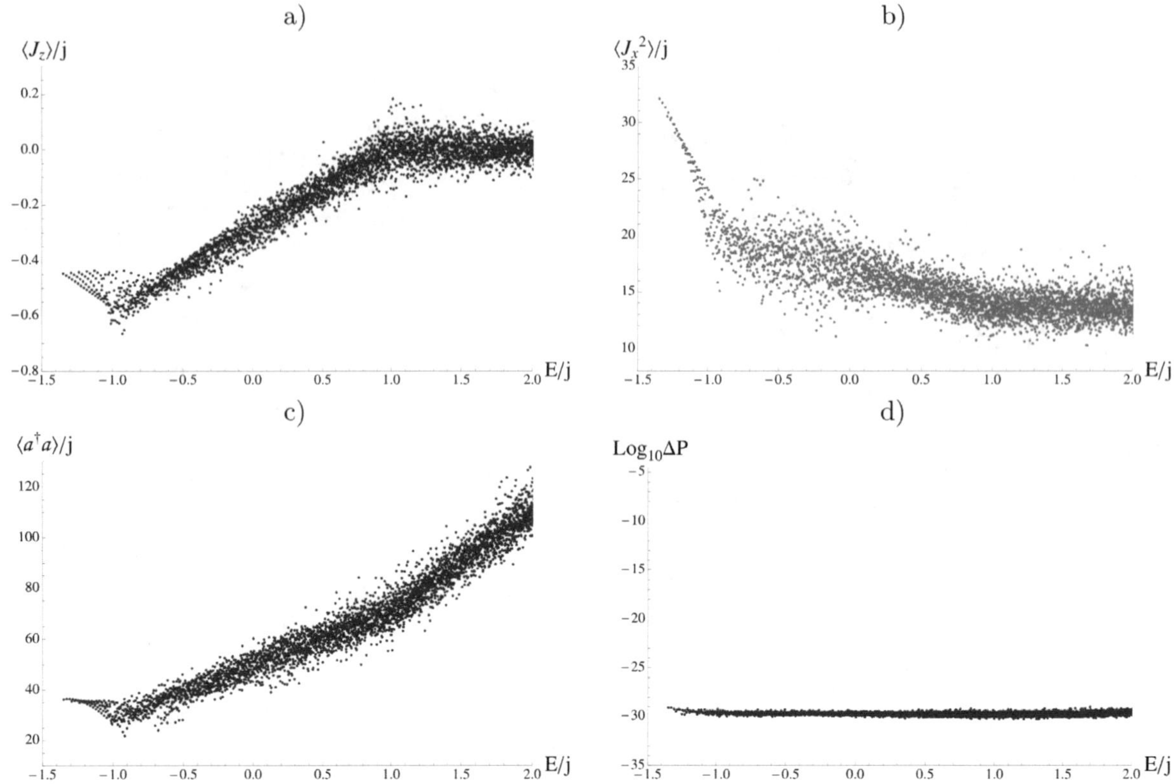

Figure 3. Peres lattices for the Dicke model. Superradiant phase $\gamma = 1.5\gamma_c$ with $N_{max} = 250$. Peres operators: J_z (a), J_x^2 (b) and n (c). The numerical precision in the wave function for each individual state is shown in (d).

Acknowledgments
We thank S. Lerma, P. Stransky and P. Cejnar for many useful and interesting conversations.This work was partially supported by CONACyT- México and DGAPA-UNAM.

Appendix A. Precision in the wave function

We express the ground state wave function as

$$|\Psi(N_{max})\rangle = \sum_{N=0}^{N_{max}} \sum_{m_x=-j}^{j} C_{N,m_x} |N; j, m_x\rangle, \tag{A.1}$$

where $|N; j, m_x\rangle$ is the extended bosonic basis [24], [10], N is the eigenvalue of the $A^\dagger A$ operator ($A = a + \frac{2\gamma}{\sqrt{N}\omega} J_x$), and m_x are the eigenvalues of J_x. C_{N,m_x} are the coefficients of the exact ground state wave function and N_{max} is the value of the truncation or cutoff in the number of displaced excitations. The probability P_N of having N excitations in the ground state is

$$P_N = |\langle N|\Psi\rangle|^2 = \sum_{m_x=-j}^{j} |C_{N,m_x}|^2 \tag{A.2}$$

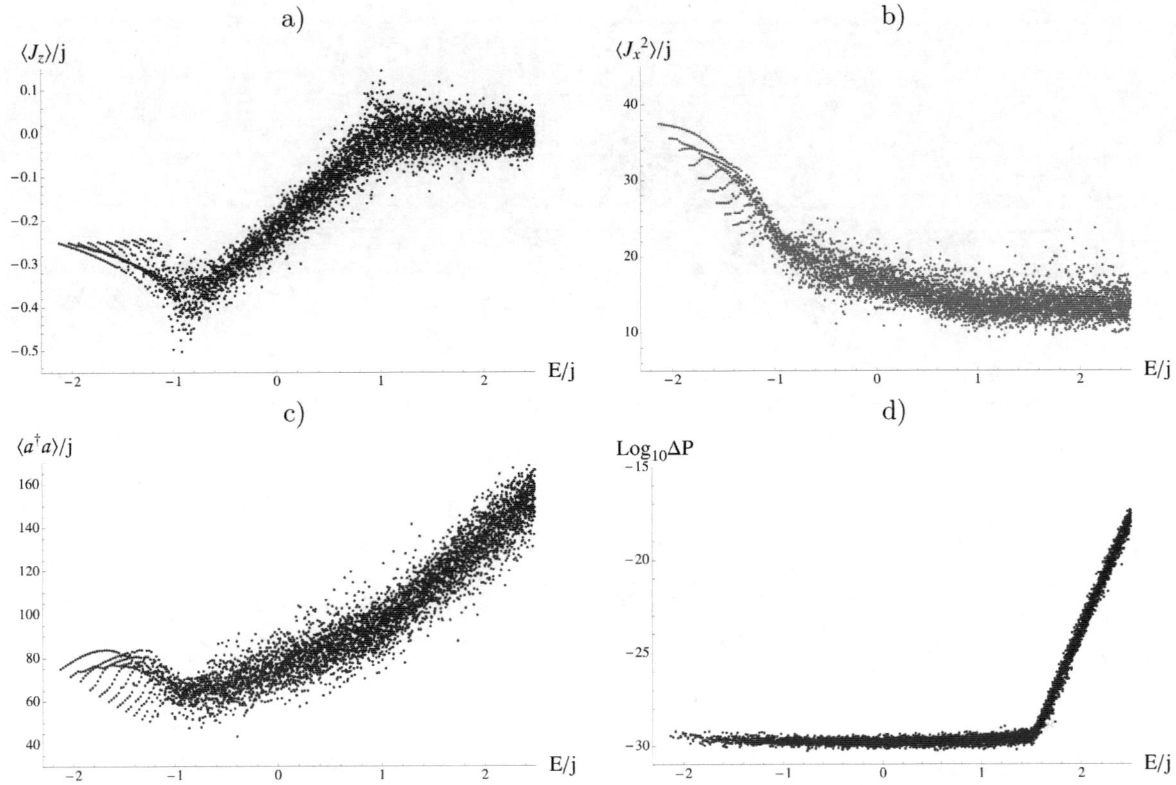

Figure 4. Peres lattices for the Dicke model. Superradiant phase $\gamma = 2.0\gamma_c$, with $N_{max} = 250$. Peres operators: J_z (a), J_x^2 (b) and n (c). The numerical precision in the wave function for each individual state is shown in (d).

We define the precision in the calculated wave function as [14]:

$$\Delta P = \sum_{m_x=-j}^{j} |C_{N_{max}+1,m_x}(N_{max}+1)|^2 . \tag{A.3}$$

By diagonalizing the Hamiltonian with several truncations, if ΔP is smaller than certain tolerance we consider that the solution has converged, being N_{max} the minimum value of the truncation necessary for obtaining the exact numerical solution.

Appendix B. Basis with well defined parity

In order to solve numerically the Dicke Hamiltonian, we use the extended coherent bosonic basis with the right parity. It can be shown the action of the parity operator (Eq. 3) over the coherent basis is:

$$\Pi|N; j, m_x\rangle = (-1)^N |N; j, -m_x\rangle \tag{B.1}$$

So, the eigenvectors of the Hamiltonian in the limit $\omega_0 \to 0$ and simultaneously of the parity operator are:

$$|N; j, m_x; \pm\rangle = \frac{1}{\sqrt{2(1+\delta_{m_x,0})}} \left(|N; j, m_x\rangle + (-1)^N |N; j, -m_x\rangle\right) \tag{B.2}$$

Journal of Physics: Conference Series **512** (2014) 012004 doi:10.1088/1742-6596/512/1/012004

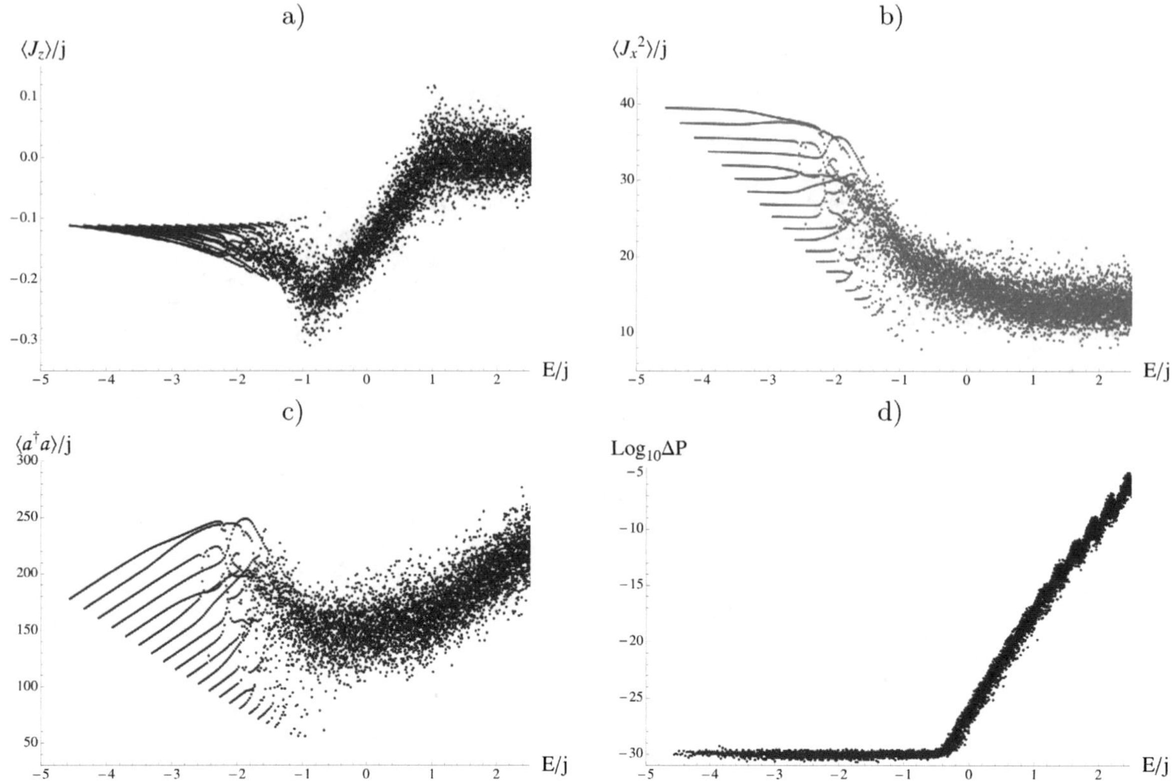

Figure 5. Peres lattices for the Dicke model. Deep superradiant phase $\gamma = 3\gamma_c$, with $N_{max} = 300$. Peres operators: J_z (a), J_x^2 (b) and n (c). The numerical precision in the wave function for each individual state is shown in (d).

Using this basis, we can obtain energy levels which are deeper in the spectrum and also we can employ level statistics techniques to analyze it.

[1] R. H. Dicke, *Phys. Rev.* **93**, 99 (1954).
[2] K. Hepp and E. H. Lieb, *Ann. Phys. (N.Y.)* **76**, 360 (1973)
[3] Y. K. Wang and F. T. Hioe, *Phys. Rev. A* **7**, 831 (1973).
[4] C. Emary and T. Brandes, *Phys. Rev. E* **67**, 066203 (2003); Phys. Rev. Lett. **90**, 044101 (2003).
[5] P. Pérez-Fernández, A. Relaño, J. M. Arias, P. Cejnar, J. Dukelsky, and J. E. García-Ramos, *Phys. Rev. E* **83**, 046208 (2011).
[6] K. Baumann, C. Guerlin, F. Brennecke, and T. Esslinger, *Nature* **464**, 1301 (2010).
[7] B. M. Garraway, Phil. Trans. R. Soc. A **369**, 1137 (2011).
[8] M. Tavis and, F. W. Cummings, Phys. Rev. **170** (2), 379 (1968).
[9] O. Castaños, R. López-Peña, E. Nahmad-Achar, J. G. Hirsch, E. López-Moreno and J. E. Vitela, Physica Scripta **79**, 065405 (2009) .
[10] M. A. Bastarrachea-Magnani and J. G. Hirsch, *Rev. Mex. Fis. S* **57** 0069 (2011) http://rmf.smf.mx/pdf/rmf-s/57/3/5730069.pdf.
[11] O. Castaños, E. Nahmad-Achar, R. López-Peña, and J. G. Hirsch, Phys. Rev. **A 83**, 051601(R) (2011) ; *ibid* Phys. Rev. **A 84**, 013819 (2011).
[12] Wallraff, A., Schuster, D. I., Blais, A., Frunzio, L., Huang, R.-S., Majer, J., Kumar, S., Girvin, S. M. and Schoelkopf, R. J. 2004 Nature **431**, 162.; Fink, J. M., Bianchetti, R., Baur, M., Gppl, M., Steffen, L., Filipp, S., Leek, P. J., Blais, A. and Wallraff, A. 2009 Phys. Rev. Lett. **103**, 083601; Niemczyk, T. et al. 2010 Nat. Phys. **6**, 772; Fink, J. M., Gppl, M., Baur, M., Bianchetti, R., Leek, P. J., Blais, A. and Wallraff, A. 2008 Nature **454**, 315.

[13] M. A. Bastarrachea-Magnani and J. G. Hirsch, *AIP Conf. Proc.* **1488** 418 (2012).
[14] M. A. Bastarrachea-Magnani and J. G. Hirsch, Journal of Physics Conf. Series in press (2013), arXiv:1312.1963.
[15] M.A. Caprio, P. Cejnar, F. Iachello, *Annals of Physics* **323** 1106 (2008).
[16] A. Relaño, J. M. Arias, J. Dukelsky, J. E. García-Ramos, and P. Pérez-Fernández, *Phys. Rev. A* **78**, 060102 (2008).
[17] P. Pérez-Fernández, P. Cejnar, J. M. Arias, J. Dukelsky, J. E. García-Ramos, and A. Relaño, *Phys. Rev. A* **83**, 033802 (2011).
[18] T. Brandes, *Phys. Rev. E* **88**, 032133 (2013).
[19] J. V. José and E. J. Saletan, *Classical Dynamics A Contemporary Approach* (Cambridge University Press, USA, 1998).
[20] H. Goldstein, C. Poole, and J. Safko, *Classical Mechanics* (Addison Wesley, USA, 2002).
[21] A. Peres, *Phys. Rev. Lett.* **53** 1711 (1984).
[22] P. Stránský, P. Hruska, and P. Cejnar, *Phys. Rev. E* **79**, 066201 (2009).
[23] S. Lerma. M. A. Bastarrachea-Magnani and J.G. Hirsch, arXiv:1312.2665 and arXiv:1312.2672.
[24] Q. H. Chen, Y. Y. Zhang, T. Liu, and K. L. Wang, *Phys. Rev. A* **78** 051801 (2008); T. Liu, Y. Y. Zhang, Q. H. Chen, and K. L. Wang, *Phys. Rev. A* **80** 165308 (2009).

Journal of Physics: Conference Series **512** (2014) 012005 doi:10.1088/1742-6596/512/1/012005

Classical dynamics of a moving mirror due to radiation pressure

L. O. Castaños[1] and R. Weder[2]

Departamento de Física Matemática, Instituto de Investigaciones en Matemáticas Aplicadas y en Sistemas, Universidad Nacional Autónoma de México, Apartado Postal 20-726, México DF 01000, México

E-mail: 1. `LOCCJ@yahoo.com` , 2. `weder@unam.mx`

Abstract. We investigate the classical dynamics of a thin moving mirror with non-zero transparency interacting with a laser by means of radiation pressure. The thin moving mirror is modelled by a delta function and it is part of a one-dimensional cavity where the other mirror is perfect and fixed at a position. We use the exact modes of the system so that the transparency and the mirror-position dependent cavity resonance frequencies are built into them. We derive the radiation pressure force from a periodic potential with period half the wavelength of the field and this allows us to give direct physical interpretations of the dynamics of the system and to obtain approximate analytic solutions. We show that within the rotating-wave-approximation the movable mirror can exhibit either a bounded or unbounded motion governed by an approaching to and a withdrawing from resonant positions where the frequency of the field is similar to one of the cavity resonance frequencies.

1. INTRODUCTION

The area of optomechanics is concerned, broadly speaking, with coupling mechanical degrees of freedom with light. The mechanical degrees of freedom could be, for example, the motion of atoms or a movable mirror. The area has attracted much attention in recent years due to the potential applications and the technological advances. The possible applications are diverse and range from optical communications [1] to studies of both classical and quantum systems [2, 3]. Some of the investigations that stand out are the possibility to explore the transition from the quantum to the classical world [3], the preparation of non-classical states of the mechanical degrees of freedom [4], and investigations of classical non-linear and possibly chaotic dynamics [5]-[13]. On the other hand, one of the technological advances that stands out is that, in some optomechanical set-ups, the mechanical degrees of freedom have been cooled to the quantum ground-state [14, 15].

One of the paradigmatic models in optomechanics consists of a one-dimensional cavity with one movable mirror. The movable mirror is a mechanical oscillator and it is coupled to the electromagnetic field by two main forces: the radiation pressure force and the thermal or bolometric force. In the former the electromagnetic field *pushes* the mirror, while the latter consists in the absorption of light that distorts and displaces the mirror. The dynamics of this system and similar ones can be studied with both classical and quantum mechanics. In this article we are concerned with the classical dynamics. We refer the reader to [2]-[4] and [16]-[21] for studies of the quantum dynamics of similar systems.

Journal of Physics: Conference Series **512** (2014) 012005 doi:10.1088/1742-6596/512/1/012005

Depending on the experimental set-up (for example, the materials composing the movable mirror) one of the two forces mentioned above can dominate and they can operate at different time scales [9, 10, 19]. Also, the radiation pressure and thermal forces can even be in opposite directions [19]. These two types of forces alone or combined give rise to complex classical non-linear dynamics that have been studied both theoretically and experimentally. For example, in the set-up analysed in [5, 6] the classical dynamics are dominated by the radiation pressure force, while in that of [7, 8] they are dominated by the thermal force. The case where both the radiation pressure and thermal forces are relevant has also been studied [9]-[12]. In particular, [7, 8] found that there is multi-stability, that is, the amplitude of the oscillations of the mirror settles into one of several attractors (see also [13] for this phenomenon), while [9, 10] found that chaotic attractors can appear.

In this article we consider a one-dimensional cavity composed of one fixed, perfect mirror and one thin, movable mirror with non-zero transparency. The movable mirror is modelled by a delta-function and interacts with the electromagnetic field of a monochromatic laser by means of radiation pressure. We start from approximate Maxwell-Newton equations valid when the velocity and acceleration of the movable mirror are small and we consider the exact modes of the complete system. The use of the exact modes allows us to derive the radiation pressure force from a periodic potential and to interpret the dynamics directly in terms of the mirror approaching or withdrawing from positions where the frequency of the field coincides with one of the cavity resonance frequencies. Moreover, we pay special attention to determining when the approximate Maxwell-Newton equations used to describe the system are valid.

The model described in the preceding paragraph but with a *fixed very thin mirror* is sometimes known as the Lang-Scully-Lamb or LSL model [22]-[29]. It was introduced to explain why the laser line is so narrow, to form a more satisfactory picture of the nature of a laser mode, and to account for radiation losses through one of the mirrors of the laser without the use of phenomenological models [22]-[29].

After this article was completed we learned about reference [30]. It uses a scattering approach to describe a mobile scatterer (a mirror or atom) interacting with the electromagnetic field. The scatterer has a dispersive dielectric constant and moves with constant velocity. Moreover, the friction force and the diffusion affecting the scatterer were derived with the rotating-wave-approximation.

The article is organized as follows. In Section II we establish the system and the model under study. In Section III we restrict to a single-mode electromagnetic field (associated with a laser) and derive the potential associated with the radiation pressure force. In Section IV we investigate the dynamics of the movable mirror. The conclusions are given in Section V.

2. THE MODEL

Consider a one dimensional cavity composed of two mirrors parallel to the yz-plane. One mirror is perfect and it is fixed at $x = 0$. The other one can move and its mid-point has position $x = q(t) > 0$ at time t. When it is at rest, we assume that the movable mirror is a linear, isotropic, non-magnetizable, and non-conducting (it has no free charges or currents) dielectric of thickness δ_{thick}. Also, there is a classical electromagnetic field polarized along the z-axis and we use the Gaussian system of units to describe it. One can then work in the Coulomb gauge and derive both the electric and magnetic fields from a vector potential $\mathbf{A}(x,t)$ of the form

$$\mathbf{A}(x,t) = A_0(x,t)\mathbf{z} . \tag{1}$$

The electric and magnetic fields are given respectively by

$$\mathbf{E}(x,t) = -\frac{1}{c}\frac{\partial A_0}{\partial t}(x,t)\mathbf{z} , \qquad \mathbf{B}(x,t) = -\frac{\partial A_0}{\partial x}(x,t)\mathbf{y} , \tag{2}$$

and the equation for the vector potential is [31]

$$\frac{\partial^2 A_0}{\partial x^2}(x,t) \;=\; \frac{\epsilon\,[x-q(t)]}{c^2}\frac{\partial^2 A_0}{\partial t^2}(x,t) \qquad (x>0,\ t\in\mathbb{R}). \tag{3}$$

The dielectric function associated with the movable mirror is ϵ.

The perfect and fixed mirror at $x=0$ imposes the condition $\mathbf{E}(x,t)=\mathbf{B}(x,t)=\mathbf{0}$ for $x<0$. Using the well-known boundary conditions for the electromagnetic field and (2) one arrives at the following boundary condition for $A_0(x,t)$:

$$\lim_{x\to 0^+}\frac{\partial A_0}{\partial t}(x,t) \;=\; 0 \qquad (t\in\mathbb{R}). \tag{4}$$

Reference [31] shows that (3) is obtained by starting from Maxwell's equations in inertial reference frames in which the movable mirror is instantaneously at rest and then using instantaneous Lorentz transformations to change back to the reference frame where the perfect mirror is fixed at $x=0$. Moreover, terms of order $\dot{q}(t)/c$, $\ddot{q}(t)/(c\omega_0)$, and higher powers of them are neglected. Here c is the speed of light in vacuum and ω_0 is the characteristic frequency of the electromagnetic field. It follows that (3) is an accurate approximation to the exact equation governing the dynamics of the field if both the velocity and acceleration of the movable mirror are small or, more precisely, if

$$\left|\frac{\dot{q}(t)}{c}\right|,\ \left|\frac{\ddot{q}(t)}{c\omega_0}\right| \;\ll\; 1 \qquad (t\in\mathbb{R})\,. \tag{5}$$

Note that these conditions are reasonable, since according to (3) one has that $A_0(x,t)$ evolves as if the mirror were instantaneously fixed at $x=q(t)$.

Reference [31] also shows that (3) and (4) are valid for general dielectric functions $\epsilon[x-q(t)]$. In all that follows we assume that the movable mirror is very thin, that is,

$$\delta_{\rm thick} \;\ll\; \lambda \;\equiv\; \frac{2\pi c}{\omega_0}\,. \tag{6}$$

Therefore, one can approximate $\epsilon\,[x-q(t)]$ by a delta function:

$$\epsilon\,[x-q(t)] \;=\; 1+4\pi\chi_0\delta\,[x-q(t)]\,. \tag{7}$$

Here χ_0 is a quantity with units of length. We note that movable mirrors satisfying (6) have already been used experimentally [32].

From (3), (7), and the law of conservation of linear momentum [33] for the complete system (electromagnetic field + dielectric mirror at $x=q(t)$ + fixed perfect mirror at $x=0$) one obtains that the equation governing the dynamics of the movable mirror is given by [31]

$$M_0\ddot{q}(t) \;=\; \lim_{\eta\to 0^+} -\frac{1}{8\pi}\left\{\left[\frac{\partial A_0}{\partial x}(q(t)+\eta,t)\right]^2 - \left[\frac{\partial A_0}{\partial x}(q(t)-\eta,t)\right]^2\right\} \qquad (t\in\mathbb{R}). \tag{8}$$

Here M_0 is the mass per unit area of the movable mirror. Note that, in accordance with (3), the force on the right-hand side of (8) corresponds to the radiation pressure exerted by the field on the mirror as if the mirror were instantaneously fixed. Also, the right-hand side of (8) is the difference of the radiation pressure exerted by the field outside de cavity minus the radiation pressure exerted by the field inside the cavity.

Since we have neglected terms of order $\dot{q}(t)/c$, $\ddot{q}(t)/(c\omega_0)$, and higher powers of them, we are only considering the leading term in the force affecting the movable mirror. We now comment on what processes are not taken into account because of this approximation. The first corrections to the force on the right-hand side of (8) are a friction force proportional to $\dot{q}(t)/c$ [30, 31] and an acceleration-dependent force proportional to $\ddot{q}(t)/(c\omega_0)$ [31]. The former leads to dissipation in the motion of the movable mirror, while the latter modifies the mass of the mirror. In addition, field amplitudes corresponding to different wave-numbers are mixed if terms proportional to $\dot{q}(t)/c$ are not neglected [30, 31]. In our case the aforementioned processes are small because we demand that conditions (5) are always satisfied.

We now introduce the exact modes of the system for fixed $q(t)$.

2.1. Modes for fixed q(t)

In this subsection we assume that the movable mirror is fixed at $q > 0$, that is, $q(t) = q > 0$ for all $t \in \mathbb{R}$. To calculate the modes associated with (3), (4), and (7) we consider the function

$$A_{0,k}(x,t) = \mathrm{Re}\left[V_k(x)e^{-i\omega t}\right] , \tag{9}$$

with $\omega = ck$, $k > 0$, and Re the real part of a complex number.

Substituting (9) in (4) one immediately obtains the following boundary condition:

$$V_k(0+) = 0 , \tag{10}$$

with

$$f(y\pm) = \lim_{x \to y^{\pm}} f(x) , \tag{11}$$

for any function $f(x)$ of a real variable x.

If one substitutes (9) in (3) one obtains the following differential equation:

$$\frac{d^2}{dx^2}V_k(x) + k^2\epsilon(x-q)V_k(x) = 0 \qquad (x > 0). \tag{12}$$

It follows from (7) (recall that we are considering $q(t) = q$) that (12) is equivalent to the following problem:

$$\begin{aligned} \frac{d^2}{dx^2}V_k(x) + k^2V_k(x) &= 0 \qquad (x > 0,\ x \neq q) , \\ V_k(q+) &= V_k(q-) , \\ \frac{dV_k}{dx}(q+) - \frac{dV_k}{dx}(q-) &= -4\pi\chi_0 k^2 V_k(q) . \end{aligned} \tag{13}$$

Reference [24] (see also [34]) solved (10) and (13) for a coordinate system where the dielectric is fixed at $x = 0$, while the perfect mirror is fixed at $x = q > 0$. One adapts the results to our coordinate system by making the translation and reflection in which x is replaced by $(q - x)$. One then concludes that they are given by

$$V_k(x,q) = \begin{cases} L_k(q)\sin(kx) & \text{if } 0 \le x \le q , \\ \sqrt{\frac{2}{\pi}}\sin\left[k\left(x-q\right) + \delta_k\left(q\right)\right] & \text{if } x > q . \end{cases} \tag{14}$$

for $k > 0$. Here one has

$$L_k\left(q\right) = \sqrt{\frac{2}{\pi}}\left[\, 1 + (4\pi\chi_0 k)^2\sin^2\left(kq\right) - (4\pi\chi_0 k)\sin\left(2kq\right)\,\right]^{-1/2} , \tag{15}$$

and

$$\begin{aligned}\sin[\delta_k(q)] &= \sqrt{\frac{\pi}{2}} L_k(q)\sin(kq) \ , \\ \cos[\delta_k(q)] &= \sqrt{\frac{\pi}{2}} L_k(q)[\,\cos(kq) - (4\pi\chi_0 k)\sin(kq)\,] \ . \end{aligned} \tag{16}$$

Observe that $V_k(x,q)$ are real valued functions.

We want to use the modes $V_k(x,q)$ to represent the plane wave of a laser approaching the movable mirror from the right. In order to do this we must eliminate the mirror position dependence in $V_k(x,q)$ in the second line of (14). To accomplish this we introduce a phase shift as follows:

$$\tilde{V}_k(x,q) = e^{i[\delta_k(q)-kq]}V_k(x,q) = \begin{cases} \frac{i}{2}L_k(q)e^{i[\delta_k(q)-kq]}(e^{-ikx}-e^{ikx}) & \text{if } 0 \le x \le q \ , \\ \frac{i}{2}\sqrt{\frac{2}{\pi}}\left\{e^{-ikx}-e^{ikx}e^{i2[\delta_k(q)-kq]}\right\} & \text{if } x > q \ . \end{cases} \tag{17}$$

Notice that (17) expresses the modes in terms of incoming and outgoing waves. A plane wave e^{-ikx} (which could be associated with a laser) with amplitude 1 comes from $+\infty$, travels to the left, and is incident on the mirror at $x = q > 0$. There is a reflected plane wave e^{ikx} travelling to the right and with amplitude given by the scattering matrix $S(k) = e^{i2[\delta_k(q)-kq]}$ with $2[\delta_k(q) - kq]$ the phase shift. Also, there is a standing wave $(e^{-ikx} - e^{ikx})$ in the region $0 \le x \le q$ produced by the plane wave coming from $+\infty$ and the perfect mirror at $x = 0$. Moreover, $L_k(q)$ is a normalization factor. The factor i has been left in (17) for two reasons. First, $\tilde{V}_k(x,q)$ reduces to $(2/\pi)^{1/2}\sin(kx)$ when $\chi_0 \to 0$ (that is, there is no movable mirror) and second, (17) is $\sqrt{2/\pi}$ times the standard physical solution in scattering in the half-line $(0,+\infty)$ with a Dirichlet boundary condition at $x = 0$. The factor $\sqrt{2/\pi}$ is necessary for the orthogonality relation below.

The modes (14) form an orthogonal set of generalized eigenfunctions [24, 34]. Adapting those relations for the $\tilde{V}_k(x,q)$ one obtains

$$\delta(k-k') = \int_0^{+\infty} dx\ \epsilon(x-q)\tilde{V}_k(x,q)^*\tilde{V}_{k'}(x,q) \ , \qquad f(x) = \int_0^{+\infty} dk\ f_k(q)\tilde{V}_k(x,q) \ , \tag{18}$$

with $f(x) \in \mathcal{L}^2\,[0,+\infty)$, $k, k' > 0$, and the kth-mode $f_k(q)$ given by

$$f_k(q) = \int_0^{+\infty} dx\ \epsilon(x-q)\tilde{V}_k(x,q)^* f(x) \ . \tag{19}$$

The transmissivity T of the dielectric is also calculated in [24]. It is given by

$$T = \left[1 + \left(\frac{4\pi\chi_0 k}{2}\right)^2\right]^{-1} \ . \tag{20}$$

for each fixed $k > 0$. Therefore, the transparency of the dielectric is small if $T \ll 1$ or, equivalently, $4\pi\chi_0 k \gg 1$.

We now list some properties of the function $L_k(q)$ for $k > 0$ fixed. For their proof we refer the reader to [35]. In all that follows $\mathbb{Z}^+$ denotes the set of non-negative integers.

(i) $L_k(q)$ is maximized for a discrete set of values q_{2n} of q with $n \in \mathbb{Z}^+$. If $4\pi\chi_0 k \gtrsim 5$, then one has to good approximation

$$kq_{2n} \simeq n\pi + \frac{1}{4\pi\chi_0 k} \ , \qquad L_k(q_{2n}) \simeq \sqrt{\frac{2}{\pi}}(4\pi\chi_0 k) \ . \tag{21}$$

(ii) $L_k(q)$ is minimized for a discrete set of values q_{2n+1} of q with $n \in \mathbb{Z}^+$. If $4\pi\chi_0 k \gtrsim 5$, then one has to good approximation

$$kq_{2n+1} \simeq \left(n+\frac{1}{2}\right)\pi + \frac{1}{4\pi\chi_0 k}\,, \qquad L_k(q_{2n+1}) \simeq \sqrt{\frac{2}{\pi}}\left(\frac{1}{4\pi\chi_0 k}\right). \tag{22}$$

(iii) q_{2n} tends to the values $n\pi/k$ corresponding to the case where the movable mirror is perfect [36] as the transparency tends to zero:

$$kq_{2n} \to n\pi \qquad \text{as} \qquad 4\pi\chi_0 k \to +\infty\,. \tag{23}$$

(iv) $L_k(q)^2$ can be approximated by a Lorentzian if the transparency of the movable mirror is small and one restricts q to an interval around q_{2n} whose endpoints are not near the minimizers $q_{2n\pm1}$, that is,

$$L_k(q)^2 \simeq \left(\frac{2}{\pi\xi^2}\right)\frac{1}{k^2(q-q_{2n})^2+\frac{1}{\xi^4}} \tag{24}$$

if $k|q-q_{2n}| \ll 3/2$ and $2/\xi \equiv 2/(4\pi\chi_0 k) \ll 1$.

(v) For fixed $k>0$, the frequency $\omega = ck$ coincides with one of the cavity resonance frequencies if and only if $q \simeq q_{2n}$ for some $n \in \mathbb{Z}^+$. If $k|q-q_{2n}| \ll 3/2$ and $2/\xi \equiv 2/(4\pi\chi_0 k) \ll 1$, then it follows form (24) that $\omega = ck$ coincides with one of the cavity resonance frequencies if and only if $k|q-q_{2n}| \leq 1/\xi^2$. Therefore, the half-width-at-half-maximum (HWHM) of the resonant position kq_{2n} is ξ^{-2} if $k|q-q_{2n}| \ll 3/2$ and $2/\xi \equiv 2/(4\pi\chi_0 k) \ll 1$.

Figure 1a illustrates $L_k(q)$ as a function of q. The Lorentzian approximation to $L_k(q)^2$ given in (24) is illustrated in figure 1b for the interval $[q_{2n} - 1/(k\xi),\ q_{2n} + 1/(k\xi)]$ with $n=4$ and $\xi = 4\pi\chi_0 k$. Notice that the agreement between both is good.

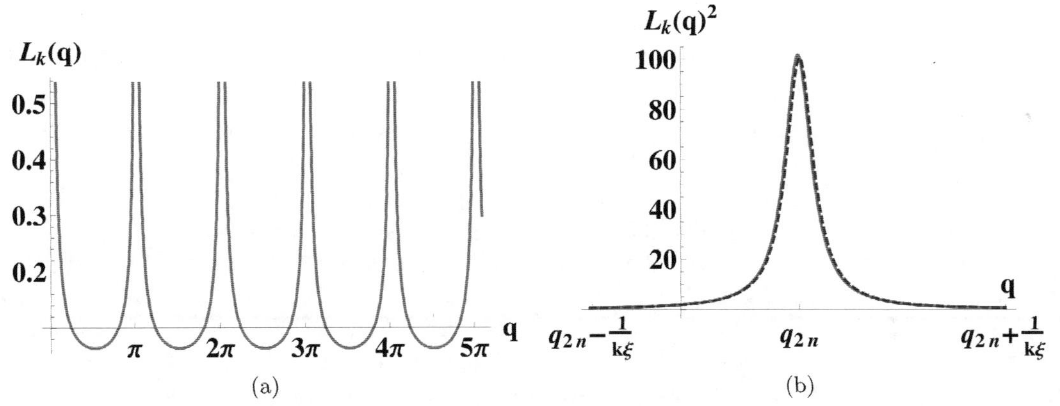

Figure 1: (Color online) Figure 1a illustrates $L_k(q)$ as a function of q, while figure 1b shows $L_k(q)^2$ (red-solid line) and the Lorentzian approximation (blue-dashed line) in (24) as functions of q. Both figures have $\chi_0 = 1$ m and $k = 1$ m^{-1}, so that $\xi = 4\pi\chi_0 k = 4\pi$. Also, figure 1b has $n=4$ with q_{2n} given in (21). For a better illustration only part of the y-axis is shown in 1a.

2.2. Expansion in terms of the instantaneous modes
We now use the instantaneous modes of the system, that is, we consider the exact modes introduced in the previous subsection with $q(t)$ replacing q. One can then expand $A_0(x,t)$ in terms of these using (18) and (19) as follows:

$$A_0(x,t) = \int_0^{+\infty} dk\, Q_k\left[t, q(t)\right] \tilde{V}_k\left[x, q(t)\right] , \quad (25)$$

with

$$Q_k\left[t, q(t)\right] = \int_0^{+\infty} dx\, \epsilon\left[x - q(t)\right] \tilde{V}_k\left[x, q(t)\right]^* A_0(x,t) . \quad (26)$$

To obtain the equations governing the evolution of the modes $Q_k\left[t, q(t)\right]$ first substitute (25) into (3). Then neglect terms proportional to $\dot{q}(t)/c$, $\left[\dot{q}(t)/c\right]^2$, and $\ddot{q}(t)/(c\omega_0)$ (ω_0 the characteristic frequency of the field) and use the orthonormalization relation in (18). One obtains

$$0 = \frac{d^2}{dt^2} Q_k\left[t, q(t)\right] + \omega_k^2 Q_k\left[t, q(t)\right] , \quad (27)$$

with $\omega_k = ck$, $k > 0$, and $t \in \mathbb{R}$. These are harmonic oscillator equations and their solution is given by

$$Q_k\left[t, q(t)\right] = g(k) e^{-i\omega_k t} + g(k)^* e^{i\omega_k t} e^{-i2\{\delta_k[q(t)] - kq(t)\}} , \quad (28)$$

for $k > 0$ and $t \in \mathbb{R}$. Here we have used that $A_0(x,t)$ must be a real quantity.

Recall that we have neglected terms proportional to $\dot{q}(t)/c$, $\left[\dot{q}(t)/c\right]^2$, and $\ddot{q}(t)/(c\omega_0)$ to obtain (27). We have done this because terms of these orders were neglected to obtain (3). Furthermore, only by dropping these does one arrive to (27) and recover the physical situation envisioned in (3): the field evolves as if the movable mirror were fixed.

3. A SINGLE-MODE FIELD

Consider a monochromatic laser on the far right. It is turned on and the plane wave associated with it propagates to the left. This plane wave is reflected partially by the movable mirror at $q(t)$ and it is completely reflected by the perfect mirror located at $x = 0$. After a transient, a standing wave defined by one of the modes in (17) is approximately formed. Therefore, one approximately has a single-mode field that is exerting radiation pressure on the movable mirror and at the same time it is driving the electromagnetic field inside the cavity. We describe the dynamics of the system after the aforementioned transient.

Notice that the field does not decay because the laser is always turned on. Moreover, the restriction to a single-mode is possible because we have neglected terms of order $\dot{q}(t)/c$. Recall from the discussion at the end of Sec. II that field amplitudes corresponding to different wave-numbers are mixed if terms of order $\dot{q}(t)/c$ are not neglected. In our case this mixing is small because we demand that $|\dot{q}(t)/c| \ll 1$, see (5).

In the rest of the article we consider the physical situation described in the preceding paragraphs and we take the field as being composed of a single-mode, that is,

$$g(k) = g_0 e^{i\phi_0} \delta\left(k - k_N^0\right) , \quad (29)$$

with $k_N^0, g_0 > 0$ and $\phi_0 \in \mathbb{R}$. In the following we take the initial phase equal to zero, that is, $\phi_0 = 0$. This corresponds to redefining the instant $t = 0$ by making a time translation.

Substituting (28) in (25) with $g(k)$ given in (29), one obtains

$$A_0(x,t) = 2g_0\cos\left\{\omega_0 t - \delta_{k_N^0}[q(t)] + k_N^0 q(t)\right\} V_{k_N^0}[x, q(t)] \ , \tag{30}$$

with $\omega_0 \equiv ck_N^0$.

To proceed we introduce a characteristic time and length of the system and then use these to express (8) in terms of non-dimensional quantities.

Define

$$\Delta = \sqrt{\frac{g_0^2\omega_0^3}{\pi^2 M_0 c^3}} \ . \tag{31}$$

Then Δ is a quantity with units $1/s$ and can be thought of as 1 over the characteristic time-scale of the movable mirror. Therefore, $\tau = \Delta t$ is the non-dimensional time. Also, we measure lengths in units of 1 over the wave-number k_N^0 of the field, that is, we take $1/k_N^0$ to be the characteristic length of the system. We note that the form of Δ is dictated by the differential equation in (8), by (30), and by choosing the characteristic length to be $1/k_N^0$.

Now define the non-dimensional quantities

$$\xi = 4\pi\chi_0 k_N^0 \ , \qquad \Omega = 2\frac{\omega_0}{\Delta} \ , \qquad x(\tau) = k_N^0 q\left(\frac{\tau}{\Delta}\right) \ . \tag{32}$$

Notice that the transmissivity T of the movable mirror given in (20) can be rewritten as

$$T = \left[1 + \left(\frac{\xi}{2}\right)^2\right]^{-1} \ . \tag{33}$$

Here we used the definition of ξ in (32). It follows that the transparency will be small if ξ is large.

One can rewrite equation (8) for the movable mirror in terms of non-dimensional quantities using the expression for $A_0(x,t)$ in (30) and the quantities defined in (32). One obtains that

$$x''(\tau) = \left[1 + \cos\left\{\Omega\tau + 2x(\tau) - 2\delta_{k_N^0}\left[\frac{x(\tau)}{k_N^0}\right]\right\}\right] f_{\mathrm{RWA}}[x(\tau)] \ . \tag{34}$$

Here we have introduced the non-dimensional force

$$f_{\mathrm{RWA}}(x) = -\frac{1}{2}\left[1 - \frac{1}{1 + \xi^2\sin^2(x) - \xi\sin(2x)}\right] \ . \tag{35}$$

Also, derivatives with respect to τ are denoted by a prime, while derivatives with respect to t are denoted by a dot.

If one uses the Taylor series of the function $1/(1+y)$ with $y = \xi^2\sin^2(x) - \xi\sin(2x)$ and one neglects terms of order ξ^3 and higher powers, then (35) is the same force as in [30] in the case where the transparency of the movable mirror is large and terms of order $\dot{q}(t)/c$ are neglected and the rotating-wave-approximation (RWA) is performed. We note that one must also change the coordinate system and Gaussian units to SI units for this comparison.

We now list a set properties of $f_{\mathrm{RWA}}(x)$ that will be used throughout the article. We refer the reader to [35] for their proof.

(i) $f_{\mathrm{RWA}}(x)$ is well defined for all $x \in \mathbb{R}$.
This follows from the following identity valid for all $x \in \mathbb{R}$:

$$1 + \xi^2\sin^2(x) - \xi\sin(2x) = \sin^2(x) + [\xi\sin(x) - \cos(x)]^2 > 0 \ . \tag{36}$$

Journal of Physics: Conference Series **512** (2014) 012005 doi:10.1088/1742-6596/512/1/012005

(ii) The maximizers of $f_{\rm RWA}(x)$ are denoted by x_{2n} with $n \in \mathbb{Z}^+$. If $\xi \gtrsim 5$, then one has to good approximation

$$x_{2n} \simeq n\pi + \frac{1}{\xi}\,, \qquad f_{\rm RWA}(x_{2n}) \simeq \frac{\xi^2}{2} \qquad (n \in \mathbb{Z}^+). \tag{37}$$

(iii) The minimizers of $f_{\rm RWA}(x)$ are denoted by x_{2n+1} with $n \in \mathbb{Z}^+$. If $\xi \gtrsim 5$, then one has to good approximation

$$x_{2n+1} \simeq \left(n + \frac{1}{2}\right)\pi + \frac{1}{\xi}\,, \qquad f_{\rm RWA}(x_{2n+1}) \simeq -\frac{1}{2} \qquad (n \in \mathbb{Z}^+). \tag{38}$$

(iv) The zeros of $f_{\rm RWA}(x)$ are denoted by x_n^* and x_n^{**} with $n \in \mathbb{Z}^+$. One has

$$x_n^* = n\pi\,, \qquad x_n^{**} \simeq n\pi + \frac{2}{\xi} \qquad (n \in \mathbb{Z}^+). \tag{39}$$

(v) One can approximate $f_{\rm RWA}(x)$ by a displaced Lorentzian

$$f_{\rm RWA}(x) \simeq -\frac{1}{2} + \frac{1}{2\xi^2}\left(\frac{1}{u^2 + \xi^{-4}}\right), \tag{40}$$

if $(2m-1)\pi/2 \le x \le (2m+1)\pi/2$, $x \ge 0$, $u = x - (m\pi + 1/\xi)$, $1 \gg 3^{-1}(u + \xi^{-1})^2$, and $\xi \gg 1$ with $m \in \mathbb{Z}^+$.

Before proceeding we delve a bit more on the non-dimensional quantities introduced in (32) and the equation (34) governing the dynamics of the movable mirror.

Recall that (3) and (8) are the equations governing the dynamics of the field and of the movable mirror and that these are accurate approximations to the exact equations governing the dynamics of the system when (5) are satisfied. These conditions say that the movable mirror has to move very slowly and that it has to be subject to very small accelerations so that the field evolves as if the movable mirror were instantaneously fixed. Therefore, the model of Sec. II assumes the existence of two very different time-scales: one very fast time-scale for the evolution of the field and one very slow time-scale for the evolution of the movable mirror. The former is given by $2\pi/\omega_0$ while the latter is given by an unknown quantity $2P_{\rm mirror} \sim 1/\Delta$ that satisfies $2\pi/\omega_0 \ll 2P_{\rm mirror}$. In non-dimensional units the previous condition takes the form

$$2\pi/\Omega = \pi\Delta/\omega_0 \ll P \equiv \Delta P_{\rm mirror}\,. \tag{41}$$

Notice that P is the non-dimensional time scale in which $x(\tau)$ changes appreciably. From (41) it follows that the cosine term in (34) oscillates very rapidly and averages to zero. Hence, one can perform the rotating-wave-approximation (RWA) to obtain the approximate equation

$$x''(\tau) = f_{\rm RWA}[x(\tau)]\,. \tag{42}$$

We shall say that the system is in the *RWA regime* or *limit of low field intensity* whenever (41) is satisfied and the RWA is performed. The name *RWA regime* comes from the fact that condition (41) was used to perform the RWA. On the other hand, the name *limit of low field intensity* comes from the fact that $2\pi/\Omega$ is proportional to g_0, see (32).

From the discussion above it follows that the RWA regime is a natural setting for the model presented in Sec. II, since the movable mirror evolves on a time-scale much larger than that of the field. These two very different and separate time-scales are easily illustrated if one takes frequencies in the optical regime [35]: $2\pi/\omega_0 \sim 10^{-14}$ and $P_{\rm mirror} \sim 1/\Delta \gtrsim 10^{-6}$.

It is important to note that (41) may be satisfied, but the RWA may not be applicable. For the RWA to be applicable one needs ω_0 several orders of magnitude larger than $1/P_{\rm mirror} \sim \Delta$ (say, 4 orders of magnitude larger) as in the previous example [35].

In all that follows we restrict to the RWA regime. We now deduce the non-dimensional radiation pressure force $f_{\rm RWA}(x)$ from a potential.

3.1. The radiation pressure potential

From (34) it follows that the non-dimensional radiation pressure force $f_{\mathrm{RWA}}(x)$ can be obtained from a non-dimensional radiation pressure potential $V_{\mathrm{RWA}}(x)$, that is,

$$f_{\mathrm{RWA}}(x) \;=\; -\frac{d}{dx}V_{\mathrm{RWA}}(x)\;. \tag{43}$$

An analytic expression for $V_{\mathrm{RWA}}(x)$ can be obtained by integrating $f_{\mathrm{RWA}}(x)$ from $x' = 0$ to $x' = x > 0$. One obtains that

$$V_{\mathrm{RWA}}(x) \;=\; \frac{x}{2} - \frac{1}{2}\tan^{-1}\left[\,(1+\xi^2)\tan(x) - \xi\,\right] - \frac{1}{2}\left[\tan^{-1}(\xi) + m\pi\,\right]\,, \tag{44}$$

if $(2m-1)\pi/2 \le x \le (2m+1)\pi/2$ and $x \ge 0$ with $m \in \mathbb{Z}^+$. Here $\tan^{-1}(\theta) \in (-\pi/2, \pi/2)$ for $\theta \in \mathbb{R}$.

We now list properties of $V_{\mathrm{RWA}}(x)$ that will be used throughout the article. For their proof we refer the reader to [35].

(i) $V_{\mathrm{RWA}}(x)$ is a periodic function of period π.
This follows from the piecewise definition of $V_{\mathrm{RWA}}(x)$ on intervals of length π and the fact that $\tan(x)$ is periodic of period π.

(ii) The maximizers of $V_{\mathrm{RWA}}(x)$ are given by x_n^* in (39). Moreover,

$$V_{\mathrm{RWA}}(x_n^*) \;=\; 0\;, \tag{45}$$

and 0 is the absolute maximum value of $V_{\mathrm{RWA}}(x)$.

(iii) The minimizers of $V_{\mathrm{RWA}}(x)$ are given by x_n^{**} in (39) and $V_{\mathrm{RWA}}(x_n^{**})$ is the absolute minimum value of $V_{\mathrm{RWA}}(x)$. Also, if $\xi \gtrsim 5$, then

$$V_{\mathrm{RWA}}(x_n^{**}) \;\simeq\; V_{\mathrm{RWA}}(2/\xi) \;\simeq\; -\tan^{-1}(\xi)\;. \tag{46}$$

(iv) For all $x \ge 0$ one has

$$-\pi/2 \;<\; V_{\mathrm{RWA}}(x) \;\le\; 0\;. \tag{47}$$

Also, $V_{\mathrm{RWA}}(2/\xi) \to -\pi/2$ as $\xi \to +\infty$.

(v) All of the wells of $V_{\mathrm{RWA}}(x)$ have non-dimensional length π, which corresponds to a length of half the wavelength of the field, that is, $\lambda_0/2 = \pi/k_N^0$.

Figure 2 illustrates $V_{\mathrm{RWA}}(x)$ for $\xi = 2, 5,$ and 50. Notice that the depth of the wells increases and that $V_{\mathrm{RWA}}(x)$ tends to a *saw tooth wave* as $\xi \to +\infty$. Therefore, $V_{\mathrm{RWA}}(x)$ can be approximated for sufficiently large ξ (say, $\xi \gtrsim 50$) as follows:

$$V_{\mathrm{RWA}}(x) \;\simeq\; V_{\mathrm{RWA}}^{\mathrm{approx}}(x) \;\equiv\; \begin{cases} 0 & \text{if } \; n\pi \le x \le n\pi + \frac{1}{\xi}\,, \\ V_{\mathrm{RWA}}\left(\frac{2}{\xi}\right) & \text{if } \; n\pi + \frac{1}{\xi} < x \le n\pi + \frac{2}{\xi}\,, \\ V_{\mathrm{RWA}}\left(\frac{2}{\xi}\right) + m_+\left[\,x - \left(n\pi + \frac{2}{\xi}\right)\,\right] & \text{if } \; n\pi + \frac{2}{\xi} \le x \le (n+1)\pi\,. \end{cases} \tag{48}$$

with $n\pi \le x \le (n+1)\pi$, $n \in \mathbb{Z}^+$, and

$$m_+ \;=\; -\frac{V_{\mathrm{RWA}}(2/\xi)}{\pi - 2/\xi}\;. \tag{49}$$

Recall that $x = (n\pi + 1/\xi)$ is (approximately) the maximizer x_{2n} of $f_{\mathrm{RWA}}(x)$, see (37).

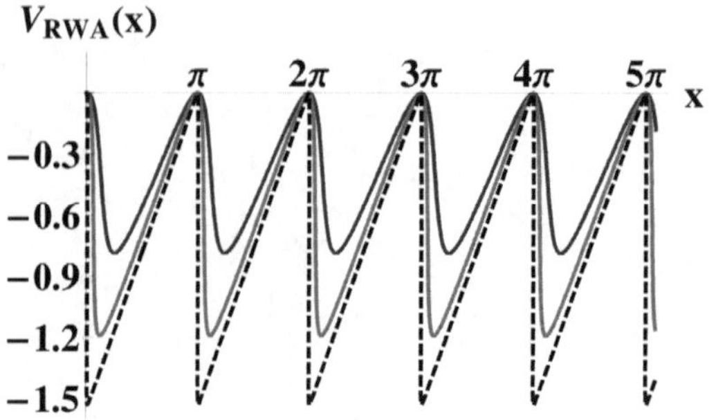

Figure 2: (Color online) The figure shows $V_{\text{RWA}}(x)$ for $\xi = 2$ (blue-solid line), 5 (red-solid line), and 50 (black-dashed line).

Figure 3a shows a close-up of one of the wells of $V_{\text{RWA}}(x)$ and $V_{\text{RWA}}^{\text{approx}}(x)$ for $\xi = 5$, while figure 3b shows the relative error $|V_{\text{RWA}}(x) - V_{\text{RWA}}^{\text{approx}}(x)| \,/\, |V_{\text{RWA}}(x)|$ for $\xi = 50$. Notice that the approximation is quite good except at thin layers at the maximizers $x = 4\pi$ and $x = 5\pi$ of $V_{\text{RWA}}(x)$ and at the maximizer $x = (4\pi + 1/\xi)$ of $f_{\text{RWA}}(x)$. These layers appear because the approximation $V_{\text{RWA}}^{\text{approx}}(x)$ does not take into account the curvature of $V_{\text{RWA}}(x)$.

Using (43) and the approximation for $V_{\text{RWA}}(x)$ given in (48) one obtains for $\xi \gg 1$ that

$$f_{\text{RWA}}(x) \simeq \begin{cases} 0 & \text{if } n\pi < x < \left(n\pi + \frac{2}{\xi}\right) \text{ and } x \neq \left(n\pi + \frac{1}{\xi}\right) , \\ -m_+ & \text{if } \left(n\pi + \frac{2}{\xi}\right) < x < (n+1)\pi . \end{cases} \tag{50}$$

It follows that, for sufficiently large ξ, $f_{\text{RWA}}(x)$ is approximately a non-dimensional piecewise constant force.

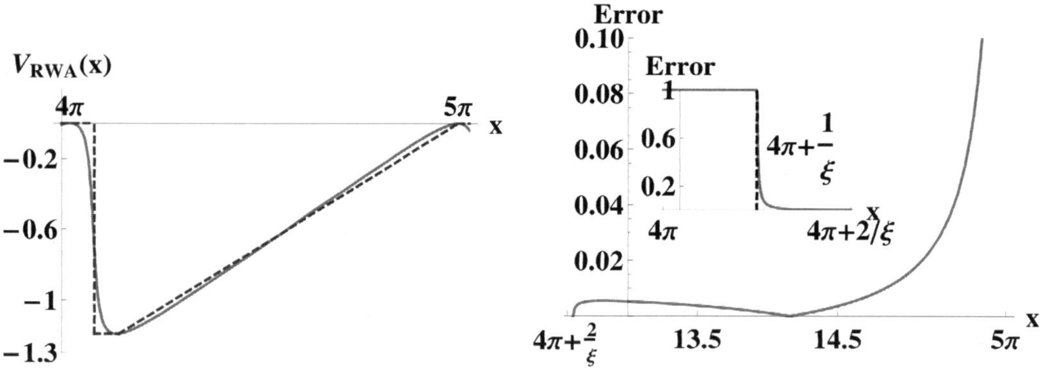

(a) (b)

Figure 3: (Color online) Figure 3a shows a close up of one of the wells of $V_{\text{RWA}}(x)$ (red-solid line) and the approximate $V_{\text{RWA}}^{\text{approx}}(x)$ (blue-dashed line) for $\xi = 5$ and the region $[4\pi, 5\pi]$. Figure 3b illustrates the relative error $|V_{\text{RWA}}(x) - V_{\text{RWA}}^{\text{approx}}(x)|/|V_{\text{RWA}}(x)|$ for $\xi = 50$ (red-solid line). The outer figure shows the region $[4\pi + 2/\xi, 5\pi]$, while the inside figure shows a close-up of the interval $[4\pi, 4\pi + 2/\xi]$ and the vertical line $x = 4\pi + 1/\xi$ (black-dashed line) indicating the (approximate) position of the maximizer of $f_{\text{RWA}}(x)$.

Journal of Physics: Conference Series **512** (2014) 012005 doi:10.1088/1742-6596/512/1/012005

4. DYNAMICS IN THE RWA

We now analyse and solve to good approximation (42). First notice from (42) that the non-dimensional energy $E\left[x(\tau), x'(\tau)\right]$ of the movable mirror is conserved. It is given by

$$E\left[x(\tau), x'(\tau)\right] \;=\; \frac{1}{2}\left[x'(\tau)\right]^2 + V_{\mathrm{RWA}}\left[x(\tau)\right] \, , \tag{51}$$

Figure 4 shows a contour plot of the energy as a function of x and x' for $\xi = 2$.

Since the mirror moves under the potential $V_{\mathrm{RWA}}(x)$ (see (45)-(47) and figure 2) and energy is conserved, one has the following facts [35]:

(i) The motion of the mirror is bounded to one of the potential wells if and only if the mirror has non-positive energy, that is, if and only if $E(x, x') \leq 0$.

(ii) If the motion of the mirror is bounded to one of the potential wells, then the movable mirror has a closed orbit in phase space, see figure 4. In each of these periodic trajectories the mirror takes on its maximum speed v_0 when it is located at one of the minimizers x_n^{**} of the potential $V_{\mathrm{RWA}}(x)$.

(iii) The maximizers $x_n^* = n\pi$ of $V_{\mathrm{RWA}}(x)$ are points of unstable equilibrium.

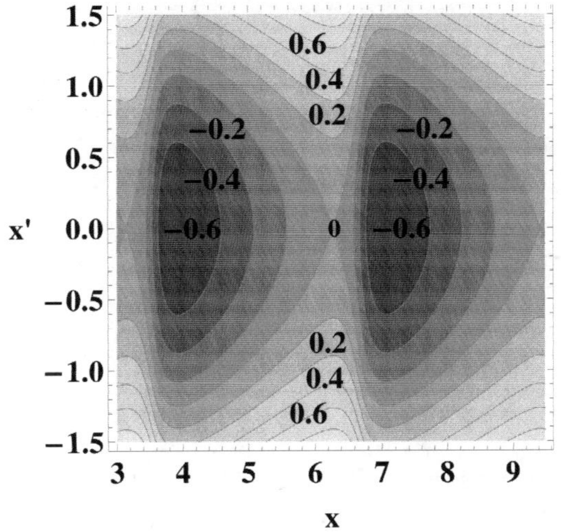

Figure 4: (Color online) The figure shows a contour plot of the non-dimensional energy $E(x, x')$ of the movable mirror as a function of x and x' for $\xi = 2$. The contours $E(x, x') = \pm 0.6$, ± 0.4, ± 0.2, 0 are shown.

The use of the exact instantaneous modes (14) of the system allows us to give a direct physical interpretation of the periodic bounded motion of the movable mirror in terms of an approaching and withdrawing from a position where the frequency of the field coincides with one of the cavity resonance frequencies. Assume that $\xi \gtrsim 5$ (that is, the transparency of the movable mirror is small), $E[x(0), x'(0)] < 0$ (that is, the movable mirror is bound to one of the potential wells), and $q(0) = q_{2n}$ for some $n \in \mathbb{Z}^+$ (that is, the movable mirror is at a position where the frequency of the field ω_0 coincides with one of the cavity resonance frequencies). Then the field inside the cavity is much larger than the field outside of it, see figures 5a and 5b. As a consequence, the radiation pressure exerted by the field inside the cavity is much larger than that exerted by the field outside of it and the mirror is pushed to the right (recall that $f_{\mathrm{RWA}}(x_{2n}) \simeq \xi^2/2$, see (37)). As the mirror withdraws from the resonant position q_{2n}, the frequency of the field becomes very different from any of the cavity resonance frequencies. Therefore, the field inside the cavity decreases (see figures 5c and 5d) and the radiation pressure exerted by the field outside of the cavity eventually surpasses that of the field inside . As a consequence the mirror slows down, stops, and starts moving to the left (recall that $f_{\rm RWA}(x_{2n+1}) \simeq -1/2$, see (38)). As the mirror now approaches the *resonant position* q_{2n}, the frequency of the field becomes similar to one of the cavity resonance frequencies and the field inside the cavity increases. Hence, the radiation pressure exerted by the field inside the cavity eventually surpasses that of the field outside (see figures 5a and 5b and recall that $f_{\rm RWA}(x_{2n}) \simeq \xi^2/2$). The mirror then slows down, stops, and starts moving to the right and the process starts over again.

Notice that the process described in the preceding paragraph can also be explained with the radiation pressure potential $V_{\rm RWA}(x)$, see figure (2) and recall from (37) and (39) that the resonant positions $x_{2n} = k_N^0 q_{2n}$ are located between consecutive maximizers x_n^* and minimizers x_n^{**} of $V_{\rm RWA}(x)$. The potential also allows one to understand the unbounded motion of the mirror. Suppose that the movable mirror has positive energy so that its motion is not bounded to one of the potential wells. As the mirror starts moving (if it moves to the right (left), then see the contours with positive energy and $x' > 0$ $(x' < 0)$ in Fig. 4), it accelerates as it approaches a minimizer and it decelerates as it approaches a maximizer of $V_{\rm RWA}(x)$. The radiation pressure is never large enough so as to decelerate the mirror to a complete stop as in the case of the bounded motion.

Now assume that $E(x, x') < 0$ and that $\xi \gg 1$ (say, $\xi \gtrsim 50$), that is, the mirror is confined to one of the potential wells and has small transparency. Then one can use $V_{\rm RWA}^{\rm approx}(x)$ in (48) to solve approximately (42). Notice that the mirror cannot approach $x = n\pi$ with this approximation. Therefore, $V_{\rm RWA}^{\rm approx}(x)$ cannot be used to describe accurately the motion of the mirror in a small layer at $E(x, x') = 0$.

If the mirror is in the n-th well, then an approximate solution of (42) is given by [35]

$$x_{\rm RWA}(\tau) = \begin{cases} \left(n\pi + \frac{2}{\xi}\right) + v_0(\tau - \tau_{2k}) - \frac{m_+}{2}(\tau - \tau_{2k})^2 & \text{if } \tau_{2k} \le \tau \le \tau_{2k+1}\,, \\ \left(n\pi + \frac{2}{\xi}\right) - v_0(\tau - \tau_{2k+1}) & \text{if } \tau_{2k+1} \le \tau \le \tau'_{2k+1}\,, \\ \left(n\pi + \frac{1}{\xi}\right) + v_0(\tau - \tau'_{2k+1}) & \text{if } \tau'_{2k+1} \le \tau \le \tau_{2(k+1)}\,. \end{cases} \tag{52}$$

Here $k \in \mathbb{Z}$ and τ_0 is such that $x(\tau_0) = n\pi + 2/\xi$ and $v_0 = (dx/d\tau)(\tau_0) > 0$. Notice that v_0 is the maximum (non-dimensional) speed of the mirror. Given $k \in \mathbb{Z}$ one has

$$\tau_{2k+1} - \tau_{2k} = \frac{2v_0}{m_+}\,, \qquad \tau_{2(k+1)} - \tau'_{2k+1} = \tau'_{2k+1} - \tau_{2k+1} = \frac{1}{\xi v_0}\,. \tag{53}$$

It follows from (52) and (53) that the periodic trajectory has period P given by

$$P = \frac{2v_0}{m_+} + \frac{2}{\xi v_0} \simeq 4v_0 + \frac{2}{\xi v_0}\,. \tag{54}$$

Notice that P depends on the maximum speed v_0 of the mirror. Moreover, its absolute minimum value $P_{\rm min}$ occurs for $v_0 = \sqrt{m_+/\xi} \simeq 1/\sqrt{2\xi}$ and is given by [35]

$$P_{\rm min} = \frac{4}{\sqrt{m_+\xi}} \simeq 4\sqrt{\frac{2}{\xi}}\,, \tag{55}$$

since $m_+ \simeq 1/2$ for $\xi \gg 1$, see (47) and (49).

Figure 6 illustrates the numerical solution $x(\tau)$ (red-solid line) of (42) and the approximate solution $x_{\rm RWA}(\tau)$ (blue-dashed line) given in (52) for $\xi = 20$ and the initial conditions

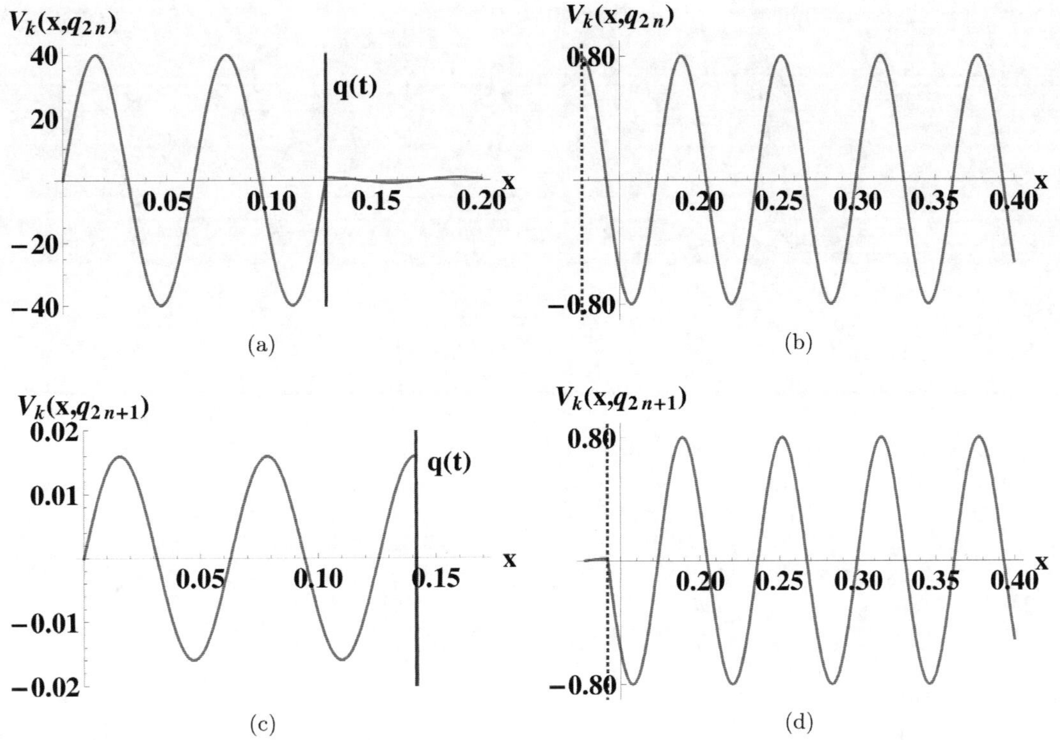

Figure 5: (Color online) Figures 5a and 5b illustrate $V_{k_N^0}(x, q_{2n})$ as a function of x. Here the position q_{2n} of the mirror is such that ω_0 coincides with one of the cavity resonance frequencies. Figures 5c and 5d illustrate $V_{k_N^0}(x, q_{2n+1})$ as a function of x. Here the position q_{2n+1} of the mirror is such that ω_0 is very different from all the cavity resonance frequencies. Figures 5a and 5c show the region inside the cavity, while figures 5b and 5d show the region outside of it. In all figures the vertical blue solid or dashed line shows the position of the movable mirror. Also, all figures have $k_N^0 = 100\ \mathrm{m}^{-1}$, $n = 4$, and $\xi = 4\pi\chi_0 k_N^0 = 50$.

$x(0) = 4\pi + 1/\xi$ and $x'(0) = 0$. Therefore, the movable mirror starts from rest at a position where the frequency ω_0 of the field coincides with one of the cavity resonance frequencies. Observe that the agreement between $x(\tau)$ and $x_{\mathrm{RWA}}(\tau)$ is very good and that (54) gives a good approximation for the period, since one can show that $P \simeq 5$.

To end this section we mention that the initial velocity of the movable mirror has to be adjusted in order to apply (52): if $\mathcal{E} \equiv E(4\pi + 1/\xi, 0)$ is the energy of the movable mirror, then according to the approximation in (48) the corresponding initial conditions for $x_{\mathrm{RWA}}(\tau)$ are $x(0) = 4\pi + 1/\xi$ and $x'(0) = x_0'(0)$ with $x_0'(0) = \sqrt{2[\mathcal{E} - V_{\mathrm{RWA}}(2/\xi)]}$.

4.1. Validity of the model

We now establish the regime of validity of the model, that is, we determine sufficient conditions for (5) to be valid and we rewrite condition (41) for the RWA regime. In the following we drop the assumption that $\xi \gg 1$ and we do not restrict the motion to be bounded.

From (51) we know that energy is conserved, that is,

$$E[x(\tau), x'(\tau)] = E_0 \qquad (\tau \in \mathbb{R}) . \tag{56}$$

Using (56) to bound the velocity of the movable mirror, (42) to bound its acceleration, (37) to bound $f_{\mathrm{RWA}}(x)$, (47) to bound $V_{\mathrm{RWA}}(x)$, and (32) to relate $x(\tau)$ and its derivatives with $q(t)$ and its derivatives, it is straightforward to show that

$$2\sqrt{2E_0 + \pi} \ll \Omega \quad \Rightarrow \quad \left|\frac{\dot{q}(t)}{c}\right| \ll 1 \qquad (t \in \mathbb{R}), \tag{57}$$

and

$$4f_{\mathrm{RWA}}(x_{2n}) \ll \Omega^2 \quad \Rightarrow \quad \left|\frac{\ddot{q}(t)}{c\omega_0}\right| \ll 1 \quad (t \in \mathbb{R}). \tag{58}$$

We now illustrate (57) and (58) using the typical values $\xi = 6.4$ (the reflectivity of the movable mirror is 0.91) and $\Omega \geq 10^9$ given in [35]. From (57) it is clear that $|\dot{q}(t)/c| \ll 1$ if the energy E_0 of the movable mirror is not too large (for example, $E_0 \leq 0$ for bounded motion). Using (37) to evaluate $f_{\mathrm{RWA}}(x_{2n})$ it follows that $4f_{\mathrm{RWA}}(x_{2n}) \simeq 81.92 \ll 10^{18} \leq \Omega^2$. Then, from (58) one obtains that $|\ddot{q}(t)/(c\omega_0)| \ll 1$ for all $t \in \mathbb{R}$. Observe that this last condition is valid for both bounded and unbounded motion. Therefore, the model presented in Sec.II is an accurate description of the physical system under study if these typical values are used.

If one considers $E_0 \leq 0$ (bounded motion) and $\xi \gg 1$ (small transparency of the movable mirror), then (using (37)) the results (57) and (58) can be reduced to the following:

$$2\xi^2 \ll \Omega^2 \quad \Rightarrow \quad \left|\frac{\ddot{q}(t)}{c\omega_0}\right|, \left|\frac{\dot{q}(t)}{c}\right| \ll 1 \quad (t \in \mathbb{R}). \tag{59}$$

Using the typical value $\Omega \geq 10^9$ given in [35] it follows that one only requires $\xi \ll 10^8$ for the model in Sec. II to be valid.

Finally, we rewrite condition (41) for the RWA regime and for bounded motion. In this case P in (41) can be taken to be the period of the periodic trajectory of the mirror. If in addition $\xi \gg 1$ (small transparency of the movable mirror), then (55) can be used to obtain a general sufficient condition for the RWA to be valid:

$$\Omega \;\gg\; \frac{\pi}{2}\sqrt{m_+\xi} \;\simeq\; \frac{\pi}{2}\sqrt{\frac{\xi}{2}}\,. \tag{60}$$

Comparing (60) with (59) it follows that the low intensity limit (or RWA regime) is a natural setting for the consideration of the model established in the first section in the case where the transparency of the movable mirror is small.

5. CONCLUSIONS

In this article we investigated the classical dynamics of a thin movable mirror with non-zero transparency as it interacts with a monochromatic laser by means of radiation pressure. The movable mirror is part of a one-dimensional cavity whose other mirror is perfect and fixed. The model used in the article describes the following physical situation: a single-mode laser on the far right is always turned on. The plane wave associated with the laser travels to the left, is partially reflected by the movable mirror at $q(t) > 0$, and it is completely reflected by the perfect, fixed mirror at $x = 0$. After a transient a standing wave is formed and the laser is responsible for maintaining it. We described the dynamics after the aforementioned transient.

Starting from approximate Maxwell-Newton equations describing the dynamics of the system when the movable mirror has small speed and is subject to small accelerations, we considered the exact instantaneous modes of the system and we established an effective equation governing the dynamics of the movable mirror. This allowed us to deduce the radiation pressure force from a periodic potential with period half the wavelength of the field.

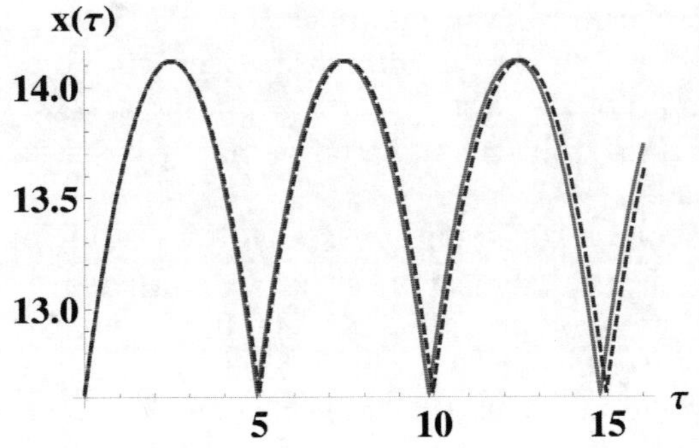

Figure 6: (Color online) The figure illustrates the solution $x(\tau)$ (red-solid line) of (42) computed numerically for $\xi = 20$ and the initial conditions $x(0) = 4\pi + 1/\xi$, $x'(0) = 0$. It also shows the approximate solution $x_{\rm RWA}(\tau)$ (blue-dashed line) in (52) corresponding to the same energy $\mathcal{E} = E(4\pi + 1/\xi, 0)$. The corresponding initial conditions for $x_{\rm RWA}(\tau)$ are $x(0) = 4\pi + 1/\xi$ and $x'(0) = x_0'(0)$ with $x_0'(0) = \sqrt{2[\mathcal{E} - V_{\rm RWA}(2/\xi)]}$.

The conditions under which the approximate equations describing the dynamics of the system were derived allowed us to identify two very different time-scales in the system: one very fast time-scale for the evolution of the field and another slow time-scale for the evolution of the movable mirror. This allowed us to perform the rotating-wave-approximation and, hence, to simplify the equation governing the dynamics of the mirror. Within this approximation the energy of the movable mirror is conserved. Combining this result with the radiation pressure potential we concluded that the motion of the movable mirror can be either bounded or unbounded. In the case of unbounded motion the mirror accelerates (decelerates) as it approaches a minimizer (maximizer) of the potential. In the case of bounded motion, the movable mirror has a periodic trajectory around one of the minimizers of the potential. The use of the exact instantaneous modes and of the potential also allowed us to interpret this periodic motion as follows. As the movable mirror approaches a position where the frequency of the field is similar to one the cavity resonance frequencies, the field inside the cavity builds up and eventually exerts a radiation pressure that surpasses that exerted by the field outside the cavity. As a consequence, the movable mirror slows down, stops, and eventually starts moving to the right. As the movable mirror withdraws form the *resonant position*, the frequency of the field becomes very different from all cavity resonance frequencies and the field inside the cavity is depleted. As a result, the field outside the cavity eventually exerts a radiation pressure that exceeds that of the field inside the cavity. The movable mirror then slows down, stops, and eventually starts moving to the left. Then the mirror approaches the resonant position once more and the process repeats itself over and over giving rise to the periodic motion.

We note that the radiation pressure potential also allowed us to obtain accurate analytic approximations of the solutions of the equations governing the dynamics of the movable mirror in the case were its transparency is small. Moreover, we established that for some typical values both the rotating-wave-approximation and the approximate Maxwell-Newton equations are valid.

We note that our approach differs from that taken in most articles because the movable mirror can deviate far from an equilibrium position and we considered the exact instantaneous modes of the system. In other words, we did not divide the modes of the system into modes of the cavity and modes outside of it as if the cavity mirrors were perfect and then coupled them through a phenomenological interaction. This allowed us to incorporate directly into the modes the transparency of the movable mirror and the positions of the mirror where the frequency of the field coincides with the cavity resonance frequencies. The result of this approach is that we were able to deduce the radiation pressure force from a potential and to give a direct physical interpretation of the dynamics of the system.

Reference [35] also investigates the case where the mirror is subject both to friction and to a harmonic oscillator potential.

It is important to note that we have only considered the leading order of the electromagnetic force acting on the movable mirror, that is, we neglected terms of order $\dot{q}(t)/c$, $\ddot{q}(t)/(c\omega_0)$, and higher powers of them. Here c is the speed of light in vacuum and ω_0 is the characteristic frequency of the field. These terms are responsible, for example, for a friction force arising from the interaction of the movable mirror with the electromagnetic field and for a mixing of field amplitudes corresponding to different wave-numbers [30, 31]. In our case these processes are small because we always demanded that $|\dot{q}(t)/c| \ll 1$ and $|\ddot{q}(t)/(c\omega_0)| \ll 1$. In future work we shall consider the terms we neglected.

Acknowledgments
We thank Pablo Barberis-Blostein, Marc Bienert, and Christian Gerard for fruitful discussions. L. O. Castaños wishes to thank the Universidad Nacional Autónoma de México for support. R. Weder is a fellow of Sistema Nacional de Investigadores. Research partially supported by CONACYT under Project CB-2008-01-99100.

References

[1] Wu M C, Solgaard O and Ford J E 2006 *J. Lightwave Technol.* **24** 4433
[2] Marquardt F and Girvin S M 2009 *Physics* **2** 40
[3] Hunger D, Camerer S, Korppi M, Jöckel A, Hänsch T W and Treutlein P 2011 *C. R. Physique* **12** 871
[4] Qian J, Clerk A A, Hammerer K and Marquardt F 2012 *Phys. Rev. Lett.* **109** 253601
[5] Carmon T, Rokhsari H, Yang L, Kippenberg T J, and Vahala K J 2005 *Phys. Rev. Lett.* **94** 223902
[6] Kippenberg T J, Rokhsari H, Carmon T, Scherer A and Vahala K J 2005 *Phys. Rev. Lett.* **95**, 033901
[7] Marquardt F, Harris J G E and Girvin S M 2006 *Phys. Rev. Lett.* **96** 103901
[8] Metzger C, Ludwig M, Neuenhahn C, Ortlieb A, Favero I, Karrai K and Marquardt F 2008 *Phys. Rev. Lett.* **101** 133903
[9] Marino F and Marin F 2011 *Phys. Rev. E* **83** 015202(R)
[10] Marino F and Marin F 2013 *Phys. Rev. E* **87** 052906
[11] Zaitsev S, Gottlieb O and Buks E 2012 *Nonlinear Dyn.* **69** 1589
[12] Zaitsev S, Pandey A K, Shtempluck O and Buks E 2011 *Phys. Rev. E* **84** 046605
[13] Blocher D, Rand R H and Zehnder A T 2013 *International Journal of Non-Linear Mechanics* **52** 119
[14] Teufel J D, Donner T, Li D, Harlow J W, Allman M S, Cicak K, Sirois A J, Whittaker J D, Lehnert K W and Simmonds R W 2011 *Nature (London)* **475** 359
[15] Chan J, Alegre T P M, Safavi-Naeini A H, Hill J T, Krause A, Groblacher S, Aspelmeyer M and Painter O 2011 *Nature (London)* **478** 89
[16] Heinrich G, Harris J G E and Marquardt F 2010 *Phys. Rev. A* **81** 011801(R)
[17] Ludwig M, Safavi-Naeini A H, Painter O and Marquardt F 2012 *Phys. Rev. Lett.* **109** 063601
[18] Ludwig M, Kubala B and Marquardt F 2008 *New Journal of Physics* **10** 095013
[19] Restrepo J, Gabelli J, Ciuti C and Favero I 2011 *C. R. Physique* **12** 860
[20] Borkje K and Girvin S M 2012 *New Journal of Physics* **14** 085016
[21] Abdi M, Bahrampour A R and Vitali D 2012 *Phys. Rev. A* **86** 043803
[22] Spencer M B and Lamb Jr W E 1972 *Phys. Rev. A* **5** 884
[23] Lang R, Scully M O and Lamb Jr W E 1973 *Phys. Rev. A* **7** 1788
[24] Baseia B and Nussenzveig H M 1984 *Optica Acta: International Journal of Optics* **31** 39
[25] Penaforte J C and Baseia B 1984 *Phys. Rev. A* **30** 1401
[26] Guedes I, Penaforte J C and Baseia B 1989 *Phys. Rev. A* **40** 2463
[27] Guedes I and Baseia B 1990 *Phys. Rev. A* **42** 6858
[28] Barnett S M and Radmore P M 1988 *Opt. Commun.* **68** 364
[29] van der Plank R W F and Suttorp L G 1996 *Phys. Rev. A* **53** 1791
[30] Xuereb A, Domokos P, Asbóth J K, Horak P and Freegarde T 2009 *Phys. Rev. A* **79** 053810
[31] Castaños L O and Weder R *Equations of a moving mirror and the electromagnetic field*, to be published.

[32] J. D. Thompson, B. M. Zwickl, A. M. Jayich, Florian Marquardt, S. M. Girvin, and J. G. E. Harris, Nature **452**, 06715 (2008).
[33] Jackson J D 1998 *Classical Electrodynamics* 3rd edition (Wiley)
[34] Nussenzveig H M 1972 *Causality and Dispersion Relations* (Academic Press)
[35] Castaños L O and Weder R *Classical dynamics of a thin moving mirror and a single-mode field*, arXiv:1312.0648 [quant-ph]
[36] Schwartz M 1987 *Principles of Electrodynamics* (Dover Publications)

Journal of Physics: Conference Series **512** (2014) 012006 doi:10.1088/1742-6596/512/1/012006

Single and collective regimes in three-level systems interacting with a one-mode electromagnetic field

O. Castaños, S. Cordero, R. López-Peña, and E. Nahmad-Achar

Instituto de Ciencias Nucleares, Universidad Nacional Autónoma de México, Apdo. Postal 70-543 México 04510 D.F.

E-mail: ocasta@nucleares.unam.mx

Abstract. A semiclassical analysis is presented to determine the quantum phase transition from single to collective regimes in three-level atoms in the presence of a radiation field. The energy surfaces of the Ξ, V, and Λ configurations are constructed by taking the expectation value with respect to U(3) coherent states that carry the totally symmetric representation, determined by the total number of atoms. The corresponding stability and equilibrium properties are calculated by means of the catastrophe theory, discovering the bifurcation and Maxwell sets. We establish that the atoms with Λ and Ξ configurations have the presence of double points, i.e., there are two independent quantum states with the same energy that can be obtained depending on the values of the dipolar strengths of the interaction between the atoms and the radiation field. Additionally, the Ξ configuration exhibits a fixed triple point.

1. Introduction

The theory of quantum electrodynamics (QED) is used to study the interaction of atomic or molecular matter with a radiation field [1]. It requires a quantum treatment of both the particles and the field. QED is one of the most successful theories of physics, as is evident from the examples of the Lamb shift, the anomalous magnetic moment of the electron, and the electrodynamic relative displacement of the ground state of the positronium. Traditionally, the Hamiltonian related to a system of non-interacting atoms in an electromagnetic field is formed by three parts:

$$H = H_a + H_f + H_{af} \,, \tag{1}$$

where the interaction term H_{af} is responsible for the exchange of energy between the radiation field and the atomic sector. When the radiation field is considered classically and fixed externally, a semiclassical method has been developed in which only the atomic and coupling parts are described within the quantum mechanics formalism [2]. In this contribution, we take into account the complete Hamiltonian within the quantum formalism.

The radiation field is usually described in terms of an infinite set of harmonic oscillators, one for each mode of radiation. The amplitude of the electric field associated to a mode of frequency ω is given by $E_0 = \sqrt{\hbar\omega/(2\epsilon_0 V)}$ with ϵ_0 the permittivity of free space, and V the quantization volume. Each atom is considered as a two-level system, and its coupling to the radiation is described by the Rabi frequency $\Omega_{af} = d_{ge}E_0/\hbar$. Here d_{ge} denotes the matrix element of the electric dipole operator of the atom between the ground and excited states. In free space the probability of photon emission per unit time is proportional to the Rabi frequency and to the number of modes per unit of frequency interval [3, 4], that is,

$$\Gamma_0 = \frac{\omega^3}{3\pi\hbar c^3} \frac{|d_{ge}|^2}{\epsilon_0} . \tag{2}$$

The probability of survival of an atom in its excited state is given by the Fermi rule $P_e(t) = e^{-\Gamma_0 t}$. This exponential law describes an irreversible decay process of the atom to its ground state, g. However, this can be changed dramatically in a high-Q cavity regime, like that of emission of radiation between mirrors. The radiation rate in a cavity of volume V is given by [3]

$$\Gamma_{cav} = \Gamma_0 \frac{Q\lambda^3}{V} , \tag{3}$$

where Q denotes the quality factor of the cavity (which measures the lifetime of a photon in the cavity), and λ is the wavelength of the radiation field. For some high-Q cavities, one can achieve storage times of a photon running from a few milliseconds to a fraction of a second. Therefore, we see that the rate of emission in free space can be increased by the ratio $\frac{Q\lambda^3}{V}$, an advantage which in [5] was exploited to enhance Γ_0 by a factor of approximately 500. This type of cavity leads to a reversible atom-field evolution [6].

A review of the dynamics of two- and three-level systems interacting with quantized cavity electromagnetic fields is given in [7]. The Jaynes-Cummings model is the coupling of a two-level system with a harmonic oscillator and it corresponds to the typical situation of QED in a cavity. This model can also describe an ion in a trap when two ionic states are coupled by laser beams with those states of the oscillator in the trap. The model was introduced in 1963 by Jaynes and Cummings as an idealization of the matter-field coupling in free space [8]. Many theoretical predictions of this model, such as the existence of collapse and revivals in the Rabi oscillations [9], the formation of macroscopic quantum states, the many experimental studies of Rydberg atoms with very large principal quantum number within single-mode cavities, and measures of entanglement associated with spin-squeezed states, have been confirmed [10].

A system of N_a non-interacting two-level atoms coupled to a quantized electromagnetic field, using the dipolar and the rotating wave approximations, was described by the Tavis-Cummings Model [11, 12], having since then an extensive use in quantum optics [10]. The investigation of the phase transitions of the system has been carried out in the thermodynamic limit [13, 14], and at zero temperature [15, 16, 17]. The quantum description of a system of N_a non-interacting atoms under the action of a one-mode electromagnetic field allowed Dicke to predict the superradiance phenomenon [18]. The Dicke model does not consider the rotating wave approximation and it has been physically realized using a QED cavity with Bose-Einstein condensates [19, 20]. The localization of the quantum phase transitions in the dipolar strength parameters was studied in [21, 22, 23], and its relation with the position of extremal values for the Rényi-Wehrl entropies in [24]. The system of three-level atoms in the Ξ and Λ configurations interacting with a one mode radiation field together with a dipole-dipole interaction between the atoms has been used to define a new concept about atomic squeezing [25, 26]. Spin variances for the V and Λ configurations of an ensemble of atoms interacting with two light fields, a coherent pump state and a squeezed vacuum as a probe, have been calculated by means of Langevin equations derived from the Bloch equations [27]. By means of the Holstein-Primakoff transformation, normal and superradiant stable states for the Λ configuration have been identified in the thermodynamic limit [28].

More recently, we obtained in the rotating wave approximation (RWA), analytical expressions of the quantum phase transitions and their order from the single to collective regimes for three-level atoms interacting with a one-mode electromagnetic field for the Ξ, Λ, and V

configurations [29]. These transitions appear reflected in the ground state energy surface $\mathcal{E}^c$ and in their corresponding total number of excitations $\mathcal{M}^c$, when plotted as functions of the dipolar coupling constants (control parameters).

In this contribution we consider three-level atoms in a one-mode electromagnetic cavity in the three possible atomic configurations Ξ, Λ, and V. We review thoroughly the procedure to define the separatrix of the system in terms of the bifurcation and Maxwell sets. This includes the use of good variational states associated to the coherent states of U(3) to describe the matter part, together with the Heisenberg-Weyl coherent states for the description of the field. The separatrix divides the single and collective regimes of the system, which can be shown through the analytic expression for the vector states together with the corresponding statistical properties of light and matter. We establish the presence of double points for the Ξ, and Λ atomic configurations; for the Ξ case we additionally find the existence of a fixed triple point. This is independent of the number of atoms of the system, and depends only on the values of the dipole coupling strengths between the three-level states.

2. Model Hamiltonian

The one atom Hamiltonian can be written in terms of a three dimensional complete set of states as follows

$$H_a = \sum_{k,j=1}^{3} (H_a)_{kj} |k\rangle\langle j| = \sum_{j=1}^{3} E_j \, |j\rangle\langle j| \, , \tag{4}$$

where in the last expression, we use the fact that the matrix element $(H_a)_{kj}$ of the Hamiltonian satisfies $(H_a)_{kj} = E_j \delta_{jk}$. This means that the set of states $\{|k\rangle\}$ with $k = 1, 2, 3$, are eigenstates of the atom.

The one-mode electromagnetic field of frequency Ω is denoted by the Hamiltonian of a harmonic oscillator $H_f = \hbar\Omega a^\dagger a$, and the coupling interaction is given by $H_{af} = -\vec{d}\cdot\vec{E}$ with $\vec{d} = e\,\vec{r}$ denoting the dipole moment operator and $\vec{E}$ the external quantum electric field at the position of the atom. In the long wavelength approximation the coupling interaction term is given by

$$H_{af} = -\frac{1}{2} \sum_{j\neq k=1}^{3} \mu_{jk} \left(|j\rangle\langle k| + |k\rangle\langle j| \right)(a + a^\dagger) \, , \tag{5}$$

where we have defined $\mu_{jk} \equiv \sqrt{\frac{2\pi\hbar\Omega}{V}} \sum_\beta (\vec{E}_0 \cdot \vec{e}_\beta \, (x_\beta)_{jk})$. Besides, we have used the fact that the parameters μ_{jk}, which are proportional to the dipolar strengths, are only different from zero when $j \neq k$. We take, without loss of generality, $\mu_{jk} = \mu_{kj}$.

Therefore the one three-level atom Hamiltonian interacting with a one-mode radiation field can be written in the form

$$H = \Omega a^\dagger a + \sum_{j=1}^{3} \omega_j \, A_{jj}^{(1)} - \frac{1}{2} \sum_{j\neq k=1}^{3} \mu_{jk} \left(A_{jk}^{(1)} + A_{kj}^{(1)} \right)(a + a^\dagger) \, , \tag{6}$$

where we have defined $A_{ij}^{(1)} = |i\rangle\langle j|$ and it is easy to prove that these operators satisfy the commutation relations of a unitary group in 3 dimensions, U(3). We use from here onwards $E_j = \hbar\,\omega_j$ and units where $\hbar = 1$. This Hamiltonian presents the phenomenon of collapses and revivals in the probabilities of finding the atom in a determined energy level in a dynamic evolution of an arbitrary state, for a sufficiently large average number of photons in the cavity ($\nu \approx 10$).

The extension of the problem to consider N_a three-level atoms is straightforward through the replacement of the matter operators as follows: $A_{jk} = \sum_s A_{jk}^{(s)}$, where $A_{jk}^{(s)}$ denotes the

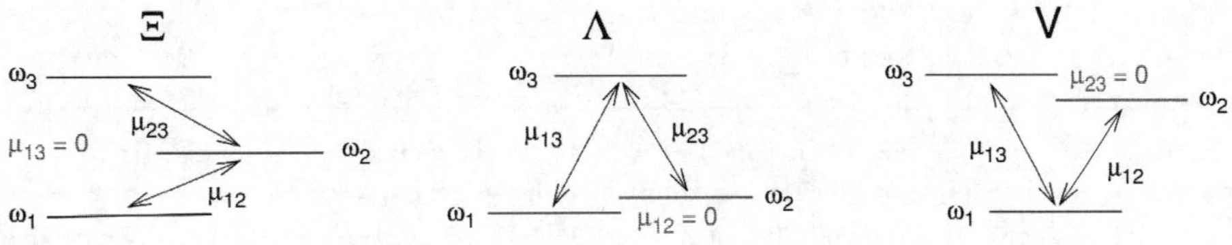

Figure 1. Atomic configurations and dipolar coupling parameters.

atomic operator of the s-th atom and satisfies again the commutation relations of a U(3) group. By means of the rotating wave approximation (RWA), that is, neglecting the terms in the Hamiltonian that do not conserve the total number of excitations, we arrive to the expression of the Hamiltonian model that is going to be used to determine the region in parameter space in which there is a sudden change in the properties of the ground state of the system,

$$\begin{aligned} H &= \Omega a^\dagger a + \omega_1 A_{11} + \omega_2 A_{22} + \omega_3 A_{33} \\ &- \frac{\mu_{12}}{\sqrt{N_a}}\left(A_{12}\,a^\dagger + A_{21}\,a\right) - \frac{\mu_{13}}{\sqrt{N_a}}\left(A_{13}\,a^\dagger + A_{31}\,a\right) - \frac{\mu_{23}}{\sqrt{N_a}}\left(A_{23}\,a^\dagger + A_{32}\,a\right). \end{aligned} \tag{7}$$

We have included the factor $\sqrt{N_a}$ in the expression of the dipolar strengths to replace the quantization volume by its corresponding density of atoms. Additionally we are considering that the energy levels obey the frequency inequalities $\omega_1 \leq \omega_2 \leq \omega_3$, in all of the atomic configurations mentioned above. It can be easily proved that the total number of excitations operator given by

$$M_\Xi = a^\dagger\, a + A_{22} + 2A_{33}\,, \quad M_\Lambda = a^\dagger\, a + A_{33}\,, \quad M_V = a^\dagger\, a + A_{22} + A_{33}\,, \tag{8}$$

commutes with respect to the corresponding Hamiltonian. In Fig.(1) we show that: for the Ξ atomic configuration one takes $\mu_{13} = 0$; for Λ atoms we consider $\mu_{12} = 0$; and for the V case, $\mu_{23} = 0$. The condition $\mu_{ij} = 0$ means that the transition from the atomic state j to the state i is forbidden.

3. Energy surface

The energy surface is defined by means of the expectation value of the Hamiltonian with respect to a variational test function. The structure of the Hamiltonian (7) leads us to choose the Heisenberg-Weyl coherent states as a good starting point for the field part and the U(3) coherent states for the matter. Major reasons for using these variational states are the facts that one can obtain analytical expressions for the expectation values of all the matter and field observables, and that they form a basis for the complete Hilbert space of the system in terms of four complex parameters: one to describe the field, α, and three associated to the matter part, γ_k, with $k = 1, 2, 3$. In this work we are going to consider only totally symmetric representations of the U(3) group; then the corresponding coherent states depend only on two complex parameters γ_2 and γ_3. Therefore, the variational state constructed from the tensor product of matter and field components can be written in the form [29]

$$\begin{aligned} |\alpha; N_a\, \gamma_2, \gamma_3\rangle &= \mathrm{e}^{-|\alpha|^2/2}\,\mathrm{e}^{\alpha a^\dagger}\,|0\rangle_F \otimes \frac{\left(b_1^\dagger + \gamma_3\, b_2^\dagger + \gamma_2\, b_3^\dagger\right)^{N_a}}{\sqrt{N_a!}\,\left(1 + |\gamma_2|^2 + |\gamma_3|^2\right)^{N_a/2}}\,|0,0,0\rangle_F\,, \\ &= \frac{\sqrt{N_a!}\,\mathrm{e}^{-|\alpha|^2/2}}{\left(1 + |\gamma_2|^2 + |\gamma_3|^2\right)^{N_a/2}} \sum_{\nu=0}^{\infty}\sum_{q=0}^{N_a}\sum_{r=0}^{q} \frac{\alpha^\nu\,\gamma_2^{N_a-q}\,\gamma_3^{q-r}}{\sqrt{\nu!\,(N_a-q)!\,(q-r)!\,r!}}\,|\nu\rangle_F\,|N_a, q, r\rangle_{GT}\,. \end{aligned} \tag{9}$$

In the last expression, we expand the variational state in terms of the Fock states for the field, $|\nu\rangle_F$ and the Gelfand-Tsetlin basis [30, 31] for the matter, $|N_a, q, r\rangle_{GT}$. Notice that $a|0\rangle = 0$ defines the vacuum state of the field and $|\alpha|^2$ measures the average number of photons. We defined $\Gamma^\dagger \equiv b_1^\dagger + \gamma_3\, b_2^\dagger + \gamma_2\, b_3^\dagger$ as a combination of three boson creation operators, this operator was normalized by asking that $[\Gamma, \Gamma^\dagger] = 1$, with $\Gamma \equiv b_1 + \gamma_3^*\, b_2 + \gamma_2^*\, b_3$.

For the totally symmetric representation, the U(3) generators can be realized in the form $A_{ij} = b_i^\dagger\, b_j$. It is straightforward to check that they satisfy the appropriate commutation relations and that the action of the weight generators A_{ii} on the GT basis is given by

$$\begin{aligned} A_{11}|N_a, q, r\rangle_{GT} &= r\,|N_a, q, r\rangle_{GT}\,, \\ A_{22}|N_a, q, r\rangle_{GT} &= (q-r)\,|N_a, q, r\rangle_{GT}\,, \\ A_{33}|N_a, q, r\rangle_{GT} &= (N_a - q)\,|N_a, q, r\rangle_{GT}\,, \end{aligned} \tag{10}$$

while for the lowering generators one has that

$$\begin{aligned} A_{12}|N_a, q, r\rangle_{GT} &= \sqrt{(q-r)(r+1)}\;|N_a, q, r+1\rangle_{GT}\,, \\ A_{13}|N_a, q, r\rangle_{GT} &= \sqrt{(N_a-q)(r+1)}\;|N_a, q+1, r+1\rangle_{GT}\,, \\ A_{23}|N_a, q, r\rangle_{GT} &= \sqrt{(N_a-q)(q-r+1)}\;|N_a, q+1, r\rangle_{GT}\,, \end{aligned} \tag{11}$$

where it is easy to check that the weight of the states in the right hand side is lower than that of the states on the left hand side. The action of the raising generators can be obtained from the previous results.

Because it is convenient to have an intensive quantity that describes the energy per particle, in units of the photon energy, the expectation value of (7) with respect to (9) is given by

$$\mathcal{E}^c(\alpha, N_a, \gamma_2, \gamma_3) = \frac{1}{N_a\,\Omega}\,\langle \alpha; N_a,\, \gamma_2,\, \gamma_3|H|\alpha; N_a,\, \gamma_2,\, \gamma_3\rangle\,. \tag{12}$$

If we rewrite the complex parameters in their polar form, i.e., $\alpha = \rho e^{i\phi}$ and $\gamma_j = \varrho_j e^{i\varphi_j}$ with $j = 2, 3$, one finds that the energy surface only depends on the angles $\vartheta_3 = \phi - \varphi_3$, $\vartheta_2 = \phi - \varphi_2$ and $\vartheta_1 = \phi - \varphi_2 + \varphi_3$. Minimizing $\mathcal{E}^c$ with respect these phases, the critical values are given by $\vartheta_{kc} = 0,\ \pi$ with $k = 1, 2, 3$, and it is not difficult to prove that a minimum is always obtained if the energy surface is rewritten as

$$\mathcal{E}^c = r^2 + \frac{\bar\omega_1 + \bar\omega_2\,\varrho_3^2 + \bar\omega_3\,\varrho_2^2}{1 + \varrho_2^2 + \varrho_3^2} - 2\,r\,\frac{\bar\mu_{12}\,\varrho_3 + \bar\mu_{13}\,\varrho_2 + \bar\mu_{23}\,\varrho_2\,\varrho_3}{1 + \varrho_2^2 + \varrho_3^2}. \tag{13}$$

where we have made the following definitions: $r \equiv \rho/\sqrt{N_a}$, $\bar\omega_k \equiv \omega_k/\Omega$ with $k = 1, 2, 3$, and $\bar\mu_{ij} \equiv |\mu_{ij}|/\Omega$. The absolute values for the dipolar strengths appear because the energy surface is invariant under the interchange $\mu_{ij} \to -\mu_{ij}$.

The critical points are determined by

$$\left(\frac{\partial}{\partial r}, \frac{\partial}{\partial \rho_2}, \frac{\partial}{\partial \rho_3}\right)\mathcal{E}^c = \vec{0}\,. \tag{14}$$

and the stability properties by its corresponding Hessian matrix,

$$\mathcal{H} = \left(\begin{array}{ccc} \frac{\partial^2\mathcal{E}^c}{\partial r^2} & \frac{\partial^2\mathcal{E}^c}{\partial r\,\partial \rho_2} & \frac{\partial^2\mathcal{E}^c}{\partial r\,\partial \rho_3} \\ \frac{\partial^2\mathcal{E}^c}{\partial \rho_2\,\partial r} & \frac{\partial^2\mathcal{E}^c}{\partial \rho_2^2} & \frac{\partial^2\mathcal{E}^c}{\partial \rho_2\,\partial \rho_3} \\ \frac{\partial^2\mathcal{E}^c}{\partial \rho_3\,\partial r} & \frac{\partial^2\mathcal{E}^c}{\partial \rho_3\,\partial \rho_2} & \frac{\partial^2\mathcal{E}^c}{\partial \rho_3^2} \end{array}\right)_{(r,\rho_2,\rho_3)\to(r_c,\rho_{2c},\rho_{3c})}\,,$$

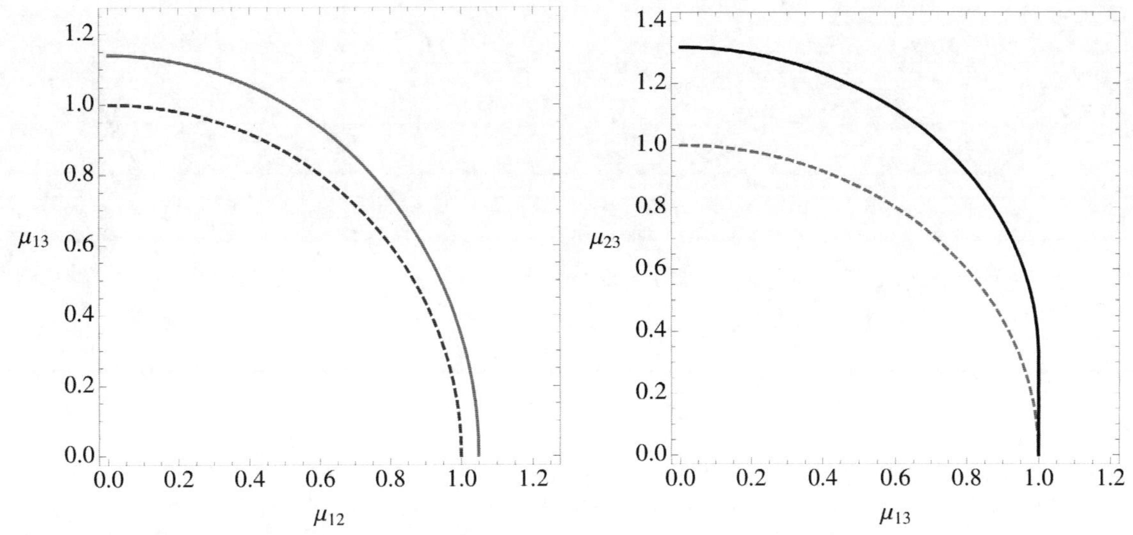

Figure 2. (Color online.) At left the bifurcation set of the V atomic configuration is shown by a continuous line for the frequencies $\bar{\omega}_{31} = 1.3$ and $\bar{\omega}_{21} = 1.1$ and with a dashed line for the double resonance case, $\bar{\omega}_{31} = 1$ and $\bar{\omega}_{21} = 1$. At right the separatrix of the Λ case is shown for the single resonance case $\bar{\omega}_{21} = 0.1$ by a continuous line, and with a dashed line when in double resonance $\bar{\omega}_{21} = 0$.

where r_c, ϱ_{2c} and ϱ_{3c} stand for the critical values of r, ϱ_2 and ϱ_3, respectively.

From the expression (14) it is immediate that the variable r takes the critical value

$$r_c = \frac{\bar{\mu}_{12}\,\varrho_{3c} + \bar{\mu}_{13}\,\varrho_{2c} + \bar{\mu}_{23}\,\varrho_{2c}\,\varrho_{3c}}{1 + \varrho_{2c}^2 + \varrho_{3c}^2}\,. \tag{15}$$

In general we are not able to have analytic solutions for ϱ_{2c} and ϱ_{3c}; however, the values $\varrho_{2c} = \varrho_{3c} = 0$ are critical points for all values of the parameters of the Hamiltonian. It is then possible to determine the eigenvalues of the Hessian matrix for each one of the atomic configurations and conclude that there are bifurcations sets for each case [32]. They are given by the locus of points

$$(\bar{\mu}_{12})_\Xi = \sqrt{\bar{\omega}_{21}}\,, \quad (\bar{\mu}_{13})_\Lambda = \sqrt{\bar{\omega}_{31}}\,, \quad \left(\frac{\bar{\mu}_{12}^2}{\bar{\omega}_{21}} + \frac{\bar{\mu}_{13}^2}{\bar{\omega}_{31}}\right)_V = 1\,, \tag{16}$$

where we have used $\bar{\omega}_{ij} = \bar{\omega}_i - \bar{\omega}_j$ to denote the energy shifts between the atomic levels i and j. For these sets the determinant of the Hessian is zero, and we note that the bifurcation sets for the Ξ and Λ configurations are independent of the value of $\bar{\mu}_{23}$. In all the bifurcation sets the energy of the system is given by $\mathcal{E}^c = \bar{\omega}_1$, and there are two degenerate critical points. These sets separate the control parameter space into a region where the total number of excitations and the average number of photons are zero, and a region where they are different from zero. The bifurcation set for the V configuration is displayed in Fig. 2 for the cases with detuning $\Delta_{21} = 0.1$ and $\Delta_{31} = 0.3$, and without detuning ($\Delta_{ij} \equiv \bar{\omega}_{ij} - 1$).

Recently [29] we found the minimum energy surface E^c as a function of the control parameters μ_{ij}. It changes its value from $E^c = 0$ (taking $\bar{\omega}_1 = 0$) to $E^c < 0$, when a transition from $M^c = 0$ (normal regime) to $M^c > 0$ (collective regime) in the total number of excitations takes place.

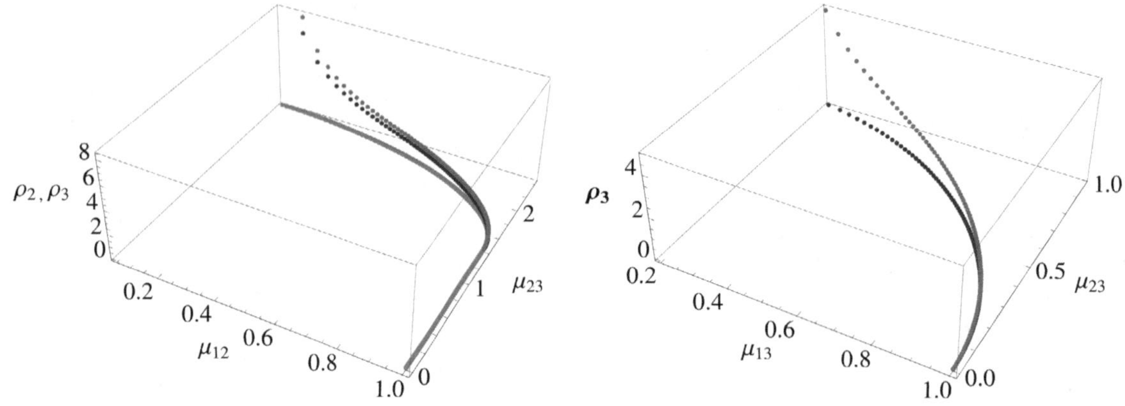

Figure 3. (Color online.) The separatrix of the three-level system is given together with the minima ρ_{2c} and ρ_{3c} as functions of the dipolar strengths, in the double resonance case. At left: the Ξ atomic configuration. At right; the Λ atomic configuration. In this last case the variable $\rho_{2c} = 0$. In both cases, the minima points yield an energy value $\mathcal{E}^c = 0$.

This leads to the existence of a separatrix in parameter space, for which we were able to propose the following ansatz:

$$\begin{aligned}
\bar\omega_{21} &= \bar\mu_{12}^2 + \left[\bar\mu_{23} - \sqrt{\bar\omega_{31}}\right]^2 \Theta\left[\bar\mu_{23} - \sqrt{\bar\omega_{31}}\right], \quad &(\Xi \text{ configuration}),\\
\bar\omega_{31} &= \bar\mu_{13}^2 + \left[\bar\mu_{23} - \sqrt{\bar\omega_{21}}\right]^2 \Theta\left[\bar\mu_{23} - \sqrt{\bar\omega_{21}}\right], \quad &(\Lambda \text{ configuration}),\\
1 &= \frac{\bar\mu_{12}^2}{\bar\omega_{21}} + \frac{\bar\mu_{13}^2}{\bar\omega_{31}}, \quad &(V \text{ configuration}).
\end{aligned} \tag{17}$$

Here, $\Theta[x]$ stands for the Heaviside theta function. From these expressions one immediately sees that for values of the dipolar strengths $\bar\mu_{23} < \sqrt{\bar\omega_{31}}$ (Ξ config.) and $\bar\mu_{23} < \sqrt{\bar\omega_{21}}$ (Λ config.) one recovers Eq. (16). The separatrix establishes two regions in parameter space: in the internal region the atoms manifest a single state behavior with a total number of excitations equal to zero, while in the external one they have a multicomponent eigenstate with a total number of excitations different from zero.

3.1. Ξ atomic configuration

The Ξ configuration forbids the transition $\omega_1 \longleftrightarrow \omega_3$, and this is introduced in the Hamiltonian by taking $\mu_{13} = 0$. Also the condition $\omega_2 \approx \omega_3/2$ is fulfilled, and the detuning should satisfy $\Delta_{21} \approx \Delta_{32}$ with $|\Delta_{ij}|$ small in order to maintain the RWA. The minima points of the energy surface for the Ξ configuration can be obtained numerically as functions of $\bar\mu_{12}$ and $\bar\mu_{23}$. They are displayed in Fig. 3 together with the separatrix. For values of $\bar\mu_{23} > \sqrt{2}$, on the locus of points of the separatrix, there are two values of the variables with the same minimum energy: $\varrho_{2c} = \varrho_{3c} = 0$, and $\varrho_{2c} \neq 0$ and $\varrho_{3c} \neq 0$. They have different values of the total number of excitations. This shows the existence of a double point, that is, two independent variational states with the same energy but different total number of excitations and average number of photons. The value $(\bar\mu_{12}, \bar\mu_{23}) = (1, \sqrt{2})$ defines a triple point in parameter space, which is fixed in the sense that it does not move when the number of particles is changed.

3.2. Λ atomic configuration

For atoms in the Λ configuration it is required that the transition $\omega_1 \longleftrightarrow \omega_2$ be forbidden, and so we take $\mu_{12} = 0$. Because of the convention $\omega_1 \le \omega_2 \le \omega_3$ used in the labeling of the energy levels, the condition $\omega_1 \approx \omega_2$ requires $\Delta_{31} - \Delta_{32} \approx 0$ with $\Delta_{31} \ge \Delta_{32}$. The minima points can be obtained analytically for the double resonant case $\Delta_{31} = \Delta_{32} = 0$. These are given by $\varrho_{2c} = \varrho_{3c} = 0$ in the normal regime, with $\bar{\mu}_{13}^2 + \bar{\mu}_{23}^2 \le 1$; while in the collective regime we have

$$\varrho_{2c} = \frac{1}{\bar{\mu}_{13}}\sqrt{\frac{(\bar{\mu}_{13}^2 + \bar{\mu}_{23}^2)\,(\bar{\mu}_{13}^2 + \bar{\mu}_{23}^2 - 1)}{\bar{\mu}_{13}^2 + \bar{\mu}_{23}^2 + 1}}, \qquad \varrho_{3c} = \frac{\bar{\mu}_{23}}{\bar{\mu}_{13}}, \tag{18}$$

where basis states with $\mathcal{M} > 0$ contribute to the ground state. In Fig. 3 the minima are plotted along the separatrix, for which $\varrho_{2c} = 0$ and $\varrho_{3c} = \frac{\sqrt{1-\bar{\mu}_{13}^2}}{\bar{\mu}_{13}}$. This separatrix is displayed in Fig. 2 (right) for the single (one detuning parameter different from zero) and double resonance (both detuning parameters equal to zero) cases.

3.3. V atomic configuration

A system of atoms in the V configuration requires $\mu_{23} = 0$, since transitions between the levels ω_3 and ω_2 are forbidden. Notice that the condition $\omega_2 \approx \omega_3$ on $\omega_1 \le \omega_2 \le \omega_3$ reads, in terms of the detuning, as $\Delta_{21} \approx \Delta_{31}$ but satisfying $\Delta_{21} \le \Delta_{31}$.

In a similar way to the Λ configuration, when the detuning parameters are equal, $\Delta_{21} = \Delta_{31}$, the problem has an analytic solution. The critical points are $\varrho_{2c} = \varrho_{3c} = 0$ for the normal regime implying an energy surface for the ground state equal to zero. For the collective regime, ϱ_{2c} and ϱ_{3c} take the values

$$\varrho_{2c} = \bar{\mu}_{13}\sqrt{\frac{\bar{\mu}_{12}^2 + \bar{\mu}_{13}^2 - 1}{(\bar{\mu}_{12}^2 + \bar{\mu}_{13}^2)\,(\bar{\mu}_{12}^2 + \bar{\mu}_{13}^2 + 1)}}, \qquad \varrho_{3c} = \bar{\mu}_{12}\sqrt{\frac{\bar{\mu}_{12}^2 + \bar{\mu}_{13}^2 - 1}{(\bar{\mu}_{12}^2 + \bar{\mu}_{13}^2)\,(\bar{\mu}_{12}^2 + \bar{\mu}_{13}^2 + 1)}}. \tag{19}$$

Notice that on the separatrix the critical points coalesce to the values $\rho_c = \rho_2 = \rho_3 = 0$, and all the expectation values of the matter and field observables are given in terms of those values.

4. Quantum solution

The Hamiltonian (7) can be diagonalized in terms of the Fock states $|\nu\rangle$ for the field part and the Gelfand-Tsetlin (GT) states for the matter part

$$|\nu;\, h_1\, h_2\, h_3;\, q_1\, q_2;\, r\rangle = |\nu\rangle \otimes \left| \begin{array}{ccccc} h_1 & & h_2 & & h_3 \\ & q_1 & & q_2 & \\ & & r & & \end{array} \right\rangle, \tag{20}$$

where the labels of each row of the GT states denote the irreducible representations (irreps) of the canonical chain of unitary groups $U(3) \supset U(2) \supset U(1)$. The quantum labels of $U(3)$ satisfy the inequalities $h_1 \ge h_2 \ge h_3 \ge 0$; the ones for $U(2)$ satisfy $q_1 \ge q_2 \ge 0$; and for $U(1)$ $r \ge 0$. The branching rules [30] are the following: $h_2 \le q_1 \le h_1$, $h_3 \le q_2 \le h_2$, and $q_2 \le r \le q_1$. The irreps of $U(3)$ are determined by means of the different partitions of the total number of atoms N_a in terms of h_1, h_2, and h_3 that satisfy the previous rules. As an example one may consider the three atoms case $N_a = 3$; the $U(3)$ irreps are $[3,0,0]$, $[2,1,0]$, and $[1,1,1]$. These partitions of $N_a = 3$ also indicate the symmetry of the particles under the action of the permutation group S_3. The first one is symmetric under the interchange of the atoms, and the last one is antisymmetric. The middle irrep has a mixed symmetry.

In this work, we want to consider only the symmetric case and then we set $h_2 = h_3 = q_2 = 0$. Therefore, the exact numerical calculation of the ground state energy may be evaluated using $h_1 = N_a$ and we can simplify the GT states notation as

$$|\, h_1\, 0\, 0; q_1\, 0;\, r\rangle_{GT} \equiv |N_a\, q_1\, r\rangle_{GT}\,. \tag{21}$$

The corresponding matrix elements of the operators A_{ij} (for this particular basis) are easily calculated in the bosonic realization $A_{ij} = b_i^\dagger\, b_j$, which can be used to evaluate the matrix elements of the Hamiltonian and to obtain numerically its eigenvalues.

For each particular atomic configuration (Ξ, Λ or V) there is an additional constant of motion M, namely the total number of excitations M_Ξ, M_Λ, and M_V given by (8). These can be gathered in a single equation as

$$\hat{M} = a^\dagger\, a + \lambda_2 A_{22} + \lambda_3 A_{33}\,, \tag{22}$$

where $(\lambda_2, \lambda_3) = (1, 2)$ for the Ξ configuration; $(\lambda_2, \lambda_3) = (0, 1)$ for the Λ case, and $(\lambda_2, \lambda_3) = (1, 1)$ for V. Then one has two constants of the motion: the total number of atoms $\hat{N} = A_{11} + A_{22} + A_{33}$, and the total number of excitations, and we can use them to write the Hamiltonian (7) in units of the photon energy and per particle, $\mathcal{H} = \frac{H}{N_a \Omega}$, as

$$\begin{aligned}\mathcal{H} &= \frac{M}{N_a} + \frac{\Delta}{N_a} A_{22} + \frac{\bar{\Delta}}{N_a} A_{33} - \frac{\bar{\mu}_{12}}{N_a^{3/2}} \left(A_{12}\, a^\dagger + A_{21}\, a\right) \\ &- \frac{\bar{\mu}_{13}}{N_a^{3/2}} \left(A_{13}\, a^\dagger + A_{31}\, a\right) - \frac{\bar{\mu}_{23}}{N_a^{3/2}} \left(A_{23}\, a^\dagger + A_{32}\, a\right),\end{aligned} \tag{23}$$

where we have taken $\bar{\omega}_1 = 0$, and defined the parameters $\Delta \equiv \bar{\omega}_{21} - \lambda_2$, and $\bar{\Delta} \equiv \bar{\omega}_{32} + \bar{\omega}_{21} - \lambda_3$. Additionally, we have replaced the eigenvalues of $\hat{N}$ and $\hat{M}$ by N_a and M, respectively.

Therefore, the bases of the different configurations can be written in the form

$$\begin{aligned}|\nu\, N_a\, q_1\, r\rangle_\Xi &\equiv |M - 2N_a + q_1 + r, N_a\, q_1\, r\rangle\,, & M - 2N_a + q_1 + r \geq 0\,,\\ |\nu\, N_a\, q_1\, r\rangle_\Lambda &\equiv |M - N_a + q_1,\, N_a\, q_1\, r\rangle\,, & M - N_a + q_1 \geq 0\,,\\ |\nu\, N_a\, q_1\, r\rangle_V &\equiv |M - N_a + r,\, N_a\, q_1\, r\rangle\,, & M - N_a + r \geq 0\,.\end{aligned} \tag{24}$$

where we have used the corresponding values of λ_2 and λ_3.

4.1. Dimensions of the Hilbert space

The Hilbert space is strongly dependent on the values of M and N_a. We start by giving the dimension for the Ξ configuration:

$$d_\Xi(N_a, M) = \begin{cases} (M/2+1)^2 & M \,\text{even}, M \leq N_a\,,\\ (M+1)\,(M+3)\,/4 & M\,\text{odd},\;\; M \leq N_a\,,\\ \{2\,N_a(2M - N_a + 1) - M(M-2) + 4\}\,/4 & M\text{even}, N_a \leq M \leq 2\,N_a\,,\\ \{2\,N_a(2M - N_a + 1) - (M-1)^2 + 4\}\,/4 & M\;\text{odd}, N_a \leq M \leq 2\,N_a\,,\\ (N_a+1)(N_a+2)/2 & M \geq 2\,N_a\,.\end{cases} \tag{25}$$

For the Λ configuration, the dimension of the Hilbert space is given by

$$d_\Lambda(N_a, M) = \begin{cases} (M+1)\,(2N_a + 2 - M)/2 & M \leq N_a\,,\\ (N_a+1)(N_a+2)/2 & M \geq N_a\,;\end{cases} \tag{26}$$

while for the V configuration one has

$$d_V(N_a, M) = \begin{cases} (M+1)(M+2)/2 & M \leq N_a\,, \\ (N_a+1)\,(N_a+2)/2 & M \geq N_a\,. \end{cases} \tag{27}$$

These relations are important for the study of the system for a small average number of photons in the cavity, because they imply a small number of total excitations M.

To find the quantum ground energy and its corresponding eigenstate, we proceed as follows. For each configuration of the atom, a given value of M, and for fixed parameters Ω, ω_1, ω_2, and ω_3 the eigenvalues and their corresponding eigenstates are evaluated numerically as functions of the control parameters μ_{ij}. This gives us the ground state energy for each corresponding total number of excitations. It is worth mentioning that, for a fixed region of values of the interaction intensity, one may estimate the maximum value of M that is required to find the minimum energy from the semi-classical calculation.

4.2. Single Regime

Next we show the single regime of the atoms for the three atomic configurations. As an example we consider $N_a = 5$ atoms; according to the dimension of the space one has a unique state for the Ξ and V configurations. For the Λ case there are two situations: in the double resonance case, $\bar{\omega}_{21} = 0$ and $\bar{\omega}_{31} = 1$, there are 6 degenerate eigenstates. If one has $\bar{\omega}_{21} \neq \delta$, with $\delta \leq 0.2$ the degeneracy is broken and the ground state of the system has all the atoms in their lowest level and zero photons. However, if $\bar{\omega}_{31} = \bar{\omega}_{32} \neq 1$, the system continues being degenerate. The degeneracy for any number of atoms is equal to $N_a + 1$, and it is present because the first and second levels of the system have the same energy; this degeneracy will be broken when a small energy difference between them is present. Assuming this approximation for the Λ configuration, in the three cases one has all the non-interacting atoms in their lowest energy level and zero photons for the radiation field.

4.3. Collective Regime

In the double resonance case, $\Delta = \bar{\Delta} = 0$, the Hamiltonian for any of the atomic configurations takes the form

$$\mathcal{H}_{DR} = \frac{M}{N_a} - \frac{\bar{\mu}_{12}\,(A_{12}\,a^\dagger + A_{21}\,a)}{N_a^{3/2}} - \frac{\bar{\mu}_{13}\,(A_{13}\,a^\dagger + A_{31}\,a)}{N_a^{3/2}} - \frac{\bar{\mu}_{23}\,(A_{23}\,a^\dagger + A_{32}\,a)}{N_a^{3/2}}\,, \tag{28}$$

where the first term is a constant of motion. For the Ξ configuration one has $\bar{\mu}_{13} = 0$ and M denotes the eigenvalue of the total excitation number M_Ξ; for the Λ type of atomic spectra $\bar{\mu}_{12} = 0$ and M is the eigenvalue of M_Λ; while for the V atomic case one considers $\bar{\mu}_{23} = 0$ and M is the eigenvalue of M_V. Therefore one has

$$\begin{aligned} \mathcal{H}_{DR\Xi} &= \frac{M}{N_a} - \frac{\bar{\mu}_{12}\,(A_{12}\,a^\dagger + A_{21}\,a)}{N_a^{3/2}} - \frac{\bar{\mu}_{23}\,(A_{23}\,a^\dagger + A_{32}\,a)}{N_a^{3/2}}\,, \\ \mathcal{H}_{DR\Lambda} &= \frac{M}{N_a} - \frac{\bar{\mu}_{13}\,(A_{13}\,a^\dagger + A_{31}\,a)}{N_a^{3/2}} - \frac{\bar{\mu}_{23}\,(A_{23}\,a^\dagger + A_{32}\,a)}{N_a^{3/2}}\,, \\ \mathcal{H}_{DRV} &= \frac{M}{N_a} - \frac{\bar{\mu}_{12}\,(A_{12}\,a^\dagger + A_{21}\,a)}{N_a^{3/2}} - \frac{\bar{\mu}_{13}\,(A_{13}\,a^\dagger + A_{31}\,a)}{N_a^{3/2}}\,. \end{aligned} \tag{29}$$

By means of the interaction picture of quantum mechanics one has the evolution equation in terms of the coupling term of the Hamiltonian. In this double resonance case the coupling term

Journal of Physics: Conference Series **512** (2014) 012006 doi:10.1088/1742-6596/512/1/012006

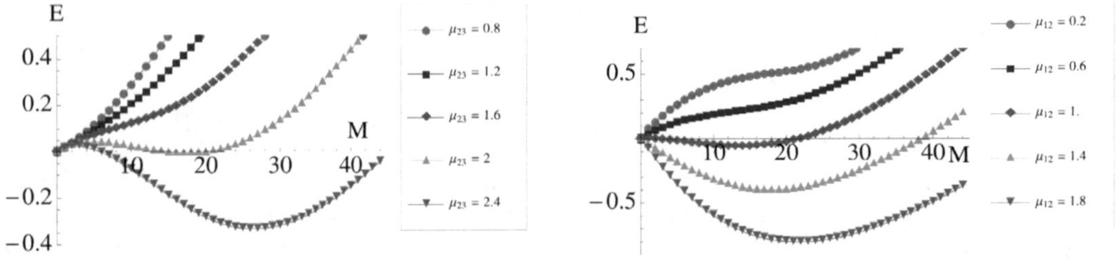

Figure 4. (Color online.) Energy of the atomic configuration Ξ as a function of the total number of excitations, M (horizontal axis). At left, the parameter value $\bar{\mu}_{12} = 0.8$ was used while at right we took $\bar{\mu}_{23} = 1.8$. In both cases the number of atoms is $N_a = 10$.

is time independent, so to find the spectra of the system one has to solve only the corresponding eigenvalue equation. For these atomic configurations, when one of the dipolar strengths are equal to zero the three-level system reduces to a two-level problem and only the levels involved change their properties.

Ξ *configuration*

For the double resonance case, together with the equality of the dipolar strengths, $\bar{\mu}_{12} = \bar{\mu}_{23}$, it is easy to prove that the system is equivalent to a two-level system with the operators for the matter given by $S_- = A_{12} + A_{23}$, $S_+ = A_{21} + A_{32}$, and $S_0 = (A_{33} - A_{11})/2$. The operator of the total number of excitations takes the form $M_\Xi = a^\dagger a + \hat{N} + 2S_0$.

In the general case, for fixed values of $\bar{\mu}_{12}$ and $\bar{\mu}_{23}$ we explore in Fig. 4 the energy eigenvalues as functions of the total number of excitations. The minimum of each curve determines the ground state of the system. These are calculated for a fixed value of the dipolar strength $\bar{\mu}_{12} = 0.8$ and for several values of $\bar{\mu}_{23}$ to see the transition from the dominance of the state with $M = 0$ to states with $M > 0$. The same was done for a fixed value of $\bar{\mu}_{23} = 1.8$ and changing $\bar{\mu}_{12}$ to see the transition from the normal regime to the collective regime. This is happening because the separatrix of the Hamiltonian model is crossed.

Λ *configuration*

In the double resonance case the dipolar strengths are associated to transitions from two degenerate levels to a third atomic level. Then, if one proposes

$$\bar{\mu}_{13} = \mu \cos\theta\,, \quad \bar{\mu}_{23} = \mu \sin\theta \tag{30}$$

together with the unitary transformation for the boson operators

$$c^\dagger = \cos\theta\, b_1^\dagger + \sin\theta\, b_2^\dagger\,, \quad c = \cos\theta\, b_1 + \sin\theta\, b_2\,, \tag{31}$$

it is straightforward to find that the system is equivalent to a two-level case with the following definitions for the matter observables: $S_+ = b_3^\dagger\, c$, $S_- = c^\dagger\, b_3$, and $S_0 = (b_3^\dagger b_3 - c^\dagger c)/2$.

In Fig. 5 we study the energy spectra as functions of the total number of excitations. The spectra are calculated for a fixed value of the dipolar strength $\mu_{13} = 0.5$ and detuning $\Delta_{32} = -0.1$, and for several values of μ_{23} to see the transition from the dominance of the state with $M = 0$ to states with $M > 0$. The same study was done in double resonance for a fixed value of $\mu_{23} = 1.0$ and changing μ_{13} to see again the transition to the collective regime.

V *configuration*

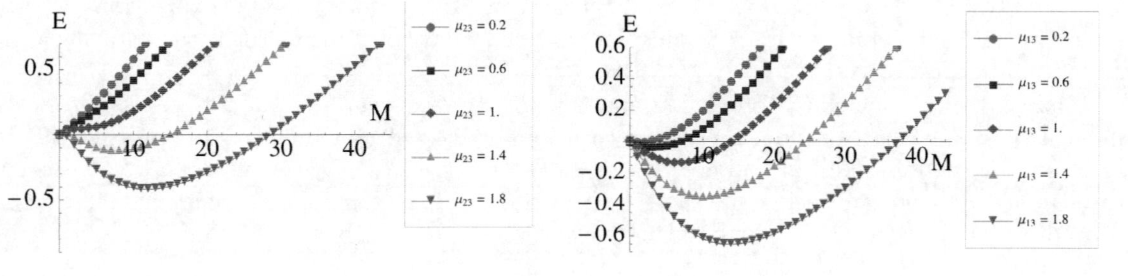

Figure 5. (Color online.) Energy of the atomic configuration Λ as function of the total number of excitations, M (horizontal axis). At left, the parameters $\bar{\mu}_{13} = 0.5$ and $\Delta_{32} = -0.1$ were used while at right we took $\bar{\mu}_{23} = 1.0$ and $\Delta_{32} = 0$. In both cases the number of atoms is $N_a = 10$.

In the double resonance case one has the transition from an atomic energy level to two degenerate levels. One can then proceed in a similar way as for the Λ configuration, defining

$$\bar{\mu}_{12} = \mu \cos\theta\,, \quad \bar{\mu}_{13} = \mu \sin\theta \tag{32}$$

together with the unitary boson transformation

$$d^\dagger = \cos\theta\, b_2^\dagger + \sin\theta\, b_3^\dagger\,, \quad d = \cos\theta\, b_2 + \sin\theta\, b_3\,. \tag{33}$$

It is straightforward to find that the system is equivalent to a two-level case with the following definitions for the matter observables $S_+ = d^\dagger\, b_1$, $S_- = b_1^\dagger\, d$, and $S_0 = (d^\dagger d - b_1^\dagger b_1)/2$.

For this case the results for the energy spectra are very similar to those displayed in Fig. 5 for the Λ configuration.

5. Conclusions

We have established the existence of single and collective regimes for three-level atoms in the Ξ, Λ, and V atomic configurations. These regions, in the dipolar parameter space, are separated by the corresponding separatrix of each configuration. The separatrix of the Ξ and Λ atomic configurations are constituted by bifurcation and Maxwell sets. In the bifurcation sector, the transitions are of second order, two critical points coalesce: one of them yields zero values for the expectation values of the total number of excitations and the photon number operator while the other produces $\langle M\rangle$ and $\langle N_{ph}\rangle$ different from zero. In the Maxwell region the quantum phase transitions are of first order, and there are double critical points yielding the same energy (see Fig.3). These double points are described by two independent states with different distribution functions for M and for the number of photons. For the Ξ configuration there is additionally a triple point, that is, three independent solutions with the same energy for any number of particles. These states have total excitation quantum numbers $M_\Xi = 0$, $M_\Xi = 1$, and $M_\Xi = 2$, noticing that the point is on the border between the bifurcation and Maxwell sets.

For the V atomic configuration we have found that the separatrix is formed by a bifurcation set, where there are quantum phase transitions of second order and two critical degenerate points. One of them implies zero values for the expectation values $\langle M\rangle$ and $\langle N_{ph}\rangle$ and the other yields expectation values of the same observables different from zero.

A connection between the double resonance case of the V and Λ configurations and two-level systems was established by means of unitary transformations. These occupy degenerate energy levels. This result allows for a check of consistency of our separatrix expressions with the localization of the quantum phase transitions in the dipolar parameter space for corresponding two-level systems. Additionally, we expect that the form of the analytic expressions for the separatrix of each configuration will only be modified by a constant factor if the rotating wave approximation is not considered.

Finally, because the Hamiltonian commutes with M, a new variational state can be proposed if we truncate the series expansion of the variational coherent states given in expression (9) by replacing the photon quantum number as $\nu \to M - \lambda_3 N_a + (\lambda_3 - \lambda_2)q + \lambda_2 r$, renormalizing the resulting expression, and using the same critical points of the variational coherent states. This procedure has been used to calculate the fluctuation of the number of photons with very good agreement with the exact quantum calculation [29].

5.1. Acknowledgments

This work was partially supported by CONACyT-México (under project 101541) and DGAPA (under projects IN101614 and IN110114).

References

[1] Milonni P W 1994 *The quantum vacuum: An introduction to quantum electrodynamics* (Academic Press).
[2] Craig D P and Thirunamachandrani T 1998 *Molecular quantum electrodynamics: An introduction to radiation molecule interactions* (Dover).
[3] Haroche S and Kleppner D 1991*Nonclassical effects in quantum optics*, ed P. Myestre P and Walls D F (AIP) p 402.
[4] Haroche S, Raimond J M 2007 *Exploring the Quantum: Atoms, Cavities and Photons* (Oxford University Press).
[5] Goy P, Raymond J M, Gross M, and Haroche S 1983 *Phys. Rev. Lett.* **50** 1903.
[6] Haroche S 1999 *Latin-American School of Physics XXXI ELAF: New perspectives of quantum mechanics*, Vol. 464, ed Hacyan S, Jáuregui R, and López-Peña R, (AIP Conference Proceedings) p 45.
[7] Yoo H J and Eberly J H 1985 *Phys. Rep.* **118** 239.
[8] Jaynes E T and Cummings F W 1963 *Proc. IEEE* Vol **51** 89.
[9] Eberly J H, Narozhny N B, and Sanchez-Mondragon J J 1980 *Phys. Rev. Lett.*, **44** 1323.
[10] *Theory of Nonclassical States of Light*, Eds. V.V. Dodonov and V.I. Man'ko (Taylor & Francis, London, 2003).
[11] Tavis M and Cummings F W 1968 *Phys. Rev.* **170** 379.
[12] Tavis M and Cummings F W 1969 *Phys. Rev.* **188** 692.
[13] Hepp K, Lieb E H 1973 *Ann. Phys.* **76** 360404.
[14] Wang YK, Hioe F T 1973 *Phys Rev A* **7** 831836.
[15] Buzek V, Orszag M, Rosko M 2005 *Phys Rev Lett.* **94**163601.
[16] Castaños O, López-Peña R, Nahmad-Achar E, Hirsch JG, López-Moreno E, Vitela J E 2009 *Phys Scr.* **79** 065405.
[17] Castaños O, Nahmad-Achar E, López-Peña R, Hirsch J G 2009 *Phys Scr.* **80** 055401.
[18] Dicke, R H 1954 *Phys. Rev.* **93** 99.
[19] Baumann K, Guerlin C, Brennecke F, and Esslinger T 2010 *Nature* **464** 1301.
[20] Nagy D, Konya G, Szirmai G, Domokos P 2010 *Phys. Rev. Lett.* **104** 130401.
[21] Castaños O, Nahmad-Achar E, López-Peña R, Hirsch J G 2011 *Phys Rev.* **A 84** 013819.
[22] Castaños O, Nahmad-Achar E, López-Peña R, Hirsch J G 2012 *Phys Rev.* **A 86** 023814.
[23] Romera E, del Real R and Calixto M 2012 *Phys. Rev.* **A 85** 053831
[24] Romera E and Nagy A 2011 *Phys. Lett.* **A 375**, 3066.
[25] Civitarese O, Reboiro M 2006 *Phys. Lett.* **A 357** 224.
[26] Civitarese O, Reboiro M, Rebón L, Tielas D 2010 *Phys. Lett.***A 374** 2117.
[27] Dantan A, Pinard M, Josse V, Nayak N, Berman P R 2003 *Phys. Rev.* **A 67** 045801.
[28] Hayn M, Emary C, Brandes T 2011 *Phys Rev A* 84 053856.
[29] Cordero S, López-Peña R, Castaños O, Nahmad-Achar E 2013 *Phys. Rev.* **A 87** 023805.
[30] Gelfand I M, Tsetlin M L. 1950 *Dokl Akad Nauk SSSR* **71** 825.
[31] Moshinsky M *Group Theory and the Many-Body Problem* (Gordon and Breach, New York, 1967).
[32] Gilmore R, *Catastrophe Theory for Scientists and Engineers*, (Wiley, New York, 1981).

Journal of Physics: Conference Series **512** (2014) 012007 doi:10.1088/1742-6596/512/1/012007

Discrete and continuous symmetries in α-cluster nuclei

Roelof Bijker

Instituto de Ciencias Nucleares, Universidad Nacional Autónoma de México, A.P. 70-543, 04510 México, D.F., México

E-mail: bijker@nucleares.unam.mx

Abstract. I discuss some properties of the algebraic cluster model which is based on the introduction of a spectrum generating algebra for the relative motion of the clusters. Particular attention is paid to the permutation symmetry of identical clusters. As an example, it is shown how the method can be used to study α-cluster configurations in ^{12}C and ^{16}O with point group symmetries, $\mathcal{D}_{3h}$ and $\mathcal{T}_d$, respectively.

1. Introduction

The binding energy per nucleon for light nuclei shows large oscillations with maxima for nuclei with $A = 4n$ and $Z = N$, indicating the importance of α-cluster phenomena in these nuclei. There exists a large literature on α cluster models to describe the spectral properties of nuclei like ^{12}C and ^{16}O [1, 2, 3, 4] (for a recent review on cluster models see [5]).

Recently, there has been a lot of renewed interest in the structure of ^{12}C with the measurement of a new lowlying 2^+ state around 10 MeV with a width of approximately 0.8 MeV [6, 7, 8] which was interpreted as a rotational excitation of the Hoyle state, thus providing the first evidence of a collective band structure built on top of the Hoyle state.

In view of the new experimental information, in this contribution I revisit an algebraic treatment of α-cluster nuclei, called the Algebraic Cluster Model [9], discuss the permutation symmetry of identical clusters, as well as two special solutions which are relevant to the α-cluster nuclei ^{12}C and ^{16}O.

2. The Algebraic Cluster Model

The Algebraic Cluster Model (ACM) is an interacting boson model to describe the relative motion of n clusters based on the spectrum generating algebra of $U(\nu + 1)$ with $\nu = 3(n - 1)$. As special cases the ACM contains the vibron model [10] for two-body problems ($n = 2$), the $U(7)$ model for three-body clusters ($n = 3$) with applications in hadronic physics [11], molecular physics [12] and nuclear physics [9], and the $U(10)$ model for four-body clusters ($n = 4$) [13].

The building blocks of the ACM consist of a dipole boson $b_k^\dagger$ for each independent relative Jacobi coordinate $k = 1, \ldots, n - 1$ and an auxiliary scalar boson $s^\dagger$ which is added under the restriction that the total number of bosons $N = n_s + n_b = n_s + \sum_k n_k$ is conserved. The bilinear products of creation and annihilation operators span the Lie algebra of $U(3n-2)$. All operators of interest, such as the Hamiltonian and electromagnetic transition operators, are expanded into elements of this algebra. This procedure leads to a compact spectrum generating algebra of $U(3n-2)$ whose model space contains the oscillator shells with $n_b = 0, 1, 2, \ldots, N$. The introduction of the scalar boson makes it possible to investigate the mixing of oscillator shells.

For n identical objects, the Hamiltonian has to be invariant under the permuation group S_n. The transformation properties under S_n of all operators in the model follow from those of the building blocks. The scalar boson $s^\dagger$ transforms as the symmetric representation $[n]$, whereas the dipole bosons $b_k^\dagger$ transform as the $n-1$ components of the mixed symmetry representation $[n-1, 1]$ [14]. Next, one can use the multiplication rules for S_n to construct physical operators with the appropriate symmetry properties. As a result, the most general one- and two-body Hamiltonian that describes the relative motion of a system of n identical clusters, is a scalar under the permutation group S_n, is rotationally invariant, conserves parity as well as the total number of bosons, is given by

$$\begin{aligned} H &= \epsilon_0\, s^\dagger \tilde{s} - \epsilon_1 \sum_k b_k^\dagger \cdot \tilde{b}_k + u_0\, s^\dagger s^\dagger\, \tilde{s}\tilde{s} - u_1 \sum_k s^\dagger b_k^\dagger \cdot \tilde{b}_k \tilde{s} + v_0 \left[\sum_k b_k^\dagger \cdot b_k^\dagger\, \tilde{s}\tilde{s} + \text{h.c.} \right] \\ &\quad + \sum_L \Big[w_L\, T_{[n]}^{\dagger(L)} \cdot \tilde{T}_{[n]}^{(L)} + x_L\, T_{[n-1,1]}^{\dagger(L)} \cdot \tilde{T}_{[n-1,1]}^{(L)} \\ &\qquad + y_L\, T_{[n-2,1,1]}^{\dagger(L)} \cdot \tilde{T}_{[n-2,1,1]}^{(L)} + z_L\, T_{[n-2,2]}^{\dagger(L)} \cdot \tilde{T}_{[n-2,2]}^{(L)} \Big] , \end{aligned} \tag{1}$$

with $\tilde{b}_{k,m} = (-1)^{1-m} b_{k,-m}$ and $\tilde{s} = s$. The coefficients ϵ_0, ϵ_1, u_0, u_1, v_0, w_L, x_L, y_L and z_L parametrize the interactions. The operators $T_{[f]}^{(L)}$ are spherical tensor operators with respect to the rotation group as well as irreducible tensor operators with respect to the permutation group

$$T_{[f]k}^{\dagger(L)} = \sum_{i,j=1}^{n-1} c_{ij}^{[f]k}\, (b_i^\dagger \times b_j^\dagger)^{(L)} . \tag{2}$$

Here L represents the angular momentum ($L = 0, 2$ for $i = j$ and $L = 0, 1, 2$ for $i \neq j$) and the coefficients $c^{[f]}$ denote the Clebsch-Gordan coefficients for S_n

$$c_{ij}^{[f]k} = \langle [n-1,1]i, [n-1,1]j | [f]k \rangle , \tag{3}$$

with

$$[n-1,1] \otimes [n-1,1] = [n] \oplus [n-1,1] \oplus [n-2,1,1] \oplus [n-2,2] \qquad \text{for } n \geq 4 . \tag{4}$$

For $n = 3$ the last term is absent. Recently, the Clebsch-Gordan coefficients of Eq. (3) were derived in closed analytic form [15], in agreement with the expressions obtained in [14].

The eigenvalues and corresponding eigenvectors are obtained numerically by diagonalizing the Hamiltonian in a coupled harmonic oscillator basis. By construction, the wave functions are characterized by the total number of bosons N, angular momentum and parity L^P, and their transformation property $[f]$ under the permutation group S_n. For the case of the harmonic oscillator the construction of states with good permutation symmetry was studied by Kramer and Moshinsky [14]. In particular, it was shown that for the n-body problem, the reduction $U(3n-3) \supset U(3) \otimes U(n-1)$ with $U(3) \supset SO(3)$ and $U(n-1) \supset O(n-1) \supset S_n$ provides a basis for an explicit construction of states with good permutation symmetry. Here $U(3)$ denotes the symmetry group for the coupled harmonic oscillator and $U(n-1)$ is the symmetry group associated with the transformation of the indices of the $n-1$ relative Jacobi vectors.

However, in the application to the ACM the number of oscillator shells can be large ($n_b = 0, 1, \ldots, N$ with $N \sim 10 - 20$), and moreover the v_0 term in Eq. (1) mixes different oscillator shells with $\Delta n_b = \pm 2$. For this reason, there is need for a general procedure to create a set of basis states with good permutation symmetry which is easy to implement. One possible solution is to generate these symmetry-adapted basis states numerically by diagonalizing S_n invariant interactions for each oscillator shell separately and to determine the permutation symmetry $[f]$ of a given basis state from the transformation properties under the interchange $P(12)$ and the cyclic permutation $P(12\cdots n)$. Since the Hamiltonian of Eq. (1) is invariant under $P(12)$, basis states with n_1 even and n_1 odd do not mix, and can therefore be treated separately. In operator form, one has

$$\left\langle \psi_{[f]k} \right| P(12) \left| \psi_{[f]k} \right\rangle = \left\langle \psi_{[f]k} \right| \mathrm{e}^{i\pi b_1^\dagger b_1} \left| \psi_{[f]k} \right\rangle . \tag{5}$$

Next, the permutation symmetry $[f]$ of the wave functions can be determined by evaluating the matrix elements of the cyclic permutation $P(12\cdots n)$

$$\left\langle \psi_{[f]k} \right| P(12\cdots n) \left| \psi_{[f]k} \right\rangle = \left\langle \psi_{[f]k} \right| \mathrm{e}^{i\pi \sum_{i=1}^{n-1} b_i^\dagger b_i} \prod_{j=1}^{n-2} \mathrm{e}^{\theta_j (b_j^\dagger b_{j+1} - b_{j+1}^\dagger b_j)} \left| \psi_{[f]k} \right\rangle , \tag{6}$$

with $\theta_j = \arctan \sqrt{j(j+2)}$. The first term depends on the total number of oscillator quanta n_b and gives rise to a factor $+1$ (-1) for states with even (odd) parity. The other terms can be interpreted as a change of oscillator coordinates, and hence their matrix elements can be expressed in terms of Talmi-Moshinksy brackets [16, 17] which are calculated with the program TMB developed by Dobeš [18]. Finally, the relative phases of the degenerate representations can be determined from the off-diagonal matrix elements

$$\left\langle \psi_{[f]k} \right| P(12\cdots n) \left| \psi_{[f]k'} \right\rangle \tag{7}$$

by requiring that they transform like the components of the degenerate representations.

This numerical technique was developed for systems of three identical clusters [11] and later generalized to four identical clusters [13].

3. Special solutions

The ACM has a rich algebraic structure, which includes both continuous and discrete symmetries. It is of general interest to study limiting cases of the Hamiltonian in which the energy spectra can be obtained in closed form. The S_n invariant ACM Hamiltonian of Eq. (1) has dynamic symmetries corresponding to the group chains

$$U(3n-2) \supset \left\{ \begin{array}{l} U(3n-3) \\ \\ SO(3n-2) \end{array} \right. . \tag{8}$$

These dynamic symmetries were studied in general for the n-body problem and were shown to correspond to the $(3n-3)$-dimensional (an)harmonic oscillator and a deformed oscillator, respectively [13].

In addition, there are several other limiting cases of S_n invariant Hamiltonians for which approximate solutions can be obtained in a semiclassical mean-field analysis and which are of interest for the application to α-cluster nuclei. For the case of three-body clusters I discuss the oblate symmetric top with three particles located at the vertices of an equilateral triangle, and for four-body clusters the spherical top with four particles at the vertices of a regular tetrahedron.

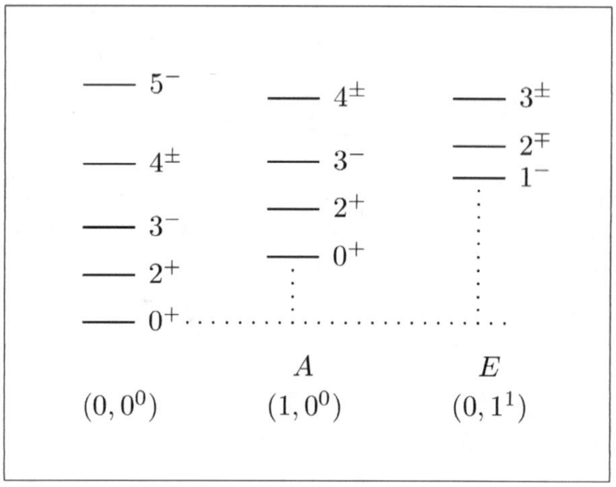

Figure 1. Schematic spectrum of an oblate symmetric top. The rotational bands are labeled by (v_1, v_2^l) (bottom). All states are symmetric under $S_3 \sim \mathcal{D}_3$.

3.1. Three-body clusters: oblate top

For the case of three identical clusters the spectrum generating algebra of the ACM is given by $U(7)$. Here I consider a Hamiltonian which is a special case of Eq. (1)

$$\begin{aligned} H_1 &= \xi_1 \,(R^2\, s^\dagger s^\dagger - b_1^\dagger \cdot b_1^\dagger - b_2^\dagger \cdot b_2^\dagger)\,(R^2\, \tilde{s}\tilde{s} - \tilde{b}_1 \cdot \tilde{b}_1 - \tilde{b}_2 \cdot \tilde{b}_2) \\ &\quad + \xi_2 \left[(b_1^\dagger \cdot b_1^\dagger - b_2^\dagger \cdot b_2^\dagger)\,(\tilde{b}_1 \cdot \tilde{b}_1 - \tilde{b}_2 \cdot \tilde{b}_2) + 4\,(b_1^\dagger \cdot b_2^\dagger)\,(\tilde{b}_2 \cdot \tilde{b}_1) \right] \\ &\quad + 2\kappa_1 \,(b_1^\dagger \times \tilde{b}_1 + b_2^\dagger \times \tilde{b}_2)^{(1)} \cdot (b_1^\dagger \times \tilde{b}_1 + b_2^\dagger \times \tilde{b}_2)^{(1)} \\ &\quad + 3\kappa_2 \,(b_1^\dagger \times \tilde{b}_2 - b_2^\dagger \times \tilde{b}_1)^{(0)} \cdot (b_2^\dagger \times \tilde{b}_1 - b_1^\dagger \times \tilde{b}_2)^{(0)} \,. \end{aligned} \tag{9}$$

For $R^2 = 0$, this Hamiltonian has $U(7) \supset U(6)$ symmetry (anharmonic oscillator), whereas for $R^2 = 1$ and $\xi_2 = 0$ it has $U(7) \supset SO(7)$ symmetry (deformed oscillator). The general case with $R^2 \neq 0$ and ξ_1, $\xi_2 > 0$ corresponds to a geometric configuration in which the α particles are located at the vertices of an equilateral triangle [9]. The corresponding point group symmetry is $\mathcal{D}_{3h}$ whose subgroup $\mathcal{D}_3$ is isomorphic to the permutation group S_3. Even though in this case the energy eigenvalues cannot be derived in closed form, an approximate expression for the energy levels can be obtained by making use of the method of intrinsic or coherent states (valid in the limit of large N) [11, 12]

$$E_1 \approx \omega_1\,(v_1 + \frac{1}{2}) + \omega_2\,(v_2 + 1) + \kappa_1\, L(L+1) + \kappa_2\,(K \pm 2l)^2 \,, \tag{10}$$

with frequencies $\omega_1 = 4NR^2\xi_1$ and $\omega_2 = 4NR^2\xi_2/(1+R^2)$. The quantum numbers have the following meaning: v_1, v_2 are vibrational quantum numbers: v_1 corresponds to a symmetric one-dimensional stretching vibration, and v_2 represents a two-dimensional bending vibration; $l = v_2, v_2-2, \ldots, 1$ or 0 is the vibrational angular momentum of the doubly degenerate vibration; L is the angular momentum, and K its projection on a body-fixed axis [9].

Fig. 1 shows the rotation-vibration spectrum of a triangular configuration with $\mathcal{D}_{3h}$ symmetry which is characterized by a series of rotational bands labeled by (v_1, v_2^l). It is assumed that the spin of the identical clusters is zero, as is relevant for the description of the ^{12}C nucleus as a

Journal of Physics: Conference Series **512** (2014) 012007 doi:10.1088/1742-6596/512/1/012007

cluster of three α particles. As a consequence, all states in Fig. 1 are symmetric under S_3. For this case, the bands with $(v_1, 0^0)$ can have angular momenta and parity $L^P = 0^+, 2^+, 3^-, 4^\pm, 5^-, \ldots$, whereas the angular momentum content of the doubly degenerate vibrations $(v_1, 1^1)$ is given by $L^P = 1^-, 2^\mp, 3^\mp, \ldots$, in agreement with Ref. [19]. The structure of the rotational bands can be considered as the fingerprint of the underlying $\mathcal{D}_{3h}$ symmetry of three identical structureless particles at the vertices of an equilateral triangle.

3.2. Four-body clusters: spherical top

As a second example, I discuss the spherical top with tetrahedral symmetry $\mathcal{T}_d$ as a special case of the ACM for four identical clusters. The spectrum generating algebra is given by $U(10)$. Let us consider a S_4 invariant Hamiltonian of the form

$$\begin{aligned} H_2 &= \xi_1 \, (R^2 \, s^\dagger s^\dagger - b_1^\dagger \cdot b_1^\dagger - b_2^\dagger \cdot b_2^\dagger - b_3^\dagger \cdot b_3^\dagger) \, (\text{h.c.}) \\ &\quad + \xi_2 \left[(-2\sqrt{2} \, b_1^\dagger \cdot b_3^\dagger + 2 b_1^\dagger \cdot b_2^\dagger) \, (\text{h.c.}) \right. \\ &\qquad\qquad \left. + \, (-2\sqrt{2} \, b_2^\dagger \cdot b_3^\dagger + (b_1^\dagger \cdot b_1^\dagger - b_2^\dagger \cdot b_2^\dagger)) \, (\text{h.c.}) \right] \\ &\quad + \xi_3 \left[(2 b_1^\dagger \cdot b_3^\dagger + 2\sqrt{2} \, b_1^\dagger \cdot b_2^\dagger) \, (\text{h.c.}) \right. \\ &\qquad\qquad + (2 b_2^\dagger \cdot b_3^\dagger + \sqrt{2} \, (b_1^\dagger \cdot b_1^\dagger - b_2^\dagger \cdot b_2^\dagger)) \, (\text{h.c.}) \\ &\qquad\qquad \left. + (b_1^\dagger \cdot b_1^\dagger + b_2^\dagger \cdot b_2^\dagger - 2 b_3^\dagger \cdot b_3^\dagger) \, (\text{h.c.}) \right] \\ &\quad + \kappa_1 \, \vec{L} \cdot \vec{L} + \kappa_2 \, (\vec{L} \cdot \vec{L} - \vec{I} \cdot \vec{I})^2 \;, \end{aligned} \tag{11}$$

where $\vec{L}$ denotes the angular momentum in coordinate space and $\vec{I}$ the angular momentum in index space. For $R^2 = 0$, this Hamiltonian has $U(10) \supset U(9)$ symmetry (anharmonic oscillator), whereas for $R^2 = 1$ and $\xi_2 = \xi_3 = 0$ it has $U(10) \supset SO(10)$ symmetry (deformed oscillator). The general case with $R^2 \neq 0$ and ξ_1, ξ_2, $\xi_3 > 0$ corresponds to a geometric configuration in which the four clusters are located at the vertices of a regular tetrahedron [13]. The corresponding point group symmetry is $\mathcal{T}_d$ which is isomorphic to the permutation group S_4. Just as in the previous case of the oblate top, also for the Hamiltonian of Eq. (11) an approximate energy formula can be derived in a semiclassical mean-field analysis by making use of the methods of intrinsic or coherent states which is valid in the limit of large N. The resulting energy spectrum is that of the vibrational and rotational excitations of a spherical top with tetrahedral symmetry

$$E_2 \approx \omega_1 (v_1 + \frac{1}{2}) + \omega_2 (v_2 + 1) + \omega_3 (v_3 + \frac{3}{2}) + \kappa_1 \, L(L+1) \;, \tag{12}$$

with frequencies $\omega_1 = 4NR^2\xi_1$, $\omega_2 = 8NR^2\xi_2/(1+R^2)$ and $\omega_3 = 8NR^2\xi_3/(1+R^2)$. Here v_1 represents the vibrational quantum number for a symmetric stretching A vibration, v_2 denotes a doubly degenerate E vibration, and v_3 a three-fold degenerate F vibration.

Whereas the angular momentum L is an exact symmetry of H_2 of Eq. (11), the angular momentum in index space I in general does not commute with the Hamiltonian. Only if $\xi_2 = \xi_3$ in Eq. (11), does I become a good quantum number. The rotational excitations of the ground state vibrational band of H_2 with $(v_1, v_2, v_3) = (0,0,0)$ are characterized by $L = I$. This property is a consequence of the fact that the operator $\vec{L} \cdot \vec{L} - \vec{I} \cdot \vec{I}$ annihilates the coherent (or intrinisic) state corresponding to the rigid equilibrium shape of a regular tetrahedron.

Fig. 2 shows the structure of the spectrum of a spherical top with tetrahedral symmetry according to the approximate energy formula of Eq. (12). The energy spectrum consists of a series of rotational bands labeled by (v_1, v_2, v_3). It is assumed that the spin of the identical clusters is zero, as is relevant for the description of the ^{16}O nucleus as a cluster of four α

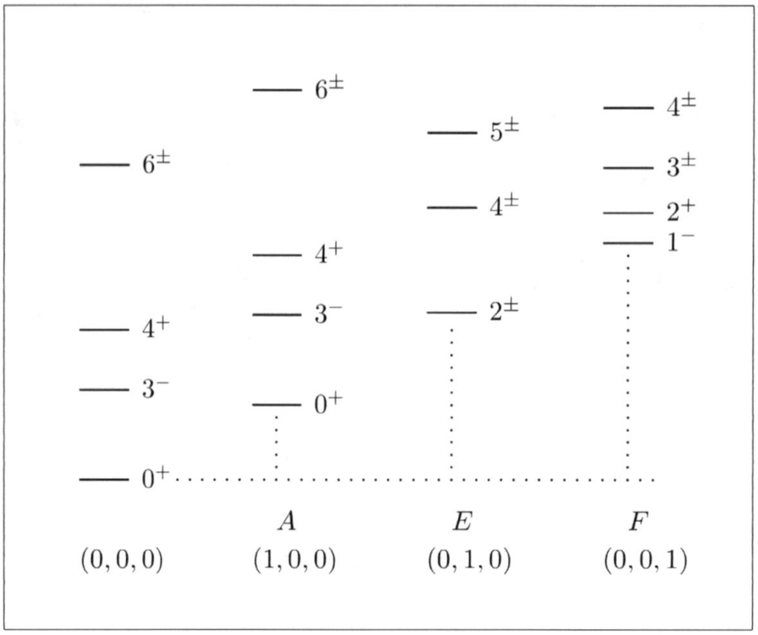

Figure 2. Schematic spectrum of a spherical top with tetrahedral symmetry. The rotational bands are labeled by (v_1, v_2, v_3) (bottom). All states are symmetric under $S_4 \sim \mathcal{T}_d$.

particles. As a consequence, all states in Fig. 2 are symmetric under S_4. For this case, the bands with $(v_1, 0, 0)$ can have angular momenta and parity $L^P = 0^+, 3^-, 4^+, 6^\pm, \ldots$, whereas the angular momentum content of the doubly degenerate E vibration $(v_1, 1, 0)$ is given by $L^P = 2^\pm, 4^\pm, 5^\pm, \ldots$, and for the triply degenerate F vibration $(v_1, 0, 1)$ by $L^P = 1^-, 2^+, 3^\pm, 4^\pm, \ldots$, in agreement with Ref. [19]. The structure of the rotational bands can be considered as the fingerprint of the underlying $\mathcal{T}_d$ tetrahedral symmetry of four identical structureless particles at the vertices of a regular tetrahedron.

4. Summary and conclusions

In this contribution, I discussed the interplay of continuous and discrete symmetries in the framework of the Algebraic Cluster Model (ACM). The ACM incorporates all vibrational and rotational degrees of freedom from the outset. In addition, the permutation symmetry of identical clusters can be taken into account in an exact manner.

It was shown that for systems with identical clusters the structure of rotational bands is determined by the underlying point-group symmetry of the geometric configuration of the α particles. For the case of three-body clusters with the three α particles located at the vertices of an equilateral triangle (oblate top), the rotational structure of the ground state band is given by the sequence $L^P = 0^+$, 2^+, 3^-, $4^\pm$, 5^-, ..., whereas for the case of four-body clusters with the four α particles located at the vertices of a regular tetrahedron (spherical top), the rotational sequence is given by $L^P = 0^+$, 3^-, 4^+, $6^\pm$, ...

If a physical system, like an α-cluster nucleus, is claimed to consist of identical structureless particles at the vertices of an equilateral triangle (as in ^{12}C) or a regular tetrahedron (as in ^{16}O), then its spectrum *must* be as shown in Figs. 1 or 2, respectively. The rotational structure provides a fingerprint of the underlying point-group symmetry of the geometric configuration of the α particles.

Acknowledgments

This work was supported in part by a grant from PAPIIT-UNAM, Mexico.

References

[1] Wheeler J A 1937 *Phys. Rev.* **52** 1083
[2] Hafstad L R and Teller E 1938 *Phys. Rev.* **54** 681
[3] Brink D M 1966 *International School of Physics: "Enrico Fermi" Course XXXVI* (New York: Academic Press)
[4] Robson D 1978 *Nucl. Phys. A* **308** 381
[5] Von Oertzen W, Freer M and Kanada-En'yo Y 2006 *Phys. Rep.* **432** 43
[6] Itoh M, *et al.* 2011 *Phys. Rev. C* **84** 054308
[7] Freer M, *et al.* 2012 *Phys. Rev. C* **86** 034320
[8] Zimmerman W R, *et al.* 2013 *Phys. Rev. Lett.* **110** 152502
[9] Bijker R and Iachello F 2000 *Phys. Rev. C* **61** 067305
Bijker R and Iachello F 2002 *Ann. Phys. (N.Y.)* **298** 334
[10] Iachello F and Levine R D 1995 *Algebraic Theory of Molecules* (Oxford: Oxford U. Press)
[11] Bijker R, Iachello F and Leviatan A 1994 *Ann. Phys. (N.Y.)* **236** 69
Bijker R, Iachello F and Leviatan A 2000 *Ann. Phys. (N.Y.)* **284** 89
[12] Bijker R, Dieperink A E L and Leviatan A 1995 *Phys. Rev. A* **52** 2786
Bijker R and Leviatan A 1998 *Few-Body Systems* **25** 89
[13] Bijker R 2010 *AIP Conference Proceedings* **1323** 28
Bijker R 2012 *J. Phys.: Conf. Ser.* **380** 012003
[14] Kramer P and Moshinsky M 1966 *Nucl. Phys.* **82** 241
[15] Doma S B and Hammad M M 2012 *J. Math. Phys.* **53** 033511
[16] I. Talmi, Helv. Phys. Acta **25**, 185 (1952).
[17] M. Moshinksy, Nucl. Phys. **13**, 104 (1959).
[18] J. Dobeš, J. Phys. A: Math. Gen. **10**, 2053 (1977).
[19] Herzberg G 1991 *Molecular Spectra and Molecular Structure. II. Infrared and Raman Spectra of Polyatomic Molecules* (Malabar Florida: Krieger)

Journal of Physics: Conference Series **512** (2014) 012008 doi:10.1088/1742-6596/512/1/012008

Symmetry breaking in noncommutative finite temperature $\lambda\phi^4$ theory with a nonuniform ground state

J. M. Hernández, C. Ramírez, M. Sánchez

Facultad de Ciencias Físico Matemáticas, Universidad Autónoma de Puebla, P.O. Box 1364, 72000 Puebla, México.

E-mail: [1]`javierh@fcfm.buap.mx`, [2]`cramirez@fcfm.buap.mx`, [3]`hanamichi2_2000@yahoo.com`

Abstract. We consider the CJT effective action at finite temperature for a noncommutative real scalar field theory, with noncommutativity among space and time variables. We study the solutions of a stripe type nonuniform background, which depends on space and time. The analysis in the first approximation shows that such solutions appear in the planar limit, but also under normal anisotropic noncommutativity. Further we show that the transition from the uniform ordered phase to the non uniform one is first order and that the critical temperature depends on the nonuniformity of the ground state.

1. Introduction

Noncommutativity in field theory has received much attention in recent years, as a means to explore high energy effects [1]. An interesting question is the finite temperature behavior and phase transitions. In particular there is a breaking of translational invariance [2], which leads to the appearance of a stripe phase, related to the UV/IR mixing [2, 3, 4]. It is known that the noncommutativity of time with space variables leads to causality problems [5]. However, at finite temperature, in the imaginary time formalism, we could consider full noncommutativity [6], which could be related to nonequilibrium processes. This type of processes has been studied in [7, 8] by means of the effective action of Cornwall, Jackiw and Tomboulis (CJT) [9]. This effective action allows to make selective summations to higher loop graphs [10], which on the other side can be seen as consistent truncations which describe dissipative processes [7]. In [11] we considered the conditions for the existence of a stripe phase, in a scalar field theory with $\lambda\phi^4$ interaction at finite temperature, with noncommutativity among all variables. We included a dependence of the stripe background on the imaginary time. The study non uniform solutions was made by means of a Rayleigh-Ritz variation of the CJT effective action. We obtained that there are consistent solutions in the planar limit, as already found in [12, 2], but also with anisotropic noncomutativity, with a temperature dependent background. Moreover there is a discontinuity from the uniform phase to the non uniform one, which points to a first order transition between these phases, thus confirming the phase transition in four dimensions conjectured in [2], corresponding to a Lifshitz point.

Journal of Physics: Conference Series **512** (2014) 012008 doi:10.1088/1742-6596/512/1/012008

2. CJT effective action

Let us consider an action of scalar fields $I(\Phi)$. The CJT [9] effective action $\Gamma(\phi, G)$ is obtained including a quadratic source term in the generating functional of the Green functions,

$$Z(J,K) = \int D\Phi \exp\left\{ i \left[I(\Phi) + \int d^4x \Phi(x) J(x) + \frac{1}{2} \int d^4x d^4y \Phi(x) K(x,y) \Phi(y) \right] \right\}, \quad (1)$$

and then doing a Legendre transformation of the generating functional of the connected diagrams

$$\begin{aligned} \Gamma(\phi, G) = W(J,K) \quad &- \int d^4x \phi(x) J(x) - \frac{1}{2} \int d^4x d^4y \phi(x) K(x,y) \phi(y) \\ &- \frac{1}{2} \int d^4x d^4y G(x,y) K(y,x). \end{aligned} \quad (2)$$

Defining

$$\frac{\delta W(J,K)}{\delta J(x)} = \phi(x), \qquad \text{and} \qquad \frac{\delta W(J,K)}{\delta K(x,y)} = \frac{1}{2} \left[\phi(x)\phi(y) + G(x,y) \right], \quad (3)$$

it turns out that $\phi(x)$ is the vacuum expectation value of $\Phi(x)$, and $G(x,y)$ is the connected two point function. The physical solutions are obtained from the stationarity conditions,

$$\frac{\delta}{\delta \phi(x)} \Gamma(\phi, G) = 0, \qquad \text{and} \qquad \frac{\delta}{\delta G(x,y)} \Gamma(\phi, G) = 0. \quad (4)$$

The resulting effective action is [9]

$$\Gamma(\phi, G) = I(\phi) + \frac{i}{2} Tr \ln G^{-1} + \frac{i}{2} Tr \left[\Delta^{-1}(\phi) G \right] + \Gamma_2(\phi, G) - \frac{i}{2} Tr(1), \quad (5)$$

where $\Gamma_2(\phi, G)$ is the 2PI diagram expansion, $iD^{-1}(x-y) = -\left(\partial^\mu \partial_\mu + m^2\right) \delta^4(x-y)$ and

$$i\Delta^{-1}(x-y) = \frac{\delta^2 I(\phi)}{\delta\phi(x)\delta\phi(y)} = -\left(\partial^\mu \partial_\mu + m^2\right) \delta^4(x-y) + \frac{\delta^2 I_{\text{int}}(\phi)}{\delta\phi(x)\delta\phi(y)}. \quad (6)$$

Further, for a $\lambda\Phi^4$ action

$$I(\Phi) = \int_{-\infty}^{\infty} d^4x \left(\frac{1}{2} \partial^\mu \Phi \partial_\mu \Phi - \frac{1}{2} m^2 \Phi^2 - \frac{\lambda}{4!} \Phi^4 \right), \quad (7)$$

Γ_2 is given to first order in λ by [9],

$$\Gamma_2(\phi, G) = -\frac{\lambda}{8} \int_{-\infty}^{\infty} d^4x G^2(x,x). \quad (8)$$

3. Noncommutative finite temperature action

Let us consider the noncommutative action for a real scalar field with a $\lambda\Phi^4$ potential,

$$I(\Phi) = \int_{-\infty}^{\infty} d^4x \left(\frac{1}{2} \partial^\mu \Phi \partial_\mu \Phi - \frac{1}{2} m^2 \Phi^2 - \frac{\lambda}{4!} \Phi * \Phi * \Phi * \Phi \right). \quad (9)$$

The noncommutative Weyl-Moyal product is given by,

$$\Phi(x) * \Psi(x) = e^{\frac{i}{2}\theta^{\mu\nu}\partial^x_\mu \partial^y_\nu} \Phi(x) \Psi(y) \Big|_{y=x}. \quad (10)$$

We are interested on symmetry breaking of noncommutative finite temperature field theory. Action (9), as well as (7), is symmetric under $\Phi \to -\Phi$. The imaginary time formulation is obtained from the relativistic field theory by setting the fields periodic in the imaginary time variable $\tau = it$, $\Phi(\tau + \beta) = \Phi(\tau)$. The imaginary time integration interval is $[0, \beta]$, and $i \int d^4x \to \int_0^\beta d\tau \int d^3x$. The fourier transform of the fields is then given by

$$\phi(x) = \frac{1}{\beta} \sum_{-\infty}^{\infty} \int \frac{d^3p}{(2\pi)^3} e^{i(\omega_n \tau + \vec{p}\vec{x})} \phi_n(p), \tag{11}$$

where $\omega_n = \frac{2\pi n}{\beta}$ is the Matsubara frequency, see e.g. [14]. Thus the free action entering into the CJT effective action will be,

$$I_0(\phi) = -\frac{1}{2\beta} \sum_{n=-\infty}^{\infty} \int \frac{d^3p}{(2\pi)^3} \left(p_n^2 + m^2\right) \phi_n(\vec{p}) \phi_{-n}(-\vec{p}), \tag{12}$$

where $p_n \equiv (\omega_n, \vec{p})$, i.e. $p_n^2 = \omega_n^2 + \vec{p}^2$. Further, if $\Theta_{\tau i} = i\theta^{0i}$ and $\Theta_{ij} = \theta^{ij}$, we write

$$\phi(x) * \phi(x) = e^{\frac{i}{2}\partial_x \wedge \partial_y} \phi(x) \phi(y)\Big|_{y=x} \equiv e^{\frac{i}{2}[\Theta_{\tau i}(\partial_\tau^x \partial_i^y - \partial_i^x \partial_\tau^y) + \Theta_{ij}\partial_i^x \partial_j^y]} \phi(x) \phi(y)\Big|_{y=x}. \tag{13}$$

Therefore, if $(\Theta_\tau)_i = \Theta_{\tau i}$, $\Theta_i = \epsilon_{ijk}\Theta_{jk}$, for the interaction action we get

$$I_{int}(\phi) = \frac{\lambda}{4!}(2\pi)^2 \prod_{k=1}^{4} \left[\sum_{n_k} \int \frac{d^3p}{(2\pi)^3} \phi_{n_k}(\vec{p}_k) \right] e^{\frac{i}{2}(p_{n1} \wedge p_{n2} + p_{n3} \wedge p_{n4})} \delta_{\sum_i n_i, 0}\, \delta^3\left(\sum_i \vec{p}_i\right), \tag{14}$$

where $p_{n1} \wedge p_{n2} = \Theta_{\tau i}(\omega_{n_1} p_{2i} - \omega_{n_2} p_{1i}) + \Theta_{ij} p_{1i} p_{2j} = \vec{\Theta}_\tau(\omega_{n_1}\vec{p}_2 - \omega_{n_2}\vec{p}_1) + \vec{\Theta}(\vec{p}_1 \times \vec{p}_2)$. The noncommutative CJT action can be obtained from (9) and (5), as follows

$$i\Gamma(\phi, G) = iI(\phi) - \frac{1}{2} Tr \ln G^{-1} + \frac{1}{2} Tr \left\{ \frac{\delta^2}{\delta\phi\delta\phi}[iI(\phi)]G \right\} + i\Gamma_2(\phi, G) + \frac{1}{2} Tr(1), \tag{15}$$

which becomes real by the continuation to imaginary time. Thus we get, including the nonplanar two loop terms,

$$\begin{aligned} i\Gamma(\phi, G) &= -\frac{1}{2\beta} \sum_l \int \frac{d^3p}{(2\pi)^3} \left(p_l^2 + m^2\right) \phi_l(\vec{p}) \phi_{-l}(-\vec{p}) \\ &+ \frac{\lambda}{4!} \left(\frac{2\pi}{\beta}\right)^3 \prod_{k=1}^{4} \left[\sum_{n_k} \int \frac{d^3p_k}{(2\pi)^3} \phi_{n_k}(\vec{p}_k) \right] e^{\frac{i}{2}(p_{n1} \wedge p_{n2} + p_{n3} \wedge p_{n4})} \delta_{\sum_i n_i, 0}\, \delta^3\left(\sum_i \vec{p}_i\right) \\ &+ \frac{1}{2} \sum_l \int d^3p \left\{ (p_l^2 + m^2) G(p_l) + \log[G(p_l)] \right\} \delta^3(\vec{0}) \\ &+ \frac{\lambda}{4\beta} \sum_{k,l} \int \frac{d^3p}{(2\pi)^3} \frac{d^3q}{(2\pi)^3} \left[\frac{1}{\beta} \phi_k(\vec{p}) \phi_{-k}(-\vec{p}) G(q_l) - \frac{(2\pi)^3}{2} G(p_k) G(q_l) \delta^3(\vec{0}) \right] \left[1 + \frac{1}{2} e^{i(p_k \wedge q_l)} \right]. \end{aligned} \tag{16}$$

4. Nonuniform solutions

In [2], following [15], the ground state field is chosen in a stripe phase, $\phi(x) = A\cos(\vec{Q}\vec{x})$. Considering the possibility of non closed systems [16], we take a dependence on the imaginary time

$$\phi(\tau, \vec{x}) = A\cos(\omega_n \tau + \vec{Q}\vec{x}). \tag{17}$$

These fields are commutative, i.e. $\phi(\tau,\vec{x}) * \phi(\tau,\vec{x}) = \phi^2(\tau,\vec{x})$. For simplicity, following [13], we keep $G(x,y) = G(x-y)$ translational invariant. The Fourier expansion coefficients of (17) are

$$\phi_l(\vec{p}) = (2\pi)^3 \frac{\beta}{2} A \left[\delta_{ln}\delta^3(\vec{p}-\vec{Q}) + \delta_{l,-n}\delta^3(\vec{p}+\vec{Q}) \right]. \tag{18}$$

Thus

$$\begin{aligned} i\Gamma(\phi,G) &= \Bigg\{ -\frac{1}{4}\beta A^2 \left(Q_n^2 + m^2 + \frac{\lambda}{16}A^2 \right) \\ &+ \frac{1}{2}\sum_l \int \frac{d^3q}{(2\pi)^3} \{ (q_l^2+m^2)G(q_l) + \log[G(q_l)] \} \\ &+ \frac{\lambda}{8}A^2 \sum_l \int \frac{d^3q}{(2\pi)^3} G(q_l)\left[1+\frac{1}{2}\cos(Q_n \wedge q_l)\right] \\ &- \frac{\lambda}{8\beta}\sum_{k,l}\int \frac{d^3p}{(2\pi)^3}\frac{d^3q}{(2\pi)^3} G(p_k)G(q_l)\left[1+\frac{1}{2}\cos(p_k\wedge q_l)\right] \Bigg\} V_3 \\ &+ \frac{1}{4}\beta A^2 \left[-\frac{5\lambda}{48}A^2 - m^2 + \frac{3\lambda}{4\beta}\sum_l \int \frac{d^3q}{(2\pi)^3} G(q_l) \right] \delta_{n,0}\delta_{\vec{Q}} V_3 + \text{const.} \end{aligned} \tag{19}$$

where $\delta_{\vec{Q}}$ vanishes if $\vec{Q} \neq \vec{0}$, and is 1 when $\vec{Q} = \vec{0}$. Further $V_3 = \int d^3x$ is the three dimensional volume. This expression is discontinuous at $\vec{Q} \neq \vec{0}$. Due to the dependence of (17) on parameters, a Rayleigh-Ritz variation with respect to A, $\vec{Q}$ and $G(p_k)$ of (19) is in order [9]. Thus

$$\begin{aligned} A\Bigg\{ \frac{\lambda}{8}A^2 + Q_n^2 + m^2 - \frac{\lambda}{2\beta}\sum_l \int \frac{d^3q}{(2\pi)^3} G(q_m)\left[1+\frac{1}{2}\cos(Q_n\wedge q_l)\right] \\ + \left[\frac{5\lambda}{24}A^2 + m^2 - \frac{3\lambda}{4\beta}\sum_l\int\frac{d^3q}{(2\pi)^3}G(q_l)\right]\delta_{n,0}\delta_{\vec{Q}} \Bigg\} = 0, \end{aligned} \tag{20}$$

$$A^2\left[-Q_i + \frac{\lambda}{2\beta}\sum_l \int \frac{d^3q}{(2\pi)^3} G(q_l)\,(2\pi l \Theta_{\tau i}T + \epsilon_{ijk}\Theta_j q_k)\sin(Q_n\wedge q_l) \right] = 0, \tag{21}$$

$$\begin{aligned} G^{-1}(p_k) + p_k^2 + m^2 + \frac{\lambda}{4}A^2\left[1+\frac{1}{2}\cos(Q_n\wedge p_k)\right] \\ -\frac{\lambda}{2\beta}\sum_l\int\frac{d^3q}{(2\pi)^3}G(q_l)\left[1+\frac{1}{2}\cos(p_k\wedge q_l)\right] + \frac{3\lambda}{8}A^2\delta_{n,0}\delta_{\vec{Q}} = 0. \end{aligned} \tag{22}$$

These equations have solutions, at least for low temperatures, only if $m^2 < 0$. Thus, in the following we suppose that this is the case and we set $\mu^2 = -m^2$. From the third equation, we see that the integrals $\int d^3q G(q_l)$ have UV divergences and in order to regularize them, a high-momentum Λ cutoff is made, and we define the (cut-off regularized) integral

$$I(p_k) = -\sum_l \int_\Lambda \frac{d^3q}{(2\pi)^3}\, G(q_l)\left[1+\frac{1}{2}\cos(p_k\wedge q_l)\right]. \tag{23}$$

For the ordered phase, at $A = 0$, (22) has the consistent, Q_n independent, solution

$$G^{-1}(p_k) = -p_k^2 + \mu^2 - \frac{\lambda}{2(2\pi)^3\beta}\, I(p_k)|_{A=0}\,. \tag{24}$$

Further, equations (20) and (22) are discontinuous at $\vec{Q} = \vec{0}$, which points to a phase transition from the uniform, constant ϕ field, to the stripe phase (17). Indeed, defining $A_0 = A|_{\phi=\text{constant}}$, $G_0(p_k) \equiv G(p_k)|_{\phi=\text{constant}}$, $G(q_l, 0) = G(q_l, Q_n)|_{\vec{Q}=\vec{0},n=0}$, where we made explicit the dependence on Q_n, and $\Delta G(p_k) = G(p_k, 0) - G_0(p_k)$, then the nonvanishing solution of (20) leads to the following equations for the discontinuity

$$A^2(0) - A_0^2 = \frac{18}{\beta} \sum_l \int_\Lambda \frac{d^3q}{(2\pi)^3} \, \Delta G(q_l), \tag{25}$$

$$G^{-1}(p_k, 0) - G_0^{-1}(p_k) = \frac{3}{2}\mu^2 - \frac{\lambda}{4\beta} \sum_l \int_\Lambda \frac{d^3q}{(2\pi)^3} \left\{ -\frac{9}{2} \, G_0(q_l) + \Delta G(q_l) \left[7 - \cos(p_k \wedge q_l)\right] \right\}. \tag{26}$$

Further, as (21) is a very complicated equation, we will look for solutions which, by consistency, are necessarily non vanishing. For $A \neq 0$, if we define the adimensional quantities, $\vec{\theta} = \vec{\Theta}\Lambda^2$, $\vec{\theta}_\tau = \vec{\Theta}_\tau \Lambda^2$, $\vec{\kappa} = \frac{\vec{Q}}{\Lambda}$, $\hat{\mu} = \frac{\mu}{\Lambda}$, $\vartheta = \frac{T}{\Lambda}$ and $r_i = \frac{q_i}{\Lambda}$ for the integration variables, (21) turns into

$$\kappa_i = -\frac{\lambda\vartheta}{2} \sum_l \int_\Lambda \frac{d^3r}{(2\pi)^3} \mathcal{G}(r_l) \left(2\pi l \vartheta \theta_{\tau i} + \epsilon_{ijk}\theta_j r_k\right) \sin\left[2\pi\vartheta\vec{\theta}_\tau(l\vec{\kappa} - n\vec{r}) + \vec{\theta}(\vec{r} \times \vec{\kappa})\right], \tag{27}$$

where $\mathcal{G}(r_l) = r^2 - \hat{\mu}^2 + \left[2(\hat{\mu}^2 - \kappa^2) - \frac{\lambda\vartheta}{\Lambda} I(\kappa_n)\right] \left\{1 + \frac{1}{2}\cos\left[2\pi\vartheta\vec{\theta}_\tau(l\vec{\kappa} - n\vec{r}) + \vec{\theta}(\vec{r} \times \vec{\kappa})\right]\right\} - \frac{\lambda\vartheta}{2\Lambda} I(r_l)$. As expected, κ_i vanish in the commutative limit. As we are interested on non vanishing solutions, we will look at the limit of the r.h.s., when $\kappa_i \to 0$. If the integral tends to zero, then $\kappa_i = 0$ is a consistent solution, otherwise, if the integral remains finite, then $\kappa_i \neq 0$. For finite values of $\vec{\theta}_\tau$ and $\vec{\theta}$, when κ_i tend to zero we get

$$\kappa_i \simeq \frac{\lambda\vartheta}{2} \sum_l \int_\Lambda \frac{d^3r}{(2\pi)^3} \frac{\epsilon_{ijk}\theta_j r_k \sin\left(2\pi n\vartheta\vec{\theta}_\tau\vec{r}\right)}{r^2 + \hat{\mu}^2 \left[1 + \cos\left(2\pi n\vartheta\vec{\theta}_\tau\vec{r}\right)\right]} + \mathcal{O}[(\lambda\vartheta)^2], \tag{28}$$

where the term proportional to $2\pi l\vartheta$ in the numerator of (27) vanishes because the integrand is an odd function. Decomposing θ_j in the numerator of (27) parallel and perpendicular to $\vec{\theta}_\tau$, $\vec{\theta} = \vec{\theta}_\parallel + \vec{\theta}_\perp$, the part of the integral with $\vec{\theta}_\parallel$ vanishes as well because the integrand is odd, hence

$$\kappa_i \simeq \frac{\lambda\vartheta}{2} \sum_l \int_\Lambda \frac{d^3r}{(2\pi)^3} \frac{\epsilon_{ijk}(\theta_\perp)_j r_k \sin\left(2\pi n\vartheta\vec{\theta}_\tau\vec{r}\right)}{r^2 + \hat{\mu}^2 \left[1 + \cos\left(2\pi n\vartheta\vec{\theta}_\tau\vec{r}\right)\right]}, \tag{29}$$

which does not vanish if $n \neq 0$, and there is a nonvanishing solution, $\kappa_i \neq 0$. The components $\vec{\theta}_\tau$ correspond to the maximal rank parametrization taken in [2] for $\Theta_{\mu\nu}$, and $\vec{\theta}_\perp$ corresponds to anisotropic noncommutativity, $\vec{\theta} \times \vec{\theta}_\tau \sim (\Theta^2)_{0i}$. Therefore for maximal rank noncommutativity there is no stripe phase, but for anisotropic noncommutativity $\kappa_i \neq 0$. In the planar limit, $\vec{\theta}_\tau$ and $\vec{\theta}$ tend to infinity, letting at the same time κ_i tend to zero, such that $\alpha_\tau = \vec{\theta}_\tau\vec{\kappa}$ and $\vec{\alpha} = \vec{\theta} \times \vec{\kappa}$ remain constant. We multiply (27) by κ_i and proceed to these limits, as follows.

If $n = 0$

$$\kappa^2 \simeq -\frac{\lambda\vartheta}{2} \sum_l \int_\Lambda \frac{d^3r}{(2\pi)^3} \frac{(2\pi l\vartheta\alpha_\tau + \vec{\alpha}\vec{r}) \sin(2\pi\vartheta l\alpha_\tau + \vec{\alpha}\vec{r})}{r^2 + \hat{\mu}^2 \left[1 + \cos(2\pi\vartheta l\alpha_\tau + \vec{\alpha}\vec{r})\right]} + \mathcal{O}[(\lambda\vartheta)^2]. \tag{30}$$

If $n \neq 0$

$$\kappa^2 \simeq -\frac{\lambda\vartheta}{2}\sum_l \int_\Lambda \frac{d^3r}{(2\pi)^3} \frac{(2\pi l\vartheta\alpha_\tau + \vec{\alpha}\vec{r})\sin\left(2\pi n\vartheta\vec{\theta}_\tau\vec{r}\right)}{r^2 + \hat{\mu}^2\left[1+\cos\left(2\pi n\vartheta\vec{\theta}_\tau\vec{r}\right)\right]} + \mathcal{O}[(\lambda\vartheta)^2]. \tag{31}$$

Both integrals differ by the argument of the trigonometric functions, which is finite when $n = 0$, and tends to infinity when $n \neq 0$. In both cases the denominator is positive. For $n = 0$, in general the integral does not vanish. For $n \neq 0$, the highly oscillating trigonometric functions in the numerator and in the denominator have same frequency and phase, and the integral does not vanish. Therefore in the limit $\vec{\theta}_\tau, \vec{\theta} \to \infty$, $\vec{\kappa}$ does not vanish, consistently with [2, 12].

Returning to the phase transition between the non uniform and the uniform phases, i.e. at $Q_i \to 0$, define $\Delta G(p_k) = G(p_k, 0) - G_0(p_k)$. Thus, from (25) and (26), it can be shown that

$$\Delta G(p_k) = -\frac{3}{2}\frac{\mu^2}{(p_k^2 + 2\mu^2)(p_k^2 + \frac{7}{2}\mu^2)} + \mathcal{O}(\lambda), \tag{32}$$

$$\frac{\lambda}{8}[A^2(0) - A_0^2] = \frac{1}{4}\mu^2 - \frac{3\lambda T}{16}\sum_l \int_\Lambda \frac{d^3q}{(2\pi)^3}\frac{q_l^2 + 8\mu^2}{(q_l^2 + 2\mu^2)(q_l^2 + \frac{7}{2}\mu^2)} + \mathcal{O}(\lambda^2), \tag{33}$$

which can be integrated to give

$$\frac{\lambda}{8\Lambda^2}[A^2(0) - A_0^2] = \frac{1}{4}\hat{\mu}^2 - \frac{3\pi\lambda\vartheta}{16}\sum_l \left\{1 - 4\sqrt{4l^2\pi^2\vartheta^2 + 2\hat{\mu}^2}\arctan\left[\frac{1}{\sqrt{4l^2\pi^2\vartheta^2 + 2\hat{\mu}^2}}\right] + \right.$$
$$\left. +3\sqrt{4l^2\pi^2\vartheta^2 + \frac{7\hat{\mu}^2}{2}}\arctan\left[\frac{1}{\sqrt{4l^2\pi^2\vartheta^2 + \frac{7\hat{\mu}^2}{2}}}\right]\right\} + \mathcal{O}(\lambda^2). \tag{34}$$

The r.h.s. of this expression is positive and does not vanish, for all temperatures below the cutoff. Therefore, there is a first order phase transition from the uniform to the non-uniform phase. This equation can be also applied to the transition from the non uniform to the disordered phase. The r.h.s. decreases as the temperature increases, but it reaches the maximum temperature (given by the cut-off) before reaching the zero value, and this phase transition is also first order.

Another question we are interested to explore is the critical temperature, at which the symmetry $\phi \to -\phi$ is restored. We have from (20) for nonvanishing A and $\vec{Q}$

$$\frac{\lambda}{8}A^2 = -Q_n^2 - \mu^2 - \frac{\lambda}{2(2\pi)^3\beta}I(Q_n), \tag{35}$$

which substituted into (22) gives

$$G^{-1}(p_k) = -p_k^2 + \mu^2 + 2\left(-\mu^2 + Q_n^2\right)\left[1 + \frac{1}{2}\cos(Q_n \wedge p_k)\right]$$
$$-\frac{\lambda}{2(2\pi)^3\beta}\left\{\left[1 + \frac{1}{2}\cos(Q_n \wedge p_k)\right]I(Q_n) - \frac{1}{2}I(p_k)\right\}. \tag{36}$$

Further, in order to see how are these solutions, we must consider the integral (23)

$$I(Q_n) = \sum_l \int d^3q \frac{1 + \frac{1}{2}\cos(Q_n \wedge q_l)}{2(\mu^2 - Q_n^2)\left[1 + \frac{1}{2}\cos(Q_n \wedge q_l)\right] + q_l^2 - \mu^2 - \frac{\lambda T}{(2\pi)^3}I_Q(q_l)}, \tag{37}$$

where

$$I_Q(q_l) = \left[1 + \frac{1}{2}\cos(Q_n \wedge q_l)\right] I(Q_n) - \frac{1}{2} I(q_l). \tag{38}$$

The denominator of the r.h.s. of (37), at zeroth order in λ is

$$\begin{aligned} G^{-1(0)}(q_l, T) &= (2\pi)^2 \left(-2n^2 \left\{1 + \frac{1}{2}\cos\left[2\pi\vec{\Theta}_\tau(n\vec{q} - l\vec{Q})T + \vec{\Theta}(\vec{Q}\times\vec{q})\right]\right\} + l^2\right) T^2 \\ &+2(\mu^2 - \vec{Q}^2)\left\{1 + \frac{1}{2}\cos\left[2\pi\vec{\Theta}_\tau(n\vec{q} - l\vec{Q})T + \vec{\Theta}(\vec{Q}\times\vec{q})\right]\right\} + \vec{q}^2 - \mu^2, \end{aligned} \tag{39}$$

which is positive for sufficiently high values of $q_l^2 = (2\pi T)^2 l^2 + \vec{q}^2$. Otherwise it can be negative, as happens with $G^{-1(0)}(0,T) = 2\left(\mu^2 - \frac{3}{2}Q_n^2\right)$ for high temperature. Eq. (39) vanishes in general at various values of the temperature, whose minimum we call T_0. Further, we consider the nonuniformity and the noncommutativity as small perturbations. We suppose also that at the temperatures we are considering, $|Q_n \wedge q_l| < \frac{\pi}{2}$, as can be verified later. Thus, if $\mu^2 > 2\vec{Q}^2$, the second row in (39) is positive, and if $l^2 < 2n^2$, the term containing T^2 will be negative. Hence under these assumptions, the temperature at which (39) vanishes

$$T^2 = \frac{2(\mu^2 - \vec{Q}^2)\left[1 + \frac{1}{2}\cos(Q_n \wedge q_l)\right] + \vec{q}^2 - \mu^2}{(2\pi)^2\left\{2n^2\left[1 + \frac{1}{2}\cos(Q_n \wedge q_l)\right] - l^2\right\}}, \tag{40}$$

will increase when $\vec{q}^2$ and l increase. Thus (40) will be minimum for $\vec{q}^2 = 0$ and $l = 0$ at

$$T_0^2 = \frac{1}{(2\pi n)^2}\left(\frac{2}{3}\mu^2 - \vec{Q}^2\right). \tag{41}$$

Hence, at least to the lowest order in λ and for temperatures below T_0, the integral (37) will be positive, and it will increase and diverge as T approaches T_0. A numerical evaluation for $n = 1$ and $l = 0$, shows that (39) decreases monotonically as the temperature rises and becomes zero at T_0. Further, for $\lambda = 0$ the r.h.s. of (35) vanishes at $T^2 = \frac{1}{(2\pi n)^2}\left(\mu^2 - \vec{Q}^2\right) > T_0$. Therefore at first order in λ, the critical temperature will be reached when the integral in the r.h.s. of (35), evaluated at $\lambda = 0$, increases its value as $T \to T_0$. Thus in this approximation, the symmetry will be restored at a temperature $T_c \lesssim T_0$. We could go to higher orders in λ, by a selective summation in (19). We consider (37) with the denominator at first order in λ. Thus

$$\begin{aligned} G^{-1(1)}(q_l, T) &= 2(\mu^2 - Q_n^2)\left[1 + \frac{1}{2}\cos(Q_n \wedge q_l)\right] + q_l^2 - \mu^2 \\ &-\frac{\lambda T}{(2\pi)^3}\left\{\left[1 + \frac{1}{2}\cos(Q_n \wedge q_l)\right] I_\Lambda^{(0)}(Q_n) - \frac{1}{2} I_\Lambda^{(0)}(q_l)\right\} + \mathcal{O}(\lambda^2), \end{aligned} \tag{42}$$

where the integrals $I_\Lambda^{(0)}(Q_n)$ and $I_\Lambda^{(0)}(q_l)$ are at zeroth order. From the previous analysis, we know that the λ independent part in (42) becomes zero at T_0, when $\vec{q} = \vec{0}$ and $l = 0$, whereas $I_\Lambda^{(0)}(Q_l)$ and $I_\Lambda^{(0)}(q_l)$ diverge at the same temperature. Thus we must consider

$$G^{-1(0)}(0,T) = 2\mu^2 - 3\vec{Q}^2 - 3(2\pi n)^2 T^2 - \frac{\lambda T}{2(2\pi)^3}\left[3I^{(0)}(Q_n) - I^{(0)}(0)\right] + \mathcal{O}(\lambda^2), \tag{43}$$

at which, when $T \to T_0$, the difference $3I^{(0)}(Q_n) - I^{(0)}(0)$ gives a positive contribution. Thus the vanishing temperature of (43) will be eventually somewhat smaller than T_0, by a correction of order λ, and the critical temperature T_c will be corrected in a similar way. This consideration could be continuated iteratively to any order in the autoenergy.

Journal of Physics: Conference Series **512** (2014) 012008 doi:10.1088/1742-6596/512/1/012008

5. Conclusions

We consider the noncommutative effective action at finite temperature, for a real scalar field theory with $\lambda\Phi^4$ interaction, in 4 dimensions, with noncommutative time. We solve to first order in λ the stationarity equations of the CJT effective action, with a stripe type ansatz for the background, which depends on temperature. The CJT effective action shows a discontinuity between the results for uniform and non uniform background, which is present also in the stationarity equations. This discontinuity originates first order phase transitions between the corresponding phases, in particular, as conjectured in [2], between the ordered uniform and non uniform phases. Non uniform solutions arise not only in the planar limit, but also in the case of anisotropic non commutativity parameters. Further, the critical temperature depends on the physical mass of the scalar field and on the wave vector of the nonuniform ground state field, hence it depends on noncommutativity implicitly, although we would expect an explicit dependence in a more detailed computation.

Acknowledgments

We thank VIEP-BUAP and CONACyT for its support.

References

[1] Szabo R J 2003 *Phys. Rept.* **378** 207
[2] Gubser S S and Sondhi S L 2001 *Nucl.Phys.* **B605** 395
[3] S. Minwalla, M. Van Raamsdonk and N. Seiberg, JHEP **0002**, 020 (2000) [hep-th/9912072].
[4] Ambjørn J and Catterall S 2002 *Phys. Lett.* **B549** 253
[5] Seiberg N, Susskind L and Toumbas N 2000 *J. High Energy Phys.* **06** 044
[6] Strelchenko A V and Vassilevich D V 2007 *Phys. Rev.* **D76** 065014
[7] Calzetta E and Hu B L 1988 *Phys. Rev.* **D37** 2878
[8] J. Berges, Nucl. Phys A **699**, 847 (2002).
[9] Cornwall J M, Jackiw R and Tomboulis E 1974 *Phys. Rev.* **D10** 2428
[10] R. Jackiw and G. Amelino-Camelia, Banff 1993, Thermal field theories and their applications, 180, [hep-ph/9311324].
[11] J.M. Hernández, C. Ramírez and M. Sánchez, Phys. Rev. D87 (2013) 125012.
[12] Chen G and Wu Y 2002 *Nucl.Phys.* **B622** 189
[13] Castorina P and Zappalà D 2003 *Phys. Rev.* **D68** 065008
Castorina P and Zappalà D 2008 *Phys. Rev.* **D77** 027703
[14] Kapusta J I and Gale C 2006 *Finite-Temperature Field Theory and Applications* (Cambridge Univ. Press, 2nd. ed.)
[15] Brazovskii S A 1975 *Zh. Eksp. Teor. Fiz.* **68** 175
[16] Calzetta E and Hu B L 1998 *Phys. Rev.* **D37** 2878
Chou K, Zu Z, Hao B and Yu L 1985 *Phys. Rept.* **118** 1

Journal of Physics: Conference Series **512** (2014) 012009 doi:10.1088/1742-6596/512/1/012009

An algebraic approach applied to the determination of the polarizability in CO_2

R. Lemus[1], M. Sánchez-Castellanos[2] and M. Carvajal[3]

[1]Instituto de Ciencias Nucleares,Universidad Nacional Autónoma de México. A. P. 70-543, Circuito Exterior, C. U. , 04510 México, D. F. , Mexico

[2]Instituto de Química, Universidad Nacional Autónoma de México, 04510 México, DF, Mexico.

[3]Departamento de Física Aplicada, Unidad Asociada al CSIC, Facultad de Ciencias Experimentales, Universidad de Huelva, Avenida de las Fuerzas Armadas, s/n 21071, Huelva, Spain.

E-mail: `renato@nucleares.unam.mx`

Abstract. A local algebraic approach to describe vibrational excitations of molecules presenting both local and normal mode behaviors is presented. This approach allows the connection with configuration space to be established. The model consists in expanding the kinetic energy as well as the potential in terms of coordinates and momenta. An algebraic representation is obtained by introducing creation and destruction bosonic operators associated with the harmonic oscillators. From the resulting Hamiltonian a local algebraic representation is obtained through a canonical transformation to local bosonic operators. Finally an anharmonization is carried out by changing the local bosonic operators to ladder operators associated with the Morse or Pöschl-Teller functions. Since the model is connected with configuration space, non linear curvilinear coordinates are contemplated. Our model is applied to the vibrational spectroscopic description of the $^{12}C^{16}O_2$ molecule. The eigenstates are tested by calculating the derivatives for the polarizability for this molecule.

1. Introduction

Algebraic methods have been used for a long time as a useful tool to describe the vibrational spectroscopic properties of molecules. Usually these methods are based on harmonic oscillator bases because of their direct correspondence with normal modes [1, 2, 3, 4, 5]. A remarkable advantage of the algebraic representation is that it is easy to recognize the polyads [6], subspaces of the basis that interact among themselves through the most important interactions. Polyad-preserving Hamiltonian are valid as long as the description is confined to the low lying region of the spectrum. As the energy increases the polyad stop being useful. In such situations a suitable strategy may consists in breaking the polyad [2, 7, 8, 9], but an alternative route to follow may consists in choosing a different basis. The appropriate approach will depend on the system under study.

Normal modes involve nuclear collective motions, and consequently a harmonic oscillator basis is reasonable when there are not strong differences in their corresponding masses. When large mass differences are present, local modes are invoked [10, 11, 12, 13, 14, 15]. The basic idea of the local models consists in expressing the Hamiltonian in terms of a zeroth order Hamiltonian corresponding of a set of non interacting oscillators coupled by kinetic and potential energy terms. Local modes are closely related with anharmonicity, a fact that explains the strong mixture of the normal basis to describe the regularities previously mentioned [16]. When local modes are present the natural space to be used correspond to a local basis identified with Morse or Pöschl-Teller functions [17, 18].

Starting in configuration space with a power expansion of the Hamiltonian in terms of normal coordinates it is possible to obtain an algebraic representation using bosonic operators in such a way that the spectroscopic parameters allow the potential energy surface (PES) to be estimated [1, 2]. However, a different scheme either local or normal may be useful for the different molecular degrees of freedom, like in methane for instance [19].

A relatively high success has been reached in modeling the local oscillators in terms of Morse functions. In particular this can be done in the framework of an $su(2)$ algebra. Algebraic local models using an $SU(2)$ basis have been extensively used in phenomenological way in the spectroscopic description of semirigid molecules [20, 21, 22, 23, 24, 25, 26, 27, 28, 29]. The creation and annihilation operators for the Morse and/or Pöschl-Teller (PT) functions were identified as generators of the $su(2)$ algebra [30, 31, 32]. This connection allowed force constants to be obtained, and consequently the PES became available to predict spectra of isotopic species [33, 34, 35, 36, 37, 38, 39, 40].

The natural way to establish the polyad is the normal model scheme. This means that a translation into the local scheme may not be possible, in particular when a strong normal behavior is present. In recent works it has been proved that it is possible to estimate the PES involving any degree of freedom using a local algebraic approach based on $su(2)$ algebras, still preserving polyads [41, 42, 43]. The basic idea consists in starting with the traditional algebraic description in terms of bosonic operators associated to symmetry adapted coordinates, and then apply a canonical transformation to bosonic operators isomorphic to bosonic local creation (annihilation) operators. Finally an anharmonization is carried out by the substitution of the local bosonic operators by ladder operators associated with the Morse/Pöschl-Teller functions. This last step is done preserving the same connection between the spectroscopic operators with the structure and force constants. This approach is unique because it allows to improve the vibrational spectroscopic description without counterpart in configuration space.

On the other hand, the natural molecular vibration description is given in terms of curvilinear coordinates, which can be expanded in terms of normal coordinates. The first approximation consists in considering only linear terms, which implies to identify the nonlinear symmetry adapted coordinates with the normal coordinates. This approximation has been taken in all the applications of the $su(2)$ algebraic approach to describe molecular vibrational excitations [39]. In this contribution we present a general perspective of the local algebraic approach based on $su(2)$ algebras to describe semi-rigid molecules. Our approach assumes the validity of the Born-Oppenheimer approximation and the convergence of the Hamiltonian expansion in terms of both nonlinear coordinates and normal coordinates. We present the study of the vibrational excitations of the specific molecule of carbon dioxide ($^{12}CO_2$) in its ground electronic state, considering non-linear effects in the expansion of the curvilinear coordinates. This study includes the analysis of transition probabilities associated with the Raman spectrum [44], which represents a relevant test for the eigenstates. The description of this spectrum is used to obtain a suitable expansion of the molecular polarizability tensor as a Taylor series in terms of nonlinear and normal coordinates. Although the analysis of this molecule has already been presented in a letter form [45], this was done strictly in the linear approximation. In this work we present the importance and generalization of our approach from a general perspective remarking the effect of the non-linear contributions.

The outline of the present work is as follows. In Section 2 we present in detail the ingredients involved in our algebraic model to obtain the eigenstates in terms of basis of local states associated with anharmonic oscillators. Section 3 is devoted to the application of our algebraic approach to the CO_2 molecule. The estimation of the derivatives of the polarizability tensor are given in Section 4. Finally, in Section 5, a summary and conclusions are presented.

2. A general local algebraic approach

We start considering the following approximation for the vibrational Hamiltonian $\hat{H}$ in terms of curvilinear internal displacement coordinates [4, 46, 47]

$$H = \frac{1}{2}\tilde{\mathbf{P}}\mathbf{G}(\mathbf{S})\mathbf{P} + V(\mathbf{S}), \tag{1}$$

where $\mathbf{S}$ and $\mathbf{P}$ are column vectors corresponding to the internal displacement coordinates and their conjugate momenta $\hat{P}_\alpha = -i\hbar\partial/\partial S_\alpha$, respectively, while the $\mathbf{G}(\mathbf{S})$ matrix establishes the connection between the internal and Cartesian coordinates. $V(\mathbf{S})$ is the Born-Oppenheimer potential. Here the kinetic energy term not involving momentum operators has been neglected. The usual approach to obtain a suitable Hamiltonian to deal with consists in expanding both the $\mathbf{G}(\mathbf{S})$ matrix and the potential $V(\mathbf{S})$ as a Taylor series around the equilibrium configuration, truncating the expansion where an adequate convergence is achieved. In this way we have for the $\mathbf{G}(\mathbf{S})$ matrix

$$G_{ij}(\mathbf{S}) = G^o_{ij}(\mathbf{S}) + \sum_\alpha \left(\frac{\partial G_{ij}}{\partial S_\alpha}\right)_o S_\alpha + \ldots, \tag{2}$$

while for the potential

$$V(\mathbf{S}) = \frac{1}{2!}\sum_{\alpha,\beta}\left(\frac{\partial^2 V}{\partial S_\alpha \partial S_\beta}\right)_o S_\alpha S_\beta + \frac{1}{3!}\sum_{\alpha,\beta,\gamma}\left(\frac{\partial^3 V}{\partial S_\alpha \partial S_\beta \partial S_\gamma}\right)_o S_\alpha S_\beta S_\gamma \ldots \tag{3}$$

Usually the expansion of the $\mathbf{G}(\mathbf{S})$ matrix elements up to linear terms and the potential up to quadratic terms are the starting point to obtain a reasonable spectroscopic description. The substitution of the expansions (2) and (3) into (1), leads to a Hamiltonian of type

$$\hat{H} = \hat{H}(\hat{\mathbf{P}}, \mathbf{S}), \tag{4}$$

involving powers of $\hat{P}_\alpha$, S_β and products. The curvilinear coordinates may be expanded in terms of linear symmetry adapted coordinates Q_i. These coordinates are isomorphic to the normal coordinates, a remarkable fact that allows us to deal with Q_i in every calculation. Hence we have the general form for the expansion

$$S_\alpha = \sum_i L^\alpha_i Q_i + \frac{1}{2!}\sum_{i,j} L^\alpha_{ij} Q_i Q_j + \frac{1}{3!}\sum_{i,j,k} L^\alpha_{ijk} Q_i Q_j Q_k \ldots . \tag{5}$$

The corresponding momenta are obtained through their definition

$$P_\alpha = \sum_i \frac{\partial Q_i}{\partial S_\alpha}\frac{\partial T}{\partial Q_i} = \sum_i \frac{\partial Q_i}{\partial S_\alpha} p_i. \tag{6}$$

The partial derivatives involved in (6) are obtained by inversion of the system of equations

$$\delta_{\alpha\beta} = \sum_i L^\alpha_i \left(\frac{\partial Q_i}{\partial S_\beta}\right) + \sum_{i,j} L^\alpha_{ij} Q_i \left(\frac{\partial Q_j}{\partial S_\beta}\right) + \frac{3}{3!}\sum_{i,j,k} L^\alpha_{ijk}\left(\frac{\partial Q_i}{\partial S_\beta}\right) Q_j Q_k \ldots, \tag{7}$$

which in turn are got by deriving (5) with respect to the coordinate S_β. This set of equations, however, provides a non linear solution for the derivatives. An additional expansion in powers of the symmetry adapted coordinates is required in order to obtain an approximation of the form

$$P_\alpha = \sum_i (M^{i,\alpha} + \sum_r M_r^{i,\alpha} Q_r + \sum_{r,s} M_{rs}^{i,\alpha} Q_r Q_s + \ldots) p_i, \tag{8}$$

where the associated quantum operators are obtained through a symmetry projection method. Therefore the substitution of (5) and (8) into (4) yields a Hamiltonian of type

$$\hat{H} = \hat{H}(\hat{\mathbf{p}}, \mathbf{Q}), \tag{9}$$

where $\hat{p}_k = -i\hbar\partial/\partial Q_k$. The form of this Hamiltonian is not appropriate to identify the polyad(s) preserving interactions. The natural scheme is the number operator representation, also known as the algebraic representation.

A suitable algebraic representation of the Hamiltonian (9) is obtained by the introduction of the bosonic operators

$$a^\dagger_{q\Gamma\gamma} = \alpha_{q\Gamma}\, Q_{q\Gamma\gamma} - \frac{i}{2\hbar\alpha_{q\Gamma}}\, p_{q\Gamma\gamma}\,, \quad a_{q\Gamma\gamma} = \alpha_{q\Gamma}\, Q_{q\Gamma\gamma} + \frac{i}{2\hbar\alpha_{q\Gamma}}\, p_{q\Gamma\gamma}\,, \tag{10}$$

where Γ and γ are irreps associated with a group chain

$$G \supset H, \tag{11}$$

q is a multiplicity index and $\alpha^{q\Gamma}$'s are functions of the reduced mass $\mu_{q\Gamma}$ and the force constant $f^q_{\Gamma\Gamma}$

$$\alpha^2_{q,\Gamma} = \frac{1}{2\hbar} f^q_{\Gamma\Gamma}\, \mu_{q\Gamma}. \tag{12}$$

With the introduction of the bosonic operators the Hamiltonian takes thus the form

$$\hat{H}_P = \hat{H}_P(\hat{a}^\dagger_{q\Gamma\gamma}, \hat{a}_{q\Gamma\gamma}), \tag{13}$$

where the subindex P means that we assume polyad preservation by neglecting all the terms mixing states of different polyad. This algebraic representation has the advantage that a polyad preserving Hamiltonian may be proposed from the outset in a precise form. The spectroscopic parameters involved in (13) are related to the structure and force constants, and consequently the PES can be estimated from a fit of experimental vibrational levels. The matrix representation of Hamiltonian (13) can be obtained in a straightforward way in the harmonic oscillator basis $|n_1 n_2, \ldots n_s\rangle$, where n_i is defined as the eigenvalue associated to the number operator $\hat{n}_{i(q\Gamma)} = \sum_\gamma \hat{a}^\dagger_{q\Gamma\gamma}\hat{a}_{q\Gamma\gamma}$. This basis can be projected to obtain a symmetry adapted basis in which the matrix representation of the Hamiltonian acquires a blocked diagonal form. An efficient and elegant approach to carry out this projection is described in Refs. [48, 49, 50, 51].

The main nondiagonal interactions that define the polyad are identified from the fundamentals and overtones of the molecule, and it is in the normal (symmetry adapted) scheme that the polyad number appears in natural form. In molecules with local mode behavior the concept of polyad has a straightforward translation. In contrast, for molecules involving strong normal behavior, like in BF_3, for instance, the polyad can not be translated into a local scheme. This is explained by the large splitting among the states coming from the same set of equivalent local oscillators as well as the appearance of redundant coordinates [38]. This fact leads to the conclusion that local mode bases cannot be proposed for these kind of molecules. Another argument in favor of this conclusion is that, in systems with strong normal behavior, starting with local oscillators necessarily leads to the indispensableness of breaking the polyad in order to obtain the correct estimation of the force constants [39, 41, 42, 43]. In spite of these arguments, we address the possibility of using a local mode approach in cases where the normal behavior is strong enough that breaking the polyad may seem compulsory.

The approach giving the answer to this question relies on applying to the bosonic operators in the Hamiltonian (13) a canonical transformation of the type

$$\hat{a}^\dagger_{q\Gamma\gamma} = \sum_i \hat{B}^i_{q\Gamma\gamma}\, c^\dagger_i, \tag{14}$$

where $\hat{c}^\dagger_i(\hat{c}_i)$ are bosonic operators, and the coefficients $B^i_{q\Gamma_\gamma}$ correspond to the linear combinations associated with the symmetry adapted local coordinates of the corresponding subspace q. The operators $\hat{c}^\dagger_i(\hat{c}_i)$ are not local operators, but an isomorphism to the true local operators $\hat{a}^\dagger_i(\hat{a}_i)$ can be established:

$$\hat{c}^\dagger_i \approx \hat{a}^\dagger_i. \tag{15}$$

The true connection between the bosonic operators $\hat{a}^\dagger_{q\Gamma\gamma}$ and the true local operators $\hat{a}^\dagger_i(\hat{a}_i)$ is given by a Bogoliubov transformation involving creation and annihilation operators, being the relation (14) true only in the limit of weak interaction among the oscillators [43]. This fact justify the proposed isomorphism (15).

The transformation (14) preserves the polyad in the Hamiltonian. Hence the substitution of (14) into (13) leads to a Hamiltonian of type $\hat{H}_P = \hat{H}_P(\hat{c}^\dagger_i, \hat{c}_i)$, which through the isomorphism (15), takes the form

$$\hat{H}_P = \hat{H}_P(\hat{a}^\dagger_i, \hat{a}_i). \tag{16}$$

The importance of this Hamiltonian is that it can be looked as a polyad preserving Hamiltonian in a local basis, an expression impossible to be obtained starting from a local scheme in coordinates, unless a local behavior was present. The Hamiltonians (13) and (16) are equivalent since they provide the same spectrum, although in different basis connected with a linear transformation. Up to this point nothing new has been proposed. However, this treatment can be improved considerable when we carry out the anharmonization procedure [38, 41, 42]

$$a^\dagger_i \to b^\dagger_i, \qquad a_i \to b_i \ . \tag{17}$$

where the creation and annihilation operators $b^\dagger_i(b_i)$ are generators of a $su(2)$ dynamical algebra [39]. Since this anharmonization has to be done in both, the Hamiltonian and the basis, the operators $b^\dagger_i(b_i)$ may be interpreted as ladder operators for a Morse or Pöschl-Teller potential eigenstates, $|\psi^j_{v_i}\rangle$, with matrix elements [34]

$$b^\dagger|\Psi^j_v\rangle = \sqrt{(v+1)(1-(v+1)/\kappa)}\, |\Psi^j_{v+1}\rangle \ , \tag{18}$$

$$b\,|\Psi^j_v\rangle = \sqrt{v(1-v/\kappa)}\, |\Psi^j_{v-1}\rangle \ , \tag{19}$$

where v is the stretching quantum number, $v = 0, 1, 2 \ldots j-1$, and $\kappa = 2j+1$ is related to the potential depth. The anharmonization (17) is valid for both stretching and bending degrees of freedom in non linear molecules [33, 34, 35, 36, 37, 38, 39, 40]. For linear molecules an anharmonization method based on the $su(3)$ algebra should be included in order to describe the degenerate bending modes [20, 41, 42].

We remark that this approach is a semiempirical method, which means that in principle we are unable to predict vibrational energy or intensity patterns for yet unknown systems. Experimental energies and transitions intensities are needed to carry out a vibrational description of spectroscopic quality, whose accuracy relies on the quality of the input data. However, our algebraic approach is connected with configuration space and predictions are possible provided the force constants as well as the dipole moments.

We should stress that the steps (15) and (17) in our algebraic approach do not have in general an analogy in configuration space, only in the local mode limit. In the next section we shall present the application of this approach to the CO_2 molecule, which present a strong normal mode behavior with degenerate (bending) modes involved.

3. Application of the algebraic approach to CO_2

The first step to study the vibrational spectroscopic description of the CO_2 molecule consists in establishing the symmetry adapted coordinates. For the stretching degrees of freedom we have

$$S_1 \equiv S_{\Sigma_g^+} = \frac{1}{\sqrt{2}}(\Delta r_1 + \Delta r_2), \tag{20}$$

$$S_3 \equiv S_{\Sigma_u^+} = \frac{1}{\sqrt{2}}(\Delta r_1 - \Delta r_2), \tag{21}$$

where $\Delta r_i = r_i - r_e$, with $i = 1, 2$, correspond to the left and right CO distances as displayed in Figure 1 with r_e being the equilibrium distance, while for the bending coordinates we consider [52, 53]

$$S_{2a} = d\ \mathbf{e}_Y \cdot \frac{\mathbf{r}_1 \times \mathbf{r}_2}{r_e^2} \quad , \quad S_{2b} = -d\ \mathbf{e}_X \cdot \frac{\mathbf{r}_1 \times \mathbf{r}_2}{r_e^2} \quad , \tag{22}$$

where $d = 1$Å is a normalization length.

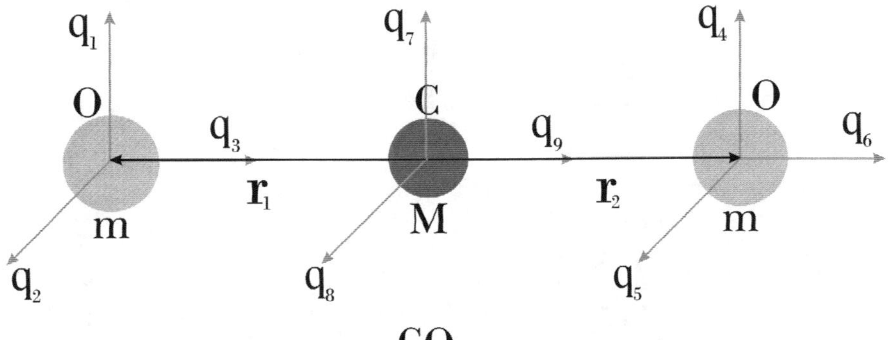

Figure 1. Mass weighted Cartesian displacement coordinates for CO_2.

For the bending modes it is convenient to introduce polar coordinates

$$S_+ = -\frac{1}{\sqrt{2}}(S_{2a} + iS_{2b}) \quad , \quad S_- = \frac{1}{\sqrt{2}}(S_{2a} - iS_{2b}) \quad , \tag{23}$$

with the corresponding conjugate momenta

$$\hat{P}_+ = -\frac{1}{\sqrt{2}}\left(\hat{P}_{2a} - i\hat{P}_{2b}\right), \quad \hat{P}_- = \frac{1}{\sqrt{2}}\left(\hat{P}_{2a} + i\hat{P}_{2b}\right) \quad . \tag{24}$$

In terms of these coordinates the expansion of the Hamiltonian equivalent to (4) may be written as

$$\hat{H} = \hat{H}_s + \hat{H}_b + \hat{H}_{sb} \quad . \tag{25}$$

with explicit expressions given in Ref.[42].

On the other hand, as expressed by (5), the curvilinear coordinates S_α may be written in terms of a linear combination of the normal coordinates. In terms of the mass weighted Cartesian displacement coordinates $q_1 = \sqrt{m_i}\Delta\xi_i$, $\xi = x, y, z$, we have the following normal coordinates in accordance with the notation of Figure 1:

$$\mathcal{Q}_1 = \frac{1}{\sqrt{2}}(q_6 - q_3), \tag{26}$$

$$\mathcal{Q}_{2a} = \sqrt{\frac{M}{2M_T}}[-(q_1 + q_4) + 2\sqrt{\frac{m}{M}}q_7], \tag{27}$$

$$\mathcal{Q}_{2b} = \sqrt{\frac{M}{2M_T}}[-(q_2 + q_5) + 2\sqrt{\frac{m}{M}}q_8], \tag{28}$$

$$\mathcal{Q}_3 = \sqrt{\frac{m}{2M_T}}[-\sqrt{\frac{M}{m}}(q_3 + q_6) + 2q_9], \tag{29}$$

where $M_T = 2m + M$. In terms of these coordinates the curvilinear coordinates (5), up to third order, take the form

$$S_1 = \frac{1}{\sqrt{m}}\mathcal{Q}_1 + \frac{1}{2\sqrt{2}}\frac{1}{r_e}\frac{M_T}{mM}\mathcal{Q}^2 - \frac{1}{4}\frac{1}{r_e^2}\frac{M_T}{mM\sqrt{m}}\mathcal{Q}_1\mathcal{Q}^2, \tag{30}$$

$$S_{2a} = \sqrt{2}\left(\frac{d}{r_e}\right)\sqrt{\frac{M_T}{mM}}\mathcal{Q}_{2a} + \left(\frac{d}{r_e^2}\right)\frac{1}{\sqrt{m}}\sqrt{\frac{M_T}{mM}}\mathcal{Q}_1\mathcal{Q}_{2a}, \tag{31}$$

$$S_{2b} = \sqrt{2}\left(\frac{d}{r_e}\right)\sqrt{\frac{M_T}{mM}}\mathcal{Q}_{2b} + \left(\frac{d}{r_e^2}\right)\frac{1}{\sqrt{m}}\sqrt{\frac{M_T}{mM}}\mathcal{Q}_1\mathcal{Q}_{2b}, \tag{32}$$

$$S_3 = \sqrt{\frac{M_T}{mM}}\mathcal{Q}_3 - \frac{1}{4r_e^2}\frac{M_T}{mM}\sqrt{\frac{M_T}{mM}}\mathcal{Q}_3\mathcal{Q}^2, \tag{33}$$

with the definition $\mathcal{Q}^2 = \mathcal{Q}_{2a}^2 + \mathcal{Q}_{2b}^2$. In order to identify the coordinates involved in the bosonic operators associates with the harmonic oscillators, we introduce the new coordinates Q_i :

$$Q_1 = \frac{1}{\sqrt{m}}\mathcal{Q}_1 \tag{34}$$

$$Q_{2a} = \sqrt{2}\sqrt{\frac{M_T}{mM}}\mathcal{Q}_{2a} \tag{35}$$

$$Q_{2b} = \sqrt{2}\sqrt{\frac{M_T}{mM}}\mathcal{Q}_{2b} \tag{36}$$

$$Q_3 = \sqrt{\frac{M_T}{mM}}\mathcal{Q}_3, \tag{37}$$

given in lenght units. In these new coordinates the expansion (30) takes thus the simplified form

$$S_1 = Q_1 + \frac{1}{4\sqrt{2}}\frac{1}{r_e}Q^2 - \frac{1}{8}\frac{1}{r_e^2}Q_1Q^2, \tag{38}$$

$$
\begin{aligned}
S_{2a} &= \left(\frac{d}{r_e}\right)Q_{2a} + \left(\frac{d}{r_e^2}\right)\frac{1}{\sqrt{2}}Q_1Q_{2a}, &(39)\\
S_{2b} &= \left(\frac{d}{r_e}\right)Q_{2b} + \left(\frac{d}{r_e^2}\right)\frac{1}{\sqrt{2}}Q_1Q_{2b}, &(40)\\
S_3 &= Q_3 - \frac{1}{8r_e^2}Q_3Q^2, &(41)
\end{aligned}
$$

where again $Q^2 = Q_{2a}^2 + Q_{2b}^2$.

Let us now turn our attention to the Hamiltonian (25). The simplest approximation corresponds to consider the first term in the expansions (38), as it was done in Refs. [41, 42, 45]. In this case the curvilinear coordinates are basically identified with the normal coordinates (34). However, in any case, an algebraic representation is obtained by introducing the bosonic creation and annihilation operators [41, 42]

$$
\begin{aligned}
a_\Gamma^\dagger &= \alpha^\Gamma Q_\Gamma - \frac{i}{2\hbar\alpha^\Gamma}\, p_\Gamma \ , \quad a_\Gamma = \alpha^\Gamma Q_\Gamma + \frac{i}{2\hbar\alpha^\Gamma}\, p_\Gamma \ ; \Gamma = \Sigma_g, \Sigma_u \ , &(42)\\
a_\pm^\dagger &= \alpha^\pm Q_\pm + \frac{i}{2\hbar\alpha^\pm}\, p_\mp \ , \quad a_\pm = -\alpha^\pm Q_\mp + \frac{i}{2\hbar\alpha^\pm}\, p_\pm \ . &(43)
\end{aligned}
$$

where α^Γ and $\alpha^\pm$ are given by

$$
\begin{aligned}
(\alpha^\Gamma)^2 &= \frac{1}{2\hbar} f_{\Gamma\Gamma}\mu_\Gamma; \qquad \Gamma = \Sigma_g, \Sigma_u, &(44)\\
(\alpha^+)^2 &= (\alpha^-)^2 = \alpha^2 = \frac{1}{2\hbar}\sqrt{\frac{f_{+-}}{g^0_{+-}}}, &(45)
\end{aligned}
$$

with $\mu_{\Sigma_g} = 1/g^o_{\Sigma_g\Sigma_g} = m$ and $\mu_{\Sigma_u} = 1/g^o_{\Sigma_u\Sigma_u} = (2/M + 1/m)^{-1}$, while for the force constants $f_{\Sigma_g\Sigma_g} = f_{rr} + f_{rr'}$ and $f_{\Sigma_u\Sigma_u} = f_{rr} - f_{rr'}$. The resulting Hamiltonian is equivalent to the one in configuration space and it is expected to be diagonalized in a harmonic oscillator basis. The same approach can be followed when considering terms of higher order in the coordinates expansion.

According to the approach previously presented in §2 we proceed to establish the algebraic polyad-preserving Hamiltonian in a local basis. This goal cannot be achieved starting from configuration space because of the strong normal behavior for the stretches. This issue is addressed by introducing the canonical transformation in the stretching coordinates

$$
a_{\Sigma_g}^\dagger = \frac{1}{\sqrt{2}}(c_1^\dagger + c_2^\dagger) \ , \quad a_{\Sigma_u}^\dagger = \frac{1}{\sqrt{2}}(c_1^\dagger - c_2^\dagger) \ , \tag{46}
$$

where $c_i^\dagger(c_i)$ are bosonic operators isomorphic to the i-th bosonic local operators $a_i^\dagger(a_i)$. Taking into account this isomorphism, the resulting Hamiltonian takes the form

$$
\begin{aligned}
\hat{H} &= \tilde{\omega}_s \sum_{k=1}^{2}\left(a_k^\dagger a_k + a_k a_k^\dagger\right) + \lambda_s \sum_{k\neq j=1}^{2} a_k^\dagger a_j + \alpha_1^s\left(\hat{n}_1^2 + \hat{n}_2^2\right)\\
&+ \alpha_2^s\left(a_1^{\dagger 2}a_2^2 + a_2^{\dagger 2}a_1^2 + 4\hat{n}_1\hat{n}_2\right) + \alpha_3^s\left(\hat{n}_1 a_2^\dagger a_1 + \hat{n}_2 a_1^\dagger a_2 + H.c.\right)\\
&+ \tilde{\omega}_b\hat{n} + \alpha_1^b\hat{n}^2 + \alpha_2^b\hat{\ell}^2 + \alpha_1^{sb}\left\{\left(a_1^\dagger + a_2^\dagger\right)a_+a_- + H.c.\right\}\\
&+ \alpha_2^{sb}\left(\hat{n}_1 + \hat{n}_2\right)\hat{n} + \alpha_3^{sb}\left(a_1^\dagger a_2 + a_2^\dagger a_1\right)\hat{n} \ , &(47)
\end{aligned}
$$

Journal of Physics: Conference Series **512** (2014) 012009 doi:10.1088/1742-6596/512/1/012009

with the definitions

$$\hat{n}_k = a_k^\dagger a_k \; ; \quad k = 1, 2 \; , \tag{48}$$
$$\hat{\ell} = a_+^\dagger a_+ - a_-^\dagger a_- = \hat{n}_+ - \hat{n}_- \; , \tag{49}$$
$$\hat{n} = a_+^\dagger a_+ + a_-^\dagger a_- = \hat{n}_+ + \hat{n}_- \; , \tag{50}$$

where $\hat{n}_k$ is the number of quanta for the k-th stretching oscillator, $\hat{n}$ is the total number of bending quanta, and $\hat{\ell}$ is the vibrational angular momentum. In Eq. (47) we have restricted the interactions to preserve the polyad number $P = 2(n_{\Sigma_g} + n_{\Sigma_u}) + (n_+ + n_-)$, which establish the subspace of states connected through the transfer of two bending quanta to one stretching quantum. In the linear approximation of (38) the expressions for the spectroscopic parameters in terms of the structure and force constants are detailed in Appendix A of Ref.[42]. The diagonalization of Eq. (47) can be carried out in the harmonic oscillator local basis

$$|n_1 n_2 n_+ n_-\rangle = \frac{1}{\sqrt{n_1! n_2! n_+! n_-!}} (\hat{a}_1^\dagger)^{n_1} (\hat{a}_2^\dagger)^{n_2} (\hat{a}_+^\dagger)^{n_+} (\hat{a}_-^\dagger)^{n_-} |0\rangle. \tag{51}$$

Instead we propose a local mode approach for the stretches in terms of Morse oscillators, together with the incorporation of anharmonic effects for the bending degrees of freedom. This goal is achieved through the introduction of the anharmonization procedure (17). Hence the operators $\{b_i^\dagger(b_i); i = 1, 2\}$ are interpreted as ladder operators for a Morse potential eigenstates with matrix elements (18).

For the bending operators $\hat{a}_\pm^\dagger(\hat{a}_\pm)$ an equivalent anharmonization procedure is performed

$$a_\pm^\dagger \to b_\pm^\dagger, \qquad a_\pm \to b_\pm \; , \tag{52}$$

but now the operators $b_\pm^\dagger(b_\pm)$ are given as a linear combination of generators of the $U(3)$ dynamical algebra [20, 21], with matrix elements [41]

$$b_\pm^\dagger |[N]; n^\ell\rangle = \sqrt{\left(\frac{n \pm \ell}{2} + 1\right)\left(1 - \frac{n}{N}\right)} \, |[N]; (n+1)^{\ell \pm 1}\rangle \; , \tag{53}$$
$$b_\pm |[N]; n^\ell\rangle = \sqrt{\left(\frac{n \pm \ell}{2}\right)\left(1 - \frac{n-1}{N}\right)} \, |[N]; (n-1)^{\ell \mp 1}\rangle \; , \tag{54}$$

where n and ℓ have the same meaning that in (48), and N is the total number of bosons (irrep of the $U(3)$ group) related with the anharmonicity. Note that the harmonic 2D oscillator matrix elements are recovered in the large N limit of Eq. (53), as expected. Hence applying the anharmonization procedures (17) and (52), yields the following Hamiltonian

$$\begin{aligned} \hat{H} &= \tilde{\omega}_s \sum_{i=1}^{2} \left(b_i^\dagger b_i + b_i b_i^\dagger\right) + \lambda_s \sum_{i \neq j = 1}^{2} b_i^\dagger b_j + \alpha_1^s \left(\hat{n}_{s,1}^2 + \hat{n}_{s,2}^2\right) \\ &+ \alpha_2^s \left(b_1^{\dagger 2} b_2^2 + b_2^{\dagger 2} b_1^2 + 4\hat{n}_{s,1}\hat{n}_{s,2}\right) + \alpha_3^s \left(\hat{n}_{s,1} b_2^\dagger b_1 + \hat{n}_{s,2} b_1^\dagger b_2 + H.c.\right) \\ &+ \tilde{\omega}_b \hat{n} + \alpha_1^b \hat{n}^2 + \alpha_2^b \hat{\ell}^2 + \alpha_1^{sb} \left\{\left(b_1^\dagger + b_2^\dagger\right) \hat{b}_+ \hat{b}_- + H.c.\right\} \\ &+ \alpha_2^{sb} \left(\hat{n}_{s,1} + \hat{n}_{s,2}\right) \hat{n} + \alpha_3^{sb} \left(b_1^\dagger b_2 + b_2^\dagger b_1\right) \hat{n}, \end{aligned} \tag{55}$$

now with the definition

$$\hat{n}_{s,i} = b_i^\dagger b_i \; , \quad i = 1, 2 \; , \tag{56}$$

Journal of Physics: Conference Series **512** (2014) 012009 doi:10.1088/1742-6596/512/1/012009

and the connection between spectroscopic parameters and force constants is assumed to be identical to the connection previously obtained making use of the harmonic oscillator basis. This resulting Hamiltonian, although not obtained from a scheme of interacting local Morse oscillators, may be interpreted as modeling three interacting oscillators: a 2D oscillator with the $U(3)$ model (bending degrees of freedom) and two 1D Morse oscillators (stretching degrees of freedom).

The matrix representation of the Hamiltonian (55) can be obtained in the $U(2) \times U(3) \times U(2)$ basis

$$|[N_s],[N];v_1v_2;n^\ell\rangle = |\psi^j_{v_1}\rangle \otimes |\psi^j_{v_2}\rangle \otimes |N;n^\ell\rangle \ , \tag{57}$$

although a symmetry adapted basis combined with a normal mode scheme is preferred [49, 50, 51]. In general the diagonalization leads to eigenvectors of the form

$$|\psi^{\Gamma\gamma}_{P,j}\rangle = \sum_{\nu_1,\nu_2,\nu_3} C^{P,j,\Gamma\gamma}_{\nu_1,\nu_2,\nu_3} |P;\nu_1,\nu_2,\nu_3;\Gamma,\gamma\rangle \tag{58}$$

with the basis set

$$|P;\nu_1,\nu_2,\nu_3;\Gamma,\gamma\rangle = \sum_{v_1,v_2,n} B^{P;\nu_1,\nu_2,\nu_3;\Gamma,\gamma}_{v_1,v_2,n} |\psi^j_{v_1}\rangle \otimes |\psi^j_{v_2}\rangle \otimes |N;n^\ell\rangle, \tag{59}$$

where P is the polyad, Γ and γ label the irreducible representations and their components associated with the chain $\mathcal{D}_{\infty h} \supset \mathcal{C}_s$ with $\mathcal{C}_s = \{E, \sigma(xz)\}$, while the set $\{\nu_1,\nu_2,\nu_3\}$ corresponds to an approximate normal-mode labeling [49]. The B coefficients in (59) come from the symmetry projection procedure [49], while the C components in (58) are provided by the diagonalization. We stress that the eigenstates (58) are obtained from an anharmonization procedure without counterpart in configuration space. Concerning the assignment of the transitions, we follow the usual labeling scheme where each state is labeled in accordance with their maximum component in the normal scheme in the expansion (58). We can appreciate that with this method the states may be also labeled in a local scheme through (59). In cases with a strong admixture of the basis, two states may carry the same labels [42, 45].

Using the linear approximation in the coordinates, in Refs. [41, 42] the vibrational spectrum of the molecule ^{12} CO_2 has been described using the Hamiltonian (55). The spectroscopic parameters were optimized with an iterative non-linear least square method [54, 55]. The fit included 101 vibrational (encompassing terms up to polyad $P = 9$) levels with a deviation of rms= 0.53 cm^{-1}. The values for the boson numbers N_s and N were found to be $N_s = 160$ and $N = 150$ [42]. From that work the value for the force constant associated with the Fermi interaction was estimated to be $f_{q_1q_aq_a} = -0.9592$ aJÅ^{-3}, in close agreement with the Chedin's result [56].

In phenomenological algebraic models a test of the quality of the wave functions is compulsory since a good energy fit does not guarantee the wave functions quality [57]. Indeed the wave functions represent a sensitive indicator as a test of the quality of the description. This situation will be discussed in detail in the next section, when we introduce the description of the Raman spectrum.

4. Polarizability derivatives

Line intensities in a Raman spectrum for a gas sample are given by the transition moments of the molecular polarizability tensor α. In the particular case of the CO_2 molecule, the non-vanishing components of α tensor have symmetries Σ_g^+ (trace) and $\Sigma_g^+ \oplus \Pi_g \oplus \Delta_g$ (anisotropy). The sharp, polarized, Q-branches in the vibrational Raman spectrum of CO_2 in the ~ 1300 cm^{-1} region are totally symmetric (Σ_g^+) transitions and their intensities are mainly due to the trace of the

Journal of Physics: Conference Series **512** (2014) 012009 doi:10.1088/1742-6596/512/1/012009

α tensor (mean-polarizability). For the trace scattering of a gas sample at thermal equilibrium, the differential cross section can be expressed in the SI system as follows [53]

$$\left(\frac{\partial\sigma}{\partial\Omega}\right)^{\text{trace}}_{i\to f} = \left(\frac{\pi}{\epsilon_0}\right)^2 \frac{(\nu_0+\nu_i-\nu_f)^4}{Z_{vib}(T)} g_{if}\,(M_{if})^2 \exp\left(-hc\nu_i/k_BT\right) \ , \tag{60}$$

where ϵ_0 is the vacuum permitivity, ν_0 is the wave number of the exciting radiation $\nu_0 = 19430$ cm^{-1}, ν_i and ν_f are the wavenumbers of initial and final states, and g_{if} is the vibrational degeneracy of the energy levels involved in the transition. $Z_{vib}(T) = \sum_j g_j e^{-\nu_j/k_BT}$ is the vibrational partition function at the temperature T of the gas sample, where g_j is the degeneracy of the j-th state with ν_j energy, and $M_{if} = \langle\nu_i|\bar{\alpha}|\nu_f\rangle$ is the transition moment of the mean molecular polarizability $\bar{\alpha}$ between vibrational states $|\nu_i\rangle$ and $|\nu_f\rangle$.

The computation of the intensity (60) involves the molecular polarizability surface, which can be expressed as a Taylor series expansion on the vibrational coordinates

$$\begin{aligned}
\bar{\alpha}_{\Sigma_g^+} &= \bar{\alpha}_0 + \left(\frac{\partial\bar{\alpha}}{\partial S_{\Sigma_g^+}}\right)_0 S_{\Sigma_g^+} + \frac{1}{2}\left(\frac{\partial^2\bar{\alpha}}{\partial S^2_{\Sigma_g^+}}\right)_0 S^2_{\Sigma_g^+} + \frac{1}{2}\left(\frac{\partial^2\bar{\alpha}}{\partial S^2_{\Sigma_u^+}}\right)_0 S^2_{\Sigma_u^+} \\
&+ \frac{1}{2}\left(\frac{\partial^2\bar{\alpha}}{\partial S^2_{2a}}\right)_0 (S^2_{2a}+S^2_{2b}) + \frac{1}{2}\left(\frac{\partial^3\bar{\alpha}}{\partial S_{\Sigma_g^+}\partial S^2_{2a}}\right)_0 S_{\Sigma_g^+}(S^2_{2a}+S^2_{2b}) \\
&+ \frac{1}{2}\left(\frac{\partial^3\bar{\alpha}}{\partial S_{\Sigma_g^+}\partial S^2_{\Sigma_u^+}}\right)_0 S_{\Sigma_g^+} S^2_{\Sigma_u^+} \ ,
\end{aligned} \tag{61}$$

where the derivatives are unknown. The last term in Eq. (61) turns out to be negligible and, consequently, it will be omitted [53]. These derivatives may be determined through *ab initio* calculations [58], but they can be also obtained from a fit to experimental transition moments [53, 59].

In the linear approximation the expansion in terms of normal coordinates takes the same form as (61), but substituting S_α by Q_α. This case was considered previously to estimate the derivatives of the polarizability as well as to calculate the Raman spectrum [45]. Here we shall proceed to investigate the nonlinearity effects in this analysis. If non linear terms are taken into account, we have for the polarizability the following expansion in terms of normal coordinates

$$\begin{aligned}
\bar{\alpha}_{\Sigma_g^+} &= \bar{\alpha}_0 + \alpha_1\Big(Q_1 + \frac{1}{4\sqrt{2}r_e}Q^2 - \frac{1}{8r_e^2}Q_1Q^2\Big) \\
&+ \frac{1}{2}\alpha_{11}\Big(Q_1^2 + \frac{1}{32r_e^2}Q^2 + \frac{1}{64r_e^4}Q_1^2Q^4 + \frac{1}{2\sqrt{2}r_e}Q_1Q^2 - \frac{1}{4r_e^2}Q_1^2Q^2 - \frac{1}{16\sqrt{2}r_e^3}Q_1Q^4\Big) \\
&+ \frac{1}{2}\alpha_{33}\Big(Q_3^2 + \frac{1}{64r_e^4}Q_3^2Q^4 - \frac{1}{4r_e^2}Q_3^2Q^2\Big) \\
&+ \alpha_{22}\Big(\frac{d^2}{4r_e^4}Q_1^2Q^2 + \frac{d^2}{\sqrt{2}r_e^3}Q^2Q_1 + \frac{d^2}{2r_e^2}Q^2\Big) \\
&+ \alpha_{122}\Big(-\frac{d^2}{32r_e^6}Q_1^3Q^4 + \Big(-\frac{1}{2}\Big)\frac{d^2}{16\sqrt{2}r_e^5}Q_1^2Q^4 + \frac{d^2}{4r_e^4}Q_1^3Q^2 + \Big(\frac{1}{2}\Big)\frac{d^2}{16r_e^4}Q_1Q^4 \\
&\quad + \frac{d^2}{\sqrt{2}\,r_e^3}Q_1^2Q^2 + \frac{d^2}{8\sqrt{2}r_e^3}Q^4 + \frac{d^2}{2r_e^2}Q_1Q^2\Big),
\end{aligned} \tag{62}$$

with the following definitions for the derivatives

$$\alpha_1 = \left(\frac{\partial\bar{\alpha}}{\partial S_{\Sigma_g^+}}\right)_0; \quad \alpha_{11} = \left(\frac{\partial^2\bar{\alpha}}{\partial S^2_{\Sigma_g^+}}\right)_0; \quad \alpha_{33} = \left(\frac{\partial^2\bar{\alpha}}{\partial S^2_{\Sigma_u^+}}\right)_0,$$

$$\alpha_{22} = \left(\frac{\partial^2\bar{\alpha}}{\partial S_{2a}^2}\right)_0; \quad \alpha_{122} = \left(\frac{\partial^3\bar{\alpha}}{\partial S_{\Sigma_g^+}\partial S_{2a}^2}\right)_0. \tag{63}$$

We now have the task of computing the matrix elements of the mean molecular polarizability $\bar{\alpha}$ between the system wavefunctions (58). The approach to follow is similar to the calculation of the matrix elements of the Hamiltonian. We first introduce an algebraic representation of $\bar{\alpha}$ introducing the bosonic operators (42). Then the cannonical transformation (46) together with the identification (15) is carried out. Finally the anharmonizations (17) and (52) are applied. In this way the matrix elements of the $\bar{\alpha}$ in the basis (57) is well defined, and the transition moments $M_{if} = |\langle\nu_i|\bar{\alpha}|\nu_f\rangle|$ can be calculated using the expansion (61) and the wavefunctions (58). The comparison of the calculated M_{if}^{calc} with those M_{if}^{exp} derived from the experimental transition intensities (60) [53, 60], allows us to obtain estimates of the molecular polarizability derivatives Eq. (63). To achieve this task the following root mean square deviation is defined [45]

$$rms = \sqrt{\sum_\beta [\log|M_\beta^{exp}| - \log|M_\beta^{calc}|]^2} \tag{64}$$

which is minimized by a suitable selection of the polarizability derivatives, using the values in Ref. [53] as initial guesses for the fitting procedure. In the previous work the subindex β run over the first six transitions presented in Table 1. In this work it was necessary to consider the first nine transitions in order to obtain convergence. At the left (first three columns) of Table 1 the experimental and calculated frequencies of the Raman transitions as well as the states involved in the transitions are displayed. At the right the experimental matrix elements M_{if}^{exp} [53, 60] and the calculated ones after the minimization of Eq. (64) are given.

The fit allows the derivatives of $\bar{\alpha}$ with respect to nonlinear symmetry adapted coordinates to be estimated. These derivatives are displayed in the left panel of Table 2. For comparison we include the experimental values as well as the derivatives obtained using the linear approximation [45]. In general, there is a good agreement between our results and the experimental polarizability derivatives, in particular for the first three derivatives α_1, α_{11} and α_{33}. For the derivative α_{122} a slightly different value is obtained, but significant discrepancies appear in the bending polarizability derivatives. While the experimental one takes the value $\alpha_{122} = 0.36$, we obtained $\alpha_{122} = -0.06$. At this point the linear approximation seems to represent a better result. However, we have to take into account that the derivatives (63) in Ref.[53] were obtained from the derivatives with respect to the dimensionless normal coordinates and consequently a much better physical comparison is expected to be obtained by calculating such derivatives. The connection between both sets of derivatives is established using the expansion (5) and the chain rule to obtain

$$\bar{\alpha}'_1 \equiv \frac{\partial\bar{\alpha}}{\partial Q_1} = \frac{1}{\sqrt{m}}\frac{\partial\bar{\alpha}}{\partial S_1} \tag{65}$$

$$\bar{\alpha}'_{11} \equiv \frac{\partial^2\bar{\alpha}}{\partial Q_1^2} = \frac{1}{m}\frac{\partial^2\bar{\alpha}}{\partial S_1^2} \tag{66}$$

$$\bar{\alpha}'_{33} \equiv \frac{\partial^2\bar{\alpha}}{\partial Q_3^2} = \frac{M_T}{mM}\frac{\partial^2\bar{\alpha}}{\partial S_3^2} \tag{67}$$

$$\bar{\alpha}'_{+-} \equiv \frac{\partial^2\bar{\alpha}}{\partial Q_{2a}^2} = 2\left(\frac{d}{r_e}\right)^2\frac{M_T}{mM}\left\{\frac{r_e}{2\sqrt{2}d^2}\frac{\partial\bar{\alpha}}{\partial S_1} + \frac{\partial^2\bar{\alpha}}{\partial S_{2a}^2}\right\} \tag{68}$$

$$\bar{\alpha}'_{1+-} \equiv \frac{\partial^3\bar{\alpha}}{\partial Q_1\partial Q_{2a}^2} = 2\left(\frac{d}{r_e}\right)^2\frac{M_T}{m\sqrt{mM}}\left\{\frac{1}{4d^2}\frac{\partial\bar{\alpha}}{\partial S_1} + \frac{r_e}{2\sqrt{2}d^2}\frac{\partial^2\bar{\alpha}}{\partial S_1^2}\right.$$
$$\left. + \Lambda_1\frac{\sqrt{2}}{r_e}\frac{\partial^2\bar{\alpha}}{\partial S_{2a}^2} + \frac{\partial^3\bar{\alpha}}{\partial S_1\partial S_{2a}^2}\right\} \tag{69}$$

The derivatives with respect to the dimenssionless normal coordinates used in Ref.[53] are obtained by multiplying for a product $\prod_i \sqrt{2}b_i$, where i runs over the coordinates involved in the derivatives, and

$$b_i^2 = \frac{h}{8\pi^2\omega_i c}, \tag{70}$$

where $\omega_i = 1/\lambda_i$ is the wave number associated with the i-th mode.

Table 1. Experimental and fitted transition moments $|M_{if}| = |\langle\nu_i|\bar{\alpha}|\nu_f\rangle|$ of the mean polarizability of CO_2. The first six transitions were involved in the fit while the last five are predictions. The labeling was chosen to be the normal mode scheme (ν_1, ν_2, ν_3).

ν (cm^{-1})^a	ν (cm^{-1})^b	$\lvert\nu_i\rangle^c \longrightarrow \lvert\nu_f\rangle$ transition	$\lvert M_{if}\rvert = \lvert\langle\nu_i\lvert\bar{\alpha}\rvert\nu_f\rangle\rvert^d$	$\lvert M_{if}\rvert = \lvert\langle\nu_i\lvert\bar{\alpha}\rvert\nu_f\rangle\rvert^e$
1285.4	1286.29	$\lvert 0;000;\Sigma_g^+\rangle \longrightarrow \lvert 2;100;\Sigma_g^+\rangle$	5.58	5.59
1388.2	1387.54	$\lvert 0;000;\Sigma_g^+\rangle \longrightarrow \lvert 2;020;\Sigma_g^+\rangle$	6.79	6.89
2548.4	2549.53	$\lvert 0;000;\Sigma_g^+\rangle \longrightarrow \lvert 4;120;\Sigma_g^+\rangle$	0.088	0.084
2671.1	2671.11	$\lvert 0;000;\Sigma_g^+\rangle \longrightarrow \lvert 4;200;\Sigma_g^+\rangle$	0.114	0.116
2797.1	2795.98	$\lvert 0;000;\Sigma_g^+\rangle \longrightarrow \lvert 4;120;\Sigma_g^+\rangle$	0.026	0.026
4673.3	4673.17	$\lvert 0;000;\Sigma_g^+\rangle \longrightarrow \lvert 4;002;\Sigma_g^+\rangle$	0.050^f	0.050
1265.1	1266.04	$\lvert 1;010;\Pi_u\rangle \longrightarrow \lvert 3;110;\Pi_u\rangle$	5.4	5.41
1409.5	1408.54	$\lvert 1;010;\Pi_u\rangle \longrightarrow \lvert 3;030;\Pi_u\rangle$	7.2	7.06
2514.1	2515.01	$\lvert 1;010;\Pi_u\rangle \longrightarrow \lvert 5;130;\Pi_u\rangle$	0.095	0.098
2671.9	2671.93	$\lvert 1;010;\Pi_u\rangle \longrightarrow \lvert 5;210;\Pi_u\rangle$	...	0.109
2833.3	2831.93	$\lvert 1;010;\Pi_u\rangle \longrightarrow \lvert 5;130;\Pi_u\rangle$	≤ 0.03	0.016

^{a} Experimental transition frequency [53].
^{b} Theoretical frequency calculated from the fitted vibrational term values of Ref. [42].
^{c} The vibrational states are labeled by the ket $|P;\nu_1,\nu_2,\nu_3;\Gamma\rangle$ where P is the polyad number, (ν_1,ν_2,ν_3) are the quantum numbers of the normal-mode representation and Γ is the symmetry of the vibrational wavefunction.
^{d} Experimental values from Ref. [53] in 10^{-42}CV^{-1}m^2 otherwise is indicated.
^{e} Fitted and predicted values of the transition moment in 10^{-42}CV^{-1}m^2 obtained from this work.
^{f} Value derived from *Chrysos et al.* [60].

In the right panel of Table 2 we compare our results with the mean polarizability derivatives with respect to dimensionless normal coordinates obtained by *Tejeda et al.* in Ref. [53]. As in the previous case, the first three derivatives are very close to the experimental, which means that the non linear effects are basically negligible. In contrast for the second derivative with respect to the bending coordinate, $\bar{\alpha}''_{+-}$, and $\bar{\alpha}''_{1+-}$, the nonlinear effects turn out to be significant. In the linear approximation the derivative $\bar{\alpha}''_{+-}$ is 31 % smaller than the experimental value of *Tejeda et al.*, but when the non linear terms are incorporated the derivative becomes closer to the expected value lowering the error to 25 %. This improvement is not reflected in the left panel. From the technical point of view this fact is explained by the appearance of the derivative $\partial\bar{\alpha}/\partial S_1$ in (68) as a consequence of nonlinearity. This term is not negligible compared with $\partial^2\bar{\alpha}/\partial S_{2a}^2$. With respect the derivative $\bar{\alpha}''_{1+-}$ the improvement is slightly better. The error decreases from 8.4 % to 1.6 %.

The results displayed in Table 2 for the derivatives show the sensitiveness of the polarizability with respect to the nonlinearity. Only the derivatives involving the bending coordinates are affected by the non linear terms. The differences between our results and the ones obtained by *Tejada et al* can be attributed to the approach followed to obtain the eigenstates. In Ref.[53] the derivatives with respect to the normal coordinates were fitted using perturbation theory through the eigenfunctions of the Chedin's results. It is worth mentioning that if the latter procedure is applied to our calculated transition moments in Table 1, the value of $\bar{\alpha}''_{+-} = 2.28 \times 10^{-42}$ C m^2 V^{-1} is obtained, very close to the value 2.12×10^{-42} C m^2 V^{-1} of our fit. This is an indication of the consistency of the methodology employed in this work.

Table 2. Derivatives of the mean polarizability of CO_2 with respect to the symmetry coordinates (left panel) and dimensionless normal coordinates (right panel).

Symmetry coordinates				Dimensionless normal coordinates			
Derivative (units) [a]	Linear [b]	Exp [c]	Nonlinear [d]	Derivative[e] $(10^{-42}CV^{-1}m^2)$	Linear [f]	Exp [d]	Nonlinear
$(\partial\bar{\alpha}/\partial S_{\Sigma_g})_0(10^{-30}\,CV^{-1}m)$	3.15	3.15	3.181	$\bar{\alpha}'_1$	12.44	12.43	12.56
$(\partial^2\bar{\alpha}/\partial S^2_{\Sigma_g})_0(10^{-20}\,CV^{-1})$	2.549	2.9	2.634	$\bar{\alpha}''_{11}$	0.398	0.45	0.411
$(\partial^2\bar{\alpha}/\partial S^2_{\Sigma_u})_0(10^{-20}\,CV^{-1})$	0.447	0.50	0.448 [g]	$\bar{\alpha}''_{33}$	0.144	0.15 [g]	0.144
$(\partial^2\bar{\alpha}/\partial S^2_{2a})_0(10^{-20}\,CV^{-1})$	0.8395	0.36	−0.060	$\bar{\alpha}''_{+-}$	1.925	2.81	2.12
$(\partial^3\bar{\alpha}/\partial S_{\Sigma_g}\partial S^2_{2a})_0(10^{-10}\,CV^{-1}m^{-1})$	−1.21	−1.7	−1.2	$\bar{\alpha}'''_{1+-}$	−0.110	−0.06	−0.061

[a] Polarizability function derivatives as defined in Eq. (61).
[b] Values obtained in this work by a fitting of experimental polarizability transition moments (see Table 1).
[c] Refs. [?, 58].
[d] Experimental values [53].
[e] Polarizability function derivatives in terms of dimensionless normal coordinates as defined in Ref. [53].
[f] Polarizability derivative values given in second column after transforming them from symmetry coordinates to dimensionless normal coordinates.
[g] Best choice of the two available experimental values of *Tejeda et al.* [53] according to the *ab initio* CCSD(T) value provided in Ref. [58].

5. Conclusions

In this work a general algebraic approach to describe the vibrational spectroscopy of molecules presenting both normal and local mode behaviors has been presented. The main feature of our approach is that the PES can be estimated for every kind of semirigid molecule using a local basis with preservation of the polyad, allowing anharmonic effects to be introduced from the outset. This approach is not analogous to any method in configuration space, since any trial to obtain a PES from a local basis necessarily leads to brake the polyad when a normal mode behavior is present. Although in this work we present an application to CO_2, the current approach could be extended to deal with more complex molecular systems presenting either a local-mode or a normal-mode behavior.

In the present work we have applied the algebraic method to the description of the CO_2 molecule. Although this molecule has already been studied in the framework of this approach, here we present the general approach in detail with the extension to include non linear coordinates. Since CO_2 molecule is linear, the local-mode description is introduced via a canonical transformation in the stretching degrees of freedom, after an anharmonization procedure, gives rise to the $U(2) \times U(3) \times U(2)$ model. The model basis set is built as the direct product of two 1D local Morse potential wavefunctions for the stretching and a 2D quasi-rigid bender wavefunction. The Morse functions are introduced as a consequence of the anharmonzation procedure in the local basis while the U(3) model appears because of the anharmonization in the 2D harmonic oscillator. Our analysis starts with a summary of the previous results obtained for the energy fits and their consequences in the estimation of the PES as well as in the quality of the eigenstates. An attempt to improve the Fermi force constant through the non linear effects fails due to their non perturbative effects at the level of approximation.

The polarizability function is expanded in terms of non linear symmetry coordinates, which through the use of the transformation (5) to normal coordinates, is transformed into an expansion in terms of normal coordinates. This allows us to obtain its algebraic representation by the introduction of the bosonic operators associated with the normal modes. In similar form to the Hamiltonian, thereafter the anharmonization procedure is applied. This makes possible to compute the Raman transition matrix elements using the algebraic eigenstates. The polarizability function derivatives with the stretching coordinates obtained agree with previous estimations based an a linear expansion. The non linear effects manifest in the calculations of the derivatives associated with the bending coordinates, which moves in the right direction to agree with the experimental results.

The eigenstates provided by the Hamiltonian parameter values in Ref. [42] along with the polarizability derivatives allow us to reproduce most of the available experimental Raman intensities. This is a stringent test for the calculated wavefunctions opening up the possibility of simulations of high temperature Raman spectra of CO_2 for combustion flames diagnostics [61].

Although in this contribution the main focus of interest was the particular molecule of CO_2, this approach is general and it can be applied to any semi-rigid molecular system either linear or nonlinear. However, it does not mean that the quality of the description would be similar, because ro-vibrational interactions and/or polyad breaking interactions may appear as fundamental ingredients.

Acknowledgments
This work is partially supported by DGAPA-UNAM under project No. IN109113.

References
[1] Califano S 1976 *Vibrational States.* Wiley, New York.
[2] Papoušek D, and Aliev M R 1982 *Molecular Vibrational-Rotational Spectra.* Elsevier, Amsterdam.
[3] Sibert III E L 1988, *Com.Phys.Comm.* **51** 149-160.
[4] Jensen P and Bunker P R Editors 2000 *Computational Molecular Spectroscopy.* John Wiley & Sons.
[5] Kellman M E 1995 *Annu.Rev.Phys.Chem.*, **66** 395.
[6] Kellman M E 1990, *J.Chem.Phys.* **93** 6630-6635.
[7] Kauppi E and Halonen L 1995 *J.Chem.Phys.* **103** 6861-6872.
[8] Kauppi E and Halonen L, *J.Chem.Phys.* **96** 2933-2941.
[9] Sánchez-Castellanos M and Lemus R 2008, *J.Phys.B* **41** 145203 (11pp).
[10] Sage M L, Jortner J 1981 *Adv.Chem.Phys.* **47**293-322.
[11] Child M S and Halonen L 1984, *Adv.Chem.Phys.* **1**1-57.
[12] L.Halonen, *Adv.Chem.Phys.* **104** (1998)41-179.
[13] Jensen P 2000, *Mol.Phys.* **98** 1253-1285.
[14] Mecke R and Ziegler R, *Z.Phys.* **101**405-417.
[15] Child M and Lawton R T 1981, *Faraday Discuss. Chem. Soc.* **71**273-289.
[16] Mills I and Robiette A G 1985, *Mol.Phys.* **56**743-765.
[17] Morse P M, *Phys.Rev.* **34** 57-64.
[18] Pöchl G and Teller E Z 1933, *Z. Phys.* **83** 143-151.
[19] Halonen L 1997,*J.Chem.Phys.* **106** (1997) 831-845
[20] Iachello F and Levine R D 1995, *Algebraic Theory of Molecules*, Oxford University Press.
[21] Oss S 1996, *Adv. Chem. Phys.*, **93** 455-650.
[22] Lemus R and Frank A 1994,*J.Chem.Phys.* **101** 8321-8332.
[23] Frank A, Lemus R, Bijker B, Príez-Bernal F and Arias J M 1996,*Ann. of Phys.* **252** 211-238.
[24] Lemus R and Frank A 2000, *J.Mol.Spectrosc.* **201**198-217.
[25] Lemus R and Frank A, *Chem. Phys.* **242** 25-35.
[26] Pérez-Bernal F, Arias J M, Frank A, Lemus R and Bijker R 1997,*J.Mol.Spect.* **184**1-11.
[27] Lemus R, Frank A, Perez-Bernal F, Bijker B and Arias J M 1997, *Phys.Rev.A.* **56** 4337-4340.
[28] Bernal R and Lemus R 2006 *J.Mol. Spectrosc.* **235** 218-234.
[29] Sako T, Aoki D, Yamanouchi K and Iachello F 2000 *J. Chem. Phys.*, **113** 6063-6069.
[30] Frank A, Lemus R, Carvajal M, Jung C and Ziemniak E 1999, *Chem. Phys. Lett.* **308** 91-98.
[31] Carvajal M, Lemus R, Frank A, Jung J and Ziemniak E 2000, *Chem. Phys.* **260**105-123.
[32] Lemus R and Bernal B 2002 , *Chem. Phys.* **283**401-417.
[33] Lemus R, Carvajal M, López J C and Frank A 2002, *J. Mol. Spectrosc.* **214** 52-68.
[34] Lemus R 2004, *J. Mol. Spectrosc.* **225** 73-92.
[35] Sánchez-Castellanos M, Amezcua-Eccius C A, Álvarez-Bajo O and Lemus R 2008, *J. Mol. Spectrosc.* **247**, 140-159.
[36] Bernal B and Lemus R 2006, *J. Mol. Spectrosc.* **235** 218-234.

[37] Sánchez-Castellanos M and Lemus R 2011, *J. Mol. Spectrosc.* **266** 1-11
[38] Amezcua-Eccus C A and Lemus R 2010, *J. Mol. Spectrosc.* **260** 36-49.
[39] Lemus R 2012, *Adv. in Phys. Chem.* Volume 2012, Article ID 593872, 38 pages. doi:10.1155/2011/593872.
[40] 'Alvarez-Bajo O, Carvajal M and Pérez-Bernal F 2012 *Chem. Phys.* **392**, 63-70.
[41] Sánchez-Castellanos M, Lemus R, Carvajal M and Pérez-Bernal F 2009, *J. Mol. Spectrosc.* **253**1-15.
[42] Sánchez-Castellanos M, Lemus R, Carvajal M and Pérez-Bernal F 2012, *Int. J. Quant. Chem.* **112**(2012) 3498-3507.
[43] Castaños O and Lemus R 2010,*Mol. Phys.***108** 597-610.
[44] Fernández J M, Punge A, Tejeda G and Montero S 2006, *J. Raman Spectrosc.* **37** 175-182.
[45] Sánchez-Castellanos M, Lemus R, Carvajal M, Pérez-Bernal F and Fernández J M 2012, *Chem. Phys. Lett.* **554** 208-213.
[46] Pickett H M 1972, *J. Chem. Phys.* **56** 1715-1723.
[47] Meyer R and Günthard H 1968, *J.Chem. Phys.* **49** 1510-1521.
[48] Chen J Q 1989, *Group Representation Theory for Physicists.* World Scientific, Singapore.
[49] Lemus R 2003 , *Mol. Phys.*, **101** 2511-2528.
[50] Álvarez-Bajo O, Lemus R, Carvajal M and Pérez-Bernal F 2011, *Mol.Phys.* **109** 797-812.
[51] Lemus R 2012, *Symmetry* **4** 667-685.
[52] Hoy A R, Mills I M and Strey G 1972, *Mol. Phys.* **24** 1265-1290.
[53] Tejeda G, Maté B and Montero S 1995, *J. Chem. Phys.* **103** 568-576.
[54] Walters L 1989, *The p-shell nuclei in a* $(0+2)\hbar\omega$ *model space*, Ph.D. Thesis. University of Rijks, Groningen (1989).
[55] Brussaard P J and Glandemans P W M 1977, *Shell-model applications in Nuclear Spectrospcopy.* North-Holland.
[56] Chedin A and Teffo J L 1984, *J. Mol. Spectrosc.* **107** 333-342.
[57] Ishikawa H 2002 *et al.*, *Chem. Phys. Lett.* **365** 57-68.
[58] Haskopoulos A and Maroulis G 2006, *Chem. Phys. Lett.* **417** 235-240.
[59] Akhmedzhanov R, Atakhodzhaev A K, and Bulanin M O 1982, *J. Mol. Struct. THEOCHEM* **89** 285-290.
[60] Chrysos M, Verzhbitskiy I A, Rachet F and Kouzov A P 2011, *J. Chem. Phys.* **134** 104310 (6pp).
[61] Lemus R, Sánchez-Castellanos M, Carvajal M, Fernández, J M and Pérez-Bernal F *Simulation of the Raman spectra of* CO_2 *: bridging the gap between experimental and the algebraic approach.* To be published.

Estimado Manolo Tena,

Muchisimas gracias por esta atencion.

deseo decirle que toda la semana he estado completamente en tension sin poder dormir por este problema. Como Usted sabe,tenia la cita con el abogado representante de la demanda y Usted me dijo que la pospusiera.

Quiero pedirle disculpas debido a que decidi asistir solo para saber lo que querian. Yo se que esto fue indevido dado el consejo que Usted me dio, pero me volvia loco solo el pensar que la propiedad de mi papa estaba ya por venderse. Tenia que saber la situacion en la que estaba el terreno. La incertidumbre me tenia en zozobra. Solo escuche lo que desaban y me retire.

Le pido disculpas y espero que comprenda y me perdone.

The Dunkl oscillator in three dimensions

Vincent X. Genest
Centre de Recherches Mathématiques, Université de Montréal, Montréal, Québec, Canada
E-mail: vincent.genest@umontreal.ca

Luc Vinet
Centre de Recherches Mathématiques, Université de Montréal, Montréal, Québec, Canada
E-mail: luc.vinet@umontreal.ca

Alexei Zhedanov
Donetsk Institute for Physics and Technology, Donetsk, Ukraine
E-mail: zhedanov@yahoo.com

Abstract. The isotropic Dunkl oscillator model in three-dimensional Euclidean space is considered. The system is shown to be maximally superintegrable and its symmetries are obtained by the Schwinger construction using the raising/lowering operators of the dynamical $sl_{-1}(2)$ algebra of the one-dimensional Dunkl oscillator. The invariance algebra generated by the constants of motion, an extension of $\mathfrak{u}(3)$ with reflections, is called the Schwinger-Dunkl algebra $sd(3)$. The system is shown to admit separation of variables in Cartesian, polar (cylindrical) and spherical coordinates and the corresponding separated solutions are expressed in terms of generalized Hermite, Laguerre and Jacobi polynomials.

1. Introduction

This paper is concerned with the study of the isotropic Dunkl oscillator model in three-dimensional Euclidean space. This model, described by a Hamiltonian involving reflection operators, will be shown to be both maximally superintegrable and exactly solvable. The constants of motion will be obtained by the Schwinger construction using the $sl_{-1}(2)$ dynamical symmetry of the parabose oscillator in one dimension. The invariance algebra generated by the symmetries will be seen to be an extension of $\mathfrak{u}(3)$ by involutions and shall be called the Schwinger-Dunkl algebra $sd(3)$. The Schrödinger equation of the system will be seen to admit separation of variables in Cartesian, polar (cylindrical) and spherical coordinates. The corresponding separated solutions will be found and expressed in terms of generalized Hermite, Laguerre and Jacobi polynomials.

1.1. Superintegrability

Let us first recall the notion of superintegrability. A quantum system with d degrees of freedom defined by a Hamiltonian H is *maximally superintegrable* if it admits $2d$ 1 algebraically

Journal of Physics: Conference Series **512** (2014) 012010 doi:10.1088/1742-6596/512/1/012010

independent symmetry operators S_i that commute with the Hamiltonian, i.e.

$$[H, S_i] = 0, \qquad 1 \leq i \leq 2d - 1,$$

where one of the symmetries is the Hamiltonian itself, e.g. $S_1 = H$. A superintegrable system is said to be of order ℓ if ℓ is the maximal degree of the constants of motion S_i, excluding the Hamiltonian, in the momentum variables. The $\ell = 1$ case is associated to symmetries of geometric nature and to Lie algebras whereas the $\ell = 2$ case is typically associated to multiseparability of the Schrödinger equation and to quadratic invariance algebras.

A substantial amount of work has been done on superintegrable systems, motivated in part by their numerous applications, exact solvability and connections with the theory of special functions. In view of these properties, the search for new superintegrable models and their characterization is of significant interest in mathematical physics. For a recent review on superintegrable systems, one can consult [9].

1.2. Dunkl oscillator models in the plane

A series of novel superintegrable models in the plane with Hamiltonians involving reflection operators have been introduced recently [4, 5, 7]. The simplest of these systems, called the Dunkl oscillator in the plane, is defined by the following Hamiltonian [4]:

$$H = -\frac{1}{2}\left[\mathcal{D}_1^2 + \mathcal{D}_2^2\right] + \frac{1}{2}\left[x_1^2 + x_2^2\right], \tag{1.1}$$

where $\mathcal{D}_i$ stands for the Dunkl derivative [3]

$$\mathcal{D}_i = \partial_i + \frac{\mu_i}{x_i}(1 - R_i), \qquad i = 1, 2, \tag{1.2}$$

with $\partial_i = \frac{\partial}{\partial x_i}$ and where R_i is the reflection operator with respect to the $x_i = 0$ axis, i.e

$$R_1 f(x_1, x_2) = f(-x_1, x_2), \qquad R_2 f(x_1, x_2) = f(x_1, -x_2).$$

The Hamiltonian (1.1) can obviously be written as

$$H = H_1 + H_2$$

where H_i, $i = 1, 2$, is the one-dimensional Dunkl oscillator Hamiltonian

$$H_i = -\frac{1}{2}\mathcal{D}_i^2 + \frac{1}{2}x_i^2. \tag{1.3}$$

In [4], the Hamiltonian (1.1) was shown to be maximally superintegrable and its two independent constants of motion, obtained by the Schwinger construction, were seen to generate a $\mathfrak{u}(2)$ algebra extended with involutions. This algebra was called the Schwinger-Dunkl algebra $sd(2)$. It was further shown that the Schrödinger equation associated to (1.1) admits separation of variables in both Cartesian and polar coordinates and the separated wavefunctions were given explicitly. Furthermore, the overlap coefficients between the separated wavefunctions were expressed in terms of the dual -1 Hahn polynomials through a correspondence with the Clebsch-Gordan problem of $sl_{-1}(2)$, a $q \to -1$ limit of the quantum algebra $sl_q(2)$ (see Section II for the definition of $sl_{-1}(2)$ and [11] for background). In [5], the representation theory of the Schwinger-Dunkl algebra $sd(2)$ was examined. The results obtained in this paper further strengthened the idea that Dunkl oscillator models are showcases for -1 orthogonal polynomials (see [6, 12]).

Generalizations of the Dunkl oscillator in the plane (1.1) were considered in [7], where the singular and the 2 : 1 anisotropic Dunkl oscillators were investigated. The two systems were shown to be superintegrable, their constants of motion were constructed and their (quadratic) invariance algebra was given. It was also shown that in some cases these models exhibit multiseparability. Here we pursue our investigations on Dunkl oscillator models by considering the three-dimensional version of the system (1.1) in Euclidean space.

1.3. The three-dimensional Dunkl oscillator

The Dunkl oscillator in three-dimensional Euclidean space is defined by the Hamiltonian

$$\mathcal{H} = -\frac{1}{2}\Big[\mathcal{D}_1^2 + \mathcal{D}_2^2 + \mathcal{D}_3^2\Big] + \frac{1}{2}\Big[x_1^2 + x_2^2 + x_3^2\Big], \tag{1.4}$$

where $\mathcal{D}_i$ is the Dunkl derivative given by (1.2). An elementary calculation shows that the square of the Dunkl derivative has the expression

$$\mathcal{D}_i^2 = \partial_i^2 + \frac{2\mu_i}{x_i}\partial_i - \frac{\mu_i}{x_i^2}(1 - R_i).$$

As is directly seen, the Hamiltonian (1.4) corresponds to the combination of three independent (commuting) one-dimensional Dunkl oscillators with Hamiltonians H_i given by (1.3). It directly follows from the formulas (1.2) and (1.4) that when $\mu_i = 0$, $i = 1, 2, 3$, the Dunkl oscillator Hamiltonian (1.4) reduces to that of the standard isotropic harmonic oscillator in three dimensions. We note that the term "isotropic" here refers to the fact that the three independent one-dimensional oscillator entering the total Hamiltonian (1.4) have the same frequency. To obtain a 3D Dunkl oscillator isotropic in the sense of being $O(3)$-invariant, one must also take $\mu_1 = \mu_2 = \mu_3$.

1.4. Outline

The paper is organized as follows. In section II, the $sl_{-1}(2)$ dynamical symmetry and the Schwinger construction are used to obtain the symmetries of the total Hamiltonian (1.4), thus establishing its superintegrability. The commutation relations satisfied by the constants of motion, which define the Schwinger-Dunkl algebra $sd(3)$, are also exhibited in two different bases. In section III, the separated solutions of the Schrödinger equation associated to (1.4) are given in Cartesian, polar and spherical coordinates. For each coordinate system, the symmetries responsible for the separation of variables are given. A short conclusion follows.

2. Superintegrability

In this section, the $sl_{-1}(2)$ dynamical algebra of the three-dimensional Dunkl oscillator is made explicit and is exploited to obtain the spectrum of the Hamiltonian. Following the Schwinger construction, the constants of motion are constructed using the raising/lowering operators of the $sl_{-1}(2)$ algebra of the one-dimensional constituents. The Schwinger-Dunkl algebra $sd(3)$ formed by the symmetries is seen to correspond to a deformation of the Lie algebra $\mathfrak{u}(3)$ by involutions and the defining relations of $sd(3)$ are given in two different bases.

2.1. Dynamical algebra and spectrum

The three-dimensional Dunkl oscillator Hamiltonian (1.4) possesses an $sl_{-1}(2)$ dynamical symmetry inherited from the one of its one-dimensional constituents H_i. This can be seen as follows. Consider the following operators:

$$A_{\pm}^{(i)} = \frac{1}{\sqrt{2}}\Big(x_i \pm \mathcal{D}_i\Big), \tag{2.1}$$

and define $A_0^{(i)} = H_i$ with H_i given by (1.3). It is easily checked that these operators, together with the reflection operator R_i, satisfy the defining relations of the $sl_{-1}(2)$ algebra which are of the form [11]

$$[A_0^{(i)}, A_{\pm}^{(i)}] = \pm A_{\pm}^{(i)}, \quad [A_0^{(i)}, R_i] = 0, \quad \{A_+^{(i)}, A_-^{(i)}\} = 2A_0^{(i)}, \quad \{A_{\pm}^{(i)}, R_i\} = 0, \tag{2.2}$$

where $[x, y] = xy - yx$ and $\{x, y\} = xy + yx$. As is seen from the above commutation relations, the operators $A_{\pm}^{(i)}$ act as raising/lowering operators for the one-dimensional Hamiltonians $A_0^{(i)} = H_i$. In the realization (2.1), the $sl_{-1}(2)$ Casimir operator

$$Q^{(i)} = A_+^{(i)} A_-^{(i)} R_i - A_0^{(i)} R_i + R_i/2,$$

is a multiple of the identity $Q^{(i)} = -\mu_i$. It thus follows that for $\mu_i > -1/2$, the operators $A_{\pm}^{(i)}$, $A_0^{(i)}$ and R_i realize the positive-discrete series of representations of $sl_{-1}(2)$ [11]. In these representations, the spectrum $\mathcal{E}^{(i)}$ of the operators $A_0^{(i)}$ is given by

$$\mathcal{E}_{n_i}^{(i)} = n_i + \mu_i + 1/2,$$

where n_i is a non-negative integer. Given that the full Hamiltonian (1.4) of the three-dimensional Dunkl oscillator is of the form $\mathcal{H} = H_1 + H_2 + H_3$, it follows that its energy eigenvalues $\mathcal{E}$ have the expression

$$\mathcal{E}_N = N + \mu_1 + \mu_2 + \mu_3 + 3/2, \tag{2.3}$$

where N is a non-negative integer. Since the integer N can be written as $N = n_1 + n_2 + n_3$, the spectrum of $\mathcal{H}$ has degeneracy

$$g_N = \sum_{n_3=0}^{N} (N - n_3 + 1) = (N + 1)(N + 2)/2,$$

at energy level $\mathcal{E}_N$. Hence the three-dimensional Dunkl oscillator exhibits the same degeneracy in its spectrum as the standard three-dimensional isotropic oscillator.

The coproduct of $sl_{-1}(2)$ (see [11]) can be used to construct the $sl_{-1}(2)$ dynamical algebra of the total Hamiltonian $\mathcal{H}$. Indeed, upon defining the operators

$$\mathcal{A}_{\pm} = A_{\pm}^{(1)} R_2 R_3 + A_{\pm}^{(2)} R_3 + A_{\pm}^{(3)}, \quad \mathcal{R} = R_1 R_2 R_3, \qquad \mathcal{A}_0 = \mathcal{H},$$

it is directly checked that one has

$$[\mathcal{A}_0, \mathcal{A}_{\pm}] = \pm \mathcal{A}_{\pm}, \quad [\mathcal{A}_0, \mathcal{R}] = 0, \quad \{\mathcal{A}_+, \mathcal{A}_-\} = 2\mathcal{A}_0, \quad \{\mathcal{A}_{\pm}, \mathcal{R}\} = 0.$$

Hence the operators $\mathcal{A}_{\pm}$ act as the raising/lowering operators for the full Hamiltonian $\mathcal{H}$ of the three-dimensional Dunkl oscillator.

2.2. Constants of motion and the Schwinger-Dunkl algebra sd(3)

Given that $\mathcal{H} = H_1 + H_2 + H_3$, it is clear that combining a raising operator for H_i with a lowering operator for H_j when $i \neq j$ will result in an operator preserving the eigenspace associated to a given energy $\mathcal{E}_N$, thus producing a constant of motion of the total Hamiltonian $\mathcal{H}$. Moreover, it is obvious that the one-dimensional components H_i commute with the total Hamiltonian $\mathcal{H}$ and hence these one-dimensional Hamiltonians can be considered as constants of the motion. Furthermore, it is observed that the total Hamiltonian commutes with the reflection operators

$$[\mathcal{H}, R_i] = 0, \qquad i = 1, 2, 3,$$

so that the reflections can be considered as symmetries. Following the Schwinger construction of $\mathfrak{u}(3)$ with standard creation/annihilation operators, we define

$$J_1 = \frac{1}{i}(A_+^{(2)}A_-^{(3)} - A_-^{(2)}A_+^{(3)}), \quad J_2 = \frac{1}{i}(A_+^{(3)}A_-^{(1)} - A_-^{(3)}A_+^{(1)}),$$
$$J_3 = \frac{1}{i}(A_+^{(1)}A_-^{(2)} - A_-^{(1)}A_+^{(2)}).$$

The operators J_i, $i = 1, 2, 3$, can be interpreted as Dunkl "rotation" generators since in terms of Dunkl derivatives, they read

$$J_1 = \frac{1}{i}(x_2\mathcal{D}_3 - x_3\mathcal{D}_2), \quad J_2 = \frac{1}{i}(x_3\mathcal{D}_1 - x_1\mathcal{D}_3), \quad J_3 = \frac{1}{i}(x_1\mathcal{D}_2 - x_2\mathcal{D}_1). \tag{2.4}$$

We also introduce the operators

$$K_1 = (A_+^{(2)}A_-^{(3)} + A_-^{(2)}A_+^{(3)}),\ K_2 = (A_+^{(3)}A_-^{(1)} + A_-^{(3)}A_+^{(1)}),\ K_3 = (A_+^{(1)}A_-^{(2)} + A_-^{(1)}A_+^{(2)}),$$

which in coordinates have the expression

$$K_1 = (x_2x_3 - \mathcal{D}_2\mathcal{D}_3), \quad K_2 = (x_3x_1 - \mathcal{D}_3\mathcal{D}_1), \quad K_3 = (x_1x_2 - \mathcal{D}_1\mathcal{D}_2). \tag{2.5}$$

It is directly checked that the operators J_i, K_i, $i = 1, 2, 3$, are symmetries of the total Dunkl oscillator Hamiltonian (1.4), that is

$$[J_i, \mathcal{H}] = [K_i, \mathcal{H}] = 0,$$

for $i = 1, 2, 3$. Upon writing

$$L_1 = H_1/2, \quad L_2 = H_2/2, \quad L_3 = H_3/2,$$

which satisfy $\mathcal{H} = 2L_1 + 2L_2 + 2L_3$, a direct computation shows that the non-zero commutation relations between the symmetries are given by

$$[J_j, J_k] = i\epsilon_{jk\ell}J_\ell(1 + 2\mu_\ell R_\ell), \qquad [K_j, K_k] = -i\epsilon_{jk\ell}J_\ell(1 + 2\mu_\ell R_\ell), \tag{2.6a}$$

where ϵ_{ijk} stands for the totally antisymmetric tensor with summation over repeated indices understood and

$$[J_j, K_k] = -i\epsilon_{jk\ell}K_\ell(1 + 2\mu_\ell R_\ell), \quad [J_j, K_j] = -i\epsilon_{jk\ell}L_k(1 + 2\mu_\ell R_\ell), \tag{2.6b}$$

for $j \neq k$. One also has

$$[J_j, L_k] = ig_{jk}K_j/2, \quad [K_j, L_k] = -ig_{jk}J_j/2, \tag{2.6c}$$

where g_{ij} are the elements of a 3×3 antisymmetric matrix with $g_{12} = g_{23} = 1$, $g_{13} = -1$. The commutation relations involving the reflection operators are of the form

$$[L_i, R_i] = [L_j, R_k] = [J_i, R_i] = [K_i, R_i] = 0, \tag{2.6d}$$
$$\{J_j, R_k\} = \{K_j, R_k\} = 0, \tag{2.6e}$$

where $j \neq k$. Hence it follows that the three-dimensional Dunkl oscillator model is maximally superintegrable. The invariance algebra is defined by the commutation and anticommutation relations (2.6) and we shall refer to this algebra as the Schwinger-Dunkl algebra $sd(3)$. It is clear from the defining relations (2.6) that the algebra $sd(3)$ corresponds to a deformation of the $\mathfrak{u}(3)$ Lie algebra by involutions; the central element here is of course the total Hamiltonian $\mathcal{H} = H_1 + H_2 + H_3$. If one takes $\mu_1 = \mu_2 = \mu_3 = 0$ in the commutation relations (2.6), one recovers the $\mathfrak{u}(3)$ symmetry algebra of the standard isotropic harmonic oscillator in three dimensions realized by the standard creation/annihilation operators.

2.3. An alternative presentation of sd(3)

The Schwinger-Dunkl algebra $sd(3)$ obtained here can be seen as a "rank two" version of the Schwinger-Dunkl algebra $sd(2)$ which has appeared in [4] as the symmetry algebra of the Dunkl oscillator in the plane. It is possible to present another basis for the symmetries of the three-dimensional Dunkl oscillator in which the $sd(2)$ algebra explicitly appears as a subalgebra of $sd(3)$. In order to define this basis, it is convenient to introduce the standard 3×3 Gell-Mann matrices [1] denoted by Λ_i, $i=1,\ldots,8$, and obeying the $\mathfrak{su}(3)$ commutation relations

$$[\Lambda_i,\Lambda_j]=if^{ijk}\Lambda_k,$$

with $f^{123}=2$, $f^{458}=f^{678}=\sqrt{3}$ and

$$f^{147}=f^{165}=f^{246}=f^{257}=f^{345}=f^{376}=1.$$

The symmetries of the three-dimensional Dunkl oscillator can be expressed in terms of these matrices as follows. One takes

$$M_j=(A_+^{(1)},A_+^{(2)},A_+^{(3)})\,\Lambda_j\,(A_-^{(1)},A_-^{(2)},A_-^{(3)})^t, \tag{2.7}$$

for $j=1,2,4,5,6,7$ and also

$$\begin{aligned} M_3&=\frac{1}{4}\left(\{A_+^{(1)},A_-^{(1)}\}-\{A_+^{(2)}A_-^{(2)}\}\right),\\ M_8&=\frac{1}{4\sqrt{3}}\Big(\{A_+^{(1)},A_-^{(1)}\}+\{A_+^{(2)},A_-^{(2)}\}-2\{A_+^{(3)},A_-^{(3)}\}\Big). \end{aligned}$$

It is straightforward to verify that the operators M_i, $i=1,\ldots,8$ commute with the Hamiltonian $\mathcal{H}$ of the three-dimensional Dunkl oscillator. Using the commutation relations (2.2) as well as the extra relation

$$[A_-^{(i)},A_+^{(j)}]=\delta_{ij}(1+2\mu_iR_i),$$

the $sd(3)$ commutation relations expressed in the basis of the symmetries M_i can easily be obtained. Consider the constants of motion

$$M_1=\frac{1}{2}\left(A_+^{(1)}A_-^{(2)}+A_-^{(1)}A_+^{(2)}\right),\quad M_2=\frac{1}{2i}\left(A_+^{(1)}A_-^{(2)}-A_-^{(1)}A_+^{(2)}\right),$$

as well as M_3. These symmetry operators satisfy the commutation relations of the Schwinger-Dunkl algebra $sd(2)$

$$[M_2,M_3]=iM_1,\quad [M_3,M_1]=iM_2, \tag{2.8a}$$

$$[M_1,M_2]=i\Big(M_3+M_3(\mu_1R_1+\mu_2R_2)-\frac{1}{3}\left(\mathcal{H}+\sqrt{3}M_8\right)(\mu_1R_1-\mu_2R_2)\Big), \tag{2.8b}$$

$$\{M_1,R_i\}=0,\quad \{M_2,R_i\}=0,\quad [M_3,R_i]=0, \tag{2.8c}$$

for $i=1,2$. In the subalgebra (2.8), the operator $\left(\mathcal{H}+\sqrt{3}M_8\right)$ is central and corresponds to the Hamiltonian of the Dunkl oscillator in the plane

$$\frac{2}{3}\left(\mathcal{H}+\sqrt{3}M_8\right)=H_1+H_2.$$

Journal of Physics: Conference Series **512** (2014) 012010 doi:10.1088/1742-6596/512/1/012010

3. Separated Solutions: Cartesian, cylindrical and spherical coordinates

In this section, the exact solutions of the time-independent Schrödinger equation

$$\mathcal{H}\Psi = \mathcal{E}\Psi, \tag{3.1}$$

associated to the three-dimensional Dunkl oscillator Hamiltonian (1.4) are obtained in Cartesian, polar (cylindrical) and spherical coordinates. The operators responsible for the separation of variables in each of these coordinate systems are given explicitly.

3.1. Cartesian coordinates

Since $\mathcal{H} = H_1 + H_2 + H_3$, where H_i, $i = 1, 2, 3$, are the one-dimensional Dunkl oscillator Hamiltonians

$$H_i = -\frac{1}{2}\mathcal{D}_i^2 + \frac{1}{2}x_i^2,$$

it is obvious that the Schrödinger equation (3.1) admits separation of variable in Cartesian coordinates $\{x_1, x_2, x_3\}$. In this coordinate system, the separated solutions are of the form

$$\Psi(x_1, x_2, x_3) = \psi(x_1)\psi(x_2)\psi(x_3),$$

where $\psi(x_i)$ are solutions of the one-dimensional Schrödinger equation

$$\left[-\frac{1}{2}\mathcal{D}_i^2 + \frac{1}{2}x_i^2\right]\psi(x_i) = \mathcal{E}^{(i)}\psi(x_i). \tag{3.2}$$

The regular solutions of (3.2) are well known [4, 10]. To obtain these solutions, one uses the fact that the reflection operator R_i commutes with the one-dimensional Hamiltonian H_i, which allows to diagonalize both operators simultaneously. For the one-dimensional problem, the two sectors corresponding to the possible eigenvalues $s_i = \pm 1$ of the reflection operator R_i can be recombined to give the following expression for the wavefunctions:

$$\psi_{n_i}(x_i) = e^{-x_i^2/2} H_{n_i}^{\mu_i}(x_i), \tag{3.3}$$

where n_i is a non-negative integer. The corresponding energy eigenvalues are

$$\mathcal{E}_{n_i}^{(i)} = n_i + \mu_i + 1/2,$$

and $H_n^{\mu}(x)$ stands for the generalized Hermite polynomials [2]

$$H_{2m+p}^{\mu}(x) = (-1)^n \sqrt{\frac{n!}{\Gamma(m+p+\mu+1/2)}}\, x^p\, L_m^{(\mu-1/2+p)}(x^2), \tag{3.4}$$

with $p \in \{0, 1\}$. In (3.4), $L_n^{(\alpha)}(x)$ are the standard Laguerre polynomials [8] and $\Gamma(x)$ is the classical Gamma function [1]. The wavefunctions (3.3) satisfy

$$R_i\psi_{n_i}(x_i) = (-1)^{n_i}\psi_{n_i}(x_i),$$

so that the eigenvalue s_i of the reflection operator R_i is given by the parity of n_i. Using the orthogonality relation of the Laguerre polynomials, one finds that the wavefunctions (3.3) satisfy the orthogonality condition

$$\int_{-\infty}^{\infty} \psi_{n_i}(x_i)\psi_{n_i'}(x_i)|x_i|^{2\mu_i}\, \mathrm{d}x_i = \delta_{n_i,n_i'}.$$

In Cartesian coordinates, the separated solution of the Schrödinger equation associated to the three-dimensional Dunkl oscillator Hamiltonian (1.4) are thus given by

$$\Psi_{n_1,n_2,n_3}(x_1,x_2,x_3) = e^{-(x_1^2+x_2^2+x_3^2)/2} H_{n_1}^{\mu_1}(x_1) H_{n_2}^{\mu_2}(x_2) H_{n_3}^{\mu_3}(x_3), \tag{3.5}$$

and the corresponding energy $\mathcal{E}$ is

$$\mathcal{E} = n_1 + n_2 + n_3 + \mu_1 + \mu_2 + \mu_3 + 3/2, \tag{3.6}$$

where n_i, $i = 1,2,3$, are non-negative integers. It is directly seen from (3.4), (3.5) and (3.6) that if one takes $\mu_i = 0$, the solutions and the spectrum of the isotropic three-dimensional oscillator in Cartesian coordinates are recovered.

3.2. Cylindrical coordinates

In cylindrical coordinates

$$x_1 = \rho\cos\varphi, \quad x_2 = \rho\sin\varphi, \quad x_3 = z.$$

the Hamiltonian (1.4) of the three-dimensional Dunkl oscillator reads

$$\mathcal{H} = \mathcal{A}_\rho + \frac{1}{\rho^2}\mathcal{B}_\varphi + \mathcal{C}_z,$$

where

$$\mathcal{A}_\rho = -\frac{1}{2}\left[\partial_\rho^2 + \frac{1}{\rho}\partial_\rho\right] - \frac{(\mu_1+\mu_2)}{\rho}\partial_\rho + \frac{1}{2}\rho^2, \tag{3.7a}$$

$$\mathcal{B}_\varphi = -\frac{1}{2}\partial_\varphi^2 + (\mu_1\tan\varphi - \mu_2\cot\varphi)\partial_\varphi + \frac{\mu_1}{2\cos^2\varphi}(1-R_1) + \frac{\mu_2}{2\sin^2\varphi}(1-R_2), \tag{3.7b}$$

$$\mathcal{C}_z = -\frac{1}{2}\partial_z^2 - \frac{\mu_3}{z}\partial_z + \frac{1}{2}z^2 + \frac{\mu_3}{2z^2}(1-R_3). \tag{3.7c}$$

The reflection operators are easily seen to have the action

$$R_1 f(\rho,\varphi,z) = f(\rho,\pi-\phi,z), \quad R_2 f(\rho,\varphi,z) = f(\rho,-\varphi,z), \quad R_3 f(\rho,\varphi,z) = f(\rho,\varphi,-z).$$

Upon taking $\Psi(\rho,\varphi,z) = P(\rho)\Phi(\varphi)\psi(z)$, one finds that (3.1) is equivalent to the system of ordinary equations

$$\mathcal{A}_\rho P(\rho) - \widetilde{\mathcal{E}}P(\rho) + \frac{k^2}{2\rho^2}P(\rho) = 0, \tag{3.8a}$$

$$\mathcal{B}_\varphi\Phi(\varphi) - \frac{k^2}{2}\Phi(\varphi) = 0, \tag{3.8b}$$

$$\mathcal{C}_z\psi(z) = \mathcal{E}^{(3)}\psi(z), \tag{3.8c}$$

where $\mathcal{E}^{(3)}$, $k^2/2$ are the separation constants and where $\widetilde{\mathcal{E}} = \mathcal{E} - \mathcal{E}^{(3)}$. The solutions to the equation are given by (3.3) and (3.4) with $\mathcal{E}^{(3)} = n_3 + \mu_3 + 1/2$. The solutions to (3.8a) and (3.8b) have been obtained in [4]. For the angular part, the solutions are labeled by the eigenvalues s_1, s_2 with $s_i = \pm 1$ of the reflection operators R_1, R_2 and read

$$\Phi_m^{(s_1,s_2)}(\varphi) = \eta_m \cos^{e_1}\varphi \sin^{e_2}\varphi\, P_{m-e_1/2-e_2/2}^{(\mu_2+e_2-1/2,\mu_1+e_1-1/2)}(\cos 2\varphi), \tag{3.9}$$

where $P_n^{(\alpha,\beta)}(x)$ are the classical Jacobi polynomials [8], η_m is a normalization factor and where e_1, e_2 are the indicator functions for the eigenvalues of the reflections R_1 and R_2, i.e.:

$$e_i = \begin{cases} 0, & \text{if } s_i = 1, \\ 1, & \text{if } s_i = -1, \end{cases}$$

for $i = 1, 2$. When $s_1 s_2 = -1$, m is a positive half-integer whereas when $s_1 s_2 = 1$, m is a non-negative integer; note also that for $m = 0$, only the $s_1 = s_2 = 1$ state exists. In all parity cases, the separation constant takes the value

$$k^2 = 4m(m + \mu_1 + \mu_2).$$

If one takes

$$\eta_m = \left[\frac{(2m + \mu_1 + \mu_2)\Gamma(m + \mu_1 + \mu_2 + \frac{e_1}{2} + \frac{e_2}{2})(m - \frac{e_1}{2} - \frac{e_2}{2})!}{2\,\Gamma(m + \mu_1 + \frac{e_1}{2} - \frac{e_2}{2} + 1/2)\,\Gamma(m + \mu_2 + \frac{e_2}{2} - \frac{e_1}{2} + 1/2)}\right]^{1/2},$$

as the normalization factor, the angular part of the separated wavefunction satisfy the orthogonality relation

$$\int_0^{2\pi} \Phi_m^{(s_1,s_2)} \Phi_{m'}^{(s_1',s_2')} \,|\cos\phi|^{2\mu_1}\, |\sin\phi|^{2\mu_2}\, \mathrm{d}\phi = \delta_{m,m'}\delta_{s_1,s_1'}\delta_{s_2,s_2'},$$

which can be deduced from the orthogonality relation satisfied by the Jacobi polynomials [8]. The radial solutions have the expression

$$P_{n_\rho}(\rho) = \left[\frac{2\,n_\rho!}{\Gamma(n_\rho + 2m + \mu_1 + \mu_2 + 1)}\right]^{1/2} e^{-\rho^2/2}\rho^{2m} L_{n_\rho}^{(2m+\mu_1+\mu_2)}(\rho^2), \tag{3.10}$$

where n_ρ is a non-negative integer and where $L_n^{(\alpha)}(x)$ are the Laguerre polynomials. They satisfy the orthogonality relation

$$\int_0^{\infty} P_{n_\rho}(\rho) P_{n_\rho'}(\rho)\, \rho^{2\mu_1+2\mu_2+1}\, \mathrm{d}\rho = \delta_{n_\rho,n_\rho'}.$$

The separated wavefunctions of the three-dimensional Dunkl oscillator in cylindrical coordinates are thus given by

$$\Psi_{n_\rho,m,n_z}(\rho,\varphi,z) = P_{n_\rho}(\rho)\Phi_m^{(s_1,s_2)}(\varphi)\psi_{n_z}(z)$$

where $P_{n_\rho}(\rho)$, $\Phi_m^{(s_1,s_2)}(\varphi)$ and $\psi_{n_z}(z)$ are given by (3.10), (3.9) and (3.3), respectively. The energy $\mathcal{E}$ is expressed as

$$\mathcal{E} = 2n_\rho + 2m + n_z + \mu_1 + \mu_2 + \mu_3 + 3/2,$$

where n_ρ, n_z are non-negative integers and where m is a non-negative integer when $s_1 s_2 = 1$ or a positive half-integer when $s_1 s_2 = -1$. In the cylindrical basis, the operators $\mathcal{C}_z$ and $\mathcal{B}_\phi$ are diagonal with eigenvalues $\mathcal{E}^{(3)}$ and $k^2/2$. A direct computation shows that one has

$$J_3^2 = 2\mathcal{B}_\varphi + 2\mu_1\mu_2(1 - R_1 R_2),$$

where J_3 is the symmetry given in (2.4). It thus follows that J_3 and H_3 are the symmetries responsible for the separation of variables in cylindrical coordinates.

3.3. Spherical coordinates
In spherical coordinates

$$x_1 = r\cos\phi\sin\theta,\quad x_2 = r\sin\phi\sin\theta,\quad x_3 = r\cos\theta,$$

the Hamiltonian (1.4) of the three-dimensional Dunkl oscillator takes the form

$$\mathcal{H} = \mathcal{M}_r + \frac{1}{r^2}\mathcal{N}_\theta + \frac{1}{r^2\sin^2\theta}\mathcal{B}_\phi,$$

where $\mathcal{B}_\phi$ is given by (3.7b) and where

$$\mathcal{M}_r = -\frac{1}{2}\partial_r^2 - \frac{(1+\mu_1+\mu_2+\mu_3)}{r}\partial_r + \frac{1}{2}r^2, \tag{3.11a}$$

$$\mathcal{N}_\theta = -\frac{1}{2}\partial_\theta^2 + (\mu_3\tan\theta - (1/2+\mu_1+\mu_2)\cot\theta)\partial_\theta + \frac{\mu_3}{2\cos^2\theta}(1-R_3). \tag{3.11b}$$

The reflection operators have the action

$$R_1 f(r,\theta,\phi) = f(r,\theta,\pi-\phi),\quad R_2 f(r,\theta,\phi) = f(r,\theta,-\phi),\quad R_3 f(r,\theta,\phi) = f(r,\pi-\theta,\phi).$$

Upon taking $\Psi(r,\theta,\phi) = R(r)\Theta(\theta)\Phi(\phi)$ in the Schrödinger equation (3.1), one finds the following system of ordinary differential equations

$$\left[\mathcal{M}_r + \left(\frac{q^2}{2r^2} - \mathcal{E}\right)\right] R(r) = 0, \tag{3.12a}$$

$$\left[\mathcal{N}_\theta + \left(\frac{k^2}{2\sin^2\theta} - \frac{q^2}{2}\right)\right]\Theta(\theta) = 0, \tag{3.12b}$$

$$\left[\mathcal{B}_\phi - \frac{k^2}{2}\right]\Phi(\phi) = 0, \tag{3.12c}$$

where $k^2/2$ and $q^2/2$ are the separation constants. It is directly seen that the azimuthal solution $\Phi(\phi)$ to (3.12c) is given by (3.9) with value of the separation constant $k^2 = 4m(m+\mu_1+\mu_2)$. The zenithal solutions $\Theta(\theta)$ are labeled by the eigenvalue $s_3 = \pm 1$ of the reflection operator R_3. One has

$$\Theta_\ell^{(s_3)}(\theta) = \iota_\ell \cos^{e_3}\theta\,\sin^{2m}\theta\, P_{\ell-e_3/2}^{(2m+\mu_1+\mu_2,\mu_3+e_3-1/2)}(\cos 2\theta), \tag{3.13}$$

where the value of the separation constant is $q^2 = 4(\ell+m)(\ell+m+\mu_1+\mu_2+\mu_3+1/2)$. When $s_3 = 1$, ℓ is a non-negative integer whereas ℓ is a positive half-integer when $s_3 = -1$. The normalization constant has the expression

$$\iota_\ell = \left[\frac{(2\ell+2m+\mu_1+\mu_2+\mu_3+1/2)\Gamma(\ell+2m+\mu_1+\mu_2+\mu_3+1/2+e_3/2)(\ell-e_3/2)!}{\Gamma(\ell+2m+\mu_1+\mu_2+1-e_3/2)\Gamma(\ell+\mu_3+1/2+e_3/2)}\right]^{1/2}.$$

The radial solutions are given by

$$R_{n_r}(r) = \left[\frac{2n_r!}{\Gamma(n_r+\alpha+1)}\right]^{1/2} e^{-r^2/2} r^{2(\ell+m)} L_{n_r}^{(\alpha)}(r^2), \tag{3.14}$$

with $\alpha = 2(\ell+m)+\mu_1+\mu_2+\mu_3+1/2$. The separated wavefunctions of the three-dimensional Dunkl oscillator in spherical coordinates thus read

$$\Psi_{n_r,\ell,m}^{(s_1,s_2,s_3)}(r,\theta,\phi) = R_{n_r}(r)\,\Phi_m^{(s_1,s_2)}(\phi)\,\Theta_\ell^{(s_3)}(\theta), \tag{3.15}$$

where $R(r)$, $\Theta(\theta)$ and $\Phi(\phi)$ are respectively given by (3.14), (3.13) and (3.9) and correspond to the total energy

$$\mathcal{E} = 2(n_r + \ell + m) + \mu_1 + \mu_2 + \mu_3 + 3/2. \tag{3.16}$$

These wavefunctions are eigenfunctions of the reflection operators R_i with eigenvalues $s_i = \pm 1$ for $i = 1, 2, 3$. When $s_3 = 1$, the quantum number ℓ takes non-negative integer values and when $s_3 = -1$, the number ℓ takes positive half-integer values. Similarly, when $s_1 s_2 = 1$, the quantum number m is a non-negative integer and when $s_1 s_2 = -1$, m is a positive half-integer. The wavefunctions satisfy the orthogonality relation

$$\int_0^\infty \int_0^\pi \int_0^{2\pi} r^{2\mu_1+2\mu_2+2\mu_3} |\sin\theta|^{2\mu_1+2\mu_2} |\cos\theta|^{2\mu_3} |\cos\phi|^{2\mu_1} |\sin\phi|^{2\mu_2}\, r^2 \sin\theta\, \mathrm{d}r\, \mathrm{d}\theta\, \mathrm{d}\phi$$
$$R_{n_r'}(r) R_{n_r}(r) \Theta_{\ell'}^{(s_3')}(\theta) \Theta_{\ell}^{(s_3)}(\theta) \Phi_{m'}^{(s_1',s_2')}(\phi) \Phi_{m}^{(s_1,s_2)}(\phi) = \delta_{n_r,n_r'} \delta_{\ell,\ell'} \delta_{s_3,s_3'} \delta_{m,m'} \delta_{s_1,s_1'} \delta_{s_2,s_2'}.$$

In analogy with the standard three-dimensional oscillator, the symmetries responsible for the separation of variables in spherical coordinates are related to the Dunkl "rotation" generators. Indeed, one has that the operator

$$J_3^2 = \left\{ \frac{1}{i}(x_1 \mathcal{D}_2 - x_2 \mathcal{D}_1) \right\}^2 = 2\mathcal{B}_\phi + 2\mu_1\mu_2(1 - R_1 R_2), \tag{3.17}$$

is diagonal on the separated wavefunction in spherical coordinates and has eigenvalues

$$J_3^2\, \Psi_{n_r,\ell,m}^{(s_1,s_2,s_3)}(r,\theta,\phi) = \left[4m(m + \mu_1 + \mu_2) + 2\mu_1\mu_2(1 - s_1 s_2)\right] \Psi_{n_r,\ell,m}^{(s_1,s_2,s_3)}(r,\theta,\phi).$$

Furthermore, a direct computations shows that the Dunkl total angular momentum operator

$$\mathbf{J}^2 = \left\{ \frac{1}{i}(x_2 \mathcal{D}_3 - x_3 \mathcal{D}_2) \right\}^2 + \left\{ \frac{1}{i}(x_3 \mathcal{D}_1 - x_1 \mathcal{D}_3) \right\}^2 + \left\{ \frac{1}{i}(x_1 \mathcal{D}_2 - x_2 \mathcal{D}_1) \right\}^2$$

has the following expression in spherical coordinates:

$$\mathbf{J}^2 = 2\left(\mathcal{N}_\theta + \frac{1}{\sin^2\theta} \mathcal{B}_\phi \right) + 2\mu_1\mu_2(1 - R_1R_2) + 2\mu_2\mu_3(1 - R_2R_3) + 2\mu_1\mu_3(1 - R_1R_3)$$
$$+ \mu_1(1 - R_1) + \mu_2(1 - R_2) + \mu_3(1 - R_3),$$

where $\mathcal{N}_\theta$ and $\mathcal{B}_\phi$ are as in (3.11b) and (3.7b). It thus follows that the separated wavefunctions in spherical coordinates $\Psi_{n_r,\ell,m}^{(s_1,s_2,s_3)}(r,\theta,\phi)$ satisfy

$$\mathbf{J}^2 \Psi_{n_r,\ell,m}^{(s_1,s_2,s_3)}(r,\theta,\phi) = \lambda_{\ell,m} \Psi_{n_r,\ell,m}^{(s_1,s_2,s_3)}(r,\theta,\phi),$$

with

$$\lambda_{\ell,m} = 4(\ell + m)(\ell + m + \mu_1 + \mu_2 + \mu_3 + 1/2) + 2\mu_1\mu_2(1 - s_1s_2) + 2\mu_1\mu_3(1 - s_1s_3)$$
$$+ 2\mu_2\mu_3(1 - s_2s_2) + \mu_1(1 - s_1) + \mu_2(1 - s_2) + \mu_3(1 - s_3).$$

4. Conclusion

In this paper, we have examined the isotropic Dunkl oscillator model in three-dimensional Euclidean space and we have shown that this system is maximally superintegrable. The symmetries of the model were exhibited and the invariance algebra they generate, called the Schwinger-Dunkl algebra $sd(3)$, has been seen to be a deformation of the Lie algebra $\mathfrak{u}(3)$ by involutions. So far, we have examined Dunkl systems with oscillator type potentials. In view of the superintegrability and importance of the Coulomb problem, the examination of the Dunkl-Coulomb problem is of considerable interest. This will be the subject of a future publication.

Acknowledgements

V.X.G. holds an Alexander-Graham-Bell fellowship from the Natural Sciences and Engineering Research Council of Canada (NSERC). The research of L.V. is supported in part by NSERC.

References

[1] G. B. Arfken, H. Weber, and F. E. Harris. *Mathematical Methods for Physicists.* Academic Press, 7 edition, 2012.

[2] T. Chihara. *An Introduction to Orthogonal Polynomials.* Dover Publications, 2011.

[3] C. F. Dunkl. Differential-difference operators associated to reflection groups. *Trans. Amer. Math. Soc.*, 311:167–183, 1989.

[4] V.X. Genest, M.E.H. Ismail, L. Vinet, and A. Zhedanov. The Dunkl oscillator in the plane: I. Superintegrability, separated wavefunctions and overlap coefficients. *J. Phys. A: Math. Theor.*, 46:145201, 2013.

[5] V.X. Genest, M.E.H. Ismail, L. Vinet, and A. Zhedanov. The Dunkl oscillator in the plane: II. Representations of the symmetry algebra. *Comm. Math. Phys. (to appear)*, 2013.

[6] V.X. Genest, L. Vinet, and A. Zhedanov. Bispectrality of the Complementary Bannai–Ito polynomials. *SIGMA*, 9:18–37, 2013.

[7] V.X. Genest, L. Vinet, and A. Zhedanov. The singular and 2:1 anisotropic Dunkl oscillators in the plane. *J. Phys. A: Math. Theor.*, 46:325201, 2013.

[8] R. Koekoek, P.A. Lesky, and R.F. Swarttouw. *Hypergeometric orthogonal polynomials and their q-analogues.* Springer, 1st edition, 2010.

[9] W. Miller, S. Post, and P. Winternitz. Classical and quantum superintegrability with applications. *J. Phys. A: Math. Theor.*, 46:423001, 2013.

[10] N. Mukunda, E. C. G. Sudarshan, J. K. Sharma, and C. L. Mehta. Representations and properties of para-Bose oscillator operators I. Energy position and momentum eigenstates. *J. Math. Phys.*, 21:2386, 1980.

[11] S. Tsujimoto, L. Vinet, and A. Zhedanov. From $sl_q(2)$ to a parabosonic Hopf algebra. *SIGMA*, 7:93–105, 2011.

[12] S. Tsujimoto, L. Vinet, and A. Zhedanov. Dunkl shift operators and Bannai–Ito polynomials. *Adv. Math.*, 229:2123–2158, 2012.

Journal of Physics: Conference Series **512** (2014) 012011 doi:10.1088/1742-6596/512/1/012011

The Racah algebra and superintegrable models

Vincent X. Genest

Centre de Recherches Mathématiques, Université de Montréal, C.P. 6128, Centre-ville station, Montréal (QC), Canada, H3C 3J7

E-mail: genestvi@crm.umontreal.ca

Luc Vinet

Centre de Recherches Mathématiques, Université de Montréal, C.P. 6128, Centre-ville station, Montréal (QC), Canada, H3C 3J7

E-mail: vinetl@crm.umontreal.ca

Alexei Zhedanov

Donetsk Institute for Physics and Technology, Donetsk 83114, Ukraine

E-mail: zhedanov@kinetic.ac.donetsk.ua

Abstract. The universal character of the Racah algebra will be illustrated by showing that it is at the center of the relations between the Racah polynomials, the recoupling of three $\mathfrak{su}(1,1)$ representations and the symmetries of the generic second-order superintegrable model on the 2-sphere.

1. Introduction

This paper offers a review of the central role that the Racah algebra plays in connection with superintegrable models [2].

1.1. Superintegrable models

A quantum system with d degrees of freedom described by a Hamiltonian H is maximally superintegrable (S.I.) if it possesses $2d-1$ algebraically independent constants of motion S_i (also called symmetries) such that:

$$[S_i, H] = 0, \qquad 1 \leq i \leq 2d-1, \tag{1.1}$$

where one of the symmetries is the Hamiltonian. Since the maximal number of symmetries that can be in involution is d, the constants of motion of a superintegrable system generate a non-Abelian algebra whose representations can in general be used to obtain an exact solution to the dynamical equations. A S.I. system is said to be of order ℓ if the maximal order of the symmetries in the momenta (apart from H) is ℓ. We shall be concerned here with second-order ($\ell = 2$) S.I. systems for which the Schrödinger equation is known to admit separation of variables and for which the symmetry algebras are quadratic.

S.I. systems, which include the classical examples of the isotropic harmonic oscillator and of the Coulomb-Kepler problem, are most interesting as models in applications and for pedagogical purposes. In particular, they form the bedrock for the analysis of symmetries and their description. Their study has helped to understand how Lie algebras, superalgebras, quantum algebras, polynomials algebras and algebras with involutions serve that purpose.

1.2. Second-order S.I. systems in 2D

The model that we shall focus on is the generic 3-parameter system on the 2-sphere. Its Hamiltonian is

$$H = J_1^2 + J_2^2 + J_3^2 + \frac{a_1}{x_1^2} + \frac{a_2}{x_2^2} + \frac{a_3}{x_3^2}, \qquad a_i = k_i^2 - 1/4, \tag{1.2}$$

where

$$x_1^2 + x_2^2 + x_3^2 = 1, \tag{1.3}$$

and where

$$J_1 = -i(x_2\partial_{x_3} - x_3\partial_{x_2}), \quad J_2 = -i(x_3\partial_{x_1} - x_1\partial_{x_3}), \quad J_3 = -i(x_1\partial_{x_2} - x_2\partial_{x_1}), \tag{1.4}$$

are the familiar angular momentum operators satisfying the commutation relations

$$[J_1, J_2] = iJ_3, \quad [J_2, J_3] = iJ_1, \quad [J_3, J_1] = iJ_2. \tag{1.5}$$

This 3-parameter system is 2nd order superintegrable [9]. Most importantly, all 2nd S.I. systems in 2D can be obtained from this model through specializations, limits and contractions [11]. The superintegrability of H is confirmed by checking that the two operators:

$$L_1 = J_1^2 + \frac{a_2 x_3^2}{x_2^2} + \frac{a_3 x_2^2}{x_3^2}, \quad L_2 = J_2^2 + \frac{a_3 x_1^2}{x_3^2} + \frac{a_1 x_3^2}{x_1^2}, \tag{1.6}$$

commute with H. Kalnins, Miller and Pogosyan [9] have examined the symmetry algebra and presented it as follows. With

$$L_3 = J_3^2 + \frac{a_1 x_2^2}{x_1^2} + \frac{a_2 x_1^2}{x_2^2}, \qquad R = [L_1, L_2], \tag{1.7}$$

one has $[H, L_3] = 0$, $H = L_1 + L_2 + L_3 + a_1 + a_2 + a_3$, and the relations (with (ijk) cyclic)

$$[L_i, R] = 4\{L_i, L_j\} - 4\{L_i, L_k\} - (8 - 16a_j)L_j + (8 - 16a_k)L_k + 8(a_j - a_k), \tag{1.8a}$$

$$\begin{aligned} R^2 = &-\frac{8}{3}\{L_1, L_2, L_3\} - \sum_{i=1}^{3}\left\{(12 - 16a_i)L_i^2 + \frac{1}{3}(16 - 176a_i)L_i + \frac{32}{3}a_i\right\} \\ &+ \frac{52}{3}(\{L_1, L_2\} + \{L_2, L_3\} + \{L_1, L_3\}) + 48(a_1a_2 + a_2a_3 + a_3a_1) \\ &- 64\, a_1a_2a_3, \end{aligned} \tag{1.8b}$$

where $\{A, B\} = AB + BA$. Remarkably, Kalnins, Miller and Post [10] have further shown that the quadratic symmmetry algebra can be realized in terms of the difference operators associated to the Racah polynomials, that these same polynomials occur as transition coefficients between bases in which L_1 or L_2 is diagonal and furthermore that contractions of representations of the symmetry algebra lead to the symmetry algebras of the other 2nd order S.I. systems and other families of orthogonal polynomials [11]. Details about these Racah polynomials [12], denoted by $R_n(\lambda(x); \alpha, \beta, \gamma, \delta)$, will be given later. Suffice it to say for now that they are defined in terms of generalized hypergeometric functions, that they are of degree n in the variable $\lambda(x) = x(x + \gamma + \delta + 1)$, that they obey a discrete/finite orthogonality relation and that they sit atop the discrete part of the Askey scheme of hypergeometric orthogonal polynomials [12].

Journal of Physics: Conference Series **512** (2014) 012011 doi:10.1088/1742-6596/512/1/012011

1.3. Objectives

In reduced form, the (quadratic) Racah algebra has three generators K_1, K_2, K_3 and the defining relations

$$\begin{aligned} [K_1, K_2] &= K_3, \\ [K_2, K_3] &= K_2^2 + \{K_1, K_2\} + dK_2 + e_1, \\ [K_3, K_1] &= K_1^2 + \{K_1, K_2\} + dK_1 + e_2, \end{aligned} \tag{1.9}$$

where d, e_1 and e_2 are real parameters. The objectives of this paper are to show that this algebra has a universal character and intimately connects the generic second-order S.I. system on S^2, the Racah polynomials and the recoupling of three $\mathfrak{su}(1,1)$ representations. Schematically, the goal is to explain the links represented on the following diagram:

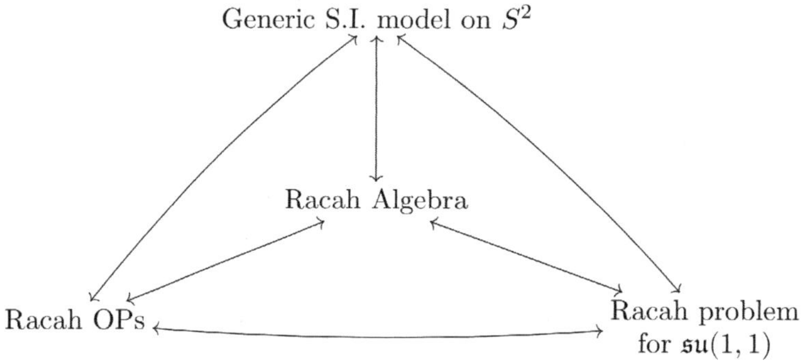

2. Warming up with a simple model

A key idea in our considerations is that 2nd order superintegrable models can be obtained by combining 1D models that are exactly solvable. This can be done in simple cases by straightforward constructions [14] and has also been realized in the R-matrix formalism [5, 13].

Consider the 2D isotropic singular oscillator

$$H = H_{x_1} + H_{x_2}, \tag{2.1}$$

where

$$H_{x_i} = -\frac{1}{2}\partial_{x_i}^2 + \frac{1}{2}\left(x_i^2 + \frac{a_i}{x_i^2}\right), \qquad a_i = k_i^2 - 1/4. \tag{2.2}$$

This is one of the four systems in the classification of second order S.I. systems in Euclidean space [1]. The spectrum of H is

$$E_N = N + (k_1 + k_2 + 1)/2, \qquad N = n_1 + n_2, \qquad n_i \in \mathbb{N}, \tag{2.3}$$

and has a $(N+1)$-fold degeneracy. It is well known that the associated Schrödinger equation separates in Cartesian and polar coordinates. To confirm that this system is maximally superintegrable, one needs to identify two independent constants of motion. This can be done using the $\mathfrak{su}(1,1)$ dynamical algebra of the one-dimensional components. Let

$$B_i^{\pm} = \frac{1}{2}\left[(x_i \mp \partial_{x_i})^2 - \frac{a_i}{x_i^2}\right], \quad i = 1, 2, \tag{2.4}$$

Journal of Physics: Conference Series **512** (2014) 012011 doi:10.1088/1742-6596/512/1/012011

it is readily verified that these operators combine with H_{x_i} to realize the $\mathfrak{su}(1,1)$ algebra since

$$[J_0, J_\pm] = \pm J_\pm, \qquad [J_+, J_-] = -2J_0, \tag{2.5}$$

with

$$J_0 = H_{x_i}/2 \qquad \text{and} \qquad J_\pm = B_i^\pm/2.$$

In the positive-discrete series of $\mathfrak{su}(1,1)$ representations thus constructed, $B_i^\pm$ act as raising and lowering operators with respect to the eigenvalues of H_{x_i} labeled by n_i. The conserved quantities of the 2D Hamiltonian $H = H_{x_1} + H_{x_2}$ are readily obtained from this observation. Indeed, combining the raising operator for H_{x_1} with the lowering operator for H_{x_2} (or vice-versa) will give an operator that leaves the total energy E_N unchanged and that commutes with H. This is a straightforward generalization of the Schwinger construction for the isotropic harmonic oscillator. The superintegrability of H is thus made manifest by exhibiting the operators

$$C^+ = B_{x_1}^+ B_{x_2}^-, \qquad C^- = B_{x_1}^- B_{x_2}^+, \tag{2.6}$$

to which we conveniently add

$$D = H_{x_1} - H_{x_2}, \tag{2.7}$$

and by noting that

$$[H, C^\pm] = 0, \qquad [H, D] = 0. \tag{2.8}$$

Defining relations for the symmetry algebra formed by the operators $C^\pm$ and D are straightforwardly obtained [14]:

$$\begin{aligned} [D, C^\pm] &= \pm 4C^\pm, \\ [C^-, C^+] &= D^3 + \alpha_1 D + \alpha_2, \end{aligned} \tag{2.9}$$

where

$$\alpha_1 = -H^2 - 2(k_1^2 + k_2^2 - 2), \qquad \alpha_2 = (2k_1^2 - 2k_2^2)H. \tag{2.10}$$

Since H is central, α_1 and α_2 can be treated as constants on eigenspaces of H. An algebraic solution of the problem is obtained by working out the appropriate representations of this algebra. A special case of (2.9) was found by Higgs in [7] as symmetry algebra of the Coulomb problem on S^2. By taking a different set of operators, it is possible to cast the relations (2.9) in a form that we would say is standard. Let

$$K_1 = \frac{1}{8}\Big(H_{x_1} - H_{x_2}\Big), \qquad K_2 = \frac{1}{8}\left(C^+ + C^- + \frac{1}{2}(D^2 - H^2)\right). \tag{2.11}$$

It is seen that K_2 can be written as

$$K_2 = \frac{1}{8}\left((x_1\partial_{x_2} - x_2\partial_{x_1})^2 - \frac{a_1 x_2^2}{x_1^2} - \frac{a_2 x_1^2}{x_2^2} - 1/2\right), \tag{2.12}$$

and is purely angular in polar coordinates. If we take

$$K_3 = [K_1, K_2] = \frac{1}{16}(C^+ - C^-), \tag{2.13}$$

Journal of Physics: Conference Series **512** (2014) 012011 doi:10.1088/1742-6596/512/1/012011

the defining relations become

$$\begin{aligned}
[K_1, K_2] &= K_3,\\
[K_2, K_3] &= \{K_1, K_2\} + \delta_1 K_1 + \delta_2,\\
[K_3, K_1] &= K_1^2 - \frac{1}{4}K_2 + \delta_3,
\end{aligned} \tag{2.14}$$

with

$$\delta_1 = -\frac{1}{4}(k_1^2 + k_2^2 - 2), \quad \delta_2 = \frac{1}{32}(k_1 - k_2)(k_1 + k_2)H, \quad \delta_3 = -\frac{1}{64}H^2. \tag{2.15}$$

This presentation allows the identification with the Hahn algebra, a special case of the "generic" Racah algebra (see next section). It is known to appear in connection with the Clebsch-Gordan problem for $\mathfrak{su}(1,1)$ [17]. This suggests a potential link between the isotropic singular oscillator in two dimensions and the Clebsch-Gordan problem for the dynamical algebra of its one-dimensional components. We shall now proceed to discuss the most general second order S.I. model in two dimensions along the lines followed in this section and shall see that the link mentioned above is not fortuitous. However before we do so, we shall introduce thoroughly the Racah algebra, present its finite-dimensional representations and go over the relations these have with Racah polynomials.

3. The Racah algebra

The Racah algebra has three generators K_1, K_2 and K_3. In the generic presentation, they obey the relations

$$\begin{aligned}
[K_1, K_2] &= K_3,\\
[K_2, K_3] &= a_2 K_2^2 + a_1\{K_1, K_2\} + c_1 K_1 + dK_2 + e_1,\\
[K_3, K_1] &= a_1 K_1^2 + a_2\{K_1, K_2\} + c_2 K_2 + dK_1 + e_2,
\end{aligned} \tag{3.1}$$

where the parameters a_1, a_2, c_1, c_2, d, e_1 and e_2 are taken to be real. This defines the most general associative quadratic algebra with two independent generators and a ladder property. To see this, let K_1, K_2 be the two independent generators and define $[K_1, K_2] = K_3$. K_1 and K_2 are assumed to be Hermitian, K_3 is thus anti-Hermitian. Consider the most general quadratic relations compatible with the hermiticity conditions:

$$\begin{aligned}
[K_2, K_3] &= a_2 K_2^2 + a_1\{K_1, K_2\} + g_1 K_1^2 + h_1 K_3^2 + c_1 K_1 + d_1 K_2 + e_1,\\
[K_3, K_1] &= a_3 K_1^2 + a_4\{K_1, K_2\} + g_2 K_2^2 + h_2 K_3^2 + c_2 K_2 + d_2 K_1 + e_2.
\end{aligned} \tag{3.2}$$

It follows from the Jacobi identity

$$[K_1, [K_2, K_3]] + [K_3, [K_1, K_2]] + [K_2, [K_3, K_1]] = 0,$$

that

$$d_1 = d_2, \quad a_3 = a_1, \quad a_4 = a_2, \quad h_1 = h_2 = 0. \tag{3.3}$$

One then requires $g_1 = g_2 = 0$ to ensure the ladder property (see later) and thus recovers (3.1). This algebra made its appearance in the work of Granovskii and Zhedanov [3] where it was used in the context of the Racah problem of $\mathfrak{su}(2)$ to derive the symmetry group of the $6j$-symbols.

It is also known as the Racah-Wilson algebra. When neither a_1 nor a_2 are zero, that is when $a_1 \cdot a_2 \neq 0$, the relations can be put in the following canonical form

$$[K_1, K_2] = K_3, \tag{3.4a}$$
$$[K_2, K_3] = K_2^2 + \{K_1, K_2\} + dK_2 + e_1, \tag{3.4b}$$
$$[K_3, K_1] = K_1^2 + \{K_1, K_2\} + dK_1 + e_2, \tag{3.4c}$$

where d, e_1 and e_2 are still real. This presentation thus retains three essential structure parameters and is arrived at by simple affine transformations of the generators $K_i \to u_i K_i + v_i$, $i = 1, 2, 3$. It is verified that this algebra has the following Casimir operator (central element)

$$\begin{aligned} Q =& \{K_1^2, K_2\} + \{K_1, K_2^2\} + K_1^2 + K_2^2 + K_3^2 \\ &+ (d+1)\{K_1, K_2\} + (2e_1 + d)K_1 + (2e_2 + d)K_2, \end{aligned} \tag{3.5}$$

which is cubic in the generators and commutes with each one of them.

4. Representations of the Racah algebra and Racah polynomials

We now wish to point out the connection between the Racah algebra and the Racah orthogonal polynomials. This can be done in at least two ways:

(i) By constructing the finite-dimensional representations of the algebra.

(ii) By realizing the algebra in terms of the operators associated to the polynomials.

We shall describe these two approaches in the following. We shall begin this section though by registering the basic definitions and properties of the Racah polynomials that we shall use.

4.1. Racah polynomials

The Racah polynomials $R_n(\lambda(x); \alpha, \beta, \gamma, \delta)$ of degree n in $\lambda(x) = x(x + \gamma + \delta + 1)$ depend on four real parameters α, β, γ, δ and are defined by the following explicit expression ($n \in \mathbb{N}$):

$$R_n(\lambda(x)) = {}_4F_3\left[\begin{matrix} -n, n+\alpha+\beta+1, -x, x+\gamma+\delta+1 \\ \alpha+1, \beta+\delta+1, \gamma+1 \end{matrix}; 1\right], \tag{4.1}$$

where ${}_pF_q$ is the generalized hypergeometric series

$${}_pF_q\left[\begin{matrix} a_1 & \cdots & a_p \\ b_1 & \cdots & b_q \end{matrix}; z\right] = \sum_{j=0}^{\infty} \frac{(a_1)_j \cdots (a_p)_j}{(b_1)_j \cdots (b_q)_j} \frac{z^j}{j!}, \tag{4.2}$$

and

$$(a)_j = a(a+1)\cdots(a+j-1).$$

The series in (4.1) truncates since $(-n)_j = 0$ for $j \geqslant n+1$. The polynomials thus defined satisfy $R_0(\lambda(x)) = 1$ and a three-term recurrence relation of the form [12]

$$\lambda(x) R_n(\lambda(x)) = A_n R_{n+1}(\lambda(x)) - (A_n + C_n) R_n(\lambda(x)) + C_n R_{n-1}(\lambda(x)), \tag{4.3}$$

with

$$\begin{aligned} A_n &= \frac{(n+\alpha+1)(n+\alpha+\beta+1)(n+\beta+\delta+1)(n+\gamma+1)}{(2n+\alpha+\beta+1)(2n+\alpha+\beta+2)}, \\ C_n &= \frac{n(n+\alpha+\beta-\gamma)(n+\alpha-\delta)(n+\beta)}{(2n+\alpha+\beta)(2n+\alpha+\beta+1)}. \end{aligned} \tag{4.4}$$

As usual it is assumed that $R_{-1}(\lambda(x)) = 0$. Like all polynomials of the Askey scheme, the Racah polynomials are bispectral: in addition to obeying the above recurrence relation, they are aso eigenfunctions of the difference equation

$$\mathcal{L}\, R_n(\lambda(x)) = n(n+\alpha+\beta+1)R_n(\lambda(x)), \tag{4.5}$$

where

$$\mathcal{L} = B(x)T^{+} + D(x)T^{-} - (B(x)+D(x))\mathbb{I}, \tag{4.6}$$

with

$$T^{\pm} f(x) = f(x \pm 1), \tag{4.7}$$

and

$$\begin{aligned} B(x) &= \frac{(x+\alpha+1)(x+\beta+\delta+1)(x+\gamma+1)(x+\gamma+\delta+1)}{(2x+\gamma+\delta+1)(2x+\gamma+\delta+2)}, \\ D(x) &= \frac{x(x-\alpha+\gamma+\delta)(x-\beta+\gamma)(x+\delta)}{(2x+\gamma+\delta)(2x+\gamma+\delta+1)}. \end{aligned} \tag{4.8}$$

Provided one of the following truncation conditions holds:

$$\alpha+1=-N, \quad \beta+\delta+1=-N, \quad \gamma+1=-N,$$

the Racah polynomials $R_n(\lambda(x))$ enjoy a finite orthogonality relation of the form

$$\sum_{x=0}^{N} w_x R_n(\lambda(x))R_m(\lambda(x)) = h_n\delta_{nm}, \tag{4.9}$$

where w_x and h_n are known explicitly. (For more details on the Racah polynomials see [12] where in particular the limit relations to other OPs of the Askey scheme are provided.)

4.2. Finite-dimensional representations

We shall now describe the finite-dimensional unitary representations of the Racah algebra and sketch how they are obtained. We take the defining relations to be in the canonical form (3.4). We begin by taking one generator, say K_1, to be diagnal on the representation space and proceed to show that the Racah algebra has a ladder property.

Let ω_p be a vector of the representation space such that

$$K_1\omega_p = \lambda_p\omega_p, \qquad p \in \mathbb{R}. \tag{4.10}$$

Suppose we look for another eigenvector $\omega_{p'}$ with eigenvalue $\lambda_{p'}$ that has the form

$$\omega_{p'} = \{\alpha(p)K_1 + \beta(p)K_2 + \gamma(p)K_3\}\,\omega_p, \tag{4.11}$$

where $\alpha(p)$, $\beta(p)$ and $\gamma(p)$ are coefficients. Imposing the eigenvalue equation

$$K_1\omega_{p'} = \lambda_{p'}\omega_{p'}, \tag{4.12}$$

using (4.10) and the commutation relations (3.4a) and (3.4c), it is seen that the eigenvalues $\lambda_{p'}$ must satisfy

$$(\lambda_{p'}-\lambda_p)^2 + (\lambda_{p'}+\lambda_p) = 0. \tag{4.13}$$

For a given λ_p, there are two solutions which we can choose to call λ_{p+1} and λ_{p-1}. Assuming that λ_p is non-degenerate and denoting by E_{λ_p} the one-dimensional eigenspace, it follows from the above considerations that a generic element of the algebra will map E_{λ_p} onto $E_{\lambda_{p-1}} \oplus E_{\lambda_p} \oplus E_{\lambda_{p+1}}$. We can thus write

$$\begin{aligned} K_1\omega_p &= \lambda_p\omega_p, \\ K_2\omega_p &= U_{p+1}\omega_{p1} + V_p\omega_p + U_p\omega_{p-1}, \\ K_3\omega_p &= [K_1, K_2]\omega_p = U_{p+1}g_{p+1}\,\omega_{p+1} - U_p g_p\,\omega_{p-1}, \end{aligned} \tag{4.14}$$

with

$$g_p = \lambda_p - \lambda_{p-1}, \tag{4.15}$$

observing that K_2 is tridiagonal and K_3 bidiagonal. Note that K_2 is self-adjoint if U_p is real. Having understood that the representations of the Racah algebra have a ladder structure, we now wish to focus on those representations that are finite-dimensional and for which the spectrum of K_1 is discrete. In the following we shall hence replace the vectors ω_p, $p \in \mathbb{R}$, by the vectors ψ_n, $n \in \mathbb{Z}$, labeled by the discrete index n. The actions (4.14) become

$$\begin{aligned} K_1\psi_n &= \lambda_n\psi_n, \\ K_2\psi_n &= U_{n+1}\psi_{n+1} + V_n\psi_n + U_n\psi_{n-1}, \\ K_3\psi_n &= U_{n+1}g_{n+1}\,\psi_{n+1} - U_n g_n\,\psi_{n-1}. \end{aligned} \tag{4.16}$$

with

$$g_n = \lambda_n - \lambda_{n-1}, \tag{4.17}$$

and there remains to determine λ_n, V_n and U_n. From (3.4c), one finds

$$\lambda_n = (\sigma - n)(n - \sigma + 1)/2, \quad g_n = \sigma - n, \tag{4.18}$$

and

$$V_n = -\frac{\lambda_n^2 + d\lambda_n + e_2}{\lambda_n}, \tag{4.19}$$

where σ is an arbitrary real parameter. We observe that the spectrum is quadratic in n. To find U_n, one uses (3.4b) that yields the following recurrence relation for U_n^2:

$$2(g_{n+3/2}U_{n+1}^2 - g_{n-1/2}U_n^2) = V_n^2 + (2\lambda_n + d)V_n + e_1. \tag{4.20}$$

Instead of solving directly (4.20), it is simpler to use the fact that the Casimir operator Q given in (3.5) is constant on irreducible representation spaces in order to find an expression for

$$2(g_{n+3/2}g_n U_{n+1}^2 + g_{n+1}g_{n-1/2}U_n^2).$$

Upon eliminating U_{n+1} with the help of this result and solving (4.20) for U_n^2 one arrives at

$$U_n^2 = \frac{\mathcal{P}(g_n^2)}{64 g_n^2 g_{n-1/2} g_{n+1/2}}, \tag{4.21}$$

where $\mathcal{P}(z)$ is the fourth degree polynomial

$$\begin{aligned}\mathcal{P}(z) = & z^4 - (4d+2)z^3 + (4d^2+4d+1+8e_2-16e_1)z^2 \\ & - 4(d^2+2e_2+4de_2+4q)z + 16e_2^2.\end{aligned} \tag{4.22}$$

In terms of the roots ξ_k^2 of $\mathcal{P}(g_n^2)$, we can write

$$U_n^2 = \frac{\prod_{k=1}^4 (g_n^2 - \xi_k^2)}{64 g_n^2 g_{n-1/2} g_{n+1/2}}. \tag{4.23}$$

For finite-dimensional representations, the index n is comprised in a finite interval $N_1 \leq n \leq N_2$, $N_1, N_2 \in \mathbb{Z}$ and the eigenvalues of K_1 are $\lambda_{N_1}, \lambda_{N_1+1}, \ldots, \lambda_{N_2}$. Setting the arbitrary parameter σ in (4.18) equal to $N_1 + \rho$ and calling $N = N_2 - N_1$, we can equivalently restrict the label n to be in the interval

$$0 \leq \sigma \leq N, \tag{4.24}$$

with the eigenvalues $\lambda_0, \lambda_1, \ldots, \lambda_N$ given by

$$\lambda_n = (\rho - n)(n - \rho + 1)/2. \tag{4.25}$$

Clearly, in a $(N+1)$-dimensional representation we must have $U_0 = U_{N+1} = 0$ so that ψ_{-1} and ψ_{N+1} cannot be reached by the actions of K_2 on ψ_0 and ψ_N, respectively. This will be observed if one of the zeros ξ_k^2, $k = 1, 2, 3, 4$, say ξ_i^2 is such that

$$\xi_i^2 = g_0^2 = \rho^2, \tag{4.26a}$$

and another say ξ_j^2, verify

$$\xi_j^2 = g_{N+1}^2 = (\rho - N - 1)^2. \tag{4.26b}$$

4.3. Connection with Racah polynomials

Having described the finite-dimensional representations of the Racah algebra in the eigenbasis of K_1, we might wonder what the picture is in the eigenbasis of K_2. It turns out to be very similar. Indeed, it is observed that the defining relations (3.4) are invariant under the exchanges

$$K_1 \leftrightarrow K_2, \qquad e_1 \leftrightarrow e_2. \tag{4.27}$$

It follows that the representation in the bases $\{\phi_s\}$ in which K_2 is diagonal

$$K_2 \phi_s = \mu_s \phi_s, \tag{4.28}$$

can be obtained from those in which K_1 is diagonal using this symmetry property. In this correspondence with the formulas of the last subsection, we make the replacement $n \to s$. It is clear that μ_s will have a from similar to λ_n say

$$\mu_s = (\nu - s)(\nu - s + 1)/2, \tag{4.29}$$

and that K_1 will be tridiagonal in the ϕ_s basis. In other words, K_1 and K_2 realize a Leonard pair [15]. We now claim that the Racah polynomials arise as overlaps between the $\{\psi_n\}$ and the $\{\phi_s\}$ bases; that is, the Racah polynomials appear as expansion coefficients between the basis in which K_1 is diagonal and the one in which K_2 is diagonal.

Let us make this more explicit. Since the bases $\{\phi_s\}$ and $\{\psi_n\}$ span isomorphic spaces we can expand the elements of one in terms of those of the other:

$$\phi_s = \sum_{n=0}^{N} W_n(s)\psi_n. \tag{4.30}$$

Let us write the coefficients $W_n(s)$ in the form

$$W_n(s) = w_0(s)P_n(\mu_s), \tag{4.31}$$

so that $P_0(\mu_s) = 1$. Using (4.28) and (4.16), it is seen upon acting with K_2 on both sides of (4.30) that the quantities $P_n(\mu_s)$ obey

$$\mu_s P_n(\mu_s) = U_{n+1}P_{n+1}(\mu_s) + V_n P_n(\mu_s) + U_n P_{n-1}(\mu_s). \tag{4.32}$$

Given the formulas for U_n and V_n, this recurrence relation is seen to coincide with that of the Racah polynomials. The zeros ξ_k^2 are related to the parameters $\alpha, \beta, \gamma, \delta$ of the polynomials $R_n(\lambda(x); \alpha, \beta, \gamma, \delta)$. If the truncations conditions are satisfied through

$$\xi_1 = \rho, \qquad \xi_4 = (\rho - N - 1), \tag{4.33}$$

(recall (4.26)) the identification is achived by the following parametrization of the roots:

$$\xi_1 = -\frac{\alpha+\beta}{2}, \quad \xi_2 = \frac{\beta-\alpha}{2} + \delta, \quad \xi_3 = \frac{\beta-\alpha}{2}, \quad \xi_4 = \gamma - \frac{\alpha+\beta}{2}. \tag{4.34}$$

4.4. The Racah algebra from the Racah polynomials

We have illustrated how the Racah polynomials can be obtained from the Racah algebra by constructing the finite-dimensional representations. Let us indicate now that conversely, the Racah algebra can be identified from the properties of the Racah polynomials. As shall be seen the Racah algebra encodes the bispectrality properties of the polynomials. These properties amount to the fact that in addition to satisfying a three-term recurrence relation (as all orthogonal polynomials must do), the Racah polynomials also obey a difference equation. Let us recall that these relations can be put in the form

$$x(x+\gamma+\delta+1)R_n(\lambda(x)) = \mathcal{M}R_n(\lambda(x)), \tag{4.35a}$$
$$\mathcal{L}R_n(\lambda(x)) = n(n+\alpha+\beta+1)R_n(\lambda(x)), \tag{4.35b}$$

where the *recurrence* operator $\mathcal{M}$ and the *difference* operator $\mathcal{L}$ are given by

$$\mathcal{M} = A_n T_n^+ + C_n T_n^- - (A_n + C_n)\mathbb{I}, \tag{4.36}$$
$$\mathcal{L} = B(x)T_x^+ + D(x)T_x^- - (B(x) + D(x))\mathbb{I}, \tag{4.37}$$

with $T_n^{\pm} f(n) = f(n \pm 1)$, $T_x^{\pm} f(x) = f(x \pm 1)$ and where A_n, C_n are given in (4.4) and $B(x)$, $D(x)$ provided by (4.8). Consider now the realization on functions of x, where $\widetilde{K}_1$ and $\widetilde{K}_2$ are the two operators occurring on the left-hand side of (4.35), i.e. $\widetilde{K}_1$ is the recurrence operator and $\widetilde{K}_2$ is the difference operator:

$$\widetilde{K}_1 = x(x+\gamma+\delta+1), \qquad K_2 = \mathcal{L}. \tag{4.38}$$

Performing the affine transformations

$$K_1 = u_1\widetilde{K}_1 + v_1, \qquad K_2 = u_2\widetilde{K}_2 + v_2, \tag{4.39}$$

one finds that K_1, K_2 verify the Racah algebra relations (3.4) with

$$\begin{aligned}
e_1 &= \frac{1}{4}\left(\frac{\alpha-\beta}{2}\right)\left(\frac{\alpha+\beta}{2}\right)\left(\frac{\alpha+\beta}{2}-\gamma\right)\left(\frac{\alpha-\beta}{2}-\delta\right),\\
e_2 &= \frac{1}{4}\left(\frac{\gamma-\delta}{2}\right)\left(\frac{\gamma+\delta}{2}\right)\left(\frac{\gamma+\delta}{2}-\alpha\right)\left(\frac{\gamma-\delta}{2}-\beta\right),\\
d &= \frac{1}{4}\left\{\left(\frac{\gamma-\delta}{2}\right)^2+\left(\frac{\gamma+\delta}{2}\right)^2+\left(\frac{\gamma+\delta}{2}-\alpha\right)^2+\left(\frac{\gamma-\delta}{2}-\beta\right)^2-2\right\}.
\end{aligned}$$

We could obviously have taken the realization on functions of n with

$$\widehat{K}_1 = \mathcal{M}, \qquad \widehat{K}_2 = n(n+\alpha+\beta+1), \tag{4.40}$$

which is bound to lead to the same algebra. The duality property of the Racah polynomials under the exchanges

$$x \leftrightarrow n, \qquad \alpha \leftrightarrow \gamma, \qquad \beta \leftrightarrow \delta, \tag{4.41}$$

which follows from (4.35) is immediately seen to correspond to the symmetry of the Racah algebra under

$$K_1 \leftrightarrow K_2, \qquad K_3 \leftrightarrow -K_3, \qquad e_1 \leftrightarrow e_2, \tag{4.42}$$

that we already observed.

5. The generic superintegrable model on S^2 and $\mathfrak{su}(1,1)$

We now return to the generic superintegrable model on S^2 with Hamiltonian (1.2) and constants of motion (1.6) and (1.7). We want to show that it is intimately connected to the $\mathfrak{su}(1,1)$ algebra.

Consider three $\mathfrak{su}(1,1)$ realizations identical to the one introduced in the discussion of the two-dimensional singular oscillator in Section 2:

$$J_0^{(i)} = \frac{1}{4}\left(-\partial_{x_i}^2 + x_2^2 + \frac{a_i}{x_i^2}\right), \quad J_\pm^{(i)} = \frac{1}{4}\left((x_i \mp \partial_{x_i})^2 - \frac{a_i}{x_i^2}\right), \tag{5.1}$$

with $a_i = k_i^2 - 1/4$ and $i = 1, 2, 3$. These provide positive discrete series representations for which the $\mathfrak{su}(1,1)$ Casimir element

$$C^{(i)} = [J_0^{(i)}]^2 - J_+^{(i)}J_-^{(i)} - J_0^{(i)}, \quad i = 1, 2, 3, \tag{5.2}$$

takes the value

$$C^{(i)} = \nu_i(\nu_i - 1), \qquad \nu_i = (k_i+1)/2. \tag{5.3}$$

These three sets of $\mathfrak{su}(1,1)$ generators can be added to produce a "fourth" realization:

$$J_0^{(4)} = J_0^{(1)} + J_0^{(2)} + J_0^{(3)}, \quad J_\pm^{(4)} = J_\pm^{(1)} + J_\pm^{(2)} + J_\pm^{(3)}. \tag{5.4}$$

Three types of Casimir operators can be distinguished in the process:

Journal of Physics: Conference Series **512** (2014) 012011 doi:10.1088/1742-6596/512/1/012011

(i) The "initial" Casimir operators (5.2)

(ii) The "intermediate" Casimir operators associated to the addition of two representations

$$C^{(ij)} = [J_0^{(i)} + J_0^{(j)}]^2 - (J_+^{(i)} + J_+^{(j)})(J_-^{(i)} + J_-^{(j)}) - (J_0^{(i)} + J_0^{(j)}), \tag{5.5}$$

with $(ij) = (12), (23), (31)$.

(iii) The "full" Casimir operator

$$C^{(4)} = [J_0^{(4)}]^2 - J_+^{(4)} J_-^{(4)} - J_0^{(4)}. \tag{5.6}$$

By a direct computation one finds that

$$C^{(ij)} = \frac{1}{4}\left\{ J_k^2 + \frac{a_i x_j^2}{x_i^2} + \frac{a_j x_i^2}{x_j^2} + a_i + a_j - 1 \right\}, \tag{5.7}$$

with (ijk) a cyclic permutation of $(1, 2, 3)$. Comparing with (1.6) and (1.7), we see that

$$L_1 = 4C^{(23)} - a_2 - a_3 + 1, \quad L_2 = 4C^{(31)} - a_3 - a_1 + 1, \quad L_3 = 4C^{(12)} - a_1 - a_2 + 1, \tag{5.8}$$

and thus observe that the constants of motion of the generic S.I. system on S^2 are basically the intermediate Casimir operators arising in the addition of three $\mathfrak{su}(1,1)$ representations. Similarly one can check that the full Casimir operator $C^{(4)}$ takes the following form when (5.1) and (5.4) are used in (5.6):

$$C^{(4)} = \frac{1}{4}\left\{ J_1^2 + J_2^2 + J_3^2 + (x_1^2 + x_2^2 + x_3^2)\left(\frac{a_1}{x_1^2} + \frac{a_2}{x_2^2} + \frac{a_3}{x_3^2}\right) - \frac{3}{4} \right\}. \tag{5.9}$$

As a consequence,

$$H = 4C^{(4)} + \frac{3}{4}, \tag{5.10}$$

if $x_1^2 + x_2^2 + x_3^2 = 1$. At this point we may ask if this restriction to S^2 can generally be ensured in the addition of the three $\mathfrak{su}(1,1)$ representations. That the answer is yes is readily seen. It is noted from (5.1) that

$$2J_0^{(i)} + J_+^{(i)} + J_-^{(i)} = x_i^2, \tag{5.11}$$

and hence that

$$S = 2J_0^{(4)} + J_+^{(4)} + J_-^{(4)} = x_1^2 + x_2^2 + x_3^2. \tag{5.12}$$

Since S commutes with $4C^{(4)} + 3/4$, it is "time-independent" and as a constant can be taken to be 1. We can thus conclude the following. The generic S.I. 3-paramater system is obtained from the addition of three $\mathfrak{su}(1,1)$ realizations. In this identification the restriction $x_1^2+x_2^2+x_3^2 = 1$ to S^2 is preserved; the Hamiltonian corresponds to the full Casimir operator $C^{(4)}$ for the addition of the three representations and the constants of motion are obviously the intermediate Casimir operators $C^{(ij)}$ which commute with $C^{(4)}$. The algebra that these intermediate Casimir operators generate is thus the symmetry algebra of the generic *S.I.* system on S^2. We shall show in the next section that the intermediate Casimir operators in the addition of three $\mathfrak{su}(1,1)$ irreducible representations generate the Racah algebra.

Note that the R-matrix approach has been used in [6] to describe the generic 3-parameter model on S^2 and construct its invariants. One proceeds via dimensional reduction from S^5 to S^2 with a Lax matrix that also involves three $\mathfrak{su}(1,1)$ elements. It is interesting to further point out that the same generic S.I. model on S^2 has been shown to correspond to one of the Krall-Sheffer classes of orthogonal polynomials in two variables [4, 16].

6. The Racah problem for $\mathfrak{su}(1,1)$ and the Racah algebra

We loosely refer to the Racah problem as the recoupling of three irreducible representations. Strictly speaking, it is about determining the unitary transformation between two canonical bases corresponding to the steps $(1 \oplus 2) \oplus 3$ and $1 \oplus (2 \oplus 3)$ which are respectively associated to the diagonalization of the intermediate Casimir operators $C^{(12)}$ and $C^{(23)}$. Our goal in this section is to comment on the algebra that $C^{(12)}$ and $C^{(23)}$ generate in the case of $\mathfrak{su}(1,1)$ (and of $\mathfrak{su}(2)$ as a matter of fact).

Consider the addition of three $\mathfrak{su}(1,1)$ representations as in (5.4) and take each initial Casimir operator $C^{(i)}$ to be a multiple of the identity:

$$C^{(i)} = \lambda_i, \qquad i = 1, 2, 3. \tag{6.1}$$

The intermediate Casimir operators $C^{(12)}$ and $C^{(23)}$ (given by (5.5)) can then be expressed as follows

$$\begin{aligned} C^{(12)} &= 2J_0^{(1)} J_0^{(2)} - (J_+^{(1)} J_-^{(2)} + J_-^{(1)} J_+^{(2)}) + \lambda_1 + \lambda_2, \\ C^{(23)} &= 2J_0^{(2)} J_0^{(3)} - (J_+^{(2)} J_-^{(3)} + J_-^{(2)} J_+^{(3)}) + \lambda_2 + \lambda_3. \end{aligned} \tag{6.2}$$

Further assume that the full Casimir operator $C^{(4)}$ which can be written in the form

$$C^{(4)} = C^{(12)} + C^{(23)} + C^{(31)} - C^{(1)} - C^{(2)} - C^{(3)}, \tag{6.3}$$

is also a multiple of the identity, i.e.

$$C^{(4)} = \lambda_4. \tag{6.4}$$

Denoting by $V^{(\lambda_i)}$ an irreducible $\mathfrak{su}(1,1)$ representation space on which the Casimir operator $C^{(i)}$ is equal to λ_i, we are thus looking at the decomposition of $V^{(\lambda_1)} \otimes V^{(\lambda_2)} \otimes V^{(\lambda_3)}$ in irreducible components $V^{(\lambda_4)}$. Consider now the algebra generated by the intermediate Casimir operators in this Racah problem. Define

$$\kappa_1 = -C^{(12)}/2 \qquad \kappa_2 = -C^{(23)}/2, \tag{6.5}$$

and let

$$\kappa_3 = [\kappa_1, \kappa_2]. \tag{6.6}$$

A direct computation in which $C^{(i)}$ is replaced by λ_i gives, remarkably, the defining relations of the Racah algebra

$$\begin{aligned} [\kappa_1, \kappa_2] &= \kappa_3, \\ [\kappa_2, \kappa_3] &= \kappa_2^2 + \{\kappa_1, \kappa_2\} + d\kappa_2 + e_1, \\ [\kappa_3, \kappa_1] &= \kappa_1^2 + \{\kappa_1, \kappa_2\} + d\kappa_1 + e_2, \end{aligned} \tag{6.7}$$

where

$$d = \frac{1}{2}(\lambda_1 + \lambda_2 + \lambda_3 + \lambda_4) \quad e_1 = \frac{1}{4}(\lambda_1 - \lambda_4)(\lambda_2 - \lambda_3) \quad e_2 = \frac{1}{4}(\lambda_1 - \lambda_2)(\lambda_4 - \lambda_3) \tag{6.8}$$

The intermediate Casimir operators in the addition of three $\mathfrak{su}(1,1)$ representations form the Racah algebra which is thus the structure behind the Racah problem for $\mathfrak{su}(1,1)$. Combining this with the identification of the intermediate Casimir operators with the constants of motion of the generic S.I. 3-parameter system on S^2, it follows that the (reduced) Racah algebra is the symmetry algebra of this model. In other words, the algebra (1.8) is isomorphic to the Racah algebra (3.4).

We recall from Section (4.3) that the Racah polynomials appear as expansion coefficients between bases for Racah algebra representation spaces in which K_1 is diagonal on the one hand and K_2 is diagonal on the other. Since as we just have seen, K_1 and K_2 can be realized by $C^{(12)}$ and $C^{(23)}$, this naturally relates to the fact that the Racah coefficients (the elements of the matrix relating the bases associated to the 2 step-wise recoupling processes) are Racah polynomials. In the context of the superintegrable model on S^2, these Racah coefficients can be connected to separation of variables. This is seen as follows. The diagonalization of

$$C^{(12)} = \frac{1}{4}(L_3 + a_1 + a_2 - 1) = J_3^2 + \frac{a_1 x_2^2}{x_1^2} + \frac{a_2 x_1^2}{x_2^2}, \tag{6.9}$$

brings the separation of variables in the usual spherical coordinates (in which x_1 and x_2 are paired):

$$x_1 = r\sin\theta\cos\phi, \quad x_2 = r\sin\theta\sin\phi, \quad x_3 = r\cos\theta, \tag{6.10}$$

while the diagonalization of

$$C^{(23)} = \frac{1}{4}(L_1 + a_2 + a_3 - 1) = J_1^2 + \frac{a_2 x_3^2}{x_2^2} + \frac{a_3 x_2^2}{x_3^2}, \tag{6.11}$$

leads to the separation of variables in another spherical coordinate system (in which x_2 and x_3 are paired) namely,

$$x_1 = r\cos\theta, \quad x_2 = r\sin\theta\cos\phi, \quad x_3 = r\sin\theta\sin\phi. \tag{6.12}$$

Hence the Racah polynomials are in this framework, the overlap coefficients between these two sets of wavefunctions of the generic 3-parameter system on S^2 which are obtained by separating the variables in the considered systems (6.10) and (6.12).

Summing up, the identification of the generic S.I. model on S^2 as the full Casimir operator in the addition of three $\mathfrak{su}(1,1)$ realizations provides a natural way of obtaining the constants of motion (as the intermediate Casimir operators) and of determining the symmetry algebra (as the Racah algebra). This intimately associates the Racah polynomials to the model on S^2. It should be said that representations with continuous spectra are found to bring the Wilson polynomials in the picture. As explained in [11], all the other second order S.I. mdels in two dimensions can be obtained from the generic system on S^2 by contractions and specializations. Correspondingly, when effected on the Racah algebra and the Racah/Wilson polynomials these operations provide the symmetry algebras of all these second order S.I. models and their tagging to orthogonal polynomials of the Askey scheme.

7. Conclusion

Let us summarize the main findings and offer perspectives. We presented the Racah algebra and its finite-dimensional representations. We proceeded to explain its universal character and the relations depicted on the diagram presented in Section I. It was shown that the Racah algebra is behind

- The generic 3-parameter superintegrable model on S^2 and hence all second order superintegrable systems in two dimensions

- The Racah problem for $\mathfrak{su}(1,1)$, that is the combination of three irreducible $\mathfrak{su}(1,1)$ representations
- The Racah polynomials that sit atop the discrete part of the Askey scheme of hypergeometric orthogonal polynomials

Looking at three and higher dimensions, the analogous connections between superintegrable models, polynomial algebras and special functions are bound to be illuminating. They are expected to feed the theory of multivariate orthogonal polynomials and their algebraic interpretations.

As an illustration of this, let us mention that Kalkins, Miller and Post have already shown that 2-variable Racah-Wilson polynomials occur in the S^3 model [8]. The construction presented here extends to any dimensions and the three-dimensional model on S^3 arises from the addition of four $\mathfrak{su}(1,1)$ representations and corresponds to the $9j$ or Fano problem for $\mathfrak{su}(1,1)$. The symmetries in this case are expected to lead to a rank 2 version of the Racah algebra which should encode the properties of the general bivariate Racah-Wilson polynomials. We intend to pursue investigations along those lines.

Acknowledgements
V.X.G. holds an Alexander-Graham-Bell fellowship from the Natural Sciences and Engineering Research Council of Canada (NSERC). The research of L.V. is supported in part by NSERC.

References

[1] J. Fris, V. Mandrosov, Ya. A. Smorodinsky, M. Uhlir, and P. Winternitz. On higher symmetries in quantum mechanics. *Phys. Lett.*, 16:354–356, 1965.
[2] V.X. Genest, L. Vinet, and A. Zhedanov. Superintegrability in two dimensions and the Racah-Wilson algebra. *ArXiv:1307.5539*, 2013.
[3] Ya. A. Granovskii and A. Zhedanov. Nature of the symmetry group of the $6j$-symbol. *Zh. Eksp. Teor. Fiz.*, 94:49–54, 1988.
[4] J. Harnad, L. Vinet, O. Yermolayeva, and A. Zhedanov. Two-dimensional Krall-Sheffer polynomials and integrable systems. *J. Phys. A: Math. Gen.*, 34:10619, 2001.
[5] J. Harnad and O. Yermolayeva. Superintegrability, Lax matrices and separation of variables. In *CRM Proc. Lecture Notes.*
[6] J. Harnad, A. Zhedanov, and O. Yermolayeva. R-Matrix Approach to the Krall-Sheffer Problem. *Proc. Inst. Math. Nat. Acad. Sc. Ukr.*, 439:314–320, 2002.
[7] P. W. Higgs. Dynamical symmetries in a spherical geometry. I. *J. Phys. A: Math. Gen.*, 12(3):309, 1979.
[8] E. G. Kalnins, W. Miller, and S. Post. Two-variable Wilson polynomials and the generic superintegrable system on the 3-sphere. *SIGMA*, 7:51–76, 2011.
[9] E.G. Kalnins, W. Miller, and G. S. Pogosyan. Superintegrability and associated polynomial solutions: Euclidean space and the sphere in two dimensions. *J. Math. Phys.*, 37:6439, 1996.
[10] E.G. Kalnins, W. Miller, and S. Post. Wilson polynomials and the generic superintegrable system on the 2-sphere. *J. Phys. A: Math. Theor.*, 40:11525–11538, 2007.
[11] E.G. Kalnins, W. Miller, and S. Post. Contractions of 2D 2nd order quantum superintegrable systems and the Askey scheme for hypergeometric orthogonal polynomials. *SIGMA*, 9:57–84, 2013.
[12] R. Koekoek, P.A. Lesky, and R.F. Swarttouw. *Hypergeometric orthogonal polynomials and their q-analogues.* Springer, 1st edition, 2010.
[13] V. Kuznetsov. Quadrics on real Riemannian spaces of constant curvature: separation of variables and connection with Gaudin magnet. *J. Math. Phys.*, 33:1992, 3240.
[14] P. Letourneau and L. Vinet. Superintegrable Systems: Polynomial Algebras and Quasi-Exactly Solvable Hamiltonians. *Ann. Phys.*, 243:144–168, 1995.
[15] P. Terwilliger. Two linear transformations each tridiagonal with respect to an eigenbasis of the other. *Lin. Alg. Appl.*, 330:149–203, 2001.
[16] L. Vinet and A. Zhedanov. Two-Dimensional Krall–Sheffer Polynomials and Quantum Systems on Spaces with Constant Curvature. *Lett. Math. Phys*, 65:83–94, 2003.
[17] A. Zhedanov. Hidden symmetry algebra and overlap coefficients for two ring-shaped potentials. *J. Phys. A: Math. Gen.*, 26(18):4633, 1993.

The theory of contractions of 2D 2nd order quantum superintegrable systems and its relation to the Askey scheme for hypergeometric orthogonal polynomials

Willard Miller, Jr.

School of Mathematics, University of Minnesota, Minneapolis, Minnesota, 55455, U.S.A.

E-mail: `miller@ima.umn.edu`

Abstract. We describe a contraction theory for 2nd order superintegrable systems, showing that all such systems in 2 dimensions are limiting cases of a single system: the generic 3-parameter potential on the 2-sphere, $S9$ in our listing. Analogously, all of the quadratic symmetry algebras of these systems can be obtained by a sequence of contractions starting from $S9$. By contracting function space realizations of irreducible representations of the $S9$ algebra (which give the structure equations for Racah/Wilson polynomials) to the other superintegrable systems one obtains the full Askey scheme of orthogonal hypergeometric polynomials.This relates the scheme directly to explicitly solvable quantum mechanical systems. Amazingly, all of these contractions of superintegrable systems with potential are uniquely induced by Wigner Lie algebra contractions of $so(3, \mathbf{C})$ and $e(2, \mathbf{C})$. The present paper concentrates on describing this intimate link between Lie algebra and superintegrable system contractions, with the detailed calculations presented elsewhere. Joint work with E. Kalnins, S. Post, E. Subag and R. Heinonen.

1. Introduction

A quantum superintegrable system is an integrable Hamiltonian system on an n-dimensional Riemannian/pseudo-Riemannian manifold with potential: $H = \Delta_n + V$, that admits $2n - 1$ algebraically independent partial differential operators commuting with H, apparently the maximum possible.

$$[H, L_j] = 0, \quad L_{2n-1} = H, \ n = 1, 2, \cdots, 2n - 1.$$

Here, Δ_n is the Laplace-Beltrami operator on the manifold. Superintegrability captures the properties of quantum Hamiltonian systems that allow the Schrödinger eigenvalue problem $H\Psi = E\Psi$ to be solved exactly, analytically and algebraically. There is a similar definition of classical superintegrable systems with Hamiltonian $\mathcal{H} = \sum g^{jk} p_j p_k + V$ on phase space with $2n - 1$ functionally independent constants of the motion $\mathcal{L}_j$ with $\mathcal{L}_{2n-1} = \mathcal{H}$ and polynomial in the momenta, definitely the maximum number possible. A system is of order K if the maximum order of the symmetry operators L_j, other than H, (or classically the maximum order of constants of the motion as polynomials) is K. For $n = 2$, $K = 2$ all systems are known. The symmetry operators of each system close under commutation (or under the Poisson bracket) to generate a quadratic algebra, and the irreducible representations of the algebra determine the eigenvalues of H and their multiplicity. Classically we get important information about the orbits through algebraic methods alone. Detailed motivation for the study of superintegrable systems, a presentation of the theory and many references can be found in [16, 13]. All the 2nd order classical and quantum superintegrable systems are limiting cases of a single system: the generic 3-parameter potential on the 2-sphere, $S9$ in our listing. Analogously all quadratic symmetry algebras of these systems are contractions of $S9$. In the quantum case this system is

$$S9: \qquad H = \Delta_2 + \frac{a_1}{s_1^2} + \frac{a_2}{s_2^2} + \frac{a_3}{s_3^2}, \qquad s_1^2 + s_2^2 + s_3^2 = 1, \tag{1}$$

$$L_1 = (s_2\partial_{s_3} - s_3\partial_{s_2})^2 + \frac{a_3 s_2^2}{s_3^2} + \frac{a_2 s_3^2}{s_2^2}, \quad L_2,\ L_3, \text{ obtained by cyclic permutation of indices,}$$

$$H = L_1 + L_2 + L_3 + a_1 + a_2 + a_3.$$

In the following sections we give brief descriptions of 1st and 2nd order 2D superintegrable systems, both free and with degenerate or nondegenerate potential. Every nonfree system is associated with a closed quadratic algebra generated by its symmetries. We state, and prove elsewhere, that a free system extends to a superintegrable system with potential if and only if its symmetries generate a closed free quadratic algebra. We point out that the theory of contractions of Lie symmetry algebras of constant curvature spaces is intimately associated with superintegrable systems of 1st order; indeed it appears to have been the motivation for the development of this theory by Wigner and Inönü. Then we show for systems on 2D constant curvature spaces how these Lie algebra contractions induce 1) contractions of the free quadratic algebras and then 2) induce contractions of the nondegenerate and degenerate quadratic algebras of systems with potential. Next we describe how the contractions of the superintegrable systems with potential can induce contractions of models of irreducible representations of the quadratic algebras through the process of 'saving' a representation. The Askey scheme for hypergeometric orthogonal polynomials emerges as a special subclass of these model contractions. We conclude with some observations.

2. 1st and 2nd order 2D superintegrable systems

1st order systems $K = 1$: In the quantum case these are the (zero-potential) Laplace-Beltrami eigenvalue equations on constant curvature spaces, such as the Euclidean Helmholtz equation $(P_1^2 + P_2^2)\Phi = -\lambda^2\Phi$ (or the Klein-Gordon equation $(P_1^2 - P_2^2)\Phi = -\lambda^2\Phi$), and the Laplace -Beltrami eigenvalue equation on the 2-sphere $(J_1^2 + J_2^2 + J_3^2)\Psi = -j(j+1)\Psi$. The first order symmetries close under commutation to form the Lie algebras $e(2,\mathbf{R})$, $e(1,1)$ or $o(3,\mathbf{R})$. The eigenspaces of these systems support differential operator models of the irreducible representations of the Lie algebras in which basis eigenfunctions are the spherical harmonics ($o(3,\mathbf{R})$),Bessel functions ($e(2,\mathbf{R})$) and more complicated special functions [1, 9].

It was exactly these 1st order systems which motivated the pioneering work of Inönü and Wigner [3] on Lie algebra contractions. While, that paper introduced Lie algebra contractions in general, the motivation and virtually all the examples were of symmetry algebras of these systems. It was shown that $o(3,\mathbf{R})$ contracts to $e(2,\mathbf{R})$. In the physical space this is accomplished by letting the radius of the sphere go to infinity, so that the surface flattens out. Under this limit the Laplace-Beltrami eigenvalue equation goes to the Helmholtz equation.

The following defines so-called natural contractions, [14], a generalization of Wigner-Inönü contractions. **Lie algebra contractions**: Let $(A; [,]_A)$, $(B; [,]_B)$ be two complex Lie algebras.

We say B is a *contraction* of A if for every $\epsilon \in (0;1]$ there exists a linear invertible map $t_\epsilon : B \to A$ such that for every $X, Y \in B$,

$$\lim_{\epsilon\to 0} t_\epsilon^{-1}[t_\epsilon X, t_\epsilon Y]_A = [X,Y]_B.$$

Thus, as $\epsilon \to 0$ the 1-parameter family of basis transformations can become nonsingular but the structure constants go to a finite limit.

Features of Wigner's contraction approach, [3, 15]:

- 'Saving' a representation. Passing through a sequence of irreducible representations of the source symmetry algebra to obtain an irreducible representation of the target algebra in the contraction limit.
- Simple models of irreducible representations. Finding models on function spaces so that the eigenfunctions of the generators are special functions.
- Limit relations between special functions, as a result of saving a model representation in the contraction limit.
- Use of the models to find expansion coefficients relating different special function bases.

Free 2nd order superintegrable systems in 2D: We will apply Wigner's ideas to 2nd order systems in 2D ($2n-1=3$). We start with the free (no potential function) case. The complex spaces with free Hamiltonians admitting at least three 2nd order symmetries (i.e., three 2nd order Killing tensors) were classified by Koenigs [11]. They are:

- The two constant curvature spaces: flat space and the complex 2-sphere. They each admit 6 linearly independent 2nd order symmetries and 3 1st order symmetries,
- The four Darboux spaces, (4 2nd order symmetries and 1 1st order symmetry):

$$ds^2 = 4x(dx^2+dy^2),\ ds^2 = \frac{x^2+1}{x^2}(dx^2+dy^2),$$

$$ds^2 = \frac{e^x+1}{e^{2x}}(dx^2+dy^2),\ ds^2 = \frac{2\cos 2x + b}{\sin^2 2x}(dx^2+dy^2),$$

- Eleven 4-parameter Koenigs spaces (3 2nd order symmetries and no 1st order symmetries). An example is

$$ds^2 = (\frac{c_1}{x^2+y^2} + \frac{c_2}{x^2} + \frac{c_3}{y^2} + c_4)(dx^2+dy^2).$$

2nd order superintegrable systems (with potential) in 2D: All such systems are known. There are 59 and each of the spaces classified by Koenigs admits at least one system. However, under the Stäckel transform, an invertible structure preserving mapping [13], the systems divide into 12 equivalence classes, each with a representative in a constant curvature space. Now the symmetry algebra is a quadratic algebra, not usually a Lie algebra, and the irreducible representations of the quantum algebra determine the eigenvalues of H and their multiplicity

There are 3 types of superintegrable systems:

(i) Nondegenerate: (3-parameter potential)

$$V(\mathbf{x}) = a_1V_{(1)}(\mathbf{x}) + a_2V_{(2)}(\mathbf{x}) + a_3V_{(3)}(\mathbf{x}) + a_4$$

(ii) Degenerate: (1-parameter potential)

$$V(\mathbf{x}) = a_1V_{(1)}(\mathbf{x}) + a_2$$

(iii) Free:

$$V = a_1.$$

Usually the trivial added constant in each potential is ignored, though it is vital for some purposes.

Nondegenerate systems ($2n-1=3$ generators): The quantum symmetry algebra generated by H, L_1, L_2 always closes under commutation. Let $R=[L_1,L_2]$ be the 3rd order commutator of the generators. Then

$$[L_j,R]=A_1^{(j)}L_1^2+A_2^{(j)}L_2^2+A_3^{(j)}H^2+A_4^{(j)}\{L_1,L_2\}+A_5^{(j)}HL_1+A_6^{(j)}HL_2$$

$$+A_7^{(j)}L_1+A_8^{(j)}L_2+A_9^{(j)}H+A_{10}^{(j)}$$

$$R^2=b_1L_1^3+b_2L_2^3+b_3H^3+b_4\{L_1^2,L_2\}+b_5\{L_1,L_2^2\}+b_6L_1L_2L_1+b_7L_2L_1L_2$$

$$+b_8H\{L_1,L_2\}+b_9HL_1^2+b_{10}HL_2^2+b_{11}H^2L_1+b_{12}H^2L_2+b_{13}L_1^2+b_{14}L_2^2+b_{15}\{L_1,L_2\}$$

$$+b_{16}HL_1+b_{17}HL_2+b_{18}H^2+b_{19}L_1+b_{20}L_2+b_{21}H+b_{22},$$

Here $\{L_j,L_k\}=L_jL_k+L_kL_j$ is the symmetrizer of L_j and L_k. This structure is an example of a *quadratic algebra*. Here the $A_i^{(\ell)}, b_j$ are constants or polynomials in the parameters a_k of the potential. The exact rules are given in [7] and [13].

Degenerate systems ($2n-1=3$): There are 4 generators: one 1st order X and 3 second order H, L_1, L_2.

$$[X,L_j]=C_1^{(j)}L_1+C_2^{(j)}L_2+C_3^{(j)}H+C_4^{(j)}X^2+C_5^{(j)},\quad j=1,2,$$

$$[L_1,L_2]=E_1\{L_1,X\}+E_2\{L_2,X\}+E_3HX+E_4X^3+E_5X,$$

Since $2n-1=3$ there must be an identity satisfied by the 4 generators. It is of 4th order:

$$c_1L_1^2+c_2L_2^2+c_3H^2+c_4\{L_1,L_2\}+c_5HL_1+c_6HL_2+c_7X^4+c_8\{X^2,L_1\}+c_9\{X^2,L_2\}$$

$$+c_{10}HX^2+c_{11}XL_1X+c_{12}XL_2X+c_{13}L_1+c_{14}L_2+c_{15}H+c_{16}X^2+c_{17}=0$$

Again the C_i, E_j, c_ℓ are constants or polynomials in the parameters a_k of the potential.

The structure of classical quadratic algebras is similar, except no symmetrizers are needed. In [8] it is shown that all of the classical and quantum structure equations for nondegenerate systems can, in fact, be derived from the equation for R^2, and all degenerate structure equations can be determined to within a constant factor from the 4th order identity.

Stäckel Equivalence Classes: There are 59 types of 2D 2nd order superintegrable systems, on a variety of manifolds but under the Stäckel transform, an invertible structure preserving mapping, they divide into 12 equivalence classes with representatives on flat space and the 2-sphere, 6 with nondegenerate 3-parameter potentials

$$\{S9, E1, E2, E3', E8, E10\}$$

and 6 with degenerate 1-parameter potentials, [13],

$$\{S3, E3, E4, E5, E6, E14\}.$$

The notation comes from [4] where all 2nd order 2D superintegrable systems on constant curvature spaces are classified.

3. Representatives of nondegenerate quantum systems

(i) $S9$: Defined in (1).

Structure equations:

$$[L_i, R] = 4\{L_i, L_k\} - 4\{L_i, L_j\} - (8 + 16a_j)L_j + (8 + 16a_k)L_k + 8(a_j - a_k),$$

$$R^2 = \frac{8}{3}\{L_1, L_2, L_3\} - (16a_1 + 12)L_1^2 - (16a_2 + 12)L_2^2 - (16a_3 + 12)L_3^2 +$$

$$\frac{52}{3}(\{L_1, L_2\} + \{L_2, L_3\} + \{L_3, L_1\}) + \frac{1}{3}(16 + 176a_1)L_1 + \frac{1}{3}(16 + 176a_2)L_2 + \frac{1}{3}(16 + 176a_3)L_3$$

$$+\frac{32}{3}(a_1 + a_2 + a_3) + 48(a_1a_2 + a_2a_3 + a_3a_1) + 64a_1a_2a_3, \quad R = [L_1, L_2].$$

(ii) $E1$ (Winternitz-Smorodinsky system)

$$H = \partial_x^2 + \partial_y^2 - \omega^2(x^2 + y^2) + \frac{b_1}{x^2} + \frac{b_2}{y^2}$$

Generators:

$$L_1 = \partial_x^2 - \omega^2 x^2 + \frac{b_1}{x^2}, \; L_2 = \partial_y^2 - \omega^2 y^2 + \frac{b_2}{y^2}, \; L_3 = (x\partial_y - y\partial_x)^2 + y^2\frac{b_1}{x^2} + x^2\frac{b_2}{y^2}$$

Structure relations:

$$[R, L_1] = 8L_1^2 - 8HL_1 - 16\omega^2 L_3 + 8\omega^2,$$

$$[R, L_3] = 8HL_3 - 8\{L_1, L_3\} + (16b_1 + 8)H - 16(b_1 + b_2 + 1)L_1,$$

$$R^2 + \frac{8}{3}\{L_1, L_1, L_3\} - 8H\{L_1, L_3\} + (16b_1 + 16b_2 + \frac{176}{3})L_1^2 - 16\omega^2 L_3^2 - (32b_1 + \frac{176}{3})HL_1$$

$$+(16b_1 + 12)H^2 + \frac{176}{3}\omega^2 L_3 + 16\omega^2(3b_1 + 3b_2 + 4b_1b_2 + \frac{2}{3}) = 0$$

(iii) $E2$

$$H = \partial_x^2 + \partial_y^2 - \omega^2(4x^2 + y^2) + bx + \frac{c}{y^2}$$

Generators:

$$L_1 = \partial_x^2 - 4\omega^2 x^2 + bx, \; L_2 = \partial_y^2 - \omega^2 y^2 + \frac{c}{y^2}, \; L_3 = \frac{1}{2}\{(x\partial_y - y\partial_x), \partial_y\} + y^2(\omega^2 x - \frac{b}{4}) + \frac{cx}{y^2}$$

Structure equations:

$$[L_1, R] + 2bL_2 - 16w^2 L_3 = 0, \; [L_3, R] + 2L_2^2 - 4L_1L_2 + 2bL_3 + \omega^2(8c + 6) = 0,$$

$$R^2 = 4L_1 \; L_2^2 + 16\omega^2 L_3^2 - 2b\{L_2, L_3\} + (12 + 16c)\omega^2 L_1 - 32w^2 L_2 - b^2(c + \frac{3}{4})$$

Here, the algebra generators are $H, L_1, L_3, \; R = [L_1, L_3]$

(iv) $E3'$

$$H = \partial_x^2 + \partial_y^2 - \omega^2(x^2+y^2) + c_1 x + c_2 y = L_1 + L_2$$

Generators:

$$L_1 = \partial_x^2 - \omega^2 x^2 + c_1 x,\ L_2 = \partial_y^2 - \omega^2 y^2 + c_2 y,\ L_3 = \partial_{xy} - \omega^2 xy + \frac{c_2 x + c_1 y}{2}$$

Structure relations:

$$[L_1, R] = 4\omega^2 L_3 - c_1 c_2,\ [L_3, R] = -2\omega^2 L_1 + 2\omega^2 L_2 + \frac{1}{2}(c_1^2 - c_2^2),$$

$$R^2 = 4\omega^2(L_3^2 - L_1 L_2) - 2c_1 c_2 L_3 + c_2^2 L_1 + c_1^2 L_2 + 4\omega^4$$

The algebra generators are $H, L_1, L_3,\ R = [L_1, L_3]$.

(v) $E10$

$$H = \partial_x^2 + \partial_y^2 + \alpha\bar{z} + \beta(z - \frac{3}{2}\bar{z}^2) + \gamma(z\bar{z} - \frac{1}{2}\bar{z}^3)$$

Generators:

$$L_1 = (\partial_x - i\partial_y)^2 + \gamma\bar{z}^2 + 2\beta\bar{z},$$

$$L_2 = 2i\{x\partial_y - y\partial_x, \partial_x - i\partial_y\} + (\partial_x + i\partial_y)^2 - 4\beta z\bar{z} - \gamma z\bar{z}^2 - 2\beta\bar{z}^3 - \frac{3}{4}\gamma\bar{z}^4 + \gamma z^2 + \alpha\bar{z}^2 + 2\alpha z$$

Structure equations:

$$[R, L_1] + 32\gamma L_1 + 32\beta^2 = 0, \quad [R, L_2] - 96L_1^2 - 64\beta H + 128\alpha L_1 - 32\gamma L_2 - 32\alpha^2,$$

$$R^2 = 64L_1^3 - 64\gamma H^2 - 128\alpha L_1^2 + 128\beta H L_1 + 32\gamma\{L_1, L_2\} - 128\alpha\beta H + 64\alpha^2 L_1 + 64\beta^2 L_2 - 256\gamma^2.$$

Here $R = [L_1, L_2],\ z = x + iy,\ \bar{z} = x - iy,$

(vi) $E8$

$$H = \partial_x^2 + \partial_y^2 + \frac{c_1 z}{\bar{z}^3} + \frac{c_2}{\bar{z}^2} + c_3 z\bar{z}$$

Generators:

$$L_1 = (\partial_x - i\partial_y)^2 - \frac{c_1}{\bar{z}^2} + c_3\bar{z}^2,\ L_2 = (x\partial_y - y\partial_x)^2 + c_1\frac{z^2}{\bar{z}^2} + c_2\frac{z}{\bar{z}}$$

Structure relations:

$$[R, L_1] = 8L_1^2 + 32c_1c_3,\ [R, L_2] = -8\{L_1, L_2\} + 8c_2 H - 16L_1,$$

$$R^2 = -\frac{16}{3}\{L_1^2, L_2\} - \frac{16}{3}L_1L_2L_1 - \frac{176}{3}L_1^2 + 16c_1H^2 + 16c_2L_1H - 64c_1c_3L_2 + 16c_3(\frac{4}{3}c_1 - c_2^2).$$

Here, $R = [L_1, L_2],\ z = x + iy,\ \bar{z} = x - iy,$

4. Representatives of degenerate systems

There are close relations between nondegenerate and degenerate systems.

- Every 1-parameter potential can be obtained from some 3-parameter potential by parameter restriction.
- It is not simply a restriction, however, because the structure of the symmetry algebra changes.
- A formally skew-adjoint 1st order symmetry appears and this induces a new 2nd order symmetry.
- Thus the restricted potential has a strictly larger symmetry algebra than is initially apparent.

We list the 6 representatives of the equivalence classes for degenerate systems:

(i) $S3$ (Higgs Oscillator)

$$H = J_1^2 + J_2^2 + J_3^2 + \frac{a}{s_3^2}$$

The system is the same as $S9$ with $a_1 = a_2 = 0$, $a_3 = a$ with the former L_2 replaced by

$$L_2 = \frac{1}{2}(J_1J_2 + J_2J_1) - \frac{as_1s_2}{s_3^2}$$

and

$$X = J_3 = s_2\partial_{s_3} - s_3\partial_{s_2}.$$

Structure relations:

$$[L_1, X] = 2L_2,\ [L_2, X] = -X^2 - 2L_1 + H - a,\ [L_1, L_2] = -(L_1X + XL_1) - (\frac{1}{2} + 2a)X,$$

$$\frac{1}{3}\left(X^2L_1 + XL_1X + L_1X^2\right) + L_1^2 + L_2^2 - HL_1 + (a + \frac{11}{12})X^2 - \frac{1}{6}H + (a - \frac{2}{3})L_1 - \frac{5a}{6} = 0.$$

(ii) $E3$ (Harmonic Oscillator)

$$H = \partial_x^2 + \partial_y^2 - \omega^2(x^2 + y^2)$$

Basis symmetries:

$$L_1 = \partial_x^2 - \omega^2x^2,\ L_3 = \partial_{xy} - \omega^2xy,\ X = x\partial_y - y\partial_x.$$

Also we set $L_2 = \partial_y^2 - \omega^2y^2 = H - L_1$.
Structure equations:

$$[L_1, X] = 2L_3,\ [L_3, X] = H - 2L_1,\ [L_1, L_3] = 2\omega^2X,$$
$$L_1^2 + L_3^2 - L_1H - \omega^2X^2 + \omega^2 = 0$$

(iii) $E4$

$$H = \partial_x^2 + \partial_y^2 + a(x + iy)$$

Basis Symmetries: (with $M = x\partial_y - y\partial_x$)

$$L_1 = \partial_x^2 + ax,\ L_2 = \frac{i}{2}\{M, X\} - \frac{a}{4}(x + iy)^2,\ X = \partial_x + i\partial_y$$

Structure equations:

$$[L_1, X] = a,\ [L_2, X] = X^2,\ [L_1, L_2] = X^3 + HX - \{L_1, X\},$$
$$X^4 - 2\left\{L_1, X^2\right\} + 2HX^2 + H^2 + 4aL_2 = 0$$

(iv) $E5$

$$H = \partial_x^2 + \partial_y^2 + ax$$

Basis symmetries: (where $M = x\partial_y - y\partial_x$)

$$L_1 = \partial_{xy} + \frac{1}{2}ay, \; L_2 = \frac{1}{2}\{M, X\} - \frac{1}{4}ay^2, \; X = \partial_y$$

Structure equations:

$$[L_1, L_2] = 2X^3 - HX, \; [L_1, X] = -\frac{a}{2}, \; [L_2, X] = L_1,$$

$$X^4 - HX^2 + L_1^2 + aL_2 = 0$$

(v) $E6$

$$H = \partial_x^2 + \partial_y^2 + \frac{a}{x^2}$$

Basis symmetries: ($M = x\partial_y - y\partial_x$)

$$L_1 = \frac{1}{2}\{M, \partial_x\} - \frac{ay}{x^2}, \; L_2 = M^2 + \frac{ay^2}{x^2}, \; X = \partial_y$$

Structure equations:

$$[L_1, L_2] = \{X, L_2\} + (2a + \frac{1}{2})X, \; [L_1, X] = H - X^2, \; [L_2, X] = 2L_1,$$

$$L_1^2 + \frac{1}{4}\{L_2, X^2\} + \frac{1}{2}XL_2X - L_2H + (a + \frac{3}{4})X^2 = 0$$

(vi) $E14$

$$H = \partial_x^2 + \partial_y^2 + \frac{b}{\overline{z}^2}$$

Basis symmetries: (with $M = x\partial_y - y\partial_x$, $z = x + iy, \overline{z} = x - iy$,)

$$X = \partial_x - i\partial_y, \; L_1 = \frac{i}{2}\{M, X\} + \frac{b}{\overline{z}}, \; L_2 = M^2 + \frac{bz}{\overline{z}}$$

Structure equations:

$$[L_1, L_2] = -\{X, L_2\} - \frac{1}{2}X, \; [X, L_1] = -X^2, \; [X, L_2] = 2L_1,$$

$$L_1^2 + XL_2X - bH - \frac{1}{4}X^2 = 0$$

5. Contractions of superintegrable systems

Suppose we have a nondegenerate quantum superintegrable system with generators H, L_1, L_2, $R = [L_1, L_2]$ and the usual structure equations, defining a quadratic algebra Q. If we make a change of basis to new generators $\tilde{H}, \tilde{L}_1, \tilde{L}_2$ and parameters $\tilde{a}_1, \tilde{a}_2, \tilde{a}_3$ such that

$$\begin{pmatrix} \tilde{L}_1 \\ \tilde{L}_2 \\ \tilde{H} \end{pmatrix} = \begin{pmatrix} A_{1,1} & A_{1,2} & A_{1,3} \\ A_{2,1} & A_{2,2} & A_{2,3} \\ 0 & 0 & A_{3,3} \end{pmatrix} \begin{pmatrix} L_1 \\ L_2 \\ H \end{pmatrix} + \begin{pmatrix} B_{1,1} & B_{1,2} & B_{1,3} \\ B_{2,1} & B_{2,2} & B_{2,3} \\ B_{3,1} & B_{3,2} & B_{3,3} \end{pmatrix} \begin{pmatrix} a_1 \\ a_2 \\ a_3 \end{pmatrix},$$

$$\begin{pmatrix} \tilde{a}_1 \\ \tilde{a}_2 \\ \tilde{a}_3 \end{pmatrix} = \begin{pmatrix} C_{1,1} & C_{1,2} & C_{1,3} \\ C_{2,1} & C_{2,2} & C_{2,3} \\ C_{3,1} & C_{3,2} & C_{3,3} \end{pmatrix} \begin{pmatrix} a_1 \\ a_2 \\ a_3 \end{pmatrix}$$

for some 3×3 constant matrices $A = (A_{i,j}), B, C$ such that $\det A \cdot \det C \neq 0$, we will have the same system with new structure equations of the same form for $\tilde{R} = [\tilde{L}_1, \tilde{L}_2]$, $[\tilde{L}_j, \tilde{R}]$, $\tilde{R}^2$, but with transformed structure constants.

- Choose a continuous 1-parameter family of basis transformation matrices $A(\epsilon), B(\epsilon), C(\epsilon,$ $0 < \epsilon \leq 1$ such that $A(1) = C(1)$ is the identity matrix, $B(1) = 0$ and $\det A(\epsilon) \neq 0$, $\det C(\epsilon) \neq 0$.
- Now suppose as $\epsilon \to 0$ the basis change becomes singular, (i.e., the limits of A, B, C either do not exist or, if they exist do not satisfy $\det A(0) \det C(0) \neq 0$) but the structure equations involving $A(\epsilon), B(\epsilon), C(\epsilon)$, go to a limit, defining a new quadratic algebra Q'.
- We call Q' a contraction of Q in analogy with Lie algebra contractions.

There is a similar definition of a contraction of a degenerate superintegrable system. Further, we say that the 2D system without potential, $H_0 = \Delta_2$, and with 3 algebraically independent second-order symmetries is a 2nd order free triplet. The possible spaces admitting free triplets are just those classified by Koenigs. Note that every nondegenerate or degenerate superintegrable system defines a free triplet, simply by setting the parameters $a_j = 0$ in the potential. Similarly, this free triplet defines a free quadratic algebra, i.e., a quadratic algebra with all $a_j = 0$. In general, a free triplet cannot be obtained as a restriction of a superintegrable system and its associated algebra does not close to a free quadratic algebra. All of these definitions extend easily to classical superintegrable systems.

We have the following closure theorems:

Theorem 1 *Closure Theorem: A free triplet (classical or quantum) extends to a superintegrable system if and only if it generates a free quadratic algebra.*

Theorem 2 *A superintegrable system, degenerate or nondegenerate, is uniquely determined by its free quadratic algebra.*

Proofs of these results will appear in [8]. The main ideas are as follows. Suppose we have a classical free triplet with basis

$$\mathcal{L}_{(s)} = \sum_{i,j=1}^{2} a^{ij}_{(s)} p_i p_j \quad a^{ij}_{(s)} = a^{ji}_{(s)}, \ s = 1, 2, 3, \ \mathcal{L}_{(3)} = \mathcal{H}_0 = \frac{p_1^2 + p_2^2}{\lambda(x, y)},$$

that determines a free nondegenerate quadratic algebra, hence a free nondegenerate superintegrable system. From the free system alone we can compute the functions A^{ij}, B^{ij}, expressed in terms of the Cartesian-like coordinates (x, y), that determine the system of equations for an additive potential

$$\begin{aligned} V_{22} &= V_{11} + A^{22} V_1 + B^{22} V_2, \\ V_{12} &= A^{12} V_1 + B^{12} V_2, \end{aligned} \tag{2}$$

These equations always admit a constant potential for a solution, but they will admit a full 4-dimensional vector space of solutions V if and only if the integrability conditions for (2) are identically satisfied. In [8] we show that the integrability conditions hold if and only if the free system generates a quadratic algebra. This is an algebraic solution for an analytic problem.

Further, if a potential function satisfies (2) then it is guaranteed that the Bertrand-Darboux integrability conditions for equations

$$W_i^{(s)} = \lambda \sum_{j=1}^{2} a_{(s)}^{ij} V_j, \quad i, s = 1, 2,$$

hold and we can compute the solutions $W^{(s)}$, $W^{(3)} = V$, unique up to additive constants, such that the constants of the motion $\mathcal{L}_{(s)} = \sum a_{(s)}^{ij} p_i p_j + W^{(s)}$ define a nondegenerate superintegrable system. This system is guaranteed to determine a nondegenerate quadratic algebra with potential whose highest order (potential-free) terms agree with the free quadratic algebra. The functions A^{ij}, B^{ij} are defined independent of the basis chosen for the free triplet although, of course, they do depend upon the particular coordinates chosen.

Similarly, there is an associated 2nd order quantum free triplet

$$L_s = \frac{1}{\lambda} \sum_{i,j=1}^{2} \partial_i(\lambda a_{(s)}^{ij} \partial_j), \; s = 1, 2, 3, \; L_3 = H_0 = \frac{1}{\lambda(\mathbf{x})}(\partial_{11} + \partial_{22}),$$

that defines a free nondegenerate quantum quadratic algebra with potential. The functions $W^{(s)}$ are the same as before.

There is an analogous construction of degenerate superintegrable systems with potential from free triplets that generate a free quadratic algebras, but are such that one generator say, $\mathcal{L}_1 = \mathcal{X}^2$, is a perfect square.

6. Lie algebra contractions

The contractions of the Lie algebras $e(2, \mathbf{C})$ and $o(3, \mathbf{C})$ have long since been classified, e.g. [17]. There are 7 nontrivial contractions of $e(2, \mathbf{C})$ and 4 of $o(3, \mathbf{C})$. However, 2 of the contractions of $e(2, \mathbf{C})$ take it to an abelian Lie algebra so are not of interest to us.

Wigner-Inonu contractions of $e(2, \mathbf{C})$:

(i) $\{\mathcal{J}', p_1', p_2'\} = \{\mathcal{J}, \; \epsilon p_1, \; \epsilon p_2\} : \; e(2, \mathbf{C})$,
coordinate implementation $x' = \frac{x}{\epsilon}, y' = \frac{y}{\epsilon}$,

(ii) $\{\mathcal{J}', p_1', p_2'\} = \{\epsilon \mathcal{J}, \; p_1, \; \epsilon p_2\}$: Heisenberg algebra,
coordinate implementation $x' = x, y' = \frac{y}{\epsilon}, \mathcal{J}' = x' p_2'$,

(iii) $\{\mathcal{J}', \; p_1' + i p_2', \; p_1' - i p_2'\} = \{\epsilon \mathcal{J}, \; \epsilon(p_1 + i p_2), \; p_1 - i p_2\}$: abelian algebra,

(iv) $\{\mathcal{J}', p_1', p_2'\} = \{\epsilon \mathcal{J}, \; p_1, \; p_2\}$: abelian algebra,

(v) $\{\mathcal{J}', \; p_1' + i p_2', \; p_1' - i p_2'\} = \{\mathcal{J}, \; \epsilon(p_1 + i p_2), \; p_1 - i p_2\} : \; e(2, \mathbf{C})$,
coordinate implementation $x' + iy' = x + iy, x' - iy' = \frac{x - iy}{\epsilon}$,

The other natural contractions of $e(2, \mathbf{C})$:

(vi) $\{\mathcal{J}', p_1', p_2'\} = \{\mathcal{J} + \frac{p_1}{\epsilon}, \; p_1, \; p_2\} : \; e(2, \mathbf{C})$,
coordinate implementation $x' = x, y' = y - \frac{1}{\epsilon}$,

(vii) $\{\mathcal{J}', p_1', p_2'\} = \{\mathcal{J} + \frac{p_1 + i p_2}{\epsilon}, \; p_1, \; p_2\} : \; e(2, \mathbf{C})$,
coordinate implementation $x' = x + \frac{i}{\epsilon}, y' = y - \frac{1}{\epsilon}$.

We use the classical realization for $o(3, \mathbf{C})$ acting on the 2-sphere, with basis $\mathcal{J}_1 = s_2 p_3 - s_3 p_2$, $\mathcal{J}_2 = s_3 p_1 - s_1 p_3$, $\mathcal{J}_3 = s_1 p_2 - s_2 p_1$, commutation relations

$$\{\mathcal{J}_2, \mathcal{J}_1\} = \mathcal{J}_3, \quad \{\mathcal{J}_3, \mathcal{J}_2\} = \mathcal{J}_1, \quad \{\mathcal{J}_1, \mathcal{J}_3\} = \mathcal{J}_2,$$

and Hamiltonian $\mathcal{H} = \mathcal{J}_1^2 + \mathcal{J}_2^2 + \mathcal{J}_3^2$. Here $s_1^2 + s_2^2 + s_3^2 = 1$ and restriction to the sphere gives $s_1p_1 + s_2p_2 + s_3p_3 = 0$.

Wigner-Inonu contractions of $o(3, \mathbf{C})$:

(i) $\{\mathcal{J}_1', \mathcal{J}_2', \mathcal{J}_3'\} = \{\epsilon\mathcal{J}_1,\ \epsilon\mathcal{J}_2,\ \mathcal{J}_3\} :\ e(2, \mathbf{C})$,
coordinate implementation $x = s_1/\epsilon, y = s_2/\epsilon, s_3 \approx 1, \mathcal{J} = \mathcal{J}_3$,

(ii) $\{\mathcal{J}_1' + i\mathcal{J}_2',\ \mathcal{J}_1' - i\mathcal{J}_2',\ \mathcal{J}_3'\} = \{\mathcal{J}_1 + i\mathcal{J}_2,\ \epsilon(\mathcal{J}_1 - i\mathcal{J}_2),\ \epsilon\mathcal{J}_3\}$: Heisenberg algebra,
coordinate implementation $s_1 = \frac{\cos\phi}{\cosh\psi},\ s_2 = \frac{\sin\phi}{\cosh\psi},\ s_3 = \frac{\sinh\psi}{\cosh\psi}$,
we set $\phi = \epsilon\theta - i\ln\sqrt{\epsilon},\ \psi = \xi\sqrt{\epsilon}$, to get
$\mathcal{J}_3' = p_\theta,\ \mathcal{J}_1' + i\mathcal{J}_2' = -i(\xi p_\theta + p_\xi),\ \mathcal{J}_1' - i\mathcal{J}_2' = -\xi p_\theta + p_\xi$,

(iii) $\{\mathcal{J}_1' + i\mathcal{J}_2',\ \mathcal{J}_1' - i\mathcal{J}_2',\ \mathcal{J}_3'\} = \{\mathcal{J}_1 + i\mathcal{J}_2,\ \epsilon(\mathcal{J}_1 - i\mathcal{J}_2),\ \mathcal{J}_3\} :\ e(2, \mathbf{C})$,
coordinate implementation $s_1 + is_2 = \epsilon z,\ s_1 - is_2 = \bar{z},\ s_3 \approx 1$,
Using $zp_z + \bar{z}p_{\bar{z}} + s_3p_{s_3} = 0$, we get $\mathcal{J}_3' = i(zp_z - \bar{z}p_{\bar{z}})$,
$\mathcal{J}_1' + i\mathcal{J}_2' = 2ip_{\bar{z}},\ \mathcal{J}_1' - i\mathcal{J}_2' = -2ip_z$,

The other natural contraction of $o(3, \mathbf{C})$:

(v) $\{\mathcal{J}_1' + i\mathcal{J}_2', \mathcal{J}_1' - i\mathcal{J}_2', \mathcal{J}_3'\} = \{\epsilon(\mathcal{J}_1 + i\mathcal{J}_2), \frac{\mathcal{J}_1 - i\mathcal{J}_2}{\epsilon}, \mathcal{J}_3\} :\ o(3, \mathbf{C})$,
coordinate implementation $s_1' = \frac{\epsilon+\epsilon^{-1}}{2}s_1 + i\frac{\epsilon-\epsilon^{-1}}{2}s_2$,
$s_2' = -i\frac{\epsilon-\epsilon^{-1}}{2}s_1 + \frac{\epsilon+\epsilon^{-1}}{2}s_2,\ s_3' = s_3$.

Note that once we choose a basis for a Lie algebra A, the structure of its enveloping algebra is uniquely determined by the structure constants. All structure relations in the enveloping algebra are continuous functions of the structure constants. Thus a contraction of one Lie algebra A to another, B induces a similar contraction of the corresponding enveloping algebras of A and B. In the case of $e(2, \mathbf{C})$ and $o(3, \mathbf{C})$, free quadratic algebras constructed in the enveloping algebras will contract to free quadratic algebras generated by the target Lie algebras, [8] We illustrate the process with several examples. In the following examples we work out all of the induced contractions for the systems $\tilde{E}1$, $\tilde{S}9$, $\tilde{S}3$ and $\tilde{E}3$ to illustrate the contraction procedure for each of these Lie algebras and for both nondegenerate and degenerate systems.

(i) $\tilde{E}1 \to \tilde{E}8$: Use $\{\mathcal{J}',\ p_1' + ip_2',\ p_1' - ip_2'\} = \{\mathcal{J},\ \epsilon(p_1 + ip_2),\ p_1 - ip_2\}$.

$$\begin{aligned}
\mathcal{L}_1 &= \mathcal{J}^2 = (\mathcal{J}')^2 = \mathcal{L}_1' \\
\mathcal{L}_2 &= p_1^2 = \frac{1}{4}\left((p_1 + ip_2) + (p_1 - ip_2)\right)^2 = \frac{1}{4}\left(\frac{(p_1' + ip_2')}{\epsilon} + (p_1' - ip_2')\right)^2 \\
&\approx \frac{(p_1' + ip_2')^2}{4\epsilon^2} = \frac{\mathcal{L}_2'}{4\epsilon^2} \\
\mathcal{H} &= (p_1 + ip_2)(p_1 - ip_2) = \frac{(p_1' + ip_2')(p_1' - ip_2')}{\epsilon} = \frac{\mathcal{H}'}{\epsilon}.
\end{aligned}$$

(ii) $\tilde{E}1 \to \tilde{E}2$: Use $\{\mathcal{J}', p_1', p_2'\} = \{\mathcal{J} + \frac{p_2}{\epsilon},\ p_1,\ p_2\}$.

$$\begin{aligned}
\mathcal{L}_1 &= \mathcal{J}^2 = (\mathcal{J}' - \frac{p_2'}{\epsilon})^2 = (\mathcal{J}'^2) - 2\frac{p_2'\mathcal{J}'}{\epsilon} + \frac{(p_2')^2}{\epsilon^2} \approx -2\frac{\mathcal{L}_2'}{\epsilon} + \frac{\mathcal{L}_1'}{\epsilon^2} \\
\mathcal{L}_2 &= p_1^2 = (p_1')^2 = \mathcal{H}' - \mathcal{L}_1',\quad \mathcal{L}_2' = \mathcal{J}'p_2' \\
\mathcal{H} &= p_1^2 + p_2^2 = (p_1')^2 + (p_2')^2 = \mathcal{H}'.
\end{aligned}$$

(iii) $\tilde{E}1 \to \tilde{E}3'$: Use $\{\mathcal{J}', p_1', p_2'\} = \{\mathcal{J} + \frac{p_1+p_2}{\epsilon},\ p_1,\ p_2\}$.

$$\begin{aligned}
\mathcal{L}_1 &= \mathcal{J}^2 = (\mathcal{J}' - \frac{p_1'+p_2'}{\epsilon})^2 = \mathcal{J}'^2 - 2\frac{\mathcal{J}'(p_1'+p_2')}{\epsilon} + \frac{{p_1'}^2 + 2p_1'p_2' + {p_2'}^2}{\epsilon^2} \\
&\approx \frac{\mathcal{L}_2'}{\epsilon^2} + \frac{\mathcal{H}'}{\epsilon^2} \\
\mathcal{L}_2 &= p_1^2 = \mathcal{H}' - \mathcal{L}_1', \quad \mathcal{L}_2' = 2p_1'p_2' \\
\mathcal{H} &= p_1^2 + p_2^2 = \mathcal{H}'.
\end{aligned}$$

(iv) $\tilde{E}1 \to \tilde{E}3'$ (alternate version). Use $\{\mathcal{J}', p_1', p_2'\} = \{\mathcal{J} + (p_1 + ip_2)/\epsilon, p_1, p_2\}$.

$$\begin{aligned}
\mathcal{L}_1' &= p_1p_2 = \frac{\epsilon^2\mathcal{L}_1 + \mathcal{H} - 2\mathcal{L}_2}{2i} \\
\mathcal{L}_2' &= p_1^2 = \mathcal{L}_2 \\
\mathcal{H}' &= p_1^2 + p_2^2 = \mathcal{H}.
\end{aligned}$$

(v) $\tilde{E}1 \to \tilde{E}1$. Use $\{\mathcal{J}', p_1', p_2'\} = \{\mathcal{J}, \epsilon p_1, \epsilon p_2\}$.

$$\begin{aligned}
\mathcal{L}_1' &= \mathcal{J}^2 = \mathcal{L}_1 \\
\mathcal{L}_2' &= \epsilon^2 p_1^2 = \epsilon^2\mathcal{L}_2 \\
\mathcal{H}' &= \epsilon^2(p_1^2 + p_2^2) = \epsilon^2\mathcal{H}.
\end{aligned}$$

(vi) $\tilde{E}_1 \to$ Heisenberg. Use $\{\mathcal{J}', p_1', p_2'\} = \{\epsilon\mathcal{J}, p_1, \epsilon p_2\}$.

$$\begin{aligned}
\mathcal{L}_1 &= \mathcal{J}^2 = \frac{\mathcal{J}'^2}{\epsilon^2} \\
\mathcal{L}_2 &= p_1^2 = {p_1'}^2 \\
\mathcal{H} &= p_1^2 + p_2^2 = {p_1'}^2 + \frac{{p_2'}^2}{\epsilon^2} = \mathcal{L}_2 + \frac{{p_2'}^2}{\epsilon^2},
\end{aligned}$$

so,

$$\begin{aligned}
\mathcal{L}_1' &= \epsilon^2\mathcal{L}_1 = \mathcal{J}'^2 = {x'}^2{p_2'}^2 \\
\mathcal{L}_2' &= \mathcal{L}_2 = {p_1'}^2 \\
\mathcal{H}' &= \epsilon^2(\mathcal{H} - \mathcal{L}_2) = {p_2'}^2.
\end{aligned}$$

Structure relations:

$$\mathcal{R} = \{\mathcal{L}_1', \mathcal{L}_2'\},\ \mathcal{R}^2 = 4{p_1'}^2{p_2'}^4 = 4\mathcal{L}_1'\mathcal{H}'^2.$$

(vii) $\tilde{S}9 \to \tilde{E}1$: Use $\{\mathcal{J}_1', \mathcal{J}_2', \mathcal{J}_3'\} = \{\epsilon\mathcal{J}_1,\ \epsilon\mathcal{J}_2,\ \mathcal{J}_3\}$.

$$\begin{aligned}
\mathcal{L}_1 &= \mathcal{J}_3^2 = \mathcal{J}^2 = \mathcal{L}_1' \\
\mathcal{L}_2 &= \mathcal{J}_1^2 \approx \frac{{p_1'}^2}{\epsilon^2} = \frac{{\mathcal{L}_2'}^2}{\epsilon^2} \\
\mathcal{H} &= \mathcal{J}_1^2 + \mathcal{J}_2^2 + \mathcal{J}_3^2 \approx \frac{{p_1'}^2 + {p_2'}^2}{\epsilon^2} = \frac{\mathcal{H}'}{\epsilon^2}
\end{aligned}$$

(viii) $\tilde{S}9 \to \tilde{S}2$: Use $\{\mathcal{J}_1'+i\mathcal{J}_2', \mathcal{J}_1'-i\mathcal{J}_2', \mathcal{J}_3'\} = \{\epsilon(\mathcal{J}_1+i\mathcal{J}_2), \frac{\mathcal{J}_1-i\mathcal{J}_2}{\epsilon}, \mathcal{J}_3\}$.

$$\begin{aligned}
\mathcal{L}_1 &= \mathcal{J}_3^2 = \mathcal{J}_3'^2 = \mathcal{L}_2' \\
\mathcal{L}_2 &= \mathcal{J}_1^2 = \frac{1}{4}((\mathcal{J}_1+i\mathcal{J}_2)+(\mathcal{J}_1-i\mathcal{J}_2))^2 = \frac{1}{4}\left(\frac{\mathcal{J}_1'+i\mathcal{J}_2'}{\epsilon}+\epsilon(\mathcal{J}_1'-i\mathcal{J}_2')\right)^2 \approx \frac{1}{4}\frac{\mathcal{L}_1'}{\epsilon^2} \\
\mathcal{H} &= (\mathcal{J}_1+i\mathcal{J}_2)(\mathcal{J}_1-i\mathcal{J}_2)+\mathcal{J}_3^2 = \mathcal{J}_1'^2+\mathcal{J}_2'^2+\mathcal{J}_3'^2 = \mathcal{H}', \quad \mathcal{L}_1' = (\mathcal{J}_1'+i\mathcal{J}_2')^2,
\end{aligned}$$

so the change of basis

$$\begin{aligned}
\mathcal{L}_1' &= 4\epsilon^2\mathcal{L}_2, \\
\mathcal{L}_2' &= \mathcal{L}_1, \\
\mathcal{H}' &= \mathcal{H}.
\end{aligned}$$

determines the contraction to $\tilde{S}2$ in the limit as $\epsilon \to 0$.

(ix) $\tilde{S}9 \to \tilde{E}8$: Use $\{\mathcal{J}_1'+i\mathcal{J}_2', \ \mathcal{J}_1'-i\mathcal{J}_2', \ \mathcal{J}_3'\} = \{\mathcal{J}_1+i\mathcal{J}_2, \ \epsilon(\mathcal{J}_1-i\mathcal{J}_2), \ \mathcal{J}_3\}$, with coordinate implementation $s_1+is_2=\epsilon z, \ s_1-is_2=\bar{z}, \ s_3\approx 1$, so $\mathcal{J}_3'=i(zp_z-\bar{z}p_{\bar{z}}), \mathcal{J}_1'+i\mathcal{J}_2'=2ip_{\bar{z}}$, $\mathcal{J}_1'-i\mathcal{J}_2'=-2ip_z$.

$$\begin{aligned}
\mathcal{L}_1 &= \mathcal{J}_3^2 = \mathcal{J}_3'^2 = \mathcal{L}_1' \\
\mathcal{L}_2 &= \mathcal{J}_1^2 = \frac{1}{4}((\mathcal{J}_1+i\mathcal{J}_2)+(\mathcal{J}_1-i\mathcal{J}_2))^2 = \frac{1}{4}\left((\mathcal{J}_1'+i\mathcal{J}_2')+\frac{\mathcal{J}_1'-i\mathcal{J}_2'}{\epsilon}\right)^2 \approx \frac{1}{4}\frac{\mathcal{L}_2'}{\epsilon^2} \\
\mathcal{H} &= (\mathcal{J}_1+i\mathcal{J}_2)(\mathcal{J}_1-i\mathcal{J}_2)+\mathcal{J}_3^2 = \frac{(\mathcal{J}_1'+i\mathcal{J}_2')(\mathcal{J}_1'-i\mathcal{J}_2')}{\epsilon}+\mathcal{J}_3'^2 = \frac{\mathcal{H}'}{\epsilon}+\mathcal{L}_1',
\end{aligned}$$

where $\mathcal{L}_2' = (\mathcal{J}_1'-i\mathcal{J}_2')^2$, so the change of basis

$$\begin{aligned}
\mathcal{L}_1' &= \mathcal{L}_1, \\
\mathcal{L}_2' &= 4\epsilon^2\mathcal{L}_2, \\
\mathcal{H}' &= \epsilon\mathcal{H},
\end{aligned}$$

determines the contraction to $\tilde{E}8$ in the limit as $\epsilon \to 0$.

(x) $\tilde{S}9 \to$ Heisenberg algebra: Use $\{\mathcal{J}_1'+i\mathcal{J}_2', \ \mathcal{J}_1'-i\mathcal{J}_2', \ \mathcal{J}_3'\} = \{\mathcal{J}_1+i\mathcal{J}_2, \ \epsilon(\mathcal{J}_1-i\mathcal{J}_2), \ \epsilon\mathcal{J}_3\}$. with coordinate implementation $s_1 = \frac{\cos\phi}{\cosh\psi}$, $s_2 = \frac{\sin\phi}{\cosh\psi}$, $s_3 = \frac{\sinh\psi}{\cosh\psi}$, and substitutions $\phi = \epsilon\theta - i\ln\sqrt{\epsilon}$, $\psi = \xi\sqrt{\epsilon}$, to get $\mathcal{J}_3' = p_\theta$, $\mathcal{J}_1'+i\mathcal{J}_2' = -i(\xi p_\theta + p_\xi)$, $\mathcal{J}_1'-i\mathcal{J}_2' = -\xi p_\theta + p_\xi$.

$$\begin{aligned}
\mathcal{L}_1 &= \mathcal{J}_3^2 = \frac{\mathcal{J}_3'^2}{\epsilon^2} = \frac{\mathcal{L}_1'}{\epsilon^2} \\
\mathcal{L}_2 &= \mathcal{J}_1^2 = \frac{1}{4}((\mathcal{J}_1+i\mathcal{J}_2)+(\mathcal{J}_1-i\mathcal{J}_2))^2 = \frac{1}{4}\left((\mathcal{J}_1'+i\mathcal{J}_2')+\frac{\mathcal{J}_1'-i\mathcal{J}_2'}{\epsilon}\right)^2 \approx \frac{1}{4}\frac{\mathcal{L}_2'}{\epsilon^2} \\
\mathcal{H} &= (\mathcal{J}_1+i\mathcal{J}_2)(\mathcal{J}_1-i\mathcal{J}_2)+\mathcal{J}_3^2 = \frac{\mathcal{J}_1'^2+\mathcal{J}_2'^2}{\epsilon}+\frac{\mathcal{J}_3'^2}{\epsilon^2},
\end{aligned}$$

where $\mathcal{L}_1' = (\mathcal{J}_1'-i\mathcal{J}_2')^2$, $\mathcal{L}_2' = \mathcal{J}_1'^2+\mathcal{J}_2'^2$. so the change of basis

$$\begin{aligned}
\mathcal{H}' &= \epsilon^2\mathcal{H}, \\
\mathcal{L}_1' &= 4\epsilon^2\mathcal{L}_2, \\
\mathcal{L}_2' &= \epsilon(\mathcal{H}-\mathcal{L}_1),
\end{aligned}$$

$\mathcal{R}'^2 = -16\mathcal{H}'\mathcal{L}_1'^2$, determines the contraction.

(xi) $\tilde{S}3 \to \tilde{E}3$. Use $\{\mathcal{J}_1', \mathcal{J}_2', \mathcal{J}_3'\} = \{\epsilon \mathcal{J}_1, \epsilon \mathcal{J}_2, \mathcal{J}_3\}$.

$$\begin{aligned}
\mathcal{X}' &= \mathcal{X}, \\
\mathcal{L}_1' &= \epsilon^2 \mathcal{L}_1, \\
\mathcal{L}_2' &= \epsilon^2 \mathcal{L}_2, \\
\mathcal{H}' &= \epsilon^2 \mathcal{H}.
\end{aligned}$$

(xii) $\tilde{S}3 \to \tilde{E}3$ (alternate contraction). Use $\{\mathcal{J}_1'+i\mathcal{J}_2', \mathcal{J}_1'-i\mathcal{J}_2', \mathcal{J}_3'\} = \{\mathcal{J}_1+i\mathcal{J}_2, \epsilon(\mathcal{J}_1-i\mathcal{J}_2), \mathcal{J}_3\}$.

$$\begin{aligned}
\mathcal{X}' &= \mathcal{X} = i(zp_z - \bar{z}p_{\bar{z}}), \\
\mathcal{L}_1' &= p_{\bar{z}}^2 = -\frac{1}{2}(\mathcal{L}_1 + i\mathcal{L}_2) + \frac{1}{4}\mathcal{H} - \frac{1}{4}\mathcal{X}^2, \\
\mathcal{L}_2' &= p_z^2 = -i\epsilon^2 \mathcal{L}_2, \\
\mathcal{H}' &= 4p_z p_{\bar{z}} = \epsilon \mathcal{H}.
\end{aligned}$$

(xiii) $\tilde{S}3 \to \tilde{S}3$. Use $\{\mathcal{J}_1' + i\mathcal{J}_2', \mathcal{J}_1' - i\mathcal{J}_2', \mathcal{J}_3'\} = \{\epsilon(\mathcal{J}_1 + i\mathcal{J}_2), (\mathcal{J}_1 - i\mathcal{J}_2)/\epsilon, \mathcal{J}_3\}$.

$$\begin{aligned}
\mathcal{X}' &= \mathcal{X} = \mathcal{J}_3, \\
\mathcal{L}_1' &= (\mathcal{J}_1' + i\mathcal{J}_2')^2 = 4i\epsilon^2 \mathcal{L}_2, \\
\mathcal{L}_2' &= (\mathcal{J}_1' - i\mathcal{J}_2')^2 = \frac{2}{\epsilon^2}(\mathcal{L}_1 - i\mathcal{L}_2 - \frac{1}{2}\mathcal{H} + \frac{1}{2}\mathcal{X}^2), \\
\mathcal{H}' &= \mathcal{J}_1'^2 + \mathcal{J}_2'^2 + \mathcal{J}_3'^2 = \mathcal{H}.
\end{aligned}$$

(xiv) $\tilde{S}3 \to$ Heisenberg. Use $\{\mathcal{J}_1' + i\mathcal{J}_2', \mathcal{J}_1' - i\mathcal{J}_2', \mathcal{J}_3'\} = \{\mathcal{J}_1 + i\mathcal{J}_2, \epsilon(\mathcal{J}_1 - i\mathcal{J}_2), \epsilon \mathcal{J}_3\}$.

$$\begin{aligned}
\mathcal{X}' &= \mathcal{J}_3' = \epsilon \mathcal{X}, \\
\mathcal{L}_1' &= (\mathcal{J}_1' + i\mathcal{J}_2')^2 = \mathcal{L}_1, \\
\mathcal{L}_2' &= (\mathcal{J}_1' - i\mathcal{J}_2')^2 = \epsilon^2 \mathcal{L}_2, \\
\mathcal{H}' &= \mathcal{J}_3'^2 = \epsilon^2 \mathcal{H}.
\end{aligned}$$

The structure relation is $\mathcal{H}' - \mathcal{X}'^2 = 0$.

(xv) $\tilde{E}3 \to \tilde{E}3$. Use $\{\mathcal{J}', p_1', p_2'\} = \{\mathcal{J}, \epsilon p_1, \epsilon p_2\}$.

$$\begin{aligned}
\mathcal{X}' &= \mathcal{J}' = \mathcal{X}, \\
\mathcal{L}_1' &= {p_2'}^2 = \epsilon^2 \mathcal{L}_1, \\
\mathcal{L}_2' &= p_1' p_2' = \epsilon^2 \mathcal{L}_2, \\
\mathcal{H}' &= {p_1'}^2 + {p_2'}^2 = \epsilon^2 \mathcal{H}.
\end{aligned}$$

(xvi) $\tilde{E}3 \to \tilde{E}3$ (alternate form contraction). Use $\{\mathcal{J}', p_1'+ip_2', p_1'-ip_2'\} = \{\mathcal{J}, \epsilon(p_1+ip_2), p_1-ip_2\}$.

$$\begin{aligned}
\mathcal{X}' &= \mathcal{J}' = \mathcal{X}, \\
\mathcal{L}_1' &= (p_1' - ip_2')^2 = 2i(\mathcal{L}_1 - \mathcal{L}_2) - \mathcal{H}, \\
\mathcal{L}_2' &= (p_1' + ip_2')^2 = 4i\epsilon^2 \mathcal{L}_2, \\
\mathcal{H}' &= {p_1'}^2 + {p_2'}^2 = \epsilon \mathcal{H}.
\end{aligned}$$

(xvii) $\tilde{E}3 \to \tilde{E}5$. Use $\{\mathcal{J}', p_1', p_2'\} = \{\mathcal{J} + \frac{p_1}{\epsilon}, p_1, p_2\}$. Take the 1st order basis for $\tilde{E}3$ as $\mathcal{X} = \mathcal{J}$, with 2nd order basis $\mathcal{X}^2, \mathcal{L}_1 = p_1^2, \mathcal{L}_2 = p_1p_2, \mathcal{H} = p_1^2 + p_2^2$.

$$\begin{aligned} \mathcal{X}' &= p_1' = \epsilon\mathcal{X}, \\ \mathcal{L}_1' &= {p_1'}^2 = \mathcal{L}_1, \\ \mathcal{L}_2' &= p_1'p_2' = \mathcal{L}_2, \\ \mathcal{H}' &= {p_1'}^2 + {p_2'}^2 = \mathcal{H}. \end{aligned}$$

However, $\mathcal{L}_1' = \mathcal{X}'^2$ so the space of 2nd order symmetries would appear to have dimension only 3. The missing 2nd order symmetry is constructed from $\mathcal{X}^2$ and $\mathcal{L}_1$:

$$\mathcal{L}_3' = \mathcal{J}'p_1' = -\frac{\epsilon}{2}(\mathcal{X}^2 - \frac{\mathcal{L}_1}{\epsilon^2}).$$

(xviii) $\tilde{E}3 \to \tilde{E}4$. Use $\{\mathcal{J}', p_1', p_2'\} = \{\mathcal{J} + \frac{p_1+ip_2}{\epsilon}, p_1, p_2\}$. Take the 1st order basis for $\tilde{E}3$ as $\mathcal{X} = \mathcal{J}$, with 2nd order basis $\mathcal{X}^2, \mathcal{L}_1 = p_2^2, \mathcal{L}_2 = p_1p_2, \mathcal{H} = p_1^2 + p_2^2$.

$$\begin{aligned} \mathcal{X}' &= p_1' + ip_2' = \epsilon\mathcal{X}, \\ \mathcal{L}_1' &= {p_2'}^2 = \mathcal{L}_1, \\ \mathcal{L}_2' &= p_1'p_2' = \mathcal{L}_2, \\ \mathcal{H}' &= {p_1'}^2 + {p_2'}^2 = \mathcal{H}. \end{aligned}$$

However, $\mathcal{H}' - 2\mathcal{L}_1' + 2i\mathcal{L}_2' = \mathcal{X}'^2$ so the space of 2nd order symmetries would appear to have dimension only 3. The missing 2nd order symmetry is constructed as

$$\mathcal{L}_3' = \mathcal{J}'(p_1' + ip_2') = -\frac{\epsilon}{2}(\mathcal{X}^2 - \frac{\mathcal{H} + 2i\mathcal{L}_2 - 2\mathcal{L}_1}{\epsilon^2}).$$

(xix) $\tilde{E}3 \to$ Heisenberg. Use $\{\mathcal{J}', p_1', p_2'\} = \{\epsilon\mathcal{J}, p_1, \epsilon p_2\}$. Take the basis as $\mathcal{X} = \mathcal{J}$,and $\mathcal{X}^2, \mathcal{L}_1 = p_1^2, \mathcal{L}_2 = p_1p_2, \mathcal{H} = p_1^2 + p_2^2$.

$$\begin{aligned} \mathcal{X}' &= \mathcal{J}' = \epsilon\mathcal{X}, \\ \mathcal{L}_1' &= {p_1'}^2 = \mathcal{L}_1, \\ \mathcal{L}_2' &= p_1'p_2' = \epsilon\mathcal{L}_2, \\ \mathcal{H}' &= {p_2'}^2 = \epsilon^2\mathcal{H}. \end{aligned}$$

The functional relation is $\mathcal{L}_1'\mathcal{H}' - {\mathcal{L}_2'}^2 = 0$.

Suppose we have a classical free triplet $\mathcal{H}^{(0)}, \mathcal{L}_1^{(0)}, \mathcal{L}_2^{(0)}$ that determines a nondegenerate quadratic algebra $Q^{(0)}$ and structure functions $A^{ij}(\mathbf{x}), B^{ij}(\mathbf{x})$ in some set of Cartesian-like coordinates (x_1, x_2). Further, suppose this system contracts to another nondegenerate system $\mathcal{H}'^{(0)}, \mathcal{L}'^{(0)}_1, \mathcal{L}'^{(0)}_2$ with quadratic algebra $Q'^{(0)}$ via the mechanism described in the preceding sections. We show here that this contraction induces a contraction of the associated nondegenerate superintegrable system $\mathcal{H} = \mathcal{H}^{(0)} + V$, $\mathcal{L}_1 = \mathcal{L}_1^{(0)} + W^{(1)}$, $\mathcal{L}_2 = \mathcal{L}_2^{(0)} + W^{(2)}$, Q to $\mathcal{H}' = \mathcal{H}'^{(0)} + V'$, $\mathcal{L}_1' = \mathcal{L}'^{(0)}_1 + W^{(1)'}$, $\mathcal{L}_2' = \mathcal{L}'^{(0)}_2 + W^{(2)'}$, Q'. The point is that in the contraction process the symmetries $\mathcal{H}'^{(0)}(\epsilon)$, $\mathcal{L}'^{(0)}_1(\epsilon)$, $\mathcal{L}'^{(0)}_2(\epsilon)$ remain continuous functions of ϵ, linearly independent as quadratic forms, and $\lim_{\epsilon\to 0}\mathcal{H}'^{(0)}(\epsilon) = \mathcal{H}'^{(0)}$, $\lim_{\epsilon\to 0}\mathcal{L}'^{(0)}_j(\epsilon) = \mathcal{L}'^{(0)}_j$. Thus the

Journal of Physics: Conference Series **512** (2014) 012012 doi:10.1088/1742-6596/512/1/012012

associated functions $A^{ij}(\epsilon), B^{ij}(\epsilon)$ will also be continuous functions of ϵ and $\lim_{\epsilon\to 0} A^{ij}(\epsilon) = A'^{ij}$, $\lim_{\epsilon\to 0} B^{ij}(\epsilon) = B'^{ij}$. Similarly, the integrability conditions for the potential equations

$$\begin{aligned} V^{(\epsilon)}_{22} &= V^{(\epsilon)}_{11} + A^{22}(\epsilon)V^{(\epsilon)}_1 + B^{22}(\epsilon)V^{(\epsilon)}_2, \\ V^{(\epsilon)}_{12} &= A^{12}(\epsilon)V^{(\epsilon)}_1 + B^{12}(\epsilon)V^{(\epsilon)}_2, \end{aligned} \tag{3}$$

will hold for each ϵ and in the limit. This means that the 4-dimensional solution space for the potentials V will deform continuously into the 4-dimensional solution space for the potentials V'. Thus the target space of solutions V' is uniquely determined by the free quadratic algebra contraction.

Example 1 *We describe the contraction of S9 to E1, including the potential terms. Recall for S9 in coordinates $x_1 = \psi, x_2 = \phi$ we have*

$$A^{12} = 0, \quad A^{22} = \frac{3\cosh^2\psi - \sinh^2\psi}{\sinh\psi\cosh\psi}, \quad B^{12} = 2\frac{\sinh\psi}{\cosh\psi},$$

$$B^{22} = -3\frac{(\cos^2\phi - \sin^2\phi)}{\sin\phi\cos\phi},$$

$$V = \frac{a_1\cosh^2\psi}{\cos^2\phi} + \frac{a_2\cosh^2\psi}{\sin^2\phi} + \frac{a_3\cosh^2\psi}{\sinh^2\psi} + a_4. \tag{4}$$

For E1 and using polar coordinates $y_1 = R, y_2 = \phi'$ where $x = e^R\cos\phi'$, $y = e^R\sin\phi'$, we have

$$A'^{12} = 0, \quad A'^{22} = -2, \quad B'^{12} = -2, \quad B'^{22} = -3\frac{(\cos^2\phi - \sin^2\phi)}{\sin\phi\cos\phi},$$

The general potential is

$$V' = b_1e^{2R} + \frac{b_2e^{-2R}}{\cos^2\phi} + \frac{b_3e^{-2R}}{\sin^2\phi} + b_4. \tag{5}$$

In terms of these coordinates the standard contraction of the sphere to flat space is expressed as $\psi \approx \frac{1}{2}\ln(\frac{1}{\epsilon}) - R$, $\phi = \phi'$. In the limit as $\epsilon \to 0$ we have

$$A^{12} \to A'^{12} = 0, \; A'^{22} \to 2 = -A'^{22}, \; B^{12} \to 2 = -B'^{12},$$

$$B^{22} = -3\frac{(\cos^2\phi - \sin^2\phi)}{\sin\phi\cos\phi} = B'^{22}.$$

The change in sign for A^{22} and B^{12} is due to the fact that y_1 corresponds to $-x_1$ whereas y_2 corresponds to x_2. In the limit the 4 dimensional space of potentials (4) must go to the 4 dimensional vector space (5). However the basis functions for the S9 potential,

$$\frac{\cosh^2\psi}{\cos^2\phi}, \; \frac{\cosh^2\psi}{\sin^2\phi}, \; \frac{\cosh^2\psi}{\sinh^2\psi}, \; 1$$

will not go to a new basis in the limit; 2 basis functions become unbounded and 2 go to a constant. There are many ways to choose an ϵ dependent basis so that the limit can be taken. One of the simplest choices of basis is

$$V^{(1)}(\epsilon) = \frac{1}{4\epsilon}(\frac{\cosh^2\psi}{\sinh^2\psi} - 1) \to e^{2R},$$

$$V^{(2)}(\epsilon) = \epsilon\frac{\cosh^2\psi}{\cos^2\phi} \to \frac{e^{-2R}}{\cos^2\phi}, \; V^{(3)}(\epsilon) = \epsilon\frac{\cosh^2\psi}{\sin^2\phi} \to \frac{e^{-2R}}{\sin^2\phi}, \; V^{(4)}(\epsilon) = 1 \to 1.$$

7. Models of superintegrable systems

- A representation of a quadratic algebra Q is a homomorphism of Q into the associative algebra of linear operators on some vector space: products go to products, commutators to commutators, etc.
- A model M is a faithful representation of Q in which the vector space is a space of polynomials in one complex variable and the action is via differential/difference operators acting on that space. We study classes of irreducible representations realized by these models.
- Suppose a quadratic algebra Q contracts to a algebra Q' via a continuous family of transformations indexed by ϵ. If we have a model M of Q we can try to "save" this representation by passing through a continuous family of models $M(\epsilon)$ of $Q(\epsilon)$ to obtain a model M' of Q'.
- There are three closely related limits tying one superintegrable system to another: 1) The pointwise coordinate limit of the source physical system to the target system. 2) The induced contraction of the source quadratic algebra to the target quadratic algebra. 3) The process of saving a representation of the target quadratic algebra by passing through a continuous family of models of representations of the source quadratic algebra, see Figure 1,
- As a byproduct of contractions to systems from $S9$ for which we save representations in the limit, we obtain the Askey Scheme for hypergeometric orthogonal polynomials. See Figure 2.

8. Hypergeometric polynomials and the Askey scheme

Recall, [2], that the Wilson polynomials are defined as

$$w_n(t^2) \equiv w_n(t^2, a, b, c, d) = (a+b)_n(a+c)_n(a+d)_n \times \tag{6}$$

$${}_4F_3\left(\begin{array}{llll} -n, & a+b+c+d+n-1, & a-t, & a+t \\ a+b, & a+c, & & a+d \end{array} ;1\right)$$

$$= (a+b)_n(a+c)_n(a+d)_n \Phi_n^{(a,b,c,d)}(t^2),$$

where $(a)_n$ is the Pochhammer symbol and ${}_4F_3(1)$ is a hypergeometric function of unit argument. The polynomial $w_n(t^2)$ is symmetric in a, b, c, d. For the finite dimensional representations the spectrum of t^2 is $\{(a+k)^2,\ k = 0, 1, \ldots, m\}$ and the orthogonal basis eigenfunctions are Racah polynomials. In the infinite dimensional case they are Wilson polynomials. They are eigenfunctions for the difference operator $\tau^*\tau$ defined via

$$\begin{aligned} \tau &= \frac{1}{2t}(E_t^{1/2} - E_t^{-1/2}), \\ \tau^* &= \frac{1}{2t}\left[(a+t)(b+t)(c+t)(d+t)E_t^{1/2} - (a-t)(b-t)(c-t)(d-t)E_t^{-1/2}\right], \end{aligned}$$

with $E_t^A F(t) = F(t+A)$.

The Askey Scheme, [12, 10], organizes the theory of hypergeometric orthogonal polynomials of one variable by exhibiting the relations such that each of these polynomials can be obtained as a sequence of pointwise limits from either the Racah polynomials in the finite dimensional case or the Wilson polynomials in the infinite dimensional case.

$$\lim_{\tau \to \infty} \Phi_n(\tau) = \Phi'_n.$$

The irreducible representations of $S9$ have a realization in terms of difference operators in 1 variable [5], exactly the structure algebra for the Wilson and Racah polynomials! By contracting these representations to obtain the representations of the quadratic symmetry algebras of the other superintegrable systems we obtain the full Askey scheme of orthogonal hypergeometric polynomials. This relationship ties the structure equations directly to physical phenomena. The full details of the contractions are given in [7]; our contribution here is to show how these contractions were induced in a natural and unique way from Lie algebra contractions which have clear physical and geometrical significance. In the following we just give some examples.

9. The $S9$ difference operator model

There is no model of the irreducible representations of the quadratic algebra $S9$ in terms of differential operators but there is a difference operator model [5]:

$$L_2 f_{n,m} = (-4t^2 - \frac{1}{2} + B_1^2 + B_3^2) f_{n,m},$$

$$L_3 f_{n,m} = (-4\tau^*\tau - 2[B_1+1][B_2+1] + \frac{1}{2}) f_{n,m},$$

$$H = L_1+L_2+L_3+\frac{3}{4}-(B_1^2+B_2^2+B_3^2) = -4(m+1)(B_1+B_2+B_3+m+1)-2(B_1B_2+B_1B_3+B_2B_3) + \frac{3}{4} - (B_1^2+B_2^2+B_3^2).$$

Here $n = 0, 1, \cdots, m$ if m is a nonnegative integer and $n = 0, 1, \cdots$ otherwise. Also

$$a_j = \frac{1}{4} - B_j^2,\ \alpha = -(B_1+B_3+1)/2 - m,\ \beta = (B_1+B_3+1)/2,$$

$$\gamma = (B_1 - B_3 + 1)/2,\ \delta = (B_1+B_3-1)/2 + B_2 + m + 2,$$

$$E^A F(t) = F(t+A),\ \tau = \frac{1}{2t}(E^{1/2} - E^{-1/2}),$$

$$\tau^* = \frac{1}{2t}\left[(\alpha+t)(\beta+t)(\gamma+t)(\delta+t)E^{1/2} - (\alpha-t)(\beta-t)(\gamma-t)(\delta-t)E^{-1/2}\right],$$

$$w_n(t^2) = (\alpha+\beta)_n(\alpha+\gamma)_n(\alpha+\delta)_n {}_4F_3\left(\begin{matrix} -n, & \alpha+\beta+\gamma+\delta+n-1, & \alpha-t, & \alpha+t \\ \alpha+\beta, & \alpha+\gamma, & \alpha+\delta & \end{matrix}; 1\right)$$

$$= (\alpha+\beta)_n(\alpha+\gamma)_n(\alpha+\delta)_n \Phi_n^{(\alpha,\beta,\gamma,\delta)}(t^2), \quad \Phi_n \equiv f_{n,m},$$

$$\tau^*\tau\Phi_n = n(n+\alpha+\beta+\gamma+\delta-1)\Phi_n,$$

where $(a)_n$ is the Pochhammer symbol and ${}_4F_3(1)$ is a hypergeometric function of unit argument. The polynomial $w_n(t^2)$ is symmetric in $\alpha, \beta, \gamma, \delta$. For the finite dimensional representations the spectrum of t^2 is $\{(\alpha+k)^2,\ k = 0, 1, \cdots, m\}$ and the orthogonal basis eigenfunctions are Racah polynomials. In the infinite dimensional case they are Wilson polynomials.

The action of L_2 and L_3 on an L_3 eigenbasis is

$$L_2 f_{n,m} = -4K(n+1,n) f_{n+1,m} - 4K(n,n) f_{n,m} - 4K(n-1,n) f_{n-1,m} + (B_1^2 + B_3^2 - \frac{1}{2}) f_{n,m},$$

$$L_3 f_{n,m} = -(4n^2 + 4n[B_1+B_2+1] + 2[B_1+1][B_2+1] - \frac{1}{2}) f_{n,m},$$

$$K(n+1,n) = \frac{(B_1+B_2+n+1)(n-m)(-B_3-m+n)(B_2+n+1)}{(B_1+B_2+2n+1)(B_1+B_2+2n+2)},$$

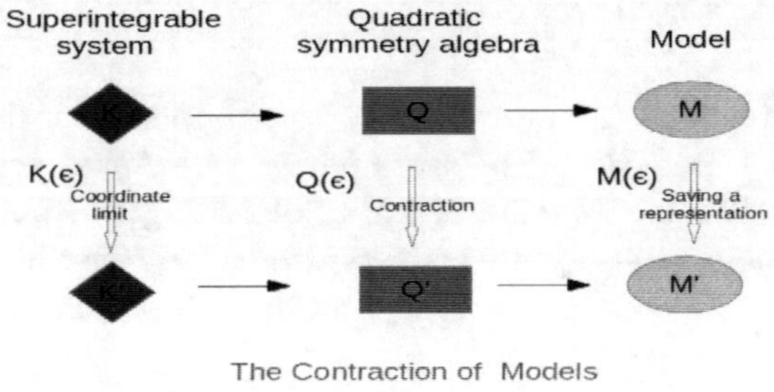

Figure 1.

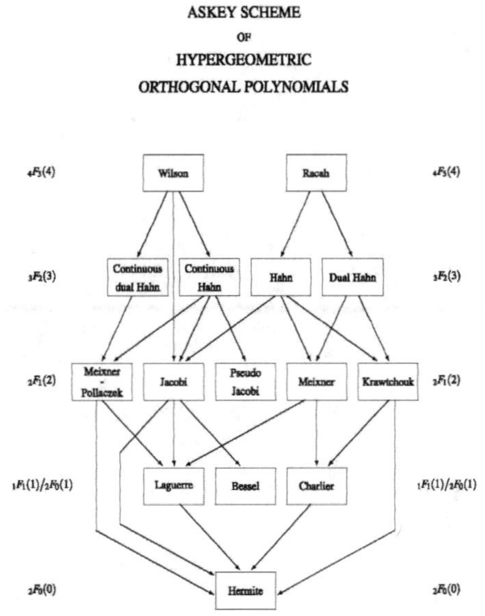

Figure 2.

$$K(n-1,n) = \frac{n(B_1+n)(B_1+B_2+B_3+m+n+1)(B_1+B_2+m+n+1)}{(B_1+B_2+2n)(B_1+B_2+2n+1)},$$

$$K(n,n) = [\frac{B_1+B_2+2m+1}{2}]^2 - K(n+1,n) - K(n-1,n),$$

We give an example showing how a contraction of one superintegrable system to another induces a similar contraction of models and recovers part of the Askey scheme. Our example is the contraction of $S9$ to $E1$. The full scheme of limits of orthogonal polynomials is recovered through sequences of contractions of superintegrable systems, starting from $S9$.

Quantum system limit:

$$H_{S9} = J_1^2 + J_2^2 + J_3^2 + \frac{a_1}{s_1^2} + \frac{a_2}{s_2^2} + \frac{a_3}{s_3^2}$$

where $J_3 = s_1\partial_{s_2} - s_2\partial_{s_1}$ and J_2, J_3 are obtained by cyclic permutations of the indices $1, 2, 3$.

$$H_{E1} = \partial_x^2 + \partial_y^2 - \omega^2(x^2+y^2) + \frac{b_1}{x^2} + \frac{b_2}{y^2}$$

In $S9$ we contract about the north pole of the unit sphere. Set

$$s_1 = \sqrt{\epsilon}x,\ s_2 = \sqrt{\epsilon}y,\ s_3 = \sqrt{1-s_1^2-s_2^2} \approx 1 - \frac{\epsilon}{2}(x^2+y^2),$$

$$a_1' = b_2 = a_1,\ a_2' = b_1 = a_2,\ a_3' = -\omega^2 = \epsilon^2 a_3,$$

in H_{S9} to get $\epsilon(H_{S9} - a_3) \to H_{E1}$ as $\epsilon \to 0$.

Quadratic algebra contraction:

$$L_1' = \epsilon L_1,\ L_2' = \epsilon L_2,\ L_3' = L_3,\ H' = \epsilon(H - a_3)$$

$$R' = \epsilon R,\ a_1' = b_2 = a_1,\ a_2' = b_1 = a_2,\ a_3' = -\omega^2 = \epsilon^2 a_3.$$

Saving a representation: We set

$$t = -x + B_3/2 + (B_1+1)/2 + m,\ B_3 = \frac{\omega}{\epsilon} \to \infty \implies$$

$$f_{n,m}' = {}_3F_2\left(\begin{matrix} -n, & B_1+B_2+n+1, & -x \\ -m, & B_2+1 & \end{matrix};1\right) = Q_n(x; B_2, B_1, m)$$

where the Q_n are Hahn polynomials. We have the model

$$L_2' f_{n,m}' = 2\omega(2x-2m-B_1-1)f_{n,m}' = -4K'(n+1,n)f_{n+1,m}' - 4K'(n,n)f_{n,m}' - 4K'(n-1,n)f_{n-1,n}',$$

$$L_3' f_{n,m}' = -\left(4n^2 + 4n[B_1+B_2+1] + 2[B_1+1][B_2+1] - \frac{1}{2}\right) f_{n,m}' =$$

$$\Big[-4(x-m)(x+B_2+1)E_x^1 + 4x(x-m-B_1-1)E_x^{-1} + 8x^2 + 4x(B_1+B_2-2m)$$

$$-4m(B_2+1) - 2(B_2+1)(B_1+1) + \frac{1}{2}\Big] f_{n,m}',$$

$$H' = L_1' + L_2' = -2\omega(2m+2+B_1+B_2).$$

Here the K' are the appropriate limits of the K as $B_3 \to \infty$.

See Figures 3 and 4 for the contraction description of the Askey Scheme.

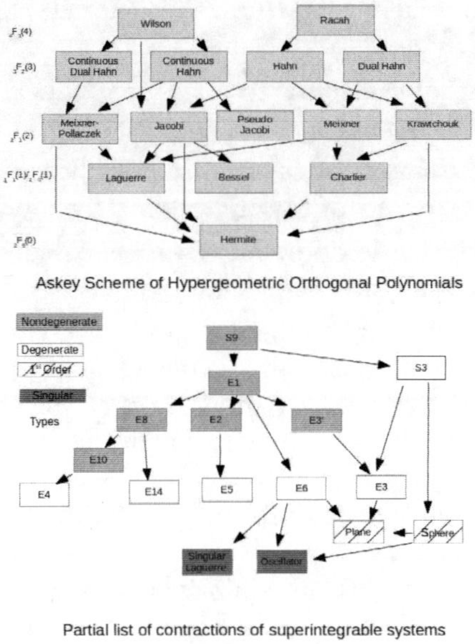

Figure 3. The Askey scheme and contractions of superintegrable systems

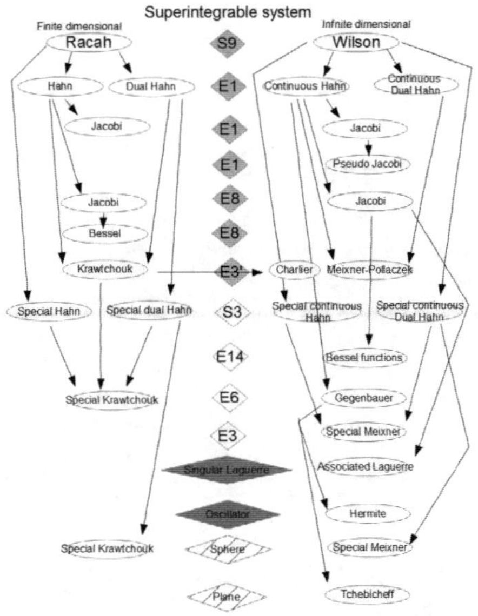

Figure 4. The Askey contraction scheme

10. Observations and conclusions

- Free quadratic algebras uniquely determine associated superintegrable systems with potential.
- A contraction of a free quadratic algebra to another uniquely determines a contraction of the associated superintegrable systems.
- For a 2D superintegrable systems on a constant curvature space these contractions can be induced by Lie algebra contractions of the underlying Lie symmetry algebra.
- Every 2D superintegrable system is obtained either as a sequence of contractions from $S9$ or is Stäckel equivalent to a system that is so obtained.
- Taking contractions step-by-step from the $S9$ model we can recover the Askey Scheme. However, the contraction method is more general. It applies to all special functions that arise from the quantum systems via separation of variables, not just polynomials of hypergeometric type, and it extends to higher dimensions [6]. The special functions arising from the models can be described as the coefficients in the expansion of one separable eigenbasis for the original quantum system in terms of another separable eigenbasis. The functions in the Askey Scheme are just those hypergeometric polynomials that arise as the expansion coefficients relating two separable eigenbases that are *both* of hypergeometric type. Thus, there are some contractions which do not fit in the Askey scheme since the physical system fails to have such a pair of separable eigenbases.
- The details of the Askey Scheme derivation can be found in [7]. The origin of the complicated multiparameter contractions was not clear in that paper. In this paper we have demonstrated that all of these contractions were uniquely induced by the contractions of the Lie algebras $e(2, \mathbf{C})$, $o(3, \mathbf{C})$. Details will follow in [8]. There are only a small number of these Lie algebra contractions and their action on physical space is well known.
- Even though 2nd order 2D nondegenerate superintegrable systems admit no group symmetry, their structure is determined completely by the underlying symmetry of constant curvature spaces.
- To extend the method to Askey-Wilson polynomials we would need to find appropriate q-quantum mechanical systems with q-symmetry algebras and we have not yet been able to do so.

Acknowledgments

This work was partially supported by a grant from the Simons Foundation (# 208754 to Willard Miller, Jr.).

References

[1] A.A.Izmest'ev, G.S.Pogosyan, A.N.Sissakian, and P.Winternitz. Contractions of lie algebras and the separation of variables. *J. Phys. A*, 29:5940–5962, 1996.

[2] G. E. Andrews, R. Askey, and R. Roy. *Special Functions. Encyclopedia of Mathematics and its Applications.* Cambridge University Press, 1999.

[3] E. Inönü and E. P. Wigner. On the contraction of groups and their representations. *Proc. Nat. Acad. Sci.*, 39:510–524, 1953.

[4] E. G. Kalnins, J.M. Kress, W. Miller Jr. , and G. S. Pogosyan. Completeness of superintegrability in two-dimensional constant curvature spaces. *J. Math Phys.*, 34:4705–472, 2001.

[5] E. G. Kalnins, W. Miller Jr., and S. Post. Wilson polynomials and the generic superintegrable system on the 2-sphere. *J. Math Phys. A*, 40:11525–11538, 2007.

[6] E. G. Kalnins, W. Miller Jr., and S. Post. Two-variable Wilson polynomials and the generic superintegrable system on the 3-sphere. *SIGMA*, 7:051, 2011.

[7] E. G. Kalnins, W. Miller, Jr, and S. Post. Contractions of 2d 2nd order quantum superintegrable systems and the askey scheme for hypergeometric orthogonal polynomials. *SIGMA*, 9:057, 28 pages, arXiv:1212.4766v1 [math–ph], 2013.

[8] E. G. Kalnins, W. Miller. Jr., E. Subag, and R. Heinenen. Contractions of 2nd order superintegrable systems in 2d. *(In preparation)*, 2013.

[9] E.G. Kalnins, Jr. W. Miller, and G.S. Pogosyan. Contractions of lie algebras and special function identities. *J. Phys. A*, 32:4709–4732, 1999.

[10] R. Koekoek, P. A. Lesky, and R. F. Swarttouw. *Hypergeometric orthogonal polynomials and their q-analogues.* Springer Monographs in Mathematics. Springer-Verlag, 2010.

[11] G. Koenigs. *Lecons sur la théorie générale des surfaces*, volume III, chapter Sur les géodésiques a intégrales quadratiques, pages 368–404. Chelsea Publishing, 1872. In book by G. Darboux.

[12] T. H. Koornwinder. Group theoretic interpretations of askey's scheme of hypergeometric orthogonal polynomials. In *Orthogonal polynomials and their applications (Segovia, 1986)*, volume 1329 of *Lecture Notes in Math.*, pages 46–72, Berlin, 1988.

[13] W Miller Jr., S. Post, and P.. Winternitz. Classical and quantum superintegrability with applications. *J. Phys. A: Math. Theor.*, 2013. A 97 page review article.

[14] E.. Saletan. Contractions of lie groups. *J. Math. Phys.*, 2:1–21, 1961.

[15] J. Talman. *Special Functions: A Group Theoretic Approach (based on the lecture notes of Eugene Wigner).* W.A. Benjamin, New York, 1968.

[16] P. Tempesta, P. Winternitz, W. Miller, and G. Pogosyan, editors. *Superintegrability in Classical and Quantum Systems*, volume 37 of *CRM Proceedings and Lecture Notes*, Providence, RI, 2004. AMS.

[17] E. Weimar-Woods. The three-dimensional real lie algebras and their contractions. *J. Math. Phys.*, 32:2028–2033, 1991.

Journal of Physics: Conference Series **512** (2014) 012013 doi:10.1088/1742-6596/512/1/012013

Canonical Transformations in Crystals

E. Sadurní

Instituto de Física, Benemérita Universidad Autónoma de Puebla, Apartado Postal J-48, 72570 Puebla, México

E-mail: `sadurni@ifuap.buap.mx`

Abstract. The space representations of linear canonical transformations were studied by Moshinsky and Quesne in 1972. For a few decades, the bilinear hamiltonian remained as the only exactly solvable representative for such problems. In this work we show that the Mello-Moshinsky equations can be solved exactly for a class of problems with discrete symmetry, leading to exact propagators for Wannier-Stark ladders in one and two dimensional crystals. We give a detailed study for a particle in a triangular lattice under the in uence of a time-dependent electric field. A more general set of Mello-Moshinsky equations for arbitrary lattices is presented.

1. Introduction

It is well known that the study of quantum-mechanical representations of canonical transformations was initiated with Bargmann's work [1], although Dirac's book [2] already gave a hint to this important subject. In 1972 the problem was revisited by Moshinsky and Quesne [3], who found that the space representations of linear canonical transformations in arbitrary dimensions were given by gaussian kernels. It was then clear that the propagator associated to any bilinear hamiltonian – in space and momentum – could be obtained explicitly with group theoretical methods, in this case the symplectic group. Notably, the harmonic oscillator with time-dependent mass and frequency belongs to this category [5].

For four decades, the bilinear hamiltonian remained as the only example for which the transformation equations – also known as the Mello-Moshinsky (MM) equations [6] – could be solved analytically in terms of elementary functions [1].

In this contribution we show that new solutions for the representations of evolution maps can be obtained in the context of tight-binding models. In particular, we shall find the propagator of a particle on a lattice under the influence of a time-dependent homogeneous field. The corresponding MM equations shall be written as recurrence relations on the *discrete* coordinates of the system, finding thereby that the kernels can be written in terms of Bessel functions or their multi-index generalizations [7].

From the physical point of view, we may also point out that the study of the evolution of wave packets in crystals has a long history of its own, which started with Bloch's work in 1928 [8]. Solid state physics and, quite recently, matter waves in lattice traps [9] serve as a motivation to our work.

[1] An amusing example involving a quadratic canonical transformation and Airy functions comes from the time-energy variables (t, H) of a particle in a linear potential [4]. Such a transformation does not connect (x, p) with (t, H) continuously, and shall not be considered in our discussions.

Journal of Physics: Conference Series **512** (2014) 012013 doi:10.1088/1742-6596/512/1/012013

In connection with canonical variables in crystals, we underscore the fact that such operators *can* be defined by a careful construction in terms of position operators and crystal translations.[2] Moreover, a proper definition of Lie derivatives has been given in [11] through generalizations of the Leibniz rule, contemplating local and non-local operators. In section 2 we take care of this point by providing the canonical and non-canonical maps that set the foundations in further developments. In section 3 we formulate the discrete MM equations for canonical transformations in one-dimensional lattices and obtain explicitly the propagator of a model hamiltonian, namely the particle in discrete space with time-dependent hopping parameters and a time-dependent homogeneous field. Section 4 deals with two-dimensional generalizations of the problem: triangular and square lattices. There, we obtain the corresponding MM equations and their solution. In section 5 we formulate the problem using tight-binding lattices in arbitrary dimensions and coordination numbers. We conclude briefly in section 6.

2. Independent vs. canonical variables in one-dimensional lattices

The study of canonical transformations as a function of time is usually based on the invariance of $[x, p]$. We show first that a canonical pair in discrete variables can be constructed. Then we propose another pair which is non-canonical, but whose commutator transforms canonically; through such a pair we will be able to perform computations easily in further sections. A schematic view of the procedure is given in figures 1 and 2.

2.1. A canonical pair in discrete space

The first question arising in the context of crystalline structures is whether one can define a momentum operator canonically conjugate to a discrete position. The concept of derivative is obviously not applicable here, but a path offers itself through the central discretization of the derivative. Let us define our units in the form $\hbar = 1$ and denote by a the lattice spacing parameter with units of length. Then we have the following example of canonical operators ξ, π in terms of the usual continuous-space operators x and $p = -i\partial/\partial x$:

$$\xi \equiv \frac{1}{2}\left\{\sec(ap), x\right\}, \qquad \pi \equiv \frac{1}{a}\sin(ap). \tag{1}$$

One can easily check that $a \to 0$ leads to an appropriate continuous limit and that the new operators ξ, π satisfy the canonical commutator

$$[\xi, \pi] = i \qquad \Longleftrightarrow \qquad [x, p] = i. \tag{2}$$

Although the map (1) is not invertible –we shall discuss this later on– its functional form guarantees that unitary transformations $UxU^\dagger, UpU^\dagger$ representing arbitrary canonical maps [3] induce canonical transformations on the pair ξ, π. In fact, the transformation rule (1) can be written as a power series of x and p:

$$\xi = \sum_{n=0}^{\infty} \frac{(ia)^{2n} E_{2n}}{2(2n)!}\left\{p^{2n}, x\right\}, \qquad \pi = \sum_{n=0}^{\infty} \frac{a^{2n}}{(2n+1)!}p^{2n+1}, \tag{3}$$

where E_{2n} denotes the Euler number. Therefore one can easily show that

$$\xi' \equiv \xi\left(UxU^\dagger, UpU^\dagger\right) = U\xi U^\dagger, \tag{4}$$

[2] Interesting possibilities were provided in [10]. In this paper we follow a different approach.

[3] We shall focus on unitary representations of canonical transformations, embracing probability conservation as our guiding principle.

$$\pi' \equiv \pi\left(UxU^{\dagger}, UpU^{\dagger}\right) = U\pi U^{\dagger}. \tag{5}$$

Moreover, the meaning of these operators can be found by their application to localized (or atomic) states. Let us denote such sates by $|n\rangle$, where $n \in \mathbf{Z}$ is the site number and $\psi_n(x) = \langle x|n\rangle$ represents the Wannier function [4] around site n. We have

$$\begin{aligned} \xi\psi_n(x) &= \sum_{m=0}^{\infty}(-)^m\left(2x-[2m+1]\,a\right)\psi_n\left(x-[2m+1]\,a\right) \\ &= \sum_{m=0}^{\infty}(-)^m\left(2x-[2m+1]\,a\right)\psi_{n+2m+1}\left(x\right), \end{aligned} \tag{6}$$

$$\begin{aligned} \pi\psi_n(x) &= \frac{\psi_n(x+a)-\psi_n(x-a)}{2ia} \\ &= \frac{\psi_{n-1}(x)-\psi_{n+1}(x)}{2ia}. \end{aligned} \tag{7}$$

By virtue of these relations and the transformations (4), (5), the study of canonical transformations in crystals through discrete variables is justified; the action of ξ and π on functions of a continuous variable x can be written explicitly as an operation on the discrete variable n, as indicated by the following vector representation of (6) and (7):

$$\xi|n\rangle = a\sum_{m=0}^{\infty}(-)^m\left[2(n-m)-1\right]|n+2m+1\rangle, \tag{8}$$

$$\pi|n\rangle = \frac{1}{2ia}|n-1\rangle - \frac{1}{2ia}|n+1\rangle. \tag{9}$$

Finally, through a map such as (1), it is fair to restrict our considerations to the invariance of the commutator $[\xi,\pi]$ rather than $[x,p]$. Thus, our first question can be answered.

2.2. A convenient pair of non-canonical variables

For practical reasons one might be interested in non-canonical variables whose transformations are more amenable; according to (6) the operator ξ acts non-locally on lattice functions, giving complicated expressions for the matrix elements $\langle n|\xi|m\rangle$. This problem was already noted in [11]. Other operators given by arbitrary functions $N(x,p)$ and $T(x,p)$ might allow a simpler analysis of canonical transformations by means of a covariant $[N,T]$. A careful explanation of this process is given in [4], where it was pointed out that the elements of U could be obtained by analyzing the transformation equations of N and T, albeit $[N,T] \neq i$. It turns out that in the case of crystals there is a particularly useful set of operators N,T of this type:

$$N = \frac{x}{a}, \qquad T = \exp\left(-iap\right), \tag{10}$$

with the algebraic properties

[4] Only one band is considered for simplicity. The band index of the Wannier function shall be omitted.

$$[N,T]=T, \qquad TT^{\dagger}=T^{\dagger}T=\mathbf{1}. \tag{11}$$

These operators have a simple interpretation in terms of lattice coordinates, as they act on atomic states in the form

$$N|n\rangle = n|n\rangle, \qquad T|n\rangle = |n+1\rangle. \tag{12}$$

By means of (10) we may infer that any unitary transformation of x, p leads again to $N' = UNU^{\dagger}$, $T' = UTU^{\dagger}$, ensuring the covariance of the commutator $[N', T'] = T'$. Furthermore, the relations in (12) show that the matrix elements of these operators are considerably simpler than those of ξ and π:

$$N_{n,m} = n\delta_{n,m}, \qquad T_{n,m} = \delta_{n,m+1}. \tag{13}$$

Finally, our study of the representations $\langle x|U|x'\rangle$ can be reduced to the analysis of $U_{n,m} \equiv \langle n|U|m\rangle$.

2.3. Non-invertibility and ambiguity groups

The transformation equations (1), (10) and the property $T^{\dagger} = 1/T$ imply that the map $(N,T) \mapsto (\xi, \pi)$ can be written explicitly as

$$\xi(N,T) = a\left\{\left(T+T^{\dagger}\right)^{-1}, N\right\} = a\sum_{m=0}^{\infty}(-)^{m}\left(2\left[N-m\right]+1\right)T^{2m+1},$$
$$\pi(T) = \frac{1}{2ia}\left(T+T^{\dagger}\right) = \frac{1}{2ia}\left(T+\frac{1}{T}\right). \tag{14}$$

This map cannot be inverted for N and T, since it is 2 to 1. Moreover, the transformation (10) relating (N,T) with the original variables (x,p), maps an infinite number of Riemann sheets to one plane. According to [4] and references cited therein, this gives rise to what is known as *ambiguity spin*. In other words, the space representation of U is multiply defined by all matrix elements of the type $\langle x|U|x'\rangle \exp\left[2\pi i q(x-x')/a\right]$, where $q \in \mathbf{Z}$. This can be seen more clearly by recognizing that in the space (x,p) all the preimages of the form

$$p_q \equiv p + \frac{2\pi q}{a} = \exp\left[-i\frac{2\pi q x}{a}\right]\, p\, \exp\left[i\frac{2\pi q x}{a}\right] \tag{15}$$

lead to the same image (N,T). An infinite dimensional matrix with spinorial indices q, q' should take care of the ambiguity and it can be proposed in the form

$$U_{n,m}^{q,q'} = \int\int dx dx' \psi_n^*(x)\psi_m(x')\langle x|U|x'\rangle \times \exp\left[i\frac{2\pi(qx-q'x')}{a}\right]\delta_{q,q'}. \tag{16}$$

Such $U_{n,m}^{q,q'}$ represents a transformation between (N,T) and (N',T') in discrete coordinates n and m, just as $\langle x|U|x'\rangle$ represents the corresponding transformation between (x,p) and (x',p') in the Hilbert space $\{\psi_n(x)\}$. The ambiguity of the crystalline group, *i.e.* $(\mathbf{Z},+)$, is therefore represented irreducibly by $\exp\left[i\frac{2\pi q(x-x')}{a}\right]\delta_{q,q'}$.

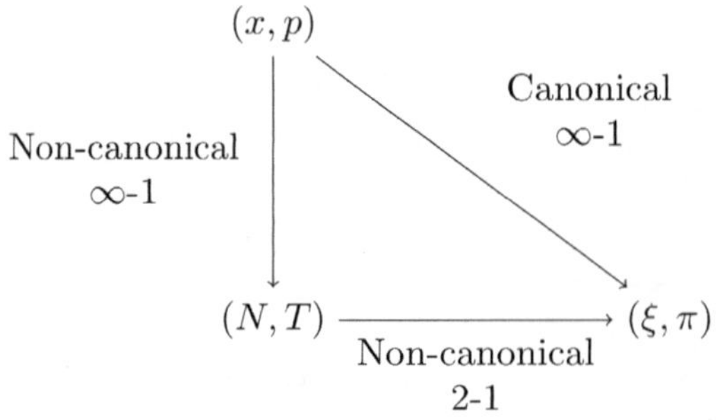

Figure 1. Commutative diagram showing the maps (1), (10) and (14).

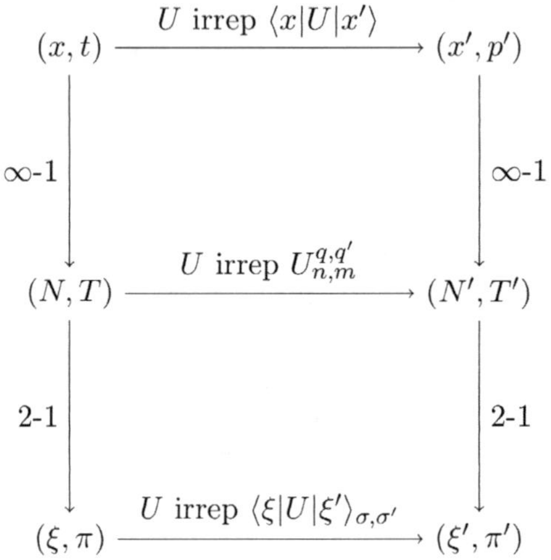

Figure 2. Diagram showing the canonical transformation U and its Hilbert space representations, for each pair of variables. In this work we study the map $N' = UNU^\dagger$ $T' = UTU^\dagger$ represented by $U^{q,q'}_{n,m}$. The third level contains extra indices $\sigma = \pm$, $\sigma' = \pm$ due to a 2-1 map.

To summarize, we have three maps relating the pairs (x, p), (ξ, π) and (N, T). The diagram in fig.1 commutes, but the morphisms are not necessarily one-to-one. The ambiguity spin comes to our rescue, and in fact it is possible to work with the representation $U^{q,q'}_{n,m}$ alone. The three levels at which the canonical transformation is represented in Hilbert space are indicated in the diagram of fig.2.

In the rest of this paper we shall work with $q = 0$. We recognize that such a choice is equivalent to a particular phase factor of the (tight-binding) functions $\psi_n(x)$; the r.h.s. of (16) shows it clearly, for the replacement $\psi_n(x) \mapsto e^{i2\pi qx/a}\psi_n(x)$ does the required job.

Ultimately, the MM equations for a canonical transformation can be written for any pair of independent variables (N, T) [4] and we shall proceed in this direction in what follows.

Journal of Physics: Conference Series **512** (2014) 012013 doi:10.1088/1742-6596/512/1/012013

3. The discrete version of the MM equations

3.1. Transformation equations with a continuous parameter

In full analogy with what is done for the harmonic oscillator and the associated symplectic group [3], let us consider a set of transformation equations $N'(N,T)$ and $T'(N,T)$ which represent the evolution of the operators (N,T) under some hamiltonian.

Since the pair (N,T) is now our main focus, it should be stressed that if $T^\dagger = 1/T$, *i.e.* if $T^\dagger$ is not independent of T, the commutator $\left[N, T^\dagger\right] = -T^\dagger$ is redundant. With this in mind, we propose the non-linear transformation

$$N' = \alpha T + \frac{\alpha^*}{T} + \beta N, \tag{17}$$

$$T' = \gamma T + \frac{\bar{\gamma}}{T} + \delta N, \tag{18}$$

where $\alpha, \gamma, \bar{\gamma}$ and δ are complex functions of a continuous parameter t (time) and β is real. The conditions $T'T'^\dagger = T'^\dagger T' = \mathbf{1}$ and $[N', T'] = T'$ lead to restrictions on such functions, *i.e.*

$$\gamma\beta - \alpha\delta = 1, \qquad \alpha^*\delta = \bar{\gamma}\beta, \qquad \delta\gamma^* = \delta^*\gamma, \qquad \delta\bar{\gamma}^* = \delta^*\bar{\gamma}. \tag{19}$$

From (17), (18) and the unitary operator U acting from the right, we have

$$UN = \alpha TU + \alpha^* T^\dagger U + \beta NU, \tag{20}$$

$$UT = \gamma TU + \bar{\gamma} T^\dagger U + \delta NU. \tag{21}$$

Finally, the matrix elements of both relations (20), (21) can be found using the atomic states $\langle n|$, $|m\rangle$ and (13) :

$$mU_{n,m} = \alpha U_{n-1,m} + \alpha^* U_{n+1,m} + \beta n U_{n,m}, \tag{22}$$

$$U_{n,m+1} = \gamma U_{n-1,m} + \bar{\gamma} U_{n+1,m} + \delta n U_{n,m}. \tag{23}$$

These are two coupled recurrence relations in the indices n, m and their solution is the unitary representation of the canonical transformation in discrete coordinates. We shall refer to it as the discrete MM equations. Specific choices of the coefficients $\alpha, \beta, \gamma, \delta, \bar{\gamma}$ yield solutions in terms of Bessel functions [12, 13]. A concrete example is given in the following section.

3.2. The Stark Hamiltonian

Suppose now that a concrete hamiltonian is given. For a particle in discrete space with a time-dependent hopping parameter α and under the influence of a time-dependent homogeneous field β, we have

$$H = \alpha(t)T + \alpha^*(t)T^\dagger + \beta(t)N. \tag{24}$$

Journal of Physics: Conference Series **512** (2014) 012013 doi:10.1088/1742-6596/512/1/012013

A canonical transformation is obtained by means of the unitary evolution, therefore we analyze the equations of motion for N and T:

$$\dot{T} = i\beta(t)T, \qquad \dot{T}^\dagger = -i\beta(t)T^\dagger, \qquad \dot{N} = i\alpha^*(t)T^\dagger - i\alpha(t)T. \tag{25}$$

The solutions of (25) yield the evolution map, which results in

$$T(t) = e^{if(t)}T(0), \qquad T^\dagger(t) = e^{-if(t)}T^\dagger(0), \qquad N(t) = F^*(t)T^\dagger + F(t)T + N(0), \tag{26}$$

where we have used the definitions $f(t) = \int^t \beta(t)dt$ and $F(t) = -\int^t i\alpha(t)e^{if(t)}dt$. The discrete MM equations corresponding to such a map are a particular case of (22) and (23) in the previous section. By computing the corresponding matrix elements, we obtain the recurrences

$$(m-n)U_{n,m} = F(t)U_{n-1,m} + F^*(t)U_{n+1,m}, \tag{27}$$

$$U_{n,m+1} = e^{if(t)}\,U_{n-1,m}, \tag{28}$$

and their solutions are explicitly given by

$$U_{n,m}(t) = e^{imf(t)}\left[\frac{F(t)}{F^*(t)}\right]^{(n-m)/2} J_{m-n}\left(2|F(t)|\right). \tag{29}$$

Here J denotes a Bessel function of the first kind; the Bessel function of the second kind Y is not allowed, as it does not fulfill the requirement $U_{n,m}(0) = \delta_{n,m}$ due to its irregularity at the origin.

4. The MM equations in two-dimensional lattices and their solution

The next non-trivial step is to increase the number of dimensions. For crystalline sheets with tight-binding hamiltonians and a time-dependent field, there are two examples that offer themselves: the square lattice and the triangular lattice. We shall see that the topological differences between these tessellations lead to different canonical maps; therefore, the corresponding propagators are different.

4.1. Square lattice immersed in an external field

Let $T_i = \exp(-ia_ip_i)$ be the translation operator along the i-th primitive vector ($i = 1$ for $\mathbf{i}$ and $i = 2$ for $\mathbf{j}$) with lattice spacings a_i for each direction. For a field with components $\beta_i(t)$, the hamiltonian of interest has the form

$$H = \alpha_1(t)T_1 + \alpha_2(t)T_2 + \text{h.c.} + \beta_1(t)N_1 + \beta_2(t)N_2. \tag{30}$$

At this point it is convenient to introduce an appropriate Hilbert space: we denote the atomic states by $|n_1, n_2\rangle$ where n_1, n_2 are integers. We have

$$N_1|n_1,n_2\rangle = n_1|n_1,n_2\rangle, \qquad N_2|n_1,n_2\rangle = n_2|n_1,n_2\rangle, \tag{31}$$

$$T_1|n_1,n_2\rangle = |n_1+1,n_2\rangle, \qquad T_2|n_1,n_2\rangle = |n_1,n_2+1\rangle. \tag{32}$$

It is important to note that H can be separated in two parts which commute: a part which contains operators with index 1 and another with index 2. The coordination number $c = 4$ makes possible a separation of variables and we expect that the resulting MM equations can be decoupled in two sets of independent equations. In fact, we have

$$(m_1 - n_1)U_{n_1,m_1;n_2,m_2} = F_1(t)U_{n_1-1,m_1;n_2,m_2} + F_1^*(t)U_{n_1+1,m_1;n_2,m_2}, \quad (33)$$
$$(m_2 - n_2)U_{n_1,m_1;n_2,m_2} = F_2(t)U_{n_1,m_1;n_2-1,m_2} + F_2^*(t)U_{n_1,m_1;n_2+1,m_2}, \quad (34)$$

$$U_{n_1,m_1+1;n_2,m_2} = e^{if_1(t)}\, U_{n_1-1,m_1;n_2,m_2}, \quad (35)$$
$$U_{n_1,m_1;n_2,m_2+1} = e^{if_2(t)}\, U_{n_1,m_1;n_2-1,m_2}, \quad (36)$$

which are easily solved by

$$U_{n_1,m_1;n_2,m_2}(t) = \prod_{j=1,2} e^{im_j f_j(t)} \left[\frac{F_j(t)}{F_j^*(t)}\right]^{(n_j-m_j)/2} J_{m_j-n_j}\left(2|F_j(t)|\right). \quad (37)$$

The parameters are given by $f_j(t) = \int^t \beta_j(t)dt$ and $F_j(t) = -\int^t i\alpha_j(t)e^{if_j(t)}dt$.

4.2. Triangular lattice immersed in an external field

This example is crucial in our understanding of lattices. Although the corresponding MM equations can be solved, we shall see that their solutions cannot be expressed as a simple product. Take primitive lattice vectors $\mathbf{a}_1 = \frac{1}{2}(\sqrt{3}\mathbf{i} + 3\mathbf{j})$, $\mathbf{a}_2 = -\sqrt{3}\mathbf{i}$ and $\mathbf{a}_3 = -\mathbf{a}_1 - \mathbf{a}_2$, such that any lattice point can be expressed as $\mathbf{A} = n_1\mathbf{a}_1 + n_2\mathbf{a}_2$. The translation operators are given by $T_i = \exp(-i\mathbf{a}_i \cdot \mathbf{p}_i)$ for each direction, with the additional property $T_3 = T_1^\dagger T_2^\dagger$. With this notation, a hamiltonian analogous to (30) but with coordination number $c = 6$ can be proposed:

$$H = \sum_{i=1}^{3} \alpha_i(t)T_i + \text{h.c.} + \beta_1(t)N_1 + \beta_2(t)N_2. \quad (38)$$

The action of the operators N_i, T_i on the atomic states $|n_1, n_2\rangle$ is, as before,

$$N_1|n_1,n_2\rangle = n_1|n_1,n_2\rangle, \qquad N_2|n_1,n_2\rangle = n_2|n_1,n_2\rangle, \quad (39)$$
$$T_1|n_1,n_2\rangle = |n_1+1,n_2\rangle, \qquad T_2|n_1,n_2\rangle = |n_1,n_2+1\rangle. \quad (40)$$

However, in the triangular case we have the additional relations

$$T_3|n_1,n_2\rangle = |n_1-1,n_2-1\rangle, \qquad [T_3, N_j] = T_3, \quad j = 1,2. \quad (41)$$

Let us define $\beta_3 = -\beta_1 - \beta_2$ and $f_i(t) = \int^t \beta_i(t)dt$ for $i = 1,2,3$. Similarly, take $F_j = -\int i\alpha_j(t)e^{if_j(t)}dt$ for $j = 1,2$, but define the third function as $F_3 = +\int i\alpha_3(t)e^{if_3(t)}dt$, *i.e.* with a reversed sign. With these definitions, the dynamical map generated by H can be written in compact form as

$$T_i(t) = e^{if_i(t)} T_i(0), \quad i = 1,2,3. \tag{42}$$

$$N_j(t) = F_j(t)T_j(0) + F_3(t)T_3(0) + \text{h.c.} + N_j(0), \quad j = 1,2. \tag{43}$$

The relation (43) differs considerably from the case of a square lattice. The discrete MM equations are now

$$\begin{aligned}(m_1 - n_1)U_{n_1,m_1;n_2,m_2} &= F_1(t)U_{n_1-1,m_1;n_2,m_2} + F_1^*(t)U_{n_1+1,m_1;n_2,m_2}, \\ &+ F_3(t)U_{n_1+1,m_1;n_2+1,m_2} + F_3^*(t)U_{n_1-1,m_1;n_2-1,m_2},\end{aligned} \tag{44}$$

$$\begin{aligned}(m_2 - n_2)U_{n_1,m_1;n_2,m_2} &= F_2(t)U_{n_1,m_1;n_2-1,m_2} + F_2^*(t)U_{n_1,m_1;n_2+1,m_2}, \\ &+ F_3(t)U_{n_1+1,m_1;n_2+1,m_2} + F_3^*(t)U_{n_1-1,m_1;n_2-1,m_2},\end{aligned} \tag{45}$$

and

$$U_{n_1,m_1+1;n_2,m_2} = e^{if_1(t)}\, U_{n_1-1,m_1;n_2,m_2}, \tag{46}$$

$$U_{n_1,m_1;n_2,m_2+1} = e^{if_2(t)}\, U_{n_1,m_1;n_2-1,m_2}, \tag{47}$$

where we have omitted the recurrence arising from $T_3(t) = e^{if_3(t)}T_3(0)$, which can be proved redundant. We may attempt a separable solution of the form

$$U_{n_1,m_1;n_2,m_2} = \exp\left[i\frac{(n_1+m_1)f_1(t)}{2} + i\frac{(n_2+m_2)f_2(t)}{2}\right] \times K_{m_1-n_1,m_2-n_2}(t), \tag{48}$$

which solves (46) and (47). Substitution of (48) in (44) and (45) gives the simplified recurrences

$$nK_{n,m} = F_1 e^{-i\frac{f_1}{2}} K_{n+1,m} + F_1^* e^{i\frac{f_1}{2}} K_{n-1,m} + F_3 e^{-i\frac{f_3}{2}} K_{n-1,m-1} + F_3^* e^{i\frac{f_3}{2}} K_{n+1,m+1}, \tag{49}$$

$$mK_{n,m} = F_2 e^{-i\frac{f_2}{2}} K_{n,m+1} + F_2^* e^{i\frac{f_2}{2}} K_{n,m-1} + F_3 e^{-i\frac{f_3}{2}} K_{n-1,m-1} + F_3^* e^{i\frac{f_3}{2}} K_{n+1,m+1}. \tag{50}$$

Therefore, we have reduced a problem of four variables n_1, n_2, m_1, m_2 to two variables $n = m_1 - n_1$ and $m = m_2 - n_2$. The two recurrence relations (49), (50) are indeed simple; they can be solved by generalizations of Bessel functions in full analogy with the recurrence (27), but now extended to two indices and three variables. One of such generalizations is of particular interest, as it was studied in [7] in a completely different context. Suppose now that the parameters in (49) are such that

$$-\arg(F_3) = \arg(F_1) + \arg(F_2). \tag{51}$$

This restriction of the parameters is not completely unnatural, as it includes the free problem in the triangular lattice and the problem of equal couplings $\alpha_1 = \alpha_2 = \alpha_3$ and constant external field $\dot{\beta}_i = 0$, among other interesting possibilities. Therefore we may use (51) to further reduce (49) and (50), arriving at the following relations

$$nG_{n,m} = |F_1| \left(G_{n+1,m} + G_{n-1,m}\right) + |F_3| \left(G_{n-1,m-1} + G_{n+1,m+1}\right), \tag{52}$$

$$mG_{n,m} = |F_2| \left(G_{n,m+1} + G_{n,m-1}\right) + |F_3| \left(G_{n-1,m-1} + G_{n+1,m+1}\right). \tag{53}$$

These relations are uniquely solved (recall our intial condition $U(0) = \mathbf{1}$) by Bessel functions of two indices, *i.e.* $G_{n,m}(t) = J^{(+,-)}_{n,m}\left(2|F_1(t)|, 2|F_2(t)|, 2|F_3(t)|\right)$, which is a special case of (14) in [7]. Thus, the full solution for our unitary operator reads

$$\begin{aligned} U_{n_1,m_1+1;n_2,m_2} &= \exp\left[i\frac{(n_1+m_1)f_1(t)}{2} + i\frac{(n_2+m_2)f_2(t)}{2}\right] \\ &\times \exp\left[i(m_1-n_1)\Phi_1 + i(m_2-n_2)\Phi_2\right] \\ &\times J^{(+,-)}_{m_1-n_1,m_2-n_2}\left(2|F_1(t)|, 2|F_2(t)|, 2|F_3(t)|\right), \end{aligned} \tag{54}$$

with the abbreviation $\Phi_j = f_j(t)/2 - \arg[F_j(t)]$.

It is important to note that the two-index, three-parameter Bessel function $J^{(+,-)}_{n,m}$ comes from an exponential generating function of the type

$$e^{[x(a-1/a)+y(b-1/b)+z(ab-1/ab)]/2} = \sum_{n,m\in\mathbf{Z}} a^n b^m J^{(+,-)}_{n,m}(x,y,z). \tag{55}$$

This simple origin of our special functions brings forth a number of properties and relations that can be obtained for our propagator (54). In particular, the partial derivatives of J with respect to x, y, z can be employed to show that (54) is indeed the Green's function of the time-dependent Schrödinger equation with hamiltonian (38). Other properties such as expansions in products of ordinary Bessel functions and ascending series in x, y, z may be helpful in the numerical evaluation of the kernel.

5. The discrete MM equations for arbitrary lattices

Now that we have solved two relevant examples, we proceed to generalize our treatment to many dimensions and coordination numbers. Although we do not solve explicitly the resulting relations for the general case, we may sketch the path towards a reduction of the problem. Consider a set of translation operators $T_i = \exp\left(-i\mathbf{a}_i \cdot \mathbf{p}\right)$ along lattice vectors $\{\mathbf{a}_i\}_{i=1,\dots,c}$ with c the coordination number. Let the primitive vectors of the lattice be the subset $\{\mathbf{a}_j\}_{j=1,\dots,d}$ with d the dimension of the space; usually we have $d \leq c$. Finally, let us denote by $\{N_j\}_{j=1,\dots,d}$ the operators of discrete positions along the primitive vectors a_j.

With this notation, consider now the hamiltonian

$$H = \sum_{i=1}^{c} \alpha_i(t) T_i + \text{h.c.} + \sum_{j=1}^{d} \beta_j(t) N_j. \tag{56}$$

The evolution map on the operators N_j, T_i can be obtained by virtue of the commutation relations

$$[N_j, T_i] = \delta_{ji} T_i, \quad \text{if} \quad i \leq d \tag{57}$$

Journal of Physics: Conference Series **512** (2014) 012013 doi:10.1088/1742-6596/512/1/012013

and

$$[N_j, T_l] = q_l T_l, \quad \text{for some} \quad l > d. \tag{58}$$

The last relation is clearly determined by the topology of the lattice, for which some of the neighbouring sites can be connected only by linear combinations of the primitive vectors and not by the primitives themselves – the triangular lattice is a clear example with $q_3 = -1$. We may encode both (57) and (58) in a function σ for simplicity:

$$[N_j, T_i] = \sigma_{ji} T_i, \quad 1 \le j \le d, \, 1 \le i \le c. \tag{59}$$

From this commutator, it is straightforward to find the Heisenberg equations of motion and extract the dynamical map. We obtain

$$T_i(t) = e^{if_i(t)} T_i(0), \tag{60}$$

$$N_j(t) = N_j(0) + \sum_{i=1}^{c} F_{ji}(t) T_i(0) + F_{ji}^*(t) T_i^\dagger(0). \tag{61}$$

The definition of parameters is now

$$f_i(t) = \sum_{j=1}^{d} \sigma_{ji} \int_0^t \beta_j(t) dt, \qquad F_{ji}(t) = -i\sigma_{ji} \int_0^t \alpha_i(t) e^{if_i(t)} dt. \tag{62}$$

At this stage it is convenient to introduce the atomic states of the lattice $|\{m_i\}_{i=1,\dots,d}\rangle$ and $\langle\{n_i\}_{i=1,\dots,d}|$. With these states, we are able to compute the matrix elements of (61) and finally write down the corresponding MM equations:

$$U_{n_1 \dots n_d; m_1 \dots m_i+1 \dots m_d} = e^{if_i(t)} U_{n_1 \dots n_i-1 \dots n_d; m_1 \dots m_d}, \tag{63}$$

$$(m_j - n_j) U_{n_1 \dots n_d; m_1 \dots m_d} = \sum_{i=1}^{c} \left[F_{ji}(t) U_{n_1 \dots n_i-1 \dots n_d; m_1 \dots m_d} + F_{ji}^*(t) U_{n_1 \dots n_i+1 \dots n_d; m_1 \dots m_d} \right], \tag{64}$$

which is a set of d recurrences. As before, we have omitted a number $c-d$ of redundant equations arising from $T_l(t), l = d, \dots, c$. Repeating our previous steps for low-dimensional problems, the first equation can be directly solved by products of the type

$$U_{n_1 \dots n_d; m_1 \dots m_d} = \exp\left(\frac{i}{2} \sum_{j=1}^{d} (n_j + m_j) f_j(t) \right) K_{m_1-n_1, \dots, m_d-n_d}(t). \tag{65}$$

The final relation for K becomes a set of d recurrences for which no precedent can be found in the literature. It seems plausible that new generalizations of Bessel functions may arise from such equations. In all, we can say that the representations of a certain type of canonical transformations in arbitrary lattices have been characterized as the solutions of a specific finite-difference equation of many indices, namely the relations (63) and (64).

6. Conclusion

Our efforts have led us to the construction of discrete representations of canonical transformations and their explicit forms by means of dynamical maps. In the first part we developed the necessary techniques in one-dimensional lattices, finding thereby that the propagator of a particle in discrete space with time-dependent site couplings and a time-modulated homogeneous field can be written in terms of Bessel functions. In full analogy with such a simple example, we went on and obtained two new propagators for two-dimensional lattices immersed in a time-dependent field. Finally, we indicated the path to follow in the case of multidimensional lattices of an arbitrary topology.

With this, it is fair to say that particles in crystals (in the basis of Wannier functions) are susceptible of phase space analysis, and that a number of relevant models can be solved explicitly in time domain. In particular, the works [12, 13] are devoted to examples which include diffraction in discrete space and the problem of pulse propagation under external fields. The resulting physical effects are related to caustic control, providing a generalization of Bloch oscillations.

Acknowledgments

I am grateful to A. Turbiner for bringing his work to my attention. Financial support from PROMEP Project 103.5/12/4367 is acknowledged.

References

[1] Bargmann V 1968 Group Representations on Hilbert Spaces of Analytic Functions *Analytic Methods of Mathematical Physics* ed P P Gilbert and R G Newton (Gordon and Breach, New York).
[2] Dirac P A M 1958 *The Principles of Quantum Mechanics* (Clarendon Press, Oxford).
[3] Moshinsky M and Quesne C 1971 *J Math Phys* **12** 1772.
[4] Moshinsky M 1981 *AIP Conf Proc* **71** 312.
[5] Grosche C and Steiner F 1998 *Handbook of Feynman Path Integrals* (Springer).
[6] Moshinsky M and Smirnov Y F 1996 *The Harmonic Oscillator in Modern Physics* (Harwood Academic Publishers, The Netherlands).
[7] Dattoli G, *et al.* 1994 *J Math Phys* **35** (7) 3636-3649.
[8] Bloch F 1928 *Z Phys* **52** 555-600.
[9] Bloch I 2005 *Nature Physics* **1** 23-30.
[10] Smirnov Y and Turbiner A 1995 *Mod Phys Lett A* **10** 1795.
[11] Levi D, Negro J and del Olmo M A 2001 *J Phys A: Math Gen* **34** 2023-2030.
[12] Sadurní E 2012 *J Phys A: Math Theor* **45** 465302.
[13] Sadurní E 2013 *J Phys A: Math Theor* **46** 135302.

Journal of Physics: Conference Series **512** (2014) 012014 doi:10.1088/1742-6596/512/1/012014

F_4 Quantum Integrable, rational and trigonometric models: space-of-orbits view

A.V. Turbiner and J.C. López Vieyra

Instituto de Ciencias Nucleares,
Universidad Nacional Autónoma de México, 04510 Mexico D.F.

E-mail: turbiner@nucleares.unam.mx, vieyra@nucleares.unam.mx

Abstract. Algebraic-rational nature of the four-dimensional, F_4-invariant integrable quantum Hamiltonians, both rational and trigonometric, is revealed and reviewed. It was shown that being written in F_4 Weyl invariants, polynomial and exponential, respectively, both similarity-transformed Hamiltonians are in algebraic form, they are quite similar the second order differential operators with polynomial coefficients; the flat metric in the Laplace-Beltrami operator has polynomial (in invariants) matrix elements. Their potentials are calculated for the first time: they are meromorphic (rational) functions with singularities at the boundaries of the configuration space. Ground state eigenfunctions are algebraic functions in a form of polynomials in some degrees. Both Hamiltonians preserve the same infinite flag of polynomial spaces with characteristic vector $(1, 2, 2, 3)$, it manifests exact solvability. A particular integral common for both models is derived. The first polynomial eigenfunctions are presented explicitly.

1. Introduction

In four dimensional Euclidian space there exist several remarkable completely integrable quantum systems originally discovered in the Hamiltonian reduction method (for a review, see [1]). These are particular cases of the Calogero-Moser-Sutherland systems and all of them are characterized by a discrete symmetry given by Weyl group W, acting on root spaces A_4, BC_4, F_4 and H_4. In such models, the potential has four possible forms (rational ($\sim 1/x^2$), trigonometric ($\sim 1/(\sin x)^2$), hyperbolic ($\sim 1/(\sinh x)^2$) and elliptic ($\sim \wp(x)$)). All these $4D$ Hamiltonians are exactly solvable in terms of Weyl-invariant variables. For rational and trigonometric (also hyperbolic) systems, exact solvability was obtained in an explicit constructive fashion in [2, 3, 4, 5]. The spectrum is found explicitly, in closed analytic form being a first/second degree polynomial in the quantum numbers for the rational/trigonometric systems respectively.

The purpose of the present paper is to revisit the F_4-invariant model in its rational and trigonometric forms showing how it looks like in the space of orbits, in the space of F_4 invariants. For the rational F_4 model it is mainly reduced to a review the results obtained in previous studies [4] (see also [6]) and their reinterpretation. In addition, we show for the first time the explicit form of the ground state eigenfunction and the original potential in terms of Weyl invariant variables. For the F_4 trigonometric model we present a more succinct solution compared to our previous analysis [4] (see also [7]), and we also present for the first time the explicit form of the ground state eigenfunction and the original potential in terms of trigonometric Weyl invariant variables. For both F_4 rational and trigonometric models the extra, particular integral [8] will be derived and it will be shown that is the same in the space-of-orbits. Eventually, it will be given a place to both models in the space-of-orbit formalism [9, 10].

2. Generalities

The F_4-Weyl-invariant Calogero-Moser-Sutherland Hamiltonian is defined as follows

$$H = -\frac{1}{2}\sum_{j=1}^{N}\frac{\partial^2}{\partial x_j^2} + \frac{1}{2}\sum_{\alpha\in\mathcal{F}_4{}^+} g_{|\alpha|}|\alpha|^2\,\mathcal{V}((\alpha\cdot x))\ , \tag{1}$$

where $\mathcal{F}_4{}^+$ is the set of positive roots, $N = 4$ is the rank of the root system (the dimension of the configuration space), and $g_{|\alpha|}$ are coupling constants. For roots of the same length the same coupling constant $g_{|\alpha|}$ is assigned. In (1) x denotes the ($N = 4$)-dimensional vector $x = (x_1, x_2, x_3, x_4)$. The potential consists of a linear superposition of the terms $\mathcal{V}((\alpha\cdot x))$ of the argument $(\alpha\cdot x)$ where $\alpha\in\mathcal{F}_4{}^+$ is a positive root (see [1]). For the rational case the four-dimensional isotropic harmonic oscillator potential is usually added to the Hamiltonian (1).

The set $\mathcal{F}_4{}^+$ consists of 24 positive roots:

$$\begin{aligned} \alpha_{\text{short}} &= e_1, e_2, e_3, e_4,\ \tfrac{1}{2}(e_1 \pm e_2 \pm e_3 \pm e_4)\,, \quad |\alpha_{\text{short}}| = 1\,, \\ \alpha_{\text{long}} &= e_i \pm e_j, \quad i > j \quad i, j = 1, \ldots 4\,, \qquad |\alpha_{\text{long}}| = \sqrt{2}\,, \end{aligned} \tag{2}$$

where e_1, e_2, e_3, e_4 are the standard orthonormal vectors in the 4-dimensional Euclidean space. From a geometrical viewpoint, the short root vectors in (2) are, in fact, the vertices of the 4-dimensional 24-cell regular polytope (see [11]), while the long roots are the vertices of its dual 24-cell. The rotational symmetry group of the 24-cell has the order 576. The full symmetry group of the 24-cell is the F_4 Weyl group (of order 1152), which is generated by reflections on the hyperplanes orthogonal to the F_4 roots.

3. F_4 Rational Model

In a four-dimensional Euclidean space with coordinates $x = (x_1, x_2, x_3, x_4)$ the Hamiltonian of the F_4 Rational Model is given by

$$\begin{aligned} H_{F_4}^{\rm rat} &= \frac{1}{2}\sum_{i=1}^{4}(-\partial_{x_i}^2 + \omega^2 x_i^2) + g_\ell \sum_{j<i=1}^{4}\left[\frac{1}{(x_i+x_j)^2} + \frac{1}{(x_i-x_j)^2}\right] \\ &+ \frac{g_s}{2}\sum_{i=1}^{4}\frac{1}{x_i^2} + 2g_s \sum_{\nu's=0,1} \frac{1}{[x_1 + (-)^{\nu_2}x_2 + (-)^{\nu_3}x_3 + (-)^{\nu_4}x_4]^2}, \end{aligned} \tag{3}$$

with $\sum \nu's$ = even. The parameters g_ℓ, g_s are coupling constants associated with *long* and *short* roots respectively, and ω is the frequency of $4D$-isotropic harmonic oscillator potential. The configuration space is the F_4-Weyl Chamber.

The ground state eigenfunction of the Hamiltonian (3) has the form

$$\psi_0(x) = (\Delta_+ \Delta_-)^\nu (\Delta\, \Delta_0)^\mu \, e^{-\frac{1}{2}\omega \sum_{i=1}^4 x_i^2} . \tag{4}$$

where

$$\Delta_+ \Delta_- = \prod_{i<j=1}^{4}(x_i \pm x_j), \quad \Delta\, \Delta_0 = \frac{1}{256}\prod_{i=1}^{4} x_i(x_1 \pm x_2 \pm x_3 \pm x_4), \tag{5}$$

are generalized Vandermonde determinants. The exponents ν, μ of the ground state eigenfunction (4) are connected to the coupling constants g_ℓ , g_s by the relations

$$g_\ell = \nu(\nu-1), \; g_s = \mu(\mu-1), \tag{6}$$

and the ground state energy is

$$E_0 = 2\omega(1 + 6\mu + 6\nu). \tag{7}$$

The eigenfunctions $\psi_n(x)$ of the Hamiltonian (3) have a factorized form

$$\psi_n(x) = \psi_0(x)\phi_n(x), \tag{8}$$

where the common factor $\psi_0(x)$ is the ground state eigenfunction (4) and n is multi-index given by four quantum numbers of the eigenstate. The functions $\phi_n(x)$ in (8) are eigenfunctions of the transformed Hamiltonian

$$\mathfrak{h}_{F_4}^{\rm rat} \equiv -2{\psi_0}^{-1}\left[H_{F_4}^{\rm (rat)} - E_0\right]\psi_0, \tag{9}$$

Then, the eigenvalue problem becomes

$$\mathfrak{h}_{F_4}^{\rm rat}\,\phi_n = \epsilon_n \phi_n ,$$

Journal of Physics: Conference Series **512** (2014) 012014 doi:10.1088/1742-6596/512/1/012014

where $\epsilon_n = -2(E_n - E_0)$.

The Hamiltonian (9) acquires an algebraic form (*i.e.* the form of a second order differential operator with polynomial coefficients) when it is written in terms of an algebraically independent set of Weyl invariant variables defined as

$$t_a^{(\Omega)} = \frac{1}{12}\sum_{k=1}^{|\Omega|}(\alpha_k, x)^a, \qquad \alpha_k \in \Omega\,, \tag{10}$$

where $a = 2, 6, 8, 12$ are the degrees of the polynomial invariants of the F_4 Weyl group, and α_k are the roots of a conveniently chosen Weyl orbit Ω. In the present study, Ω was chosen as the orbit generated by the *highest root* [1] $e_3 + e_4$ with size $|\Omega| = 24$. As a matter of fact, this orbit consists of all (positive and negative) long roots. The factor 1/12 in the definition (10) was introduced for convenience.

The fact that the invariants of given degrees (10) are defined up to polynomials in invariants of lower degree was used in [6] to obtain a particular set of invariants of fixed degrees τ_a, $(a{=}2,6,8,12)$ leading to a simple algebraic form of the Hamiltonian. These invariants are related to orbital ones (10) in the following manner

$$\begin{aligned}
\tau_2 &= t_2^{(\Omega)}\,, \quad \tau_6 = \frac{1}{12}t_6^{(\Omega)} - \frac{1}{12}\big(t_2^{(\Omega)}\big)^3\,,\\
\tau_8 &= \frac{1}{80}t_8^{(\Omega)} - \frac{1}{30}t_2^{(\Omega)}t_6^{(\Omega)} + \frac{1}{48}\big(t_2^{(\Omega)}\big)^4\,,\\
\tau_{12} &= \frac{1}{720}t_{12}^{(\Omega)} - \frac{5}{288}\big(t_2^{(\Omega)}\big)^2 t_8^{(\Omega)} + \frac{1}{27}\big(t_2^{(\Omega)}\big)^3 t_6^{(\Omega)} - \frac{29}{1440}\big(t_2^{(\Omega)}\big)^6 - \frac{1}{1080}\big(t_6^{(\Omega)}\big)^2\,,
\end{aligned} \tag{11}$$

and are given explicitly in Appendix A. In terms of the $\tau_2, \tau_6, \tau_8, \tau_{12}$, the Hamiltonian (9)

$$\mathfrak{h}^{\rm rat}_{F_4}(\tau) = \mathcal{A}_{\mu\eta}\frac{\partial^2}{\partial\tau_\mu\partial\tau_\eta} + \mathcal{B}_\mu\frac{\partial}{\partial\tau_\mu}\,, \tag{12}$$

takes the algebraic form where the coefficient functions $\mathcal{A}_{\mu,\eta}, \mathcal{B}_\mu$ are polynomial in τ_a,

$$\begin{aligned}
\mathfrak{h}^{\rm rat}_{F_4}(\tau) =& 4\tau_2\frac{\partial^2}{\partial\tau_2^2} + \frac{2}{3}\tau_2(\tau_2\tau_6 + 10\tau_8)\frac{\partial^2}{\partial\tau_6^2} + 2(\tau_2\tau_{12} + 2\tau_6\tau_8)\frac{\partial^2}{\partial\tau_8^2} + 6\tau_8(\tau_2\tau_{12} + 2\tau_6\tau_8)\frac{\partial^2}{\partial\tau_{12}^2}\\
&+ 24\tau_6\frac{\partial^2}{\partial\tau_2\partial\tau_6} + 32\tau_8\frac{\partial^2}{\partial\tau_2\partial\tau_8} + 48\tau_{12}\frac{\partial^2}{\partial\tau_2\partial\tau_{12}} + \frac{8}{3}({\tau_2}^2\tau_8 + 6\tau_{12})\frac{\partial^2}{\partial\tau_6\partial\tau_8}\\
&+ 4({\tau_2}^2\tau_{12} + 8{\tau_8}^2)\frac{\partial^2}{\partial\tau_6\tau_{12}} + 4(2\tau_2{\tau_8}^2 + 3\tau_6\tau_{12})\frac{\partial^2}{\partial\tau_8\tau_{12}}\\
&- 4[\omega\tau_2 - 2(6\nu + 6\mu + 1)]\frac{\partial}{\partial\tau_2} - [12\omega\tau_6 - {\tau_2}^2(4\nu + 2\mu + 1)]\frac{\partial}{\partial\tau_6}\\
&- 4[4\omega\tau_8 - \tau_6(1 + 3\nu)]\frac{\partial}{\partial\tau_8} - 4[6\omega\tau_{12} - \tau_2\tau_8(2 + 3\nu)]\frac{\partial}{\partial\tau_{12}}\,.
\end{aligned} \tag{13}$$

[1] Since the root $e_1 + e_2 \in \Omega(e_3 + e_4)$, the orbit actually coincides with the orbit used in the previous studies[4, 6].

This algebraic Hamiltonian has infinitely many invariant polynomial subspaces of the form

$$\mathcal{P}_n^{(1,2,2,3)} = \langle \tau_2^{p_1}\tau_6^{p_3}\tau_8^{p_4}\tau_{12}^{p_6} |\ 0 \le 1p_1 + 2p_3 + 2p_4 + 3p_6 \le n\rangle\,, \quad n = 0, 1, 2, \ldots \tag{14}$$

which are marked by the characteristic vector $(1, 2, 2, 3)$ of minimal possible *gradings* for the invariants $\tau_2, \tau_6, \tau_8, \tau_{12}$, respectively. Incidentally, this characteristic vector coincides with the F_4 highest root among short roots when written in the basis of simple roots. Interestingly, the invariant subspaces (14) form an infinite flag

$$\mathcal{P}_0^{(1,2,2,3)} \subset \mathcal{P}_1^{(1,2,2,3)} \subset \mathcal{P}_2^{(1,2,2,3)} \subset \ldots \subset \mathcal{P}_n^{(1,2,2,3)} \subset \ldots\,, \tag{15}$$

which we call *minimal flag*. Since the algebraic Hamiltonian $\mathfrak{h}_{F_4}^{\rm rat}(\tau)$ (τ stands for $\tau_2, \tau_6, \tau_8, \tau_{12}$) preserves this infinite flag (15), thus it is exactly solvable (for a discussion see e.g. [6]). The eigenfunctions from the spaces $\mathcal{P}_n^{(1,2,2,3)}$, $n = 0, 1, 2$ are presented in Appendix B.

There exists an algebra of differential operators for which the space $\mathcal{P}_n^{(1,2,2,3)}$ is the finite-dimensional irreducible representation space [4]. This algebra is called $f^{(4)}$. This algebra is infinite-dimensional but finitely-generated. The rational F_4 Hamiltonian in algebraic form $\mathfrak{h}_{F_4}^{\rm rat}(\tau)$ can be rewritten in terms of the generators of this algebra (see [4]).

3.1. Flat Metric of the F_4-rational Model.

Gauge rotated Hamiltonian $\mathfrak{h}_{F_4}^{\rm rat}$ (9) in algebraic form (12) can be written as

$$\mathfrak{h}_{F_4}^{\rm rat} \ = \ \Delta_g \ + \ \mathcal{C}^b \frac{\partial}{\partial \tau_b}\,, \tag{16}$$

where the Laplace-Beltrami operator

$$\Delta_g \ = \ \frac{1}{g^{1/2}}\frac{\partial}{\partial \tau_a}\left(g^{ab}g^{1/2}\right)\frac{\partial}{\partial \tau_b} \ = \ g^{ab}\frac{\partial^2}{\partial \tau_a \partial \tau_b} \ + \ g^b\frac{\partial}{\partial \tau_b}\,,\ g^b \equiv \frac{1}{g^{1/2}}\frac{\partial}{\partial \tau_a}\left(g^{ab}g^{1/2}\right) \tag{17}$$

with $a, b = 2, 6, 8, 12$ and g is determinant. The symmetric in $a \to b$ metric, $g^{ab} = g^{ba}$ has polynomial in τ matrix elements

$$g^{ab} \ = \ \left|\begin{array}{cccc} 4\tau_2 & 12\tau_6 & 16\tau_8 & 24\tau_{12} \\ & \frac{2}{3}\tau_2(\tau_2\tau_6 + 10\tau_8) & \frac{4}{3}(\tau_2{}^2\tau_8 + 6\tau_{12}) & 2(\tau_2{}^2\tau_{12} + 8\tau_8{}^2) \\ & & 2(\tau_2\tau_{12} + 2\tau_6\tau_8) & 2(2\tau_2\tau_8{}^2 + 3\tau_6\tau_{12}) \\ & & & 6\tau_8(\tau_2\tau_{12} + 2\tau_6\tau_8) \end{array}\right| \ . \tag{18}$$

Journal of Physics: Conference Series **512** (2014) 012014 doi:10.1088/1742-6596/512/1/012014

This metric has a special property: any component g^b of the vector $\vec{g}$ in (17) is also polynomial in τ

$$\vec{g} \;=\; (8, \tau_2^2, \tau_6, 2\tau_2\tau_6)\;.$$

The same property has the vector $\mathcal{C}^b$ (see (16)) which characterizes the interaction in (3)

$$-\vec{\mathcal{C}} \;=\; \Big(4[\omega\tau_2 - 12(\nu+\mu)],\; 2[6\omega\tau_6 - (2\nu+\mu){\tau_2}^2],\; 4[4\omega\tau_8 - 3\nu\tau_6],\; 12[2\omega\tau_{12} - \nu\tau_2\tau_8]\Big)$$

3.2. Spectrum of the F_4-rational Model.
Since the invariants $\tau_2, \tau_6, \tau_8, \tau_{12}$ have a natural ordering given by their degrees, the Hamiltonian (13) has a triangular form in the basis of monomials $\tau_2^{n_1}\tau_6^{n_2}\tau_8^{n_3}\tau_{12}^{n_4}$ The spectrum is easily obtained:

$$\epsilon_{n_1,n_2,n_3,n_4} = -4\omega(2n_1 + 6n_2 + 8n_3 + 12n_4)\;,\; n_{1,2,3,4} = 0, 1, 2, \dots\;. \tag{19}$$

The spectrum (19) does not depend on couplings g_ℓ, g_s and is equidistant with a degeneracy corresponding to the partitions of integer $N = 2n_1 + 6n_2 + 8n_3 + 12n_4$. So, the interactions in (3) have the only effect of modifying the degeneracy of the energies corresponding to the 4-dimensional isotropic harmonic oscillator part of the Hamiltonian (3).

3.3. Ground state eigenfunction and Potential in τ variables
The ground state eigenfunction (4) and the original potential can be written in terms of the invariant variables (11). After straightforward but sometimes rather tedious calculations we find that

$$(\Delta_+\,\Delta_-)^2 \;=\; 64\,(-3\,\tau_{12}^2 + 4\,\tau_8^3) \equiv 64\;P_1(\tau)\;,$$

$$\begin{aligned}(\Delta_0\,\Delta)^2 =& \frac{1}{4096}(-192\,\tau_{12}^2 + 256\,\tau_8^3 + 144\,\tau_6^2\,\tau_{12} - 27\,\tau_6^4 - 192\,\tau_2\,\tau_6\,\tau_8^2\\ &+ 48\,\tau_2^2\,\tau_8\,\tau_{12} + 30\,\tau_2^2\,\tau_6^2\,\tau_8 - 12\tau_2^3\,\tau_6\,\tau_{12} + \frac{1}{2}\tau_2^3\,\tau_6^3\\ &+ \tau_2^4\,\tau_8^2 - \frac{1}{2}\tau_2^5\,\tau_6\,\tau_8 + \frac{1}{6}\tau_2^6\,\tau_{12}) \equiv \frac{1}{2^{12}}P_2(\tau)\;,\end{aligned}$$

hence, the unnormalized ground state eigenfunction (4) is

$$\psi_0(\tau) \;=\; P_1(\tau)^{\nu/2}P_2(\tau)^{\mu/2}\,e^{-\frac{1}{2}\omega\tau_2}\;, \tag{20}$$

(cf. (4)).

The potential $V_{F_4}^{\rm rat}$ can be represented as

$$V_{F_4}^{\rm rat} = \omega^2 V_\omega + g_\ell V_\nu + g_s V_\mu\;, \tag{21}$$

where

$$V_\omega \;=\; \frac{\tau_2}{2}\,, \tag{22}$$

$$V_\nu \;=\; -\,9\,\frac{(\tau_2\,\tau_{12} - 2\,\tau_6\,\tau_8)\,\tau_8}{P_1(\tau)}\,, \tag{23}$$

$$V_\mu \;=\; -\,\frac{1}{8}\,\frac{({\tau_2}^4 - 48\,\tau_2\,\tau_6 + 192\,\tau_8)\left({\tau_2}^3\tau_8 - 3\,\tau_2({\tau_6}^2 + 8\,\tau_{12}) + 48\,\tau_6\,\tau_8\right)}{P_2(\tau)}\,. \tag{24}$$

Thus, the original potential (3) is a rational (meromorphic) function in τ-variables. From the expressions (23) and (24), it is clear that the singularities of the potential are defined by the zeroes of the (ground state) factors $P_1(\tau), P_2(\tau)$ (cf.(5)), and therefore they define the boundaries of the configuration space.

3.4. sl(2) quasi-exactly-solvable F_4-invariant rational model

It can be shown [6] that the polynomial generalization of the potential $V_{F_4}^{(\mathrm{rat})}$ (21),

$$V_{F_4}^{(\mathrm{qes})}(\tau) \;=\; V_{F_4}^{(\mathrm{rat})} + a^2\tau_2^3 + 2a\omega\tau_2^2 + 2a[2k - \gamma + 3(4\mu + 4\nu + 1)]\tau_2 + \frac{2\gamma[\gamma - 12\mu - 12\nu - 1)]}{\tau_2}\,,$$

where a, γ are real parameters, k is non-negative integer, μ, ν are given in (6), leads to the so-called $sl(2)$ quasi-exactly-solvable F_4-invariant rational Hamiltonian

$$H_{F_4}^{(\mathrm{qes})} \;=\; -\Delta_g(\tau) + V_{F_4}^{(\mathrm{qes})}(\tau)\,.$$

This Hamiltonian has a single finite-dimensional invariant subspace $\mathcal{P}_k^{(1,2,2,3)}$ (see (14)). Hence, $(k+1)$ eigenfunctions have the form

$$\Psi_k^{(\mathrm{qes})} \;=\; (\Delta_-\Delta_+)^\nu\,(\Delta_0\Delta)^\mu\,\tau_2^\gamma P_k(\tau_2)\;\exp\left[-\omega\tau_2 - \frac{a}{4}\tau_2^2\right] \tag{25}$$

$$=\; \tau_2^\gamma P_k(\tau_2)\;\exp\left(-\frac{a}{4}\tau_2^2\right)\Psi_0 \;\equiv\; P_k(\tau_2)\Psi_0^{(\mathrm{qes})}\,,$$

where $P_k(\tau_2)$ is a kth degree polynomial, and can be found algebraically. It can be easily checked that

$$\mathfrak{h}_{F_4}^{(\mathrm{qes})}(\tau) \;=\; (\Psi_0^{(\mathrm{qes})})^{-1}\left[H_{F_4}^{(\mathrm{qes})} - E_0\right]\Psi_0^{(\mathrm{qes})}\,,$$

is algebraic operator.

3.5. Particular integral
Let us consider a certain Euler-Cartan operator in τ-variables

$$\mathcal{J}^0 \;=\; \tau_1\frac{\partial}{\partial\tau_1} + 2\tau_2\frac{\partial}{\partial\tau_2} + 2\tau_3\frac{\partial}{\partial\tau_3} + 3\tau_4\frac{\partial}{\partial\tau_4}\ .$$

and form

$$i_{par}^{(k)}(\tau) \;=\; \prod_{j=0}^{k}(\mathcal{J}^0 - j)\ . \tag{26}$$

It is easy to check that

$$i_{par}^{(k)}(\tau)\ :\ \mathcal{P}_k^{(1,2,2,3)}\ \mapsto\ 0\ .$$

So, $\mathcal{P}_k^{(1,2,2,3)}$ is the space of zero modes for $i_{par}^{(k)}(\tau)$. Hence, this operator commutes with both $\mathfrak{h}_{F_4}^{\rm rat}$ and $\mathfrak{h}_{F_4}^{\rm qes}$,

$$[\mathfrak{h}_{F_4}^{\rm rat}\ (\mathfrak{h}_{F_4}^{\rm qes})\ ,\ i_{par}^{(k)}(\tau)] :\ \mathcal{P}_k^{(1,2,2,3)}\ \mapsto\ 0\ .$$

The operator $i_{par}^{(k)}$ is called a particular integral.

4. F_4 Trigonometric Model

The Hamiltonian of the quantum F_4 trigonometric model is given by [2]

$$\begin{aligned} H_{F_4}^{\rm trig} \;=\; & -\frac{1}{2}\sum_{k=1}^{4}\frac{\partial^2}{\partial x_k^2} + g_l\frac{\beta^2}{4}\sum_{1\le j<i\le 4}\left(\frac{1}{\sin^2\frac{\beta}{2}(x_i-x_j)} + \frac{1}{\sin^2\frac{\beta}{2}(x_i+x_j)}\right) \\ & + g_s\frac{\beta^2}{8}\sum_{i=1}^{4}\frac{1}{\sin^2\frac{\beta}{2}x_i} + g_s\frac{\beta^2}{8}\sum_{\nu_j=0,1}\frac{1}{\sin^2\frac{\beta}{4}\Big(x_1+(-)^{\nu_2}x_2\ +\ (-)^{\nu_3}x_3+(-)^{\nu_4}x_4\Big)}\ , \end{aligned} \tag{27}$$

where g_s, g_ℓ are coupling constants assigned to the potential terms associated with short and long roots respectively. The configuration space is the F_4-Weyl Alcove. The Hamiltonian (27) is invariant with respect to F_4-Weyl group transformations. The parameter β (the inverse of the period) is a parameter introduced for convenience in such a way that

$$\lim_{\beta\to 0} H_{F_4}^{\rm trig} \to H_{F_4}^{\rm rat}(\omega=0)\,,$$

i.e. the rational model (without the harmonic oscillator term) is reproduced. If $\beta\to i\beta$, the Hamiltonian (27) becomes the Hamiltonian of the F_4-invariant hyperbolic model.

The ground state eigenfunction of the Hamiltonian (27) is given by

$$\psi_0(x) = (\Delta_+\,\Delta_-)^{\nu}\,(\Delta\,\Delta_0)^{\mu}\ , \tag{28}$$

[2] In previous studies [4, 7] the analysis of the F_4-trigonometric Hamiltonian was done in the dual root space.

Journal of Physics: Conference Series **512** (2014) 012014 doi:10.1088/1742-6596/512/1/012014

where

$$\Delta_+ \Delta_- = \prod_{\alpha_{\text{long}} \in \mathcal{F}_4{}^+} \sin \frac{\beta}{2}(\alpha_{\text{long}} \cdot x)\,, \tag{29}$$

$$\Delta\,\Delta_0 = \prod_{\alpha_{\text{short}} \in \mathcal{F}_4{}^+} \sin \frac{\beta}{2}(\alpha_{\text{short}} \cdot x)\,. \tag{30}$$

and the exponents μ, ν in the ground state eigenfunction (28) are connected to the coupling constants g_s, g_ℓ by the relations

$$g_\ell = \nu(\nu-1) \quad , \quad g_s = \mu(\mu-1)\,.$$

The ground state eigenvalue is then

$$E_0 = \frac{1}{2}\beta^2 \left(14\,\nu^2 + 18\,\nu\,\mu + 7\,\mu^2\right)\,,$$

which can be written in terms of the *deformed Weyl vector*[3]

$$\varrho = \frac{1}{2}\left(\mu \sum_{\alpha_{\text{short}} \in \mathcal{F}_4^+} \alpha_{\text{short}} + \nu \sum_{\alpha_{\text{long}} \in \mathcal{F}_4^+} \alpha_{\text{long}}\right)\,, \tag{31}$$

as

$$E_0 = \frac{1}{2}\beta^2 \varrho^2\,.$$

All eigenfunctions $\psi_n(x)$ of the Hamiltonian operator (27) have a factorized form

$$\psi_n(x) = \psi_0(x)\,\phi_n(x)\,.$$

The functions $\phi_n(x)$ are eigenfunctions of the gauge rotated Hamiltonian operator defined by

$$\mathfrak{h}_{F_4}^{\text{trig}} = -\frac{2}{\beta^2}\psi_0^{-1}\left[H_{F_4}^{\text{trig}} - E_0\right]\psi_0\,. \tag{32}$$

Then, the eigenvalue problem becomes

$$\mathfrak{h}_{F_4}^{\text{trig}}\,\phi_n = \epsilon_n \phi_n\,,$$

where $\epsilon_n = -\frac{2}{\beta^2}(E_n - E_0)$.

Let us introduce the Weyl-invariant trigonometric coordinates (see [7])

$$\tau_a \equiv \sum_{\omega \in \Omega_a} e^{i\,\beta(\omega \cdot x)}\,, \qquad a = 1, 2, 3, 4 \tag{33}$$

[3] The standard definition for the Weyl vector is $\rho = \frac{1}{2}\sum_{\alpha \in \mathcal{R}^+} \alpha$.

where Ω_a is the Weyl orbit generated by the fundamental weight ϖ_a: [4]

$$\varpi_1 = e_4, \quad \varpi_2 = e_3 + e_4, \quad \varpi_3 = \frac{1}{2}(e_1 + e_2 + e_3 + 3e_4), \quad \varpi_4 = e_2 + e_3 + 2e_4. \tag{34}$$

The corresponding orbit sizes are $|\Omega_a| = 24, 24, 96, 96$ for $a = 1, 2, 3, 4$ respectively. The explicit form of variables $\tau_a, a = 1, 2, 3, 4$ is given at Appendix C.

Changing variables $\{x\} \to \{\tau\}$ we obtain the gauge-rotated Hamiltonian (32)

$$\mathfrak{h}_{F_4}^{\rm trig} = \sum_{a,b=1}^{4} \mathcal{A}_{ab} \frac{\partial^2}{\partial \tau_a \, \partial \tau_b} + \sum_{a=1}^{4} \mathcal{B}_a \frac{\partial}{\partial \tau_a} , \tag{35}$$

with coefficient functions

$$\begin{aligned}
\mathcal{A}_{1,1} &= -\tau_1{}^2 + 12\,\tau_1 + 6\,\tau_2 + \tau_3 + 48 \ , \\
\mathcal{A}_{1,2} &= -\tau_1\tau_2 + 12\,\tau_1 + 3\,\tau_3 \ , \\
\mathcal{A}_{1,3} &= 12\,\tau_1{}^2 + 4\,\tau_1\tau_2 - 3/2\,\tau_1\tau_3 - 96\,\tau_1 - 42\,\tau_2 - 24\,\tau_3 + 3/2\,\tau_4 - 288 \ , \\
\mathcal{A}_{1,4} &= 4\,\tau_1\tau_2 - 2\,\tau_1\tau_4 + 2\,\tau_2\tau_3 - 48\,\tau_1 - 12\,\tau_3 \ , \\
\mathcal{A}_{2,2} &= 12\,\tau_1{}^2 - 2\,\tau_2{}^2 - 96\,\tau_1 - 48\,\tau_2 - 24\,\tau_3 + 2\,\tau_4 - 192 \ , \\
\mathcal{A}_{2,3} &= -24\,\tau_1{}^2 - 4\,\tau_1\tau_2 + 3\,\tau_1\tau_3 - 2\,\tau_2\tau_3 + 240\,\tau_1 + 108\,\tau_2 + 60\,\tau_3 - 9\,\tau_4 + 576 \ , \\
\mathcal{A}_{2,4} &= -24\,\tau_1{}^3 - 4\,\tau_1{}^2\tau_2 + 96\,\tau_1{}^2 + 104\,\tau_1\tau_2 + 72\,\tau_1\tau_3 - 6\,\tau_1\tau_4 + 12\,\tau_2{}^2 + 8\,\tau_2\tau_3 \\
&\quad -3\,\tau_2\tau_4 + 3\,\tau_3{}^2 + 1536\,\tau_1 + 480\,\tau_2 + 288\,\tau_3 - 48\,\tau_4 + 2304 \ , \\
\mathcal{A}_{3,3} &= 12\,\tau_1^3 + 4\,\tau_1{}^2\tau_2 - 96\,\tau_1{}^2 - 60\,\tau_1\tau_2 - 36\,\tau_1\tau_3 + \tau_1\tau_4 - 4\,\tau_2\tau_3 - 3\,\tau_3{}^2 \\
&\quad -384\,\tau_1 - 48\,\tau_2 - 48\,\tau_3 + 12\,\tau_4 \ , \\
\mathcal{A}_{3,4} &= -16\,\tau_1{}^2\tau_2 + 2\,\tau_1\tau_2\tau_3 + 96\,\tau_1{}^2 + 144\,\tau_1\tau_2 - 12\,\tau_1\tau_3 - 8\,\tau_1\tau_4 + 72\,\tau_2{}^2 \\
&\quad +32\,\tau_2\tau_3 - 6\,\tau_2\tau_4 - 4\,\tau_3\tau_4 - 960\,\tau_1 - 48\,\tau_2 - 240\,\tau_3 + 36\,\tau_4 - 2304 \ , \\
\mathcal{A}_{4,4} &= 9216 + 2880\,\tau_1\tau_2 + 576\,\tau_1\tau_3 + 512\,\tau_2\tau_3 + 16\,\tau_1\tau_4 - 24\,\tau_2\tau_4 - 96\,\tau_1{}^2\tau_2 \\
&\quad +16\,\tau_3\tau_4 + 2\,\tau_2\tau_3{}^2 - 8\,\tau_1{}^2\tau_4 + 48\,\tau_1\tau_2{}^2 - 16\,\tau_1{}^3\tau_2 + 7680\,\tau_1 + 6144\,\tau_2 + 1152\,\tau_3 \\
&\quad +96\,\tau_4 + 1344\,\tau_1{}^2 + 864\,\tau_2{}^2 - 192\,\tau_1{}^3 + 24\,\tau_3{}^2 - 6\,\tau_4{}^2 + 48\,\tau_1\tau_2\tau_3 - 4\,\tau_1\tau_2\tau_4 \ ,
\end{aligned}$$

and

$$\begin{aligned}
\mathcal{B}_1 &= -\tau_1 - 24\,\mu - (5\,\mu + 6\,\nu)\,\tau_1 \ , \\
\mathcal{B}_2 &= -2\,\tau_2 - 48\,\nu - 6\,\mu\,\tau_1 - (6\,\mu + 10\,\nu)\,\tau_2 \ , \\
\mathcal{B}_3 &= -3\,\tau_3 - 24\,(\,\mu + \,\nu)\,\tau_1 - 12\,\mu\,\tau_2 - 3\,(\,3\mu + \,4\nu)\,\tau_3 \ , \\
\mathcal{B}_4 &= -6\,\tau_4 + 576\,\nu + 24\,(\mu + \,8\nu)\,\tau_1 - 24\,(\,\mu - 4\,\nu)\,\tau_2 + 48\,\nu\,\tau_3 - 6\,(\,2\mu + \,3\nu)\,\tau_4 \\
&\quad -24\,\nu\,\tau_1{}^2 - 4\,\mu\,\tau_1\tau_2 \ .
\end{aligned}$$

[4] Bourbaki numbering: $\varpi_1 = e_1 + e_2$, $\varpi_2 = 2e_1 + e_2 + e_3$, $\varpi_3 = \frac{1}{2}(3e_1 + e_2 + e_3 + e_4)$, $\varpi_4 = e_1$, (see [12]).

Journal of Physics: Conference Series **512** (2014) 012014 doi:10.1088/1742-6596/512/1/012014

4.1. Flat Metric of the F_4-trigonometric Model.
It is evident that gauge rotated Hamiltonian $\mathfrak{h}_{F_4}^{\rm trig}$ (32) in algebraic form (35) can be written as

$$\beta^2\, \mathfrak{h}_{F_4}^{\rm trig} \ = \ \Delta_g \ + \ \mathcal{C}^b \frac{\partial}{\partial \tau_b}\ , \tag{36}$$

where the Laplace-Beltrami operator

$$\Delta_g \ = \ \frac{1}{g^{1/2}}\frac{\partial}{\partial \tau_a}\,(g^{ab} g^{1/2})\,\frac{\partial}{\partial \tau_b} \ = \ g^{ab}\frac{\partial^2}{\partial \tau_a \partial \tau_b} \ + \ g^b \frac{\partial}{\partial \tau_b}\ ,\ g^b \equiv \frac{1}{g^{1/2}}\frac{\partial}{\partial \tau_a}\,(g^{ab} g^{1/2})\ , \tag{37}$$

with $a, b = 1, 2, 3, 4$ and g is determinant. The symmetric (in $a \leftrightarrow b$) metric, $g^{ab} = g^{ba}$ has polynomial in τ matrix elements

$$g^{ab} \ = \ \left|\begin{array}{cccc} \mathcal{A}_{1,1} & \mathcal{A}_{1,2} & \mathcal{A}_{1,3} & \mathcal{A}_{1,4} \\ & \mathcal{A}_{2,2} & \mathcal{A}_{2,3} & \mathcal{A}_{2,4} \\ & & \mathcal{A}_{3,3} & \mathcal{A}_{3,4} \\ & & & \mathcal{A}_{4,4} \end{array}\right| \ . \tag{38}$$

This flat metric has a special property: any component g^b of the vector $\vec{g}$ in (37) is also polynomial in τ

$$-\vec{g} \ = \ (\tau_1,\ 2\tau_2,\ 3\tau_3,\ 6\tau_4)\ .$$

The same property has the vector $\mathcal{C}^b$ (see (36)) which characterizes the interaction in (27)

$$-\vec{\mathcal{C}} \ = \ \Big(24\mu + (5\mu+6\nu)\tau_1,\ 48\nu + 6\mu\tau_1 + 2(3\mu+5\nu)\tau_2,\ 24(\mu+\nu)\tau_1 + 12\mu\tau_2 + 3(3\mu+4\nu)\tau_3,$$

$$-576\nu - 24(\mu+8\nu)\tau_1 + 24(\mu-4\nu)\tau_2 - 48\nu\tau_3 + 6(2\mu+3\nu)\tau_4 + 24\nu{\tau_1}^2 + 4\mu\tau_1\tau_2\Big)\ .$$

4.2. Ground state eigenfunction and potential of F_4-trigonometric model in τ-variables
The ground state eigenfunction (28) and the original potential in (27) can be written in terms of the invariant variables (33). After straightforward but very cumbersome calculations we find that

$$(\Delta_+\Delta_-)^2 = \left(\frac{1}{2}\right)^{24} P_1(\tau)\,, \qquad (\Delta\Delta_0)^2 = \left(\frac{1}{2}\right)^{24} P_2(\tau)\,,$$

where

$$
\begin{aligned}
P_1(\tau) \;=\; & -1728\,\tau_1{}^6 - 1728\,\tau_1{}^5\tau_2 - 432\,\tau_1{}^4\tau_2{}^2 + 32\,\tau_1{}^3\tau_2{}^3 + 16\,\tau_1{}^2\tau_2{}^4 + 20736\,\tau_1{}^5 \\
& +34560\,\tau_1{}^4\tau_2 + 10368\,\tau_1{}^4\tau_3 - 864\,\tau_1{}^4\tau_4 + 18432\,\tau_1{}^3\tau_2{}^2 + 8640\,\tau_1{}^3\tau_2\tau_3 \\
& -576\,\tau_1{}^3\tau_2\tau_4 + 432\,\tau_1{}^3\tau_3{}^2 + 2976\,\tau_1{}^2\tau_2{}^3 + 1728\,\tau_1{}^2\tau_2{}^2\tau_3 - 72\,\tau_1{}^2\tau_2{}^2\tau_4 \\
& +216\,\tau_1{}^2\tau_2\tau_3{}^2 - 224\,\tau_1\tau_2{}^4 - 96\,\tau_1\tau_2{}^3\tau_3 + 8\,\tau_1\tau_2{}^3\tau_4 - 64\,\tau_2{}^5 - 32\,\tau_2{}^4\tau_3 \\
& -4\,\tau_2{}^3\tau_3{}^2 + 103680\,\tau_1{}^4 + 6912\,\tau_1{}^3\tau_2 - 34560\,\tau_1{}^3\tau_3 + 1728\,\tau_1{}^3\tau_4 \\
& -88128\,\tau_1{}^2\tau_2{}^2 - 79488\,\tau_1{}^2\tau_2\tau_3 + 5184\,\tau_1{}^2\tau_2\tau_4 - 18144\,\tau_1{}^2\tau_3{}^2 + 2592\,\tau_1{}^2\tau_3\tau_4 \\
& -108\,\tau_1{}^2\tau_4{}^2 - 45888\,\tau_1\tau_2{}^3 - 43200\,\tau_1\tau_2{}^2\tau_3 + 2592\,\tau_1\tau_2{}^2\tau_4 - 13392\,\tau_1\tau_2\tau_3{}^2 \\
& +1296\,\tau_1\tau_2\tau_3\tau_4 - 36\,\tau_1\tau_2\tau_4{}^2 - 1296\,\tau_1\tau_3{}^3 + 108\,\tau_1\tau_3{}^2\tau_4 - 6384\,\tau_2{}^4 - 6592\,\tau_2{}^3\tau_3 \\
& +328\,\tau_2{}^3\tau_4 - 2520\,\tau_2{}^2\tau_3{}^2 + 144\,\tau_2{}^2\tau_3\tau_4 + \tau_2{}^2\tau_4{}^2 - 432\,\tau_2\tau_3{}^3 + 18\,\tau_2\tau_3{}^2\tau_4 \\
& -27\,\tau_3{}^4 - 774144\,\tau_1{}^3 - 1465344\,\tau_1{}^2\tau_2 - 663552\,\tau_1{}^2\tau_3 + 62208\,\tau_1{}^2\tau_4 \\
& -787968\,\tau_1\tau_2{}^2 - 566784\,\tau_1\tau_2\tau_3 + 48384\,\tau_1\tau_2\tau_4 - 103680\,\tau_1\tau_3{}^2 + 17280\,\tau_1\tau_3\tau_4 \\
& -864\,\tau_1\tau_4{}^2 - 129024\,\tau_2{}^3 - 119808\,\tau_2{}^2\tau_3 + 9024\,\tau_2{}^2\tau_4 - 36288\,\tau_2\tau_3{}^2 + 4608\,\tau_2\tau_3\tau_4 \\
& -192\,\tau_2\tau_4{}^2 - 3456\,\tau_3{}^3 + 432\,\tau_3{}^2\tau_4 - 4\,\tau_4{}^3 - 5308416\,\tau_1{}^2 - 4866048\,\tau_1\tau_2 \\
& -1990656\,\tau_1\tau_3 + 221184\,\tau_1\tau_4 - 1096704\,\tau_2{}^2 - 774144\,\tau_2\tau_3 + 78336\,\tau_2\tau_4 \\
& -138240\,\tau_3{}^2 + 27648\,\tau_3\tau_4 - 1728\,\tau_4{}^2 - 10616832\,\tau_1 - 4423680\,\tau_2 - 1769472\,\tau_3 \\
& +221184\,\tau_4 - 7077888\ ,
\end{aligned}
$$

$$
\begin{aligned}
P_2(\tau) \;=\; & -16\,\tau_1{}^5 + 48\,\tau_1{}^3\tau_2 + 112\,\tau_1{}^3\tau_3 - 4\,\tau_1{}^3\tau_4 + \tau_1{}^2\tau_3{}^2 + 4608\,\tau_1{}^3 + 1728\,\tau_1{}^2\tau_2 \\
& +384\,\tau_1{}^2\tau_3 - 144\,\tau_1{}^2\tau_4 - 216\,\tau_1\tau_2\tau_3 - 192\,\tau_1\tau_3{}^2 + 18\,\tau_1\tau_3\tau_4 - 4\,\tau_3{}^3 - 18432\,\tau_1{}^2 \\
& -20736\,\tau_1\tau_2 - 14976\,\tau_1\tau_3 + 1728\,\tau_1\tau_4 - 3888\,\tau_2{}^2 - 5184\,\tau_2\tau_3 + 648\,\tau_2\tau_4 - 1728\,\tau_3{}^2 \\
& +432\,\tau_3\tau_4 - 27\,\tau_4{}^2 - 110592\,\tau_1 - 41472\,\tau_2 - 27648\,\tau_3 + 3456\,\tau_4 - 110592\ .
\end{aligned}
$$

Thus, the unnormalized ground state eigenfunction (28) becomes

$$
\psi_0(\tau) \;=\; P_1(\tau)^{\nu/2}\, P_2(\tau)^{\mu/2}\ , \tag{39}
$$

(cf. (28)).

It is evident that the original potential of F_4-invariant trigonometric model (27) can be written in terms of τ variables. The simplest way to do it is to invert the gauge rotation (32) with the ground state eigenfunction (39) of the algebraic operator (35). After quite sophisticated technically, tedious calculations we get

$$
V_{F_4}^{\rm trig}(\tau) = -g_\ell \beta^2 \frac{\mathcal{N}_1(\tau)}{P_1(\tau)} - \frac{g_s\beta^2}{2}\frac{\mathcal{N}_2(\tau)}{P_2(\tau)}\ , \tag{40}
$$

where

$$\begin{aligned}\mathcal{N}_1 = & (1728\,\tau_1^6 + 1728\,\tau_1^5\tau_2 + 432\,\tau_1^4\tau_2^2 - 8\,\tau_1^3\tau_2^3 - 8\,\tau_1^2\tau_2^4 - 20736\,\tau_1^5 \\ & -34560\,\tau_1^4\tau_2 - 10368\,\tau_1^4\tau_3 + 864\,\tau_1^4\tau_4 - 19296\,\tau_1^3\tau_2^2 - 8640\,\tau_1^3\tau_2\tau_3 \\ & +504\,\tau_1^3\tau_2\tau_4 - 432\,\tau_1^3\tau_3^2 - 3456\,\tau_1^2\tau_2^3 - 1728\,\tau_1^2\tau_2^2\tau_3 + 60\,\tau_1^2\tau_2^2\tau_4 \\ & -216\,\tau_1^2\tau_2\tau_3^2 + 88\,\tau_1\tau_2^4 + 24\,\tau_1\tau_2^3\tau_3 - 2\,\tau_1\tau_2^3\tau_4 + 48\,\tau_2^5 + 16\,\tau_2^4\tau_3 \\ & +\tau_2^3\tau_3^2 - 103680\,\tau_1^4 + 34560\,\tau_1^3\tau_3 + 96192\,\tau_1^2\tau_2^2 + 79488\,\tau_1^2\tau_2\tau_3 \\ & -4176\,\tau_1^2\tau_2\tau_4 + 18144\,\tau_1^2\tau_3^2 - 2592\,\tau_1^2\tau_3\tau_4 + 72\,\tau_1^2\tau_4^2 + 50016\,\tau_1\tau_2^3 \\ & +45792\,\tau_1\tau_2^2\tau_3 - 2496\,\tau_1\tau_2^2\tau_4 + 13392\,\tau_1\tau_2\tau_3^2 - 1080\,\tau_1\tau_2\tau_3\tau_4 + 18\,\tau_1\tau_2\tau_4^2 \\ & +1296\,\tau_1\tau_3^3 - 108\,\tau_1\tau_3^2\tau_4 + 6912\,\tau_2^4 + 7072\,\tau_2^3\tau_3 - 352\,\tau_2^3\tau_4 + 2628\,\tau_2^2\tau_3^2 \\ & -120\,\tau_2^2\tau_3\tau_4 + 432\,\tau_2\tau_3^3 - 9\,\tau_2\tau_3^2\tau_4 + 27\,\tau_3^4 + 774144\,\tau_1^3 + 1423872\,\tau_1^2\tau_2 \\ & +663552\,\tau_1^2\tau_3 - 72576\,\tau_1^2\tau_4 + 785664\,\tau_1\tau_2^2 + 546048\,\tau_1\tau_2\tau_3 - 52992\,\tau_1\tau_2\tau_4 \\ & +103680\,\tau_1\tau_3^2 - 22464\,\tau_1\tau_3\tau_4 + 1584\,\tau_1\tau_4^2 + 134208\,\tau_2^3 + 120960\,\tau_2^2\tau_3 \\ & -9936\,\tau_2^2\tau_4 + 35424\,\tau_2\tau_3^2 - 5184\,\tau_2\tau_3\tau_4 + 396\,\tau_2\tau_4^2 + 3456\,\tau_3^3 - 648\,\tau_3^2\tau_4 \\ & +72\,\tau_3\tau_4^2 + \tau_4^3 + 5308416\,\tau_1^2 + 4534272\,\tau_1\tau_2 + 1990656\,\tau_1\tau_3 - 304128\,\tau_1\tau_4 \\ & +1022976\,\tau_2^2 + 718848\,\tau_2\tau_3 - 96768\,\tau_2\tau_4 + 138240\,\tau_3^2 - 41472\,\tau_3\tau_4 + 4032\,\tau_4^2 \\ & +10616832\,\tau_1 + 3981312\,\tau_2 + 1769472\,\tau_3 - 331776\,\tau_4 + 7077888)\ ,\end{aligned}$$

and

$$\begin{aligned}\mathcal{N}_2 = & (12\,\tau_1^5 - 4\,\tau_1^4\tau_2 + 288\,\tau_1^4 + 84\,\tau_1^3\tau_2 - 100\,\tau_1^3\tau_3 + \tau_1^3\tau_4 + 24\,\tau_1^2\tau_2\tau_3 - 9216\,\tau_1^3 \\ & -3600\,\tau_1^2\tau_2 - 1584\,\tau_1^2\tau_3 + 252\,\tau_1^2\tau_4 - 180\,\tau_1\tau_2\tau_3 + 180\,\tau_1\tau_3^2 - 9\,\tau_1\tau_3\tau_4 \\ & -36\,\tau_2\tau_3^2 + \tau_3^3 + 34560\,\tau_1^2 + 31104\,\tau_1\tau_2 + 25920\,\tau_1\tau_3 - 2592\,\tau_1\tau_4 + 3888\,\tau_2^2 \\ & +7776\,\tau_2\tau_3 - 648\,\tau_2\tau_4 + 3168\,\tau_3^2 - 648\,\tau_3\tau_4 + 27\,\tau_4^2 + 165888\,\tau_1 + 41472\,\tau_2 \\ & +41472\,\tau_3 - 3456\,\tau_4 + 110592)\ .\end{aligned}$$

Then, the F_4-trigonometric potential written in τ-variables is a rational (meromorphic) function! From the structure of the potential given in (40) it is clear that the singularities of the potential are defined by the zeroes of the (ground state) factors $P_1(\tau), P_2(\tau)$, and therefore they lie on the boundaries of the configuration space.

4.3. Exact Solvability (Invariant subspaces)

The Hamiltonian (35) has an infinite number of invariant polynomial subspaces

$$\mathcal{P}_n^{(p_1,p_2,p_3,p_4)} \equiv \{\tau_1^{n_1}\tau_2^{n_2}\tau_3^{n_3}\tau_4^{n_4} \mid 0 \ \le p_1\,n_1 + p_2\,n_2 + p_3\,n_3 + p_4\,n_4 \le n\}\,, \quad p_i, n_i \in \mathbb{N}$$

which are labeled by certain characteristic vectors

$$(p_1, p_2, p_3, p_4)\,.$$

For fixed (p_1, p_2, p_3, p_4) such invariant subspaces form a *flag*:

$$\mathcal{P}_0 \subset \mathcal{P}_1 \subset \ldots \subset \mathcal{P}_k \subset \ldots$$

where, for simplicity in notations, we have denoted $\mathcal{P}_n^{(p_1,p_2,p_3,p_4)} = \mathcal{P}_n$. The characteristic vector corresponding to the *minimal flag* is:

$$(1, 2, 2, 3)\,,$$

which coincides with the minimal flag found for the rational case. There are other flags, which are invariant under the action of the Hamiltonian (35), are associated with the characteristic vectors $(2, 2, 3, 4)$, $(2, 3, 4, 6)$, $(2, 4, 4, 6)$ etc. Among these flags there are two special flags, the first of which is characterized by the components[5] of the Weyl-vector (with respect to simple roots):

$$\rho = (8, 11, 15, 21)\,,$$

and a second flag with characteristic vector defined by the components of the co-Weyl vector :

$$\rho^\vee = (11, 16, 21, 30)\,.$$

4.4. Spectrum of the F_4-trigonometric model

In order to find the spectrum of the F_4-trigonometric model we should introduce some relevant concepts.

The *weight lattice* $L(W)$ is defined as the $\mathbb{Z}$-span of the set of fundamental weights (34) (denoted here as $\{w_a, a = 1 \ldots 4\}$). The cone of *dominant weights* $L_+(W)$ contains the lattice points with nonnegative integer coordinates:

$$L_+ = \left\{ \mathbf{n} = \sum_{a=1}^{4} n_a w_a = (n_1, n_2, n_3, n_4) \mid n_a \geq 0 \right\}. \tag{41}$$

We can define a basis set of monomials

$$\{\tau^{\mathbf{n}} = \tau_1^{n_1} \tau_2^{n_2} \tau_3^{n_3} \tau_4^{n_4} | \quad \mathbf{n} \in L_+\}\,,$$

which is ordered according to the relation

$$|\mathbf{n}|^2 > |\mathbf{m}|^2 \Rightarrow \tau_{\mathbf{n}} > \tau_{\mathbf{m}} \quad \text{for} \quad \mathbf{n}, \mathbf{m} \in L_+\,.$$

The Hamiltonian (35) is triangular in this basis. Any eigenfunction can be marked by a dominant weight $\mathbf{p}$ being of the form

$$\varphi_{\mathbf{p}} = \tau^{\mathbf{p}} + \sum_{|\mathbf{m}|^2 < |\mathbf{p}|^2} c_{\mathbf{m}} \tau^{\mathbf{m}} \tag{42}$$

[5] arranged by increasing value, *i.e.* following a non-standard Dynkin ordering of the simple roots.

with energy (see [13]):

$$-\epsilon_{\mathbf{p}} = \beta^2\Big((\mathbf{p},\mathbf{p}) + 2(\mathbf{p},\varrho)\Big)\,, \tag{43}$$

where ϱ is the deformed Weyl vector (31).

4.5. Eigenfunctions

We consider the minimal flag of polynomial subspaces $\mathcal{P}_n^{(1,2,2,3)}$, where

$$\mathcal{P}_n^{(1,2,2,3)} \equiv \{\tau_1^{p_1}\tau_2^{p_2}\tau_3^{p_3}\tau_4^{p_4}|0 \le 1\,p_1 + 2\,p_2 + 2\,p_3 + 3\,p_4 \le n\}\,,\quad n = 0,1,2,3\ldots \tag{44}$$

The ordering of monomials $\tau^{\mathbf{p}}$ is given by the norm of the associated dominant weight $\mathbf{p} = (p_1, p_2, p_3, p_4)$, *i.e.* by the formula

$$|\mathbf{p}|^2 \;=\; 3\,{p_3}^2 + 8\,p_3p_4 + 6\,{p_4}^2 + 6\,p_2p_4 + 2\,{p_2}^2 + 4\,p_2p_3 + {p_1}^2 + 4\,p_1p_4 + 3\,p_1p_3 + 2\,p_1p_2\,.$$

On the other side, the grading of the monomials $\tau^{\mathbf{p}}$ is defined as $1\,p_1 + 2\,p_2 + 2\,p_3 + 3\,p_4$ (for the minimal flag). Eigenfunctions are labeled by the coordinates of leading dominant weight in (42) with an extra super-index labeling the grading n of invariant subspace (44) to which the eigenfunction belongs. The eigenfunctions from the spaces $\mathcal{P}_n^{(1,2,2,3)}$, $n = 0,1,2$ are presented in Appendix D.

5. Conclusion

We revealed the algebraic-rational nature of the F_4-invariant, rational and trigonometric models in the space of orbits. It is shown that the kinetic energy operator is the Laplace-Beltrami operator with metric with polynomial entries of zero Riemann tensor and the potential is ratio of two polynomials. We have obtained algebraic forms for the F_4-invariant Hamiltonians, both rational and trigonometric, in terms of the Weyl invariants, polynomial and exponential, respectively. They are second order differential operators with polynomial coefficients. The both Hamiltonians preserve the same infinite flag of polynomial spaces marked by the characteristic vector $(1, 2, 2, 3)$, indicating exact solvability. It turns out that for the trigonometric case a solution (formulas) for the Hamiltonian defined in the standard root space looks easier than one obtained in previous studies [4, 7] where the Hamiltonian was defined in the dual root space. It allows to unify the description of the F_4-trigonometric model under the same systematic description as it was done for the $G_2, E_{6,7,8}$ trigonometric models (see [7]). Explicit examples of first polynomial eigenfunctions were presented for the rational and trigonometric models. We showed for the first time that both F_4-invariant rational and trigonometric models have common particular integral in the space of orbits, which annihilates the invariant subspace.

Journal of Physics: Conference Series **512** (2014) 012014 doi:10.1088/1742-6596/512/1/012014

Acknowledgments
This work was supported in part by the University Program FENOMEC, by the PAPIIT grant **IN109512** and CONACyT grant **166189** (Mexico).

Appendix A. *The Weyl-invariant variables τ for the F_4 rational model.*

$$
\begin{aligned}
\tau_2 &= {x_1}^2 + {x_2}^2 + {x_3}^2 + {x_4}^2 \ , \qquad \text{(A.1)}\\
\tau_6 &= \frac{1}{6}\left(x_1^4x_2^2 + {x_1}^4{x_3}^2 + x_1^4{x_4}^2 + x_1^2x_2^4 - 3\,x_1^2x_2^2x_3^2 - 3\,x_1^2x_2^2x_4^2\right.\\
&\quad + x_1^2x_3^4 - 3\,x_1^2x_3^2x_4^2 + x_1^2x_4^4 + {x_2}^4{x_3}^2 + {x_2}^4{x_4}^2 + {x_2}^2{x_3}^4\\
&\quad \left.-3\,x_2^2x_3^2x_4^2 + x_2^2x_4^4 + x_3^4x_4^2 + x_3^2x_4^4\right) \ ,\\
\tau_8 &= \frac{1}{12}\left({x_1}^4{x_2}^4 - {x_1}^4{x_2}^2{x_3}^2 - {x_1}^4{x_2}^2{x_4}^2 + {x_1}^4{x_3}^4 - {x_1}^4{x_3}^2{x_4}^2 + {x_1}^4{x_4}^4 - {x_1}^2{x_2}^4{x_3}^2\right.\\
&\quad -{x_1}^2{x_2}^4{x_4}^2 - {x_1}^2{x_2}^2{x_3}^4 + 6\,{x_1}^2{x_2}^2{x_3}^2{x_4}^2 - {x_1}^2{x_2}^2{x_4}^4 - {x_1}^2{x_3}^4{x_4}^2 - {x_1}^2{x_3}^2{x_4}^4\\
&\quad \left.+{x_2}^4{x_3}^4 - {x_2}^4{x_3}^2{x_4}^2 + {x_2}^4{x_4}^4 - {x_2}^2{x_3}^4{x_4}^2 - {x_2}^2{x_3}^2{x_4}^4 + {x_3}^4{x_4}^4\right) \ ,\\
\tau_{12} &= \frac{1}{72}\left({x_1}^2{x_2}^2 + {x_1}^2{x_3}^2 - 2\,{x_1}^2{x_4}^2 - 2\,{x_2}^2{x_3}^2 + {x_2}^2{x_4}^2 + {x_3}^2{x_4}^2\right)\\
&\quad \left(x_1^2x_2^2 - 2\,x_1^2x_3^2 + x_1^2x_4^2 + x_2^2x_3^2 - 2\,x_2^2x_4^2 + x_3^2x_4^2\right)\\
&\quad \left(2\,x_1^2x_2^2 - x_1^2x_3^2 - x_1^2x_4^2 - x_2^2x_3^2 - x_2^2x_4^2 + 2\,x_3^2x_4^2\right) \ .
\end{aligned}
$$

Appendix B. *Lowest Eigenfunctions of the F_4-rational Model.*

Here the eigenfunctions and eigenvalues corresponding to the spaces $\mathcal{P}_n^{(1,2,2,3)}$ for $n = 0, 1, 2$ are presented in the explicit form:

- $n = 0$ (one eigenstate)

$$
\begin{aligned}
\phi_0 &= 1 \ ,\\
\epsilon_0 &= 0 \ .
\end{aligned}
$$

- $n = 1$ (one eigenstate)

$$
\begin{aligned}
\phi_1 &= \tau_2 - \frac{2}{\omega}(6\mu + 6\nu + 1) \ ,\\
\epsilon_1 &= -4\omega \ .
\end{aligned}
$$

- $n = 2$ (three eigenstates)

$$
\begin{aligned}
\phi_2^{(1)} &= \tau_2^2 - \frac{6}{\omega}(4\mu + 4\nu + 1)\tau_2 + \frac{6}{\omega^2}(4\mu + 4\nu + 1)(6\mu + 6\nu + 1) , \\
\epsilon_2^{(1)} &= -8\omega , \\
\phi_2^{(2)} &= \tau_6 - \frac{1}{4\omega}(2\mu + 4\nu + 1)\tau_2^2 + \frac{3}{4\omega^2}(2\mu + 4\nu + 1)(4\mu + 4\nu + 1)\tau_2 \\
&\quad - \frac{1}{2\omega^3}(2\mu + 4\nu + 1)(6\mu + 6\nu + 1)(4\mu + 4\nu + 1) , \\
\epsilon_2^{(2)} &= -12\omega , \\
\phi_2^{(3)} &= \tau_8 - \frac{1}{\omega}(3\nu + 1)\tau_6 + \frac{1}{8\omega^2}(3\nu + 1)(2\mu + 4\nu + 1)\tau_2^2 \\
&\quad - \frac{1}{4\omega^3}(3\nu + 1)(2\mu + 4\nu + 1)(4\mu + 4\nu + 1)\tau_2 \\
&\quad + \frac{1}{8\omega^4}(3\nu + 1)(2\mu + 4\nu + 1)(6\mu + 6\nu + 1)(4\mu + 4\nu + 1) , \\
\epsilon_2^{(3)} &= -16\omega .
\end{aligned}
$$

Appendix C. *The Weyl-invariant variables τ for the F_4 trigonometric model.*

$$\begin{aligned}\frac{\tau_1}{2} &= 8\cos(\frac{\beta}{2}x_1)\cos(\frac{\beta}{2}x_2)\cos(\frac{\beta}{2}x_3)\cos(\frac{\beta}{2}x_4)\\ &+\cos(\beta x_1)+\cos(\beta x_2)+\cos(\beta x_3)+\cos(\beta x_4)\end{aligned}\tag{C.1}$$

$$\begin{aligned}\frac{\tau_2}{4} &= \cos(\beta x_1)\cos(\beta x_2)+\cos(\beta x_1)\cos(\beta x_3)+\cos(\beta x_1)\cos(\beta x_4)\\ &+\cos(\beta x_2)\cos(\beta x_3)+\cos(\beta x_2)\cos(\beta x_4)+\cos(\beta x_3)\cos(\beta x_4)\end{aligned}\tag{C.2}$$

$$\begin{aligned}\frac{\tau_3}{8} &= 2\cos(\frac{\beta}{2}x_1)\cos(\frac{\beta}{2}x_2)\cos(\frac{\beta}{2}x_3)\cos(3\frac{\beta}{2}x_4)\\ &+2\cos(\frac{\beta}{2}x_1)\cos(\frac{\beta}{2}x_2)\cos(3\frac{\beta}{2}x_3)\cos(\frac{\beta}{2}x_4)\\ &+2\cos(\frac{\beta}{2}x_1)\cos(3\frac{\beta}{2}x_2)\cos(\frac{\beta}{2}x_3)\cos(\frac{\beta}{2}x_4)\\ &+2\cos(3\frac{\beta}{2}x_1)\cos(\frac{\beta}{2}x_2)\cos(\frac{\beta}{2}x_3)\cos(\frac{\beta}{2}x_4)\\ &+\cos(\beta x_1)\cos(\beta x_2)\cos(\beta x_3)+\cos(\beta x_1)\cos(\beta x_2)\cos(\beta x_4)\\ &+\cos(\beta x_1)\cos(\beta x_3)\cos(\beta x_4)+\cos(\beta x_2)\cos(\beta x_3)\cos(\beta x_4)\end{aligned}\tag{C.3}$$

$$\begin{aligned}\frac{\tau_4}{16} &= \cos^2(\beta x_1)\cos(\beta x_2)\cos(\beta x_3)+\cos^2(\beta x_1)\cos(\beta x_2)\cos(\beta x_4)\\ &+\cos^2(\beta x_1)\cos(\beta x_3)\cos(\beta x_4)+\cos(\beta x_1)\cos^2(\beta x_2)\cos(\beta x_3)\\ &+\cos(\beta x_1)\cos^2(\beta x_2)\cos(\beta x_4)+\cos(\beta x_1)\cos(\beta x_2)\cos^2(\beta x_3)\\ &+\cos(\beta x_1)\cos(\beta x_2)\cos^2(\beta x_4)+\cos(\beta x_1)\cos^2(\beta x_3)\cos(\beta x_4)\\ &+\cos(\beta x_1)\cos(\beta x_3)\cos^2(\beta x_4)+\cos^2(\beta x_2)\cos(\beta x_3)\cos(\beta x_4)\\ &+\cos(\beta x_2)\cos^2(\beta x_3)\cos(\beta x_4)+\cos(\beta x_2)\cos(\beta x_3)\cos^2(\beta x_4)\\ &-\cos(\beta x_1)\cos(\beta x_2)-\cos(\beta x_1)\cos(\beta x_3)-\cos(\beta x_1)\cos(\beta x_4)\\ &-\cos(\beta x_2)\cos(\beta x_3)-\cos(\beta x_2)\cos(\beta x_4)-\cos(\beta x_3)\cos(\beta x_4)\end{aligned}\tag{C.4}$$

Appendix D. *Lowest Eigenfunctions of the F_4-trigonometric Model.*

Here, the eigenfunctions and eigenvalues corresponding to the spaces $\mathcal{P}_n^{(1,2,2,3)}$ for $n = 0, 1, 2$ are presented in the explicit form:

- $n = 0$. Ground state:

$$\begin{aligned}\varphi^{(0)}_{[0,0,0,0]} &= 1\,,\\ \varepsilon^{(0)}_{[0,0,0,0]} &= 0\,.\end{aligned}$$

- $n = 1$. One eigenstate:

$$\text{monomials: } [1, \tau_1], \quad |\mathbf{p}|^2 : [0, 1], \quad \text{gradings: } [0, 1]$$

$$\begin{aligned}
\varphi^{(1)}_{[1,0,0,0]} &= 1 + \frac{1}{24}\frac{(1 + 6\,\nu + 5\,\mu)}{\mu}\tau_1\,, \\
\varepsilon^{(1)}_{[1,0,0,0]} &= -(1 + 6\,\nu + 5\,\mu)\,,
\end{aligned}$$

- $n = 2$. Three eigenstates:

(i) monomials: $[1, \tau_1, \tau_2]$, $|\mathbf{p}|^2 : [0, 1, 2]$, gradings:$[0, 1, 2]$

$$\begin{aligned}
\varphi^{(2)}_{[0,1,0,0]} &= 24\,\frac{(3\,\mu^2 + \mu\,\nu + 4\,\nu^2 + \nu)}{(4\,\nu + \mu + 1)\,(1 + 5\,\nu + 3\,\mu)} + 6\,\frac{\mu}{4\,\nu + \mu + 1}\tau_1 + \tau_2\,, \\
\varepsilon^{(2)}_{[0,1,0,0]} &= -2(1 + 5\,\nu + 3\,\mu)\,.
\end{aligned}$$

(ii) monomials: $[1, \tau_1, \tau_2, \tau_3]$, $|\mathbf{p}|^2 : [0, 1, 2, 3]$, gradings:$[0, 1, 2, 2]$

$$\begin{aligned}
\varphi^{(2)}_{[0,0,1,0]} &= 8\,\frac{(6\,\mu^2 + 9\,\mu\,\nu + \mu + 8\,\nu^2 + 3\,\nu)}{(3\,\nu + 2\,\mu + 1)\,(1 + 4\,\nu + 3\,\mu)} + \frac{(6\,\mu^2 + 5\,\mu\,\nu + \mu + 2\,\nu^2 + \nu)}{\mu\,(3\,\nu + 2\,\mu + 1)}\tau_1 \\
&\quad + \tau_2 + \frac{1}{12}\frac{(3\,\mu + 1 + 2\,\nu)}{\mu}\tau_3\,, \\
\varepsilon^{(2)}_{[0,0,1,0]} &= -3(1 + 4\nu + 3\mu)\,.
\end{aligned}$$

(iii) monomials: $[1, \tau_1, \tau_2, \tau_3, \tau_1^2]$, $|\mathbf{p}|^2 : [0, 1, 2, 3, 4]$, gradings:$[0, 1, 2, 2, 2]$

$$\begin{aligned}
\varphi^{(2)}_{[2,0,0,0]} &= -8\,\frac{(8\,\mu^3 - 15\,\mu^2 + 4\,\mu^2\nu - 32\,\mu\,\nu - 11\,\mu - 3 - 24\,\nu^2 - 18\,\nu)}{(5\,\mu + 3 + 6\,\nu)\,(2 + 6\,\nu + 5\,\mu)} \\
&\quad -2\,\frac{(8\,\mu^3 - 9\,\mu^2 + 4\,\mu^2\nu - 10\,\mu\,\nu - 9\,\mu - 6\,\nu - 2 - 4\,\nu^2)}{(5\,\mu + 3 + 6\,\nu)\,(1 + 3\,\mu)}\tau_1 + \tau_2 \\
&\quad + \frac{1}{3}\frac{(2\,\mu + 1 + \nu)}{1 + 3\,\mu}\tau_3 - \frac{1}{6}\frac{(2\,\mu^2 + \mu\,\nu + 3\,\mu + 1 + \nu)}{1 + 3\,\mu}{\tau_1}^2\,, \\
\varepsilon^{(2)}_{[2,0,0,0]} &= -2(2 + 6\,\nu + 5\,\mu)\,.
\end{aligned}$$

References

[1] M.A. Olshanetsky and A.M. Perelomov, *Quantum integrable systems related to Lie algebras*, *Phys. Repts.* **94** (1983) 313-393 .

[2] W. Rühl and A. Turbiner, *Exact solvability of the Calogero and Sutherland models*, *Mod. Phys. Lett.* **A10** (1995) 2213-2222 .

[3] L. Brink, A. Turbiner and N. Wyllard, *Hidden Algebras of the (super) Calogero and Sutherland models*, *Journ. Math. Phys.***39** (1998) 1285-1315 .

[4] K.G. Boreskov, J.C. Lopez V. and A.V. Turbiner, *Solvability of F_4 integrable system*, *Int.Journ.Mod.Phys.* **A16** (2001) 4769-4801 .

[5] Garcia M.A.G. and Turbiner A.V., *The quantum H_4 Integrable System*, *Mod.Phys.Lett.* **A26** (2011), 433-447, `math-ph arXiv:1011.2127`

[6] K.G. Boreskov, A.V. Turbiner and J.C. López Vieyra, *Solvability of the Hamiltonians related to exceptional root spaces: rational case*,
Comm.Math.Phys. **260** (2005) 17-44 , `hep-th/0407204` pp.51 (full version).

[7] K.G. Boreskov, A.V. Turbiner and J.C. López Vieyra, *Sutherland-type trigonometric models, trigonometric invariants, and multivariate polynomials*,
Contemporary Mathematics **471**, (2008) 15-31,
eprint arXiv:0805.0770

[8] A.V. Turbiner,
Particular Integrability and (Quasi)-exact-solvability,
Journal of Physics **A 46** (2013) 025203 (9pp)

[9] Turbiner A.V., *From quantum A_N (Calogero) to H_4 (rational) model*, *SIGMA* **7** (2011), 071, 20 pages, `math-ph arXiv: 1106.5017`

[10] Turbiner A.V., *From quantum A_N (Sutherland) to E_8 trigonometric model*, *SIGMA* **9** (2013) 003, 25 pp, `math-ph arXiv: 1210.4515`

[11] H.S.M. Coxeter, *Regular Polytopes*, Methuen and Co. LTD London (1948)

[12] N. Bourbaki, *Lie groups and Lie algebras, Ch.4-6*, Springer, Berlin, 2002

[13] S.P. Khastgir, A.J. Pocklington, R. Sasaki, *Quantum Calogero-Moser Models: Integrability for all Root Systems*, *J.Phys.* **A33** (2000) 9033-9064

Journal of Physics: Conference Series **512** (2014) 012015 doi:10.1088/1742-6596/512/1/012015

Representations of ℓ-conformal Galilei algebra and hierarchy of invariant equation

N Aizawa[1], **Y Kimura**[1], **J Segar**[2]

[1] Department of Mathematics and Information Sciences, Graduate School of Science, Osaka Prefecture University, Nakamozu Campus, Sakai, Osaka 599-8531, Japan

[2] Department of Physics, Ramakrishna Mission Vivekananda College, Mylapore, Chennai 600 004, India

E-mail: aizawa@mi.s.osakafu-u.ac.jp, segar@rkmvc.ac.in

Abstract. The ℓ-conformal Galilei algebra, denoted by $\mathfrak{g}_\ell(d)$, is a particular non-semisimple Lie algebra specified by a positive integer d and a spin value ℓ. The algebra $\mathfrak{g}_\ell(d)$ admits central extensions. We study lowest weight representations, in particular Verma modules, of $\mathfrak{g}_\ell(d)$ with the central extensions for $d = 1, 2$. We give a classification of irreducible modules over $d = 1$ algebras and a condition of the Verma modules over $d = 2$ algebras being reducible. As an application of the representation theory, hierarchies of differential equations are derived. The Lie group generated by $\mathfrak{g}_\ell(d)$ with the central extension is a kinematical symmetry of the differential equations.

1. Introduction

By ℓ-conformal Galilei algebra we mean a particular class of non-semisimple Lie algebras. Each Lie algebra in the class is specified by a pair of numbers (d, ℓ). The possible values of d and ℓ are given by

$$d = 1,\ 2,\ 3,\ \ldots, \qquad \ell = \frac{1}{2},\ 1,\ \frac{3}{2},\ 2,\ \ldots$$

The Lie algebra specified by a given pair (d, ℓ) has the generators

$$D,\ H,\ C,\ M_{ij} = -M_{ji},\ P_i^{(n)}, \qquad i,j = 1,2,\ldots,d, \quad n = 0,1,\cdots,2\ell$$

and the generators satisfy the following nonvanishing commutation relations:

$$\begin{aligned}
&[D,H] = 2H, \qquad [D,C] = -2C, \qquad [C,H] = D, \\
&[M_{ij}, M_{k\ell}] = -\delta_{ik}M_{j\ell} - \delta_{j\ell}M_{ik} + \delta_{i\ell}M_{jk} + \delta_{jk}M_{i\ell}, \\
&[H, P_i^{(n)}] = -nP_i^{(n-1)}, \qquad [D, P_i^{(n)}] = 2(\ell - n)P_i^{(n)}, \\
&[C, P_i^{(n)}] = (2\ell - n)P_i^{(n+1)}, \qquad [M_{ij}, P_k^{(n)}] = -\delta_{ik}P_j^{(n)} + \delta_{jk}P_i^{(n)}.
\end{aligned} \tag{1}$$

This Lie algebra was introduced as a nonrelativistic conformal algebra in [1, 2]. Indeed one may see that it has the subalgebra $\langle D, H, C\rangle \simeq sl(2,\mathbb{R}) \simeq so(2,1)$ which is the conformal algebra in $(1+1)$ dimensional spacetime. The commutative subalgebra $\langle P_i^{(n)}\rangle$ carries the spin ℓ

representation of $sl(2,\mathbb{R})$ subalgebra and the vector representation of $so(d)$ subalgebra generated by M_{ij}. The parameter d indicates the dimension of space on which the generators act as infinitesimal transformations of the spacetime when a particular realization is considered. We denote the algebra specified by (d,ℓ) by $\mathfrak{g}_\ell(d)$.

It is known that $\mathfrak{g}_\ell(d)$ has two distinct type of central extensions according to the values of d and ℓ :
(i) mass extension existing for any d and $\ell \in \mathbb{Z}_{>0} + \frac{1}{2}$

$$[P_i^{(m)}, P_j^{(n)}] = \delta_{ij}\, \delta_{m+n,2\ell}\, I_m M, \qquad I_m = (-1)^{m+\ell+\frac{1}{2}}(2\ell-m)!\, m!. \tag{2}$$

(ii) exotic extension existing only for $d=2$ and $\ell \in \mathbb{Z}_{>0}$

$$[P_i^{(m)}, P_j^{(n)}] = \epsilon_{ij}\, \delta_{m+n,2\ell}\, I_m \Theta, \qquad I_m = (-1)^{m}(2\ell-m)!\, m!, \tag{3}$$

where ϵ_{ij} is the antisymmetric tensor with $\epsilon_{12}=1$. Simple explanation of the existence of two distinct central extensions is found in [9]. The Schrödinger algebra considered by Niederer corresponds to $\mathfrak{g}_{1/2}(d)$ with the mass central extension [3]. The exotic extension was first found for $\ell=1$ in the study of classical mechanics having higher order time derivatives [10, 11, 12].

The purpose of the present work is to develop a representation theory of $\mathfrak{g}_\ell(d)$ with the central extensions for $d=1,2$. Namely, we shall investigate representations of $\mathfrak{g}_\ell(2)$ for all possible values of ℓ but those of $\mathfrak{g}_\ell(1)$ for half-integer values of ℓ. As an application of the representation theory we derive partial differential equations which have kinematical invariance of the group generated by $\mathfrak{g}_\ell(d)$. The choice of particular values of d is motivated by the observation that $\mathfrak{g}_\ell(d)$ has more complicated structure for $d>2$ as the spatial rotation becomes non-abelian. There are a couple of reasons of focusing on the central extensions: as shown in [5], in the case of $\mathfrak{g}_{1/2}(1)$ without the mass central extension, the representation theory is more involved and the resulting differential equations are not of physical importance. In many physical applications of $\mathfrak{g}_\ell(d)$ the central extensions play an important roles (see for example [11, 13, 14]).

The present work is an extension of the previous works [5, 6, 7, 8, 16]. In [5, 6] a classification of the irreducible lowest weight representations of $\mathfrak{g}_{1/2}(d)$ $(d=1,2,3)$ and differential equations having kinematical symmetries generated by the algebras are given. Such differential equations for $\mathfrak{g}_{1/2}(d)$ with any d have been obtained in [7, 8]. On the other hand [16] classifies the irreducible highest weight representations of $\mathfrak{g}_1(2)$ with the exotic central extension. All those works show that one may apply the techniques such as Verma modules, singular vectors to investigate irreducible representations of $\mathfrak{g}_\ell(d)$. They also showed that differential equations having the group generated by $\mathfrak{g}_\ell(d)$ as kinematical symmetry may be obtained by the method developed for semisimple Lie groups [4].

We organize this paper as follows. In the next section we provide an explicit formula of singular vectors in the Verma modules over $\mathfrak{g}_\ell(1)$ with the mass central extension. This is used to give a classification of irreducible lowest weight modules of the algebras. We then employ the method of [4] to derive differential equations, such that kinematical symmetries of the equations are given by the Lie group generated by $\mathfrak{g}_\ell(1)$ with the mass central extension. In §3 we study reducibility of Verma modules over $\mathfrak{g}_\ell(2)$ with the mass or the exotic central extensions by explicit construction of singular vectors. By the result we derive differential equations having kinematical symmetries generated by $\mathfrak{g}_\ell(2)$ with the central extensions. We close the paper with some remarks in §4.

2. Representation theory of $d=1$ conformal Galilei algebra and invariant equations

2.1. Classification of irreducible representations

We start our investigations with $\mathfrak{g}_\ell(1)$ with the mass central extension, i.e., ℓ being a positive half-integer. The algebra, denoted simply by $\mathfrak{g}$, has no generators of spatial rotations and

Journal of Physics: Conference Series **512** (2014) 012015 doi:10.1088/1742-6596/512/1/012015

does not need an index for space coordinates. Namely, one may write the generators of $\mathfrak{g}$ as $D, H, C, P^{(n)}, M$ with $n = 0, 1, \ldots, 2\ell$. Their nonvanishing commutators are given by (1) and (2). Our aim in this subsection is to give a classification of the lowest weight modules of $\mathfrak{g}$. This will be done by a way analogous to the case of semisimple Lie algebras. That is, we introduce the Verma module over $\mathfrak{g}$ which is in a sense the largest lowest weight module for a given set of lowest weights. Then use the theorem that any lowest weight module over $\mathfrak{g}$ with the same set of lowest weight is isomorphic to a quotient of the Verma module [15].

To define the Verma modules we make the following vector space decomposition of $\mathfrak{g}$:

$$\begin{aligned} \mathfrak{g}^+ &= \langle\, H, P^{(0)}, P^{(1)}, \cdots, P^{(\ell-\frac{1}{2})} \,\rangle, \\ \mathfrak{g}^0 &= \langle\, D, M \,\rangle, \\ \mathfrak{g}^- &= \langle\, C, P^{(\ell+\frac{1}{2})}, P^{(\ell+\frac{3}{2})}, \cdots, P^{(2\ell)} \,\rangle. \end{aligned} \tag{4}$$

Then one may see that $[\mathfrak{g}^0, \mathfrak{g}^{\pm}] \subset \mathfrak{g}^{\pm}$, that is, this is analogues to the triangular decomposition of semisimple Lie algebras. Suppose that there exists a lowest weight vector defined by

$$\begin{aligned} &D\,|\delta,\mu\rangle = -\delta\,|\delta,\mu\rangle\,, \quad M\,|\delta,\mu\rangle = -\mu\,|\delta,\mu\rangle\,, \\ &X\,|\delta,\mu\rangle = 0, \quad {}^{\forall}X \in \mathfrak{g}^- \end{aligned} \tag{5}$$

A Verma module $V^{\delta,\mu}$ over $\mathfrak{g}$ with the lowest weights $(-\delta,-\mu)$ is defined by $V^{\delta,\mu} = U(\mathfrak{g}^+)\,|\delta,\mu\rangle$ where $U(\mathfrak{g}^+)$ is the universal enveloping algebra of $\mathfrak{g}^+$. More precisely, one may defined $V^{\delta,\mu}$ as the vector space with the following basis:

$$V^{\delta,\mu} = \left\{ H^h \prod_{j=0}^{\ell-\frac{1}{2}} P_{\ell-\frac{1}{2}-j}^{k_j}\,|\delta,\mu\rangle \;\middle|\; h, k_0, k_1, \cdots, k_{\ell-\frac{1}{2}} \in \mathbb{Z}_{\geq 0} \right\}. \tag{6}$$

The vector space $V^{\delta,\mu}$ carries a representation of $\mathfrak{g}$ specified by the lowest weights $(-\delta,-\mu)$. In general, the representation is not irreducible (reducible). Reducibility of $V^{\delta,\mu}$ is detected by singular vectors. A singular vector $|v_q\rangle$ is a vector having the same property as $|\delta,\mu\rangle$ but different eigenvalues. If the Verma module has a singular vector, then the representation is reducible, since the space $I^{\delta,\mu} = U(\mathfrak{g}^+)\,|v_q\rangle$ is obviously an invariant subspace of $V^{\delta,\mu}$. To obtain an irreducible module we consider a quotient module $V^{\delta,\mu}/I^{\delta,\mu}$. If the quotient module has no singular vectors then the quotient module is irreducible. If the module has singular vectors we take a quotient again and repeat the same procedure until we have an irreducible module. In this way one may arrive at a complete classification of irreducible lowest weight modules over $\mathfrak{g}$. Here we give only the results of this procedure. One may find the proofs in [17] (In [17] highest weight Verma modules are considered. It is easy to convert it to the lowest weight modules).

Proposition 1 *If* $2\delta - 2(q-1) + (\ell + \frac{1}{2})^2 = 0$ $q \in \mathbb{Z}_{>0}$ *and* $\mu \neq 0$, *then* $V^{\delta,\mu}$ *has precisely one singular vector:*

$$|v_q\rangle = \mathcal{S}^q\,|0\rangle\,, \quad \mathcal{S} = 2((\ell - \tfrac{1}{2}))^2 \mu H + (P^{(\ell-\frac{1}{2})})^2 \tag{7}$$

Namely, $|v_q\rangle$ *satisfies the relations:*

$$\begin{aligned} &D\,|v_q\rangle = (2q - \delta)\,|v_q\rangle\,, \quad M\,|v_q\rangle = -\mu\,|v_q\rangle\,, \\ &X\,|v_q\rangle = 0, \quad {}^{\forall}X \in \mathfrak{g}^- \end{aligned} \tag{8}$$

Theorem 2 *The irreducible lowest weight modules of* $d = 1$ ℓ*-CGA with the mass central extension are listed as follows* $(\mu \neq 0)$:

- $V^{\delta,\mu}$ *if* $2\delta - 2(q-1) + (\ell + \frac{1}{2})^2 \neq 0$,
- $V^{\delta,\mu}/I^{\delta,\mu}$ *if* $2\delta - 2(q-1) + (\ell + \frac{1}{2})^2 = 0$,

where $I^{\delta,\mu} = U(\mathfrak{g}^+)\,|v_q\rangle$ *and* $q \in \mathbb{Z}_{>0}$. *All modules are infinite dimensional.*

2.2. Kinematically invariant differential equations

In this subsection we derive differential equations invariant under the group generated by $\mathfrak{g}$. Here the *invariance* means the kinematical symmetry in the sense of [3]. This is done by the method developed for real connected semisimple Lie groups [4]. We first summarize the recipe for deriving such differential equations.

Let g be a Lie algebra and $\mathsf{G} = \exp(\mathsf{g})$ be the group generated by g. Suppose that g admits a triangular-like decomposition such as (4): $\mathsf{g} = \mathsf{g}^+ \oplus \mathsf{g}^0 \oplus \mathsf{g}^-$. Then the Lie group G also has the corresponding decomposition: $\mathsf{G} = \mathsf{G}^+\mathsf{G}^0\mathsf{G}^-$ where $\mathsf{G}^\pm = \exp(\mathsf{g}^\pm)$ and $\mathsf{G}^0 = \exp(\mathsf{g}^0)$. Consider a Verma module V^Λ over g with the lowest weight vector $|0\rangle$ and the lowest weight Λ :

$$\begin{aligned} &V^\Lambda = U(\mathsf{g}^+)\,|0\rangle\,, \\ &X\,|0\rangle = 0, \qquad X \in \mathsf{g}^-, \\ &X\,|0\rangle = \Lambda(X)\,|0\rangle\,, \qquad X \in \mathsf{g}^0 \end{aligned} \tag{9}$$

We assume that the Verma module has a singular vector $|v_s\rangle = \mathcal{P}\,|0\rangle\,, \mathcal{P} \in U(\mathsf{g}^+)$.

Consider a C^∞ function on G having the property of *right covariance*:

$$f(gg^0g^-) = e^{\Lambda(X)}f(g), \quad {}^\forall g \in \mathsf{G},\ {}^\forall g^0 = e^X \in \mathsf{G}^0,\ {}^\forall g^- \in \mathsf{G}^- \tag{10}$$

Thus the function $f(g)$ is actually a function on the coset $\mathsf{G}/\mathsf{G}^0\mathsf{G}^-$. Now consider the space C^Λ of right covariant functions on G. We introduce the right action of g on C^Λ by the standard formula:

$$\pi_R(X)f(g) = \frac{d}{d\tau}f(ge^{\tau X})\bigg|_{\tau=0}, \quad X \in \mathsf{g},\ g \in \mathsf{G} \tag{11}$$

Then it is immediate to verify by the right covariance (10) that $f(g)$ has the properties of lowest weight vector:

$$\begin{aligned} &\pi_R(X)f(g) = 0, \qquad X \in \mathsf{g}^- \\ &\pi_R(X)f(g) = \Lambda(X)f(g), \qquad X \in \mathsf{g}^0 \end{aligned} \tag{12}$$

Thus one may have a realization of the Verma module V^Λ in terms of C^Λ.

One may consider the left regular representation of G on C^Λ defined by

$$(T^\Lambda(g)f)(g') = f(g^{-1}g'). \tag{13}$$

We remark that we take the same lowest weights for g and G. Recalling that a singular vector induces an another lowest weight representation with the different lowest weight Λ', it is not difficult to see that the singular vector in the representation π_R gives an intertwining operator of two representations of G :

$$\pi_R(\mathcal{P})T^\Lambda = T^{\Lambda'}\pi_R(\mathcal{P}). \tag{14}$$

As will be seen later $\pi_R(\mathcal{P})$ is, in general, a differential operator. If $\pi_R(\mathcal{P})$ has a kernel, i.e.,

$$\pi_R(\mathcal{P})\psi = 0, \tag{15}$$

then we have a differential equation. Furthermore the solution ψ to the differential equation (15) is transformed to another solution by the left action of G because of the relation (14). Thus equation (15) is the desired equation.

Now we apply the recipe to $\mathfrak{g}$, i.e., $\mathfrak{g}_\ell(1)$ with the mass central extension. Parameterizing an element of $G^+ = \exp(\mathfrak{g}^+)$ as $g^+ = \exp(tH + \sum_{j=0}^{\ell-\frac{1}{2}} x_jP^{(j)})$ the right action of H and $P^{(k)}$ becomes differential operators:

$$\pi_R(H) = \frac{\partial}{\partial t} + \sum_{j=1}^{\ell-\frac{1}{2}} jx_j\frac{\partial}{\partial x_{j-1}}, \qquad \pi_R(P^{(k)}) = \frac{\partial}{\partial x_k}. \tag{16}$$

By (7) and (15) we have proved the following:

Journal of Physics: Conference Series **512** (2014) 012015 doi:10.1088/1742-6596/512/1/012015

Proposition 3 *The Lie group generated by* $\mathfrak{g}$ *is a kinematical symmetry of the following hierarchy of partial differential equations:*

$$\left(a_\ell \mu \Big(\frac{\partial}{\partial t} + \sum_{j=1}^{\ell-\frac{1}{2}} j x_j \frac{\partial}{\partial x_{j-1}} \Big) + \frac{\partial^2}{\partial^2 x_{\ell-\frac{1}{2}}} \right)^q \psi(t, x_i) = 0, \qquad q \in \mathbb{Z}_{>0} \tag{17}$$

where $a_\ell = 2((\ell - \frac{1}{2})!)^2$.

For $\ell = 1/2$ (17) recovers a hierarchy of heat/Schrödinger equations in one space dimension obtained in [5, 7, 8]:

$$\left(2\mu \frac{\partial}{\partial t} + \frac{\partial^2}{\partial x^2} \right)^q \psi(t, x) = 0.$$

The central charge μ is interpreted as (imaginary) mass. For higher values of ℓ we observe an interesting deviation from the heat/Schrödinger equation and the obtained equations are highly nontrivial. As an illustration we show the hierarchy of equations for $\ell = 3/2$ and $5/2$:

$$\left(2\mu \Big(\frac{\partial}{\partial t} + x_1 \frac{\partial}{\partial x_0} \Big) + \frac{\partial^2}{\partial x_1^2} \right)^q \psi(t, x_0, x_1) = 0, \quad \ell = 3/2$$

$$\left(8\mu \Big(\frac{\partial}{\partial t} + x_1 \frac{\partial}{\partial x_0} + 2x_2 \frac{\partial}{\partial x_1} \Big) + \frac{\partial^2}{\partial x_2^2} \right)^q \psi(t, x_0, x_1, x_2) = 0, \quad \ell = 5/2$$

3. Representation theory of $d = 2$ conformal Galilei algebra and invariant equations

In this section we study $\mathfrak{g}_\ell(2)$ with the mass or the exotic central extensions. We give a condition for reducibility of Verma modules and differential equations having kinematical invariance. Here we present our main results without proofs. The proofs and other details are found in the separate publication [18].

3.1. Mass central extension

The algebra $\mathfrak{g}_\ell(2)$ with the mass central extension, we denote it simply by $\hat{\mathfrak{g}}$, has the following generators:

$$D,\ H,\ C,\ M_{12},\ P_i^{(n)}, M, \qquad i, j = 1, 2, \quad n = 0, 1, \cdots, 2\ell$$

where ℓ is a positive half-integer. We make the following changes of generators:

$$P_{\pm}^{(n)} = P_1^{(n)} \pm i P_2^{(n)}, \qquad J = -i M_{12}, \qquad M = 2M, \tag{18}$$

then their nonvanishing commutators are given by

$$\begin{aligned}
&[D, H] = 2H, \qquad [D, C] = -2C, \qquad [C, H] = D,\\
&[H, P_{\pm}^{(n)}] = -n P_{\pm}^{(n-1)}, \qquad [D, P_{\pm}^{(n)}] = 2(\ell - n) P_{\pm}^{(n)},\\
&[C, P_{\pm}^{(n)}] = (2\ell - n) P_{\pm}^{(n+1)}, \qquad [J, P_{\pm}^{(n)}] = \pm P_{\pm}^{(n)},\\
&[P_{\pm}^{(m)}, P_{\mp}^{(n)}] = \delta_{m+n, 2\ell}\, I_m M,
\end{aligned} \tag{19}$$

where I_m is given by (2). We introduce a vector space decomposition of $\hat{\mathfrak{g}}$:

$$\begin{aligned}
&\hat{\mathfrak{g}}^+ = \langle\ H,\ P_{\pm}^{(n)}\ \rangle, \quad n = 0, 1, \cdots, \ell - \frac{1}{2}\\
&\hat{\mathfrak{g}}^0 = \langle\ D,\ J,\ M\ \rangle,\\
&\hat{\mathfrak{g}}^- = \langle\ C,\ P_{\pm}^{(n)}\ \rangle, \quad n = \ell + \frac{1}{2}, \cdots, 2\ell
\end{aligned} \tag{20}$$

This corresponds to the triangular decomposition of semisimple Lie algebras.

Suppose the existence of the lowest weight vector $|\delta, r, \mu\rangle$ defined by

$$\begin{aligned} D\,|\delta,r,\mu\rangle &= -\delta\,|\delta,r,\mu\rangle\,, \quad & J\,|\delta,r,\mu\rangle &= -r\,|\delta,r,\mu\rangle\,,\\ M\,|\delta,r,\mu\rangle &= -\mu\,|\delta,r,\mu\rangle\,, \quad & X\,|\delta,r,\mu\rangle &= 0, \ {}^{\forall}X \in \hat{\mathfrak{g}}^- \end{aligned} \tag{21}$$

One may build a Verma module $V^{\delta,r,\mu}$ on $|\delta, r, \mu\rangle$ whose basis is given by

$$H^k \prod_{n=0}^{\ell-\frac{1}{2}} (P_+^{(n)})^{a_n} (P_-^{(n)})^{b_n}\,|\delta,r,\mu\rangle\,, \tag{22}$$

where k, a_n, b_n $(n = 0, 1, \ldots, \ell - \frac{1}{2})$ are non-negative integers. The Verma module $V^{\delta,r,\mu}$ is not always irreducible as indicated in the following proposition.

Proposition 4 *The Verma module $V^{\delta,r,\mu}$ has a singular vector if $\delta - q + (\ell + \frac{1}{2})^2 + 1 = 0$ for a positive integer q. It is given by*

$$|v_q\rangle = (a_\ell \mu H + P_+^{(\ell-\frac{1}{2})} P_-^{(\ell-\frac{1}{2})})^q\,|\delta,r,\mu\rangle\,, \quad \alpha_\ell = \left(\left(\ell - \frac{1}{2}\right)!\right)^2 \tag{23}$$

and satisfies the relations

$$\begin{aligned} &D\,|v_q\rangle = (2q-\delta)\,|v_q\rangle\,, \quad J\,|v_q\rangle = -r\,|v_q\rangle\,, \quad M\,|v_q\rangle = -\mu\,|v_q\rangle\,,\\ &X\,|v_q\rangle = 0, \quad {}^{\forall}X \in \hat{\mathfrak{g}}^- \end{aligned} \tag{24}$$

The existence of the singular vectors (24) allows us to derive differential equations such that their kinematical symmetry is given by the Lie group $\exp(\hat{\mathfrak{g}})$ as was done in §2.2. Let us parametrize an element $g \in \exp(\hat{\mathfrak{g}}^+)$ as

$$g = \exp(tH)\exp\left(\sum_{n=0}^{\ell-\frac{1}{2}} (x_n P_+^{(n)} + y_n P_-^{(n)})\right), \tag{25}$$

then the right action of elements in $\hat{\mathfrak{g}}^+$ on a right covariant function $f(g)$ with $g \in \exp(\hat{\mathfrak{g}})$ becomes differential operators:

$$\begin{aligned} \pi_R(H) &= \frac{\partial}{\partial t} + \sum_{n=1}^{\ell-\frac{1}{2}} n\left(x_n \frac{\partial}{\partial x_{n-1}} + y_n \frac{\partial}{\partial y_{n-1}}\right),\\ \pi_R(P_+^{(n)}) &= \frac{\partial}{\partial x_n}, \qquad \pi_R(P_-^{(n)}) = \frac{\partial}{\partial y_n}. \end{aligned} \tag{26}$$

By this and Proposition 4 we have prove the following:

Proposition 5 *The Lie group generated by $\hat{\mathfrak{g}}$ is a kinematical symmetry of the following hierarchy of partial differential equations:*

$$\left[a_\ell \mu \left(\frac{\partial}{\partial t} + \sum_{n=1}^{\ell-\frac{1}{2}} n\left(x_n \frac{\partial}{\partial x_{n-1}} + y_n \frac{\partial}{\partial y_{n-1}}\right)\right) + \frac{\partial^2}{\partial x_{\ell-\frac{1}{2}} \partial y_{\ell-\frac{1}{2}}}\right]^q \psi(t, x_i, y_i) = 0, \tag{27}$$

where $q \in \mathbb{Z}_{>0}$.

For $\ell = 1/2$ (27) yields the following form:

$$\left(\mu\frac{\partial}{\partial t} + \frac{\partial^2}{\partial x \partial y}\right)^q \psi(t,x,y) = 0.$$

By an appropriate change of variables, this recovers a hierarchy of heat/Schrödinger equations in two space dimension obtained in [6, 7, 8].

3.2. Exotic central extension

We denote $\mathfrak{g}_\ell(2)$ with the exotic central extension by $\check{\mathfrak{g}}$. It has the generators

$$D,\ H,\ C,\ M_{12},\ P_i^{(n)}, \Theta, \qquad i,j=1,2, \quad n=0,1,\cdots,2\ell$$

where ℓ is a positive integer. We denote the central element of $\check{\mathfrak{g}}$ by Θ to make clear the difference from the case of $\hat{\mathfrak{g}}$. We redefined the generators as

$$P_\pm^{(n)} = P_1^{(n)} \pm iP_2^{(n)}, \qquad J=-iM_{12}, \qquad \Theta=-2i\Theta, \tag{28}$$

then the nonvanishing commutators of $\check{\mathfrak{g}}$ are identical to (19) except the central extension. The central extension for $\check{\mathfrak{g}}$ in terms of new generators is given by

$$[P_\pm^{(m)}, P_\mp^{(n)}] = \pm\delta_{m+n,2\ell} I_m \Theta \tag{29}$$

with I_m defined in (3).

We make a triangular like decomposition of $\check{\mathfrak{g}}$:

$$\begin{aligned} \check{\mathfrak{g}}^+ &= \langle\ H,\ P_+^{(\ell)},\ P_\pm^{(n)}\ \rangle, \quad n=0,1,\cdots,\ell-1 \\ \check{\mathfrak{g}}^0 &= \langle\ D,\ J,\ \Theta\ \rangle, \\ \check{\mathfrak{g}}^- &= \langle\ C,\ P_-^{(\ell)},\ P_\pm^{(n)}\ \rangle, \quad n=\ell+1,\ell+2,\cdots,2\ell \end{aligned} \tag{30}$$

The lowest weight vector $|\delta,r,\theta\rangle$ is defined as usual:

$$\begin{aligned} &D\,|\delta,r,\theta\rangle = -\delta\,|\delta,r,\theta\rangle\,, \qquad J\,|\delta,r,\theta\rangle = -r\,|\delta,r,\theta\rangle\,, \\ &\Theta\,|\delta,r,\theta\rangle = \theta\,|\delta,r,\theta\rangle\,, \qquad X\,|\delta,r,\theta\rangle = 0 \text{ for } {}^\forall X \in \check{\mathfrak{g}}^-. \end{aligned} \tag{31}$$

We construct a Verma module $V^{\delta,r,\theta}$ over $\check{\mathfrak{g}}$ by repeated applications of an element of $\check{\mathfrak{g}}^+$ on $|\delta,r,\theta\rangle$. A basis of $V^{\delta,r,\theta}$ is given by

$$H^h (P_+^{(\ell)})^{a_\ell} \prod_{n=0}^{\ell-1} (P_+^{(n)})^{a_n} (P_-^{(n)})^{b_n}\,|\delta,r,\mu\rangle\,, \tag{32}$$

where $h, a_n\ (n=0,1,\ldots,\ell), b_n\ (n=0,1,\ldots,\ell-1)$ are non-negative integers. The Verma module $V^{\delta,r,\theta}$ is not always irreducible as $\mathfrak{g}$ or $\hat{\mathfrak{g}}$.

Proposition 6 *The Verma module $V^{\delta,r,\theta}$ has a singular vector if $\delta-q+\ell(\ell+1)+1=0$ for a positive integer q. It is given by*

$$|v_q\rangle = (\alpha_\ell \theta H + (-1)^\ell P_-^{(\ell-1)} P_+^{(\ell)})^q\,|\delta,r,\theta\rangle\,, \quad \alpha_\ell = \ell!(\ell-1)! \tag{33}$$

and satisfies the relations

$$\begin{aligned} &D\,|v_q\rangle = (2q-\delta)\,|v_q\rangle\,, \quad J\,|v_q\rangle = -r\,|v_q\rangle\,, \quad \Theta\,|v_q\rangle = \theta\,|v_q\rangle\,, \\ &X\,|v_q\rangle = 0, \quad {}^\forall X \in \check{\mathfrak{g}}^- \end{aligned} \tag{34}$$

This generalize the result for $\ell=1$ investigated in [16].

Now let us derive differential equations having kinematical symmetries generated by $\check{\mathfrak{g}}$. We parametrize an element $g \in \exp(\check{\mathfrak{g}}^+)$ as

$$g = e^{tH} \exp\left(\sum_{n=0}^{\ell-1} (x_n P_+^{(n)} + y_n P_-^{(n)}) + x_\ell P_+^{(\ell)}\right). \tag{35}$$

Then the right action of elements in $\breve{\mathfrak{g}}^+$ on a right covariant function $f(g)$ with $g \in \exp(\breve{\mathfrak{g}})$ becomes differential operators:

$$\pi_R(H) = \frac{\partial}{\partial t} + \sum_{n=0}^{\ell-1}(n+1)x_{n+1}\frac{\partial}{\partial x_n} + \sum_{n=0}^{\ell-2}(n+1)y_{n+1}\frac{\partial}{\partial y_n},$$
$$\pi_R(P_+^{(n)}) = \frac{\partial}{\partial x_n}, \qquad \pi_R(P_-^{(n)}) = \frac{\partial}{\partial y_n}. \tag{36}$$

By this and Proposition 6 we have prove the following:

Proposition 7 *The Lie group generated by* $\hat{\mathfrak{g}}$ *is a kinematical symmetry of the following hierarchy of partial differential equations:*

$$\left[\alpha_\ell\theta\left(\frac{\partial}{\partial t} + \sum_{n=1}^{\ell} nx_n\frac{\partial}{\partial x_{n-1}} + \sum_{n=1}^{\ell-1} ny_n\frac{\partial}{\partial y_{n-1}}\right) + (-1)^\ell\frac{\partial^2}{\partial y_{\ell-1}\partial x_\ell}\right]^q \psi = 0, \tag{37}$$

where $q \in \mathbb{Z}_{>0}$.

4. Concluding remarks

We have studied the lowest weight representations of ℓ-conformal Galilei algebras with central extensions for $d = 1, 2$. For $d = 1$ algebras we gave a classification of all irreducible lowest weight modules, however, the result for $d = 2$ algebras is partial in the sense that a full classification of irreducible modules has not been done yet. In fact, our knowledge on irreducible modules of $\mathfrak{g}_\ell(d)$ is very limited. We summarize in Table 1 the works on lowest (highest) weight representations of $\mathfrak{g}_\ell(d)$ have been done so far. In Table 1 'full classification' lists the algebras for which a classification of irreducible lowest (highest) weight modules has been given. On the other hand 'partial result' means that it has been shown that some Verma modules are reducible because of the existence of singular vectors, but reducibility of other Verma modules are still open. The algebras are indicated by a pair (d, ℓ). Recalling that d takes any positive integers and ℓ takes any spin values, one can see that, what has been done so far is very little compared with the whole family of $\mathfrak{g}_\ell(d)$. There are many things to be done in order to achieve a complete classification of irreducible modules of $\mathfrak{g}_\ell(d)$. However the method of the classification employed in [5, 6, 16, 17], which is useful for lower values of d, will not be efficient for higher values of d. We need to develop an alternate method.

Table 1. Works on irreducible modules of $\mathfrak{g}_\ell(d)$.

full classification	partial result
(1,any half-integer)	$(2,\ell \geq \frac{3}{2})$
$(2,\frac{1}{2})$	$(d \geq 4, \frac{1}{2})$
(2,1)	
$(3,\frac{1}{2})$	

In the present work we also discuss differential equations having kinematical symmetries generated by $\mathfrak{g}_\ell(d)$. The differential equations for $\ell = 1/2$ are a hierarchy of free Schrödinger equations. However physical implication of other equations is still an open problem. Because of the method of construction, we obtained linear differential equations having kinematical symmetries. It may be interesting to consider differential equations having full Lie symmetries generated by $\mathfrak{g}_\ell(d)$ along the line of [19, 20, 21, 22, 23].

Acknowledgments
N.A. is supported by a grants-in-aid from JSPS (Contract No.23540154).

References
[1] Negro J, del Olmo M A and Rodríguez-Marco A 1997 *J. Math. Phys.* **38** 3786
[2] Negro J, del Olmo M A and Rodríguez-Marco A 1997 *J. Math. Phys.* **38** 3810
[3] Niederer U 1972 *Helv. Phys. Acta* **45** 802
[4] Dobrev V K 1988 *Rep. Math. Phys.* **25** 159
[5] Dobrev V K, Doebner H-D and Mrugalla Ch 1997 *Rep. Math. Phys.* **39** 201
[6] Mrugalla Ch 1997 Ph.D Thesis, TU Clausthal
[7] Aizawa N, Dobrev V K and Doebner H-D 2002 *Proc. Second Int. Symposium on Quantum Theory and Symmetries* ed E Kapuścik and A Horzela (World Scientific) p 222
[8] Aizawa N, Dobrev V K, Doebner H-D and Stoimenov S 2008 *Proc. VII Int. Workshop on Lie Theory and Its Appl. in Phys.* ed H-D Doebner and V K Dobrev (Sofia: Heron Press) p 372
[9] Martelli D and Tachikawa Y 2010 *J. High Energy Phys.* **JHEP05(2010)** 091
[10] Stichel P C and Zakrzewski W J 2004 *Ann. Phys., NY* **310** 158
[11] Lukierski J, Stichel P C and Zakrzewski W J 2006 *Phys. Lett.* A **357** 1
[12] Lukierski J, Stichel P C and Zakrzewski W J 2007 *Phys. Lett.* B **650** 203
[13] Son D T 2009 *Phys. Rev.* **D78** 046003
[14] Balasubramanian K and McGreevy J 2008 *Phys. Rev. Lett.* **101** 061601
[15] Dixmier J 1996 *Enveloping Algebras* (AMS)
[16] Aizawa N and Isaac P S 2011 *J. Phys. A:Math. Theor.* **44** 035401
[17] Aizawa N, Isaac P S and Kimura Y 2012 *Int. J. Math.* **23** 1250118
[18] Aizawa N, Kimura Y and Segar J 2013 *J. Phys. A: Math. Theor.* **46** 405204
[19] Ovsiannikov L V 1980 *The Group Analysis of Differential Equations* (Academic Press)
[20] Olver P 1986 *Applications of Lie Groups to Differential Equations* (Springer)
[21] Fushchich W and Cherniha R 1989 Ukrainian Math. J. **41** 1161
[22] Rideau G and Winternitz P 1993 J. Math. Phys. **34** 558
[23] Fushchych W I and Cherniha R M 1995 *J. Phys. A:Math. Gen.* **28** 5569

Journal of Physics: Conference Series **512** (2014) 012016 doi:10.1088/1742-6596/512/1/012016

Cotangent bundle over all the compact Hermitian symmetric spaces and projective superspace

Masato Arai

Fukushima National College of Technology, Iwaki, Fukushima 970-8034, Japan

Institute of Experimental and Applied Physics, Czech Technical University in Prague, Horská 22, 128 00 Prague 2, Czech Republic

E-mail: masato.arai@fukushima-nct.ac.jp

Abstract. We construct the $\mathcal{N}=2$ supersymmetric nonlinear sigma model on the cotangent bundle over the compact Hermitian symmetric space $E_7/E_6 \times U(1)$ by using the projective superspace formalism which is an off-shell superfield formulation in four-dimensional space-time. We also give a simple formula giving the hyper-Kähler potential of the cotangent bundle over all the compact Hermitian symmetric spaces.

1. Introduction

Supersymmetry (SUSY) has intimate relations to the complex geometry [1]. In particular, the target space of SUSY nonlinear sigma models (NLSMs) possessing 8 supercharges must be hyper-Käbler manifold [2]. It means that SUSY is a powerful tool to construct a new hyper-Käbler manifold. In other words, a new hyper-K̈haler manifold can be obtained through a new SUSY NLSM with 8 supercharges. One of the most convenient formulations to construct SUSY NLSMs with 8 supercharges is the projective superspace formalism [3, 4, 5, 6] being an off-shell superfield formulation in theories with 8 supercharges [1].

The projective superspace formalism is a powerful devise to construct $\mathcal{N}=2$ SUSY NLSMs. With the use of this formalism, first the SUSY NLSMs on the tangent bundles over the compact and non-compact classical Hermitian symmetric as well as, using the generalized Legendre transform [4], the cotangent bundles corresponding to the hyperkähler metrics have been constructed [9, 10, 11, 12]. Developed the results in [9, 10, 11, 12], the cotangent bundle over the compact type of $E_6/SO(10) \times U(1)$ have been also constructed [13]. After these works, the closed form expression of the cotangent bundle over any compact Hermitian symmetric space was obtained [14], which gives a form of the cotangent bundle presented in a matrix form. We elaborated on this result to provide a new closed formula for cotangent bundle action, and applied it to construction of the $\mathcal{N}=2$ SUSY NLSM on the cotangent bundle over the compact $E_7/E_6 \times U(1)$ [15]. In this proceeding, we review our work based on [15].

[1] There is another $\mathcal{N}=2$ off-shell superfield formulation called the harmonic superspace formalism [7, 8].

2. $\mathcal{N} = 2$ sigma models in the projective superspace

We start with a family of four-dimensional $\mathcal{N} = 2$ off-shell supersymmetric nonlinear sigma models that are described in ordinary $\mathcal{N} = 1$ superspace by the action:

$$S[\Upsilon, \breve{\Upsilon}] = \frac{1}{2\pi \mathrm{i}} \oint \frac{\mathrm{d}\zeta}{\zeta} \int \mathrm{d}^8 z \, K(\Upsilon^I(\zeta), \breve{\Upsilon}^{\bar{J}}(\zeta)) \, , \quad z^M = (x^\mu, \theta_\alpha, \bar{\theta}^{\dot{\alpha}}), \tag{2.1}$$

where $\mu = 0, 1, 2, 3$ and $\alpha, \dot{\alpha} = 1, 2$. The complex coordinate ζ which is called the projective coordinate parameterizes $SU(2)_R/U(1)$ where $SU(2)_R$ is the internal symmetry. Action in the projective superspace is written by using contour integral over ζ, and reality conditions are imposed using complex conjugation of ζ composed with the antipodal map. The action is formulated in terms of the so-called polar multiplet [4, 5, 6], one of $\mathcal{N} = 2$ multiplets living in the projective superspace. The polar multiplet is described by the so-called arctic superfield $\Upsilon(\zeta)$ and antarctic superfield $\breve{\Upsilon}(\zeta)$ that are generated by an infinite set of ordinary $\mathcal{N} = 1$ superfields:

$$\Upsilon(\zeta) = \sum_{n=0}^{\infty} \Upsilon_n \zeta^n = \Phi + \Sigma \, \zeta + O(\zeta^2) \, , \qquad \breve{\Upsilon}(\zeta) = \sum_{n=0}^{\infty} \bar{\Upsilon}_n (-\zeta)^{-n} \, , \tag{2.2}$$

where $\breve{\Upsilon}$ is the conjugation of Υ under the composition of complex conjugation with the antipodal map on the Riemann sphere, $\bar{\zeta} \to -1/\zeta$. Here Φ is chiral, Σ complex linear,

$$\bar{D}_{\dot{\alpha}} \Phi = 0 \, , \qquad \bar{D}^2 \Sigma = 0 \, , \tag{2.3}$$

and the remaining component superfields are unconstrained complex superfields. The above theory is obtained as a minimal $\mathcal{N} = 2$ extension of the general four-dimensional $\mathcal{N} = 1$ supersymmetric nonlinear sigma model [1]

$$S[\Phi, \bar{\Phi}] = \int \mathrm{d}^8 z \, K(\Phi^I, \bar{\Phi}^{\bar{J}}) \, , \tag{2.4}$$

where K is the Kähler potential of a Kähler manifold $\mathcal{M}$.

To describe the theory in terms of the physical superfields Φ and Σ only, all the auxiliary superfields have to be eliminated with the aid of the corresponding algebraic equations of motion

$$\oint \frac{\mathrm{d}\zeta}{\zeta} \, \zeta^n \, \frac{\partial K(\Upsilon, \breve{\Upsilon})}{\partial \Upsilon^I} = \oint \frac{\mathrm{d}\zeta}{\zeta} \, \zeta^{-n} \, \frac{\partial K(\Upsilon, \breve{\Upsilon})}{\partial \breve{\Upsilon}^{\bar{I}}} = 0 \, , \qquad n \geq 2 \, . \tag{2.5}$$

Let $\Upsilon_*(\zeta) \equiv \Upsilon_*(\zeta; \Phi, \bar{\Phi}, \Sigma, \bar{\Sigma})$ denote a unique solution subject to the initial conditions

$$\Upsilon_*(0) = \Phi \, , \qquad \dot{\Upsilon}_*(0) = \Sigma \, . \tag{2.6}$$

For a general Kähler manifold $\mathcal{M}$, the auxiliary superfields $\Upsilon_2, \Upsilon_3, \dots$, and their conjugates, can be eliminated only perturbatively. Their elimination can be carried out using the ansatz [16]

$$\Upsilon^I_n = \sum_{p=0}^{\infty} G^I{}_{J_1 \dots J_{n+p} \bar{L}_1 \dots \bar{L}_p}(\Phi, \bar{\Phi}) \, \Sigma^{J_1} \dots \Sigma^{J_{n+p}} \, \bar{\Sigma}^{\bar{L}_1} \dots \bar{\Sigma}^{\bar{L}_p} \, , \qquad n \geq 2 \, . \tag{2.7}$$

Upon elimination of the auxiliary superfields, the action (2.1) should take the form [9, 10]

$$S_{\mathrm{tb}}[\Phi, \Sigma] = \int \mathrm{d}^8 z \Big\{ K(\Phi, \bar{\Phi}) + \mathcal{L}(\Phi, \bar{\Phi}, \Sigma, \bar{\Sigma}) \Big\} \, , \tag{2.8}$$

where

$$\mathcal{L}(\Phi,\bar{\Phi},\Sigma,\bar{\Sigma}) = \sum_{n=1}^{\infty} \mathcal{L}_{I_1\cdots I_n \bar{J}_1\cdots\bar{J}_n}(\Phi,\bar{\Phi})\Sigma^{I_1}\ldots\Sigma^{I_n}\bar{\Sigma}^{\bar{J}_1}\ldots\bar{\Sigma}^{\bar{J}_n}. \tag{2.9}$$

The first term in the expansion (2.9) is given as $\mathcal{L}_{I\bar{J}} = -g_{I\bar{J}}(\Phi,\bar{\Phi})$ and the tensors $\mathcal{L}_{I_1\cdots I_n\bar{J}_1\cdots\bar{J}_n}$ for $n>1$ are functions of the Riemann curvature $R_{I\bar{J}K\bar{L}}(\Phi,\bar{\Phi})$ and its covariant derivatives. Eq. (2.8) is written by the base manifold coordinate Φ and the tangent space coordinate Σ. Therefore this is the tangent bundle action.

The complex linear tangent variables Σ's in (2.8) can be dualized into chiral one-forms, in accordance with the generalized Legendre transform [4]. To construct a dual formulation, we consider the first-order action

$$S = \int \mathrm{d}^8 z \left\{ K(\Phi,\bar{\Phi}) + \mathcal{L}(\Phi,\bar{\Phi},\Sigma,\bar{\Sigma}) + \Psi_I\,\Sigma^I + \bar{\Psi}_{\bar{I}}\bar{\Sigma}^{\bar{I}} \right\}, \tag{2.10}$$

where the tangent vector Σ^I is now a complex unconstrained superfield, while the one-form Ψ is chiral, $\bar{D}_{\dot{\alpha}}\Psi_I = 0$. Integrating out Σ's and their conjugate should give the cotangent-bundle action

$$S_{\rm ctb}[\Phi,\Psi] \;=\; \int \mathrm{d}^8 z \left\{ K(\Phi,\bar{\Phi}) + \mathcal{H}(\Phi,\bar{\Phi},\Psi,\bar{\Psi}) \right\}, \tag{2.11}$$

where

$$\mathcal{H}(\Phi,\bar{\Phi},\Psi,\bar{\Psi}) = \sum_{n=1}^{\infty} \mathcal{H}^{I_1\cdots I_n\bar{J}_1\cdots\bar{J}_n}(\Phi,\bar{\Phi})\Psi_{I_1}\ldots\Psi_{I_n}\bar{\Psi}_{\bar{J}_1}\ldots\bar{\Psi}_{\bar{J}_n}, \tag{2.12}$$

with $\mathcal{H}^{I\bar{J}}(\Phi,\bar{\Phi}) = g^{I\bar{J}}(\Phi,\bar{\Phi})$. The variables (Φ^I,Ψ_J) parametrize the cotangent bundle $T^*\mathcal{M}$ of the Kähler manifold $\mathcal{M}$. The action (2.11) is written by chiral superfields only, so the action gives hyperkähler potential.

3. The tangent bundle

In what follows, we consider the case when $\mathcal{M}$ is a Hermitian symmetric space:

$$\nabla_L R_{I_1\bar{J}_1 I_2\bar{J}_2} = \bar{\nabla}_{\bar{L}} R_{I_1\bar{J}_1 I_2\bar{J}_2} = 0\,. \tag{3.1}$$

Then, the algebraic equations of motion (2.5) are known to be equivalent to the holomorphic geodesic equation (with complex evolution parameter) [9, 10]

$$\frac{\mathrm{d}^2\Upsilon^I_*(\zeta)}{\mathrm{d}\zeta^2} + \Gamma^I{}_{JK}\Big(\Upsilon_*(\zeta),\bar{\Phi}\Big)\,\frac{\mathrm{d}\Upsilon^J_*(\zeta)}{\mathrm{d}\zeta}\,\frac{\mathrm{d}\Upsilon^K_*(\zeta)}{\mathrm{d}\zeta} = 0\,, \tag{3.2}$$

under the same initial conditions (2.6). Here $\Gamma^I{}_{JK}(\Phi,\bar{\Phi})$ are the Christoffel symbols for the Kähler metric $g_{I\bar{J}}(\Phi,\bar{\Phi}) = \partial_I\partial_{\bar{J}}K(\Phi,\bar{\Phi})$. In particular, we have

$$\Upsilon^I_2 = -\frac{1}{2}\Gamma^I{}_{JK}(\Phi,\bar{\Phi})\,\Sigma^J\Sigma^K\,. \tag{3.3}$$

According to the principles of projective superspace [4, 5], the action (2.1) is invariant under the $\mathcal{N}=2$ supersymmetry transformations

$$\delta\Upsilon^I(\zeta) = \mathrm{i}\;\; \varepsilon^\alpha_i Q^i_\alpha + \bar{\varepsilon}^i_{\dot{\alpha}}\bar{Q}^{\dot{\alpha}}_i\;\; \Upsilon^I(\zeta), \tag{3.4}$$

when $\Upsilon^I(\zeta)$ is viewed as a $\mathcal{N}=2$ superfield. Here $i=1,2$ is the $SU(2)_R$ index. However, since the action is given in $\mathcal{N}=1$ superspace, it is only the $\mathcal{N}=1$ supersymmetry which is manifestly realized. The second hidden supersymmetry can be shown to act on the physical superfields Φ and Σ as follows (see, e.g. [6])

$$\delta\Phi^I = \bar{\varepsilon}_{\dot\alpha}\bar{D}^{\dot\alpha}\Sigma^I\,, \qquad \delta\Sigma^I = -\varepsilon^\alpha D_\alpha\Phi^I + \bar{\varepsilon}_{\dot\alpha}\bar{D}^{\dot\alpha}\Upsilon_2^I\,. \tag{3.5}$$

Upon elimination of the auxiliary superfields, the action (2.8), which is associated with the Hermitian symmetric space $\mathcal{M}$, is invariant under

$$\delta\Phi^I = \bar{\varepsilon}_{\dot\alpha}\bar{D}^{\dot\alpha}\Sigma^I\,, \qquad \delta\Sigma^I = -\varepsilon^\alpha D_\alpha\Phi^I - \frac{1}{2}\bar{\varepsilon}_{\dot\alpha}\bar{D}^{\dot\alpha}\Big\{\Gamma^I{}_{JK}(\Phi,\bar\Phi)\,\Sigma^J\Sigma^K\Big\}\,. \tag{3.6}$$

The invariance of the action (2.8) under (3.6) requires to satisfy the the following first-order differential equation

$$\frac{1}{2}R_{K\bar{J}L}{}^I\,\frac{\partial\mathcal{L}}{\partial\Sigma^I}\,\Sigma^K\Sigma^L + \frac{\partial\mathcal{L}}{\partial\bar\Sigma^{\bar J}} + g_{I\bar J}\,\Sigma^I = 0\,. \tag{3.7}$$

The solution to the equation (3.7) was first presented in [13] in the form

$$\mathcal{L} = -\frac{\mathrm{e}^{R_{\Sigma,\bar\Sigma}}-1}{R_{\Sigma,\bar\Sigma}}\,|\Sigma|^2\,, \qquad |\Sigma|^2 := g_{I\bar J}\Sigma^I\bar\Sigma^{\bar J}\,. \tag{3.8}$$

An alternative form of the solution was obtained in [14]:

$$\mathcal{L}(\Phi,\bar\Phi,\Sigma,\bar\Sigma) = -\frac{1}{2}\mathbf{\Sigma}^T\mathbf{g}\frac{\ln(\mathbf{1}+\mathbf{R}_{\Sigma,\bar\Sigma})}{\mathbf{R}_{\Sigma,\bar\Sigma}}\mathbf{\Sigma}\,, \tag{3.9}$$

where

$$\mathbf{\Sigma} = \begin{pmatrix}\Sigma^I\\ \bar\Sigma^{\bar I}\end{pmatrix}\,, \qquad \mathbf{g} = \begin{pmatrix}0 & g_{I\bar J}\\ g_{\bar I J} & 0\end{pmatrix}\,, \tag{3.10}$$

$$\mathbf{R}_{\Sigma,\bar\Sigma} = \begin{pmatrix}0 & (R_\Sigma)^I{}_{\bar J}\\ (R_{\bar\Sigma})^{\bar I}{}_J & \end{pmatrix} = \begin{pmatrix}0 & \frac{1}{2}R_K{}^I{}_{L\bar J}\Sigma^K\Sigma^L\\ \frac{1}{2}R_{\bar K}{}^{\bar I}{}_{\bar L J}\bar\Sigma^{\bar K}\bar\Sigma^{\bar L} & \end{pmatrix}\,. \tag{3.11}$$

We shall rewrite (3.9) using the following differential operators

$$R_{\Sigma,\bar\Sigma} = -(R_\Sigma)^I{}_{\bar J}\bar\Sigma^{\bar J}\frac{\partial}{\partial\Sigma^I}, \qquad \bar R_{\Sigma,\bar\Sigma} = -(R_{\bar\Sigma})^{\bar I}{}_J\Sigma^J\frac{\partial}{\partial\bar\Sigma^{\bar I}}\,. \tag{3.12}$$

These operators were originally introduced in [13]. They satisfy the relation

$$R_{\Sigma,\bar\Sigma}\mathcal{L}^{(n)} = \bar R_{\Sigma,\bar\Sigma}\mathcal{L}^{(n)}\,. \tag{3.13}$$

Performing the Taylor expansion for (3.9), we have

$$\mathcal{L} = \sum_{n=1}^{\infty} c_n F^{(n)}\,, \qquad c_n = \frac{(-1)^n}{n}\,, \tag{3.14}$$

where the functions $F^{(n)}$ given by

$$F^{(2k+2)} = \Sigma^I g_{I\bar J}\Big(\Big(R_{\bar\Sigma}R_\Sigma\Big)^k\Big)^{\bar J}{}_{\bar K}(R_{\bar\Sigma})^{\bar K}{}_L\Sigma^L\,, \quad k=0,1,2\ldots \tag{3.15}$$

$$F^{(2k+1)} = \Sigma^I g_{I\bar J}\Big(\Big(R_{\bar\Sigma}R_\Sigma\Big)^k\Big)^{\bar J}{}_{\bar K}\bar\Sigma^{\bar K}\,, \quad k=0,1,2\ldots \tag{3.16}$$

satisfy relations

$$\Sigma^I \frac{\partial}{\partial \Sigma^I} F^{(n)} = \bar{\Sigma}^{\bar{I}} \frac{\partial}{\partial \bar{\Sigma}^{\bar{I}}} F^{(n)} = nF^{(n)} , \tag{3.17}$$

$$\frac{\partial}{\partial \Sigma^I} F^{(2k+2)} = (2k+2) g_{I\bar{J}} \left((R_{\bar{\Sigma}} R_{\Sigma})^k R_{\bar{\Sigma}} \right)^{\bar{J}}{}_K \Sigma^K, \tag{3.18}$$

$$\frac{\partial}{\partial \Sigma^I} F^{(2k+1)} = (2k+1) g_{I\bar{J}} \left((R_{\bar{\Sigma}} R_{\Sigma})^k \right)^{\bar{J}}{}_{\bar{K}} \bar{\Sigma}^{\bar{K}}. \tag{3.19}$$

Using these identities, one can easily prove

$$F^{(n+1)} = \frac{(-R_{\Sigma,\bar{\Sigma}})^n}{n!} |\Sigma|^2 . \tag{3.20}$$

Putting this into (3.14), we arrive at

$$\mathcal{L} = -\sum_{n=1}^{\infty} \frac{(R_{\Sigma,\bar{\Sigma}})^{n-1}}{n!} |\Sigma|^2 = -\int_0^1 \mathrm{d}t\, \mathrm{e}^{tR_{\Sigma,\bar{\Sigma}}} |\Sigma|^2 = -\frac{\mathrm{e}^{R_{\Sigma,\bar{\Sigma}}} - 1}{R_{\Sigma,\bar{\Sigma}}} |\Sigma|^2 . \tag{3.21}$$

This form coincides with (3.8).

4. The cotangent bundle

We start with (2.10) to construct a dual formulation [13]. The action can be shown to be invariant under the following supersymmetry transformations

$$\begin{aligned} \delta\Phi^I &= \frac{1}{2}\bar{D}^2\{\overline{\varepsilon\theta}\,\Sigma^I\} , \\ \delta\Sigma^I &= -\varepsilon^\alpha D_\alpha \Phi^I - \frac{1}{2}\bar{\varepsilon}_{\dot{\alpha}} \bar{D}^{\dot{\alpha}} \Big\{ \Gamma^I{}_{JK}(\Phi,\bar{\Phi})\, \Sigma^J \Sigma^K \Big\} - \frac{1}{2}\overline{\varepsilon\theta}\, \Gamma^I{}_{JK}(\Phi,\bar{\Phi})\, \Sigma^J \bar{D}^2 \Sigma^K , \\ \delta\Psi_I &= -\frac{1}{2}\bar{D}^2\Big\{\overline{\varepsilon\theta}\, K_I(\Phi,\bar{\Phi})\Big\} + \frac{1}{2}\bar{D}^2\Big\{\overline{\varepsilon\theta}\, \Gamma^K_{IJ}(\Phi,\bar{\Phi})\, \Sigma^J\Big\} \Psi_K . \end{aligned} \tag{4.22}$$

These transformations induce the supersymmetry transformations making the action (2.11) invariant

$$\begin{aligned} \delta\Phi^I &= \frac{1}{2}\bar{D}^2\{\overline{\varepsilon\theta}\,\Sigma^I(\Phi,\bar{\Phi},\Psi,\bar{\Psi})\} , \\ \delta\Psi_I &= -\frac{1}{2}\bar{D}^2\Big\{\overline{\varepsilon\theta}\, K_I(\Phi,\bar{\Phi})\Big\} + \frac{1}{2}\bar{D}^2\Big\{\overline{\varepsilon\theta}\, \Gamma^K_{IJ}(\Phi,\bar{\Phi})\, \Sigma^J(\Phi,\bar{\Phi},\Psi,\bar{\Psi})\Big\} \Psi_K , \end{aligned} \tag{4.23}$$

with

$$\Sigma^I(\Phi,\bar{\Phi},\Psi,\bar{\Psi}) = \frac{\partial}{\partial \Psi_I} \mathcal{H}(\Phi,\bar{\Phi},\Psi,\bar{\Psi}) \equiv \mathcal{H}^I . \tag{4.24}$$

The requirement of invariance under such transformations can be shown to be equivalent to the following nonlinear equation on $\mathcal{H}$ [13]:

$$\mathcal{H}^I g_{I\bar{J}} - \frac{1}{2} \mathcal{H}^K \mathcal{H}^L R_{K\bar{J}L}{}^I \Psi_I = \bar{\Psi}_{\bar{J}} . \tag{4.25}$$

This equation also follows directly from (3.7) using the definition of the Ψ's, or, as a consequence of the superspace Legendre transform.

The closed form expression of $\mathcal{H}$ for any Hermitian symmetric space was obtained as a solution of (4.25), which is given by [14]

$$\mathcal{H}(\Phi,\bar{\Phi},\Psi,\bar{\Psi}) = \frac{1}{2}\mathbf{\Psi}^T\mathbf{g}^{-1}\mathcal{F}(-\mathbf{R}_{\Psi,\bar{\Psi}})\mathbf{\Psi}\,, \tag{4.26}$$

where

$$\mathbf{\Psi} = \begin{pmatrix}\Psi_I\\ \bar{\Psi}_{\bar{I}}\end{pmatrix}, \qquad \mathbf{g}^{-1} = \begin{pmatrix}0 & g^{I\bar{J}}\\ g^{\bar{I}J} & 0\end{pmatrix}, \tag{4.27}$$

$$\mathbf{R}_{\Psi,\bar{\Psi}} = \begin{pmatrix}0 & (R_\Psi)_I{}^{\bar{J}}\\ (R_{\bar{\Psi}})_{\bar{I}}{}^J & 0\end{pmatrix} = \begin{pmatrix}0 & \frac{1}{2}R_I{}^{K\bar{J}L}\Psi_K\Psi_L\\ \frac{1}{2}R_{\bar{I}}{}^{\bar{K}J\bar{L}}\bar{\Psi}_{\bar{K}}\bar{\Psi}_{\bar{L}} & 0\end{pmatrix}, \tag{4.28}$$

and

$$\mathcal{F}(x) = \frac{1}{x}\left[\sqrt{1+4x} - 1 - \ln\left(\frac{1+\sqrt{1+4x}}{2}\right)\right]. \tag{4.29}$$

We rewrite (4.26) to a convenient form for our purpose. In a similar way to the tangent bundle case, we introduce the following differential operators

$$R_{\Psi,\bar{\Psi}} = -(R_\Psi)_I{}^{\bar{J}}\bar{\Psi}_{\bar{J}}\frac{\partial}{\partial\Psi_I}, \quad \bar{R}_{\Psi,\bar{\Psi}} = -(R_{\bar{\Psi}})_{\bar{I}}{}^J\Psi_J\frac{\partial}{\partial\bar{\Psi}_{\bar{I}}}. \tag{4.30}$$

They satisfy

$$R_{\Psi,\bar{\Psi}}\mathcal{H}^{(n)} = \bar{R}_{\Psi,\bar{\Psi}}\mathcal{H}^{(n)}. \tag{4.31}$$

Performing the Taylor expansion for (4.26), we have

$$\mathcal{H} = \sum_{n=1}^{\infty} c_n G^{(n)}, \qquad c_n = \frac{(-1)^{n-1}\mathcal{F}^{(n-1)}(0)}{(n-1)!}, \tag{4.32}$$

where the terms $G^{(n)}$ are given by

$$G^{(2k+2)} = \Psi_I g^{I\bar{J}}\Big(\big(R_{\bar{\Psi}}R_\Psi\big)^k\Big)_{\bar{J}}{}^{\bar{K}}(R_{\bar{\Psi}})_{\bar{K}}{}^L\Psi_L, \quad k = 0,1,2\ldots \tag{4.33}$$

$$G^{(2k+1)} = \Psi_I g^{I\bar{J}}\Big(\big(R_{\bar{\Psi}}R_\Psi\big)^k\Big)_{\bar{J}}{}^{\bar{K}}\bar{\Psi}_{\bar{K}}, \quad k = 0,1,2\ldots \tag{4.34}$$

and satisfy following relations

$$\Psi_I\frac{\partial}{\partial\Psi_I}G^{(n)} = \bar{\Psi}_{\bar{I}}\frac{\partial}{\partial\bar{\Psi}_{\bar{I}}}G^{(n)} = nG^{(n)}, \tag{4.35}$$

$$\frac{\partial}{\partial\Psi_I}G^{(2k+2)} = (2k+2)g^{I\bar{J}}\Big(\big(R_{\bar{\Psi}}R_\Psi\big)^k\Big)_{\bar{J}}{}^{\bar{K}}(R_{\bar{\Psi}})_{\bar{K}}{}^L\Psi_L, \tag{4.36}$$

$$\frac{\partial}{\partial\Psi_I}G^{(2k+1)} = (2k+1)g^{I\bar{J}}\Big(\big(R_{\bar{\Psi}}R_\Psi\big)^k\Big)_{\bar{J}}{}^{\bar{K}}\bar{\Psi}_{\bar{K}}. \tag{4.37}$$

By using the above identities, one can show

$$G^{(n+1)} = \frac{(-R_{\Psi,\bar{\Psi}})^n}{n!}|\Psi|^2, \qquad |\Psi|^2 := g^{I\bar{J}}\Psi_I\bar{\Psi}_{\bar{J}}. \tag{4.38}$$

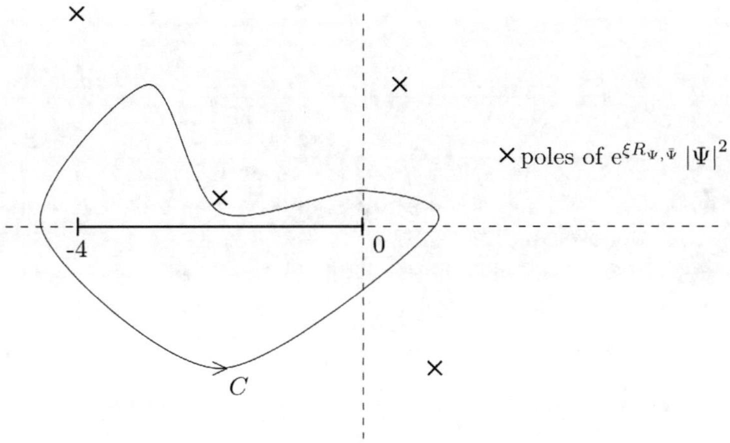

Figure 1. The contour of integration in eq. (4.42).

Putting together (4.32) and (4.38) we find

$$\mathcal{H} = \sum_{n=0}^{\infty} \frac{\mathcal{F}^{(n)}(0)}{n!} \frac{R^n_{\Psi,\bar{\Psi}}}{n!} |\Psi|^2 . \tag{4.39}$$

Now we use of the formula,

$$\frac{x^n}{n!} = \oint_C \frac{d\xi}{2\pi i} \frac{e^{\xi x}}{\xi^{n+1}} , \tag{4.40}$$

where, at this point, contour C can be any closed loop containing the origin with positive (counterclockwise) orientation. Using this identity we obtain

$$\mathcal{H} = \sum_{n=0}^{\infty} \frac{\mathcal{F}^{(n)}(0)}{n!} \oint_C \frac{d\xi}{2\pi i} \frac{e^{\xi R_{\Psi,\bar{\Psi}}}}{\xi^{n+1}} |\Psi|^2 , \tag{4.41}$$

where now the contour C must be chosen such that it lies in the region where the factor $e^{\xi R_{\Psi,\bar{\Psi}}}|\Psi|^2$ is analytic. With this in mind we can rewrite eq. (4.41) as

$$\mathcal{H} = \oint_C \frac{d\xi}{2\pi i} \frac{\mathcal{F}(1/\xi)}{\xi} e^{\xi R_{\Psi,\bar{\Psi}}} |\Psi|^2 . \tag{4.42}$$

The function $\mathcal{F}(1/\xi)/\xi$ gives a branch cut between -4 and 0. Therefore, the contour has to be chosen so that it does not cross the branch cut because the $\mathcal{H}$ is real and regular. Combining both requirements we see that the contour C in addition to avoiding the branch cut must be also bounded by the poles of the factor $e^{\xi R_{\Psi,\bar{\Psi}}}|\Psi|^2$ (see fig. 1). In the subsequent section, we derive the tangent bundle as well as the cotangent bundle over $E_7/[E_6 \times U(1)]$ by using (3.21) and (4.42).

5. The compact Hermitian symmetric space $E_7/(E_6 \times U(1))$

Locally, the compact Hermitian symmetric space $E_7/(E_6 \times U(1))$ can be described by complex variables Φ^i transforming in the **27** representation of the E_6 and their conjugates [17, 18].

$$\Phi^i , \qquad \bar{\Phi}_i := (\Phi^i)^* , \qquad i = 1, \ldots, 27 . \tag{5.1}$$

Journal of Physics: Conference Series **512** (2014) 012016 doi:10.1088/1742-6596/512/1/012016

The Kähler potential is

$$K(\Phi,\bar{\Phi}) = \ln\left(1 + \Phi^i\bar{\Phi}_i + \frac{1}{4}|\Gamma_{ijk}\Phi^j\Phi^k|^2 + \frac{1}{36}|\Gamma_{ijk}\Phi^i\Phi^j\Phi^k|^2\right). \tag{5.2}$$

Here we have used the notation, $|\Gamma_{ijk}\Phi^j\Phi^k|^2 = (\Gamma_{ijk}\Phi^j\Phi^k)(\Gamma^{ilm}\bar{\Phi}_l\bar{\Phi}_m)$ and $|\Gamma_{ijk}\Phi^i\Phi^j\Phi^k|^2 = (\Gamma_{ijk}\Phi^i\Phi^j\Phi^k)(\Gamma^{lmn}\bar{\Phi}_l\bar{\Phi}_m\bar{\Phi}_n)$, where Γ_{ijk} is a rank-3 symmetric tensor covariant under the E_6. Its complex conjugate is denoted by Γ^{ijk}.

Products of these tensors satisfy an identity [19]

$$\Gamma_{ijk}\Gamma^{ijl} = 10\delta^l_k\,. \tag{5.3}$$

And

$$\Gamma_{ijk}\left(\Gamma^{ilm}\Gamma^{jnp} + \Gamma^{ilp}\Gamma^{jnm} + \Gamma^{iln}\Gamma^{jmp}\right) = \delta^l_k\Gamma^{mnp} + \delta^p_k\Gamma^{mnl} + \delta^n_k\Gamma^{mlp} + \delta^m_k\Gamma^{lnp}\,, \tag{5.4}$$

which is called the Springer relation [20].

Let us calculate the tangent bundle Lagrangian by using (3.21). In our notation, the first-order differential operator defined in (3.12) is given by

$$R_{\Sigma,\bar{\Sigma}} = -\frac{1}{2}\Sigma^i\bar{\Sigma}_j\Sigma^k R_{i\ k}^{\ j\ l}(g^{-1})_l^{\ m}\frac{\partial}{\partial\Sigma^m}\,, \tag{5.5}$$

where $(g^{-1})_i^{\ j} = (g_j^{\ i})^{-1}$ is the inverse metric of $g_i^{\ j}$, that is $g_i^{\ k}(g^{-1})_k^{\ j} = \delta_i^{\ j}$. Since we are considering a symmetric space, it is actually sufficient to carry out the calculations at a particular point, say at $\Phi = 0$. The Riemann tensor at $\Phi = 0$ is given as

$$\begin{aligned} R_{i\ k}^{\ j\ l}\Big|_{\Phi=0} &= \partial_k\partial^l g_i^{\ j} - (g^{-1})_m^{\ n}\partial_n g_i^{\ j}\partial^m g_k^{\ l}\Big|_{\Phi=0} \\ &= -\delta_i^{\ j}\delta_k^{\ l} + \Gamma_{mik}\Gamma^{mlj} - \delta_k^{\ j}\delta_i^{\ l}\,. \end{aligned} \tag{5.6}$$

Substituting this into (5.5), we have

$$R_{\Sigma,\bar{\Sigma}} = xD - \frac{1}{2}y\partial_x - \frac{1}{3}z\partial_y\,, \qquad D := x\partial_x + y\partial_y + z\partial_z\,, \tag{5.7}$$

where we have used (5.4) and invariant quantities under the E_6 action

$$x := \Sigma^i\bar{\Sigma}_i\,, \tag{5.8}$$

$$y := (\Gamma_{ijk}\Sigma^j\Sigma^k)(\Gamma^{ilm}\bar{\Sigma}_l\bar{\Sigma}_m)\,, \tag{5.9}$$

$$z := (\Gamma_{ijk}\Sigma^i\Sigma^j\Sigma^k)(\Gamma^{lmn}\bar{\Sigma}_l\bar{\Sigma}_m\bar{\Sigma}_n)\,. \tag{5.10}$$

Using the Baker-Campbell-Hausdorff expansion formula, one can show, that

$$e^{tR_{\Sigma,\bar{\Sigma}}} = e^{txD}e^{-\frac{t}{2}y\partial_x}e^{-\frac{t}{3}z\partial_y}e^{\frac{t^2}{12}z\partial_x}e^{-\frac{t^2}{4}yD}e^{-\frac{t^3}{18}zD}\,. \tag{5.11}$$

By straight-forward application of each exponential we obtain

$$e^{tR_{\Sigma,\bar{\Sigma}}}x = -\partial_t\ln\Omega(t;x,y,z)\,, \qquad \Omega(t;x,y,z) := 1 - tx + \frac{t^2}{4}y - \frac{t^3}{36}z\,. \tag{5.12}$$

Plugging (5.12) into (3.21) we get the tangent bundle action at $\Phi = 0$.

$$\mathcal{L} = \ln\left(1 - x + \frac{1}{4}y - \frac{1}{36}z\right). \tag{5.13}$$

This result can be extended to an arbitrary point Φ of the base manifold by replacing

$$x \to g_i{}^j \Sigma^i \bar{\Sigma}_j\,, \tag{5.14}$$

$$\frac{1}{4}y \to \frac{1}{2}(g_i{}^j \Sigma^i \bar{\Sigma}_j)^2 + \frac{1}{4}R_i{}^j{}_k{}^l \Sigma^i \bar{\Sigma}_j \Sigma^k \bar{\Sigma}_l\,, \tag{5.15}$$

$$\begin{aligned} -\frac{1}{36}z \to & -\frac{1}{6}(g_i{}^j \Sigma^i \bar{\Sigma}_j)^3 - \frac{1}{4}(g_i{}^j \Sigma^i \bar{\Sigma}_j)(R_k{}^m{}_l{}^n \Sigma^k \bar{\Sigma}_m \Sigma^l \bar{\Sigma}_n) \\ & - \frac{1}{12}|g_i{}^j R_j{}^k{}_l{}^m \bar{\Sigma}_k \Sigma^l \bar{\Sigma}_m|^2\,. \end{aligned} \tag{5.16}$$

Let us turn to the cotangent bundle action. Again we restrict the calculations to the origin of the base manifold $\Phi = 0$. Defining E_6 invariant quantities in terms of the cotangent vector

$$\tilde{x} := \Psi_i \bar{\Psi}^i\,, \tag{5.17}$$

$$\tilde{y} := (\Gamma^{ijk}\Psi_j\Psi_k)(\Gamma_{ilm}\bar{\Psi}^l\bar{\Psi}^m)\,, \tag{5.18}$$

$$\tilde{z} := (\Gamma^{ijk}\Psi_i\Psi_j\Psi_k)(\Gamma_{lmn}\bar{\Psi}^l\bar{\Psi}^m\bar{\Psi}^n)\,, \tag{5.19}$$

we obtain the differential operator (4.30) of the form

$$R_{\Psi,\bar{\Psi}} = \tilde{x}\tilde{D} - \frac{1}{2}\tilde{y}\partial_{\tilde{x}} - \frac{1}{3}\tilde{z}\partial_{\tilde{y}}\,, \quad \tilde{D} := \tilde{x}\partial_{\tilde{x}} + \tilde{y}\partial_{\tilde{y}} + \tilde{z}\partial_{\tilde{z}}\,. \tag{5.20}$$

Repeating the same calculation as below (5.11), we find

$$e^{\xi R_{\Psi,\bar{\Psi}}}\tilde{x} = -\partial_\xi \ln \Omega(\xi; \tilde{x}, \tilde{y}, \tilde{z})\,, \tag{5.21}$$

where the function Ω is given in (5.12). Eq. (4.42) with the above leads to

$$\mathcal{H} = -\oint_C \frac{\mathrm{d}\xi}{2\pi \mathrm{i}} \frac{\mathcal{F}(1/\xi)}{\xi} \frac{-\tilde{x} + \frac{\xi}{2}\tilde{y} - \frac{\xi^2}{12}\tilde{z}}{1 - \xi\tilde{x} + \frac{\xi^2}{4}\tilde{y} - \frac{\xi^3}{36}\tilde{z}}\,. \tag{5.22}$$

Note that the factor in (4.41) $e^{\xi R_{\Psi,\bar{\Psi}}}|\Psi|^2$ produces poles, given by the roots of the cubic equation:

$$1 - \xi x + \frac{\xi^2}{4}y - \frac{\xi^3}{36}z = \frac{z}{36}(\xi_1 - \xi)(\xi_2 - \xi)(\xi_3 - \xi) = 0\,. \tag{5.23}$$

By construction, these poles are not inside the contour which only encircles the branch cut between the origin $\xi = 0$ and $\xi = -4$. Since the function $\mathcal{F}(1/\xi)/\xi$ has no singularity at infinity in the ξ-plane, the contour can be equivalently respected as encircling the poles ξ_1, ξ_2 and ξ_3 in the opposite direction. Thus, we can apply the Residue theorem, picking additional minus sign from the orientation of contour (which kills another minus given by eq. (5.21)) to obtain

$$\mathcal{H} = \frac{\mathcal{F}(1/\xi_1)}{\xi_1} + \frac{\mathcal{F}(1/\xi_2)}{\xi_2} + \frac{\mathcal{F}(1/\xi_3)}{\xi_3}\,. \tag{5.24}$$

The explicit form of ξ_1, ξ_2 and ξ_3 are given by

$$\begin{aligned}\xi_1 &= \frac{3\tilde{y}}{\tilde{z}} + \frac{2^{1/3}(-81\tilde{y}^2 + 108\tilde{x}\tilde{z})}{3\tilde{z}(-1458\tilde{y}^3 + 2916\tilde{x}\tilde{y}\tilde{z} - 972\tilde{z}^2 + A)^{1/3}} \\ &\quad - \frac{(-1458\tilde{y}^3 + 2916\tilde{x}\tilde{y}\tilde{z} - 972\tilde{z}^2 + A)^{1/3}}{3 \cdot 2^{1/3}\tilde{z}}, \qquad (5.25)\\ \xi_2 &= \frac{3\tilde{y}}{\tilde{z}} - \frac{(1 + i\sqrt{3})(-81\tilde{y}^2 + 108\tilde{x}\tilde{z})}{3 \cdot 2^{2/3}\tilde{z}(-1458\tilde{y}^3 + 2916\tilde{x}\tilde{y}\tilde{z} - 972\tilde{z}^2 + A^{1/3})} \\ &\quad + \frac{1 - i\sqrt{3}}{6 \cdot 2^{1/3}\tilde{z}}(-1458\tilde{y}^3 + 2916\tilde{x}\tilde{y}\tilde{z} - 972\tilde{z}^2 + A^{1/3}), \qquad (5.26)\\ \xi_3 &= \frac{3\tilde{y}}{\tilde{z}} - \frac{(1 - i\sqrt{3})(-81\tilde{y}^2 + 108\tilde{x}\tilde{z})}{3 \cdot 2^{2/3}\tilde{z}(-1458\tilde{y}^3 + 2916\tilde{x}\tilde{y}\tilde{z} - 972\tilde{z}^2 + A^{1/3})} \\ &\quad + \frac{1 + i\sqrt{3}}{6 \cdot 2^{1/3}\tilde{z}}(-1458\tilde{y}^3 + 2916\tilde{x}\tilde{y}\tilde{z} - 972\tilde{z}^2 + A^{1/3}), \qquad (5.27)\end{aligned}$$

where $A = \sqrt{4(-81\tilde{y}^2 + 108\tilde{x}\tilde{z})^3 + (-1458\tilde{y}^3 + 2916\tilde{x}\tilde{y}\tilde{z} - 972\tilde{z}^2)^2}$. The result at an arbitrary point of Φ can be obtained by the following replacements

$$\begin{aligned}\tilde{x} &\to (g^{-1})_i{}^j \Psi_j \bar{\Psi}^i, \qquad (5.28)\\ \frac{1}{4}\tilde{y} &\to \frac{1}{2}((g^{-1})_i{}^j \Psi_j \bar{\Psi}^i)^2 + \frac{1}{4}\tilde{R}_i{}^j{}_k{}^l \bar{\Psi}^i \Psi_j \bar{\Psi}^k \Psi_l, \qquad (5.29)\\ -\frac{1}{36}\tilde{z} &\to -\frac{1}{6}((g^{-1})_i{}^j \Psi_j \bar{\Psi}^j)^3 - \frac{1}{4}((g^{-1})_i{}^j \Psi_j \bar{\Psi}^i)(\tilde{R}_k{}^l{}_m{}^n \bar{\Psi}^k \Psi_l \bar{\Psi}^m \Psi_n) \\ &\quad - \frac{1}{12}|(g^{-1})_i{}^j \tilde{R}_j{}^k{}_l{}^m \Psi_k \bar{\Psi}^l \Psi_m|^2, \qquad (5.30)\end{aligned}$$

where $\tilde{R}_i{}^j{}_k{}^l = (g^{-1})_i{}^m (g^{-1})_n^j (g^{-1})_k{}^p (g^{-1})_q^l R_m{}^n{}_p{}^q$.

6. Generalization

Our reformulation suggests an alternative expression for the cotangent bundle over any compact Hermitian symmetric space. In order to explain it, let us look at the equation (5.22). We can rewrite it as follows

$$\mathcal{H} = -\oint_C \frac{\mathrm{d}\xi}{2\pi \mathrm{i}} \frac{\mathcal{F}(1/\xi)}{\xi} \partial_\xi \ln \Omega(\xi; \tilde{x}, \tilde{y}, \tilde{z}). \qquad (6.31)$$

It is easy to see that Ω is related to the Kähler potential $K(\Phi, \bar{\Phi})$ by uniform rescaling of coordinates

$$\ln \Omega(\xi; \tilde{x}, \tilde{y}, \tilde{z}) = K(\Phi \to -\sqrt{\xi}\Psi, \bar{\Phi} \to \sqrt{\xi}\bar{\Psi}) := K_\xi(\Psi, \bar{\Psi}). \qquad (6.32)$$

Thus, we see that the cotangent bundle action for any Hermitian symmetric space is given by

$$\mathcal{H} = -\oint_C \frac{\mathrm{d}\xi}{2\pi \mathrm{i}} \frac{\mathcal{F}(1/\xi)}{\xi} \partial_\xi K_\xi(\Psi, \bar{\Psi}), \qquad (6.33)$$

where, as in the $E_7/(E_6 \times U(1))$ case, the contour C must encircle the branch cut of $\mathcal{F}(1/\xi)/\xi$ and it must be bounded by the poles of $\partial_\xi K_\xi$, in the same manner as depicted in fig. 1. Since for all Hermitian symmetric spaces the Ω is a finite-order polynomial in ξ, we may write

$$\Omega \sim \Pi_i(\xi - \xi_i(\Psi, \bar{\Psi})), \qquad (6.34)$$

where ξ_i are the solutions of characteristic equation $\Omega = 0$. Since we are interested in (derivative of) logarithm of the above quantity, the constant of proportionality is actually not important. Thus, we are led to

$$\partial_\xi K_\xi = \sum_\xi \frac{1}{\xi - \xi_i} . \tag{6.35}$$

Substituting this back into (6.33) and using the Residue theorem, we arrive at

$$\mathcal{H} = -\sum_i \oint_C \frac{\mathrm{d}\xi}{2\pi \mathrm{i}} \frac{\mathcal{F}(1/\xi)}{\xi} \frac{1}{\xi - \xi_i} = \sum_i \frac{\mathcal{F}(1/\xi_i)}{\xi_i} , \tag{6.36}$$

where minus sing is absorbed because the contour C encircles poles in the clockwise direction. We checked that the results obtained by this formula for all the compact Hermitian symmetric spaces perfectly coincide with the previous results in [9, 10, 12, 13].

7. Conclusions

In this proceeding we have reviewed how to construct the closed formulas for tangent and cotangent bundle action over the compact Hermitian symmetric space $E_7/(E_6 \times U(1))$, starting from the result in [14]. This particular case allowed us to establish the correspondence between a cotangent bundle action and poles of derivatives of the Kähler potential under uniform rescaling of coordinates. Based on this correspondence we have present a general formula for a cotangent bundle action for any compact Hermitian symmetric space in Eq. (6.36). In the former methods in [9, 10, 12, 13], the Legendre transform is used to obtain a cotangent bundle action. This procedure, although in principle applicable to the $E_7/(E_6 \times U(1))$ case, is not straightforward and one often must find a suitable ansatz to solve involved algebraic equations. Our simple formulae of the cotangent bundle action obtained from the result in [14] avoids such a complication and clearly demonstrates the reduction of work in constructing of a cotangent bundle action.

Acknowledgments

The work of the authors is supported in part by the Research Program MSM6840770029, by the project of International Cooperation ATLAS-CERN of the Ministry of Education, Youth and Sports of the Czech Republic, and by Japan Society for the Promotion of Science (JSPS) and Academy of Sciences of the Czech Republic (ASCR) under the Japan - Czech Republic Research Cooperative Program.

References

[1] B. Zumino *Phys. Lett.* B 1979 **87** 203
[2] L. Alvarez-Gaumé and D. Z. Freedman 1981 *Commun. Math. Phys.* **80** 443
[3] A. Karlhede, U. Lindström and M. Roček *Phys. Lett.* B 1984 **147** 297
[4] U. Lindström and M. Roček *Commun. Math. Phys.* 1988 **115** 21
[5] U. Lindström and M. Roček *Commun. Math. Phys.* 1990 **128** 191
[6] F. Gonzalez-Rey, U. Lindström, M. Roček, S. Wiles and R. von Unge *Nucl. Phys.* B 1998 **516** 426 (*preprint* hep-th/9710250)
[7] A. Galperin, E. Ivanov, S. Kalitsyn, V. Ogievetsky and E. Sokatchev *Class. Quant. Grav.* 1984 **1** 469
[8] A. S. Galperin, E. A. Ivanov, V. I. Ogievetsky and E. S. Sokatchev 2001 *Harmonic Superspace* (UK: Univ. Pr.) p 306
[9] S. J. Gates Jr. and S. M. Kuzenko *Nucl. Phys.* B 1999 **543** 122 (*Preprint* hep-th/9810137)
[10] S. J. Gates Jr. and S. M. Kuzenko *Fortsch. Phys.* 2000 **48** 115 (*Preprint* hep-th/9903013)
[11] M. Arai and M. Nitta *Nucl. Phys.* B 2006 **745** 208 (*Preprint* hep-th/0602277)
[12] M. Arai, S. M. Kuzenko and U. Lindström *J. High Energy Phys.* JHEP02(2007)100 (*Preprint* hep-th/0612174)
[13] M. Arai, S. M. Kuzenko and U. Lindström *J. High Energy Phys.* JHEP12(2007)008 (*Preprint* arXiv:0709.2633)
[14] S. M. Kuzenko and J. Novak *J. High Energy Phys.* JHEP12(2008)072 (*Preprint* arXiv:0811.0218)

[15] M. Arai and F. Blaschke *J. High Energy Phys.* JHEP02(2013)045 (*Preprint* arXiv:1211.1537)
[16] S. M. Kuzenko and W. D. Linch *J. High Energy Phys.* JHEP02(2006)038 (*Preprint* hep-th/0507176)
[17] F. Delduc and G. Valent *Nucl. Phys.* B 1985 **253** 494
[18] K. Higashijima and M. Nitta *Prog. Theor. Phys.* 2000 **103** 635 (*Preprint* hep-th/9911139)
[19] T. W. Kephart and M. T. Vaughn *Annals Phys.* 1983 **145** 162
[20] T. A. Springer *Proc. Kon. Ned. Akad. Wet.* A 1962 **65** 259

Journal of Physics: Conference Series **512** (2014) 012017 doi:10.1088/1742-6596/512/1/012017

Hamiltonians with two-ladder spectra and solutions to the Painlevé IV equation

David Bermudez[1,2] **David . Fernández C.**[1]

[1]Departamento de Física, Cinvestav, AP 14-740, 07000 México DF, Mexico

[2]Department of Physics of Complex Systems, Weizmann Institute of Science, Rehovot 76100, Israel

E-mail: `david.bermudez@weizmann.ac.il`, `david@fis.cinvestav.mx`

Abstract. Some supersymmetric partners of the harmonic oscillator Hamiltonian possess third-order differential ladder operators and, thus, they realize the second-order polynomial Heisenberg algebras. The corresponding spectra consist of an infinite ladder of energy levels plus a finite one that can be placed at any position with respect to the infinite ladder. Both of them have equally spaced energy levels. Departing from these Hamiltonians several families of real and complex solutions to the Painlevé IV equation with real parameters will be generated.

1. Introduction

The interest in analyzing systematically the links which can be established between supersymmetric quantum mechanics (SUSY QM) and nonlinear differential equations is increasing over time [1–3]. In fact, there is a well known connection between the SUSY partners of the free particle and solutions to the KdV equation [4]. One of the aims in this paper is to discuss the existence of an additional link, now between SUSY QM and Painlevé IV (PIV) equation. By exploiting this fact, a procedure for generating solutions of the PIV equation can be easily designed [2]. Up to our knowledge, the first people who established the link between first-order SUSY QM and PIV equation were Veselov and Shabat [5], Dubov, Eleonsky and Kulagin [6], Adler [7]. This was further explored for higher-order SUSY by Andrianov, Cannata, Ioffe and Nishnianidze [8], Fernández, Negro and Nieto [9], Carballo, Fernández, Negro and Nieto [10], Mateo and Negro [11], Bermudez and Fernández [1–3, 12] (see also [13]).

In order to perform this task, we have organized the paper as follows. In sections 2 and 3 we will introduce the SUSY QM and the second-order polynomial Heisenberg algebras respectively. In section 4 we will apply the SUSY QM to the harmonic oscillator while in section 5 we will establish the corresponding connection with Painlevé IV equation. In addition, several solutions to this equation, real and complex, will be derived. Our conclusions are contained in section 6.

2. SUSY QM

Let us consider the following intertwining relations [14–21]:

$$H_{i+1}A_{i+1}^{+} = A_{i+1}^{+}H_i, \tag{1}$$

$$A_{i+1}^{+} = \frac{1}{\sqrt{2}}\left[-\frac{d}{dx} + \alpha_{i+1}(x,\epsilon_{i+1})\right], \tag{2}$$

$$H_i = -\frac{1}{2}\frac{d^2}{dx^2} + V_i(x), \quad i = 0, \dots, k. \tag{3}$$

Thus, the following equations are satisfied:

$$\alpha'_{i+1}(x, \epsilon_{i+1}) + \alpha^2_{i+1}(x, \epsilon_{i+1}) = 2[V_i(x) - \epsilon_{i+1}], \tag{4}$$
$$V_{i+1}(x) = V_i(x) - \alpha'_{i+1}(x, \epsilon_{i+1}). \tag{5}$$

The key point in this treatment is to realize that the solutions to the Riccati equation (4) can be obtained by algebraic means in terms of those of the Riccati equation for the previous system [18, 19], namely,

$$\alpha_{i+1}(x, \epsilon_{i+1}) = -\alpha_i(x, \epsilon_i) - \frac{2(\epsilon_i - \epsilon_{i+1})}{\alpha_i(x, \epsilon_i) - \alpha_i(x, \epsilon_{i+1})}, \tag{6}$$

which implies that $\alpha_{i+1}(x, \epsilon_{i+1})$ is determined either from $i+1$ solutions $\alpha_1(x, \epsilon_j)$ of the initial Riccati equation

$$\alpha'_1(x, \epsilon_j) + \alpha^2_1(x, \epsilon_j) = 2[V_0(x) - \epsilon_j], \quad j = 1, \dots, i+1, \tag{7}$$

or from $i+1$ solutions u_j of the associated Schrödinger equation

$$-\frac{1}{2}u''_j + V_0(x)u_j = \epsilon_j u_j, \quad j = 1, \dots, i+1. \tag{8}$$

The initial and final Hamiltonians, H_0 and H_k, satisfy the following intertwining relations

$$H_k B_k^+ = B_k^+ H_0, \qquad H_0 B_k = B_k H_k, \tag{9}$$
$$B_k = A_1 \dots A_k, \qquad B_k^+ = A_k^+ \dots A_1^+. \tag{10}$$

In the same way, the initial and final potentials V_0, V_k are linked through

$$V_k(x) = V_0(x) - \{\ln[W(u_1, \dots, u_k)]\}'', \tag{11}$$

where $W(u_1, \dots, u_k)$ denotes the Wronskian of the k solutions $\{u_1(x), \dots, u_k(x)\}$. In addition, the eigenfunctions of H_k are given by

$$\psi_n^{(k)} = [(E_n - \epsilon_1)\dots(E_n - \epsilon_k)]^{-\frac{1}{2}} \, B_k^+ \psi_n, \qquad E_n, \tag{12}$$
$$\psi_{\epsilon_j}^{(k)} \propto \frac{W(u_1, \dots, u_{j-1}, u_{j+1}, \dots, u_k)}{W(u_1, \dots, u_k)}, \qquad \epsilon_j. \tag{13}$$

Thus, given the potential $V_0(x)$ and k solutions $\{u_1(x), \dots, u_k(x)\}$ of the initial stationary Schrödinger equation associated with the factorization energies $\{\epsilon_1, \dots, \epsilon_k\}$, the new potential $V_k(x)$ of equation (11) is essentially determined, with eigenfunctions and eigenvalues given in equations (12-13).

3. Second-order polynomial Heisenberg algebras

The m-th order polynomial Heisenberg algebras (PHA) are deformations of the Heisenberg-Weyl algebra of kind [10] (see also [22, 23]):

$$[H, \mathcal{L}^\pm] = \pm\mathcal{L}^\pm, \tag{14}$$
$$[\mathcal{L}^-, \mathcal{L}^+] \equiv Q_{m+1}(H+1) - Q_{m+1}(H) = P_m(H), \tag{15}$$
$$Q_{m+1}(H) = \mathcal{L}^+\mathcal{L}^- = \prod_{i=1}^{m+1}(H - \mathcal{E}_i). \tag{16}$$

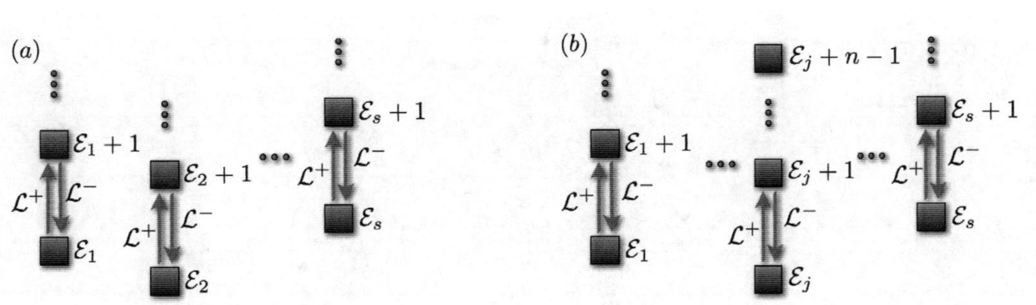

Figure 1. The spectrum of systems ruled by polynomial Heisenberg algebras can have either s infinite ladders (a) or $s-1$ infinite plus a finite one if equation (19) is satisfied (b).

One of the simplest realizations is a differential one, in which H is a one-dimensional Schrödinger Hamiltonian

$$H = -\frac{1}{2}\frac{d^2}{dx^2} + V(x), \tag{17}$$

and $\mathcal{L}^{\pm}$ are differential ladder operators of order $(m+1)$-th.

The spectral properties of systems ruled by PHA depend on the number s of physical eigenstates of H which also belong to the kernel of $\mathcal{L}^-$:

$$\mathcal{L}^-\psi_{\mathcal{E}_i} = 0, \quad H\psi_{\mathcal{E}_i} = \mathcal{E}_i\psi_{\mathcal{E}_i}, \quad i = 1, \ldots, s. \tag{18}$$

Departing from them, s physical ladders of H can be constructed through the action of $\mathcal{L}^+$, see figure 1(a). If, however, it happens that

$$\left(\mathcal{L}^+\right)^{n-1}\psi_{\mathcal{E}_j} \neq 0, \quad \left(\mathcal{L}^+\right)^{n}\psi_{\mathcal{E}_j} = 0 \quad \Rightarrow \quad \mathcal{E}_k = \mathcal{E}_j + n, \tag{19}$$

then the j-th ladder starts from $\mathcal{E}_j$ but ends at $\mathcal{E}_j + n - 1$, as shown in figure 1(b).

In this work we are particularly interested in second-order PHA, with $m = 2$, for which:

$$\begin{aligned} Q_3(H) &= (H-\mathcal{E}_1)(H-\mathcal{E}_2)(H-\mathcal{E}_3), && (20)\\ P_2(H) &= 3H^2 + [3 - 2(\mathcal{E}_1+\mathcal{E}_2+\mathcal{E}_3)]H && (21)\\ &\quad +1 - (\mathcal{E}_1+\mathcal{E}_2+\mathcal{E}_3) + \mathcal{E}_1\mathcal{E}_2 + \mathcal{E}_1\mathcal{E}_3 + \mathcal{E}_2\mathcal{E}_3. && (22)\end{aligned}$$

In this case our differential ladder operators $\mathcal{L}^{\pm}$ are of third order, and the corresponding Hamiltonian could have up to three independent physical ladders starting from $\mathcal{E}_1$, $\mathcal{E}_2$ and $\mathcal{E}_3$.

In order to identify the corresponding systems, let us take $\mathcal{L}^{\pm}$ such that [8–10]:

$$\begin{aligned} &\mathcal{L}^+ = I_1^+ I_2^+, && (23)\\ &I_1^+ = \frac{1}{\sqrt{2}}\left[-\frac{d}{dx} + f(x)\right], \quad I_2^+ = \frac{1}{2}\left[\frac{d^2}{dx^2} + g(x)\frac{d}{dx} + h(x)\right], && (24)\\ &HI_1^+ = I_1^+(H_a + 1), \quad H_a I_2^+ = I_2^+ H. && (25)\end{aligned}$$

By using the standard expressions for first and second-order SUSY QM we obtain

$$\begin{aligned} &-f' + f^2 = 2(V - \mathcal{E}_3), && (26)\\ &V_a = V + f' - 1 = V + g', && (27)\\ &\frac{g''}{2g} - \left(\frac{g'}{2g}\right)^2 - g' + \frac{g^2}{4} + \frac{(\mathcal{E}_1 - \mathcal{E}_2)^2}{g^2} + \mathcal{E}_1 + \mathcal{E}_2 - 2 = 2V, && (28)\\ &h = -\frac{g'}{2} + \frac{g^2}{2} - 2V + \mathcal{E}_1 + \mathcal{E}_2 - 2. && (29)\end{aligned}$$

Journal of Physics: Conference Series **512** (2014) 012017 doi:10.1088/1742-6596/512/1/012017

By decoupling this system it turns out that

$$f(x) = x + g(x), \tag{30}$$

$$h(x) = -x^2 + \frac{g'}{2} - \frac{g^2}{2} - 2xg + a, \tag{31}$$

$$V(x) = \frac{x^2}{2} - \frac{g'}{2} + \frac{g^2}{2} + xg + \mathcal{E}_3 - \frac{1}{2}, \tag{32}$$

where g satisfies

$$g'' = \frac{g'^2}{2g} + \frac{3}{2}g^3 + 4xg^2 + 2\left(x^2 - a\right)g + \frac{b}{g}, \tag{33}$$

which is the Painlevé IV equation with parameters ($\Delta = \mathcal{E}_1 - \mathcal{E}_2$):

$$a = \mathcal{E}_1 + \mathcal{E}_2 - 2\mathcal{E}_3 - 1, \qquad b = -2\Delta^2. \tag{34}$$

The extremal states of our system turn out to be

$$\psi_{\mathcal{E}_1} \propto \left(\frac{g'}{2g} - \frac{g}{2} - \frac{\Delta}{g} - x\right)\exp\left[\int\left(\frac{g'}{2g} + \frac{g}{2} - \frac{\Delta}{g}\right)\mathrm{d}x\right], \tag{35}$$

$$\psi_{\mathcal{E}_2} \propto \left(\frac{g'}{2g} - \frac{g}{2} + \frac{\Delta}{g} - x\right)\exp\left[\int\left(\frac{g'}{2g} + \frac{g}{2} + \frac{\Delta}{g}\right)\mathrm{d}x\right], \tag{36}$$

$$\psi_{\mathcal{E}_3} \propto \exp\left(-\frac{x^2}{2} - \int g\,\mathrm{d}x\right). \tag{37}$$

In particular, it is important to note that

$$g(x) = -x - \{\ln[\psi_{\mathcal{E}_3}(x)]\}'. \tag{38}$$

This means that, if we would know systems ruled by second-order PHA, particularly its extremal states, we could find solutions to the PIV equation. This idea will be explored further in section 5.

4. Harmonic oscillator SUSY partners

In order to apply the SUSY techniques, we need the solution of the Schrödinger equation for $V_0(x) = x^2/2$ and arbitrary ϵ, which is given by [24]:

$$u = e^{-\frac{x^2}{2}}\left[{}_1F_1\left(\frac{1-2\epsilon}{4}, \frac{1}{2}; x^2\right) + 2x\nu\frac{\Gamma(\frac{3-2\epsilon}{4})}{\Gamma(\frac{1-2\epsilon}{4})}\,{}_1F_1\left(\frac{3-2\epsilon}{4}, \frac{3}{2}; x^2\right)\right]. \tag{39}$$

Let us perform now a k-th order SUSY transformation by using $\{u_1, \ldots, u_k\}$ such that $\epsilon_k < \ldots < \epsilon_1 < 1/2$, $|\nu_j| < 1$ for odd j, $|\nu_j| > 1$ for even j. Thus, k new levels are created so that

$$\mathrm{Sp}(H_k) = \{\epsilon_j, E_n, j = k, \ldots, 1, n = 0, 1, \ldots\}. \tag{40}$$

Let us note that there are differential ladder operators for H_k of order $2k+1$:

$$L_k^- = B_k^+ a B_k, \quad L_k^+ = B_k^+ a^+ B_k. \tag{41}$$

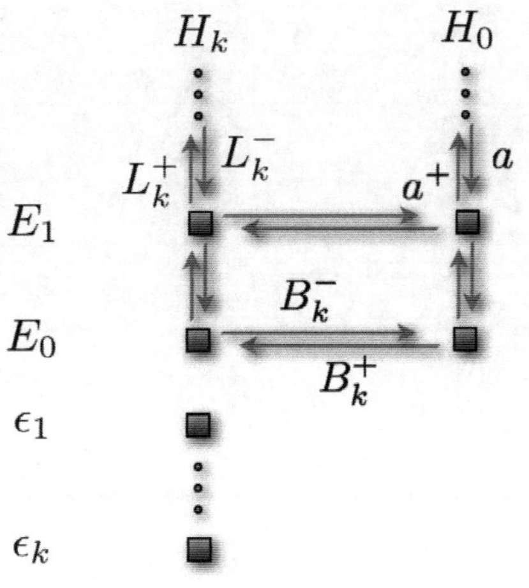

Figure 2. Diagram representing the two SUSY partner Hamiltonians H_0 and H_k and their associated ladder operators a^+, a and $L_k^{\pm}$, respectively.

Since

$$Q_{2k+1}(H_k) = \left(H_k - \frac{1}{2}\right) \prod_{i=1}^{k} (H_k - \epsilon_i - 1)\,(H_k - \epsilon_i)\,, \tag{42}$$

it is seen that they generate a PHA of order $2k$, namely:

$$[L_k^-, L_k^+] = P_{2k}(H_k). \tag{43}$$

Moreover, since to every root ϵ_j of $Q_{2k+1}(H_k)$ there is associated other one $\epsilon_j + 1$, we can say that $\mathrm{Sp}(H_k)$ contains k one-step ladders, each one of them starting and ending at $\epsilon_j, j = 1, \ldots, k$. In addition, there is an infinite ladder starting from $E_0 = 1/2$ (see diagram in figure 2).

5. Solutions of PI equation through SUSY QM

As we saw in the previous section, $L_1^{\pm}$ are third-order differential ladder operators, but $L_k^{\pm}$ are of order greater than three for $k > 1$. Thus the question arises: is it possible to reduce to three the order of these ladder operators for $k > 1$? If so, we could obtain systems which perhaps would supply us with solutions to the PIV equation [25]. The answer turns out to be positive, and it is contained in the following theorem [1, 2].

Theorem. Suppose that H_k is generated by k connected seed solutions

$$u_j = a^{j-1}u_1, \quad \epsilon_j = \epsilon_1 - (j-1), \quad j = 1, \ldots, k, \tag{44}$$

$u_1(x)$ being an independent seed solution in the form given in equation (39) such that $\epsilon_1 < 1/2$ and $|\nu_1| < 1$. Thus

$$L_k^+ = P_{k-1}(H_k) l_k^+, \tag{45}$$

where $P_{k-1}(H_k) = (H_k - \epsilon_1) \ldots (H_k - \epsilon_{k-1})$ and l_k^+ turns out to be a third-order differential ladder operator such that

$$[H_k, l_k^+] = l_k^+, \tag{46}$$

$$l_k^+ l_k^- = (H_k - \epsilon_k)\left(H_k - \tfrac{1}{2}\right)(H_k - \epsilon_1 - 1). \tag{47}$$

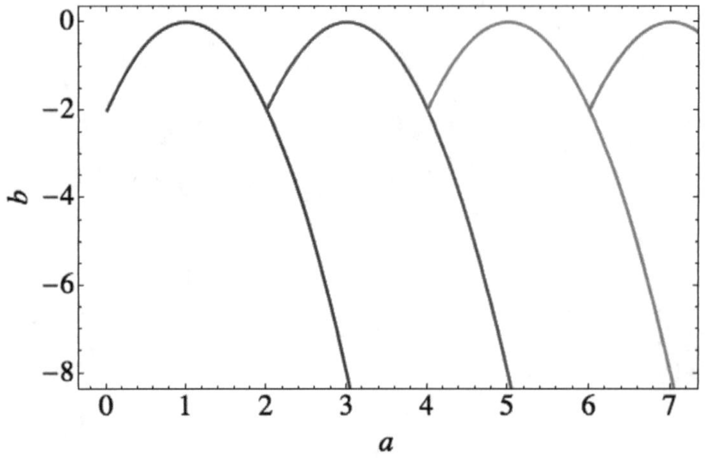

Figure 3. Parameter space where we can obtain real non-singular solutions to the PIV equation.

Let us note that the operators $l_k^{\pm}$ already connect the levels $\{\epsilon_j, j = 1, \dots, k\}$, which form a finite ladder of lenght k starting from $\epsilon_k = \epsilon_1 - (k-1)$ and finishing at ϵ_1. Moreover, the operator l_k^+ annihilates only the eigenstate associated to ϵ_1 while the operator l_k^- annihilates the following three extremal states:

$$\psi_{\mathcal{E}_1} \propto B_k^+ e^{-x^2/2}, \qquad \mathcal{E}_1 = 1/2, \tag{48}$$

$$\psi_{\mathcal{E}_2} \propto B_k^+ a^+ u_1, \qquad \mathcal{E}_2 = \epsilon_1 + 1, \tag{49}$$

$$\psi_{\mathcal{E}_3} \propto \frac{W(u_1, \dots, u_{k-1})}{W(u_1, \dots, u_k)}, \qquad \mathcal{E}_3 = \epsilon_k = \epsilon_1 - (k-1). \tag{50}$$

From equation (38), it is clear now that departing from the third extremal state we can generate non-singular real solutions to the PIV equation. Indeed, the potential and the corresponding solution to the PIV equation are given by:

$$V(x) = \frac{x^2}{2} - [\ln W(u_1, \dots, u_k)]'', \tag{51}$$

$$g(x) = \begin{cases} -x + [\ln u_1]' & \text{for } k = 1 \\ -x - \left[\ln \dfrac{u_1}{W(u_1, u_2)}\right]' & \text{for } k = 2 \\ -x - \left[\ln \dfrac{W(u_1, \dots, u_{k-1})}{W(u_1, \dots, u_k)}\right]' & \text{for } k > 2 \end{cases} \tag{52}$$

and the connection between the parameters ϵ_1, k of the SUSY transformation and a, b of the PIV equation is as follows

$$a = -\epsilon_1 + 2k - \frac{3}{2}, \qquad b = -2\left(\epsilon_1 + \frac{1}{2}\right)^2. \tag{53}$$

The curves on the parameter space a-b where we have real non-singular solutions to the PIV equation are shown in figure 3. An example of the SUSY partner potential of the oscillator as well as the corresponding PIV solution is given in figure 4.

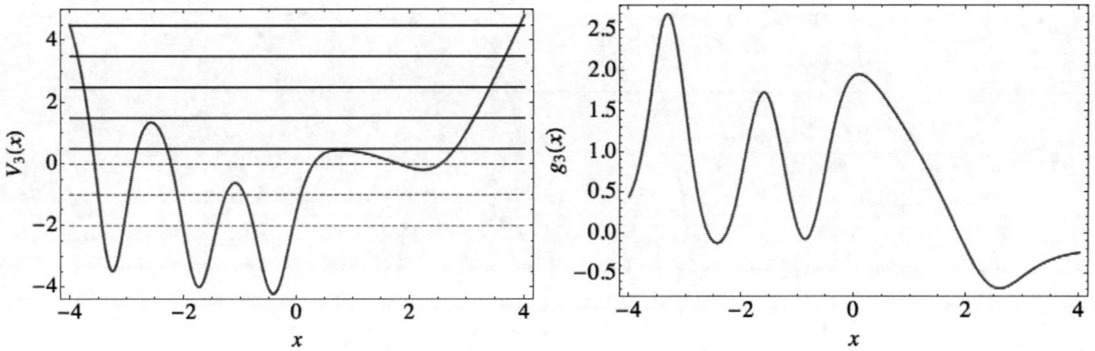

Figure 4. Third-order SUSY partner potential of the oscillator (left) and PIV solution (right) for $\epsilon_1 = 0, \nu_1 = 0.99$.

Let us stress that the restrictions $\epsilon_1 < E_0$, $|\nu_1| < 1$ were needed to produce a non-singular potential $V_k(x)$ and its PIV solution, i.e., the full finite ladder of H_k is placed below E_0. From the viewpoint of spectral design it would be important to surpass this restriction so that the finite ladder of H_k could be placed above E_0 (at least partially). This can be done by using complex transformation functions u_1 but associated to $\epsilon_1 \in \mathbb{R}$ [12, 26]. The price to pay is that the generated potentials will now be complex, but with real energy spectra. The main change with respect to our previous treatment is that the employed complex seed solution reads now

$$u_1(x) = e^{-x^2/2}\left[{}_1F_1\left(\frac{1-2\epsilon_1}{4}, \frac{1}{2}; x^2\right) + \quad x\,{}_1F_1\left(\frac{3-2\epsilon_1}{4}, \frac{3}{2}; x^2\right)\right], \tag{54}$$

where $\quad = \lambda + i\kappa$ ($\lambda, \kappa \in \mathbb{R}$) and $\epsilon_1 \in \mathbb{R}$, which leads once again to a Hamiltonian having third-order differential ladder operators.

The extremal states become once again those given in equations (48-50). The PIV solution and its associated parameters are

$$g_k(x) = -x - \{\ln[\psi_{\mathcal{E}_3}(x)]\}' = -x - \left\{\ln\left[\frac{W(u_1,\ldots,u_{k-1})}{W(u_1,\ldots,u_k)}\right]\right\}', \tag{55}$$

$$a_i = -\epsilon_1 + 2k - \tfrac{3}{2}, \quad b_i = -2\left(\epsilon_1 + \tfrac{1}{2}\right)^2. \tag{56}$$

Note that the cyclic permutations of the indexes in the extremal states and associated energies leads now to two additional non-singular solutions of the PIV equation:

$$g_k(x) = -x - \left\{\ln\left[B_k^+ e^{-x^2/2}\right]\right\}', \quad a_{ii} = 2\epsilon_1 - k, \quad b_{ii} = -2k^2, \tag{57}$$

$$g_k(x) = -x - \left\{\ln\left[B_k^+ a^+ u_1\right]\right\}', \quad a_{iii} = -\epsilon_1 - k - \tfrac{3}{2}, \quad b_{iii} = -2(\epsilon_1 - k + \tfrac{1}{2})^2. \tag{58}$$

In the parameter space of the PIV equation now we get additional curves on which we have non-singular (complex) solutions (see figure 5). An example of this kind of complex solutions can be seen in figure 6.

6. Conclusions

In this work we have shown that the general Schrödinger Hamiltonians ruled by second-order polynomial Heisenberg algebras, i.e., having third-order differential ladder operators, are linked to the PIV equation. Moreover, a prescription for generating solutions of the PIV equation

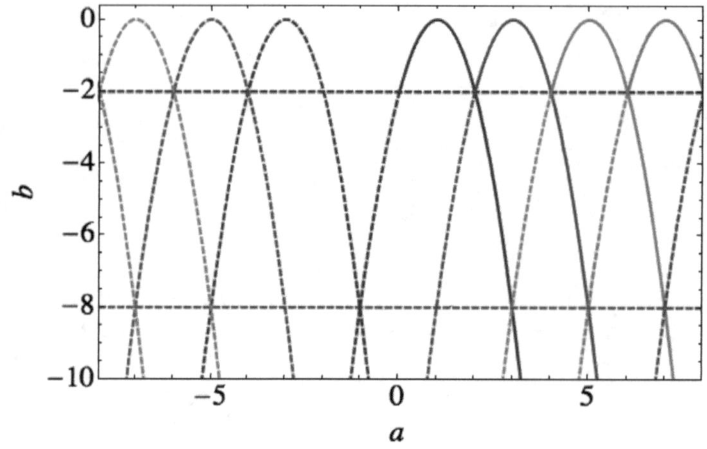

Figure 5. Parameter space where we can obtain complex non-singular solutions to the PIV equation.

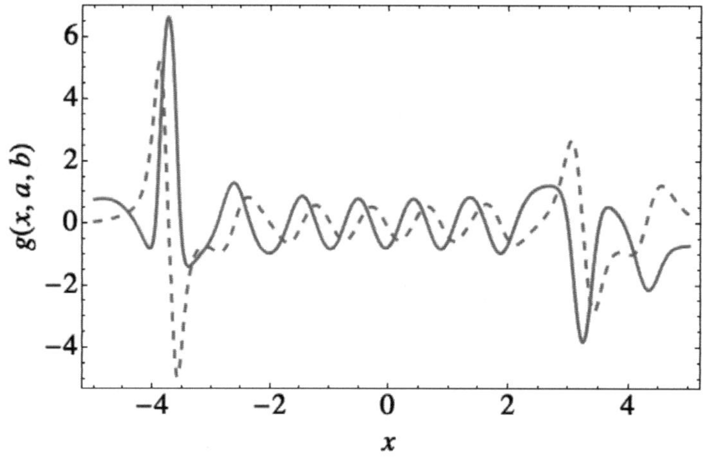

Figure 6. Real and imaginary parts (solid and dashed lines) of a complex solution to PIV for $a_{ii} = 12$, $b_{ii} = -8$ ($k = 2$, $\epsilon_1 = 7$, $\lambda = \kappa = 1$)

departing from the SUSY partners of the harmonic oscillator has been proposed. The associated potentials have two physical ladders of equally spaced energy levels: an infinite one starting from 1 2 and a finite one, placed either completely below 1 2 for real transformation functions or at any arbitrary position on the real energy axis for complex ones, both for $\epsilon_1 \in \mathbb{R}$.

Acknowledgement
The authors acknowledge the financial support of Conacyt (Mexico) project 152574. DB also acknowledges the Conacyt fellowship 207672.

References

[1] D. Bermudez, D.J. Fernández. Supersymmetric quantum mechanics and Painlevé IV equation, *SIGMA* **7** (2011) 025 (14 pages).

[2] D. Bermudez. Polynomial Heisenberg algebras and Painlevé equations, PhD thesis, Cinvestav, Mexico (2013).

[3] D. Bermudez, D.J. Fernández. Supersymmetric quantum mechanics and Painlevé equations, arXiv:1311.0647.

[4] V.B. Matveev. Generalized Wronskian formula for solutions of the KdV equations: first applications. *Phys. Lett. A* **166** (1992) 205 208.

[5] A.P. Veselov, A.B. Shabat. Dressing Chains and the Spectral Theory of the Schrödinger Operator, *Funct. Ann. App.* **27** (1993) 81 96.

[6] S.Y. Dubov, V.M. Eleonsky, N.E. Kulagin. On the equidistant spectra of anharmonic oscillators, *Chaos* **4** (1994) 47 53.

[7] V.E. Adler. Nonlinear chains and Painlevé equations, *Physica D* **73** (1994) 335 351.

[8] A.A. Andrianov, F. Cannata, M.V. Ioffe, D. Nishnianidze. Systems with higher-order shape invariance: spectral and algebraic properties, *Phys. Lett. A* **266** (2000) 341 349.

[9] D.J. Fernández, J. Negro, L.M. Nieto. Elementary systems with partial finite ladder spectra *Phys. Lett. A* **324** (2004) 139 144.

[10] J.M. Carballo, D.J. Fernández, J. Negro, L.M. Nieto. Polynomial Heisenberg algebras, *J. Phys. A: Math. Gen.* **37** (2004) 10349 10362.

[11] J. Mateo, J. Negro. Third order differential ladder operators and supersymmetric quantum mechanics, *J. Phys. A: Math. Theor.* **41** (2008) 045204 (28 pages).

[12] D. Bermudez, D.J. Fernández. Non-hermitian Hamiltonians and the Painlevé IV equation with real parameters, *Phys. Lett. A* **375** (2011) 2974 2978.

[13] R. Willox, J. Hietarinta. Painlevé equations from Darboux chains: I. P_{III}-P_V, *J. Phys. A: Math. Gen.* **36** (2003) 10615 10635.

[14] A.A. Andrianov, M.V. Ioffe, V. Spiridonov. Higher-derivative supersymmetry and the Witten index, *Phys. Lett. A* **174** (1993) 273 279.

[15] A.A. Andrianov, M.V. Ioffe, F. Cannata, J.P. Dedonder. Second order derivative supersymmetry, $q-$deformations and the scattering problem, *Int. J. Mod. Phys. A* **10** (1995) 2683 2702.

[16] V.G. Bagrov, B.F. Samsonov. Darboux transformation of the Schrödinger equation, *Phys. Part. Nucl.* **28** (1997) 374 397.

[17] B. Bagchi, A. Ganguly, D. Bhaumik, A. Mitra. Higher derivative supersymmetry, a modified Crum-Darboux transformation and coherent state, *Mod. Phys. Lett. A* **14** (1999) 27 34.

[18] B. Mielnik, O. Rosas-Ortiz. Factorization: little or great algorithm?, *J. Phys. A: Math. Gen.* **37** (2004) 10007 10035.

[19] D.J. Fernández, N. Fernández-García. Higher-order supersymmetric quantum mechanics, *AIP Conf. Proc.* **744** (2005) 236 273.

[20] D.J. Fernández. Supersymmetric quantum mechanics, *AIP Conf. Proc.* **1287** (2010) 3 36.

[21] A.A. Andrianov, M.V. Ioffe. Nonlinear supersymmetric quantum mechanics: concepts and realizations, *J. Phys. A: Math. Theor.* **45** (2012) 503001 (62 pages).

[22] I. Marquette. Superintegrability with third order integrals of motion, cubic algebras, and supersymmetric quantum mechanics. II. Painlevé transcendent potentials, *J. Math. Phys.* **50** (2009) 095202 (18 pages).

[23] I. Marquette, C. Quesne. Two-step rational extensions of the harmonic oscillator: exceptional orthogonal polynomials and ladder operators, *J. Phys. A: Math. Theor.* **46** (2013) 155201 (14 pages).

[24] G. Junker, P. Roy. Conditionally exactly solvable potentials: a supersymmetric construction method, *Ann. Phys.* **270** (1998) 155 177.

[25] A.P. Bassom, P.A. Clarkson, A.C. Hicks. Bäcklund-transformations and solution hierarchies for the 4th Painlevé equation *Stud. Appl. Math.* **95** (1995) 1 71.

[26] A.A. Andrianov, M.V. Ioffe, F. Cannata, J.P. Dedonder. SUSY quantum mechanics with complex superpotentials and real energy spectra, *Int. J. Mod. Phys. A* **14** (1999) 2675 2688.

Linearized coherent states for Hamiltonian systems with two equidistant ladder spectra

D Bermudez[1,2], A Contreras-Astorga[1] and DJ Fernández C[1]

[1]Departamento de Física, Cinvestav, A.P. 14-740, 07000 México D.F., Mexico
[2]Department of Physics of Complex Systems, Weizmann Institute of Science, Rehovot 76100, Israel

E-mail: david.bermudez@weizmann.ac.il, acontreras@fis.cinvestav.mx, david@fis.cinvestav.mx

Abstract. A simple way to construct exactly solvable Hamiltonians whose spectra contain two equidistant ladders, one finite and another infinite, appears when applying supersymmetric quantum mechanics to the harmonic oscillator. Some of those supersymmetric partners have third order differential ladder operators, although the order of the transformation is higher than one. In this work the linearized coherent states for these specific Hamiltonians are studied. To each SUSY partner Hamiltonian corresponds two families of linearized coherent states: one inside the subspace associated with the isospectral part of the spectrum and another one in the finite subspace generated by the states inserted through the SUSY technique.

1. Introduction

In the beginning of quantum mechanics, Erwin Schrödinger was interested in finding a connection between the new science and classical mechanics [1]. With this interest in mind, a set of quantum states that restored what he considered as the classical behavior for the position operator of a quantum system was developed. In the last fifty years, the study of these states has taken an important place in quantum physics, since they can be used to establish an essential link between quantum and semi-classical behaviors of a system [2–8]. The name given to them is *coherent states* (CS), and it was coined by Glauber when studying electromagnetic correlation functions [9, 10].

For the harmonic oscillator, the standard CS are expressed as

$$\psi_z = e^{-|z|^2/2} \sum_{n=0}^{\infty} \frac{z^n}{\sqrt{n!}} \phi_n, \tag{1}$$

where ϕ_n are the normalized eigenstates of the harmonic oscillator Hamiltonian with eigenvalues $E_n = n + 1/2$ (in dimensionless units) and $z \in \mathbb{C}$. Some important properties of the standard coherent states are the following: they are eigenstates of the standard annihilation operator a; they appear by acting the displacement operator $D(z) = \exp(za^+ - z^* a^-)$ onto the ground state; they minimize the Heisenberg uncertainty relation for the position and momentum operators; the resolution of the identity operator is satisfied, namely,

$$\mathbb{1} = \frac{1}{\pi} \int_{\mathbb{C}} |z\rangle\langle z| \mathrm{d}z. \tag{2}$$

These four properties have been used as definitions to look for the coherent states for systems different from the harmonic oscillator.

In this work we will study Hamiltonians with two equidistant ladder spectra, one finite and another infinite, which are obtained through supersymmetric quantum mechanics (SUSY QM) departing from the harmonic oscillator. These Hamiltonians possess third order differential ladder operator, and through them we will obtain two sets of coherent states, one inside the subspace generated by the eigenfunctions associated with the infinite ladder and another one having as basis the eigenfunctions of the finite ladder. The two sets of CS will be generated as displaced versions of the corresponding extremal states, and it will be shown that for both it is valid the resolution of the identity in the respective subspace.

The structure of this paper is the following: in section 2 the construction of Hamiltonians whose spectra consist of two equidistant ladders is done. In the third section the linearized coherent states for these systems are obtained and some of their properties will be studied. The section 4 contains the calculation of some important mean values in the limit of the harmonic oscillator. Our conclusions shall be presented in the last section.

2. Supersymmetric quantum mechanics and harmonic oscillator

The supersymmetric quantum mechanics, which is closely related to the factorization method and the intertwining technique, is a procedure to find exactly solvable Hamiltonians. It departs from a given Hamiltonian H_0, whose orthogonal set of eigenvectors is known, and looks for another Hamiltonian H_k, whose eigenfunctions are going to be obtained. Both Hamiltonians will have a similar spectrum, differing at most in a finite number of energy levels that can be chosen at will. A more detailed information about this method can be found in [11–15].

The two intertwined Hamiltonians take the form

$$H_0 = -\frac{1}{2}\frac{\mathrm{d}^2}{\mathrm{d}x^2} + V_0(x), \qquad H_k = -\frac{1}{2}\frac{\mathrm{d}^2}{\mathrm{d}x^2} + V_k(x). \tag{3}$$

In order to apply this technique, let us suppose the existence of a k-th order differential operator $B_k^\dagger$ that intertwines the previous Schrödinger Hamiltonians in the way

$$H_k B_k^\dagger = B_k^\dagger H_0. \tag{4}$$

Since $B_k^\dagger$ is a k-th order differential operator, this version of the technique is referred as k-SUSY. In the standard terminology it is also said that the potentials $V_0(x)$ and $V_k(x)$ are SUSY partners.

It can be shown that the new potential has the form

$$V_k(x) = V_0(x) - \frac{\mathrm{d}^2}{\mathrm{d}x^2}\ln W\left[u(x,\epsilon_0),\ldots,u(x,\epsilon_{k-1})\right], \tag{5}$$

where the k functions $u(x,\epsilon_j)$, $j = 0,\ldots,k-1$ are known as transformations functions and fulfill the initial Schrödinger equation

$$H_0 u(x,\epsilon) = \epsilon u(x,\epsilon). \tag{6}$$

The parameters ϵ_j are called factorization energies, and they do not necessarily belong to the spectrum of H_0, i.e., the transformations functions do not always satisfy the same boundary conditions as the eigenfunctions of H_0 do.

Journal of Physics: Conference Series **512** (2014) 012018 doi:10.1088/1742-6596/512/1/012018

The intertwining operator can be found directly from the intertwining relation (4); for the purposes of this paper it is enough to write down its formal expression:

$$B_k^\dagger = \frac{1}{W(u_{\epsilon_0}, u_{\epsilon_1}, \ldots u_{\epsilon_{k-1}})} \begin{vmatrix} u_{\epsilon_0} & u_{\epsilon_1} & \ldots & 1 \\ u'_{\epsilon_0} & u'_{\epsilon_1} & \ldots & \mathrm{d}/\mathrm{d}x \\ \vdots & \vdots & \ddots & \vdots \\ u_{\epsilon_0}^{(k)} & u_{\epsilon_1}^{(k)} & \ldots & \mathrm{d}^{(k)}/\mathrm{d}x^{(k)} \end{vmatrix}. \tag{7}$$

The intertwining relation (4) ensures that if ϕ_n is an eigenfunction of H_0 with eigenvalue E_n then $\psi_n^k \propto B_k^\dagger \phi_n$ will be an eigenfunction of H_k with the same eigenvalue. One could ask now if $\{\psi_n^k,\ n = 0, 1, 2, \ldots\}$ is a complete orthogonal set. In order to answer, let us suppose the existence of other vectors $\psi_{\epsilon_j}^k$ which are orthogonal to each one of the previous set, i.e.,

$$\langle \psi_{\epsilon_j}^k | \psi_n^k \rangle \propto \langle \psi_{\epsilon_j}^k | B_k^\dagger \phi_n \rangle = \langle B_j \psi_{\epsilon_j} | \phi_n \rangle = 0 \quad \forall \quad n. \tag{8}$$

Since $\{\phi_n,\ n = 0, 1, 2, \ldots\}$ is a complete orthogonal set, it turns out that,

$$B_k \psi_{\epsilon_j}^k = 0, \tag{9}$$

i.e., in order to complete the orthogonal set of eigenfunctions of H_k we need to analyze also the kernel of $B_k \equiv (B_k^\dagger)^\dagger$. Since B_k is a k-th order differential operator, its kernel is a subspace of dimension at most k. The different choices of the transformation functions give families of Hamiltonians whose spectra differ from the original one in a finite numbers of energy levels.

In order to generate exactly solvable Hamiltonians whose spectra are composed of two equidistant ladders, we will apply this technique to the harmonic oscillator: it turns out that one of the ladders is infinite, composed of the same energy levels of the harmonic oscillator; the second ladder is a finite one, formed by the new levels inserted by the SUSY technique. To perform the transformation we need to find the general solution $u(x, \epsilon)$ of the Schrödinger equation for the potential $V_0(x) = x^2/2$ with a factorization energy ϵ, which is given by

$$u(x,\epsilon) = e^{-x^2/2}\left[{}_1F_1\left(\frac{1-2\epsilon}{4}, \frac{1}{2}, x^2\right) + 2\nu x \frac{\Gamma(\frac{3-2\epsilon}{4})}{\Gamma(\frac{1-2\epsilon}{4})}\, {}_1F_1\left(\frac{3-2\epsilon}{4}, \frac{3}{2}; x^2\right)\right], \tag{10}$$

where ν is a real arbitrary constant and ${}_1F_1(a, c, x)$ is the confluent hypergeometric function. Suppose that $\epsilon \leq E_0$, then it is known that for $|\nu| < 1$ the solution will not have zeroes but for $|\nu| > 1$ it will have one zero in some point on the real line. In order to generate a non singular potential $V_k(x)$ with k new levels, the factorization energies have to be chosen as $\epsilon_0 < \epsilon_1 < \cdots < \epsilon_{k-1} < E_0$ with $|\nu_{k-j}| < 1$ for j odd and $|\nu_{k-j}| > 1$ for j even, $j = 1, 2, \ldots, k$. The spectrum of the corresponding Hamiltonian H_k will be

$$\mathrm{Sp}\,[H_k] = \{\epsilon_0, \epsilon_1, \ldots, \epsilon_{k-1}, E_0, E_1, E_2, \ldots\}. \tag{11}$$

It is important to define the following operators

$$L_k^+ = B_k^\dagger a^+ B_k, \qquad L_k^- = B_k^\dagger a^- B_k, \tag{12}$$

where $a^\pm$ are the standard creation and annihilation operators of the harmonic oscillator. These are $(2k + 1)$-th order differential operators that obey the following commutation relations

$$H_k, L_k^\pm = \pm L_k^\pm, \tag{13}$$

i.e. they are ladder operators for the Hamiltonian H_k.

Since the eigenfunctions associated with the inserted levels belong to the Kernel of B_k, they cannot be reached by applying $L_k^{\pm}$ onto vectors which are outside this subspace. Thus, the Hilbert space $\mathcal{H}$ is naturally separated in two orthogonal subspaces, one generated by the vectors ψ_n^k (called $\mathcal{H}_{\text{iso}}$) and other generated by the eigenfunctions associated with the new levels (named $\mathcal{H}_{\text{new}}$).

There are some special cases when third order differential ladder operators for H_k can be found, which will connect now the eigenfunctions of H_k belonging to $\mathcal{H}_{\text{new}}$. These specific cases are of major interest of this work, and they must fulfill the conditions contained in the following theorem:

Factorization theorem

Suppose that the k-th order SUSY partner H_k of the harmonic oscillator Hamiltonian H_0 is generated by k transformation functions $u(x,\epsilon_j), j = 0,\ldots,k-1$, which are connected by the standard annihilation operator in the way:

$$u(x,\epsilon_{k-j-1}) = (a^-)^j u(x,\epsilon_{k-1}), \quad \epsilon_{k-j-1} = \epsilon_{k-1} - j, \tag{14}$$

where $u(x,\epsilon_{k-1})$ is a nodeless solution of the stationary Schrödinger equation associated to H_0, given by equation (10) with $\epsilon_{k-1} < 1/2$ and $|\nu_{k-1}| < 1$. Therefore, the natural ladder operator $L_k^+ \equiv B_k^\dagger a^+ B_k$ of H_k, which is of $(2k+1)$-th order, can be factorized in the form

$$L_k^+ = P_{k-1}(H_k) l_k^+, \tag{15}$$

where $P_{k-1}(H_k) = (H_k - \epsilon_1)\ldots(H_k - \epsilon_{k-1})$ is a polynomial of degree $k-1$ in H_k, l_k^+ is a third-order differential ladder operator such that $\left[H_k, l_k^+\right] = l_k^+$ and

$$l_k^+ l_k^- = \left(H_k - \frac{1}{2}\right)(H_k - \epsilon_0)(H_k - \epsilon_{k-1} - 1). \tag{16}$$

The proof of this theorem can be found in [16] (see also [17]). It implies that a subset of Hamiltonians H_k, beside having their natural $(2k+1)$-th order differential ladder operators also have third order ones. These SUSY partner Hamiltonians have an equidistant ladder with k steps below the ground state energy $E_0 = 1/2$ whose separation is the same as for the harmonic oscillator. Also, the transformation functions are not longer arbitrary: once the first one $u(x,\epsilon_{k-1})$ is chosen, all the others become fixed by the theorem. As a result, the only parameters that can be varied are the number of levels k to be inserted, the gap $E_0 - \epsilon_{k-1}$ between the ladders (with the restriction $\epsilon_{k-1} < E_0$), and the real parameter ν_{k-1} of the transformation function $u(x,\epsilon_{k-1})$ (with $|\nu_{k-1}| < 1$).

3. Linearized coherent states

Once we have identified the Hamiltonians H_k whose spectra are composed of two equidistant ladder and used the factorization theorem to obtain the corresponding third order differential ladder operators, we proceed now to do a linearization process by defining new ladder operators that have a simpler action on the eigenvectors of H_k. With the last ladder operators we will find two sets of coherent states, one inside $\mathcal{H}_{\text{iso}}$ and other in $\mathcal{H}_{\text{new}}$.

The linearized ladder operators $\ell_k^{\pm}$ are defined departing from the third order ones $l_k^{\pm}$ as

$$\ell_k^+ \equiv \sigma(H_k) l_k^+, \qquad \ell_k^- \equiv \sigma(H_k + 1) l_k^-, \tag{17}$$

with

$$\sigma(H_k) = [(H_k - \epsilon_0)(H_k - \epsilon_0 - k)]^{-1/2}, \tag{18}$$

where we will take the positive square root. Their action onto the basis of $\mathcal{H}_{\rm iso}$ and $\mathcal{H}_{\rm new}$ will be given below, and it will help us to make the coherent state construction.

Note that this kind of linearization process has been applied previously to the generic SUSY partners of the harmonic oscillator in order to obtain several sets of CS by using different annihilation operators [18–21].

3.1. Coherent states in the subspace $\mathcal{H}_{iso}$
The action of the new ladder operators $\ell_k^\pm$ on the eigenfunctions of H_k in the subspace $\mathcal{H}_{\rm iso}$ is given by

$$\ell_k^- \psi_n^k = \sqrt{n}\psi_{n-1}^k, \qquad \ell_k^+ \psi_n^k = \sqrt{n+1}\psi_{n+1}^k. \tag{19}$$

Furthermore, we can easily show that the operators $\{\ell_k^+, \ell_k^-, H_k\}$ obey the following commutation rules inside $\mathcal{H}_{\rm iso}$:

$$[\ell_k^-, \ell_k^+] = \mathbb{1}|_{\mathcal{H}_{\rm iso}}, \quad [H_k, \ell_k^\pm] = \pm \ell_k^\pm. \tag{20}$$

Equations (20) mean that the linearized ladder operators satisfy a Heisenberg-Weyl algebra on $\mathcal{H}_{\rm iso}$.

Now, let us generate a set of linearized CS using an analogue of the displacement operator, defined as

$$\mathcal{D}(z) = \exp\left(-\frac{1}{2}|z|^2\right) \exp\left(z\ell_k^+\right) \exp\left(-z^*\ell_k^-\right). \tag{21}$$

Then, the CS are constructed by applying the last operator onto the extremal state ψ_0^k, which leads to

$$\psi_{z_{\rm iso}}^k = \mathcal{D}(z)\psi_0^k = \exp\left(-\frac{|z|^2}{2}\right) \sum_{n=0}^{\infty} \frac{z^n}{\sqrt{n!}} \psi_n^k, \tag{22}$$

i.e., we obtain a similar expression as for the CS of the harmonic oscillator of equation (1). The difference rely in the states which are involved in the summation: for the harmonic oscillator they are the eigenstates of H_0, while here they are the eigenstates of H_k in $\mathcal{H}_{\rm iso}$. Note that the coherent states of equation (22) are already normalized.

Following the same procedure as for the harmonic oscillator [22], it can be shown that for the family of coherent states $\psi_{z_{\rm iso}}^k$ it is valid the standard resolution of the identity inside this subspace

$$\frac{1}{\pi} \int_{\mathbb{C}} |z_{\rm iso}^k\rangle\langle z_{\rm iso}^k| \, {\rm d\,Re}(z) \, {\rm d\,Im}(z) = \mathbb{1}|_{\mathcal{H}_{\rm iso}}. \tag{23}$$

3.2. Coherent states in the subspace $\mathcal{H}_{new}$
It is straightforward to calculate now the actions of the linearized ladder operators $\ell_k^\pm$ onto the basis vectors $\left\{\psi_{\epsilon_j}^k, j = 0, \ldots, k-1\right\}$ of $\mathcal{H}_{\rm new}$:

$$\ell_k^+ \psi_{\epsilon_j}^k = (1 - \delta_{j,k-1})\sqrt{\epsilon_{j+1} - E_0}\, \psi_{\epsilon_{j+1}}^k, \qquad \ell_k^- \psi_{\epsilon_j}^k = (1 - \delta_{j,0})\sqrt{\epsilon_j - E_0}\, \psi_{\epsilon_{j-1}}^k, \tag{24}$$

which come from the definition of $\ell_k^\pm$ in equations (17). The first term on the right hand side of equations (24) represent the fact that ℓ_k^- annihilates the eigenstate $\psi_{\epsilon_0}^k$, while ℓ_k^+ annihilates $\psi_{\epsilon_{k-1}}^k$. Despite the fact that the operators $\ell_k^\pm$ do not satisfy the Heisenberg-Weyl algebra in the subspace $\mathcal{H}_{\rm new}$, we will use the same displacement operator of equation (21), and we will apply it now to the ground state $\psi_{\epsilon_0}^k$ in order to obtain

$$\psi_{z_{\rm new}}^k = C_z' D(z)\psi_{\epsilon_0}^k = C_z \sum_{j=0}^{k-1} \frac{(iz)^j}{j!} \sqrt{\frac{1}{\Gamma(E_0 - \epsilon_0 - j)}}\, \psi_{\epsilon_j}^k. \tag{25}$$

Without lost of generality, the normalization constant C_z will be chosen as real and positive so that

$$C_z = \left[\sum_{j=0}^{k-1} \frac{|z|^{2j}}{(j!)^2} \frac{1}{\Gamma(E_0 - \epsilon_0 - j)}\right]^{-1/2}. \tag{26}$$

It can be shown that this set satisfies the resolution of the identity in the subspace $\mathcal{H}_{\rm new}$, namely,

$$\int_{\mathbb{C}} |z^k_{\rm new}\rangle\langle z^k_{\rm new}|\mu(z)\mathrm{d}z = \mathbb{1}|_{\mathcal{H}_{\rm new}}, \tag{27}$$

where $\mu(z)$ is a positive definite function given by

$$\mu(z) = \frac{\Gamma^2(E_0 + 1 - \epsilon_0)}{\pi C_z^2} U(E_0 + 1 - \epsilon_0, 1; |z|^2), \tag{28}$$

with $U(a, c; z)$ being the logarithmic solution of the confluent hypergeometric equation (for more details see the appendix).

4. Harmonic oscillator limit

If the parameters of the transformation function $u(x, \epsilon_{k-1})$ are chosen as $\epsilon_{k-1} = -1/2$ and $\nu_{k-1} = 0$, the SUSY partner potential becomes $V_k(x) = x^2/2 - k$, i.e. we obtain again the harmonic oscillator but with the energy origin displaced. The corresponding third order ladder operators are now

$$l_k^- = a^-(H_k - 1/2), \qquad l_k^+ = (H_k - 1/2)a^+. \tag{29}$$

It can be seen that l_k^- annihilates the vectors $\psi_0^k = \phi_k$ and $\psi_{\epsilon_0}^k = \phi_0$ while l_k^+ annihilates $\psi_{\epsilon_{k-1}} = \phi_{k-1}$, as expected. The operators X and P can be expressed in terms of the standard creation and annihilation operators and with that expressions the Heisenberg uncertainty relation for a coherent state can be evaluated.

4.1. Subspace $\mathcal{H}_{iso}$

A direct calculation leads now to the followings mean values:

$$\langle X\rangle_z = \sqrt{2}\mathrm{Re}(z)e^{-|z|^2} \sum_{n=0}^{\infty} \frac{|z|^{2n}}{n!} \sqrt{\frac{n+1+k}{n+1}}, \tag{30}$$

$$\langle P\rangle_z = \sqrt{2}\mathrm{Im}(z)e^{-|z|^2} \sum_{n=0}^{\infty} \frac{|z|^{2n}}{n!} \sqrt{\frac{n+1+k}{n+1}}, \tag{31}$$

$$\langle X^2\rangle_z = |z|^2 + k + 1/2 + \mathrm{Re}(z^2)e^{-|z|^2} \sum_{n=0}^{\infty} \frac{|z|^{2n}}{n!} \sqrt{\frac{(n+1+k)(n+2+k)}{(n+1)(n+2)}}, \tag{32}$$

$$\langle P^2\rangle_z = |z|^2 + k + 1/2 - \mathrm{Re}(z^2)e^{-|z|^2} \sum_{n=0}^{\infty} \frac{|z|^{2n}}{n!} \sqrt{\frac{(n+1+k)(n+2+k)}{(n+1)(n+2)}}. \tag{33}$$

For $k = 0$, i.e., if we do not make any SUSY transformation at all, the subspace $\mathcal{H}_{\rm iso}$ becomes the complete Hilbert space and we recover the mean values for the standard coherent states. On the other hand, for $k > 1$ it turns out that $(\Delta X)_z(\Delta P)_z$ approaches $1/2$ for large $|z|$ (see an example in figure 1). In addition, for $z = 0$ it is obtained that $\psi_z = \psi_0^k = \phi_k$, i.e., the coherent state becomes the k-th excited state of the harmonic oscillator, for which $\Delta X\Delta P = k + 1/2$.

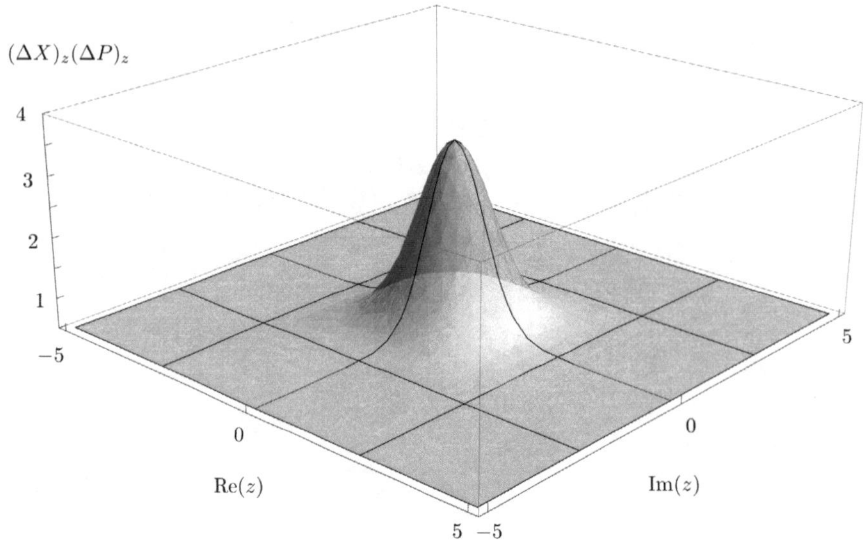

Figure 1. Heisenberg uncertainty relation for the linearized CS (considering just the first 100 terms) in the harmonic oscillator limit with $k = 3$ inside the subspace $\mathcal{H}_{\text{iso}}$.

4.2. *Subspace* $\mathcal{H}_{new}$

For this finite subspace, the corresponding mean values are now

$$\langle X\rangle_z = \sqrt{2}|C_z|^2\mathrm{Re}(iz)\sum_{j=0}^{k-2}\frac{|z|^{2j}}{(j!)^2(k-1-j)!}\frac{1}{\sqrt{(j+1)(k-1-j)}}, \tag{34}$$

$$\langle P\rangle_z = \sqrt{2}|C_z|^2\mathrm{Im}(iz)\sum_{j=0}^{k-2}\frac{|z|^{2j}}{(j!)^2(k-1-j)!}\frac{1}{\sqrt{(j+1)(k-1-j)}}, \tag{35}$$

$$\langle X^2\rangle_z = |C_z|^2\sum_{j=1}^{k-1}\frac{|z|^2 j}{(j!)^2}\frac{j}{(k-1-j)!}+\frac{1}{2}- $$
$$ - |C_z|^2\mathrm{Re}(z^2)\sum_{j=0}^{k-3}\frac{|z|^{2j}}{(j!)^2(k-3-j)!}\frac{1}{\sqrt{(j+1)(j+2)(k-1-j)(k-2-j)}}, \tag{36}$$

$$\langle P^2\rangle_z = |C_z|^2\sum_{j=1}^{k-1}\frac{|z|^2 j}{(j!)^2}\frac{j}{(k-1-j)!}+\frac{1}{2}+ $$
$$ + |C_z|^2\mathrm{Re}(z^2)\sum_{j=0}^{k-3}\frac{|z|^{2j}}{(j!)^2(k-3-j)!}\frac{1}{\sqrt{(j+1)(j+2)(k-1-j)(k-2-j)}}. \tag{37}$$

In figure 2 it can be seen an example of the Heisenberg uncertainty relation as a function of z. For $z = 0$ the corresponding CS reduces to the ground state, minimizing then the Heisenberg uncertainty relation; for large $|z|$ it approaches to $k - 1/2$, which is the value of the Heisenberg uncertainty relation for the $(k-1)$-th excited state, precisely the one with the highest energy inside this subspace.

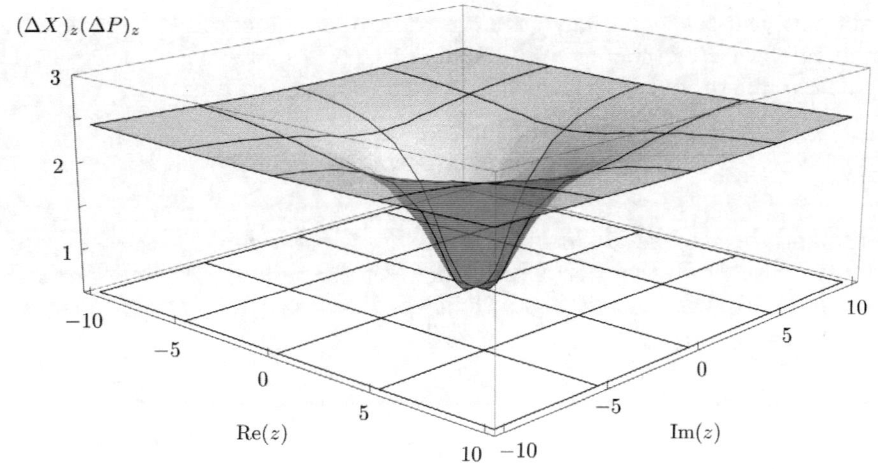

Figure 2. Heisenberg uncertainty relation for the linearized CS in the harmonic oscillator limit with $k = 3$ inside the subspace $\mathcal{H}_{\text{new}}$.

5. Concluding remarks

In this work we have studied the linearized coherent states for the k-SUSY partners of the harmonic oscillator which posses third order differential ladder operators. These operators were first linearized, and then the coherent states were calculated by applying the corresponding displacement operator onto the extremal states. As a result, two families of coherent states were generated: one inside the subspace isospectral to the harmonic oscillator and another one in the subspace generated by the eigenfunctions associated to the inserted energy levels. Both families of coherent states admit a resolution of the identity in the corresponding subspace. Since the harmonic oscillator is a shape invariance potential, then the Heisenberg uncertainty relation for the operators X and P can be simply analyzed in the case when the k-SUSY partner potential becomes the initial one.

Acknowledgements

The authors acknowledge the support of Conacyt, project No 152574. DB and ACA also acknowledge the Conacyt Ph.D. scholarships 219665 and 173288 respectively.

Appendix

In order to guarantee the resolution of the identity in the subspace $\mathcal{H}_{\text{new}}$, i.e.,

$$\int_{\mathbb{C}} |z^k_{\text{new}}\rangle\langle z^k_{\text{new}}|\mu(z)\mathrm{d}z = \mathbb{1}|_{\mathcal{H}_{\text{new}}}, \tag{38}$$

we have to find a positive definite function $\mu(z)$ which will make valid equation (38). We follow the work of Sixdeniers and Penson [23], where they have solved a similar problem. If we substitute the expression for $|z^k_{\text{new}}\rangle$ given by equation (25), write z in polar coordinates and integrate the angular one, we obtain

$$\mathbb{1}|_{\mathcal{H}_{\text{new}}} = 2\pi \sum_{j=0}^{k-1} \frac{|\epsilon^k_j\rangle\langle\epsilon^k_j|}{(j!)^2\Gamma(E_0-\epsilon_0-j)} \int_0^\infty C_z^2 r^{2j+1}\mu(r)\mathrm{d}r. \tag{39}$$

In order to simplify the last equation, let us introduce the function $f(r) = \pi C_z^2 \mu(r)$, then let us make a change of variable $r^2 = x$ and of index $j = s - 1$ in such a way that the following equation must be fulfilled

$$\int_0^\infty x^{s-1} f(x)\mathrm{d}x = \Gamma^2(s)\Gamma(E_0 + 1 - \epsilon_0 - s) \equiv \mathcal{M}[f(x); s]. \tag{40}$$

Now we need to find the *inverse Mellin transform* $\mathcal{M}^{-1}[\Gamma^2(s)\Gamma(E_0+1-\epsilon_0-s); x]$. It is possible to find several inverse Mellin transforms in tables, for example in Erdélyi's book [24]. In this case, the function $f(x)$ turns out to be a Meijer G function with $m = 2$, $n = 1$, $p = 1$, $q = 2$, $a_1 = \epsilon_0 - E_0$, $b_1 = b_2 = 0$, i.e.,

$$f(r) = G_{1\,2}^{2\,1}\left({}^{\epsilon_0 - E_0}_{0,0} \middle| r^2 \right). \tag{41}$$

Even more, in Erdélyi's book on transcendental functions [25] one can find some expressions for the Meijer G function in terms of other special functions, in particular of the Whittaker function $W_{\kappa,\mu}(z)$, which in turn can be written in terms of the logarithmic solution of the confluent hypergeometric equation $U(a, c; z)$ [26]. Then we have

$$f(r) = \Gamma^2(E_0 + 1 - \epsilon_0)U(E_0 + 1 - \epsilon_0, 1; r^2). \tag{42}$$

We still need to prove the positiveness of $\mu(z)$. To accomplish this, we use the integral representation of $U(a, c; z)$:

$$U(a, c; z) = \frac{1}{\Gamma(a)} \int_0^\infty e^{-zt} t^{a-1}(1+t)^{b-a-1}\mathrm{d}t, \tag{43}$$

and thus the measure $\mu(r)$ can be written as

$$\mu(r) = \frac{\Gamma(E_0 + 1 - \epsilon_0)}{\pi C_z^2} \int_0^\infty e^{-r^2 t} t^{E_0-\epsilon_0}(1+t)^{\epsilon_0 - E_0 - 1}\mathrm{d}t. \tag{44}$$

Besides, taking into account that $E_0 > \epsilon_0$, and that the domain of r and t is $[0, \infty)$ we can conclude that we have found, at least, one positive definite measure, i.e, this CS admit a resolution of the identity in the subspace $\mathcal{H}_{\text{new}}$.

References

[1] Schrödinger E 1928 *Collected Papers on Wave Mechanics* (London: Chelsea Publishing Company)
[2] Klauder J 1963 *J. Math. Phys.* **4** 1955–58
[3] Klauder J 1963 *J. Math. Phys.* **4** 1958–73
[4] Nieto M and Simmons L 1979 *Phys. Rev.* D **20** 1321–31
[5] Perelomov A 1986 *Generalized Coherent States and their Applications* (Belin: Springer-Verlag)
[6] Gazeau J and Klauder J 1999 *J. Phys. A: Math. Gen.* **32** 123–32
[7] Ali S, Antoine J and Gazeau J 2000 *Coherent States, Wavelets and their Generalizations* (New York: Springer-Verlag)
[8] Geloun J B, Hnybida J and Klauder J 2012 *J. Phys. A: Math. Theor.* **45** 085301
[9] Glauber R 1963 *Phys. Rev.* **130** 2529–39
[10] Glauber R 1963 *Phys. Rev.* **131** 2766–88
[11] Mielnik B 1984 *J. Math. Phys.* **25** 3387–89
[12] Cooper F, Khare A and Sukhatme U 1995 *Physics Reports* **251** 267–385
[13] Bagrov V and Samsonov B 1997 *Pramana* **49** 563–80
[14] Fernandez C D J and Fernandez-Garcia N 2004 *AIP Conf. Proc.* **744** 236–73
[15] Mielnik B and Rosas-Ortiz O 2004 *J. Phys. A: Math. and Gen.* **37** 10007
[16] Bermudez D and Fernández D 2011 *SIGMA* **7** 025

[17] Bermudez D and Fernández C D J 2013 Supersymmetric quantum mechanics and Painleve equations (*Preprint* `math-ph /1311.0647`)
[18] Fernández D, Hussin V and Nieto L 1994 *J. Phys. A: Math. Gen.* **27** 3547–64
[19] Fernández D, Nieto L and Rosas-Ortiz O 1995 *J. Phys. A: Math. Gen.* **28** 2693–708
[20] Fernández D and Hussin V 1999 *J. Phys. A: Math. Gen.* **32** 3603–19
[21] Fernández D, Hussin V and Rosas-Ortiz O 2007 *J. Math. Phys. A: Math. Theor.* **40** 6491–511
[22] Cohen-Tannoudji C, Diu B and Laloë F 1977 *Quantum Mechanics* (Paris: Wiley)
[23] Sixdeniers J and Penson K 2000 *J. Phys. A: Math. Gen.* **33** 2907–16
[24] Erdélyi A 1954 *Tables of Integral Transforms* (New York: McGraw–Hill Company)
[25] Erdélyi A 1953 *Higher Trascendental Functions* (New York: McGraw–Hill Company)
[26] Abramowitz M and Stegun I 1972 *Handbook of Mathematical Functions with Formulas, Graphs and Mathematical Tables* (New York: Dover)

Journal of Physics: Conference Series **512** (2014) 012019 doi:10.1088/1742-6596/512/1/012019

The influence of abstract group theory on molecular symmetry

L. L. Boyle
Vine Cottage, Patrixbourne, Kent CT4 5BX, UK

E-mail: Laurence.Boyle@gmail.com

Abstract. It is shown that the identification of the double groups required for treating molecular problems involving an odd number of electrons can be solved using abstract group theory without consideration of explicit matrices as in Pauli's discussion using spin matrices. This presupposes that the character tables are arrays of invariant numbers which alone are sufficient to specify the relevant properties of the spin systems. Consideration of subduction from infinite groups such as SU(2) is thus avoided and the criteria used to identify abstract groups as suitable double groups for the odd-electron spin problem include the need for subduction from finite symmetry groups of higher order as well as the requirement to produce closed shells of electrons in which no two electrons can have the same quantum numbers. Explicit calculations are provided for groups of orders 1, 2 and 4 and these provide the underlying principles for any molecular symmetry group. Novel aspects include the consideration of the separable four-fold degeneracies which occur in molecules having a single four-fold axis of symmetry. It is also shown that the treatment required for threefold symmetries is consistent with the principles derived from the study by abstract group theory of 1-, 2- and 4-fold symmetries.

1. Introduction

Abstract groups are described by presentations. These provide information about the order of the generators of the group and the relationships between them. A character table and an automorphism group can be defined for each abstract group. The following are some simple examples of presentations:-

1. The cyclic groups of order n. These can be expressed in terms of a single generator, A, such that $A^n = E$ where E is the unit element of the group. A is said to be of order n where n is the smallest power of A which generates E.
2. The direct product group of order mn. Such a group can be expressed in terms of 2 generators, A and B, such that $A^m = E$ and $B^n = E$. The elements A and B have to commute, *i.e.* BA=AB. In some notations this would be written [A,B] = E.
3. The dihedral group of order $2n$. This may be presented as $\{A^n = E;\ B^2 = E;\ BA = A^{-1}B\}$ so that the commutator $[A,B] = A^{-1}B^{-1}AB = A^{n-2}$.

The above examples are the simplest possible and it should be recorded that in more complicated groups such as the icosahedral rotation group the choice of generators to produce elements labelled by words which contain the powers of the generators in alphabetical order should not be regarded as completely free if one also aspires to mechanise the process of identifying the product of any two elements by use of the relationships given between the generators. In some cases analysis of the product will unfortunately lead to an infinite sequence of generators.

2. Some aspects of character tables

2.1 *The non-uniqueness of character tables*

Every abstract group has a character table which is a number-theoretical array. Not all character tables are different, however, but it will be demonstrated that the irreducible representations of two different groups which have identical arrays of numbers as their character tables may well differ in some of their derived properties.

To illustrate this point we may study the simplest case which happens to occur with groups of order 8. The dihedral group $\mathbf{D_4} := \{A^4=B^2=E;\ BA=A^3B\}$ and the quaternion group $\mathbf{Q} := \{A^4=B^4=E;\ B^2=A^2;\ BA=A^3B\}$ have the same character table:-

D₄	E	A^2	A, A^3	B, A^2B	AB, A^3B
Q	E	A^2	A, A^3	B, A^2B	AB, A^3B
Γ_1	1	1	1	1	1
Γ_2	1	1	1	−1	−1
Γ_3	1	1	−1	1	−1
Γ_4	1	1	−1	−1	1
Γ_5	2	−2	0	0	0

The conjugacy-class structures are identical but the order structures are not. Apart from the identity, $\mathbf{D_4}$ has five elements of order 2 and two of order 4 while $\mathbf{Q}$ has one element of order 2 and six of order 4. The differences produced in the squares of the elements, ε_i , are reflected in the calculation of the symmetric and antisymmetric squares of the 2-dimensional irreducible representation, Γ_5 , of each group. The character of the symmetric square for a representation, Γ, and an element, ε_i , is defined as $[\Gamma^2] = \frac{1}{2}[\{\chi(\varepsilon_i)\}^2 + \chi(\varepsilon_i^2)]$ while for the antisymmetric square, $\{\Gamma^2\}$, it is defined as $\frac{1}{2}[\{\chi(\varepsilon_i)\}^2 - \chi(\varepsilon_i^2)]$. The sum of these two components is the square Γ^2 and both the symmetric and antisymmetric parts are reducible representations of the group.

When we calculate the symmetric and antisymmetric squares of the 2-dimensional irreducible representation, Γ_5, of each group we obtain the following results:-

Group **D₄**

ε_i	E	A^2	A, A^3	B, A^2B	AB, A^3B	
ε_i^2	E	E	A^2	E	E	
Γ_5	2	−2	0	0	0	
$[\Gamma_5^2]$	3	3	−1	1	1	$= \Gamma_1+\Gamma_3+\Gamma_4$
$\{\Gamma_5^2\}$	1	1	1	−1	−1	$= \Gamma_2$

Group Q

ε_i	E	A^2	A, A^3	B, A^2B	AB, A^3B	
ε_i^2	E	E	A^2	A^2	A^2	
Γ_5	2	−2	0	0	0	
$[\Gamma_5^2]$	3	3	−1	−1	−1	$= \Gamma_2+\Gamma_3+\Gamma_4$
$\{\Gamma_5^2\}$	1	1	1	1	1	$= \Gamma_1$

It is to be noticed that the totally-symmetric representation, Γ_1 , occurs in the symmetric square of Γ_5 of $\mathbf{D_4}$ but in the antisymmetric square of Γ_5 of $\mathbf{Q}$.

2.2 *Single- and double-valued representations*

When, as in these two cases, a group possesses an invariant subgroup of index 2 (*i.e.* a halving subgroup) the concept of single and double value can be defined. In such cases the single-valued representations have the same character for the commuting twofold element as for the identity while the double-valued representations have minus the character for the identity as the character for the commuting twofold element. This leads to the statement that the products of two single-valued representations or two double-valued representations are single-valued while the product of a single-valued representation with a double-valued representation is double-valued.

In the above two examples, all characters of Γ_5 happened to be real numbers. According to the definition above, the irreducible representations Γ_5 of both $\mathbf{D_4}$ and **Q** can both be regarded as double-valued. The fundamental difference, however, is that Γ_5 of $\mathbf{D_4}$ is such that the totally-symmetric representation, Γ_1, occurs in the symmetric part of its square while in the case of Γ_5 of **Q**, Γ_1 occurs in the antisymmetric part of its square.

2.3 *Complex representations*

Another type of representation is characterised by having some characters which are complex numbers. For this type the above formulæ for the square and its parts cannot be used without further information. This can be illustrated with reference to the tetrahedral rotation group, **T**, which is isomorphic to the alternating group, $\mathbf{A_4}$. **T** may be presented in terms of three generators, A, B and C, such that $\{A^2=B^2=C^3=E;\ BA=AB;\ CA=ABC;\ CB=AC\}$. The character table is:-

T	E	A, B, AB	C, AC, BC, ABC	C^2, BC^2, ABC^2, AC^2
Γ_1	1	1	1	1
Γ_2	1	1	ω	ω^2
Γ_2^*	1	1	ω^2	ω
Γ_3	3	−1	0	0

where ω and ω^2 are the complex cube roots of unity. If one simply squares the characters of either of the complex representations the other is obtained and this will not lead to a useful implementation in quantum mechanics. However, if one takes the real, reducible, representation $\Gamma_2+\Gamma_2^*$ and applies the formulæ above the direct product will be $2\Gamma_1+\Gamma_2+\Gamma_2^*$ with the totally-symmetric representation, Γ_1, occurring in both the symmetric, $\Gamma_1+\Gamma_2+\Gamma_2^*$, and the antisymmetric, Γ_1, parts. That Γ_1 should appear twice is logical since $\Gamma_2+\Gamma_2^*$ consists of two representations. Γ_2 and Γ_2^* are normally described as a complex-conjugate pair of representations because, for each element their characters are related by complex conjugation. In another terminology the reducible representation $\Gamma_2+\Gamma_2^*$ would be said to be a separably-degenerate representation.

A more abstract description derives from the group-theoretical concept of the inverse automorphism rather than the algebraic concept of complex conjugation. This automorphism has the effect of replacing each element by its inverse. In this example the class {C, AC, BC, ABC} will be interchanged with the class {C^2, BC^2, ABC^2, AC^2} and the characters are replaced by their complex conjugates. The invariance of the character table to this automorphism is assured by permuting Γ_2 and Γ_2^* at the same time as interchanging the two classes of threefold elements.

Another example of a group with complex representations is the cyclic group of four elements, $\mathbf{C_4}$. The presentation is $\{A^4=E\}$ and the character table may be written as follows:-

$\mathbf{C_4}$	E	A^2	A	A^3
Γ_1	1	1	1	1
Γ_2	1	1	−1	−1
Γ_3	1	−1	i	−i
Γ_3^*	1	−1	−i	i

Unlike **T** this group has an invariant subgroup of order 2, {E, A^2}, and the characters for the twofold element, A^2, are ± those for the identity. Γ_1 and Γ_2 could therefore be regarded as single-valued representations and Γ_3 and Γ_3^* as double-valued representations. It should be stressed, however, that this is not a necessary distinction: in applications where double-valued representations are not needed as such, all four representations of $\mathbf{C_4}$ can be used as ordinary

Journal of Physics: Conference Series **512** (2014) 012019 doi:10.1088/1742-6596/512/1/012019

representations. The above character table of $\mathbf{C_4}$ illustrates one further point which is true for groups where the terms single-valued and double-valued are applicable. The weight of any double-valued representations, defined as the sum of the squares of their dimensions, has to equal the weight of the single-valued representations. In the case of $\mathbf{C_4}$ the weights of both single- and double-valued representations are $1^2+1^2=2$. In the case of **T**, however, the weight of the Γ_2 and Γ_2^* representations is $1^2+1^2=2$ while that of the Γ_1 and Γ_3 representations is $1^2+3^2=10$. In such a case the criterion of having an invariant subgroup of order two is not satisfied and so the abstract group **T** could not be used in a context where double-valued representations are required.

3. The problem of electron spin

In order to explain the Stern–Gerlach effect and the anomalous Zeeman effect it was necessary to introduce the concept of spin. Unlike electronic orbital functions, the one-electron spin-functions transform as double-valued representations of a so-called "double group" which has double the order of the group used to describe the orbital functions. Unlike orbital functions, however, the spin functions of degeneracy $2n$ have to be such that the fully-antisymmetrised $2n$-th power of the representation they span contains the totally-symmetric representation[1]. The problem is to identify what abstract groups could satisfy these requirements and hence provide information from their structure about the transformation properties of spins and spin-orbitals. Rather than assume that the special unitary group of all 2×2 matrices is involved and then reduce the symmetry to the often finite symmetry involved in molecules, the approach taken here is the converse, *viz.* to go from low-order groups to groups of higher symmetry without any mention of matrices.

4. Construction of double groups of abstract groups suitable for electron-spin problems

The principles will be illustrated by constructing the double groups of the two abstract groups of orders 1, 2 and 4. These must be groups of orders 2, 4 and 8 respectively. There are just five distinct abstract groups of order 8 which can be enumerated as:-

Order	Group	Presentation
2	$\mathbf{C_2}$	$\{A^2=E\}$
4	$\mathbf{C_4}$	$\{A^4=E\}$
	$\mathbf{C_2\times C_2}$	$\{A^2=B^2=E;\ BA=AB\}$
8	$\mathbf{C_8}$	$\{A^8=E\}$
	$\mathbf{C_4\times C_2}$	$\{A^4=B^2=E;\ BA=AB\}$
	$\mathbf{D_4}$	$\{A^4=B^2=E;\ BA=A^3B\}$
	$\mathbf{Q}$	$\{A^4=B^4=E;\ B^2=A^2;\ BA=A^3B\}$
	$\mathbf{C_2\times C_2\times C_2}$	$\{A^2=B^2=C^2=E;\ BA=AB;\ CA=AC;\ CB=BC\}$

The presentations are each given in terms of a minimum number of generators: this is noteworthy for the quaternion group where a more symmetric set is obtained using a third generator. The reason for using minimal sets is that a double group will then contain no more than one additional generator relative to the single group.

It will also be useful to enumerate the abstract halving subgroups in each case:-

Order	Group	Halving subgroups
2	$\mathbf{C_2}$	$\mathbf{C_1}\{E\}$
4	$\mathbf{C_4}$	$\mathbf{C_2}\{A^2=E\}$
	$\mathbf{C_2\times C_2}$	$\mathbf{C_2}\{A^2=E\}$
8	$\mathbf{C_8}$	$\mathbf{C_4}\{A^4=E\}$
	$\mathbf{C_4\times C_2}$	$\mathbf{C_4}\{A^4=E\};\ \mathbf{C_2\times C_2}\{A^2=B^2=E;\ BA=AB\}$
	$\mathbf{D_4}$	$\mathbf{C_4}\{A^4=E\};\ \mathbf{C_2\times C_2}\{A^2=B^2=E;\ BA=AB\}$
	$\mathbf{Q}$	$\mathbf{C_4}\{A^4=E\};\ \mathbf{C_2\times C_2}\{A^2=B^2=E;\ BA=AB\}$
	$\mathbf{C_2\times C_2\times C_2}$	$\mathbf{C_2\times C_2}\{A^2=B^2=E;\ BA=AB\}$

Journal of Physics: Conference Series **512** (2014) 012019 doi:10.1088/1742-6596/512/1/012019

4.1 *The double group of the group of order one*

This case is trivial but does serve to illustrate that real representations can serve as the symmetries of electronic spin-orbitals. The double group of $\mathbf{C_1}$ has to be $\mathbf{C_2}$ and the Γ_2 representation of $\mathbf{C_2}$ is the only available double-valued representation. Taking a pair of these the symmetric square will be $3\Gamma_1$ and the antisymmetric square will be Γ_1. $\mathbf{C_2}$ is therefore an acceptable spin double group for $\mathbf{C_1}$.

4.2 *The double group of the group of order two*

Both abstract groups of order four are candidates to be the double group of the abstract group of order two. Superimposing the character tables illustrates this point well. If $\mathbf{C_4}$ is to be the double group, the table below shows that Γ_3 and Γ_3* behave as double-valued representations for $\mathbf{C_2}$. Further the antisymmetric square of this complex-conjugate pair is the totally-symmetric representation so $\mathbf{C_4}$ is a suitable double group for studying the spin-½ problem.

	$\mathbf{C_4}$	E	A^2	A	A^3
$\mathbf{C_2}$		E		A	
Γ_1	Γ_1	1	1	1	1
Γ_2	Γ_2	1	1	−1	−1
	Γ_3	1	−1	i	−i
	Γ_3*	1	−1	−i	i

Turning now to Klein's four-group as a possible double group of $\mathbf{C_2}$ the character tables can be matched up as

	$\mathbf{C_2 \times C_2}$	E	B	A	AB
$\mathbf{C_2}$		E		A	
Γ_1	Γ_1	1	1	1	1
Γ_2	Γ_3	1	1	−1	−1
	Γ_2	1	−1	1	−1
	Γ_4	1	−1	−1	1

so that Γ_2 and Γ_4 of $\mathbf{C_2 \times C_2}$ can serve as double-valued representations. Even though they are real and independently have no antisymmetric squares, taken as the pair $\Gamma_2+\Gamma_4$ they have Γ_3 as their antisymmetric square and so are not suitable for studying the spin problem. This means that only the cyclic group of twice the order can serve as the spin double group.

4.3 *The double group of the cyclic group of order four*

Cyclic groups are 1-generator groups by definition hence the double group of a cyclic group must be presentable in terms of not more than two generators: $\mathbf{C_2 \times C_2 \times C_2}$ cannot therefore be regarded as a candidate. Further, $\mathbf{D_4}$ and $\mathbf{Q}$ cannot be candidates because their character tables do not contain any complex numbers and hence their single-valued representations cannot be mapped onto the representations of the single group, $\mathbf{C_4}$. In the case of $\mathbf{C_8}$ the only twofold element, A^4, is in the invariant subgroup of order 2 and so the character tables map as:-

	$\mathbf{C_8}$	E	A^4	A^2	A^6	A	A^5	A^3	A^7
$\mathbf{C_4}$		E		A^2		A		A^3	
Γ_1	Γ_1	1	1	1	1	1	1	1	1
Γ_2	Γ_2	1	1	1	1	−1	−1	−1	−1
Γ_3	Γ_3	1	1	−1	−1	i	i	−i	−i
Γ_3*	Γ_3*	1	1	−1	−1	−i	−i	i	i
	Γ_4	1	−1	i	−i	$\sqrt{i}$	$-\sqrt{i}$	$i\sqrt{i}$	$-i\sqrt{i}$
	Γ_4*	1	−1	−i	i	$-i\sqrt{i}$	$i\sqrt{i}$	$-\sqrt{i}$	$\sqrt{i}$
	Γ_5	1	−1	i	−i	$-\sqrt{i}$	$\sqrt{i}$	$-i\sqrt{i}$	$i\sqrt{i}$
	Γ_5*	1	−1	−i	i	$i\sqrt{i}$	$-i\sqrt{i}$	$\sqrt{i}$	$-\sqrt{i}$

where $\sqrt{i}=(i+1)/\sqrt{2}$ and $i\sqrt{i}=(i-1)/\sqrt{2}$. This case is unusually interesting as it merits further examination. If one takes the antisymmetric square of either of the complex-conjugate pairs of double-valued representations one obtains Γ_1. If one takes the pair $\Gamma_4 + \Gamma_5$ one obtains the complex, single-valued, representation Γ_3* for the antisymmetric square while Γ_4* + Γ_5* produce its complex conjugate Γ_3. However, if one takes either the pair Γ_4 + Γ_5* or the pair Γ_4* + Γ_5 one obtains Γ_2 . This unexpected result is best understood by a consideration of the automorphisms of the group **C₈**. These form a group of four elements isomorphic with Klein's four-group and generated by the inverse transformation, $A^n \rightarrow A^{7n}$, (which produces complex conjugation in the irreducible representations) and the transformation produced by replacing the generator A by A^3. The product of these two transformations is equivalent to replacing A^n by A^{5n}. The effect on the representations can be summarised in the following table:-

A^n	A^{7n}	A^{3n}	A^{5n}
Γ_3	Γ_3*	Γ_3	Γ_3*
Γ_3*	Γ_3	Γ_3*	Γ_3
Γ_4	Γ_4*	Γ_5*	Γ_5
Γ_4*	Γ_4	Γ_5	Γ_5*
Γ_5	Γ_5*	Γ_4*	Γ_4
Γ_5*	Γ_5	Γ_4	Γ_4*

The character table is invariant to the same transformations as those of the elements of the group. Mathematically the four irreducible representations {Γ_3 , Γ_3*, Γ_4 , Γ_4*} are distinct 1-dimensional representations which form an orbit under the action of the automorphism group of the group, **C₈** . Physically they form a separable quadruply-degenerate state so that they would all be associated with states of the same energy. In the presence of a linear magnetic field, however, the quadruple degeneracy is lifted into two pairs of complex-conjugate double degeneracies.

To consider **C₄×C₂** as a potential double group of **C₄** the character tables may be matched as follows:-

C₄	**C₄×C₂**	E	B	A^2	A^2B	A	AB	A^3	A^3B
		E		A^2		A		A^3	
Γ_1	Γ_1	1	1	1	1	1	1	1	1
Γ_2	Γ_2	1	1	1	1	−1	−1	−1	−1
Γ_3	Γ_3	1	1	−1	−1	i	i	−i	−i
Γ_3*	Γ_3*	1	1	−1	−1	−i	−i	i	i
	Γ_4	1	−1	1	−1	1	−1	1	−1
	Γ_5	1	−1	1	−1	−1	1	−1	1
	Γ_6	1	−1	−1	1	i	−i	−i	i
	Γ_6*	1	−1	−1	1	−i	i	i	−i

The relevant invariant subgroup of order 2 is {E, B}. Γ_4 , Γ_5 , Γ_6 and Γ_6* are then the double-valued representations as they have characters for B which are minus those for E. The problem is, however, that the real double-valued representations, Γ_4 and Γ_5 , cannot be used to construct states which have an antisymmetric square equal to Γ_1. Hence they cannot represent spin-orbitals involving spin-½ functions even though the complex-conjugate pair, Γ_6 and Γ_6*, satisfy this criterion.

The spin double group of the cyclic group of order 4 must therefore be the cyclic group of order 8.

4.4 *The double group of Klein's four-group,* **C₂×C₂**

Since Klein's four-group is a two-generator group, the cyclic group **C₈** can immediately be ruled out as a possible double group. The elementary group **C₂×C₂×C₂** and the dihedral group $\mathbf{D_4}$ are both ineligible as they only contain representations which cannot be used to produce an antisymmetric square containing the identity representation.

This leaves just $\mathbf{C_4{\times}C_2}$ and **Q** and we shall see that both are possible structures for double groups of Klein's four-group. To justify $\mathbf{C_4{\times}C_2}$ it is helpful to rearrange the rows and columns of the character table without changing the labelling of the representations of $\mathbf{C_4{\times}C_2}$ above to produce:-

	$\mathbf{C_4{\times}C_2}$	E	A^2	A	A^3	B	A^2B	AB	A^3B
$\mathbf{C_2{\times}C_2}$		E		A		B		AB	
Γ_1	Γ_1	1	1	1	1	1	1	1	1
Γ_2	Γ_4	1	1	1	1	−1	−1	−1	−1
Γ_3	Γ_2	1	1	−1	−1	1	1	−1	−1
Γ_4	Γ_5	1	1	−1	−1	−1	−1	1	1
	Γ_3	1	−1	i	−i	1	−1	i	−i
	Γ_3*	1	−1	−i	i	1	−1	−i	i
	Γ_6	1	−1	i	−i	−1	1	−i	i
	Γ_6*	1	−1	−i	i	−1	1	i	−i

from which it can be seen that the four complex representations of $\mathbf{C_4{\times}C_2}$ now serve as double-valued representations of the double group of Klein's four-group. Further, the antisymmetric squares of the sum of either of the complex-conjugate pairs is the totally-symmetric representation, Γ_1 , and hence $\mathbf{C_4{\times}C_2}$ appears to be a suitable double group for application to the electron-spin problem.

When the corresponding table for **Q** is constructed the doubly-degenerate representation Γ_5 of **Q** has all the required properties and so **Q** is also a possible structure for the spin double group of Klein's four-group.

	Q	E	A^2	A, A^3	B, A^2B	AB, A^3B
$\mathbf{C_2{\times}C_2}$		E		A	B	AB
Γ_1	Γ_1	1	1	1	1	1
Γ_2	Γ_2	1	1	1	−1	−1
Γ_3	Γ_3	1	1	−1	1	−1
Γ_4	Γ_4	1	1	−1	−1	1
	Γ_5	2	−2	0	0	0

Comparison of the conjugacy-class structures provides an elegant demonstration of the 2:1 homomorphism between the double group and the single group as well as the identification of Γ_5 of **Q** as the double-valued representation.

To decide which of the two possible groups to use as the spin double group we have to consider the relationship between the double groups of $\mathbf{C_2}$ and $\mathbf{C_4}$. On subduction from the double group of $\mathbf{C_4}$ to that of $\mathbf{C_2}$ we would have to choose which of the generators of $\mathbf{C_4{\times}C_2}$ was to remain a four-fold element and which was to become a two-fold element. But it has been seen that any two-fold generator of $\mathbf{C_2}$ becomes a four-fold generator in the double group of $\mathbf{C_2}$ which must be a subgroup of the double group of $\mathbf{C_4}$.

In considerations of spin-free molecular symmetry there are just three types of two-fold element, *viz.* space inversion, two-fold rotations and reflections. Space inversion is treated separately for this purpose as it is assumed that the spin problems are independent of whether we use a right-handed or a left-handed set of coördinates. Hence space inversion is included by factorising a symmetry group containing space inversion into a $\mathbf{C_2}$ group consisting of the identity and space inversion and a group which may contain rotations, reflections and combinations of these. Symmetry groups containing space inversion are thus best considered as direct product groups and the space inversion operation is taken to commute with all elements of the double group of the inversion-free part. Further, spin functions are taken to be invariant to space inversion, *i.e.* they are *gerade* or g-type.

This leaves the quaternion group **Q** as the only acceptable spin double group of Klein's four-group.

5. Construction of spin orbitals for the double group of C_4

This example is sufficiently interesting to merit inclusion here as it provides the key to the general problem and has probably not been published elsewhere. We know very little about electronic spin functions except that they are characterised by two quantum numbers, S and S_z, which are related to the eigenvalues of the spin operators $\hat{S}^2$ and $\hat{S}_z$ by the familiar relationships $\hat{S}^2\sigma(S,S_z) = S(S+1)\sigma(S,S_z)$ and $\hat{S}_z\sigma(S,S_z) = S_z\sigma(S,S_z)$. Converting functions with given values of S and S_z to those with neighbouring values of S_z is effected by the ladder operators $\hat{S}_+$ and $\hat{S}_-$ which have the effects $\hat{S}_\pm\sigma(S,S_z) = \{(S\mp S_z)(S\pm S_z+1)\}^{1/2}\sigma(S,S_z\pm1)$. It is usual to abbreviate $\sigma(\frac{1}{2},\frac{1}{2})$ as α and $\sigma(\frac{1}{2},-\frac{1}{2})$ as β.

For group-theoretical analysis the effects of the symmetry operators on the spin functions are required. This is given in the case of one-dimensional representations by the character tables. Thus the generator A of $\mathbf{C_8}$ acting on the spin function α produces α multiplied by one of the four eighth roots of unity. Which of the four is chosen is entirely a matter of convention and the customary choice is to take that which is the lowest power of $\exp(2\pi i)$. In this case it will be $\exp(\pi i/4)=\sqrt{i}$. Correspondingly the effect of A on β must produce the complex conjugate $-i\sqrt{i}\beta$. These definitions assign the 1-electron spin functions to the Γ_4 and Γ_4^* representations respectively.

Now suppose that we have two electrons labelled 1 and 2 with spin functions $\alpha(1)$, $\alpha(2)$, $\beta(1)$ and $\beta(2)$. The antisymmetric square of $\Gamma_4+\Gamma_4^*$ is Γ_1 so that $\alpha(1)\beta(2)-\beta(1)\alpha(2)$ is totally-symmetric and of S_z-value zero. The symmetric square produces the un-normalised functions $\alpha(1)\alpha(2)$, $\alpha(1)\beta(2)+\beta(1)\alpha(2)$ and $\beta(1)\beta(2)$ for the triplet (*i.e.* S=1) state with S_z values +1, 0 and −1 respectively. These states are of symmetries Γ_3, Γ_1 and Γ_3^* respectively. These are single-valued representations as is to be expected for states involving an even number of electrons.

To find spin functions spanning the double-valued representations Γ_5 and Γ_5^* it is necessary to consider functions for 3 or more electrons[2]. The four un-normalised three-electron functions having $S=\frac{3}{2}$ and $S_z = \frac{3}{2}, \frac{1}{2}, -\frac{1}{2}$ and $-\frac{3}{2}$ are respectively $\alpha(1)\alpha(2)\alpha(3)$, $\alpha(1)\alpha(2)\beta(3)+\alpha(1)\beta(2)\alpha(3)+\beta(1)\alpha(2)\alpha(3)$, $\beta(1)\beta(2)\alpha(3)+\beta(1)\alpha(2)\beta(3)+\alpha(1)\beta(2)\beta(3)$ and $\beta(1)\beta(2)\beta(3)$ and span the irreducible representations Γ_5^*. Γ_4, Γ_4^* and Γ_5. To show that these four spin functions form a "closed shell" requires that the determinantally-antisymmetric fourth power of $\{\Gamma_4, \Gamma_4^*, \Gamma_5, \Gamma_5^*\}$ contains the identity representation which is the case.

6. Three-fold symmetry

The other chain of symmetry groups which is important in crystallographic symmetries although not exclusively fir free molecules is that deriving from the ascent $\mathbf{C_1}\rightarrow\mathbf{C_3}$. In this case the double group of $\mathbf{C_3}$ will be a group of order 6. There are just two abstract groups of this order. The central extension produces the cyclic group $\mathbf{C_6}$ defined by $\{A^6=E\}$ while a splitting extension produces the dihedral group $\mathbf{D_3}$ defined by $\{A^3=B^2=E;\ BA=A^2B\}$. The dihedral group of six elements cannot be chosen as the spin double group because it has no complex representations which can be matched onto the single-valued complex representations Γ_2 and Γ_3 of $\mathbf{C_3}$.

The character tables of $\mathbf{C_3}$ and $\mathbf{C_6}$ may be matched as

	$\mathbf{C_6}$	E	A^3	A	A^4	A^2	A^5
$\mathbf{C_3}$		E		A		A^2	
Γ_1	Γ_1	1	1	1	1	1	1
Γ_2	Γ_2	1	1	ω^*	ω^*	ω	ω
Γ_3	Γ_3	1	1	ω	ω	ω^*	ω^*
	Γ_4	1	−1	−1	1	1	−1
	Γ_5	1	−1	$-\omega^*$	ω^*	ω	$-\omega$
	Γ_6	1	−1	$-\omega$	ω	ω^*	$-\omega^*$

where $\omega=\exp(2\pi i/3)=(-1+i\sqrt{3})/2$ and its complex conjugate, ω^*, are the complex cube roots of unity as in section 2.3 above. Antisymmetric squares can only be formed from the complex-conjugate pairs $\Gamma_2+\Gamma_3$ and $\Gamma_5+\Gamma_6$. In both cases the antisymmetric square will be Γ_1 .

If Γ_5 rather than Γ_6 of $\mathbf{C_3}$ is chosen as the symmetry of the α-spin function, the pattern established in section 5 above can be followed. The effect of the operator A of the double group is to multiply the α-spin function by $\exp(\pi i/3)=-\omega^*$ and the β-spin function, which will have the symmetry of Γ_6 , by its inverse, $\exp(-\pi i/3)=-\omega$.

7. References

Boyle L L (1972) *Int. J. Quantum Chem.* **6** 725–46.
Bennett J H and Boyle L L (1982) *Molec. Phys* . **45** 1279–84.

Journal of Physics: Conference Series **512** (2014) 012020 doi:10.1088/1742-6596/512/1/012020

Classification of Invariant Differential Operators for Non-Compact Lie Algebras via Parabolic Relations[1]

V.K. Dobrev

Institute for Nuclear Research and Nuclear Energy,
Bulgarian Academy of Sciences,
72 Tsarigradsko Chaussee, 1784 Sofia, Bulgaria

E-mail: dobrev@inrne.bas.bg

Abstract. In the present paper we review the progress of the project of classification and construction of invariant differential operators for non-compact semisimple Lie groups. Our starting points is the class of algebras, which we called earlier 'conformal Lie algebras' (CLA), which have very similar properties to the conformal algebras of Minkowski space-time, though our aim is to go beyond this class in a natural way. For this we introduced recently the new notion of *parabolic relation* between two non-compact semisimple Lie algebras $\mathcal{G}$ and $\mathcal{G}'$ that have the same complexification and possess maximal parabolic subalgebras with the same complexification. Thus, we consider the exceptional algebra $E_{7(7)}$ which is parabolically related to the CLA $E_{7(-25)}$. Other interesting examples are the orthogonal algebras $so(p,q)$ all of which are parabolically related to the conformal algebra $so(n,2)$ with $p+q=n+2$, the parabolic subalgebras including the Lorentz subalgebra $so(n-1,1)$ and its analogs $so(p-1,q-1)$. Further we consider the algebras $sl(2n,\mathbb{R})$ and for $n=2k$ the algebras $su^*(4k)$ which are parabolically related to the CLA $su(n,n)$. Further we consider the algebras $sp(r,r)$ which are parabolically related to the CLA $sp(2r,\mathbb{R})$. We consider also $E_{6(6)}$ and $E_{6(2)}$ which are parabolically related to the hermitian symmetric case $E_{6(-14)}$,

1. Introduction

Invariant differential operators play very important role in the description of physical symmetries - starting from the early occurrences in the Maxwell, d'Allembert, Dirac, equations, to the latest applications of (super-)differential operators in conformal field theory, supergravity and string theory (for reviews, cf. e.g., [1], [2]). Thus, it is important for the applications in physics to study systematically such operators. For more relevant references cf., e.g., [3–60], and others throughout the text.

In a recent paper [61] we started the systematic explicit construction of invariant differential operators. We gave an explicit description of the building blocks, namely, the parabolic subgroups and subalgebras from which the necessary representations are induced. Thus we have set the stage for study of different non-compact groups.

Since the study and description of detailed classification should be done group by group we

[1] Talk at the VIII International Symposium "Quantum Theory and Symmetries", Mexico City, August 5-9, 2013.

had to decide which groups to study. One first choice would be non-compact groups that have discrete series of representations. By the Harish-Chandra criterion [62] these are groups where holds:

$$\mathrm{rank}\, G = \mathrm{rank}\ K,$$

where K is the maximal compact subgroup of the non-compact group G. Another formulation is to say that the Lie algebra $\mathcal{G}$ of G has a compact Cartan subalgebra.

Example: the groups $SO(p,q)$ have discrete series, except when both p,q are odd numbers.

This class is rather big, thus, we decided to consider a subclass, namely, the class of Hermitian symmetric spaces. The practical criterion is that in these cases, the maximal compact subalgebra $\mathcal{K}$ is of the form:

$$\mathcal{K} \ = \ so(2) \oplus \mathcal{K}' \tag{1}$$

The Lie algebras from this class are:

$$so(n,2), \quad sp(n,R), \quad su(m,n), \quad so^*(2n), \quad E_{6(-14)}\,, \quad E_{7(-25)} \tag{2}$$

These groups/algebras have highest/lowest weight representations, and relatedly *holomorphic* discrete series representations.

The most widely used of these algebras are the conformal algebras so(n,2) in n-dimensional Minkowski space-time. In that case, there is a maximal Bruhat decomposition [63]: that has direct physical meaning:

$$\begin{aligned} so(n,2) \ &= \ \mathcal{M} \oplus \mathcal{A} \oplus \mathcal{N} \oplus \tilde{\mathcal{N}}\ , \\ \mathcal{M} \ &= \ so(n-1,1)\ , \quad \dim \mathcal{A} = 1, \quad \dim \mathcal{N} = \dim \tilde{\mathcal{N}} = n \end{aligned} \tag{3}$$

where $so(n-1,1)$ is the Lorentz algebra of n-dimensional Minkowski space-time, the subalgebra $\mathcal{A} \ = \ so(1,1)$ represents the dilatations, the conjugated subalgebras $\mathcal{N}$, $\tilde{\mathcal{N}}$ are the algebras of translations, and special conformal transformations, both being isomorphic to n-dimensional Minkowski space-time.

The subalgebra $\mathcal{P} = \mathcal{M} \oplus \mathcal{A} \oplus \mathcal{N}$ ($\cong \mathcal{M} \oplus \mathcal{A} \oplus \tilde{\mathcal{N}}$) is a maximal parabolic subalgebra.

There are other special features which are important. In particular, the complexification of the maximal compact subgroup is isomorphic to the complexification of the first two factors of the Bruhat decomposition:

$$\mathcal{K}^{\mathbb{C}} \ = \ so(n,\mathbb{C}) \oplus so(2,\mathbb{C}) \ \cong \ so(n-1,1)^{\mathbb{C}} \oplus so(1,1)^{\mathbb{C}} = \mathcal{M}^{\mathbb{C}} \oplus \mathcal{A}^{\mathbb{C}}\ . \tag{4}$$

In particular, the coincidence of the complexification of the semi-simple subalgebras:

$$\mathcal{K}'^{\mathbb{C}} \ = \ \mathcal{M}^{\mathbb{C}} \tag{$*$}$$

means that the sets of finite-dimensional (nonunitary) representations of $\mathcal{M}$ are in 1-to-1 correspondence with the finite-dimensional (unitary) representations of $\mathcal{K}'$. The latter leads to the fact that the corresponding induced representations are representations of finite $\mathcal{K}$-type [62].

It turns out that some of the hermitian-symmetric algebras share the above-mentioned special properties of $so(n,2)$. This subclass consists of:

$$so(n,2), \quad sp(n,\mathbb{R}), \quad su(n,n), \quad so^*(4n), \quad E_{7(-25)} \tag{5}$$

the corresponding analogs of Minkowski space-time V being:

$$\mathbb{R}^{n-1,1}, \quad \mathrm{Sym}(n,\mathbb{R}), \quad \mathrm{Herm}(n,\mathbb{C}), \quad \mathrm{Herm}(n,\mathbb{Q}), \quad \mathrm{Herm}(3,\mathbb{O}) \tag{6}$$

Journal of Physics: Conference Series **512** (2014) 012020 doi:10.1088/1742-6596/512/1/012020

In view of applications to physics, we proposed to call these algebras 'conformal Lie algebras', (or groups).

The corresponding groups are also called '*Hermitian symmetric spaces of tube type*' [64]. The same class was identified from different considerations in [65] called there '*conformal groups of simple Jordan algebras*'. In fact, the relation between Jordan algebras and division algebras was known long time ago. Our class was identified from still different considerations also in [66] where they were called '*simple space-time symmetries generalizing conformal symmetry*'.

We have started the study of the above class in the framework of the present approach in the cases: $so(n,2)$, $su(n,n)$, $sp(n,\mathbb{R})$, $E_{7(-25)}$, [67], [68], [69], [70], resp., and we have considered also the algebra $E_{6(-14)}$, [71].

Lately, we discovered an efficient way to extend our considerations beyond this class introducing the notion of 'parabolically related non-compact semisimple Lie algebras' [72].

• *Definition:* Let $\mathcal{G}, \mathcal{G}'$ be two non-compact semisimple Lie algebras with the same complexification $\mathcal{G}^{\mathbb{C}} \cong \mathcal{G}'^{\mathbb{C}}$. We call them parabolically related if they have parabolic subalgebras $\mathcal{P} = \mathcal{M} \oplus \mathcal{A} \oplus \mathcal{N}$, $\mathcal{P}' = \mathcal{M}' \oplus \mathcal{A}' \oplus \mathcal{N}'$, such that: $\mathcal{M}^{\mathbb{C}} \cong \mathcal{M}'^{\mathbb{C}}$ $(\Rightarrow \mathcal{P}^{\mathbb{C}} \cong \mathcal{P}'^{\mathbb{C}})$.◊

Certainly, there are many such parabolic relationships for any given algebra $\mathcal{G}$. Furthermore, two algebras $\mathcal{G}, \mathcal{G}'$ may be parabolically related via different parabolic subalgebras.

We summarize the algebras parabolically related to conformal Lie algebras with maximal parabolics fulfilling $(*)$ in the following table:

Table of conformal Lie algebras (CLA) $\mathcal{G}$ with $\mathcal{M}$-factor fulfilling $(*)$ and the corresponding parabolically related algebras $\mathcal{G}'$

$\mathcal{G}$	$\mathcal{K}'$	$\mathcal{M}$ $\dim V$	$\mathcal{G}'$	$\mathcal{M}'$
$so(n,2)$ $n \geq 3$	$so(n)$	$so(n-1,1)$ n	$so(p,q)$, $p+q=n+2$; $sl(4,\mathbb{R})$, $n=4$	$so(p-1,q-1)$ $sl(2,\mathbb{R}) \oplus sl(2,\mathbb{R})$
$su(n,n)$ $n \geq 3$	$su(n) \oplus su(n)$	$sl(n,\mathbb{C})_{\mathbb{R}}$ n^2	$sl(2n,\mathbb{R})$ $su^*(2n)$, $n=2k$	$sl(n,\mathbb{R}) \oplus sl(n,\mathbb{R})$ $su^*(2k) \oplus su^*(2k)$
$sp(2r,\mathbb{R})$ $\mathrm{rank} = 2r \geq 4$	$su(2r)$	$sl(2r,\mathbb{R})$ $r(2r+1)$	$sp(r,r)$	$su^*(2r)$
$so^*(4n)$ $n \geq 3$	$su(2n)$	$su^*(2n)$ $n(2n-1)$	$so(2n,2n)$	$sl(2n,\mathbb{R})$
$E_{7(-25)}$	e_6	$E_{6(-26)}$ 27	$E_{7(7)}$	$E_{6(6)}$

where we display only the semisimple part $\mathcal{K}'$ of $\mathcal{K}$; $sl(n,\mathbb{C})_{\mathbb{R}}$ denotes $sl(n,\mathbb{C})$ as a real Lie algebra, (thus, $(sl(n,\mathbb{C})_{\mathbb{R}})^{\mathbb{C}} = sl(n,\mathbb{C}) \oplus sl(n,\mathbb{C})$); e_6 denotes the compact real form of E_6; and we have imposed restrictions to avoid coincidences or degeneracies due to well known isomorphisms: $so(1,2) \cong sp(1,\mathbb{R}) \cong su(1,1)$, $so(2,2) \cong so(1,2) \oplus so(1,2)$, $su(2,2) \cong so(4,2)$, $sp(2,\mathbb{R}) \cong so(3,2)$, $so^*(4) \cong so(3) \oplus so(2,1)$, $so^*(8) \cong so(6,2)$.

After this extended introduction we give the outline of the paper. In Section 2 we give the preliminaries, actually recalling and adapting facts from [61]. In Section 3 we consider the case of the pseudo-orthogonal algebras $so(p,q)$ which are parabolically related to the conformal algebra $so(n,2)$ for $p+q=n+2$. In Section 4 we consider the CLA $su(n,n)$ and the parabolically related $sl(2n,\mathbb{R})$, and for $n=2k$: $su^*(4k)$. In Section 5 we consider the CLA $sp(n)$ and - for $n=2r$ - the parabolically related $sp(r,r)$. In Section 6 we consider the CLA $E_{7(-25)}$ and the parabolically related $E_{7(7)}$. In Section 7 we consider the hermitian symmetric case $E_{6(-14)}$ and the parabolically related $E_{6(6)}$ and $E_{6(2)}$.

2. Preliminaries

Let G be a semisimple non-compact Lie group, and K a maximal compact subgroup of G. Then we have an *Iwasawa decomposition* $G = KA_0N_0$, where A_0 is Abelian simply connected vector subgroup of G, N_0 is a nilpotent simply connected subgroup of G preserved by the action of A_0. Further, let M_0 be the centralizer of A_0 in K. Then the subgroup $P_0 = M_0A_0N_0$ is a *minimal parabolic subgroup* of G. A *parabolic subgroup* $P = M'A'N'$ is any subgroup of G which contains a minimal parabolic subgroup.

Further, let $\mathcal{G}, \mathcal{K}, \mathcal{P}, \mathcal{M}, \mathcal{A}, \mathcal{N}$ denote the Lie algebras of G, K, P, M, A, N, resp.

For our purposes we need to restrict to *maximal parabolic subgroups* $P = MAN$, i.e. rank$A = 1$, resp. to *maximal parabolic subalgebras* $\mathcal{P} = \mathcal{M} \oplus \mathcal{A} \oplus \mathcal{N}$ with $\dim \mathcal{A} = 1$.

Let ν be a (non-unitary) character of A, $\nu \in \mathcal{A}^*$, parameterized by a real number d, called the *conformal weight* or energy.

Further, let μ fix a discrete series representation D^μ of M on the Hilbert space V_μ, or the finite-dimensional (non-unitary) representation of M with the same Casimirs.

We call the induced representation $\chi = \mathrm{Ind}_P^G(\mu \otimes \nu \otimes 1)$ an *elementary representation* of G [73]. (These are called *generalized principal series representations* (or *limits thereof*) in [74].) Their spaces of functions are:

$$\mathcal{C}_\chi = \{\mathcal{F} \in C^\infty(G, V_\mu) \mid \mathcal{F}(gman) = e^{-\nu(H)} \cdot D^\mu(m^{-1})\,\mathcal{F}(g)\} \tag{7}$$

where $a = \exp(H) \in A'$, $H \in \mathcal{A}'$, $m \in M'$, $n \in N'$. The representation action is the left regular action:

$$(\mathcal{T}^\chi(g)\mathcal{F})(g') = \mathcal{F}(g^{-1}g'), \quad g, g' \in G. \tag{8}$$

• An important ingredient in our considerations are the *highest/lowest weight representations* of $\mathcal{G}^{\mathbb{C}}$. These can be realized as (factor-modules of) Verma modules V^Λ over $\mathcal{G}^{\mathbb{C}}$, where $\Lambda \in (\mathcal{H}^{\mathbb{C}})^*$, $\mathcal{H}^{\mathbb{C}}$ is a Cartan subalgebra of $\mathcal{G}^{\mathbb{C}}$, weight $\Lambda = \Lambda(\chi)$ is determined uniquely from χ [75].

Actually, since our ERs may be induced from finite-dimensional representations of $\mathcal{M}$ (or their limits) the Verma modules are always reducible. Thus, it is more convenient to use *generalized Verma modules* $\tilde{V}^\Lambda$ such that the role of the highest/lowest weight vector v_0 is taken by the (finite-dimensional) space $V_\mu\, v_0$. For the generalized Verma modules (GVMs) the reducibility is controlled only by the value of the conformal weight d. Relatedly, for the intertwining differential operators only the reducibility w.r.t. non-compact roots is essential.

• Another main ingredient of our approach is as follows. We group the (reducible) ERs with the same Casimirs in sets called *multiplets* [75]. The multiplet corresponding to fixed values of the Casimirs may be depicted as a connected graph, the vertices of which correspond to the reducible ERs and the lines (arrows) between the vertices correspond to intertwining operators. The explicit parametrization of the multiplets and of their ERs is important for understanding of the situation. The notion of multiplets was introduced in [76], [77] and applied to representations of $SO_o(p,q)$ and $SU(2,2)$, resp., induced from their minimal parabolic subalgebras. Then it was applied to the conformal superalgebra [78], to infinite-dimensional (super-)algebras [79], to quantum groups [80]. (For other applications we refer to [81].)

In fact, the multiplets contain explicitly all the data necessary to construct the intertwining differential operators. Actually, the data for each intertwining differential operator consists of the pair (β, m), where β is a (non-compact) positive root of $\mathcal{G}^{\mathbb{C}}$, $m \in \mathbb{N}$, such that the BGG Verma module reducibility condition (for highest weight modules) is fulfilled:

$$(\Lambda + \rho, \beta^\vee) = m, \quad \beta^\vee \equiv 2\beta/(\beta, \beta) \tag{9}$$

ρ is half the sum of the positive roots of $\mathcal{G}^{\mathbb{C}}$. When the above holds then the Verma module with shifted weight $V^{\Lambda-m\beta}$ (or $\tilde{V}^{\Lambda-m\beta}$ for GVM and β non-compact) is embedded in the Verma module V^Λ (or $\tilde{V}^\Lambda$). This embedding is realized by a singular vector v_s determined by a polynomial $\mathcal{P}_{m,\beta}(\mathcal{G}^-)$ in the universal enveloping algebra $(U(\mathcal{G}_-))\, v_0$, $\mathcal{G}^-$ is the subalgebra of $\mathcal{G}^{\mathbb{C}}$ generated by the negative root generators [82]. More explicitly, [75], $v^s_{m,\beta} = \mathcal{P}_{m,\beta}\, v_0$ (or $v^s_{m,\beta} = \mathcal{P}_{m,\beta}\, V_\mu\, v_0$ for GVMs). Then there exists [75] an intertwining differential operator

$$\mathcal{D}_{m,\beta} \;:\; \mathcal{C}_{\chi(\Lambda)} \;\longrightarrow\; \mathcal{C}_{\chi(\Lambda-m\beta)} \tag{10}$$

given explicitly by:

$$\mathcal{D}_{m,\beta} \;=\; \mathcal{P}_{m,\beta}(\widehat{\mathcal{G}^-}) \tag{11}$$

where $\widehat{\mathcal{G}^-}$ denotes the right action on the functions $\mathcal{F}$.

In most of these situations the invariant operator $\mathcal{D}_{m,\beta}$ has a non-trivial invariant kernel in which a subrepresentation of $\mathcal{G}$ is realized. Thus, studying the equations with trivial RHS:

$$\mathcal{D}_{m,\beta}\, f \;=\; 0\;, \qquad f \in \mathcal{C}_{\chi(\Lambda)}\;, \tag{12}$$

is also very important. For example, in many physical applications in the case of first order differential operators, i.e., for $m = m_\beta = 1$, these equations are called conservation laws, and the elements $f \in \ker \mathcal{D}_{m,\beta}$ are called conserved currents.

The above construction works also for the subsingular vectors v_{ssv} of Verma modules. Such a vector is also expressed by a polynomial $\mathcal{P}_{ssv}(\mathcal{G}^-)$ in the universal enveloping algebra: $v^s_{ssv} = \mathcal{P}_{ssv}(\mathcal{G}^-)\, v_0$, cf. [83]. Thus, there exists a *conditionally invariant differential operator* given explicitly by: $\mathcal{D}_{ssv} = \mathcal{P}_{ssv}(\widehat{\mathcal{G}^-})$, and a *conditionally invariant differential equation*, for many more details, see [83]. (Note that these operators (equations) are not of first order.)

Below in our exposition we shall use the so-called Dynkin labels:

$$m_i \;\equiv\; (\Lambda+\rho, \alpha_i^\vee)\;, \quad i = 1,\ldots,n, \tag{13}$$

where $\Lambda = \Lambda(\chi)$, ρ is half the sum of the positive roots of $\mathcal{G}^{\mathbb{C}}$.

We shall use also the so-called Harish-Chandra parameters:

$$m_\beta \equiv (\Lambda+\rho, \beta)\;, \tag{14}$$

where β is any positive root of $\mathcal{G}^{\mathbb{C}}$. These parameters are redundant, since they are expressed in terms of the Dynkin labels, however, some statements are best formulated in their terms. (Clearly, both the Dynkin labels and Harish-Chandra parameters have their origin in the BGG reducibility condition (9).)

3. Conformal algebras $so(n,2)$ and parabolically related

Let $\mathcal{G} = so(n,2)$, $n > 2$. We label the signature of the ERs of $\mathcal{G}$ as follows:

$$\begin{aligned} \chi \;&=\; \{\, n_1\,, \ldots, n_{\tilde{h}}\,; c\,\}\,, \quad n_j \in \mathbb{Z}/2\,, \quad c = d - \tfrac{n}{2}\,, \quad \tilde{h} \equiv [\tfrac{n}{2}], \\ &|n_1| < n_2 < \cdots < n_{\tilde{h}}\,, \quad n \text{ even}\,, \\ &0 < n_1 < n_2 < \cdots < n_{\tilde{h}}\,, \quad n \text{ odd}\,, \end{aligned} \tag{15}$$

where the last entry of χ labels the characters of $\mathcal{A}$, and the first $\tilde{h}$ entries are labels of the finite-dimensional nonunitary irreps of $\mathcal{M} \cong so(n-1,1)$.

The reason to use the parameter c instead of d is that the parametrization of the ERs in the multiplets is given in a simple intuitive way (cf. [84], [67]):

$$\begin{aligned}
\chi_1^\pm &= \{\epsilon n_1\,,\ldots,\, n_{\tilde h}\,;\, \pm n_{\tilde h+1}\}\ , \quad n_{\tilde h} < n_{\tilde h+1}\ , \\
\chi_2^\pm &= \{\epsilon n_1\,,\ldots,\, n_{\tilde h-1}\,,\, n_{\tilde h+1}\,;\, \pm n_{\tilde h}\} \\
\chi_3^\pm &= \{\epsilon n_1,\ldots, n_{\tilde h-2}, n_{\tilde h}, n_{\tilde h+1}\,;\, \pm n_{\tilde h-1}\} \\
&\ldots \\
\chi_{\tilde h}^\pm &= \{\epsilon n_1\,, n_3\,,\ldots,\, n_{\tilde h}\,,\, n_{\tilde h+1}\,;\, \pm n_2\} \\
\chi_{\tilde h+1}^\pm &= \{\epsilon n_2\,,\ldots,\, n_{\tilde h}\,,\, n_{\tilde h+1}\,;\, \pm n_1\} \\
\epsilon &= \begin{cases} \pm, & n\ even \\ 1, & n\ odd \end{cases}
\end{aligned} \tag{16}$$

Further, we denote by $\tilde{C}_i^\pm$ the representation space with signature $\chi_i^\pm$.

The number of ERs in the corresponding multiplets is equal to:

$$|W(\mathcal{G}^{\mathbb{C}},\mathcal{H}^{\mathbb{C}})|\,/\,|W(\mathcal{M}^{\mathbb{C}},\mathcal{H}_m^{\mathbb{C}})| \;=\; 2(1+\tilde h) \tag{17}$$

where $\mathcal{H}^{\mathbb{C}}$, $\mathcal{H}_m^{\mathbb{C}}$ are Cartan subalgebras of $\mathcal{G}^{\mathbb{C}}$, $\mathcal{M}^{\mathbb{C}}$, resp. This formula is valid for the main multiplets of all conformal Lie algebras.

We show some examples of diagrams of invariant differential operators for the conformal groups $so(5,1)$, resp. $so(4,2)$, in 4-dimensional Euclidean, resp. Minkowski, space-time. In Fig. 1. we show the simplest example for the most common using well known operators. In Fig. 2. we show the same example but using the group-theoretical parity splitting of the electromagnetic current, cf. [85]. In Fig. 3. we show the general classification for $so(5,1)$ given in [85]. These diagrams are valid also for $so(4,2)$ [86] and for $so(3,3) \cong sl(4,\mathbb{R})$ [72].

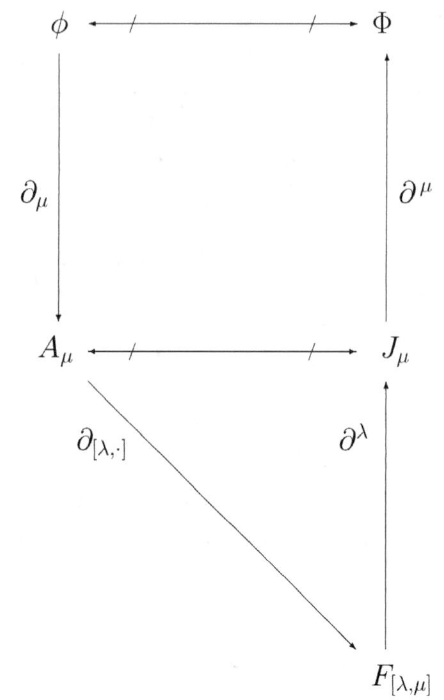

Fig. 1. Simplest example of diagram with conformal invariant operators (arrows are differential operators, dashed arrows are integral operators)

$\partial_\mu = \frac{\partial}{\partial x_\mu}$, A_μ electromagnetic potential, $\partial_\mu \phi = A_\mu$

F electromagnetic field, $\partial_{[\lambda} A_{\mu]} = \partial_\lambda A_\mu - \partial_\mu A_\lambda = F_{\lambda\mu}$

J_μ electromagnetic current, $\partial^\lambda F_{\lambda\mu} = J_\mu$, $\partial^\mu J_\mu = \Phi$

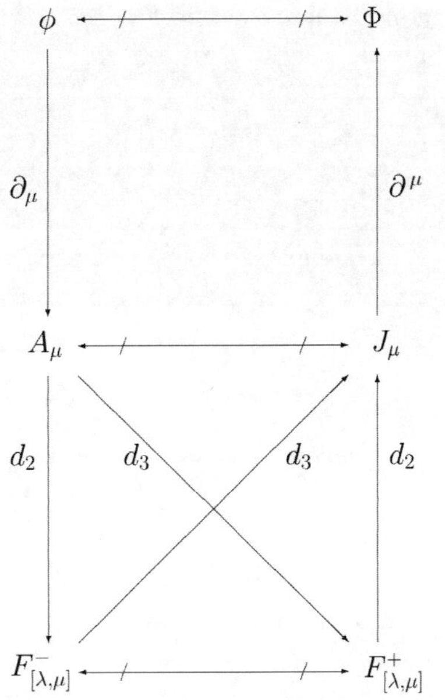

Fig. 2. More precise showing of the simplest example,
$F = F^+ \oplus F^-$ shows the parity splitting of the electromagnetic field,
d_2, d_3 linear invariant operators

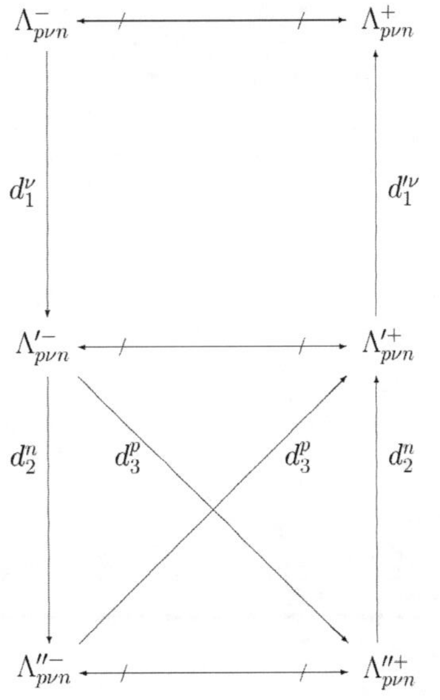

Fig. 3. The general classification of invariant differential operators valid for
$so(4,2)$, $so(5,1)$ and $so(3,3) \cong sl(4,\mathbb{R})$.
p, ν, n are three natural numbers, the shown simplest case is when $p = \nu = n = 1$,
d_1^ν is a linear differential operator of order ν, similarly $d_1'^\nu$, d_2^n, d_3^p

Next in Fig. 4. we show the general even case $so(p,q)$, $p+q=2h+2$-even, [84], [67], while in Fig. 5. we show an alternative view of the same case:

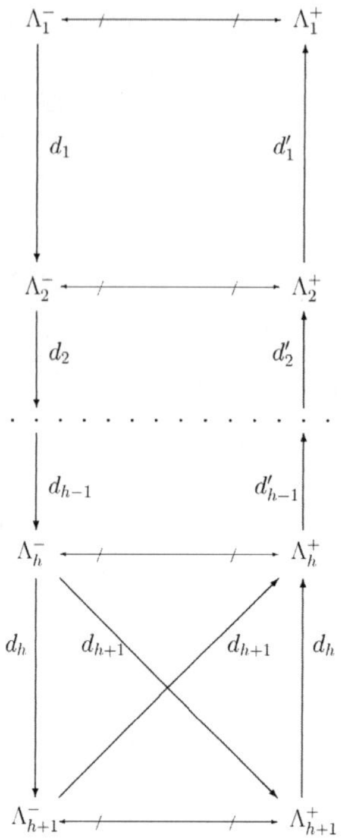

Fig. 4. The general classification of invariant differential operators in $2h$-dimensional space-time. By parabolic relation the diagram above is valid for all algebras $so(p,q)$, $p+q=2h+2$, even.

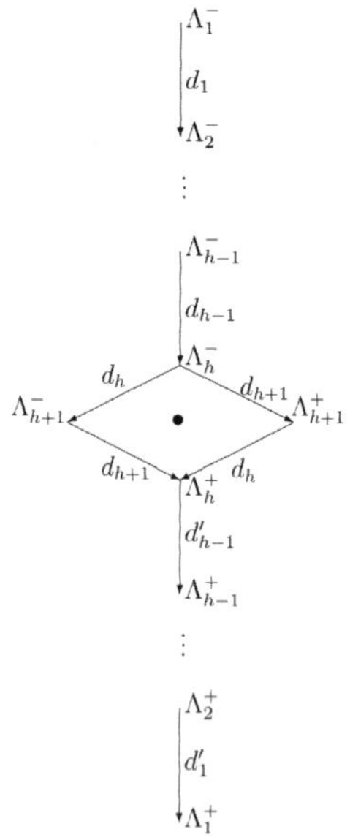

Fig. 5. Alternative showing of the case $so(p,q)$, $p+q=2h+2$, showing only the differential operators, while the integral operators are assumed as symmetry w.r.t. the bullet in the centre.

Next in Fig. 6. we show the general odd case $so(p,q)$, $p+q=2h+3$-odd, [84], [67], while in Fig. 7. we show an alternative view of the same case:

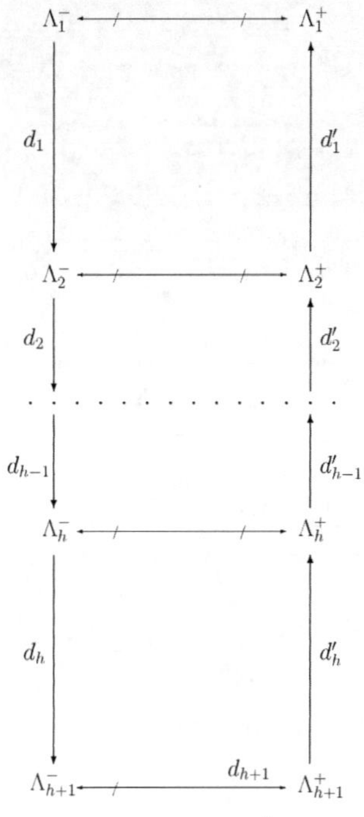

Fig. 6. The general classification of invariant differential operators in $2h+1$ dimensional space-time. By parabolic relation the diagram above is valid for all algebras $so(p,q)$, $p+q=2h+3$, odd.

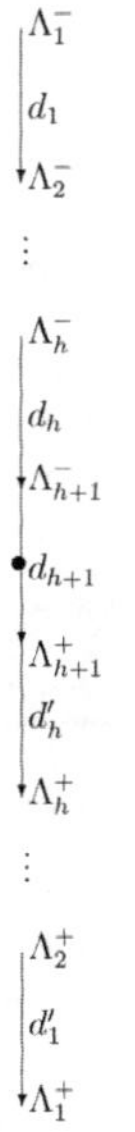

Fig. 7. Alternative showing of the case $so(p,q)$, $p+q=2h+3$, showing only the differential operators, while the integral operators are assumed as symmetry w.r.t. the bullet in the centre.

Journal of Physics: Conference Series **512** (2014) 012020 doi:10.1088/1742-6596/512/1/012020

The ERs in the multiplet are related by intertwining integral and differential operators. The integral operators were introduced by Knapp and Stein [87]. They correspond to elements of the restricted Weyl group of $\mathcal{G}$. These operators intertwine the pairs $\tilde{C}_i^{\pm}$

$$G_i^{\pm} \ : \ \tilde{C}_i^{\mp} \longrightarrow \tilde{C}_i^{\pm} \ , \quad i \ = \ 1, \ldots, 1 + \tilde{h} \tag{18}$$

The intertwining differential operators correspond to non-compact positive roots of the root system of $so(n+2, \mathbb{C})$, cf. [75]. [In the current context, compact roots of $so(n+2, \mathbb{C})$ are those that are roots also of the subalgebra $so(n, \mathbb{C})$, the rest of the roots are non-compact.] The degrees of these intertwining differential operators are given just by the differences of the c entries [84]:

$$\begin{aligned} &\deg d_i = \deg d'_i = n_{\tilde{h}+2-i} - n_{\tilde{h}+1-i} \, , \qquad i = 1, \ldots, \tilde{h} \, , \quad \forall n \\ &\deg d_{\tilde{h}+1} = n_2 + n_1 \, , \quad n \text{ even} \end{aligned} \tag{19}$$

where d'_h is omitted from the first line for $(p+q)$ even.

Matters are arranged so that in every multiplet only the ER with signature χ_1^- contains a finite-dimensional nonunitary subrepresentation in a subspace $\mathcal{E}$. The latter corresponds to the finite-dimensional unitary irrep of $so(n+2)$ with signature $\{n_1, \ldots, n_{\tilde{h}}, n_{\tilde{h}+1}\}$. The subspace $\mathcal{E}$ is annihilated by the operator G_1^+, and is the image of the operator G_1^-.

Although the diagrams are valid for arbitrary $so(p,q)$ $(p+q \geq 5)$ the contents is very different. We comment only on the ER with signature χ_1^+. In all cases it contains an UIR of $so(p,q)$ realized on an invariant subspace $\mathcal{D}$ of the ER χ_1^+. That subspace is annihilated by the operator G_1^-, and is the image of the operator G_1^+. (Other ERs contain more UIRs.)

If $pq \in 2\mathbb{N}$ the mentioned UIR is a discrete series representation. (Other ERs contain more discrete series UIRs.)

And if $q = 2$ the invariant subspace $\mathcal{D}$ is the direct sum of two subspaces $\mathcal{D} = \mathcal{D}^+ \oplus \mathcal{D}^-$, in which are realized a *holomorphic discrete series representation* and its conjugate *anti-holomorphic discrete series representation*, resp. Note that the corresponding lowest weight GVM is infinitesimally equivalent only to the holomorphic discrete series, while the conjugate highest weight GVM is infinitesimally equivalent to the anti-holomorphic discrete series.

Note that the $\deg d_i$, $\deg d'_i$, are Harish-Chandra parameters corresponding to the non-compact positive roots of $so(n+2, \mathbb{C})$. From these, only $\deg d_1$ corresponds to a simple root, i.e., is a Dynkin label.

Above we considered $so(n,2)$ for $n > 2$. The case $n = 2$ is reduced to $n = 1$ since $so(2,2) \cong so(1,2) \oplus so(1,2)$. The case $so(1,2)$ is special and must be treated separately. But in fact, it is contained in what we presented already. In that case the multiplets contain only two ERs which may be depicted by the top pair $\chi_1^{\pm}$ in the pictures that we presented. And they have the properties that we described for $so(n,2)$ with $n > 2$. The case $so(1,2)$ was given already in 1946-7 independently by Gel'fand et al [88] and Bargmann [89].

4. The Lie algebra $su(n,n)$ and parabolically related

Let $\mathcal{G} = su(n,n)$, $n \geq 2$. The maximal compact subgroup is $\mathcal{K} \cong u(1) \oplus su(n) \oplus su(n)$, while $\mathcal{M} = sl(n, \mathbb{C})_{\mathbb{R}}$. The number of ERs in the corresponding multiplets is equal to

$$|W(\mathcal{G}^{\mathbb{C}}, \mathcal{H}^{\mathbb{C}})| \, / \, |W(\mathcal{M}^{\mathbb{C}}, \mathcal{H}_m^{\mathbb{C}})| \ = \ \binom{2n}{n}$$

The signature of the ERs of $\mathcal{G}$ is:

$$\chi \quad = \quad \{\, n_1\,, \ldots, \, n_{n-1}\,, \, n_{n+1}\, \ldots, \, n_{2n-1}\,; \, c\,\}\,, \quad n_j \in \mathbb{N}\,, \quad c = d-n$$

The Knapp–Stein restricted Weyl reflection is given by:

$$\begin{aligned} G_{KS} \,:\, \mathcal{C}_\chi \,\longrightarrow\, \mathcal{C}_{\chi'}\,, \quad & \chi' = \{(n_1,\ldots,n_{n-1},n_{n+1},\ldots,n_{2n-1})^*;-c\} \\ (n_1,\ldots,n_{n-1},n_{n+1},\ldots,n_{2n-1})^* \;\doteq\; & (n_{n+1},\ldots,n_{2n-1},n_1,\ldots,n_{n-1}) \end{aligned}$$

Below in Fig. 8 and in Fig. 9 we give the diagrams for $su(n,n)$ for $n = 3, 4$, [68]. (The case $n = 2$ is already considered since $su(2,2) \cong so(4,2)$.) These are diagrams also for the parabolically related $sl(2n,\mathbb{R})$, and for $n = 2k$ these are diagrams also for the parabolically related $su^*(4k)$, [72].

We use the following conventions. Each intertwining differential operator is represented by an arrow accompanied by a symbol $i_{j...k}$ encoding the root $\beta_{j...k}$ and the number $m_{\beta_{j...k}}$ which is involved in the BGG criterion.

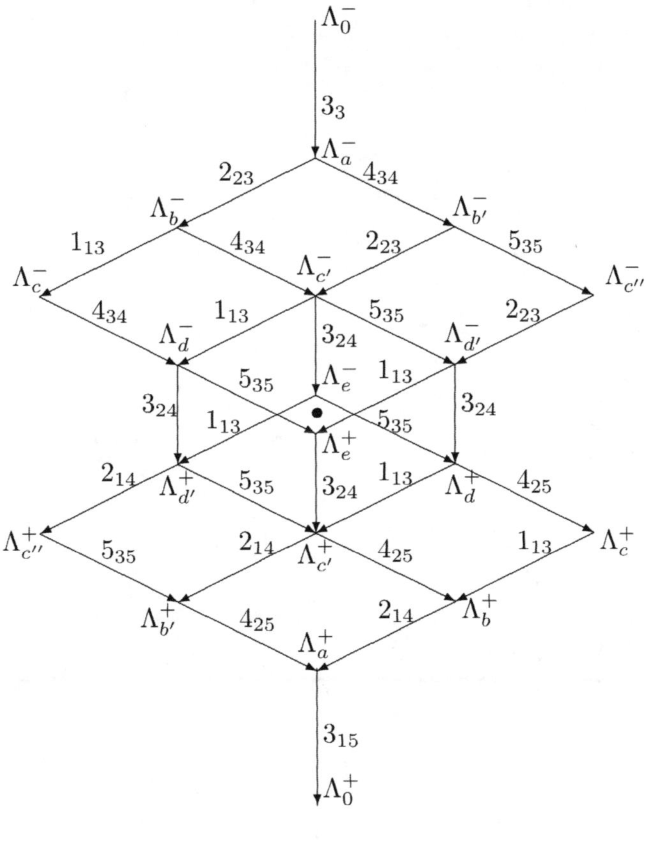

Fig. 8. Pseudo-unitary symmetry $su(3,3)$

The pseudo-unitary symmetry $su(n,n)$ is similar to conformal symmetry in n^2 dimensional space, for $n = 2$ coincides with conformal 4-dimensional case. By parabolic relation the $su(3,3)$ diagram above is valid also for $sl(6,R)$.

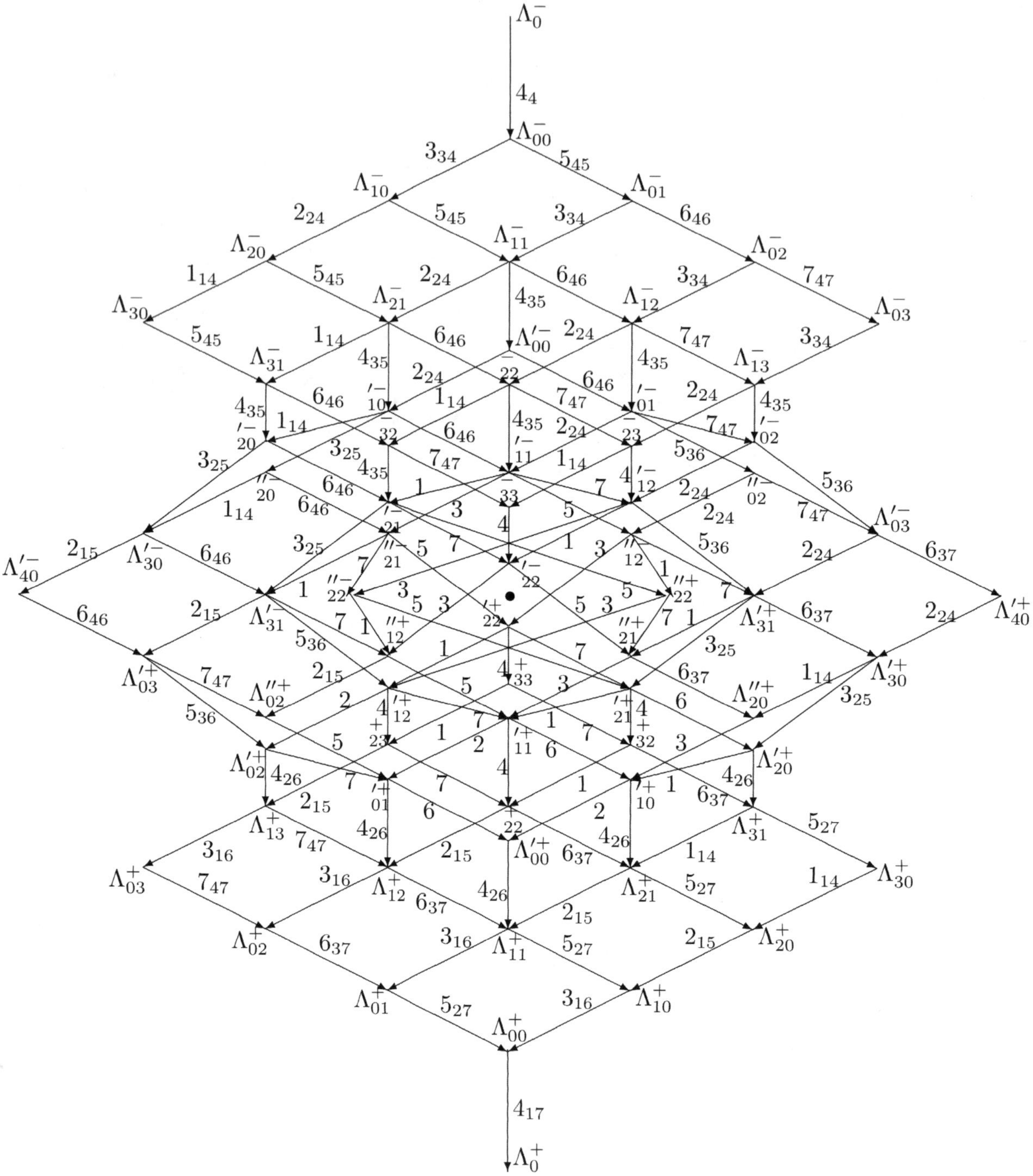

Fig. 9. Pseudo-unitary symmetry in 16-dimensional space.

By parabolic relation the $su(4,4)$ diagram above is valid also for $sl(8,R)$ and $su^*(8)$.

5. The Lie algebras $sp(n,\mathbb{R})$ and $sp(\frac{n}{2},\frac{n}{2})$ (n–even)

Let $n \geq 2$. Let $\mathcal{G} = sp(n,\mathbb{R})$, the split real form of $sp(n,\mathbb{C}) = \mathcal{G}^{\mathbb{C}}$. The maximal compact subgroup is $\mathcal{K} \cong u(1) \oplus su(n)$, while $\mathcal{M} = sl(n,\mathbb{R})$. The number of ERs in the corresponding multiplets is:

$$|W(\mathcal{G}^{\mathbb{C}},\mathcal{H}^{\mathbb{C}})| \,/\, |W(\mathcal{M}^{\mathbb{C}},\mathcal{H}^{\mathbb{C}}_m)| \;=\; 2^n$$

The signature of the ERs of $\mathcal{G}$ is:

$$\chi \;=\; \{\, n_1 \,, \ldots ,\, n_{n-1} \,;\, c \,\} \,, \qquad n_j \in \mathbb{N} \,,$$

The Knapp-Stein Weyl reflection acts as follows:

$$G_{KS} \;:\; \mathcal{C}_\chi \;\longrightarrow\; \mathcal{C}_{\chi'} \,, \chi' \;=\; \{\, (n_1, \ldots, n_{n-1})^* \,;\, -c \,\} \,,$$
$$(n_1, \ldots, n_{n-1})^* \;\doteq\; (n_{n-1}, \ldots, n_1)$$

Below in Fig. 10, Fig. 11, Fig. 12 and Fig. 13 we give pictorially the multiplets for $sp(n,\mathbb{R})$ for $n = 3,4,5,6$, [69]. (The case $n = 2$ is already considered since $sp(2,\mathbb{R}) \cong so(3,2)$.) For $n = 2r$ these are also multiplets for $sp(r,r)$, $r = 1,2,3$, [72]. (The case $n = 2, r = 1$ is already considered due to $sp(1,1) \cong so(4,1)$ and the parabolic relation between $so(3,2)$ and $so(4,1)$.)

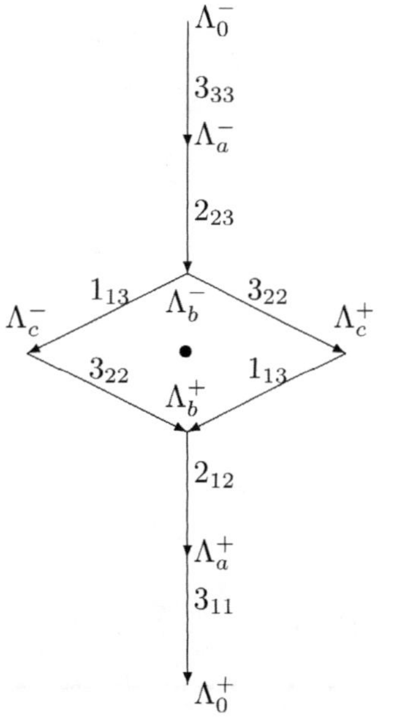

Fig. 10. Symplectic symmetry $sp(3,R)$
with diagram coinciding with 6-dimensional conformal case

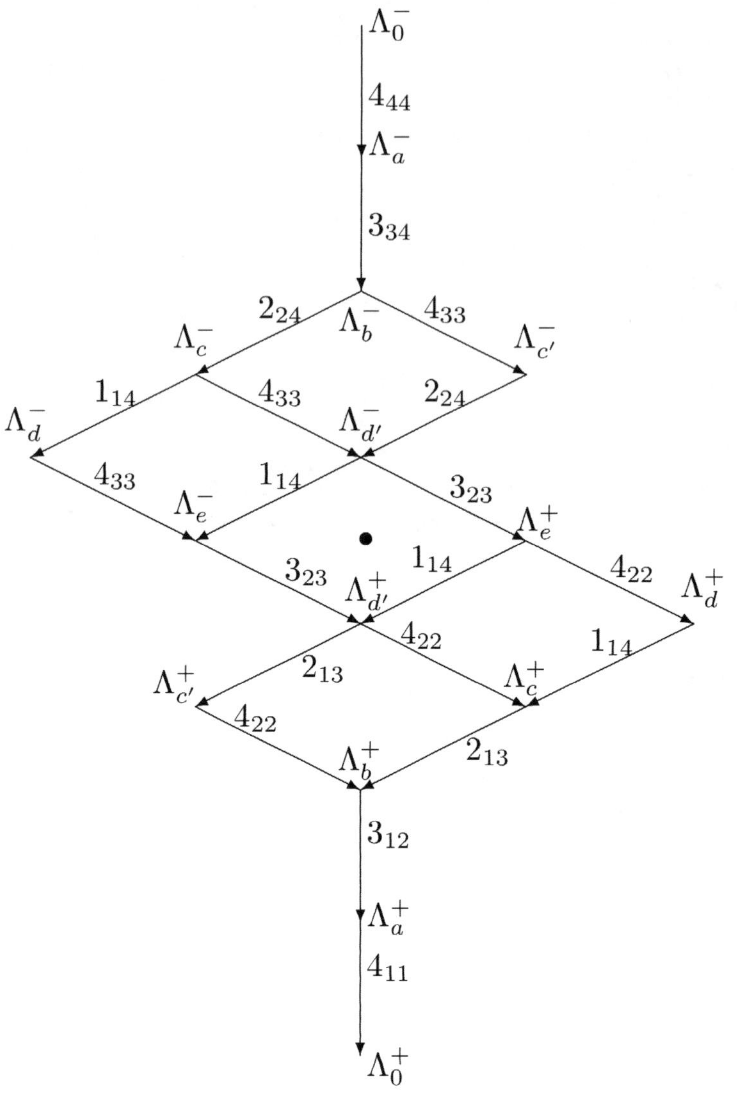

Fig. 11. Main multiplets for $sp(4,\mathbb{R})$ and $sp(2,2)$.

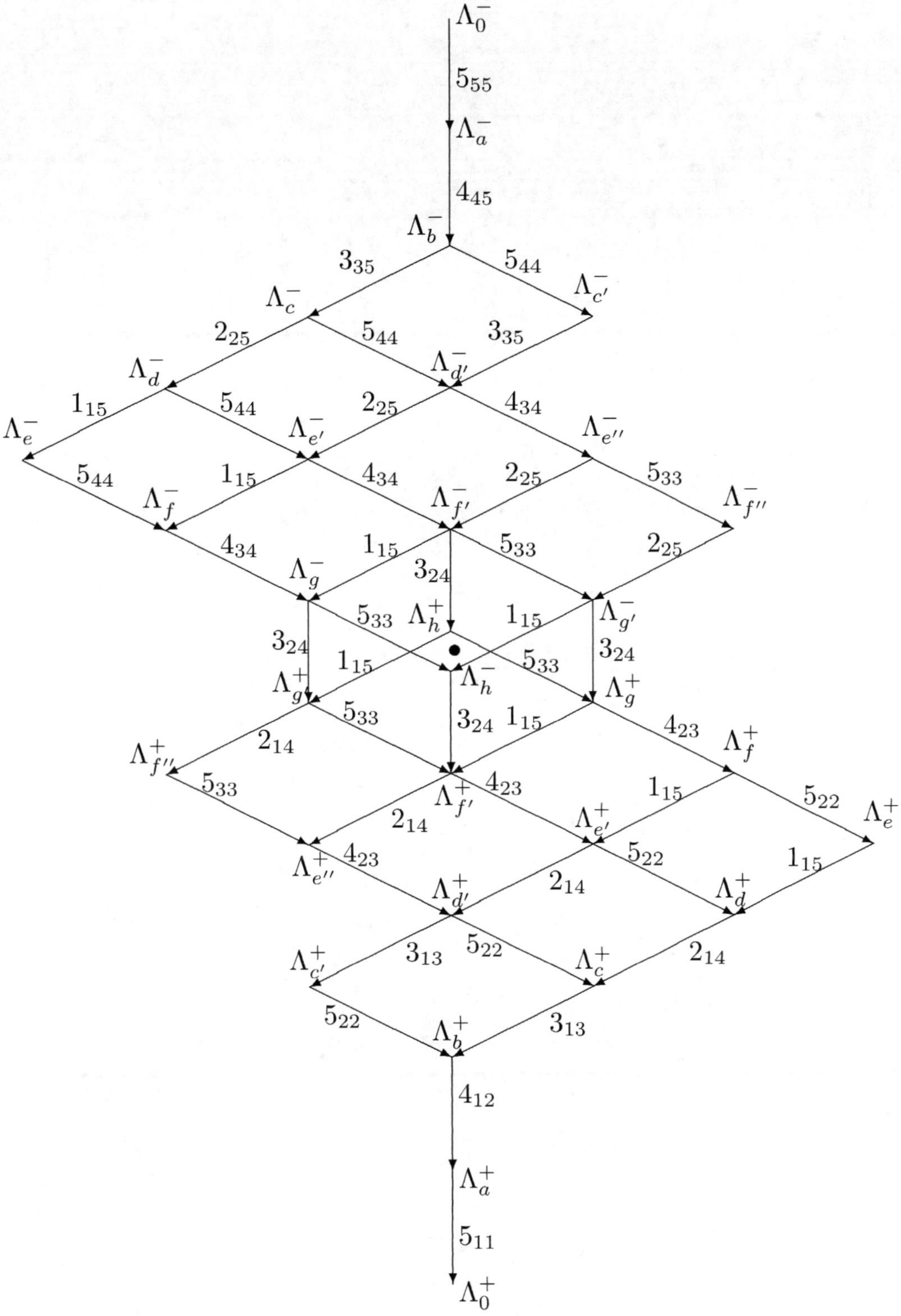

Fig. 12. Main multiplets for $sp(5, I\!R)$

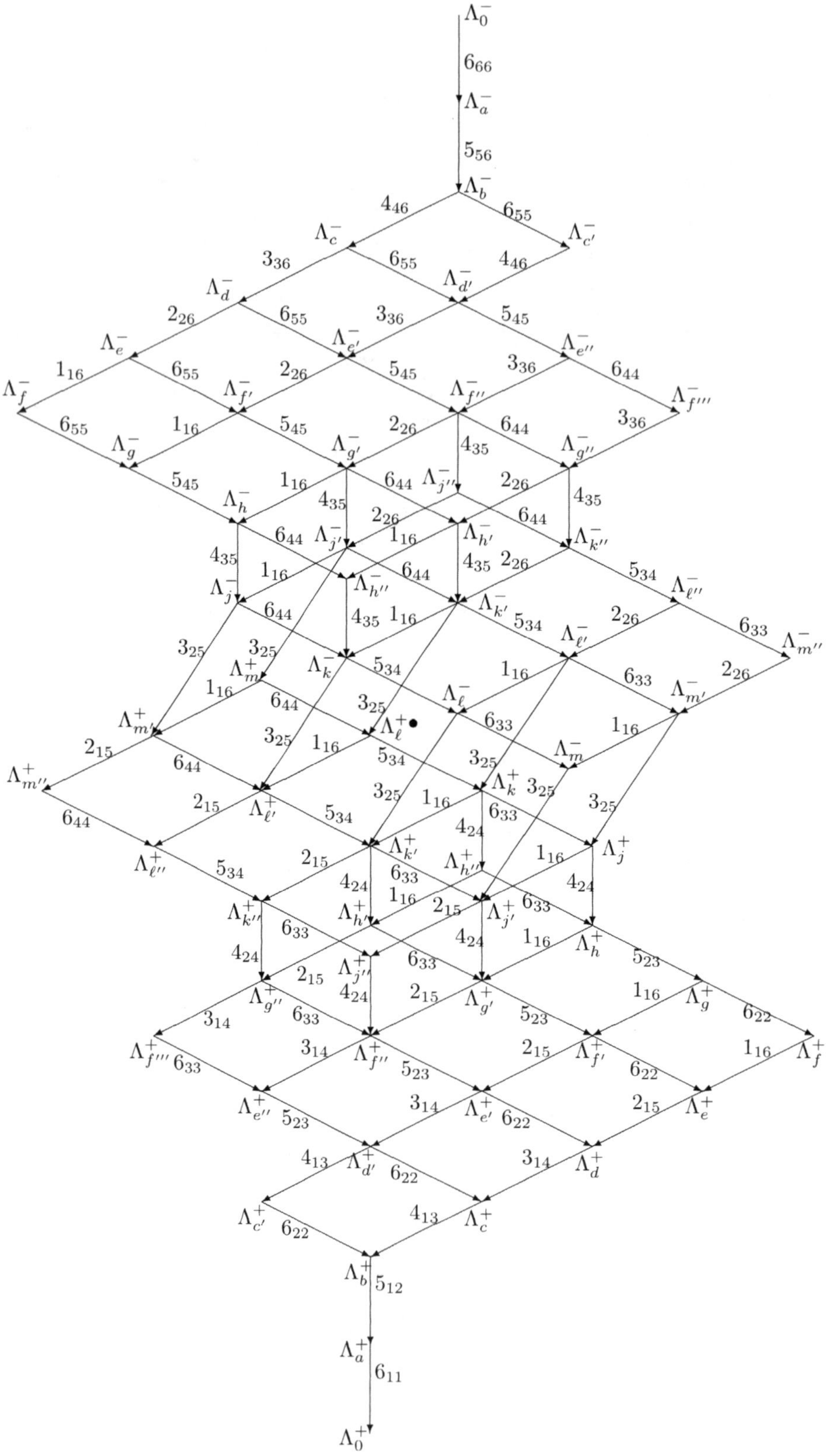

Fig. 13. Main multiplets for $sp(6,I\!R)$ and $sp(3,3)$.

6. The Lie algebras $E_{7(-25)}$ and $E_{7(7)}$

Let $\mathcal{G} = E_{7(-25)}$. The maximal compact subgroup is $\mathcal{K} \cong e_6 \oplus so(2)$, while $\mathcal{M} \cong E_{6(-6)}$.

The Satake diagram [90] is:

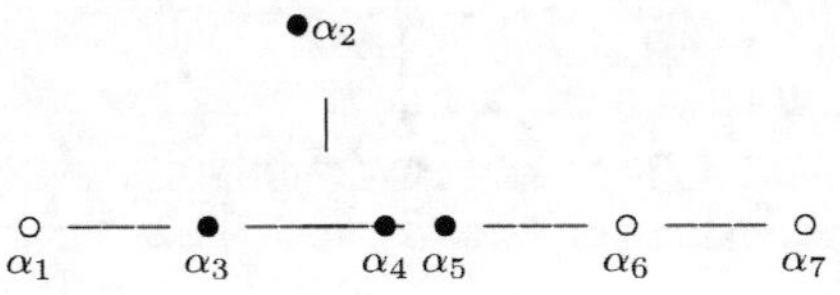

The signatures of the ERs of $\mathcal{G}$ are:

$$\chi = \{ n_1 , \ldots , n_6 ; c \} , \qquad n_j \in \mathbb{N} .$$

expressed through the Dynkin labels:

$$n_i = m_i , \qquad c = -\tfrac{1}{2}(m_{\tilde\alpha} + m_7) = -\tfrac{1}{2}(2m_1 + 2m_2 + 3m_3 + 4m_4 + 3m_5 + 2m_6 + 2m_7)$$

The same signatures can be used for the parabolically related exceptional Lie algebra $E_{7(7)}$ (with $\mathcal{M}$-factor $E_{6(6)}$).

The noncompact roots of the complex algebra E_7 are:

$$\begin{aligned}
&\alpha_7 , \quad \alpha_{17} , \quad \ldots , \quad \alpha_{67} , \\
&\alpha_{1,37} , \ \alpha_{2,47} , \ \alpha_{17,4} , \ \alpha_{27,4} , \\
&\alpha_{17,34} , \ \alpha_{17,35} , \ \alpha_{17,36} , \ \alpha_{17,45} , \ \alpha_{17,46} , \\
&\alpha_{27,45} , \ \alpha_{27,46} , \\
&\alpha_{17,25,4} , \ \alpha_{17,26,4} , \ \alpha_{17,35,4} , \ \alpha_{17,36,4} , \\
&\alpha_{17,26,45} , \ \alpha_{17,36,45} , \\
&\alpha_{17,26,35,4} , \ \alpha_{17,26,45,4} , \\
&\alpha_{17,16,35,4} = \tilde\alpha ,
\end{aligned}$$

given through the simple roots α_i :

$$\begin{aligned}
\alpha_{ij} &= \alpha_i + \alpha_{i+1} + \cdots + \alpha_j , \ i < j , \\
\alpha_{ij,k} = \alpha_{k,ij} &= \alpha_i + \alpha_{i+1} + \cdots + \alpha_j + \alpha_k , \quad i < j , \qquad \text{etc.}
\end{aligned}$$

The multiplets of the main type are in 1-to-1 correspondence with the finite-dimensional irreps of E_7, i.e., they will be labelled by the seven positive Dynkin labels $m_i \in \mathbb{N}$.

The number of ERs in the corresponding multiplets is equal to

$$|W(\mathcal{G}^{\mathbb{C}}, \mathcal{H}^{\mathbb{C}})| \,/\, |W(\mathcal{K}^{\mathbb{C}}, \mathcal{H}^{\mathbb{C}})| = 56$$

The multiplets are given in Fig. 14, [70, 72].

The Knapp-Stein operators $G^{\pm}_{\chi}$ act pictorially as reflection w.r.t. the bullet intertwining each $\mathcal{T}^-_{\chi}$ member with the corresponding $\mathcal{T}^+_{\chi}$ member.

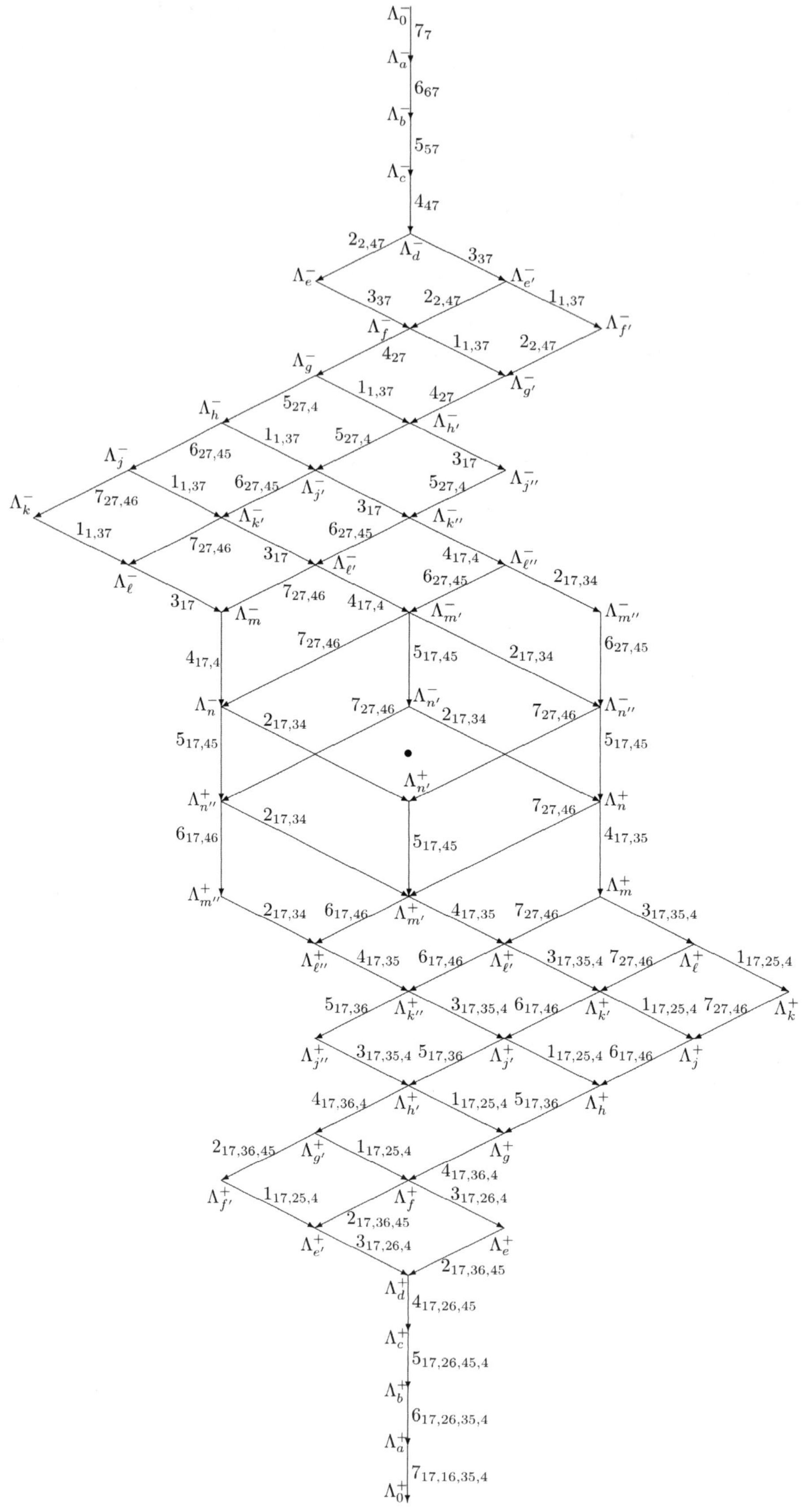

Fig. 14. Main multiplets for $E_{7(-25)}$ and $E_{7(7)}$.

7. The Lie algebras $E_{6(-14)}$, $E_{6(6)}$ and $E_{6(2)}$

Let $\mathcal{G} = E_{6(-14)}$. The maximal compact subalgebra is $\mathcal{K} \cong so(10)\oplus so(2)$, while $\mathcal{M} \cong su(5,1)$.

The Satake diagram [90] is:

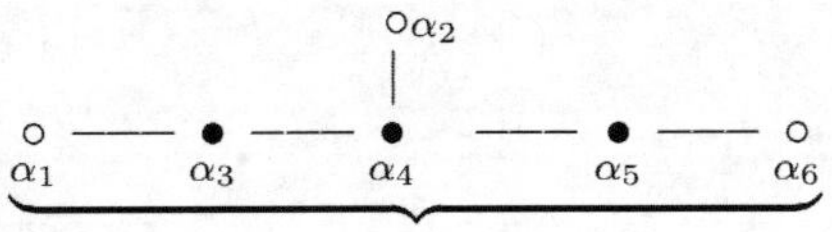

(20)

The signature of the ERs of $\mathcal{G}$ is:

$$\chi = \{\, n_1 \,, n_3 \,, n_4 \,, n_5 \,, n_6 \,;\, c\} \,, \qquad c = d - \tfrac{11}{2} \,. \tag{21}$$

expressed through the Dynkin labels as:

$$n_i = m_i \,, \qquad -c \;=\; \tfrac{1}{2} m_{\tilde{\alpha}} \;=\; \tfrac{1}{2}(m_1 + 2m_2 + 2m_3 + 3m_4 + 2m_5 + m_6) \tag{22}$$

The same signatures can be used for the parabolically related exceptional Lie algebras $E_{6(6)}$ and $E_{6(2)}$ with $\mathcal{M}$–factors $sl(6,\mathbb{R})$ and $su(3,3)$, resp.

Further, we need the noncompact roots of the complex algebra E_6 :

$$\begin{array}{l} \alpha_2 \,,\ \alpha_{14} \,,\ \alpha_{15} \,,\ \alpha_{16} \,,\ \alpha_{24} \,,\ \alpha_{25} \,,\ \alpha_{26} \\ \alpha_{2,4} \,,\ \alpha_{2,45} \,,\ \alpha_{2,46} \,,\ \alpha_{25,4} \,,\ \alpha_{15,4} \,,\ \alpha_{26,4} \\ \alpha_{16,4} \,,\ \alpha_{15,34} \,,\ \alpha_{26,45} \,,\ \alpha_{16,34} \,,\ \alpha_{16,45} \\ \alpha_{16,35} \,,\ \alpha_{16,35,4} \,,\ \alpha_{16,25,4} \;=\; \tilde{\alpha} \end{array} \tag{23}$$

The multiplets of the main type are in 1-to-1 correspondence with the finite-dimensional irreps of $\mathcal{G}$, i.e., they will be labelled by the six positive Dynkin labels $m_i \in \mathbb{N}$.

Since these algebras do not belong to the class of conformal Lie algebras (CLA) the number of ERs/GVMs in the multiplet is not given by formula (17). It turns out that each such multiplet contains 70 ERs/GVMs - see Fig. 15, [71, 72]. Another difference with the CLA class is that pictorially the the Knapp-Stein operators $G^{\pm}_{\chi}$ act as reflection w.r.t. the dotted line separating the $\mathcal{T}^{-}_{\chi}$ members from the $\mathcal{T}^{+}_{\chi}$ members (and not as reflection w.r.t. a central dot (bullet) as in the CLA cases).

Note that there are five cases when the embeddings correspond to the highest root $\tilde{\alpha}$: $V^{\Lambda^-} \longrightarrow V^{\Lambda^+}$, $\Lambda^+ = \Lambda^- - m_{\tilde{\alpha}}\,\tilde{\alpha}$. In these five cases the weights are denoted as: $\Lambda^{\pm}_{k''}$, $\Lambda^{\pm}_{k'}$, $\Lambda^{\pm}_{\tilde{k}}$, $\Lambda^{\pm}_{k}$, $\Lambda^{\pm}_{k^o}$, then: $m_{\tilde{\alpha}} = m_1, m_3, m_4, m_5, m_6$, resp. Thus, their action coincides with the action of the Knapp-Stein operators G^{+}_{χ} which in the above five cases degenerate to differential operators as we discussed for $so(3,2)$.

Note that the figure has the standard E_6 symmetry, namely, conjugation exchanging indices $1 \longleftrightarrow 6$, $3 \longleftrightarrow 5$.

Journal of Physics: Conference Series **512** (2014) 012020 doi:10.1088/1742-6596/512/1/012020

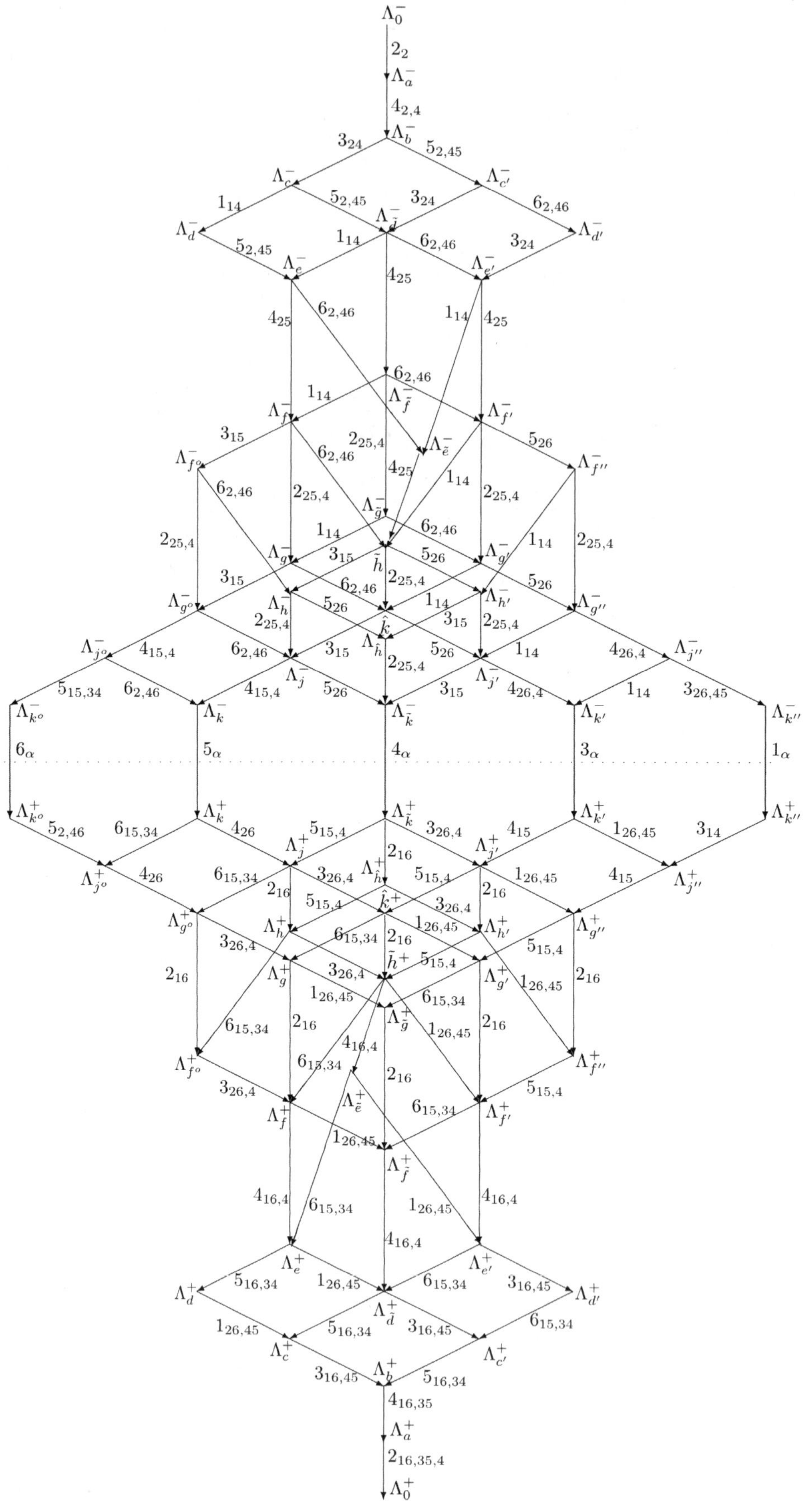

Fig. 15. Main multiplets for $E_{6(-14)}$, $E_{6(6)}$ and $E_{6(2)}$.

Acknowledgments

It is a pleasure to thank the organizers of the VIII International Symposium "Quantum Theory and Symmetries", and in especially Kurt Bernardo Wolf, for the hospitality. The author has received partial support from COST action MP-1210.

References

[1] Maldacena J 2005 Large N Field Theories, String Theory and Gravity, in: *Lectures on Quantum Gravity*, Series of the Centro De Estudios Cientficos, eds A Gomberoff and D. Marolf (New York: Springer) p. 91-150.

[2] Terning J 2005 *Modern Supersymmetry: Dynamics and Duality*, International Series of Monographs on Physics # 132 (Oxford University Press).

[3] Harish-Chandra 1966 Ann. Math. **116** 1-111.

[4] Bernstein I N, Gel'fand I M and Gel'fand S I 1971 Funkts. Anal. Prilozh. **5** (1) 1-9; English translation: Funct. Anal. Appl. **5** 1-8.

[5] Warner G 1972 *Harmonic Analysis on Semi-Simple Lie Groups I* (Berlin: Springer).

[6] Langlands R P 1988 *On the classification of irreducible representations of real algebraic groups*, Math. Surveys and Monographs, Vol. 31 (AMS), (first as IAS Princeton preprint (1973)).

[7] Ferrara S, Wess J and Zumino B 1974 Phys. Lett. **B51** 239; Ferrara S and Zumino B 1974 Nucl. Phys. **B79** 413; Ferrara S and Zumino B 1975 Nucl. Phys. **B87** 207.

[8] Zhelobenko D P 1974 *Harmonic Analysis on Semisimple Complex Lie Groups* (Moscow: Nauka, in Russian).

[9] Kostant B 1975 in: Lecture Notes in Math., Vol. 466, eds A Dold and B Eckmann (Berlin: Springer) p. 101-128.

[10] Sokatchev E 1975 Nucl. Phys. **B99** 96; Sokatchev E 1986 Phys. Lett. **B169** 209; Sokatchev E 1987 Class. Quant. Grav. **4** 237.

[11] Freedman D Z, van Nieuwenhuizen P and Ferrara S 1976 Phys. Rev. **D13** 3214; Ferrara S and van Nieuwenhuizen P 1978 Phys. Lett. **B74** 333.

[12] Wolf J 1976 *Unitary Representations of Maximal Parabolic Subgroups of the Classical Groups*, Memoirs Amer. Math. Soc. 180 (AMS).

[13] Ademollo M, Brink L, D'Adda A, D'Auria R, Napolitano E, Sciuto S, Del Giudice E, Di Vecchia P, Ferrara S, *et al.* 1976 Phys. Lett. **B62** 105; Nucl. Phys. **B111** 77.

[14] Fayet P and Ferrara S 1977 Phys. Rept. **32** 249.

[15] Wolf J 1979 *Classification and Fourier inversion for parabolic subgroups with square integrable nilradical*, Memoirs Amer. Math. Soc. 225 (AMS).

[16] Knapp A W and Zuckerman G J 1977 in: Lecture Notes in Math., Vol. 587 (Berlin: Springer) p. 138-159; Knapp A W and Zuckerman G J 1982 Ann. Math. **116** 389-501.

[17] Dobrev V K, Mack G, Petkova V B, Petrova S G and Todorov I T 1976 Rept. Math. Phys. **9** 219-246; Dobrev V K, Petkova V B, Petrova S G and Todorov I T 1976 Phys. Rev. **D13** 887-912.

[18] Ogievetsky V and Sokatchev E 1977 Nucl. Phys. **B124** 309; Ogievetsky V and Sokatchev E 1978 Phys. Lett. **B79** 222.

[19] Cremmer E, Scherk J and Ferrara S 1978 Phys. Lett. **B74** 61; Cremmer E, Ferrara S, Girardello L and Van Proeyen A 1982, Phys. Lett. **B116** 231.

[20] Speh B and Vogan D 1980 Acta Math. **145** 227-299.

[21] Vogan D 1981 *Representations of Real Reductive Lie Groups*, Progr. Math., Vol. 15 (Boston-Basel-Stuttgart: Birkhäuser).

[22] Enright T, Howe R and Wallach N 1983 in: *Representations of Reductive Groups*, ed P Trombi (Boston: Birkhäuser) p. 97-143.

[23] Galperin A, Ivanov E, Kalitsyn S, Ogievetsky V and Sokatchev E 1984 Class. Quant. Grav. **1** 469; Galperin A, Ivanov E, Ogievetsky V and Sokatchev E 1986 Commun. Math. Phys. **103** 515; Galperin A and Sokatchev E 1992 Phys. Rev. **D46** 714.

[24] Dobrev V K and Petkova V B 1985 Phys. Lett. **B162** 127-132; Dobrev V K and Petkova V B 1987 Fortsch. Phys. **35** 537-572.

[25] Dobrev V K 1987 Phys. Lett. **B186** 43-51.

[26] Delduc F and Sokatchev E 1992 Class. Quant. Grav. **9** 361; Delduc F, Galperin A and Sokatchev E 1992 Nucl. Phys. **B368** 143; Delduc F, Galperin A, Howe P S and Sokatchev E 1993 Phys. Rev. **D47** 578.

[27] Truini P and Varadarajan V S 1993 Rev. Math. Phys. **5** 363; in: "Symmetries in Science VI", B Gruber (ed), (New York: Plenum Press), p. 731; in: CGTMP Salamanka 92 Proc. Anales de Física, Monografias, eds M.A. del Olmo et al (Madrid: CIEMAT/RSEF) Vol. I, p. 208.

[28] Jakobsen H P 1986 in: Lect. Notes in Phys., Vol. 261 (Berlin: Springer) p. 253-265.
[29] Kac V G and Wakimoto M 1994 in: "Lie Theory and Geometry", Progr. Math., vol. 123 (Boston: Birkhäuser) p. 415-456.
[30] Kobayashi T S 1994 Inv. Math. **117** 181-205.
[31] Witten E 1994 Int. J. Mod. Phys. **A9** 4783-4800; Witten E 2004 in: 'From Fields to Stings: Circumnavigating Theoretical Physics', eds M Shifman et al (Singapore: World Scientific) vol. 2, p. 1173-1200; Witten E 2007 preprint arXiv:0712.0157.
[32] Argyres P C, Presser M, Seiberg N and Witten E 1996 Nucl. Phys. **B461** 71-84.
[33] Ferrara S, Harvey J A, Strominger A and Vafa C 1995 Phys. Lett. **B361** 59; Ferrara S, Kallosh R and Strominger A 1995 Phys. Rev. **D52** 5412; Ferrara S and Kallosh R 1996 Phys. Rev. **D54** p. 1514 & p. 1525.
[34] Ceresole A, D'Auria R and Ferrara S 1996, Nucl. Phys. Proc. Suppl. **46** 67; Ceresole A, Dall'Agata G, D'Auria R and Ferrara S 2000 Phys. Rev. **D61** 066001.
[35] Antoniadis I, Ferrara S and Taylor T R 1996 Nucl. Phys. **B460** 489; Antoniadis I, Ferrara S, Minasian R and Narain K S 1997 Nucl. Phys. **B507** 571.
[36] Branson T P, Olafsson G and Orsted B 1996 J. Funct. Anal. **135** 163-205.
[37] Andrianopoli L, Bertolini M, Ceresole A, D'Auria R, Ferrara S, Fre P and Magri T 1997, J. Geom. Phys. **23** 111; Andrianopoli L, D'Auria R and Ferrara S 1998 Int. J. Mod. Phys. A **13** 431; Andrianopoli L, Ferrara S, Sokatchev E and Zupnik B 2000 Adv. Theor. Math. Phys. **4** 1149.
[38] Ferrara S and Maldacena J M 1998 Class. Quant. Grav. **15** 749.
[39] Ferrara S and Fronsdal C 1998 Class. Quant. Grav. **15** 2153; Ferrara S, Fronsdal C and Zaffaroni A 1998 Nucl. Phys. **B532** 153; Ferrara S and Gunaydin M 1998 Int. J. Mod. Phys. **A13** 2075.
[40] Howe P S, Sokatchev E and West P C 1998 Phys. Lett. **B444** 341.
[41] Aharony O, Gubser S S, Maldacena J, Ooguri H and Oz Y 2000 Phys. Rept. **323** 184-386.
[42] Eden B, Howe P S, Schubert C, Sokatchev E and West P C 1999 Nucl. Phys. **B557** 355; Phys. Lett. **B466** 20; Eden B, Schubert C and Sokatchev E 2000 Phys. Lett. **B482** 309; Eden B, Petkou A C, Schubert C and Sokatchev E 2001 Nucl. Phys. **B607** 191.
[43] Dolan L, Nappi C R and Witten E 2001 JHEP 0110:016.
[44] Arutyunov G, Eden B, Petkou A C and Sokatchev E 2002 Nucl. Phys. **B620** 380.
[45] Knapp A W 2002 *Lie Groups Beyond an Introduction*, 2nd ed., Progr. Math., vol. 140 (Boston-Basel-Stuttgart: Birkhäuser).
[46] Kac V, Roan S S and Wakimoto M 2003 Comm. Math. Phys. **241** 307-342.
[47] Ferrara S and Sokatchev E 2000 JHEP 0005:038; Lett. Math. Phys. **52** 247-262; Int. J. Mod. Phys. **B14** 2315-2333.
[48] Kostant B 2004 Inv. Math. **158** 181-226.
[49] Baur K and Wallach N 2005 Represent. Theory, **9** 1-29.
[50] Gannon T and Vasudevan M 2005 JHEP 0507:035.
[51] Carmeli C, Cassinelli G, Toigo A and Varadarajan V S 2006 Comm. Math. Phys. **263** 217.
[52] Duff M J and Ferrara S 2007 Phys. Rev. **D76** 124023.
[53] Faraggi A E, Kounnas C and Rizos J 2007 Nucl. Phys. **B774** 208-231.
[54] Kinney J, Maldacena J, Minwalla S and Raju S 2007 Comm. Math. Phys. **275** 209-254.
[55] Gurrieri G, Lukas L and Micu A 2007 JHEP 0712:081.
[56] Hofman D M and Maldacena J 2008 JHEP 0805:012.
[57] Bernardoni F, Cacciatori S L, Cerchiai B L and Scotti A 2008 J. Math. Phys. **49** 012107.
[58] Kallosh R and Soroush M 2008 Nucl. Phys. **B801** 25-44; Kallosh R and Kugo T 2009 JHEP 0901:072.
[59] Mizoguchi S 2008 JHEP 0811:022
[60] Ferrara S, Kallosh R and Marrani A 2012 JHEP 1206:074.
[61] Dobrev V K 2008 Rev. Math. Phys. **20** 407-449.
[62] Harish-Chandra 1955 Am. J. Math. **77** 743-777; Harish-Chandra 1956 Am. J. Math. **78** 1-41.
[63] Bruhat F 1956 Bull. Soc. Math. France, **84** 97-205.
[64] Faraut J and Korányi A 1994 *Analysis on Symmetric Cones*, Oxford Mathematical Monographs, (Oxford: Clarendon Press).
[65] Gunaydin M 1993 Mod. Phys. Lett. **A8** 1407-1416.
[66] Mack G and de Riese M 2007 J. Math. Phys. **48** 052304.
[67] Dobrev V K 2008 J. Phys. **A41** 425206.
[68] Dobrev V K 2013 Physics of Atomic Nuclei, **76**, No 8, 983-990.
[69] Dobrev V K 2013 in: Proceedings of the 9. International Workshop *Lie Theory and Its Applications in Physics*, (Varna, Bulgaria, June 2011), "Springer Proceedings in Mathematics and Statistics" Vol. 36 (Tokyo-Heidelberg: Springer) p. 311-335 (preprint CERN-PH-TH/2012-143).

[70] Dobrev V K 2009 J. Phys. **A42** 285203.
[71] Dobrev V K 2009 Proceedings, eds B Dragovich and Z Rakic, (Institute of Physics, Belgrade, 2009) p. 95-124.
[72] Dobrev V K 2013 J. High Energy Phys. 1302:015 (preprint CERN-PH-TH/2012-215).
[73] Dobrev V K, Mack G, Petkova V B, Petrova S G and Todorov I T 1977 *Harmonic Analysis on the n-Dimensional Lorentz Group and Its Applications to Conformal Quantum Field Theory*, Lecture Notes in Physics, Vol. 63 (Berlin: Springer).
[74] Knapp A W 1986 *Representation Theory of Semisimple Groups (An Overview Based on Examples)*, (Princeton Univ. Press).
[75] Dobrev V K 1988 Rept. Math. Phys. **25** 159-181 (first as ICTP Trieste preprint IC/86/393 (1986)).
[76] Dobrev V K 1985 Lett. Math. Phys. **9** 205-211.
[77] Dobrev V K 1985 J. Math. Phys. **26** 235-251.
[78] Dobrev V K and Petkova V B 1985 Lett. Math. Phys. **9** 287-298.
[79] Dobrev V K 1986 Lett. Math. Phys. **11** 225-234 (first as ICTP Trieste preprint IC/85/9 (1985); Dobrev V K 1986 in: Proc. XIII Int. Conf. Diff.-Geom. Meth. Theor. Phys., Shumen (1984), eds H-D Doebner and T D Palev, (Singapore: World Sci) p. 348-370; Dobrev V K 1986 in: Proc. Int. Symp. Topol. Geom. Methods Field Theory, Espoo (1986), eds J Hietarinta et al, (Singapore: World Sci) p. 93-102.
[80] Dobrev V K 1991 in: Proceedings Int. Group Theory Conf. (St. Andrews, 1989), eds C M Campbell et al, Vol. 1, London Math. Soc. Lecture Note Series 159 (Cambridge University Press) p. 87-104.
[81] Dobrev V K 1991 Lett. Math. Phys. **22** 251-266; Dobrev V K 1994 J. Phys. **A27** 4841; Dobrev V K 2007 Phys. Part. Nucl. **38** 564-609; Dobrev V K and Ganchev A Ch 1988 Mod. Phys. Lett. **A3** 127-137; Dobrev V K and Moylan P J 1993 Phys. Lett. **B315** 292-298.
[82] Dixmier J 1977 *Enveloping Algebras*, (New York: North Holland).
[83] Dobrev V K 1995 J. Phys. **A28** 7135-7155; Dobrev V K 1997 in: Proceedings, Symposium "Symmetries in Science IX", Bregenz, Austria, (August 1996), eds B Gruber and M Ramek, (New York: Plenum Press) pp. 47-80.
[84] Dobrev V K 1996 Suppl. Rendiconti Circolo Matematici di Palermo, Serie II, Numero 43 15-56.
[85] Dobrev V K and Petkova V B 1978 Rept. Math. Phys. **13** 233-277.
[86] Petkova V B and Sotkov G M 1984 Lett. Math. Phys. **8** 217-226.
[87] Knapp A W and Stein E M 1971 Ann. Math. **93** 489-578; Knapp A W and Stein E M 1980 Inv. Math. **60** 9-84.
[88] Gelfand I M and Naimark M A 1946 Acad. Sci. USSR. J. Phys. **10** 93-94.
[89] Bargmann V 1947 Annals Math. **48** 568-640.
[90] Satake I 1960 Ann. Math. **71** 77-110.

A note on the geometry of the quantum states of a two-level atom in external radiation fields

Marco Enríquez, Rosa Reyes and Oscar Rosas-Ortiz
Physics Department, Cinvestav, A.P. 14-740, 07000 México D.F., Mexico
E-mail: rmreyes@fis.cinvestav.mx

Abstract. We discuss some of the geometric properties of the state spaces in Quantum Mechanics with emphasis in the two-level systems case. Then the dynamics of a two-level atom affected by radiation fields is revisited in both the semiclassical and quantum approaches. In the calculations we use Hubbard operators to reduce the matrix operations into simple relations of subscripts. The time-evolution of the energy states of the atom is associated to trajectories either on the 2-sphere or in the three-dimensional unit ball according to the classical or quantum approach we are using.

1. Introduction

The geometry of state spaces is a subtle subject in quantum theories [1–6] that find applications in diverse branches of contemporary physics as the dynamical control of quantum systems [7–15] and quantum information [16], among others. Quantum states ψ cannot always be represented by vectors $|\psi\rangle$ in a Hilbert space $\mathcal{H}$ [1]; the most realistic approach uses density operators ρ to represent the state of a quantum system $\mathcal{S}$ since this last is usually prepared in a statistical ensemble of pure states $\{p_k, \psi_k\}$ better than in a given pure state ψ_r [17]. This situation provides an indication of the geometry involved since the density operator is expressed as a convex combination of orthogonal projectors $\pi_k = |\psi_k\rangle\langle\psi_k|$ that map the Hilbert space $\mathcal{H}$ into a one-dimensional subspace Span$\{|\psi_k\rangle\}$ each one. The projectors can be visualized as filters characterized by certain *absorption coefficients* that are interpreted as the transition probabilities between pure states and define the geometry of the state space [1]. In this context, the pure states are represented by filters that perform the finest selection possible (*minimal filters*) and correspond to the orthogonal projectors π_k, so that the state space is the *convex hull* of all the minimal filters π_k associated to a given physical system $\mathcal{S}$. Using only vectors in a Hilbert space is not enough to represent all the possible quantum states, indeed the involved geometric structure is strongly limited in this case. It follows that "the physical reality can be too complex in order to fit in any Hilbert space" [1].

The simplest space of quantum states is associated to any two-level system (*qubit*) like the polarization of light or the spin-state of a silver atom, and this is just a three-dimensional ball of unit radius. Such a *convex body* is large enough to carry the canonical transformations that characterize the dynamics of the two-level system [1, 16]. The points on the surface (i.e. the points on the 2-sphere S^2) are in a one-to-one correspondence with the pure states of any qubit. A line segment $\overline{ab}$ connecting two arbitrary points a and b on S^2 is a convex set that represents the space of states of some classical bit whenever a and b are antipodal. Any interior point of

$\overline{ab}$ is a statistical ensemble of the pure states (*extreme points*) a and b. The entire state space of a qubit (that is, the three-dimensional ball) is therefore the intersection of all the convex sets $\overline{ab}$ that can be formed by taking arbitrary pairs $a, b \in S^2$.

Using a geometric approach the dynamics of the two-level pure states is reduced to rotations and reflections of the 2-sphere [12]. This condition is useful to determine the (classical) magnetic fields that must be applied on a spin-1/2 particle in order to control the time-evolution of its quantum state [9–11]. For instance, the particle can be compelled to evolve cyclically [9, 11] in a process that is known as *evolution loop* [7]. It is also possible to visualize experimental setups for the steering eigen-energy state to a destination without net adiabatic transitions [10]. The qubit pure states are also useful in the constructing of mesoscopic *kittens* as the appropriate linear combinations of $j = 1/2$ angular momentum eigenkets. The kittens manifest quantum interference effects analogous to those occurring for linear combinations of coherent light waves and are distinguishable by collecting them in *kitten-hoods* according to their invariance under azimuthal rotations [14].

In this communication we analyze the geometry of the state space associated with a two-level atom in external radiation fields. We shall address two different situations. A semiclassical approach is first used by considering that the energy of the atom is quantized while the field is treated as a classical entity. The state space of the entire system (atom+field) is just the 2-sphere since only qubit pure states are involved. The time-evolution of the qubit describes trajectories on S^2 that are parameterized by the field variables and can be reduced to rotations and reflections of the sphere as stated above. The population of the atom energy states is inverted in time according to the initial conditions and describes a geodesic connecting the north and south poles of S^2 (Rabi oscillations) if the atom and the field are in resonance. Non-resonant cases give rise to evolution loops that are represented as circles on the sphere. In a second step we use the quantum approach in which both the atom energies and the field are quantized. The entire system is in a pure state that is represented by vectors in the Kronecker product of the Hilbert spaces of the atom energy and field states. Here the population invertion of the atom energy states depends on the photon statistical distribution of the field: the well known Rabi oscillations are recovered for fields in a Fock state and the collapses and revivals phenomenon for fields in a Glauber state. The geometry of the atom energy states is then analyzed after the partial trace of the density operator ρ of the entire system over the field states $\rho_{at} = \text{Tr}_f \rho$. At the time t, the reduced density operator $\rho_{at}(t)$ is a point of the three-dimensional ball, so that the time-evolution of the atom energy states describes trajectories in the interior of the ball.

The organization of the paper is as follows. In Section 2 we present a short survey on the geometry of quantum states by emphasizing the case of two-level systems (Section 2.1). In Section 3 we analyze the interaction between the atom and the radiation field in the X-representation. That is, we use the operators $X^{i,j}$ that cause a transition from the pure state $|\psi_j\rangle$ to the pure state $|\psi_i\rangle$ of either the atom or the field. These operators satisfy a definite set of algebraic properties and are called after Hubbard (for a recent review see [18] and references quoted therein). The X-representation reduces the matrix operations we are going to deal with into simple relations of subscripts. The well known semiclassical result of the Rabi oscillations describing the atomic transitions in time is recovered in Section 3.1. There, the atomic population invertion is connected to circles on the 2-sphere described by the time-evolution of the qubit pure states. The X-representation of the quantum treatment is given in Section 3.2; the phenomenon of collapses and revivals of the atomic transitions is recovered by assuming that the field is in a Glauber state. The analysis of $\rho_{at}(t)$ shows that the reduced state of the atom is an ensemble of pure states that describes trajectories in the interior of the three dimensional ball as it evolves in time (Section 4). We close the paper with some final remarks and conclusions.

2. A survey to the geometry of quantum states

In quantum theory the pure states ψ are represented by regular kets $|\psi\rangle$ in a Hilbert space $\mathcal{H}$. These last are vectors of finite norm ($\|\varphi\|^2 = \langle\varphi|\varphi\rangle < +\infty$) that can be normalized to unity, $\langle\varphi|\varphi\rangle = 1$. On the other hand, one of the simplest devices of measuring in quantum mechanics is a filter checking for a particle to be in a given (*idler*) state $|\varphi\rangle$. The action of the device is represented by the operator $P_\varphi = |\varphi\rangle\langle\varphi|$ producing the orthonormal projection of $\mathcal{H}$ onto the one-dimensional subspace Span$\{|\varphi\rangle\} \subset \mathcal{H}$. For an arbitrary (*signal*) state $|\psi\rangle$ the action of P_φ is given by $P_\varphi|\psi\rangle = \Gamma|\varphi\rangle$, with $\Gamma = \langle\varphi|\psi\rangle \in \mathbb{C}$. Thus, the testing gives rise to one of the following results (see e.g. [2, 5]):

a) If the idler and signal kets are parallel ($|\varphi\rangle \propto |\psi\rangle$) then the answer is affirmative.
b) If the kets are orthogonal ($|\varphi\rangle \perp |\psi\rangle$), the output of the checking is negative.
c) If $|\psi\rangle = \alpha|\varphi\rangle + \beta|\eta\rangle$ with $|\varphi\rangle \perp |\eta\rangle$, the result is uncertain and depends on the coefficient $\alpha = \Gamma$. Here $|\alpha|^2 = |\langle\varphi|\psi\rangle|^2$ is the *transition probability* connected to the possibility that the system will pass the $|\varphi\rangle$-checking.

The one-dimensional space Span$\{|\varphi\rangle\}$ corresponds to a whole bunch of proportional regular kets called a *ray* and denoted $[\varphi]$. Hence, the kets representing pure states are not unique since two normalized kets that differ only by a phase factor are elements of the same equivalence class

$$[\varphi] =: \{|\psi\rangle \in \mathcal{H} : |\psi\rangle = e^{i\theta}|\varphi\rangle, e^{i\theta} \in U(1)\}, \tag{1}$$

and lead to the same probabilistic predictions of the theory (i.e., $||\psi|| = ||\varphi||$) [2, 5, 14]. In general, whenever two normalized kets $|\psi_1\rangle$ and $|\psi_2\rangle$ represent two possible pure states of a system, their linear combination $|\psi\rangle = \alpha_1|\psi_1\rangle + \alpha_2|\psi_2\rangle$, with $\alpha_1, \alpha_2 \in \mathbb{C}$, represents a pure state as well.

A quantum system, however, may not be in a pure state. In practice, the quantum state of most of the physical systems is a *mixed state* ρ rather than a given pure state $|\psi\rangle \in \mathcal{H}$. The state ρ is a statistical ensemble[1] $\{p_k, \psi_k\}_{k\in\mathcal{I}}$, $\mathcal{I} \subseteq \mathbb{Z}^+$, where each pure state $|\psi_k\rangle$ occurs with probability $0 \leq p_k \leq 1$ and $\{|\psi_k\rangle\}_{k\in\mathcal{I}}$ is an orthonormal set. As a mathematical object, ρ is not a vector in the Hilbert space $\mathcal{H}$ but a linear operator $\rho : \mathcal{H} \to \mathcal{H}$ that satisfies

i) $\rho = \rho^\dagger$ (Hermiticity)
ii) $\langle\psi|\rho|\psi\rangle \geq 0$ for all $|\psi\rangle \in \mathcal{H}$ (positive definiteness)
iii) $\mathrm{Tr}\rho = 1$ (normalizability)

Thereby, the probabilities p_k and the pure states $|\psi_k\rangle$ are respectively the eigenvalues and eigenvectors of ρ which, in canonical form, is written as

$$\rho = \sum_{k\in\mathcal{I}} p_k|\psi_k\rangle\langle\psi_k|, \quad \mathcal{I} \subseteq \mathbb{Z}^+. \tag{2}$$

Assuming that we know how to prepare the pure states $|\psi_k\rangle$, we can use a classical random number generator to produce the number k with probability p_k. Every time we get a number k from the generator a copy of the system is prepared in the state $|\psi_k\rangle$. The ensemble (2) is then obtained by repeating at will such a procedure [19]. In this case p_k represents the probability

[1] We assume discrete basis for the systems we are interested in, the transition to the continuous case is straightforward.

of being prepared in the state $|\psi_k\rangle$ as well as the probability of being found in that state[2]. A mixed state ρ represents also an imprecise knowledge or inaccuracy in the preparation process of the pure states [19] (see also [20]).

Properties (ii) and (iii) give rise to the conditions

$$\sum_{k\in\mathcal{I}} p_k = 1, \quad 0 \leq p_k \leq 1. \tag{3}$$

If $p_k = 1$ for a given $k \in \mathcal{I}$ then (2) is reduced to the orthogonal projector $\rho_k := |\psi_k\rangle\langle\psi_k|$ that fulfills (i–iii) and besides is such that $\text{Tr}\rho_k^2 = 1$. In this case ρ_k as well as $|\psi_k\rangle$ represent the same pure state ψ_k. Therefore, we realize that (2)–(3) correspond to a *convex combination* of the pure states ρ_k, that is,

$$\rho = \sum_{k\in\mathcal{I}} p_k\rho_k, \quad \sum_{k\in\mathcal{I}} p_k = 1, \quad 0 \leq p_k \leq 1. \tag{4}$$

In the case of a nontrivial convex combination (i.e., $0 < p_k < 1$ for at least two different values of $k \in \mathcal{I}$) one immediately shows that $\text{Tr}\rho^2 < 1$. Then, for any quantum state we have

$$\text{Tr}\rho^2 \leq 1, \tag{5}$$

where the equality holds only for pure states. Considering all the possible combinations (4) we see that the collection of all the density operators ρ associated to the ensemble $\{p_k, \psi_k\}_{k\in\mathcal{I}}$ is just a *convex set* with the pure states ρ_k as *extreme points*[3] [1, 3, 6, 16].

2.1. The geometry of two-level systems

Let us consider an atom with just two levels of energy. The excited $|+\rangle$ and ground $|-\rangle$ states are orthogonal

$$\langle +|+\rangle = \langle -|-\rangle = 1, \qquad \langle +|-\rangle = \langle -|+\rangle = 0, \tag{6}$$

and will be written in their simplest representation

$$|+\rangle = \begin{pmatrix} 1 \\ 0 \end{pmatrix}, \qquad |-\rangle = \begin{pmatrix} 0 \\ 1 \end{pmatrix}. \tag{7}$$

An arbitrary (pure) energy state of the atom $|\psi\rangle$ is then written as a normalized linear combination in the Hilbert space $\mathcal{H}_a = \text{Span}\{|+\rangle, |-\rangle\}$, that is

$$|\psi\rangle = c_1|+\rangle + c_2|-\rangle, \quad |c_1|^2 + |c_2|^2 = 1, \quad c_1, c_2 \in \mathbb{C}. \tag{8}$$

The 2-tuples (7) are associated to the spectral decomposition and matrix representation of the Pauli operator σ_3 as follows

$$\sigma_3 = |+\rangle\langle +| - |-\rangle\langle -| = \begin{pmatrix} 1 & 0 \\ 0 & -1 \end{pmatrix}. \tag{9}$$

[2] In a more general situation, an arbitrary set of pure states $\{\phi_k\}_{k\in\mathcal{I}'}$ need not be orthogonal in order to define a statistical ensemble $\rho' = \{p'_k, \phi_k\}_{k\in\mathcal{I}'}$. If the kets $|\phi_k\rangle$ are not orthogonal then the probabilities p'_k of 'being prepared in' and p_k of 'being found in' the state $|\phi_k\rangle$ are not equal since, according to the point (c) indicated above, a system prepared in the state $|\phi_k\rangle$ with $k \neq \ell$ can be found (with nonzero probability) also in the state $|\phi_\ell\rangle$. Moreover, if the set $\{\phi_k\}_{k\in\mathcal{I}'}$ is not orthogonal, the probabilities p'_k are not the eigenvalues of ρ' [19].

[3] The points of a convex set that cannot be expressed as a convex combination of other points in the set are called *extreme points*.

This last together with the *creation* and *annihilation* spin-1/2 operators

$$\sigma_+ = \begin{pmatrix} 0 & 1 \\ 0 & 0 \end{pmatrix}, \quad \sigma_- = \begin{pmatrix} 0 & 0 \\ 1 & 0 \end{pmatrix}, \tag{10}$$

are the generators of the $su(2)$ Lie algebra[4]

$$[\sigma_3, \sigma_\pm] = 2\sigma_\pm, \quad [\sigma_+, \sigma_-] = \sigma_3, \tag{11}$$

and satisfy the following relations

$$\sigma_+^2 = \sigma_-^2 = 0, \quad \sigma_+^\dagger = \sigma_-, \quad \sigma_\pm|\pm\rangle = 0, \quad \sigma_\pm|\mp\rangle = |\pm\rangle. \tag{12}$$

In this representation the observable of the energy is the Hamiltonian

$$H_a = \frac{1}{2}\hbar\omega_a\sigma_3, \tag{13}$$

with ω_a the atomic transition frequency. Another observable of interest is the density operator ρ which in the representation (9) acquires the general form

$$\rho = \begin{pmatrix} \rho_{11} & \rho_{12} \\ \rho_{21} & \rho_{22} \end{pmatrix} = \frac{1}{2}\left(\mathbb{I} + \vec{\tau}\cdot\vec{\sigma}\right), \quad \vec{\tau} = (\tau_1, \tau_2, \tau_3), \quad \vec{\sigma} = (\sigma_1, \sigma_2, \sigma_3). \tag{14}$$

Indeed, this operator is (i) Hermitian if $\rho_{11}, \rho_{22} \in \mathbb{R}$ and $\rho_{12} = \overline{\rho}_{21}$ (ii) definite positive if $|\rho_{12}| = |\rho_{21}| \leq \sqrt{\rho_{11}\rho_{22}}$, and (iii) normalized if $\rho_{11} + \rho_{22} = 1$. Considering these three properties one can verify that the elements of the *Bloch vector* $\vec{\tau}$ satisfy the equation of a three-dimensional ball of unit radius $B[0,1]$,

$$\mathrm{Tr}\rho_a^2 = \tau_1^2 + \tau_2^2 + \tau_3^2 \leq 1, \tag{15}$$

and are given by

$$\tau_1 = \rho_{12} + \rho_{21} = 2\mathrm{Re}(\rho_{12}), \quad \tau_2 = i(\rho_{12} - \rho_{21}) = 2\mathrm{Im}(\rho_{12}), \quad \tau_3 = \rho_{11} - \rho_{22}. \tag{16}$$

It is illustrative to notice that the Bloch vector $\vec{\tau}$ is also associated to the expectation value of $\vec{\sigma}$ calculated in the basis of pure states $|\pm\rangle$; namely $\tau_k = \langle\sigma_k\rangle = \mathrm{Tr}(\rho\sigma_k)$ for $k = 1,2,3$. Thus, $\mathbf{n} = \langle\vec{\sigma}\rangle = \mathrm{Tr}(\rho\vec{\sigma})$ is a point on the 2-sphere $S^2 \subset B[0,1]$ since $\vec{\tau} = \mathbf{n}$ in Eq. (15) is in this case such that $||\mathbf{n}|| = 1$. This representation of the energy pure states of the two-level atom leads to a fibre-bundle formulation of the qubit dynamics. The *Hopf fibring* [21] (see also [22]) of the 3-sphere S^3 over the base space S^2 with fibre S^1 yields the geometry of qubit pure states

$$S^1 - S^3 \xrightarrow{\pi} S^2.$$

Equations (16) correspond to the *Hopf mapping* π, the 3-sphere is the space of rays (1) belonging to the 2-dimensional Hilbert space $\mathcal{H}_a = \mathrm{Span}\{|+\rangle, |-\rangle\}$, and the fibre S^1 defines the rays in terms of the phases $e^{i\theta} \in U(1)$. All the points on S^2 represent energy pure states of the atom, any inner point of $B[0,1]$, on the other hand, represents a mixed state of $|\pm\rangle$.

[4] In terms of the Pauli matrices σ_3, $\sigma_1 = \sigma_+ + \sigma_-$, and $\sigma_2 = i(\sigma_- - \sigma_+)$, the $su(2)$ algebra reads $[\sigma_i, \sigma_j] = 2\epsilon_{ijk}\sigma_k$, where $i,j,k = 1,2,3$ and ϵ_{ijk} is the total anti-symmetric Levi-Civita tensor.

3. Atom-Field interaction

In this section we are interested in the interaction between a two-level atom and a radiation field. To simplify the calculations we shall use the X-representation of operators that is obtained by considering the Hubbard operators in their simplest form [18]. That is, we shall use n-square matrices $X^{i,j}$ which have entry 1 in position (i, j) and zero in all other entries at the time that they fulfill the multiplication rule

$$X^{i,j}X^{k,m} = \delta_{jk}X^{i,m}, \tag{17}$$

and have the properties

$$(X^{i,j})^\dagger = X^{j,i}, \quad \sum_k X^{k,k} = \mathbb{I}, \quad [X^{i,j}, X^{k,m}]_\pm = \delta_{jk}X^{i,m} \pm \delta_{mi}X^{k,j}, \tag{18}$$

with $\mathbb{I}$ the identity matrix and $[A, B]_\pm$ standing for either the commutator $(-)$ or the anticommutator $(+)$ of A and B.

In X-representation any operator A acting on $\mathcal{H}_a = \text{Span}\{|+\rangle, |-\rangle\}$ is expressed as a linear combination

$$A = \sum_{k,j=1}^{2} a_{k,j}X^{i,j}, \quad a_{k,j} \in \mathbb{C}, \tag{19}$$

where the coefficients $a_{k,j}$ are defined accordingly. For instance, the $su(2)$ generators (9) and (10) read as

$$\sigma_3 = X^{1,1} - X^{2,2}, \quad \sigma_+ = X^{1,2}, \quad \sigma_- = X^{2,1}, \tag{20}$$

while the Hamiltonian (13) and the identity $\mathbb{I}_2$ are written in the form

$$H_a = \frac{1}{2}\hbar\omega_a(X^{1,1} - X^{2,2}), \quad \mathbb{I}_2 = X^{1,1} + X^{2,2}. \tag{21}$$

3.1. Classical fields

As a first approach let us consider the situation in which the (electric) radiation field can be described in a purely classical manner. Considering the case of single mode we have

$$\vec{E}(t) = \mathcal{E}\left(e^{-i\omega_f t} + e^{i\omega_f t}\right)\mathbf{e}_f. \tag{22}$$

Here $\mathcal{E}$, ω_f and $\mathbf{e}_f$ are respectively the amplitude, the frequency and the polarization of the field. The most general form of the interaction Hamiltonian H_{Icl} is in this case given by

$$H_{Icl} = \sum_{k,j=1}^{2} \lambda_{k,j}X^{i,j}. \tag{23}$$

The coefficients $\lambda_{i,j}$ of this last combination are characterized by the field parameters [23, 24]. Using the dipole approximation one immediately gets $\lambda_{1,1} = \lambda_{2,2} = 0$ and $\lambda_{1,2} = \lambda_{2,1} = 2\hbar\gamma\cos(\omega_f t)$, where $\gamma = (\frac{p}{\hbar}\mathcal{E})\mathbf{e}_f \cdot \mathbf{e}_p$ stands for the classical *Rabi frequency* and p, $\mathbf{e}_p$, are respectively the "size" and orientation of the atom dipole momentum $\vec{p}$ that defines the approximation. From (20) the interaction Hamiltonian (23) acquires the usual form

$$H_{Icl} = \hbar\gamma(\sigma_+ + \sigma_-). \tag{24}$$

Now we use the Hamiltonian (13) to introduce a rotating frame description by means of the unitary transformation

$$|\psi\rangle_R = e^{iH_a t/\hbar}|\psi\rangle. \tag{25}$$

Then H_a is invariant under such a transformation while H_{Icl} is mapped into the expression

$$\widetilde{H}_{Icl} = H_{Icl} + 2\hbar\gamma(\sigma_+ e^{i2\omega_f t})_{sim}, \tag{26}$$

with $A_{sim} = (A + A^\dagger)/2$. In the rotating wave approximation we can omit the second term in the last expression so that $\widetilde{H}_{Icl} = H_{Icl}$. Finally, let us take $\frac{\hbar}{2}\omega_f$ and $\hbar\omega_a$ respectively as the zero and unit of energy to write the Hamiltonian of the entire system (atom+field) in the conventional form (see e.g. [23–26]):

$$H_{cl} = H_a + H_{Icl} = \frac{\Delta}{2}\sigma_3 + g(\sigma_+ + \sigma_-), \qquad \Delta = 1 - \frac{\omega_f}{\omega_a}, \quad g = \frac{\gamma}{\omega_a}. \tag{27}$$

The (dimensionless) parameter Δ stands for the *detuning* between the radiation field and the atomic transition. The time-evolution operator $U_{cl}(t) = e^{iH_{cl}t}$ is easily achieved

$$U_{cl}(t) = c_+(t)X^{1,1} + c_+^*(t)X^{2,2} + 2\left(c_-^*(t)X^{1,2}\right)_{sim} \tag{28}$$

where z^* is the complex conjugate of $z \in \mathbb{C}$ and

$$c_+(t) = \cos\Omega t - i\frac{\Delta}{2\Omega}\sin\Omega t, \qquad c_-(t) = i\frac{g}{\Omega}\sin\Omega t, \tag{29}$$

with $\Omega = \sqrt{\Delta^2/4 + g^2}$ a generalized Rabi frequency (this is reduced to the usual Rabi frequency when $\Delta = 0$). To get some insights on the time-evolution of the system let us assume that the atom is initially in the pure state $|\psi(0)\rangle_R = |+\rangle$. One gets

$$|\psi(t)\rangle_R = U(t)|\psi(0)\rangle_R = c_+(t)|+\rangle + c_-(t)|-\rangle, \tag{30}$$

so that the atomic population is inverted according to the rule

$$\langle\sigma_3(t)\rangle = |c_+(t)|^2 - |c_-(t)|^2 = \left(\frac{g}{\Omega}\right)^2\cos(2\Omega t) + \left(\frac{\Delta}{2\Omega}\right)^2. \tag{31}$$

At resonance ($\Delta = 0$) the well known Rabi oscillations between $|+\rangle$ and $|-\rangle$ are recovered

$$\langle\sigma_3(t)\rangle = \cos(2gt). \tag{32}$$

Figure 1 shows the behavior of (31) for different values of the detuning. If $\Delta = 0$ the atomic population is inverted from $|+\rangle$ to $|-\rangle$ and vice versa at the (dimensionless) Rabi frequency $g = \gamma/\omega_a$. Out of resonance ($\Delta \neq 0$) the initial atomic population $|c_+(t)|^2$ periodically decreases its value and recovers this at the generalized Rabi frequency Ω. In this case $|c_+(t)|^2$ is never zero so that the atom is not able to arrive at the state $|-\rangle$ in any time. Indeed this evolves from $|+\rangle$ to a linear superposition of $|+\rangle$ and $|-\rangle$, and after a while determined by Ω the atom goes back to the state $|+\rangle$ (see the cases $\Delta = g$ and $\Delta = 4g$ in Figure 1).

The pure state (30) can be represented also by a projector with the properties i)–iii) of a density operator we have discussed in the previous section, this reads as

$$\rho(t) = |\psi(t)\rangle\langle\psi(t)| = |c_+|^2X^{1,1} + |c_-|^2X^{2,2} + 2\left(c_+\bar{c}_-X^{1,2}\right)_{sim}, \tag{33}$$

where the time-dependence of the coefficients has been assumed as implicit. The Bloch vector associated to the state (33) is normalized ($\vec{r} = \mathbf{n}$) and evolves in time by describing a trajectory on the 2-sphere S^2 that is parameterized as follows

$$\begin{gathered} n_1(t) = 2\mathrm{Re}(c_+\bar{c}_-) = -\frac{g\Delta}{\Omega^2}\sin^2(\Omega t), \quad n_2(t) = 2\mathrm{Im}(c_+\bar{c}_-) = -\frac{g}{\Omega}\sin(2\Omega t), \\ n_3(t) = |c_+|^2 - |c_-|^2 = \left(\frac{g}{\Omega}\right)^2\cos(2\Omega t) + \left(\frac{\Delta}{2\Omega}\right)^2. \end{gathered} \tag{34}$$

Journal of Physics: Conference Series **512** (2014) 012021 doi:10.1088/1742-6596/512/1/012021

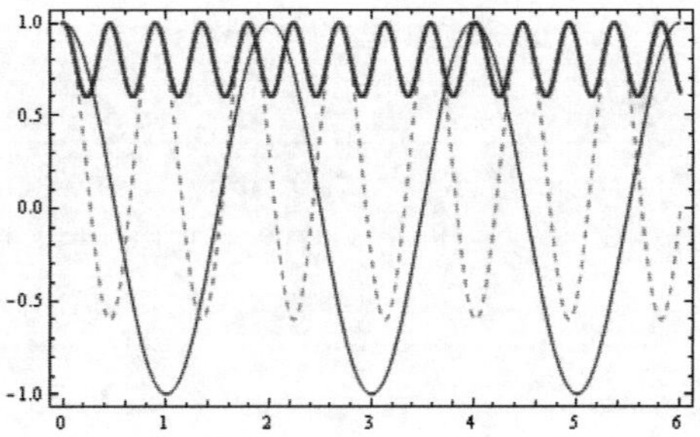

Figure 1. The atomic population invertion obtained in the semiclassical approach (31). If the atomic transitions are in resonance with the field (black curve) one recovers the well known Rabi oscillations between $|+\rangle$ and $|-\rangle$. For other values of the detuning ($\Delta = g$ in red, $\Delta = 4g$ in blue) an atom in the initial state $|+\rangle$ has not the possibility to evolve into $|-\rangle$ with certainty.

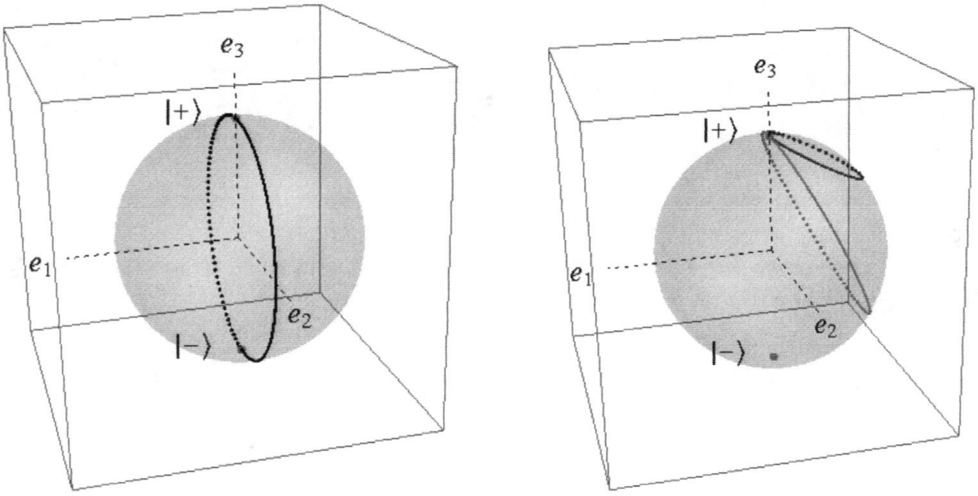

Figure 2. In the semiclassical approach the state space of an atom in a single mode radiation field is just a 2-sphere because only the pure states of the atom are involved. At resonance, the qubit describes a geodesic on the yz-plane (left) while other values of the detuning give rise to circles on the sphere that have the north pole as common point. The paths for $\Delta = g$ (red) and $\Delta = 4g$ (blue) have been depicted here (right) for comparison with Figure 1. In all cases the north pole is the initial state and the system comes back to this point in a period $T_n = n\pi/\Omega, n \in \mathbb{Z}$.

In general, the Bloch vector $\mathbf{n}$ describes circles on S^2 that pass over the north pole (N) periodically in time; the greater the value of Δ the smaller the diameter of the circle (see Figure 2). The points P integrating these circles are connected to N by a chord $\overline{NP}$ that fulfills $0 \leq \overline{NP} \leq 2$, with 2 the diameter of the 2-sphere. Such paths are also the geometric representation of the atomic population invertion (31) as this evolves in time. At resonance $\Delta = 0$, the circle corresponds to a geodesic passing over the south pole (S) in the yz-plane. In the vicinity of resonance the circles are defined over a plane which is almost parallel to the yz-one and pass over points that are close to the south pole. Thus, the atom evolves from $|+\rangle$ to a linear superposition (30) that is dominated by $|-\rangle$ at a half a period $T/2 = \frac{\pi}{2\Omega} \approx \frac{\pi}{4g}$ for $\Delta \approx 0$. Other values of Δ produce different linear superpositions at $T/2$. For instance, in Figure 2 we see that the chord $\overline{NP}$ defined by the points P belonging to the circle for $\Delta = 2g$ is smaller than the chord obtained for $\Delta = g/2$. This means that the influence of $|+\rangle$ in the superposition (30) increases as the value of Δ. For arbitrary initial states written as the linear superposition of the north and south poles the trajectories are the same as those described above. This is just necessary to consider that any point on the sphere can be transformed into N, and vice versa, by the appropriate set of rigid rotations.

3.2. quantized fields

If the description of the radiation field requires the inclusion of quantum properties then the electric vector (22) is substituted by the vectorial operator

$$\vec{\hat{E}}(t) = \mathcal{E}\left(\hat{a}e^{-i\omega_f t} + \hat{a}^\dagger e^{i\omega_f t}\right)\mathbf{e}_f, \tag{35}$$

where $\hat{a}$ and $\hat{a}^\dagger$ are the boson annihilation and creation operators fulfilling the photon algebra

$$[\hat{a}, \hat{a}^\dagger] = \mathbb{I}_f, \quad [\hat{N}, \hat{a}] = -\hat{a}, \quad [\hat{N}, \hat{a}^\dagger] = \hat{a}^\dagger, \tag{36}$$

with $\hat{N} = \hat{a}^\dagger \hat{a}$ the number operator and $\mathbb{I}_f$ the identity operator in the Hilbert space $\mathcal{H}_f = \text{Span}\{|n\rangle\}_{n\in\mathbb{Z}^+}$. The action of the above operators on the Fock vectors $|n\rangle$, $n \in \mathbb{Z}^+$, is as follows

$$\hat{a}|n\rangle = \sqrt{n}|n-1\rangle, \quad \hat{a}^\dagger|n\rangle = \sqrt{n+1}|n+1\rangle, \quad \hat{N}|n\rangle = n|n\rangle. \tag{37}$$

For simplicity we shall omit the "hat" in the representing of operators ($\hat{A} \leftrightarrow A$) while we use the number 1 instead of $\mathbb{I}_f$ whenever there will be no confusion in notation. The Hamiltonian of the field is in the Fock's representation given by

$$H_f = \hbar\omega_f\left(N + \frac{1}{2}\right), \tag{38}$$

so that $H_f|n\rangle = \hbar\omega_f(n+\frac{1}{2})|n\rangle$. Thus, $\mathcal{H}_f$ is the state space spanned by the energy eigenvectors of the field. The entire quantum system $\mathcal{S} = \mathcal{S}_a + \mathcal{S}_f$ is a composite of the atom $\mathcal{S}_a$ and the field $\mathcal{S}_f$, so that the energy observables (21) and (38) should be promoted to act on the state space of $\mathcal{S}$. We have

$$H_a \to H_a \otimes \mathbb{I}_f, \qquad H_f \to \mathbb{I}_2 \otimes H_f, \tag{39}$$

with $\otimes$ the Kronecker product of operators [18]. Then H_a and H_f are the energy observables of the atom alone and the field alone. Accordingly, the free of interaction Hamiltonian reads as

$$H_0 = H_a \otimes \mathbb{I}_f + \mathbb{I}_a \otimes H_f = \begin{pmatrix} \frac{\hbar\omega_a}{2}\Delta + \hbar\omega_f(N+1) & 0 \\ 0 & -\frac{\hbar\omega_a}{2}\Delta + \hbar\omega_f N \end{pmatrix}. \tag{40}$$

Notice that the matrix elements of H_0 are in the Fock's representation. The eigenvectors and eigenvalues of this operator are defined as follows

$$\begin{aligned} H_0|+,n\rangle &= \hbar\left[\omega_f(n+1) + \tfrac{\omega_a}{2}\Delta\right]|+,n\rangle, \\ H_0|-,n+1\rangle &= \hbar\left[\omega_f(n+1) - \tfrac{\omega_a}{2}\Delta\right]|-,n+1\rangle, \end{aligned} \tag{41}$$

where

$$|+,n\rangle \equiv |+\rangle \otimes |n\rangle = \begin{pmatrix} |n\rangle \\ |\emptyset\rangle \end{pmatrix}, \quad |-,n+1\rangle \equiv |-\rangle \otimes |n+1\rangle = \begin{pmatrix} |\emptyset\rangle \\ |n+1\rangle \end{pmatrix}, \tag{42}$$

with $|\emptyset\rangle$ the null vector in $\mathcal{H}_f$. If the interaction between the atom and the field is considered then, in a first approximation, we can write

$$H_I = \sum_{i,j=1}^{2} \sum_{k,\ell=0}^{+\infty} \lambda_{i,j;k,\ell} X^{i,j} \otimes X_f^{k,\ell}, \tag{43}$$

where $X_f^{k,\ell}$ is an infinite square matrix that has entry 1 in position (k, ℓ) and zero in all other entries [18]. This last kind of matrices is useful in the constructing of the X-representation of the boson operators [23, 24]:

$$a = \sqrt{N} \sum_{n=0}^{+\infty} X_f^{n,n+1}, \quad a^\dagger = \sqrt{N} \sum_{n=0}^{+\infty} X_f^{n+1,n}. \tag{44}$$

Using the same approximations as in Section 3.1 one arrives at the simplest form of the interaction Hamiltonian

$$H_I = \gamma(\sigma_+ \otimes a + \sigma_- \otimes a^\dagger) = \begin{pmatrix} 0 & \gamma a \\ \gamma a^\dagger & 0 \end{pmatrix}. \tag{45}$$

Therefore, the Hamiltonian of the entire system in the rotating frame is just the one appearing in the Jaynes-Cummings model [27] (see also [28]):

$$H = \frac{\Delta}{2} \sigma_3 \otimes \mathbb{I}_f + H_I, \tag{46}$$

where $\hbar\omega_f/2$ and $\hbar\omega_a$ are respectively the zero and the unit of the energy. The time-evolution operator is in this case [18, 23, 24]:

$$U(t) = e^{iHt} = C_+(N+1,t) X^{1,1} + C_+^\dagger(N,t) X^{2,2} + 2\left(C_-^\dagger(N,t) X^{1,2}\right)_{sim} \tag{47}$$

where the operators $C_\pm(N,t)$ are defined in the Fock's representation

$$C_+(N,t) = \cos(\Omega_N t) - \frac{i\Delta}{2\Omega_N} \sin(\Omega_N t), \quad C_-(N,t) = a^\dagger \left(\frac{ig}{\Omega_{N+1}}\right) \sin(\Omega_{N+1} t), \tag{48}$$

with $\Omega_N = \sqrt{\Delta^2/4 + gN}$. Thus, they act on the number states $|n\rangle$ as follows

$$C_+(N,t)|n\rangle = c_+(n,t)|n\rangle, \quad C_-(N,t)|n\rangle = c_-(n,t)|n+1\rangle. \tag{49}$$

The $c_\pm(n,t)$ functions are given by

$$c_+(n,t) = \cos(\Omega_n t) - \frac{i\Delta}{2\Omega_n} \sin(\Omega_n t), \quad c_-(n,t) = \sqrt{n+1} \left(\frac{ig}{\Omega_{n+1}}\right) \sin(\Omega_{n+1} t). \tag{50}$$

Let us assume that $|\psi(0)\rangle_R = |+, n\rangle$ is the initial sate of the entire system. The time-evolved state is then of the form

$$|\psi(t)\rangle_R = U(t)|\psi(0)\rangle_R = c_+(n+1,t)|+,n\rangle + c_-(n,t)|-,n+1\rangle, \tag{51}$$

and the atomic population invertion as a function of time is

$$\langle \sigma(t) \rangle_n = \left(\frac{g\sqrt{n+1}}{\Omega_{n+1}}\right)^2 \cos(2\Omega_{n+1} t) + \left(\frac{\Delta}{2\Omega_{n+1}}\right)^2. \tag{52}$$

Journal of Physics: Conference Series **512** (2014) 012021 doi:10.1088/1742-6596/512/1/012021

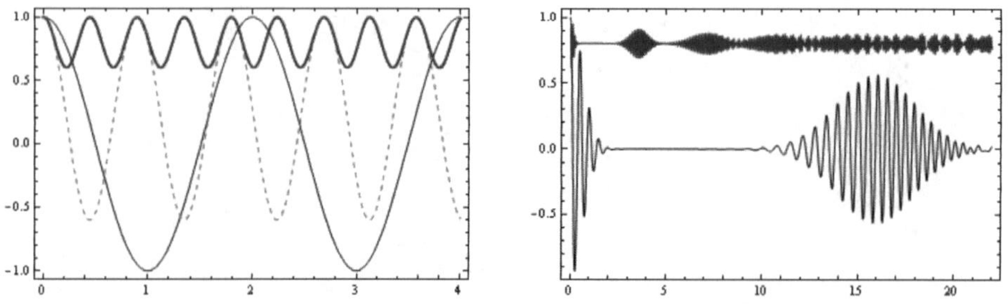

Figure 3. The atomic population invertion obtained in the quantum approach for the field in the Fock state $|n\rangle$ (Eq. (52), left) and in the Glauber state with $\overline{n} = 15$ (Eq. (56), right) for different values of the detuning ($\Delta = 0$ in black, $\Delta = g$ in red and $\Delta = 4g$ in blue).

In Figure 3 we can see the dependence of the population invertion (52) on the detuning Δ. At resonance ($\Delta = 0$) the atomic population oscillates in time from $|+\rangle$ to $|-\rangle$ while the field transits from the Fock state $|n\rangle$ to $|n{+}1\rangle$ at the frequency Ω_n (notice this occurs even for the zero photons $n = 0$ case, as expected). The situation illustrated in Figure 3 is quite similar to that presented in Figure 1 (except in the zero photons case) because in both cases we have analyzed the time-evolution of the pure states of the atom+field system. In the quantum approach this is given by the Kronecker product state $|+, n\rangle$ while in the classical approach we have the simple pure state $|+\rangle$. In the next section we shall discuss on the main differences between the semiclassical and the quantum approaches when one looks for the dynamics of the atom states alone.

If the field is not in a definite state of the energy one can consider the initial state

$$|\psi(0)\rangle = \sum_{k=0}^{\infty} \alpha_k |+, k\rangle, \quad \text{with} \quad \sum_{k=0}^{\infty} |\alpha_k|^2 = 1. \tag{53}$$

The atomic population invertion is in this case

$$\langle\sigma_3\rangle = \sum_{k=0}^{\infty} |\alpha_k|^2 \langle\sigma_3\rangle_k. \tag{54}$$

For instance, if (53) is a Glauber state then

$$|\alpha_k|^2 = \frac{e^{-\overline{n}}\, \overline{n}^k}{k!}, \qquad \overline{n} = |\alpha|^2, \quad \alpha \in \mathbb{C}, \tag{55}$$

with $\overline{n}$ standing for the mean photon number. At resonance the atomic population invertion (54) yields

$$\langle\sigma_3\rangle = e^{-\overline{n}} \sum_{k=0}^{\infty} \frac{e^{-\overline{n}}\, \overline{n}^k}{k!} \cos(2gt). \tag{56}$$

The Figure 3 shows the behavior of the atomic population invertion in the cases (52) and (56) for different values of the involved parameters. There, we can appreciate that the time evolution of the atom energy states depends on the statistical distribution of the photons that integrate the radiation bath. The collapses and revivals phenomenon is a direct consequence of the classical properties associated to the Glauber quantum states.

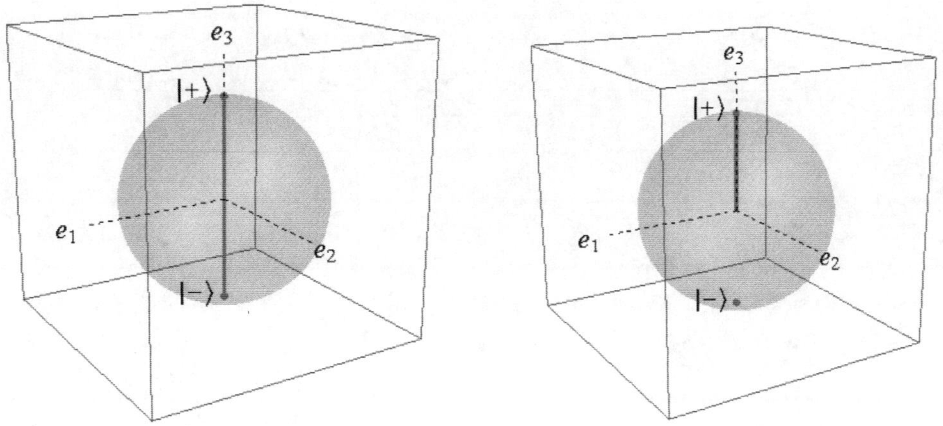

Figure 4. If the atom is initially in the energy state $|+\rangle$ and the field in the Fock state $|n\rangle$ then the entire system (atom+field) is in the pure state (51) and describes the time-evolution presented in Figure 3. If one focuses on the atomic part only, then the reduced state ρ_c in (57) describes the segment lines illustrated at the left for $\Delta = 0$ and at the right for $\Delta = 2g\sqrt{n+1}$ and $n = 2$. Notice that in this last case the total chaos $(0,0,0)$ is a part of the involved segment line.

4. Mixed states

The pure state (51) can be represented also by the projector $\rho(t) = |\psi(t)\rangle_R\langle\psi(t)|$, which is the density operator associated to the entire system and fulfills $\text{Tr}\rho^2(t) = 1$. We are interested in the description of the atom alone, so that we use the partial trace over the field states to get

$$\rho_a(t) = \text{Tr}_f\rho(t) = |c_+(n+1,t)|^2 X^{1,1} + |c_-(n,t)|^2 X^{2,2}. \tag{57}$$

This last operator is such that $\text{Tr}\rho_a^2(t) \leq 1$. Therefore (57) is the mixed state associated to the ensemble of pure states $X^{1,1} \leftrightarrow |+\rangle$ and $X^{2,2} \leftrightarrow |-\rangle$, with probabilities $|c_+(n+1,t)|^2$ and $|c_-(n,t)|^2$ respectively. Moreover, $\rho_a(t)$ defines a convex set which is just the chord connecting the north and the south poles of the three-dimensional unit ball (see Figure 4). This assertion is verified by calculating the Bloch vector $\vec{\tau}$, one finds that only the third component is different from zero,

$$\tau_3 = \left(\frac{g\sqrt{n+1}}{\Omega_{n+1}}\right)^2 \cos(2\Omega_{n+1}t) + \left(\frac{\Delta}{2\Omega_{n+1}}\right)^2. \tag{58}$$

The parametrization(58) describes a segment of line on the z-axis which is contained in the chord $\overline{NS}$ connecting any of the circles passing over the north and the south poles of $B[0,1]$. At resonance the segment of line is just $\overline{NS}$. The point of "total chaos" $(0,0,0)$ is reached whenever

$$\cos(2\Omega_{n+1}t) = -\frac{\Delta^2}{4g^2(n+1)}, \tag{59}$$

is fulfilled, so that $\Delta \leq 2g\sqrt{n+1}$ (in Figure 4 we show the case when the identity of this last expression is fulfilled with $n = 2$).

For arbitrary states of the field energy (53) the atom state is off-diagonal

$$\rho_a = \sum_{ij=1}^{2} \rho_{ij} X^{i,j}, \tag{60}$$

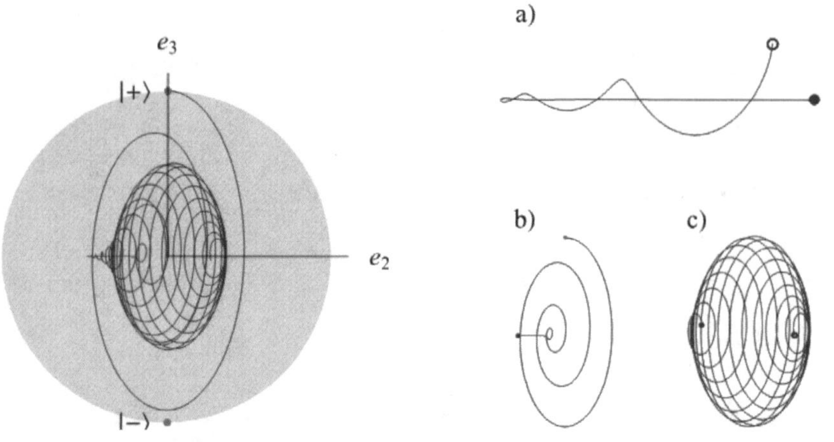

Figure 5. If the atom is in the initial state $|+\rangle$ and the field is in a Glauber state (55) the atomic reduced state ρ_c describes trajectories in the interior of the three-dimensional ball (left). The plot is at resonance with $\overline{n} = 15$. The time-evolution depicted here is in the interval $[0, 9.9\pi]$, in Ω units. At the right we show the partial intervals (b) $[0, 4\pi]$ (a) $[2.6\pi, 5\pi]$ and (c) $[6\pi, 9.9\pi]$, in all cases a red disk means pure state, a blue circle is for initial state and a blue disk denotes final state. All the path is described on the disk defined by the intersection of the yz-plane and the ball $B[0,1]$.

with coefficients (the time dependence is implicit)

$$\rho_{11} = \sum_{k=0}^{+\infty} |\alpha_k \, c_+(k+1)|^2, \quad \rho_{12} = \rho_{21}^* = \sum_{k=0}^{+\infty} \alpha_{k+1}\alpha_k^* \, c_+(k+2)\, c_-^*(k), \quad \rho_{22} = \sum_{k=0}^{+\infty} |\alpha_k \, c_-(k)|^2. \quad (61)$$

Using (16) we can evaluate the trajectories described by the Bloch vector in the three-dimensional unit ball. The Figure 5 shows the situation obtained when the atom is initially in the pure state $|+\rangle$ and the field is in a (pure) Glauber state defined by $\overline{n} = 15$. There, we can see that the atom state departures from the north pole of the ball (pure state $|+\rangle$) at $t = 0$ and goes inside that convex body for arbitrary values of $t \neq 0$. In the picture we have assumed that the atomic transitions and the field are in resonance so that all the path is depicted in the disk defined by the intersection of the yz-plane and the ball $B[0,1]$. Other values of the detuning $\Delta \neq 0$ give rise to paths that are more elaborated in the interior of the ball. Similar results can be found in [28].

5. Conclusions

We have studied the interaction of a single qubit with a single mode radiation field both in the semiclassical and quantum approaches. The geometry of the related state space has been discussed and the time-evolution of the energy states of the atom are found to describe trajectories in the three-dimensional ball of unit radius. In the semiclassical treatment only the pure states $|\pm\rangle$ of the atomic energy are involved so the geometric representation is on the points of the 2-sphere. At resonance with the radiation bath the energy state of the atom describes a geodesic on the yz-plane that passes over the north and the south poles. For other values of the detuning Δ, the paths followed by the energy state are circles over the sphere that has the north pole as common point. In the quantum case we have shown that the entire system (atom+field) is in a pure state that evolves in time depending on the statistical photon distribution of the radiation bath. If the field is in a Fock state $|n\rangle$ and the atom in an energy state $|+\rangle$ then the state of the entire system oscillates according to the Rabi frequency and depicts closed paths on the 2-sphere. The field in Glauber states leads to the collapses and revivals of the atomic population invertion that are well known in the literature. The reduced state of the atom, on the other hand, is just a mixed state and describes trajectories in the interior of the ball. In summary, we have analyzed here the geometry of the space state associated to a single qubit. The analysis of the case when the atom is in a mixed state as initial condition is going to be published elsewhere.

References

[1] Mielnik B., Geometry of quantum states, *Comm. Math. Phys.* **9** (1968) 55-80
[2] Mielnik B, Theory of filters, *Comm. Math. Phys.* **15** (1969) 1
[3] Mielnik B., Generalized quantum mechanics, *Comm. Math. Phys.* **37** (1974) 221
[4] Brody D.C. and Hughston L.P., Geometric quantum mechanics, *J. Geom. Phys.* **38** (2001) 19
[5] Mielnik B. and Rosas-Ortiz O., *Quantum Mechanical Laws*, in Fundamentals of Physics [Morán-López J.L. Ed.], Encyclopedia of Life Support Systems (EOLSS), Developed under the Auspices of the UNESCO, Eolss Publishers, Oxford, United Kingdom (2007/Rev. 2009)
[6] Mielnik B., Convex Geometry: A Travel to the Limits of Our Knowledge, in *Geometric Methods in Physics*, Springer Basel, 2013, p. 253
[7] Mielnik B., Evolution loops, *J. Math. Phys.* **27** (1986) 2290
[8] Fernández D. and Mielnik B., Controlling quantum motion, *J. Math. Phys.* **35** (1994) 2083
[9] Fernández D. and Rosas-Ortiz O., Inverse techniques and evolution of spin-1/2 systems, *Phys. Lett. A.* **236** (1997) 275
[10] Emmanouilidou A. and Zhao X.G., Steering an eigenstate to a destination, *Phys. Rev. Lett.* **85** (2000) 1626
[11] Fernández D. and Rosas-Ortiz O., Evolution loops and spin-1/2 systems, in *Coherent States, Quantization and Gravity*, Schlichenmaier M *et al* (Eds), Warsaw University Press 2001, p. 81
[12] Rosas-Ortiz O., Quantum control of two-level systems, *IOP Conf. Proc.* **185** (2004) 485
[13] Rezakhani A.T. and Zanardi P., General setting for a geometric phase of mixed states under an arbitrary non unitary evolution, *Phys. Rev. A* **73** (2006) 012107
[14] Enríquez-Flores M. and Rosas-Ortiz O., Atomic Schrödinger cat-like states, *AIP Conf. Proc.* **1287** (2010) 74
[15] Mielnik B., Quantum operations: technical or fundamental challenge?, e-print arXiv:1305.3664
[16] Bengtsson I. and Życzkowski K., *Geometry of Quantum States. An Introduction to Quantum Entanglement*, Cambridge University Press, 2006
[17] von Neumann J, *Mathematical foundations of quantum mechanics*, Princeton University Press, 1955
[18] Enríquez M. and Rosas-Ortiz O., The Kronecker product in terms of Hubbard operators and the Clebsch-Gordan decomposition of $SU(2) \times SU(2)$, *Ann. Phys.* **339** (2013) 218
[19] Thaller B., *Advanced Visual Quantum Mechanics*, Springer-Verlag, 2005
[20] Schlosshauer M., *Decoherence and the Quantum-to-Classical Transition*, Springer-Verlag, 2007
[21] Hopf H., Über die Abbildungen der dreidimensionalen Sphäre auf die Kugelfiche, *Math. Ann.* **104** (1931) 637
[22] Nakahara M., *Topology, Geometry and Physics*, IOP Publishing, 2003
[23] Enríquez M. and Rosas-Ortiz O., Spectral design to control the time-evolution of a two-level atom in single mode radiation fields, preprint Cinvestav (2013)
[24] Enríquez M. and Rosas-Ortiz O., Dynamics of two atoms in a double Jaynes-Cummings model, preprint Cinvestav (2013)
[25] Gerry C. and Knight P. L, *Introductory Quantum Optics*, Cambridge University Press, 2005
[26] Haroche S. and Raimond J. M., *Exploring the Quantum: Atoms, Cavities and Photons*, Oxford Graduate Texts, 2006
[27] Jaynes T. and Cummings F.W., Comparison of quantum and semiclassical radiation theories with application to the beam maser, *Proc. IEEE* **51** (1963) 89
[28] Gea-Banacloche J., A new look at the Jaynes-Cummings model for large fields: Bloch sphere evolution and detuning effects. *Optics communications* **88** (1992) 531

Time-evolution of entangled bipartite atomic systems in quantized radiation fields

Marco Enríquez, Claudia Quintana and Oscar Rosas-Ortiz
Physics Department, Cinvestav, A.P. 14-740, 07000 México D.F., Mexico
E-mail: lquintana@fis.cinvestav.mx

Abstract. We analyze the time-evolution of a pair of atoms that are initially entangled and are each of them in an independent isolated QED cavity. The cavities are assumed to contain a radiation bath of a given number of photons. The *sudden death* and recovering of entanglement are studied in terms of the concurrence introduced by Hill and Wootters. This last gives the value 1 for maximal entanglement and 0 for unentangled systems. We find that the entanglement of the bipartite atomic systems depends on the number of photons in the cavities when the radiation bath is assumed to be in a Fock state.

1. Introduction

Entanglement is a concept appearing in physics after the celebrated discussions deliberated by some of the quantum theory founder fathers (see e.g. [1–7] for a historical account on the matter). The term was introduced in 1935 by Schrödinger [8–10] to describe what occurs with our knowledge of two systems that are separated after they were interacting for a while, and from which we had maximal knowledge before the interaction [8]. According to Schrödinger, after the interaction, the systems can no longer be described in the same way as before. He accepted that, as a result of the interaction, instead of two isolated systems there is just a single composite system and therefore any change to one subsystem would affect the other, no mater the distance between them. In his own words "I would not call that *one* but rather *the* characteristic trait of quantum mechanics, the one that enforces its entire departure from classical lines of thought" [9]. Considering the *psi* function as an expectation catalog, it results that one information is *sharp* while the other must be *blurred* (uncertainty principle) [11]. Wondering whether the blurring can affect also the macroscopic systems, Schrödinger illustrated a "ridiculous" situation in which a cat entangled to an unstable atom enters into a superposed state of "being alive" and "being dead" [8]. This humorous tale is nowadays known as the *Schrödinger's cat* problem and has been the subject of polemic discussions over the years [5] (see also [7, 11–13]). However, the results obtained from interaction-free measurements [14, 15] are an indication that entanglement is indeed a fundamental property of the quantum systems. The quantum electrodynamics (QED) cavity experiments [16, 17], on the other hand, illustrate the actuality of the superposed atomic energy states and the field quantization [16], as well as the reality of decisive *though experiments* just as the observation of a single photon in a cavity [17]. The latter results are a consequence of the entanglement between the energy states of the atom and the radiation bath in the cavity.

In general, the characterization of the differences between pure, mixed and maximally entangled states is a timely issue in contemporary physics; consider for example that the entangled states represent a fundamental resource in quantum information, teleportation and computing [18–21]. Interestingly, the entangled mixed states could make a big difference in the usefulness of quantum correlations [22, 23]. The pairs of correlated photons produced from parametric down conversion, for instance, can be efficiently concentrated by simultaneously increasing both the purity and the degree of entanglement [22]. Besides, the mixed states can be more entangled in terms of the relative entropy of entanglement than the pure states if the appropriate measurements are done [23]. A fundamental question is then to what extent is a given quantum state entangled? The answer seems to be not unique [21, 24–26], though the *concurrence* of two qubits introduced by Hill and Wootters represents a good option [27, 28] (some generalizations can be found in e.g. [29–32]). Concurrence is a measure of the degree of entanglement that takes the value 1 for maximal entanglement and 0 for completely unentangled systems. This has been successfully applied in combination of the Jaynes-Cummings (JC) model [33] to the study of the entanglement of a pair of two-level atoms in QED cavities [34–39].

The interaction between a resonant cavity mode and a two-level atom can be analyzed in detail by using the JC model. Introduced in 1963, this model has been extended to be applicable in a diversity of situations (see e.g. [34–39] and references quoted therein). For instance, a "double JC" model consisting of a pair of two-level atoms, each one in its own perfect one-mode near-resonant cavity and isolated from the other atom and cavity, has been used to investigate the involved entanglement dynamics in [34]. By tracing over the cavity modes one arrives at a mixed state of the atoms, the entanglement of which can be conveniently measured by the Hill and Wooters concurrence. The results included in [34] assume that initially both cavities are prepared in the vacuum state and the atoms are prepared in a pure entangled state. Therefore, there is never more than one photon in the cavities. Using this model the so called *sudden death effect* is found as a result of the time-evolution of the system [34] (see also [35]). Although these results were immediately generalized to the case of six pairwise concurrences [36], the constraint of having no more than one photon in the cavities was preserved; similar results can be found in [37]. The case of an imperfect matching of the atoms to the cavity fields has been also analyzed in connection to the entanglement evolution [38]. A different situation has been reported in [39] where the case of a pair of two-level atoms are embedded in a common bath. There, it is found that using a 'hierarchy equation method' the rotating-wave approximation is not necessary to analyze the entanglement of the atoms. Indeed, the case of two photons in the cavities can be studied by including the counter-rotating-wave terms. It is also remarkable that the concurrence oscillates in time even though the initial state of the atoms is separable and the fields in cavities are in the vacuum state [39]. In almost all the previous works the vacuum is included as the initial condition of either one or both of the fields in the cavities as a manner of simplification in the calculations. The more elaborated initial condition includes no more than one photon in each of the cavities. A natural question arises: is the initial number of photons in each of the cavities playing a relevant role in the time-evolution of the atoms' entanglement?

In this contribution we are interested in a double JC model consisting of a pair of atoms, each one in its own cavity containing a given initial number of photons and isolated from the other atom and cavity. The initial number of photons in the cavities can be either the same or substantially different and will be used as a tuning parameter. Our interest is focused on the time-evolution of the entanglement between the two atoms as a function of such a parameter. With this aim we are going to use the Hill and Wooters concurrence. As we shall see, the results show that the initial number of photons is fundamental in defining the concurrence. The paper is organized as follows. In Section 2 we first make a minimal account of the mathematics associated to the concepts of entanglement and concurrence. As we are interested in the interaction between atoms and photons, in Section 3 we use the Jaynes-Cummings model to determine the Hamiltonian of the atom+field system and solve the related Schrödinger equation (Section 3.1). These results are used in Section 3.2 to get the solution of the bipartite atomic

Journal of Physics: Conference Series **512** (2014) 012022 doi:10.1088/1742-6596/512/1/012022

problem and to calculate the concurrence of the two entangled atoms when they are located in different isolated cavities. Some conclusions are given at the very end of this work.

2. Entanglement and Concurrence

Consider a bipartite system $\mathcal{S} = \mathcal{S}_1 + \mathcal{S}_2$ which is a composition of a pair of two-level atoms $\mathcal{S}_1$ and $\mathcal{S}_2$. Let us label the excited and ground pure states of each of these atoms as $|+\rangle_i$ and $|-\rangle_i$, $i = 1, 2$, respectively. The related two-dimensional Hilbert spaces are then written as

$$\mathcal{H}_1 = \text{Span}\{|+\rangle_1, |-\rangle_1\}, \quad \mathcal{H}_2 = \text{Span}\{|+\rangle_2, |-\rangle_2\}. \tag{1}$$

The Hilbert space of the entire system atom+atom is then 4-dimensional and this is integrated by all the possible linear combinations of the Kronecker product between the basis elements of $\mathcal{H}_1$ and those of $\mathcal{H}_2$, that is

$$\mathcal{H} = \text{Span}\{|+\rangle_1 \otimes |+\rangle_2, |+\rangle_1 \otimes |-\rangle_2, |-\rangle_1 \otimes |+\rangle_2, |-\rangle_1 \otimes |-\rangle_2\}. \tag{2}$$

The general properties of the Kronecker product between vectors and/or operators can be consulted in [40] and references quoted therein. From now on we adhere to the simplified notation

$$\begin{aligned} |+\rangle_1 \otimes |+\rangle_2 \equiv |+,+\rangle = |\varphi_1\rangle, \quad |+\rangle_1 \otimes |-\rangle_2 \equiv |+,-\rangle = |\varphi_2\rangle, \\ |-\rangle_1 \otimes |+\rangle_2 \equiv |-,+\rangle = |\varphi_3\rangle, \quad |-\rangle_1 \otimes |-\rangle_2 \equiv |-,-\rangle = |\varphi_4\rangle. \end{aligned} \tag{3}$$

Any pure state of the atom+atom system is represented as a linear combination of the basis (3). However, it is well known that not all these combinations can be factorized as the $\otimes$-product of a vector in $\mathcal{H}_1$ with a vector in $\mathcal{H}_2$. For instance, the Bell state

$$|\beta\rangle = \frac{1}{\sqrt{2}}\left(|+,-\rangle + |-,+\rangle\right) \equiv \frac{1}{\sqrt{2}}\left(|\varphi_2\rangle + |\varphi_3\rangle\right), \tag{4}$$

is such that there are not $|\phi^{(1)}\rangle \in \mathcal{H}_1$ and $|\phi^{(2)}\rangle \in \mathcal{H}_2$ fulfilling

$$|\beta\rangle = |\phi^{(1)}\rangle \otimes |\phi^{(2)}\rangle. \tag{5}$$

The statement is easily verified by taking $\mathcal{H}_1 \ni |\phi^{(1)}\rangle = \alpha_1|+\rangle + \alpha_2|-\rangle$ and $\mathcal{H}_2 \ni |\phi^{(2)}\rangle = \gamma_1|+\rangle + \gamma_2|-\rangle$, with the α's and γ's to be determined such that (5) holds. Then, one arrives at the system

$$\alpha_1\gamma_1 = \alpha_2\gamma_2 = 0, \quad \alpha_1\gamma_2 = \alpha_2\gamma_1 = \frac{1}{\sqrt{2}}, \tag{6}$$

that admits not solution. In contraposition to the basis vectors (3) which are defined as the Kronecker products of the bases of the subsystems, the vectors that cannot be factorized just like (4) encode information about the correlation of the subsystems. Namely, $|\beta\rangle$ represents a state of the atom+atom system that has $\frac{1}{2}$ probability of being found in the configuration $|+,-\rangle$ and $\frac{1}{2}$ probability of being found in $|-,+\rangle$. If the measurement device carries out the first result then we know with certainty that the atom $\mathcal{S}_1$ is in the excited state while the atom $\mathcal{S}_2$ is in the ground state. The second result leads to the opposite distribution of atomic states. The main difference between the pure state $|+,-\rangle$ (equivalently $|-,+\rangle$) and the Bell vector (4) is that in the first case we know all about the system: $\mathcal{S}_1$ is in $|+\rangle$ and $\mathcal{S}_2$ in $|-\rangle$ (just because either we have 'prepared them in' or we have 'measured them in' those states). The Bell state $|\beta\rangle$, on the other hand, represents the situation in which we know only that the subsystems are in orthogonal states with the configurations and probabilities indicated above, but we are not able to say any more before the measurement. In both cases we deal with pure states which are represented by vectors in $\mathcal{H}$; the correlation between the subsystems is encoded in the orthogonality of their states. The usual mathematical definition of *entangled pure states* says that these are the states of a multipartite system that cannot be factorized as the Kronecker product of the subsystems' pure states. All other pure states are named *separable.*

The bipartite atomic system $\mathcal{S} = \mathcal{S}_1 + \mathcal{S}_2$, however, may not be in a pure state. In general, it would be in a statistical ensemble of pure states $\{p_k, \varphi_k\}_{k=1}^4$, where each pure state φ_k occurs with probability $0 \leq p_k \leq 1$ and can be represented either by $|\varphi_k\rangle \in \mathcal{H}$ or by the orthogonal projector $\rho_k = |\varphi_k\rangle\langle\varphi_k|$. In this case, the *mixed state* $\{p_k, \varphi_k\}_{k=1}^4$ can be expressed as the convex combination

$$\rho = \sum_{k=1}^{4} p_k \rho_k, \quad \sum_{k=1}^{4} p_k = 1, \quad 0 \leq p_k \leq 1. \tag{7}$$

It is easy to verify that ρ_k as well as ρ fulfill the conditions to be density operators, that is they are (i) Hermitian (ii) positive definite and (iii) normalized operators (i.e., $\mathrm{Tr}A = 1$). If $p_k = 1$ for a given $k \in \{1, 2, 3, 4\}$ then (7) is reduced to the orthogonal projector ρ_k that fulfills $\mathrm{Tr}\rho_k^2 = 1$, this last verifies that $|\varphi_k\rangle$ as well as ρ_k represent the same pure state φ_k. In the case of a nontrivial combination (i.e., $0 < p_k < 1$ for at least two different values of k) one shows easily that $\mathrm{Tr}\rho^2 < 1$. Thus, nontrivial convex combinations (7) represent mixed states of the bipartite atomic system. The intersection of all the possible convex combinations (7) that can be formed with the pure states ρ_k as *extreme points* is the *convex hull* of the set $\{\rho_k\}_{k=1}^4$ and represents the state space of the bipartite atomic system we are dealing with. As a convex body, this state space can be visualized as a tetrahedron in which different geometric regions can be identified with different physical properties [21] (see also [41, 42]). Notice that the density operator ρ so constructed is such that the probabilities p_k and the pure states $|\varphi_k\rangle$ are the solutions of the eigenvalue equation $\rho|\varphi_k\rangle = p_k|\varphi_k\rangle$. That is, (7) is the spectral decomposition of ρ. Then one says that the mixed state ρ is *separable.*

In a more general situation the density operator associated to the bipartite atomic state is not necessarily in its canonical (diagonal) form but it reads [40]:

$$\rho = \sum_{i,j=1}^{4} \lambda_{i,j} X_4^{i,j}, \quad \lambda_{i,j} \in \mathbb{C}, \tag{8}$$

where the operators $X_4^{i,j}$ produce the transition from the pure state $|\varphi_j\rangle$ to the pure state $|\varphi_i\rangle$. The simplest representation of $X_4^{i,j}$ is a square matrix of order 4 that has entry 1 in position (i, j) and zero in all other entries. These operators are named after Hubbard; their properties and applications can be consulted in [40] and references quoted therein. Remark that only the diagonal Hubbard operators are connected to pure states: $X_4^{k,k} \equiv \rho_k$. The λ-coefficients in (8) are in general different from zero and are defined in correspondence to the external fields that are applied on the bipartite atomic system [41]. When (8) cannot be reduced to the spectral decomposition (7) it is said that the state ρ is *entangled.*

In what follows we assume we can prepare entangled bipartite atomic states by producing an interaction (somehow) between them. After the interaction, the entire system is in the state

$$|\psi_\alpha\rangle = \cos\alpha|+,-\rangle + \sin\alpha|-,+\rangle, \quad \alpha \in [0, \pi/2]. \tag{9}$$

Here the parameter α is a primary manifestation of entanglement since $\alpha = 0$ produces the factorized state $|+,-\rangle$ while $\alpha = \pi/2$ leads to $|-,+\rangle$. The equally weighted entangled state (4) is obtained for $\alpha = \pi/4$. Then $|\psi_{\alpha=0}\rangle = |+,-\rangle$ and $|\psi_{\alpha=\pi/2}\rangle = |-,+\rangle$ are not entangled

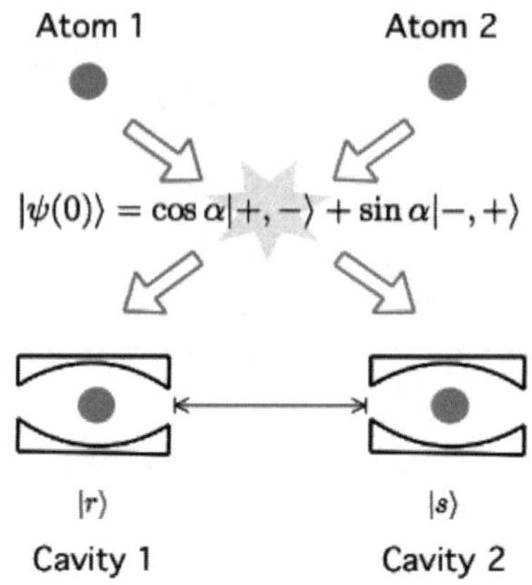

Figure 1. Schematic representation of the preparation of entangled bipartite atomic states. In a first step a pair of two-level atoms collide in such a manner that their energy states become correlated. The state of the atom+atom system is given by Equation (9). Then each of the atoms is placed in isolated cavities where a radiation bath is present. We assume that the radiation in the cavities is quantized and described by the Fock states $|r\rangle$ and $|s\rangle$ respectively.

states while $|\psi_{\alpha=\pi/4}\rangle = |\beta\rangle$ is a maximally entangled state of the bipartite atomic system. Other intermediary states can be prepared. For example, the linear combinations

$$\begin{aligned}|\psi_{\alpha=1\pi/20}\rangle &= \sin\left(\tfrac{9}{20}\pi\right)|+,-\rangle + \cos\left(\tfrac{9}{20}\pi\right)|-,+\rangle \equiv |\phi_+\rangle,\\ |\psi_{\alpha=9\pi/20}\rangle &= \cos\left(\tfrac{9}{20}\pi\right)|+,-\rangle + \sin\left(\tfrac{9}{20}\pi\right)|-,+\rangle \equiv |\phi_-\rangle,\end{aligned} \tag{10}$$

correspond to situations in which the system is prepared to be found in a specific configuration after the measuring of the state in the majority of the times. Namely, $|\phi_+\rangle$ represents a system having a probability of $\sin^2\left(\frac{9}{20}\pi\right) \approx 0.976$ to be found in the state $|+,-\rangle$, and a probability of $\cos^2\left(\frac{9}{20}\pi\right) \approx 0.024$ to be found in the state $|-,+\rangle$. If the system is in the linear combination $|\phi_-\rangle$ then the probabilities are interchanged. Once we have the atoms in the state (9) we place each of them in isolated cavities where a radiation bath is present. The radiation in cavities is quantized and prepared in the Fock states $|r\rangle$ and $|s\rangle$, so that the atoms interact with r and s photons respectively. The situation is depicted in Figure 1. Our problem is to determine the time-evolution of the atoms in the cavities and to evaluate the degree of entanglement as a function of the time.

Following [27, 28], given a two-qubit density operator ρ, the concurrence $C(\rho)$ is calculated from the expression

$$C(\rho) = \max\left\{0, \sqrt{\lambda_1} - \sqrt{\lambda_2} - \sqrt{\lambda_3} - \sqrt{\lambda_4}\right\}, \tag{11}$$

where the λ's are sorted in decreasing order and correspond to the eigenvalues of the matrix $\rho(\sigma_2 \otimes \sigma_2)\overline{\rho}(\sigma_2 \otimes \sigma_2)$, with $\overline{\rho}$ the complex conjugation of the matrix elements of ρ. The spin flip operator $\sigma_2 \otimes \sigma_2$ preserves entangled states while cancels unentangled states[1]. As a manner of example, the concurrence associated to the pure state

$$|\varphi\rangle = \sum_{k=1}^{4} a_k|\varphi_k\rangle, \quad a_k \in \mathbb{C}, \tag{12}$$

[1] For instance, the action of the spin flip on the factorized vector $|+,-\rangle$ gives $\langle +,-|\,(\sigma_2 \otimes \sigma_2|+,-\rangle) = 0$.

reads as follows [24]:

$$C\left(|\varphi\rangle\langle\varphi|\right) = 2|a_1a_4 - a_2a_3|. \tag{13}$$

Setting $a_1 = a_4 = 0$, and $a_2 = a_3 = 1/\sqrt{2}$, the Bell state (4) is recovered and the concurrence yields the maximum value $C\left(|\varphi\rangle\langle\varphi|\right) = 1$. If the state (12) is separable, the coefficients satisfy $a_1a_4 - a_2a_3 = 0$ and the concurrence is in this case equal to zero. To get the state (9) we make $a_2 = \cos\alpha$, $a_3 = \sin\alpha$, and $a_1 = a_4 = 0$. Then the concurrence is $C\left(|\psi_\alpha\rangle\langle\psi_\alpha|\right) = 2|\sin 2\alpha|$, which is maximum for the equally weighted entangled state $|\psi_{\alpha=\pi/4}\rangle = |\beta\rangle$.

3. The Jaynes-Cummings Model

In this section we consider the interaction of a two-level atom with a single mode of quantized radiation through the dipolar interaction term. The approach is a milestone introduced in quantum optics by Jaynes and Cummings [33]. We start with the atomic Hamiltonian given by

$$\widetilde{H}_a = \frac{\hbar\omega_a}{2}\sigma_3, \tag{14}$$

where ω_a is the transition frequency and σ_3 is the Pauli matrix

$$\sigma_3 = \begin{pmatrix} 1 & 0 \\ 0 & -1 \end{pmatrix}. \tag{15}$$

This last together with the spin-1/2 ladder operators

$$\sigma_+ = \begin{pmatrix} 0 & 1 \\ 0 & 0 \end{pmatrix}, \quad \sigma_- = \begin{pmatrix} 0 & 0 \\ 1 & 0 \end{pmatrix}, \tag{16}$$

close the $su(2)$ algebra. The operators (15) and (16) are in their simplest representation for which

$$|+\rangle = \begin{pmatrix} 1 \\ 0 \end{pmatrix}, \qquad |-\rangle = \begin{pmatrix} 0 \\ 1 \end{pmatrix}. \tag{17}$$

are orthonormal eigenvectors of σ_3 and span the state space of the atom $\mathcal{H}_a = \text{Span}\{|+\rangle, |-\rangle\}$. It is straightforward to verify that the following properties hold

$$\sigma_+^2 = \sigma_-^2 = 0, \quad \sigma_+^\dagger = \sigma_-, \quad \sigma_\pm|\pm\rangle = 0, \quad \sigma_\pm|\mp\rangle = |\pm\rangle. \tag{18}$$

On the other hand, a single mode radiation field can be written in terms of the boson ladder operators

$$\vec{E}(t) = \mathcal{E}\left(ae^{-i\omega_f t} + a^\dagger e^{i\omega_f t}\right)\mathbf{e}_f, \tag{19}$$

where ω_f and $\mathbf{e}_f$ are respectively the frequency and polarization of the field. The observable of the energy is the oscillator's Hamiltonian

$$\widetilde{H}_f = \hbar\omega_f(N + 1/2). \tag{20}$$

Hereafter we use the Fock's representation for the operators associated to the field if there is no confusion. That is, the number 1/2 in (20) should read $(1/2)\mathbb{I}_f$, with $\mathbb{I}_f$ the identity operator defined on the Hilbert space $\mathcal{H}_f = \text{Span}\{|n\rangle\}_{n\geq 0}$, where the nth Fock state $|n\rangle$ is given by

$$|n\rangle = \frac{1}{\sqrt{n!}}(a^\dagger)^n|0\rangle, \quad n \in \mathbb{Z}^+. \tag{21}$$

The ladder operators a, $a^\dagger$, and the number operator N satisfy the boson algebra

$$[a, a^\dagger] = 1, \quad [N, a] = -a, \quad [N, a^\dagger] = a^\dagger, \tag{22}$$

and are defined in the Fock's representation. That is

$$a|n\rangle = \sqrt{n}|n-1\rangle, \quad a^\dagger|n\rangle = \sqrt{n+1}|n+1\rangle, \quad N|n\rangle = n|n\rangle. \tag{23}$$

The energy of the entire system for no interaction between the atom and the field reads:

$$\widetilde{H}_0 = \hbar\omega_f \mathbb{I}_a \otimes \left(N + \frac{1}{2}\right) + \frac{\hbar\omega_a}{2}\sigma_3 \otimes \mathbb{I}_f = \left(\begin{array}{cc} \hbar\omega_f(N+1) + \frac{\hbar\Delta}{2} & 0 \\ 0 & \hbar\omega_f N - \frac{\hbar\Delta}{2} \end{array}\right), \tag{24}$$

with $\mathbb{I}_a$ the identity operator in the state space of the atom, $\Delta := \omega_a - \omega_f$ the *detuning* between the radiation field and the atomic transition. For each $n \geq 1$ the Hamiltonian (24) has two eigenvectors

$$|+, n\rangle := |+\rangle \otimes |n\rangle = \left(\begin{array}{c} |n\rangle \\ |\emptyset\rangle \end{array}\right), \qquad |-, n+1\rangle := |-\rangle \otimes |n+1\rangle = \left(\begin{array}{c} |\emptyset\rangle \\ |n+1\rangle \end{array}\right), \tag{25}$$

where $|\emptyset\rangle$ is the null vector in $\mathcal{H}_f$. The corresponding eigenvalues are given by

$$\begin{aligned} \widetilde{H}_0|+, n\rangle &= \hbar\left[\omega_f(n+1) + \tfrac{\Delta}{2}\right]|+, n\rangle, \\ \widetilde{H}_0|-, n+1\rangle &= \hbar\left[\omega_f(n+1) - \tfrac{\Delta}{2}\right]|-, n+1\rangle. \end{aligned} \tag{26}$$

The vector $|-, 0\rangle := |-\rangle \otimes |0\rangle$ is the eigenstate belonging to the eigenvalue 0. Therefore, the Hilbert space $\mathcal{H}$ of the atom+field system splits into a direct sum of subspaces $\Xi_n = \text{Span}\{|+, n\rangle, |-, n+1\rangle\}$, $n = 0, 1, 2\ldots$ In addition, the one-dimensional space $\Xi_\star = \text{Span}\ \{|-, 0\rangle\}$ is required to make the Hilbert space $\mathcal{H}$ complete [41, 42]. Thus, we have

$$\mathcal{H} = \Xi_\star \oplus \Xi_0 \oplus \Xi_1 \oplus \Xi_2 \oplus \ldots = \bigoplus_{n=\star,0}^{\infty} \Xi_n. \tag{27}$$

Two vectors belonging to different subspaces Ξ_m and Ξ_n $(m, n = 0, 1, 2, \ldots)$ are orthogonal

$$\langle +, m|+, n\rangle = \langle -, m+1|-, n+1\rangle = \delta_{m,n}, \qquad \langle +, m|-, n+1\rangle = \langle -, m+1|+, n\rangle = 0. \tag{28}$$

Remark as well that $|-, 0\rangle$ is orthogonal to any Ξ_n, $n = 0, 1, 2, \ldots$.

If now the interaction is allowed, assume that this is ruled by the dipolar term $\widetilde{H}_I = \vec{p} \cdot \vec{E}$ which, in the rotating wave approximation is given by

$$\widetilde{H}_I = \hbar\lambda(\sigma_+ \otimes a + \sigma_- \otimes a^\dagger), \tag{29}$$

where λ is the *coupling constant*. At resonance ($\Delta = 0$) the Hamiltonian of the whole system $\widetilde{H} = \widetilde{H}_0 + \widetilde{H}_I$ can be rewritten in dimensionless form $H = \widetilde{H}/(\hbar\omega)$ to read

$$H = H_0 + H_I, \quad H_0 = \mathbb{I}_a \otimes \left(N + \frac{1}{2}\right) + \frac{1}{2}\sigma_3 \otimes \mathbb{I}_f, \quad H_I = \gamma(\sigma_+ a + \sigma_- a^\dagger), \quad \gamma = \lambda/\omega. \tag{30}$$

Notice that $[H_0, H_I] = 0$, so the basis (25) is common to H_0 and H_I.

3.1. Dynamical evolution

The Schrödinger equation

$$i\frac{\partial}{\partial t}|\psi(t)\rangle = H|\psi(t)\rangle, \tag{31}$$

with H defined in (30) and t the dimensionless time-parameter, can be solved by time-translations of the initial vector state $|\psi(t)\rangle = U(t)|\psi(0)\rangle$. In this case the time evolution operator $U(t)$ is given by

$$U(t) = e^{iHt} = e^{i(H_0+H_I)t} = e^{iH_0t}e^{iH_It}. \tag{32}$$

We are considering the time-evolution of a given definite state of the energy, so that the operator e^{iH_0t} leads to a global phase in the solution, which will be not considered. Then

$$U(t) = e^{iH_It} = \begin{pmatrix} \cos(t\gamma\sqrt{aa^\dagger}) & -\frac{i}{\sqrt{aa^\dagger}}\sin(t\gamma\sqrt{aa^\dagger})a \\ -\frac{i}{\sqrt{a^\dagger a}}\sin(t\gamma\sqrt{a^\dagger a})a^\dagger & \cos(t\gamma\sqrt{a^\dagger a}) \end{pmatrix}. \tag{33}$$

Consider the initial situation in which the atom is in the excited state $|+\rangle$ and the cavity is filled with k photons. The time-evolved state of the composite system is

$$|\psi(t)\rangle = U(t)|+,k\rangle = \cos(\gamma t\sqrt{k+1})|+,k\rangle - i\sin(\gamma t\sqrt{k+1})|-,k+1\rangle. \tag{34}$$

The calculation of the corresponding atomic population invertion yields

$$\langle\sigma_3(t)\rangle = \cos(2\gamma t\sqrt{k+1}), \tag{35}$$

so that whole system oscillates in time between the states $|+,k\rangle$ and $|-,k+1\rangle$ according to the Rabi frequency $2\gamma\sqrt{k+1}$.

3.2. Two atoms dynamics and entanglement

The Hamiltonian of two non-interacting atoms that are placed in separate cavities is of the form

$$\begin{gathered} H = H_1 + H_2, \\ H_1 = H_{0;1} + H_{I;1}, \quad H_{0;1} = N_1 + \tfrac{1}{2} + \tfrac{1}{2}\sigma_3, \quad H_{I;1} = \gamma_1(\sigma_+ a + \sigma_- a^\dagger), \\ H_2 = H_{0;2} + H_{I;2}, \quad H_{0;2} = N_2 + \tfrac{1}{2} + \tfrac{1}{2}s_3, \quad H_{I;2} = \gamma_2(s_+ b + s_- b^\dagger), \end{gathered} \tag{36}$$

where N_i, $i = 1,2$, is the number operator in the ith cavity. The boson ladder operators $b, b^\dagger$, and the atomic $su(2)$ generators $\{s_3, s_\pm\}$ are associated to the second cavity. Using the Kronecker product properties [41, 42] one arrives at the Hilbert space $\mathcal{H}$ of the entire system

$$\mathcal{H} = \bigoplus_{n,m} \Xi_n^{(1)} \otimes \Xi_m^{(2)}. \tag{37}$$

Given n and m each subspace is of the form

$$\Xi_n^{(1)} \otimes \Xi_m^{(2)} = \text{Span}\{|+,n\rangle\otimes|+,m\rangle, |+,n\rangle\otimes|-,m+1\rangle, |-,n+1\rangle\otimes|+,m\rangle, |-,n+1\rangle\otimes|-,m+1\rangle\}.$$

We consider the α-entangled state (9) as the initial condition for the bipartite atomic system with r and s photons in cavities 1 and 2, respectively. Thus, at time $t = 0$ the state of the whole system is

$$|\psi(0)\rangle = \cos\alpha|+, r\rangle \otimes |-, s\rangle + \sin\alpha|-, r\rangle \otimes |+, s\rangle. \tag{38}$$

Since the two atoms do not interact, the time evolution operator is given as follows

$$U(t) = e^{iHt} = e^{iH_{I;1}t} \otimes e^{iH_{I;2}t}. \tag{39}$$

Hence, the state at time t reads

$$\begin{aligned}
|\psi(t)\rangle &= U(t)|\psi(0)\rangle \\
&= \cos\alpha \left[-i\cos(\gamma t\sqrt{r+1})\sin(\gamma t\sqrt{s})|+, r\rangle \otimes |+, s-1\rangle \right.\\
&\quad + \cos(\gamma t\sqrt{r+1})\cos(\gamma t\sqrt{s})|+, r\rangle \otimes |-, s\rangle \\
&\quad - \sin(\gamma t\sqrt{r+1})\sin(\gamma t\sqrt{s})|-, r+1\rangle \otimes |+, s-1\rangle \\
&\quad \left. -i\sin(\gamma t\sqrt{r+1})\cos(\gamma t\sqrt{s})|-, r+1\rangle \otimes |-, s\rangle\right] \qquad (40)\\
&+ \sin\alpha \left[-i\cos(\gamma t\sqrt{s+1})\sin(\gamma t\sqrt{r})|+, r-1\rangle \otimes |+, s\rangle \right.\\
&\quad - \sin(\gamma t\sqrt{r})\sin(\gamma t\sqrt{s+1})|+, r-1\rangle \otimes |-, s+1\rangle \\
&\quad + \cos(\gamma t\sqrt{r})\cos(\gamma t\sqrt{s+1})|-, r\rangle \otimes |+, s\rangle \\
&\quad \left. -i\cos(\gamma t\sqrt{r})\sin(\gamma t\sqrt{s+1})|-, r\rangle \otimes |-, s+1\rangle\right].
\end{aligned}$$

The density operator ρ of the whole system is the pure state $\rho = |\psi(t)\rangle\langle\psi(t)|$. In order to obtain information about the bipartite atomic system we calculate $\rho_{atoms} = Tr_{f_1,f_2}\rho$ to get

$$\rho_{atoms} = \begin{pmatrix} \Omega_{11} & 0 & 0 & 0 \\ 0 & \Omega_{22} & \Omega_{23} & 0 \\ 0 & \Omega_{32} & \Omega_{33} & 0 \\ 0 & 0 & 0 & \Omega_{44} \end{pmatrix}, \tag{41}$$

where

$$\begin{aligned}
\Omega_{11} &= \cos^2\alpha\cos^2(\gamma t\sqrt{r+1})\sin^2(\gamma t\sqrt{s}) + \sin^2\alpha\sin^2(\gamma t\sqrt{r})\cos^2(\gamma t\sqrt{s+1}) \\
\Omega_{22} &= \cos^2\alpha\cos^2(\gamma t\sqrt{r+1})\cos^2(\gamma t\sqrt{s}) + \sin^2\alpha\sin^2(\gamma t\sqrt{s+1})\sin^2(\gamma t\sqrt{r}) \\
\Omega_{33} &= \cos^2\alpha\sin^2(\gamma t\sqrt{r+1})\sin^2(\gamma t\sqrt{s}) + \sin^2\alpha\cos^2(\gamma t\sqrt{s+1})\cos^2(\gamma t\sqrt{r}) \\
\Omega_{44} &= \cos^2\alpha\sin^2(\gamma t\sqrt{r+1})\cos^2(\gamma t\sqrt{s}) + \sin^2\alpha\sin^2(\gamma t\sqrt{s+1})\cos^2(\gamma t\sqrt{r}) \\
\Omega_{23} &= \Omega_{32} = \cos\alpha\sin\alpha\cos(\gamma t\sqrt{r+1})\cos(\gamma t\sqrt{r})\cos(\gamma t\sqrt{s+1})\cos(\gamma t\sqrt{s}).
\end{aligned}$$

The eigenvalues of the matrix $\rho_{atoms}(\sigma_2 \otimes \sigma_2)\overline{\rho}_{atoms}(\sigma_2 \otimes \sigma_2)$ read as follows

$$\begin{aligned}
\lambda_1 &= \lambda_4 = \Omega_{11}\Omega_{44}, \\
\lambda_2 &= \left(\Omega_{23} + \sqrt{\Omega_{22}\Omega_{33}}\right)^2, \\
\lambda_3 &= \ \Omega_{23} - \sqrt{\Omega_{22}\Omega_{33}}\ ^2.
\end{aligned} \tag{42}$$

Finally, we use (11) to calculate the concurrence of the bipartite atomic system.

Figure 2 shows the behaviour of the concurrence as a function of time for the equally weighted entangled state (4), i.e., $|\psi_{\alpha=\pi/4}\rangle = |\beta\rangle$, and for the "$-$" weighted state (10), considered these last as different initial conditions for the bipartite atomic state. The state $|\phi_-\rangle$ corresponds to the situation in which the atom+atom system has the probability $\cos^2\left(\frac{9}{20}\pi\right) \approx 0.024$ to be in the state $|+,-\rangle$ and $\sin^2\left(\frac{9}{20}\pi\right) \approx 0.976$ to be in the state $|-,+\rangle$. That is, the repetition of measurement will give $|-,+\rangle$ in the majority of the occasions the system is asked about what state is it in. Remark that in this case the correlation obtained from the interaction process (see Figure 1) is present because the atoms are unavoidably in orthogonal states but we have now more information about the system: This is most probable to find it in the state $|-,+\rangle$ rather than in the state $|+,-\rangle$. In this context, we say that the state $|\phi_-\rangle$ encodes a weak entanglement between the components of the entire system as compared to the Bell state $|\beta\rangle$ which is the common reference of a maximally entangled state. In Figure 2 we can appreciate that the concurrence for the Bell state (4) departures from a maximally entangled condition (where concurrence is equal to one) and after a while this goes to the unentangled condition (i.e., this is equal to zero). The process is reverted after the same time-interval and the concurrence is again equal to 1. If $|\phi_-\rangle$ is the initial condition then we see that the process is quite similar except that this starts in a lower value of concurrence because the entanglement is not perfect here.

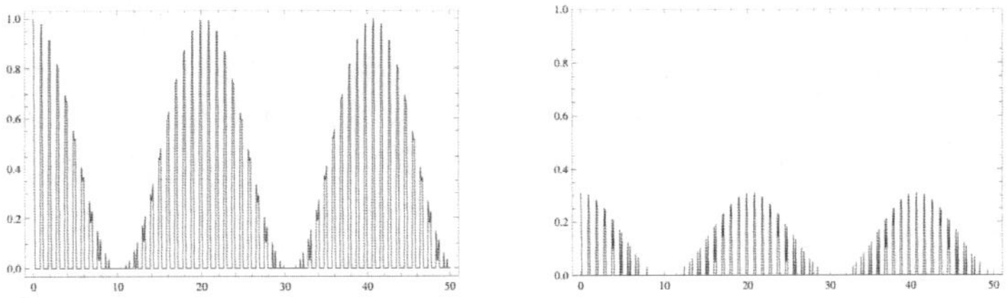

Figure 2. The time-evolution of the concurrence associated to the a bipartite atomic state $|\psi_\alpha\rangle$ defined in (9) when the radiation bath in both cavities includes 10 photons and $\alpha = \pi/4$ (left), $\alpha = 9\pi/20$ (right).

The Figure 3 shows the superposition of the concurrences discussed above. Here we can say that the time-evolution of entanglement is quite similar in both cases, no matter how perfect the entanglement was at the very beginning. Moreover, we can see that the maxima of both of these curves are centered around the same times. This is because in both cases the pair of cavities has exactly the same number of photons. Thus, it is the number of photons which determines the profile of the entanglement time-evolution. The statement is supported with the fact that the frequency Ω is a function of the number operator. Besides, it must be clear that the probability of atomic transitions increases with the number of photons that integrate the radiation bath. This is because the atom has at its disposal a bigger variety of options to capture a given photon in the cavity when the bath has more photons than having only a few of them. It is interesting that not only the frequency but also the manner in which the entanglement suffers the *sudden death* are dominated by the number of photons, as this is shown in Figure 4.

4. Conclusions

The time-evolution of the energy states of a pair of entangled atoms has been studied when the atoms are in isolated QED cavities each one. The concurrence of these systems has been calculated in order to measure entanglement. In contrast with other approaches already reported (see e.g. [34–39]), the present work provides an analysis of the bipartite atomic concurrence when

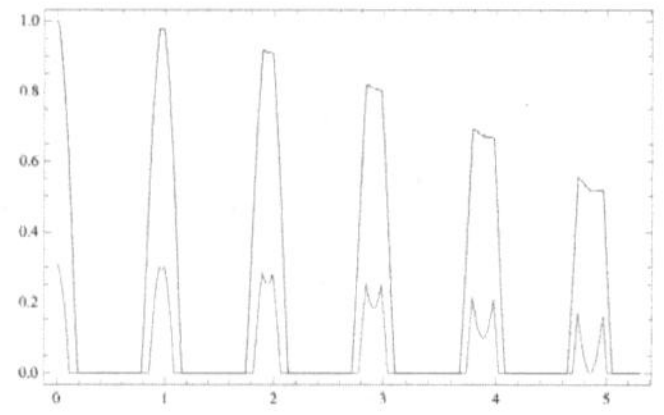

Figure 3. The curves representing concurrence in Figure 2 are superposed for comparison. The red curve is for $\alpha = 9\pi/20$ and the blue one for $\alpha = \pi/4$. Notice that the maxima are centered around the same times in both curves.

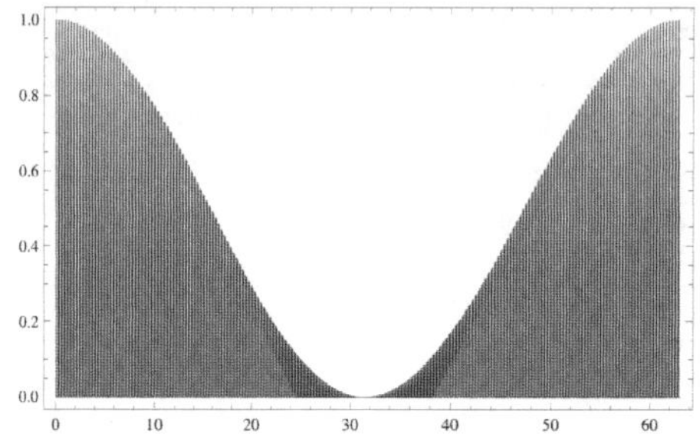

Figure 4. Time-evolution of the concurrence for a bipartite atomic entangled state which is initially in the Bell state $|\psi_{\alpha=\pi/4}\rangle$ defined in (4). The cavities include a radiation field of 100 photons each one. Compare the times of the *sudden death* and the recovering of entanglement with those reported in Figures 2 and 3.

the initial field state in each of the cavities is not longer the vacuum or the single photon state but an arbitrary Fock state. We have found that the entanglement of the atoms depends on the initial number of photons that integrate the radiation baths in the cavities. For the same initial number of photons in both cavities, the frequencies of sudden death and recovering of entanglement are well defined, no matter how perfect the entanglement between the atoms was at the very beginning. The intervals of time in which the entanglement is missing are shorter as bigger is the initial number of photons in the cavities. Some other questions as the dependence of the concurrence with the statistical distribution of the photons in the cavities are going to be discussed elsewhere.

Acknowledgments

The support of CONACyT project 152574 and IPN project SIP-SNIC-2011/04 is acknowledged. M.E. and C.Q. gratefully acknowledge the funding received through the CONACyT Scholarship.

References

[1] Jammer M., *The philosophy of quantum mechanics*, John Wiley, 1974
[2] Primas H., *Chemistry, quantum mechanics and reductionism*, Springer-Verlag, 1983
[3] Jammer M., *The conceptual development of quantum mechanics*, American Institute of Physics, 1989
[4] Mehra J., *The golden age of theoretical physics* Vols. 1 and 2, World Scientific, 2001
[5] Aczel A., *Entanglement*, Plume, 2013
[6] Gilder L., *The age of entanglement. When quantum physics was reborn*, Knopf, 2008
[7] Kumar M, *Quantum. Einstein, Bohr, and the great debate about the nature of reality*, W.W. Norton & Company, First American Edition, 2010
[8] Schrödinger E., Die gegenwärtige Situation in der Quantenmechanik, *Die Naturwissenschaften* **23** (1935)

807-812 (English translation in Wheeler J. and Zurek W., *Quantum theory and measurement*, Princeton University Press, 1983)

[9] Schrödinger E., Discussion of Probability Relations between Separated Systems, *Mathematical Proceedings of the Cambridge Philosophical Society* **31** (1935) 555-563.

[10] Schrödinger E., Probability Relations Between Separated Systems, *Mathematical Proceedings of the Cambridge Philosophical Society* **32** (1936) 446-452.

[11] Mielnik B. and Rosas-Ortiz O., *Quantum Mechanical Laws*, in Fundamentals of Physics [Morán-López J.L. Ed.], Encyclopedia of Life Support Systems (EOLSS), Developed under the Auspices of the UNESCO, Eolss Publishers, Oxford, United Kingdom (2007/Rev. 2009)

[12] Gribbin J., *In search of Schrödinger's cat: Quantum physics and reality*, Bantam Books, 1984

[13] Leggett A.J., Testing the limits of quantum mechanics: motivation, state of play, prospects, *J. Phys. Condes. Matter* **14** (2002) R415-51

[14] Elitzur A.C. and Vaidman L., Quantum mechanical interaction-free measurements, *Found. Phys.* **23** (1993) 987-997

[15] Kwiat P., Winfurter H. and Zeilinger A., Quantum seeing in the dark, *Sci. Am.* **275** (1996) 52-58

[16] Brune M., Schmidth-Kaler F., Maali A., et. al., Quantum Rabi oscillations; a direct test of field quantization in a cavity, *Phys. Rev. Lett.* **76** (1996) 1800-1803

[17] Haroche S., Entanglement, decoherence and quantum/classical boundary, *Phys. Today* **51** (1998) 36-42

[18] Bouwmeester D., Ekert A. and Zeilinger A. (Eds.), *The physics of quantum information: quantum cryptography, quantum teleportation, quantum computation*, Springer, 2000

[19] Zeilinger A, *Dance of the photons. From Einstein to quantum teleportation*, Farrar, Straus and Giroux, 2010

[20] Nielsen M.A. and Chuang I.L., *Quantum Computation and Quantum Information*, Cambridge University Press, 2000

[21] Bengtsson I. and Życzkowski K., *Geometry of Quantum States. An Introduction to Quantum Entanglement*, Cambridge University Press, 2006

[22] Peters N.A., Altepeter J.B., Branning D., et. al., Maximally entangled mixed states: Creation and concentration, *Phys. Rev. Lett.* **92** (2004) 133601

[23] Horst B., Bartkiewicz K. and Miranowicz A., Two-qubit mixed states more entangled than pure states: Comparison of the relative entropy of entanglement for a given nonlocality, *Phys. Rev. A* **87** (2013) 042108

[24] Audrestsch J., *Entangled Systems*, Wiley, 2005

[25] Plenio M. and Virmani S., An introduction to entanglement measures, *Quant. Inf. Comp.* **7**(2007) 001-051

[26] Horodecki R. *et al*, Quantum entanglement, *Rev. Mod. Phys.* **81** (2009) 865

[27] Hill S. and Wootters W.K., Entanglement of a pair of quantum bits, *Phys. Rev. Lett.*, **78** (1997) 5022

[28] Wootters W.K., Entanglement of formation of an arbitrary state of two qubits, *Phys. Rev. Lett.*, **80** (1998) 2245

[29] Ulmann A., Fidelity and concurrence of conjugated states, *Rep. Rev. A* **62** (2000) 032307

[30] Wootters W.K., Entanglement of formation and concurrence, *Quant. Inf. Comp.* **1**(2007) 27-44

[31] Runtga P., Buzek V., Caves C.M., et. al., Universal state inversion and concurrence in arbitrary dimensions, *Phys. Rev. A* **64** (2001) 042315

[32] Mintert F., Kuś M. and Buchleitner A., Concurrence of mixed bipartite quantum states in arbitrary dimensiona, *Phys. Rev. Lett.* **92** (2004) 167902

[33] Jaynnes T., Cummings F.W., Comparison of quantum and semiclassical radiation theories with application to the beam maser, *Proc. IEEE* **51** (1963) 89

[34] Yönaç M., Yu T. and Eberly J.H., Sudden death of entanglement of two Jaynes-Cummings atoms, *J. Phys. B: At. Mol. Opt. Phys.* **39** (2006) S621

[35] Yu T. and Eberly J.H., Sudden death of entanglement, *Science.*, **323** (2009) 598

[36] Yönaç, M., Yu T. and Eberly J.H., Pairwise concurrence dynamics: a four-qubit model, *J. Phys. B: At. Mol. Opt. Phys.*, **40** (2007) S45

[37] Sainz I. and Björk G., Entanglement invariant for the double Jaynes-Cummings model, *Phys. Rev. A.* **76** (2007) 042313

[38] Chan S., Reid M. D. and Ficek Z., Entanglement evolution of two remote and non-identical JaynesCummings atoms.*J. Phys. B: At. Mol. Opt. Phys.* **42** (2009) 065507

[39] Ma J., Sun Z., Wang X. and Nori F., Entanglement dynamics of two qubits in a common bath, *Phys. Rev. A.* **85** (2012) 062323

[40] Enríquez M. and Rosas-Ortiz O., The Kronecker product in terms of Hubbard operators and the Clebsch-Gordan decomposition of $SU(2) \times SU(2)$, *Ann. Phys.* **339** (2013) 218

[41] Enríquez M. and Rosas-Ortiz O., Dynamics of two atoms in a double Jaynes-Cummings model, preprint Cinvestav (2013)

[42] Enríquez M. and Rosas-Ortiz O., Spectral design to control the time evolution of a two-level atom in single-mode radiation fields, preprint Cinvestav (2013)

Bound states at exceptional points in the continuum

N. Fernández-García[1], E. Hernández[2], A. Jáuregui[3] and A. Mondragón[2]

[1]Sección de Estudios de Posgrado e Investigación, UPIITA-IPN, Av. IPN 2508, 07340, México, D.F. México
[2]Instituto de Física, Universidad Nacional Autónoma de México, Apdo. Postal 20-364, 01000 México D.F. México
[3]Departamento de Física, Universidad de Sonora, Apdo. Postal 1626, Hermosillo, Sonora, México

E-mail: `mondra@fisica.unam.mx`

Abstract. In this work, an example of exceptional points in the continuous spectrum of a Hamiltonian of von Neumann-Wigner type is presented and discussed. Remarkably, these exceptional points are not associated with a double pole in the scattering matrix but with a double pole in the normalization factor of the Jost eigenfunctions normalized to unit flux. At the exceptional points the two unnormalized Jost eigenfunctions are no longer linearly independent but coalesce to give rise to two Jordan cycles of generalized bound state energy eigenfunctions in the continuum and a Jordan block representation of the Hamiltonian. The regular scattering eigenfunction vanishes at the exceptional points and the irregular scattering eigenfunction has a double pole at the exceptional points. The scattering matrix is a regular analytical function of the wave number k for all k real including the exceptional points.

1. Introduction.

The readings in a measuring device are real numbers which, according to Quantum Mechanics, correspond to the points in the spectrum of an operator that represents an observable. This physical condition is stated in mathematical form by demanding that the spectrum of an operator representing an observable should be real [1]. When the observable is represented by a self-adjoint operator the condition is automatically satisfied. But the reality of the spectrum of an operator does not necessarily mean that the operator is selfadjoint [2, 3]. This non-equivalence of selfadjointness and the reality of the spectrum of an operator was made evident by the discovery and subsequent discussion of a large class of non-Hermitian PT-symmetric Schrödinger Hamiltonian operators with a complex valued potential term but with real energy eigenvalues [4, 5, 6]. Even in the case of a radial Schrödinger Hamiltonian with a real potential term, the reality of the energy spectrum does not necessarily mean that the Hamiltonian is self-adjoint. In many cases, the spectrum of a real non-self-adjoint Hamiltonian is real but differs from the spectrum of self-adjoint ones in two essential features. These are,

(i) the possible presence of exceptional points
(ii) the possible presence of spectral singularities.

Exceptional points already appear in the finite dimensional case of non-Hermitian Hamiltonian matrices depending on a set of control parameters [7, 8, 9, 10, 11, 12, 13], whereas spectral singularities are characteristic features of Hamiltonians having a continuous energy spectrum [14, 15, 16, 17, 18, 19, 20]. Hence, they are not possible for finite dimensional operators. Exceptional points in the real continuous spectrum of a Schrödinger Hamiltonian with a real potential have received much less attention than in the finite dimensional case [21, 22]. With the purpose of clarifying the topological nature of the exceptional points in the real, continuos energy spectrum of a quantum system, in the following we will present and discuss an example of exceptional points in a Hamiltonian with a real potential of von Neumann-Wigner type. This potential is generated from the eigenfunctions of a free particle Hamiltonian by means of a four times iterated Darboux transformation when the transformation functions are degenerated in the continuum of eigenfunctions of the free particle Hamiltonian [23, 24].

This paper is organized as follows: In section 2, we give a brief description of bound states in the continuum through the mathematical formalism of Darboux transformations. In section 3, by means of a four times iterated and completely degenerated Darboux transformation, we generate an $H[4]$ Hamiltonian which has two exceptional points in its real and continuous degenerate spectrum. In section 4, we compute the regular and irregular scattering solutions of $H[4]$. In section 5, we show that at the exceptional points $k = \pm q$, the Jost solutions coalesce, giving rise to a Jordan chain of generalized eigenfunctions of bound states. In section 6, we show that the Hamiltonian has a Jordan block matrix representation and a Jordan cycle of generalized eigenfunctions. Section 7 is devoted to show that the scattering solution vanishes at the exceptional points $k = \pm q$, while the irregular scattering solution has a double pole at $k = \pm q$. In section 8, we calculate the scattering matrix $S(k)$ and we show that at the exceptional points $k = \pm q$, it is a regular function of k, thus has no poles. We end with a brief summary and conclusions.

2. Bound states embedded in the continuum.

A variety of exactly solvable Hamiltonians of von Neumann-Wigner type that support a bound state in the continuum may be constructed from the known energy eigenfunctions of some initial Hamiltonian $\hat{H}_0$ by means of the Darboux transformation [23].

Let us consider an initial Hamiltonian $\hat{H}_0$ and its eigenfunction $\phi_\lambda(r)$ with eigenvalue E_λ, the Darboux transformation of the solution ϕ_λ is defined by

$$\phi_\lambda(r) \to \psi_\lambda[1](r) = \frac{W(\phi_\mu, \phi_\lambda)}{\phi_\mu(r)}, \tag{1}$$

where $W(\phi_\mu, \phi_\lambda)$ is the usual Wronskian determinant.

The potential function in $\hat{H}_0$, V_0 transform into $V[1]$

$$V_0 \to V[1](r) = V_0 - 2\frac{d^2}{dr^2} \ln \phi_\mu(r). \tag{2}$$

Then, the function $\psi_\lambda[1](r)$ satisfies the Schrödinger radial equation

$$H[1](r)\psi_\lambda[1](r) = E_\lambda \psi_\lambda[1](r), \tag{3}$$

with the new radial Hamiltonian,

$$H[1] = -\frac{d^2}{dr^2} + V[1](r). \tag{4}$$

The fixed eigenfunction $\phi_\mu(r)$ of the initial Hamiltonian $\hat{H}_0$ is known as the transformation function.

The Darboux transformation may be iterated N times and the resulting potential $V[N](r)$ and the corresponding eigenfunctions $\Psi[N]$ may still be expressed in terms of the eigenfunctions of the initial Hamiltonian $\hat{H}_0$. This generalization is known in the literature as the theorem of Crum [26] which states that the function

$$\Psi_E[N](r) = \frac{W(\phi_1, \phi_2, ...\phi_N, \phi_E)}{W(\phi_1, \phi_2, ...\phi_N)}, \tag{5}$$

is an eigenfunction of the radial Schrödinger equation with the potential

$$V[N] = V_0 - 2\frac{d^2}{dr^2} \ln W(\phi_1, \phi_2, ...\phi_N), \tag{6}$$

and eigenvalue E. In these expressions, $W(\phi_1, \phi_2, ...\phi_N)$ and $W(\phi_1, \phi_2, ...\phi_N, \phi_E)$ are the Wronskians of N eigenfuctions $\phi_k(r)$ of the initial Hamiltonian $\hat{H}_0$ corresponding to the eigenvalues E_k, $k = 1, 2, ..., N$ and ϕ_E is an eigenfunction of $\hat{H}_0$ with eigenvalue E. Moreover, the Hamiltonians $\hat{H}_0$ and H with potentials V_0 and $V[N]$ have the same spectrum, except for the eigenvalues $E_1, E_2, ..., E_N$. This constraint on the spectrum makes itself apparent in the time evolution of the system.

In this work, we will be interested in a special kind of potential $V[N]$ obtained from the N-times iterated Darboux transformation, when the transformation functions are degenerated in a continuum of eigenvalues of $\hat{H}_0$.

The explicit form that the transformation takes in this case is obtained by taking all the transformation functions very close in energy

$$\begin{aligned} \phi_1 &= \phi(r, q), \\ \phi_k &= \phi(r, q + \alpha_k) \qquad k = 2, 3, ..., N \end{aligned} \tag{7}$$

with $\alpha_k << q$ and $\alpha_k < \alpha_{k+1}$ for all k.

Next, the transformation functions ϕ_k are expanded in a power series in α_k and the resulting expressions are substituted for the function $\phi_k(r)$ in eqs.(5) and (6). By making use of the properties of the Wronskians in the limit when each α_k vanishes, we finally get

$$\psi_E[N] = \frac{W_2(\phi, \partial_q\phi, ..., \partial_q^{N-1}\phi; \phi_E)}{W_1(\phi, \partial_q\phi, ..., \partial_q^{N-1}\phi)} \tag{8}$$

and

$$V[N] = V_0 - 2\frac{\partial^2}{\partial r^2} \ln W(\phi, \partial_q\phi, ...\partial_q^{N-1}\phi), \tag{9}$$

the notation $\partial_q\phi$ is shorthand for $(\partial\phi/\partial q)$.

When the transformation function $\phi_q(r)$ is properly chosen, the iterated Darboux transformation with degeneracy may be used to construct potentials of von Neumann-Wigner type which support bound states in the continuum [27].

3. Jost solutions of the Hamiltonian $H[4]$.

In this section, by means of a four times iterated and completely degenerated Darboux transformation, we will generate a Hamiltonian, $H[4]$, which has two exceptional points in its spectrum.

The transformation function will be a generic solution of the Hamiltonian $\hat{H}_0$ of a free particle ($V_0 = 0$) with eigenvalue $E_q = q^2$,

$$\phi(r) = \sin(qr + \delta(q)), \tag{10}$$

the phase shift $\delta(q)$ that appears in this expression is a smooth odd function of the wave number q.

Then, the potential $V[4]$ is obtained from equation (9) as

$$V[4] = -2\frac{1}{W_1^2(r)}\Big(W_1''(r)W_1(r) - W_1'^2(r)\Big), \tag{11}$$

where, $W_1'(r) = \partial W_1(r)/\partial r$ and

$$\begin{aligned} W_1(q,r) &= W_1(\phi, \partial_q\phi, \partial_q^{(2)}\phi, \partial_q^{(3)}\phi) \\ &= 16q^4\gamma^4 - 12q^2\gamma^2 + 8q^4\gamma_2\gamma - 12q^4\gamma_1^2 + 24q^2(q\gamma_1\gamma + \gamma^2)\cos 2\theta \\ &+ 3\sin^2 2\theta + (16q^3\gamma^3 - 12q\gamma - 12q^2\gamma_1 - 4\gamma^3\gamma_2)\sin 2\theta, \end{aligned} \tag{12}$$

the functions $\theta(r), \gamma(r), \gamma_1(q)$ and $\gamma_2(q)$ are

$$\theta(r) = qr + \delta(q), \quad \gamma(r) = \partial_q\theta = r + \gamma_0 \tag{13}$$

and

$$\gamma_1 = \partial_q^{(2)}\theta \quad \text{and} \quad \gamma_2 = \partial_q^{(3)}\theta, \tag{14}$$

γ_0 is a function of q but it is independent of r.

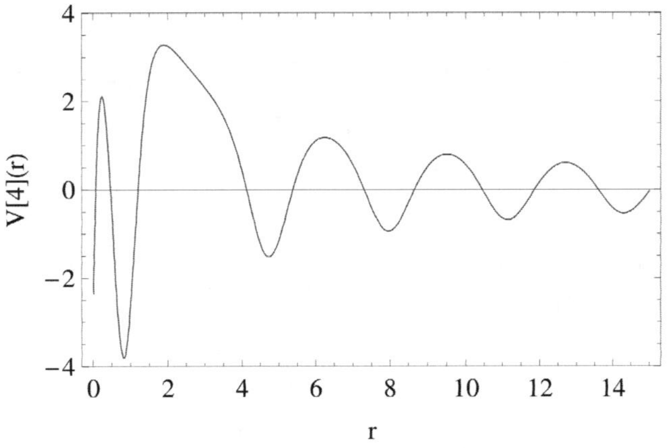

Figure 1. The graph shows the von Neumann-Wigner potential $V[4]$ as function of r for the values of the parameters in $V[4]$ fixed at $q = 1$, $\delta(q) = \pi/6$, $\gamma_0 = 0.6$, $\gamma_1 = 0.214$ and $\gamma_2 = 0.8$. This potential is such that there are exceptional points in the continuum of the spectrum of $H[4]$ at $k = \pm q$.

Explicit expressions for $W_1(qr)$ and the potential $V[4](r)$, as functions of r, are obtained when the explicit expressions for $\theta(r)$, $\gamma(r)$, γ_1 and γ_2 as functions of r are inserted in equations (11) and (12). A graphical representation of $V[4]$ as a function of r is shown in figure 1.

The asymptotic behaviour of $V[4]$ for r large is readily determined from the asymptotic behaviour of $W_1(q,r)$ for r large. Since γ is a linear function of r, the dominant term in the expression (12) for $W_1(q,r)$ is the term with the highest power of γ. Hence,

$$V[4] = 8q\frac{\sin 2(qr + \delta(q))}{r} + O(r^{-2}), \qquad \text{for } r \text{ large.} \tag{15}$$

Journal of Physics: Conference Series **512** (2014) 012023 doi:10.1088/1742-6596/512/1/012023

It follows from here that $V[4]$ is a potential of von Neumann-Wigner type with parameters $a = 8q$ and $b = 2q$. Hence, $V[4]$ is able to support a bound state in the continuum that belongs to the energy eigenvalue $b^2/4 = q^2$ independently of the numerical values of the parameters γ_0, γ_1 and γ_2.

The two independent unnormalyzed Jost functions of the Hamiltonian $H[4]$, that belong to the energy eigenvalue $E = k^2$ and behave as outgoing and incoming spherical waves for large values of r, are obtained from equation (8),

$$\psi_k^{\pm}[4] = \frac{W_2(\phi, \partial_q\phi, ..., \partial_q^{(3)}\phi; e^{\pm ikr})}{W_1(\phi, \partial_q\phi, ..., \partial_q^{(3)}\phi)} := f^{\pm}(k, r), \tag{16}$$

and take the form

$$f^{\pm}(k, r) = \frac{1}{W_1(q, r)} w^{\pm}(k, r) e^{\pm ikr}, \tag{17}$$

here, and in the following, $f^{\pm}(k, r)$ denotes the unnormalized Jost solutions of $H[4]$.

The reduced Wronskian $w^{\pm}(k, r)$ is a complex function of the arguments k, q and r,

$$w^{\pm}(k, r) = u(k, r) \pm iv(k, r). \tag{18}$$

A straightforward computation of $W_1(\phi, \partial_q\phi, \partial_q^{(2)}\phi, \partial_q^{(3)}\phi)$ and $w^{\pm}(k, r)$ allows us to find the following explicit expressions for the functions $u(k, r)$ and $v(k, r)$

$$\begin{aligned} u(k, r) &= 16q^4 \left(k^2 - q^2\right)^2 \gamma^4 - 12q^2 \left(k^4 + 6q^2k^2 + q^4\right) \gamma^2 \\ &\quad +8\gamma_2 q^4 \left(k^2 - q^2\right)^2 \gamma - 12\gamma_1^2 q^4 \left(k^2 - q^2\right)^2 \\ &\quad +24q^2 \left[\left(k^4 - 4q^2k^2 - q^4\right) \gamma^2 + q\gamma_1 \left(k^4 - q^4\right) \gamma\right] \cos 2\theta \\ &\quad + \left[16q^3 \left(k^4 - q^4\right) \gamma^3 - 12q \left(k^4 - 4q^2k^2 - q^4\right) \gamma \right. \\ &\quad \left. -4\gamma_2 q^3 \left(k^4 - q^4\right) - 12\gamma_1 q^2 \left(k^4 - 4q^2k^2 - q^4\right)\right] \sin 2\theta \\ &\quad +3 \left(k^4 + 6q^2k^2 + q^4\right) \sin^2 2\theta \end{aligned} \tag{19}$$

and

$$\begin{aligned} v(k, r) &= 64q^4k(k^2 - q^2)\gamma^3 - 24q^2k(k^2 + q^2)\gamma + 8\gamma_2 q^4k(k^2 - q^2) \\ &\quad -48\gamma_1 q^5k + [32q^4k(k^2 - q^2)\gamma^3 + 24q^2k(k^2 + q^2)\gamma \\ &\quad -8\gamma_2 q^4k(k^2 - q^2) + 48\gamma_1 q^5k] \cos 2\theta + [96q^5k\gamma^2 \\ &\quad -48\gamma_1 q^4k(k^2 - q^2)\gamma - 12qk(k^2 + q^2)] \sin 2\theta \\ &\quad +6qk(k^2 + q^2) \sin 4\theta. \end{aligned} \tag{20}$$

The asymptotic behaviour of the eigenfunction $f^{\pm}(k, r)$ for large values of r is dominated by the highest power of γ.

From equations (12), (19) and (20), we get

$$W_1(q, r) \approx 16q^4r^4[1 + O(r^{-1})], \tag{21}$$

$$u(k, r) \approx 16q^4(k^2 - q^2)^2r^4[1 + O(r^{-1})] \tag{22}$$

and

$$v(k, r) \approx 32q^4k(k^2 - q^2)(2 + \cos 2\theta)r^3[1 + O(r^{-1})]. \tag{23}$$

Hence, for large values of r, we obtain

$$f^{\pm}(k, r) = (k^2 - q^2)^2[1 + O(r^{-1})]e^{\pm ikr}, \tag{24}$$

the factor $(k^2-q^2)^2$ is the flux of probability current at infinity of the unnormalized Jost solutions $f^{\pm}(k,r)$.

Therefore, the two Jost solutions of $H[4]$, normalized to unit flux at infinity, are

$$\begin{aligned} F^{\pm}(k,r) &= \frac{f^{\pm}(k,r)}{(k^2-q^2)^2} \\ &= \frac{1}{(k^2-q^2)^2}\frac{1}{W_1(q,r)}\Big[u(k,r) \pm iv(k,r)\Big]e^{\pm ikr}, \qquad k \neq q. \end{aligned} \tag{25}$$

4. Scattering eigenfunctions and scattering matrix.

The scattering solution $\psi_s(k,r)$ expressed in terms of the Jost solutions is

$$\psi_s(k,r) = \frac{i}{2}\Big[F^-(k,r) - S(k)F^+(k,r)\Big], \qquad k \neq q, \tag{26}$$

where the scattering matrix $S(k)$ is

$$S(k) = \frac{f^-(k,0)}{f^+(k,0)} = \frac{u(k,0) - iv(k,0)}{u(k,0) + iv(k,0)}, \tag{27}$$

where the functions $u(k,0)$ and $v(k,0)$ are given by eqs.(19) and (20) with $r = 0$.

This expression may also be written in the following form

$$\psi_s(k,r) = \frac{1}{(k^2-q^2)^2}\frac{1}{W_1(q,r)}e^{i\Delta(k)}\Big[u(k,r)\sin(kr+\Delta(k)) + v(k,r)\cos(kr+\Delta(k))\Big], \tag{28}$$

where the phase $\Delta(k)$ is given by

$$\Delta(k) = -arctan\Big(\frac{v(k,0)}{u(k,0)}\Big). \tag{29}$$

The regular scattering solution $\psi_s(k,r)$ goes to zero at the origen of coordinates and behaves as $\sin(kr + \Delta(k))$ for large values of r. A graphical representation of the regular Jost solution is shown in the figure 2.

The irregular scattering solution is

$$\psi_{is}(k,r) = \frac{1}{2}[F^-(k,r) + e^{2i\Delta(k)}F^+(k,r)], \tag{30}$$

therefore

$$\psi_{is}(k,r) = \frac{1}{(k^2-q^2)^2}\frac{1}{W_1(q,r)}e^{i\Delta(k)}\{u(k,r)\cos(kr+\Delta(k)) - v(k,r)\sin(kr+\Delta(k))\}, \tag{31}$$

where $u(k,r)$ and $v(k,r)$ are given in equations (17) and (18). The figure 3 shows the behaviour of the irregular scattering solution $\psi_{is}(k,r)$, this solution is divergent at the origen and it has an oscillatory behaviour for large values of r.

5. Exceptional points of $H[4]$ and singularities of the normalized Jost solutions

In the general case, the Hamiltonian $H[4]$ has a continuous spectrum extending from zero to infinity and to each point in this spectrum belong two linearly independent energy eigenfunctions. Clearly, the continuous spectrum of $H[4]$ is doubly degenerate. If the wave number k is allowed to take complex values, the region of analyticity of the energy eigenfunctions as functions of k is

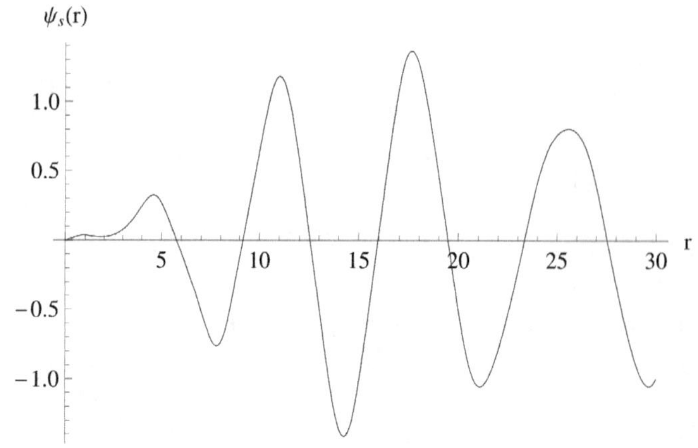

Figure 2. The graph shows the scattering solution $\psi_s(k,r)$ as function of r for $k = 0.8$ and the parameters in $V[4]$, fixed at the values $q = 1$, $\delta(q) = \pi/6$, $\gamma_0 = 0.6$, $\gamma_1 = 0.214$ and $\gamma_2 = 0.8$.

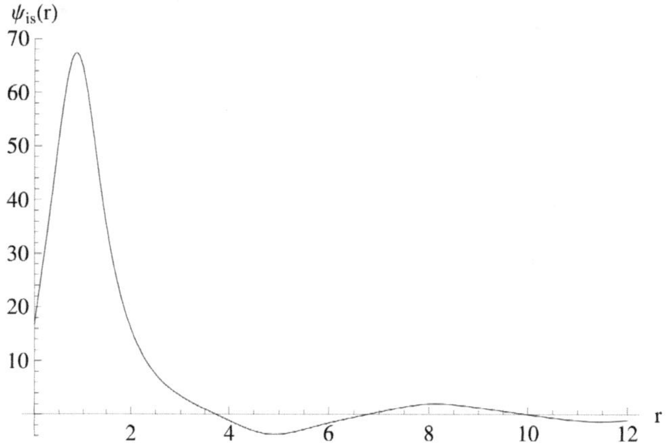

Figure 3. The graph shows the irregular scattering solution $\psi_{is}(k,r)$ as function of r for $k = 0.8$ and the values of the parameters in $V[4]$, fixed at $q = 1$, $\delta(q) = \pi/6$, $\gamma_0 = 0.6$, $\gamma_1 = 0.214$ and $\gamma_2 = 0.8$.

extended to all points in the complex k-plane. When $k \neq \pm q$, the energy eigenfunctions are the normalized, Jost solutions $F^{\pm}(k,r)$ which behave at infinity as incoming and outgoing waves. In this section, it will be shown that at the points $k = \pm q$, the two unnormalized Jost solutions are no longer linearly independent but coalesce to give rise to a Jordan chain of generalized bound state eigenfunctions of lenght 2.

From equations (17), (18) and $k > 0$, a straightforward computation of the Wronskian of the Jost solutions gives

$$W(f^+(k,r), f^-(k,r)) = -2i\Big[\frac{W(u,v)}{u^2+v^2} + k\Big] f^+(k,r) f^-(k,r), \tag{32}$$

the Wronskian of the functions $u(k,r)$ and $v(k,r)$ is readily computed from eqs.(19) and (20), we get

$$W(u,v) = -q(u^2+v^2), \tag{33}$$

substitution of this result in the previous equation gives

$$W(f^+(k,r), f^-(k,r)) = -2if^+(k,r)f^-(k,r)(k-q), \quad k>0. \tag{34}$$

Now, since the potential $V[4]$ is an even function of q, time reversal invariance of the Hamiltonian plus boundary conditions implies that [28],

$$f^+(-k,r) = +f^-(k,r), \tag{35}$$

for k real. In this case, we obtain

$$W(f^+(k,r), f^-(k,r)) = +2if^+(k,r)f^-(k,r)(k+q), \quad k<0. \tag{36}$$

Since the product of the unnormalized Jost solutions is non-vanishing for all k,

$$f^+(k,r)f^-(k,r) > 0, \tag{37}$$

the two Jost solutions are linearly independent for $k \neq \pm q$.

The points $k = \pm q$ are real non-vanishing zeroes of the Wronskian of the two Jost solutions of $H[4]$. This property identifies the points $k = \pm q$ as exceptional points in the spectrum of the Hamiltonian $H[4]$.

The unnormalized Jost solutions of $H[4]$, $f^{\pm}(k,r)$, given in eqs.(17-20), are entire functions of k for all k. Since the flux of probability current at infinity of these solutions is $(k^2-q^2)^2$, the Jost solutions $F^{\pm}(k,r)$ of $H[4]$ normalized to unit flux at infinity have double poles as function of k at $k = \pm q$, see eq.(25). In order to uncover the exceptional points of $H[4]$, we separate the singular from the regular parts of the normalized Jost solutions $F^{\pm}(k,r)$ as functions of k[29]

$$\begin{aligned}
\frac{f^{\pm}(k,r)}{(k^2-q^2)^2} &= \frac{1}{4q^2}\frac{f^{\pm}(q,r)}{(k-q)^2} + \frac{\frac{1}{4q^2}\left[\left(\frac{\partial f^{\pm}(k,r)}{\partial k}\right)_q - \frac{1}{q}f^{\pm}(q,r)\right]}{(k-q)} + h^{\pm}_{>}(k,r) \\
&+ \frac{1}{4q^2}\frac{f^{\pm}(-q,r)}{(k+q)^2} + \frac{\frac{1}{4q^2}\left[\left(\frac{\partial f^{\pm}(k,r)}{\partial k}\right)_{-q} + \frac{1}{q}f^{\pm}(-q,r)\right]}{(k+q)} \\
&+ h^{\pm}_{<}(k,r),
\end{aligned} \tag{38}$$

where

$$h^{\pm}_{\gtrless}(k,r) = \frac{1}{4q^2}\left[2g^{\pm}_{\gtrless}(k,r) \mp \frac{k}{q}g^{\pm}_{\gtrless}(k,r) \mp \frac{1}{q}\left(\frac{\partial f^{\pm}(k,r)}{\partial k}\right)_{\pm q}\right] \tag{39}$$

and

$$g^{\pm}_{\gtrless}(k,r) := \left[f^{\pm}(k,r) - \left(f^{\pm}(\pm q,r) + \left(\frac{\partial f^{\pm}(k,r)}{\partial k}\right)_{\pm q}(k\mp q)\right)\right]\frac{1}{(k\mp q)^2}. \tag{40}$$

Now, let us define a more convenient notation

$$\psi_B(q,r) = \frac{f^{\pm}(q,r)}{4q^2}e^{\pm i\delta(q)} \tag{41}$$

and

$$\chi^{\pm}_B(q,r) = \frac{1}{4q^2}\left[\left(\frac{\partial f^{\pm}(k,r)}{\partial k}\right)_{k=q} - \frac{1}{q}f^{\pm}(q,r)\right]e^{\pm i\delta(q)}. \tag{42}$$

Then, substitution of expressions (41- 42) in equation (25) allows us to write the normalized Jost solutions of $H[4]$ as the sum of singular and regular parts

$$F^{\pm}(k,r) = \frac{\psi_B(q,r)e^{\mp i\delta(q)}}{(k-q)^2} + \frac{\chi^{\pm}_B(q,r)e^{\mp i\delta(q)}}{(k-q)} + \frac{\psi_B(-q,r)e^{\pm i\delta(q)}}{(k+q)^2}$$

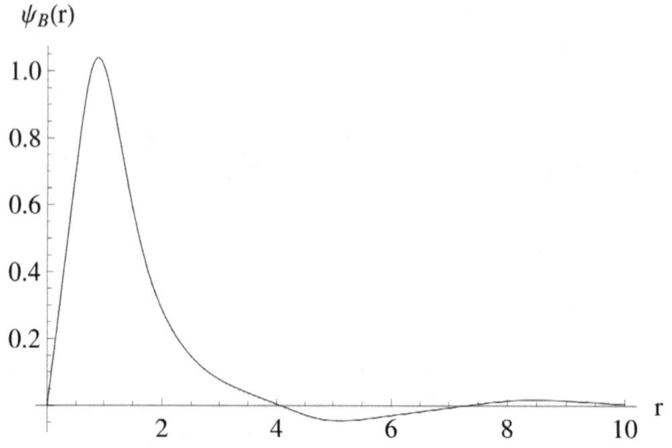

Figure 4. The graph shows the bound state eigenfuction $\psi_B(q,r)$ as a function of r computed for $q = 1$ and the values of the parameters in $V[4]$ fixed at $\delta = \pi/6$, $\gamma_0 = 0.6$, $\gamma_1 = 0.214$ and $\gamma_2 = 0.8$

$$+ \quad \frac{\chi_B^{\pm}(-q,r)e^{\pm i\delta(q)}}{(k+q)} + h^{\pm}(k,r), \tag{43}$$

where

$$h^{\pm}(k,r) = h_>^{\pm}(k,r) + h_<^{\pm}(k,r). \tag{44}$$

Explicit expressions for the functions $\psi_B(q,r)$ and $\chi_B^{\pm}(q,r)$ as functions of r, are readily obtained from the definitions (41), (42) and equations (12) and (17-20)

$$\psi_B(q,r) = \frac{24q^2}{W_1(q,r)}\{-2q^2\gamma^2\cos\theta + [q\gamma + q^2\gamma_1]\sin\theta + \sin^2\theta\cos\theta\} \tag{45}$$

and

$$\chi_B^{\pm}(q,r) = \chi_B(q,r) \mp i\gamma_0\psi_B(q,r), \tag{46}$$

where

$$\begin{aligned}\chi_B(q,r) &= \frac{8q}{W_1(q,r)}[-2q^3\gamma^3\sin\theta - 3q^2\gamma^2\cos\theta + 3q\gamma\sin^3\theta \\ &- \gamma_2 q^3\sin\theta + 3\gamma_1\gamma q^3\cos\theta + 3\sin^2\theta\cos\theta].\end{aligned} \tag{47}$$

Since $W_1(q,r) \sim 16(qr)^4[1 + O(r^{-1})]$, the generalized eigenfunctions $\psi_B(q,r)$ and $\chi_B^{\pm}(q,r)$ vanish as r^{-3} for r large, and, hence they are quadratically integrable. Furthermore, $\psi_B(q,r)$ is an even function of q and $\chi_B(q,r)$ is an odd function of q.

Now, it is straightforward to verify that the functions $\psi_B(q,r)$ and $\chi_B(q,r)$ are the elements of a Jordan cycle of lenght two of generalized eigenfunctions of $H[4]$.

From the radial Schrödinger equation satisfied by $F^{\pm}(k,r)$

$$H[4]F^{\pm}(k,r) = k^2F^{\pm}(k,r), \tag{48}$$

and the expansion of $F^{\pm}(k,r)$ in singularities, equation (43), we get

$$\left(1 + \left(\frac{k-q}{k+q}\right)^2\right)H[4]\psi_B(q,r) + (k-q)H[4]\left(\chi_B^{\pm}(q,r) + \frac{k-q}{k+q}\chi_B^{\pm}(-q,r)\right)$$

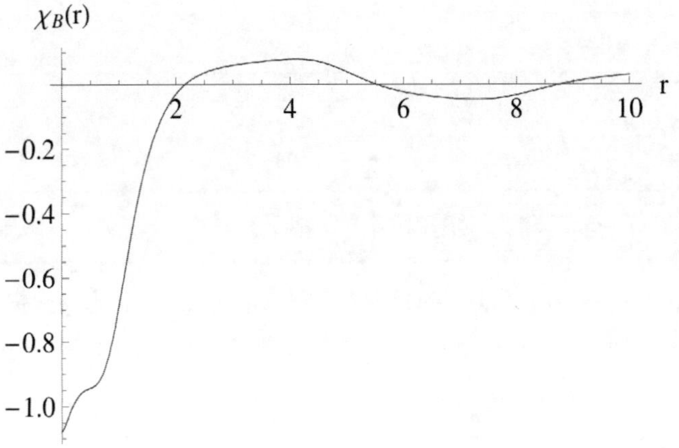

Figure 5. The generalized eigenfunction $\chi_B(q,r)$, as a function of r computed $q = 1$ and the values of the parameters in $V[4]$ fixed at $\delta = \pi/6$, $\gamma_0 = 0.6$, $\gamma_1 = 0.214$ and $\gamma_2 = 0.8$

$$+(k-q)^2 e^{\pm i\delta(q)} H[4]h^{\pm}(k,r) = k^2\Big(1+\Big(\frac{k-q}{k+q}\Big)^2\Big)\psi_B(q,r)$$
$$+k^2(k-q)\Big(\chi_B^{\pm}(q,r)+\frac{k-q}{k+q}\chi_B^{\pm}(-q,r)\Big)+k^2(k-q)^2h^{\pm}(k,r)e^{\pm i\delta(q)}. \tag{49}$$

Taking the limit when $k \to q$, we obtain

$$H[4]\psi_B(q,r) = q^2\psi_B(q,r). \tag{50}$$

Substitution of this result in (49) gives

$$H[4]\Big(\chi_B^{\pm}(q,r)+\frac{k-q}{k+q}\chi_B^{\pm}(-q,r)\Big)+(k-q)e^{\pm i\delta(q)}H[4]h^{\pm}(k,r) =$$
$$2\frac{k^2+q^2}{k+q}\psi_B(q,r)+k^2\Big(\chi_B^{\pm}(q,r)+\frac{k-q}{k+q}\chi_B^{\pm}(-q,r)\Big)+$$
$$k^2(k-q)e^{\pm i\delta(q)}h^{\pm}(k,r), \tag{51}$$

again, taking the limit when $k \to q$, and using (47) and (50) we get

$$H[4]\chi_B(q,r) = q^2\chi_B(q,r) + 2q\psi_B(q,r). \tag{52}$$

Graphical representations of the generalized eigenfunctions $\psi_B(q,r)$ and $\chi_B(q,r)$ as functions of r are shown in Figs. 4 and 5.

6. Jordan cycle of generalized eigenfunctions and Jordan block representation of $H[4]$.

The generalized eigenfunctions $\psi_B(q,r)$ and $\chi_B(q,r)$ belong to the same point $k = q$ in the spectrum of $H[4]$ and satisfy the coupled equations (50 and 52), which identify $\psi_B(q,r)$ and $\chi_B(q,r)$ as the elements of a Jordan cycle of length two of generalized eigenfunctions of $H[4]$ associated with a Jordan block representation of $\mathbf{H}[4]$ at the exceptional point.

In order to make this property evident, it will be convenient to write equations (50) and (52) in matrix form,

$$\Big[H[4]\mathbf{1_{2\times 2}}\ \Psi_B(q,r) = \mathcal{H}_B(q)\Psi_B(q,r), \tag{53}$$

where

$$\Psi_B(q,r) = \begin{pmatrix} \psi_B(q,r) \\ \chi_B(q,r) \end{pmatrix} \tag{54}$$

and

$$\mathcal{H}_B(q) = \begin{pmatrix} q^2 & 0 \\ 2q & q^2 \end{pmatrix}. \tag{55}$$

From (53), it is evident that the matrix $\mathcal{H}_B(q)$ is a matrix representation of the Hamiltonian $H[4]$ in the two dimensional functional space spanned by the generalized eigenfunctions $\{\psi_B(q,r), \chi_B(q,r)\}$.

(i) Let us recall that the domain $\mathcal{D}(\mathbf{H}[4])$ of the Hamialtonian $\mathbf{H}[4]$ is a rigged Hilbert space of continuous complex functions of the variables (k,r) with continuous first and second derivatives with respect to r, with r in the semi-infinite straight line $0 \leq r < \infty$. The two dimensional subspace of functions spanned by the generalized eigenfunctions $\{\psi_B(q,r), \chi_B(q,r)\}$ is in the domain $\mathcal{D}(\mathbf{H}[4])$ of $\mathbf{H}[4]$.

(ii) The matrix $\mathcal{H}_B(q)$ is a Jordan block[30] of 2×2 and can not be brought to diagonal form by means of a similarity transformation with a unitary operator [31]. Hence, also the Hamiltonian $\mathbf{H}[4]$ can not be diagonalized by means of a similarity transformation with a unitary operator.

(iii) The real non-symmetric matrix $\mathcal{H}_B(q)$ is not Hermitian

$$\mathcal{H}_B(q) \neq \mathcal{H}_B^\dagger(q). \tag{56}$$

Therefore, the system of coupled equations (50) and (52) plus boundary conditions is not self-adjoint. This lack of self-adjointness of $\mathcal{H}_B(q)$ implies that also the operator $\mathbf{H}[4]$ on the left hand side of eq.(53), when acting in the functional space spanned by the generalized eigenfunctions, is also not self-adjoint, that is, non-Hermitian.

7. Scattering solutions.

In this section it will be shown that the regular scattering solution $\psi_s(k,r)$ vanishes at $k = \pm q$, and the irregular Jost solution $\psi_{is}(k,r)$ has a double pole at the exceptional points $k = \pm q$.

The regular scattering solution of $H[4]$ is obtained from the Jost solutions, eqs. (25-27), as

$$\psi_s(k,r) = \frac{i}{2}\frac{1}{(k^2-q^2)^2}\frac{1}{W_1(q,r)}\frac{1}{w^+(k,0)}\Big[w^+(k,0)w^-(k,r)e^{-ikr} - w^-(k,0)w^+(k,r)e^{ikr}\Big], \tag{57}$$

where

$$\frac{i}{2}\frac{1}{w^+(k,0)}\Big[w^+(k,0)w^-(k,r)e^{-ikr} - w^-(k,0)w^+(k,r)e^{ikr}\Big] = \\ e^{i\Delta}\Big[u(k,r)\sin(kr+\Delta) + v(k,r)\cos(kr+\Delta)\Big], \tag{58}$$

explicit expressions for $u(k,r)$ and $v(k,r)$ as functions of (k,r) are given in equations (19-20).

In order to make explicit the properties of the scattering solutions as functions of k, for k close to the exceptional points at $k = \pm q$, we will write the reduced Wronskian $w^\pm(k,r)$ as an expansion in powers of $(k-q)$

$$w^\pm(k,r) = u(k,r) \pm iv(k,r) = e^{\mp i\theta(q,r)} \sum_{\ell=0}^{4} w_\ell^\pm(q,r)(k-q)^\ell, \tag{59}$$

explicit expressions for the coefficients $w_\ell^{\pm}(q,r)$, as functions of their arguments are given in the appendix.

Then, the term in square brackets in the right hand side of eq.(57) may also be expressed as an expansion in powers of $(k-q)$

$$\begin{aligned}&\left[w^+(k,0)w^-(k,r)e^{-ikr} - w^-(k,0)w^+(k,r)e^{ikr}\right] = \\ &\sum_{\ell=0}^{4}\sum_{m=0}^{4}\left[w_\ell^+(q,0)w_m^-(q,r)e^{-i(k-q)r} - w_\ell^-(q,0)w_m^+(q,r)e^{i(k-q)r}\right](k-q)^{(\ell+m)}. \end{aligned} \tag{60}$$

Now, from (A1) and (A2) in the appendix, the first three terms in the summation in the right hand side of eq.(60) are proportional to $w_0^+(q,0)$ and / or $w_0^-(q,0)$, but

$$w_0^+(q,r) = w_0^-(q,r) = 4q^2 W_1(q,r)\psi_B(q,r), \tag{61}$$

since $\psi_B(q,0)$, vanishes the first three terms in the summation in the right hand side of (60) also vanish.

It follows that

$$\begin{aligned}\psi_s(k,r) &= \frac{i}{2}\frac{(k-q)}{(k+q)^2 W_1(q,r)}\frac{1}{w^+(q.0)}\sum_{\ell=3}^{8}\sum_{m=0}^{\ell}\left[w_{\ell-m}^+(q,0)w_m^-(q,r)e^{-i(k-q)r}\right. \\ &- \left. w_{\ell-m}^-(q,0)w_m^+(q,r)e^{i(k-q)r}\right](k-q)^{\ell-3}, \end{aligned} \tag{62}$$

with the restriction $w_\ell^{\pm}(q,r) = 0$, for $\ell \geq 5$.

Therefore, at the excepcional point, the regular scattering solution also vanishes,

$$\psi_s(q,r) = 0. \tag{63}$$

The energy eigenfunctions at the exceptional point are the generalized bound state eigenfunctions $\psi_B(q,r)$ and $\chi_B(q,r)$.

Let us now turn our atention to the irregular scattering solution, equation (31),

$$\psi_{is}(k,r) = \frac{1}{(k^2-q^2)^2}\frac{1}{W_1(q,r)}e^{i\Delta(k)}\{u(k,r)\cos(kr+\Delta(k)) - v(k,r)\sin(kr+\Delta(k))\}. \tag{64}$$

In this case, we have

$$\begin{aligned}&e^{i\Delta(k)}\left[u(k,r)\cos(kr+\Delta(k)) - v(k,r)\sin(kr+\Delta(k))\right] = \\ &\frac{1}{2}\frac{1}{w^+(k,0)}\left[w^+(k,0)w^-(k,r)e^{-ikr} + w^-(k,0)w^+(k,r)e^{ikr}\right]. \end{aligned} \tag{65}$$

Notice that, when the expression (59) is substituted for $w^{\pm}(k,r)$ in (65) we obtain an expansion of the right hand side of (65) in powers of $(k-q)$ similar to (60), but now the terms proportional to $w_0^{\pm}(q,0)w_0^{\pm}(q,r)$ and $w_0^{\pm}(q,0)w_1^{\pm}(q,r)$ and $w_1^{\pm}(q,0)w_1^{\pm}(q,r)$ are added. Hence, the singular terms in the expressions (43) for the Jost solutions add instead of cancelling which makes $\psi_{is}(k,r)$ singular at the exceptional points. Hence, the irregular scattering solution $\psi_{is}(k,r)$, as a function of k, has a pole of second order at the exceptional points.

Substitution of the expressions (59-60) for $w^{\pm}(k,r)$ in (65) and, multiplying the result times $(k-q)^2$ and taking the limit, $k \to q$ gives

$$\lim_{k\to q}(k-q)^2\,\psi_{is}(k,r) = \psi_B(q,r). \tag{66}$$

Therefore, the bound state eigenfunction $\psi_B(q,r)$ is the residue of second order of the irregular scattering solution $\psi_{is}(k,r)$ at $k=q$.

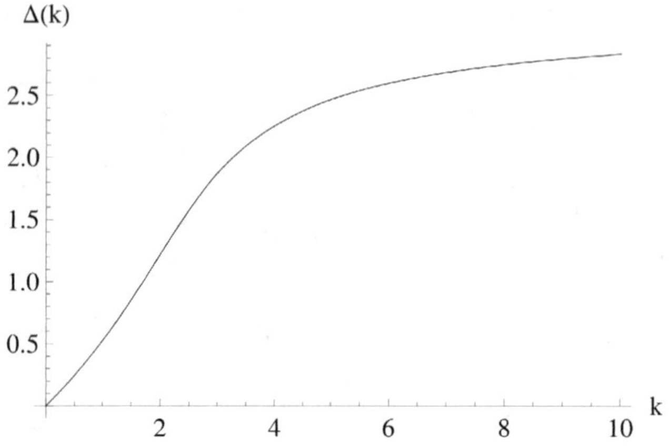

Figure 6. The phase shift $\Delta(k)$ as a function of the wave number k computed with the parameters in $V[4]$ fixed at the value $\gamma_0 = 0.61$, $\gamma_1 = 0.214$ and $\gamma_2 = 0.8$.

8. The scattering matrix $S(k)$ in the limit $k \to q$ and the differential cross section.

The scattering matrix $S(k)$ is, by definition,

$$S(k) = \frac{F^-(k,0)}{F^+(k,0)} = \frac{u(k,0) - iv(k,0)}{u(k,0) + iv(k,0)}, \tag{67}$$

in this expression $F^{\pm}(k,0)$ are the normalized Jost solutions for incoming and outgoing waves and $u(k,0) \pm iv(k,0)$ is the reduced Wronskian defined in eq. (18), and given in explicit form as a function of the arguments (k,r) in eqs. (19) and (20).

The reduced Wronskian, evaluated at the origin of coordinates, may be written as

$$u(k,0) + iv(k,0) = \mathcal{N}(k)e^{-i\delta(q)}, \tag{68}$$

where

$$\mathcal{N}(k) = \sum_{\ell=0}^{4} w_\ell^+(q,0)(k-q)^\ell . \tag{69}$$

Thus, the scattering matrix may be written as

$$S(k) = \frac{\mathcal{N}^*(k)}{\mathcal{N}(k)} e^{i2\delta(q)} . \tag{70}$$

Now, at $k = q$, only the first term in the expansion (69) constributes

$$\mathcal{N}(q) = w_0(q,0). \tag{71}$$

Since $w_0(q,0)$ is real,

$$\left.\frac{\mathcal{N}^*(k)}{\mathcal{N}(k)}\right|_{k=q} = 1 . \tag{72}$$

Therefore, the scattering matrix evaluated at the exceptional point is finite and takes the form

$$\lim_{k \to q} S(k) = e^{i2\delta(q)} . \tag{73}$$

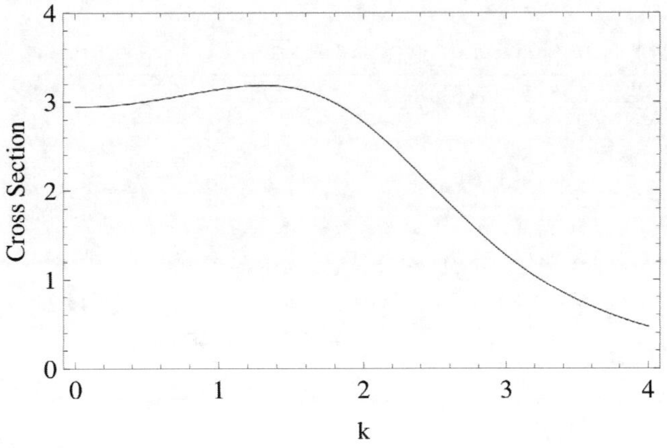

Figure 7. The cross section $\sigma(k)$ as a function of the wave number k computed with the parameter in $V[4]$ fixed at the values $\gamma_0 = 0.6$, $\gamma_1 = 0.214$ and $\gamma_2 = 0.8$

We have shown that $S(k)$ as function of k, is finite at the exceptional points, that is, it does not have a singularity, pole or otherwise, at $k = \pm q$. Hence, the bound state generalized eigenfunction are not associated with a singularity of the S(k) matrix.

The cross section is

$$\sigma(k) = \frac{4\pi}{k^2} \sin^2 \Delta(k). \tag{74}$$

The graphical representation of phase $\Delta(k)$ and the cross section are given in the figures 6 and 7.

9. Summary and Conclusions

The von Neumann-Wigner type Hamiltonian $H[4]$ discussed in this paper and the free particle Hamiltonian H_0 are isospectral. In the general case, to each point in this continuous spectrum, correspond two linearly independent Jost solutions which behave at infinity as incoming and outgoing waves. Clearly in both cases, this continuous spectrum is doubly degenerate. However, here we have shown that in the continuous spectrum of $H[4]$ there are two exceptional points at wave numbers $k = \pm q$, such that at these points the two unnormalized Jost eigenfunction of $H[4]$ are linearly dependent and coalesce to give rise to Jordan cycles of length two of generalized quadratically integrable bound state eigenfunctions and a Jordan block representation of the Hamiltonian $\mathbf{H}[4]$. At the exceptional points the regular scattering function of $H[4]$ vanishes, while the irregular scattering solution has a double pole with a second order residue equal to the bound state eigenfunction in the continuum and a Jordan block representation of the Hamiltonian. Most remarkably is the fact that the scattering matrix $S(k)$ is a regular function of k at the exceptional points, that is, the Jordan cycle of generalized bound states eigenfunctions of $H[4]$ in the continuum is not associated with a pole of the scattering matrix.

10. Acknowledgements

This work was partially supported by CONACyT México under Contract No. 132059 and by DGAPA-UNAM Contract No. PAPIIT: IN113712. N. Fernández-García acknowledges the financial support of the Instituto Politécnico Nacional (IPN), grants SIP 20130847 and SIP-SNI-2011/08

Appendix A.

The dependence of $S(k)$ on k is readily made evident by expanding the reduced Wronskian defined in eq.(18) in powers of $(k-q)$

$$w^{\pm}(k,r) = \mathrm{e}^{\mp i\theta} \sum_{\ell=0}^{4} w_{\ell}^{\pm}(q,r)(k-q)^{\ell}, \tag{A.1}$$

the coefficients of the powers of $(k-q)$ in this expansion are

$$w_0^{\pm}(q,r) = 4q^2 W_1(q,r)\psi_B(q,r) \tag{A.2}$$

$$w_1^{\pm}(q,r) = 4qW_1(q,r)\Big[\psi_B(q,r) + q\chi_B(q,r) \mp iq\gamma\psi_B(q,r)\Big] \tag{A.3}$$

$$\begin{aligned} w_2^{\pm}(q,r) &= -2W_1(q,r)\psi_B(q,r)\cos^2\theta - 3qW_1(q,r)\chi_B(q,r) \\ &+ 16q^2\Big\{4q^4\gamma^4 - 3q^2\gamma^2 - 3q^4\gamma_1^2 + 2q^4\gamma\gamma_2 + 3\sin^2\theta\cos^2\theta\Big\}\cos\theta \\ &\mp i2W_1(q,r)\Big\{(2q\gamma + q^2\gamma_1)\psi_B(q,r) + 2q^2\gamma\chi_B(q,r)\Big\} \end{aligned} \tag{A.4}$$

$$\begin{aligned} w_3^{\pm}(q,r) = \; & 8q\left\{\left[3\sin 2\theta\sin\theta - 6(q\gamma)\sin\theta\left(3\cos^2\theta - 1\right)\right.\right. \\ & +6(q\gamma)^2\cos\theta(2\cos 2\theta - 1) + 4(q\gamma)^3\sin\theta\cos 2\theta \\ & +8(q\gamma)^4\cos\theta - 6\left(\gamma_1 q^2\right)\sin 2\theta\cos\theta - 6\left(\gamma_1 q^2\right)^2\cos\theta \\ & -2\left(\gamma_2 q^3\right)\sin\theta\left(\cos^2\theta + 1\right) + 12\left(q^3\gamma_1\gamma\right)\cos^3\theta \\ & \left.+4\left(\gamma_2\gamma q^4\right)\cos\theta\right] \pm i\left[-18(q\gamma)\sin^2\theta\cos\theta\right. \\ & +6(q\gamma)^2(2\cos 2\theta - 1)\sin\theta + 4(q\gamma)^3\cos\theta\left(2\sin^2\theta + 3\right) \\ & +8(q\gamma)^4\sin\theta - 6\left(\gamma_1 q^2\right)\sin 2\theta\sin\theta - 6\left(\gamma_1 q^2\right)^2\sin\theta \\ & -12\left(q^3\gamma_1\gamma\right)\sin^3\theta - 2\left(\gamma_2 q^3\right)\sin^2\theta\cos\theta \\ & \left.\left.+4\left(\gamma_2\gamma q^4\right)\sin\theta\right]\right\}, \end{aligned} \tag{A.5}$$

and

$$\begin{aligned} w_4^{\pm}(q,r) = \; & \Big[3\sin^2 2\theta - 12(q\gamma)\sin 2\theta + 12(q\gamma)^2(2\cos 2\theta - 1) \\ & +16(q\gamma)^3\sin 2\theta + 16(q\gamma)^4 - 12\left(\gamma_1 q^2\right)\sin 2\theta - 12\left(\gamma_1 q^2\right)^2 \\ & +24\left(q^3\gamma_1\gamma\right)\cos 2\theta - 4\left(\gamma_2 q^3\right)\sin 2\theta + 8\left(\gamma_2\gamma q^4\right)\Big]\cos\theta \\ & \pm i\Big[3\sin^2 2\theta - 12(q\gamma)\sin 2\theta + 12(q\gamma)^2(2\cos 2\theta - 1) \\ & +16(q\gamma)^3\sin 2\theta + 16(q\gamma)^4 - 12\left(\gamma_1 q^2\right)\sin 2\theta - 12\left(\gamma_1 q^2\right)^2 \\ & +24\left(q^3\gamma_1\gamma\right)\cos 2\theta - 4\left(\gamma_2 q^3\right)\sin 2\theta + 8\left(\gamma_2\gamma q^4\right)\Big]\sin\theta\,. \end{aligned} \tag{A.6}$$

An expansion of the reduced wronskian, $w^{\pm}(k,r)$, about the point $k=-q$ is readily obtained from eqs. (A.1) and (35). From time reversal invariance of the Hamiltonian plus boundary conditions, we get

$$w^{\pm}(k,r) = w^{\mp}(-k,r). \tag{A.7}$$

Hence, from (A.1)

$$w^{\pm}(k,r) = e^{\mp i\theta(q)} \sum_{\ell=0}^{4} (-1)^{\ell} w_{\ell}^{\pm}(q,r)(k+q)^{\ell}, \tag{A.8}$$

the coefficents $w^{\pm}_{\ell}(q,r)$ are explicitly given in eqs.(A.2 - A.6) as entire functions of q. Their domain of analiticity is trivially extended to negative values of the argument q just by changing q by $-q$ in (A.2 - A.6). Then it may be verified that

$$(-1)^{\ell} w^{\pm}_{\ell}(q,r) = w^{\pm}_{\ell}(-q,r). \tag{A.9}$$

Therefore,

$$w^{\pm}(k,r) = e^{\pm i\theta(-q)} \sum_{\ell=0}^{4} (-1)^{\ell} w^{\pm}_{\ell}(|q|,r)(k+q)^{\ell}, \quad k<0. \tag{A.10}$$

References

[1] von Neumann, J 1996 *Mathematical Foundations of Quantum Mechanics* (Princeton, N Y: Princeton University Press)
[2] Mostafazadeh A 2010 *Physica Scripta* **82**, 038110
[3] Fülöp T 2007 *Sigma* **3** 107
[4] Bender C M 2007 *Rep. Progr. Phys.* **70** 947
[5] Mostafazadeh A 2010 *International Journal of Geometric Methods in Modern Physics* **7** 1101
[6] Cannata F, Ioffe M V, Nishnianidze D N 2007 *Phys. Lett. A* **369** 9
[7] Heiss W D 2012 *J. Phys. A: Math. Theor.* **45** 444016
[8] Mondragón, A., and Hernández, E. (1993) *J. Phys. A: Math. and Gen.* **26** 5595
[9] Cavalli S, De Fazio D 2011 *Theor. Chem. Acc.* **129** 141
[10] Félix-Beltrán O, Gómez-Bock M, Hernández E, Mondrágon A and Mondrágon M 2011 *Int. J. Theor. Phys.* **50** 2291
[11] Thilagam A 2012 *J. Phys. A: Math. Theor.* **45** 444031
[12] Dietz B, Harney H L, Kirillov O N, Miski-Oglu M, Richter A and Schäfer (2011) *Phys. Rev. Lett.* **106** 150403
[13] Gilary I, Mailybaev A A, Moiseyev N 2013 *Phys. Rev. A* **88** 010102(R)
[14] Kurasov P 2007 *Math. Proc. Camb. Phil. Soc.* **142** 161
[15] Mostafazadeh A 2009 *Phys. Rev. Lett.* **102** 220202;
Mostafazadeh A 2009 *Phys. Rev. A* **80** 032711;
Mostafazadeh A 2011 *Phys. Rev. A* **84** 023809
[16] Mostafazadeh A and Mehri-Dehnavi H 2009 *J. Phys. A: Math. Theor.* **42** 125303
[17] Gusein S H Guseinov 2009 **Pramana 73** 587
[18] Longhi S 2009 *Phys. Rev. B* **80** 165125
[19] Andrianov A A, Cannata F and Sokolov V A 2010 *J. Math. Phys.* **51** 052104
[20] Samsonov B F 2013 *Phil. Trans R Soc A* **371** 20120044
[21] Andrianov A A and Sokolov A V 2011 *Sigma* **7** 111;
Sokolov A V 2011 *Sigma* **7** 112
[22] Sokolov A V, Andrianov A A and Cannata F 2006 *J. Phys. A: Math. Gen.* **39** SI 32 10207
[23] Matveev V B and Salle M A 1991 *Darboux Transformations and Solitons* (Berlin Heidelberg: Springer-Verlag)
[24] Mostafazadeh Ali 2002 *Nucl. Phys. B* **640** 419; *arXiv:math-ph/*0203041v2 7 May
[25] von Neumann J and Wigner E.P. (1929) *Physik Z* **30** 467
[26] Crum M M 1955 *Quart. J. Math. Oxford Ser.* **6** (2) 127
[27] Stahlhofen A A 1995 *Phys. Rev. A* **51** 934;
Stahlhofen A A 1996 *J. Phys. A: Math. Gen.* **29** L581
[28] Newton R G (1982) *Scattering Theory of Wave and Particles* (New York Heidelberg Berlin: Springer-Verlag) Chapter XVII
[29] Fernández-García N, Hernández E, Jáuregui A and Mondragón A 2013 *J. Phys. A: Math. Theor.* **46** 175302
[30] Hernández E, Jáuregui A and Mondragón A 2003 *Phys. Rev. A* **67** 022721
[31] Lancaster P and Tismenetsky (1985) *The Theory of Matrices* (2nd. ed Academic Press: New York)

Journal of Physics: Conference Series **512** (2014) 012024 doi:10.1088/1742-6596/512/1/012024

Supersymmetric partners of the truncated harmonic oscillator

David J. Fernández C.[1] **and VS Morales-Salgado**[2]

Departamento de Física, Cinvestav A.P. 14-740, *07000 México D.F. Mexico*

E-mail: [1]`david@fis.cinvestav.mx`, [2]`vmorales@fis.cinvestav.mx`

Abstract. First and second order supersymmetric transformations are applied to the truncated harmonic oscillator to generate new Hamiltonians with known spectra. We also study the effect of these transformations on the eigenfunctions of the initial Hamiltonian. Finally the link between first and the second order supersymmetric partners of the truncated harmonic oscillator which possess third-order differential ladder operators with the Painlevé IV equation is used to obtain several solutions of this non-linear second-order differential equation.

1. Introduction

When generating quantum mechanical potentials with known spectra Supersymmetric Quantum Mechanics (SUSY QM) has gained a major effectiveness [1–13]. It is well known that the supersymmetric partners of the harmonic oscillator provide explicit realizations of the polynomial Heisenberg algebras (PHA) [7, 14–16]. These are deformations of the oscillator algebra, where the differential ladder operators are of order $m+1$ and the commutator between them is a polynomial of order m in the Hamiltonian.

It is remarkable that systems characterized by second order PHA have been connected to the Painlevé IV (PIV) equation [14, 17–21]. Conversely, if a Hamiltonian having third order differential ladder operators and their extremal states are found, thus solutions to the PIV equation can be generated in a simple way.

By means of this technique, plenty of non-singular solutions to the PIV equation have been derived [11, 22, 23]. Now we will start making a systematic analysis of the singular solutions by allowing the existence of one fixed singularity, which for simplicity will be placed at the origin. Our treatment is based on the harmonic oscillator with an infinite potential barrier at $x = 0$, (see e.g. [4, 24]), which we will call truncated harmonic oscillator. We are mainly interested in transformations which reproduce again the singularity present in the initial potential. We shall describe also the induced spectral modifications and the second order PHA characterizing the new Hamiltonians, which will naturally lead to solutions to the PIV equation.

In order to achieve our goals, in Section 2 we will review briefly the SUSY QM and the way in which Hamiltonians being intertwined with the harmonic oscillator realize the second order PHA, connecting them later with the PIV equation and some of its solutions. In Section 3 we will study the truncated harmonic oscillator and we will apply to it the first and second order SUSY techniques. In Section 4 we will obtain several solutions to the PIV equation, either non-singular or with a singularity at $x = 0$, by using the extremal states of the SUSY partner

Hamiltonians of the truncated harmonic oscillator. Finally, in Section 5 we will emphasize the original results contained in this paper as well as our conclusions.

2. Supersymmetric Quantum Mechanics

Supersymmetric quantum mechanical systems are characterized by three hermitian operators: a supersymmetric Hamiltonian $H_{\rm ss}$ and two supercharges Q_1, Q_2, satisfying the following supersymmetry algebra with two generators [25]:

$$[H_{\rm ss}, Q_i] = 0, \qquad \{Q_i, Q_j\} = \delta_{ij} H_{\rm ss}, \quad i, j = 1, 2, \tag{1}$$

where $[F, G]$ and $\{F, G\}$ are the commutator and anticommutator of F and G respectively.

The simplest way to realize this algebra is to suppose that a pair of quantum Hamiltonians H and $\tilde{H}$ with real potentials obey the intertwining relations [26]

$$\tilde{H} A^+ = A^+ H \quad \Leftrightarrow \quad H A = A \tilde{H}, \tag{2}$$

where

$$H = -\frac{1}{2}\frac{\mathrm{d}^2}{\mathrm{d}x^2} + V(x), \quad \tilde{H} = -\frac{1}{2}\frac{\mathrm{d}^2}{\mathrm{d}x^2} + \tilde{V}(x),$$

and operators A^+ and A are differential intertwining operators of order k which satisfy

$$AA^+ = \prod_{i=1}^{k}(H - \epsilon_i), \qquad A^+A = \prod_{i=1}^{k}(\tilde{H} - \epsilon_i), \qquad \epsilon_i \in \mathbb{R}. \tag{3}$$

For the actual realization of (1) let us choose

$$Q_1 = \frac{Q^+ + Q^-}{\sqrt{2}}, \quad Q_2 = \frac{Q^+ - Q^-}{\mathrm{i}\sqrt{2}}, \quad Q^+ = \begin{pmatrix} 0 & A^+ \\ 0 & 0 \end{pmatrix}, \quad Q^- = \begin{pmatrix} 0 & 0 \\ A & 0 \end{pmatrix},$$

so that

$$H_{\rm ss} = \begin{pmatrix} \tilde{H} - \epsilon_1 & 0 \\ 0 & H - \epsilon_1 \end{pmatrix} \cdots \begin{pmatrix} \tilde{H} - \epsilon_k & 0 \\ 0 & H - \epsilon_k \end{pmatrix}, \quad i = 1, ..., k.$$

Since A^+ and A are of k-th order, this representation is known as k-SUSY QM.

2.1. 1-SUSY

Let the operators A^+ and A be of first order (in a system of units such that $\hbar = m = 1$), i.e.,

$$A^+ = \frac{1}{\sqrt{2}}\left[-\frac{\mathrm{d}}{\mathrm{d}x} + \alpha(x)\right], \qquad A = \frac{1}{\sqrt{2}}\left[\frac{\mathrm{d}}{\mathrm{d}x} + \alpha(x)\right], \tag{4}$$

where $\alpha(x)$ is a real function of x. By plugging these expressions in the intertwining relations (2) and then substituting $\alpha = [\ln(u)]' = u'/u$ it turns out that u must satisfy the stationary Schrödinger equation $-\frac{1}{2}u'' + Vu = \epsilon u$, where ϵ is a real constant called *factorization energy*. In addition, we obtain the following expression for the new potential:

$$\tilde{V} = V - \alpha' = V - [\ln(u)]''. \tag{5}$$

Hence, if we choose a nodeless *seed solution* u of the stationary Schrödinger equation (also called *transformation function*) associated to a given factorization energy ϵ, then the intertwining operators A^+, A, and the new Hamiltonian $\tilde{H}$ become completely determined. Moreover, departing from the normalized eigenfunctions $\psi_n(x)$ of H associated to the eigenvalues E_n, the corresponding ones $\phi_n(x)$ of $\tilde{H}$ are typically found through

$$\phi_n(x) = \frac{A^+\psi_n(x)}{\sqrt{E_n - \epsilon}}. \tag{6}$$

An additional eigenfunction $\phi_\epsilon(x) \propto 1/u(x)$ of $\tilde{H}$, associated to ϵ, could exist. If $\phi_\epsilon(x)$ satisfies the given boundary conditions then ϵ must be joined to the set of eigenvalues of $\tilde{H}$.

2.2. 2-SUSY

Let us suppose now that the intertwining operators A^+ and A are of second order [26–31], i.e.,

$$A^+ = \frac{1}{2}\left[\frac{d^2}{dx^2} - \eta(x)\frac{d}{dx} + \gamma(x)\right], \qquad A = \frac{1}{2}\left[\frac{d^2}{dx^2} + \eta(x)\frac{d}{dx} + \eta'(x) + \gamma(x)\right]. \tag{7}$$

A similar treatment as for the previous subsection yields

$$\tilde{V} = V - \eta' = V - [\ln W(u_1, u_2)]''. \tag{8}$$

where $W(u_1, u_2) = u_1 u_2' - u_1' u_2$ is the Wronskian of two seed solutions $u_{1,2}$ of the stationary Schrödinger equation associated to $\epsilon_{1,2}$ respectively.

In this fashion, if $u_{1,2}$ are choosen such that $W(u_1, u_2)$ is nodeless inside the domain of $V(x)$, it turns out that A^+, A, and $\tilde{H}$ become once again completely determined. Moreover, the eigenfunctions $\phi_n(x)$ of $\tilde{H}$ associated to the eigenvalues E_n become obtained typically from those $\psi_n(x)$ of H through the standard expression:

$$\phi_n(x) = \frac{A^+\psi_n(x)}{\sqrt{(E_n - \epsilon_1)(E_n - \epsilon_2)}}. \tag{9}$$

Two extra eigenfunctions $\phi_{\epsilon_1}(x) \propto \frac{u_2}{W(u_1,u_2)}$, $\phi_{\epsilon_2}(x) \propto \frac{u_1}{W(u_1,u_2)}$ of $\tilde{H}$, associated to the eigenvalues $\epsilon_{1,2}$, could exist [26]. If $\phi_{\epsilon_{1,2}}(x)$ satisfy the given boundary conditions then $\epsilon_{1,2}$ must be included in the spectrum of $\tilde{H}$.

2.3. Polynomial Heisenberg Algebras

The polynomial Heisenberg algebras (PHA) can be seen as deformations of the harmonic oscillator algebra, for which two standard commutation relations remain $[\mathbb{H}, L^\pm] = \pm L^\pm$ and the third one defines the deformation

$$[L^-, L^+] \equiv N(\mathbb{H} + 1) - N(\mathbb{H}) = P_m(\mathbb{H}), \tag{10}$$

where $N(\mathbb{H}) \equiv L^+L^-$ is a polynomial of degree $m+1$ in the Hamiltonian $\mathbb{H}$ factorized as,

$$N(\mathbb{H}) = \prod_{i=1}^{m+1}(\mathbb{H} - \varepsilon_i), \tag{11}$$

so that $P_m(\mathbb{H})$ becomes of degree m [16].

Let us realize these PHA by expressing the commutation relation which involves $\mathbb{H}$ and L^+ in the standard intertwining form:

$$(\mathbb{H} - 1)L^+ = L^+\mathbb{H}.$$

A comparison with (2) makes it natural to identify $H = \mathbb{H}$, $\tilde{H} = \mathbb{H} - 1$, $A^+ = L^+$, $A = L^-$, $k = m+1$ and $\epsilon_i = \varepsilon_i - 1$. Thus, (3) automatically leads to the commutation relation of (10).

Let us consider now a function $\phi(x)$ in the kernel of L^-, which in turn meets

$$N(\mathbb{H})\phi = L^+L^-\phi = 0.$$

Since the kernel of L^- is invariant under the action of $\mathbb{H}$, we can choose as the linearly independent functions generating this subspace the solutions of the stationary Schrödinger equation for $\mathbb{H}$ associated to ε_i, $\mathbb{H}\phi_{\varepsilon_i} = \varepsilon_i\phi_{\varepsilon_i}$.

Looking for the more general systems ruled by PHA with $m = 0, 1$ we arrive to the harmonic oscillator and effective 'radial' oscillator potentials (which have ladder operators of first and second orders respectively). On the other hand, for $m = 2$ (third order ladder operators which will be denoted by $l^{\pm}$) the corresponding potential turns out to be determined by a function which satisfies the PIV equation [7]. In order to see this explicitly, let us assume that $\mathbb{H} = -\frac{1}{2}\frac{\mathrm{d}^2}{\mathrm{d}x^2} + \mathbb{V}(x)$, and $l^+ = I_1^+ I_2^+$, $l^- = I_2^- I_1^-$, where

$$I_1^+ = \frac{1}{\sqrt{2}}\left[-\frac{d}{dx} + f(x)\right], \qquad I_2^+ = \frac{1}{2}\left[\frac{d^2}{dx^2} + g(x)\frac{d}{dx} + h(x)\right].$$

The previous factorized expressions for $l^{\pm}$ are useful since it is employed an auxiliar Hamiltonian $H_{\rm a} = -\frac{1}{2}\frac{\mathrm{d}^2}{\mathrm{d}x^2} + V_{\rm a}(x)$ which is intertwined with $\mathbb{H}$ as follows: $(\mathbb{H} - 1)I_1^+ = I_1^+ H_{\rm a}$, $H_{\rm a} I_2^+ = I_2^+ \mathbb{H}$. By using then the formulae obtained for 1-SUSY and 2-SUSY and after several calculations we arrive to the following final result:

$$\mathbb{V} = \frac{x^2}{2} - \frac{g'}{2} + \frac{g^2}{2} + xg + \varepsilon_1 - \frac{1}{2}, \tag{12}$$

where $g(x)$ satisfies the Painlevé IV equation,

$$g'' = \frac{(g')^2}{2g} + \frac{3}{2}g^3 + 4xg^2 + 2(x^2 - a)g + \frac{b}{g},$$

with parameters $a = \varepsilon_2 + \varepsilon_3 - 2\varepsilon_1 - 1$, $b = -2(\varepsilon_2 - \varepsilon_3)^2$.

Therefore, given a solution g of the PIV equation with parameters $a = \varepsilon_2 + \varepsilon_3 - 2\varepsilon_1 - 1$, $b = -2(\varepsilon_2 - \varepsilon_3)^2$, $\varepsilon_{1,2,3} \in \mathbb{R}$, we can construct a system obeying a second-order PHA, characterized by the potential in (12).

Let us recall that ε_i, $i = 1, 2, 3$ are the three roots involved in (11) for $m = 2$, which at the same time coincide with the energies for the three extremal ϕ_{ε_i} of $\mathbb{H}$, i.e., $l^- \phi_{\varepsilon_i} = 0$, $i = 1, 2, 3$. We can obtain a simple analytic expressions for one of the extremal states [7]:

$$\phi_{\varepsilon_1} \propto \exp\left(-\frac{x^2}{2} - \int g\mathrm{d}x\right). \tag{13}$$

This implies that if we find a system ruled by second-order PHA, in particular its extremal states, then we can build solutions to the Painlevé IV equation as long as the extremal state is not identically null. In order to see this, let us rewrite the expression for the extremal state ϕ_{ε_1} of (13) in the form:

$$g(x) = -x - [\ln \phi_{\varepsilon_1}]', \tag{14}$$

i.e., a solution $g(x)$ to the PIV equation in terms of the extremal state ϕ_{ε_1} of $\mathbb{H}$ has been found.

2.4. Harmonic Oscillator

When the k-SUSY technique is applied to the harmonic oscillator Hamiltonian

$$H = -\frac{1}{2}\frac{\mathrm{d}^2}{\mathrm{d}x^2} + \frac{x^2}{2},$$

k new levels below the ground state energy $E_0 = 1/2$ of the oscillator can be created at the positions defined by the *factorization energies* ϵ_j, $j = 1, .., k$ involved in (3) [15, 26]. Supposing

Journal of Physics: Conference Series **512** (2014) 012024 doi:10.1088/1742-6596/512/1/012024

that this happens, then the eigenfunctions $\phi_n(x)$ of the new Hamiltonian $\tilde{H}$, associated to the eigenvalues $E_n = n + 1/2$ of the initial one H, are given by a generalization of (6) and (9):

$$\phi_n(x) = \frac{A^+\psi_n(x)}{\sqrt{(E_n - \epsilon_1)...(E_n - \epsilon_k)}}.$$

Furthermore, the eigenfunctions ϕ_{ϵ_j} associated to the new levels ϵ_j can be written as

$$\phi_{\epsilon_j} \propto \frac{W(u_1, .., u_{j-1}, u_{j+1}, ..., u_k)}{W(u_1, .., u_k)}, \quad j = 1, .., k, \tag{15}$$

where $W(u_1, .., u_k)$ is the Wronskian of the k seed solutions u_j, $j = 1, .., k$ used to implement the transformation, which satisfy $Hu_j = \epsilon_j u_j$. Up to a constant factor, the general solution to this equation with $V(x) = \frac{x^2}{2}$ and ϵ arbitrary is given by

$$u(x) = \mathrm{e}^{-x^2/2}\left[{}_1\mathrm{F}_1\left(\frac{1-2\epsilon}{4}; \frac{1}{2}; x^2\right) + 2\nu\frac{\Gamma(\frac{3-2\epsilon}{4})}{\Gamma(\frac{1-2\epsilon}{4})}\, x\, {}_1\mathrm{F}_1\left(\frac{3-2\epsilon}{4}; \frac{3}{2}; x^2\right)\right].$$

Thus, each u_j takes this form with ϵ substituted by ϵ_j and ν by ν_j. For this transformation not to be singular $W(u_1, .., u_k)$ must not have zeros in the real axis. For simplicity let us assume from now on that $\epsilon_k < \epsilon_{k-1} < ... < \epsilon_1 < E_0 = 1/2$. With this ordering $W(u_1, ..., u_k)$ will not have zeros if $|\nu_j| < 1$ for j odd and $|\nu_j| > 1$ for j even, and thus the new potential

$$\tilde{V}(x) = \frac{x^2}{2} - [\ln W(u_1, ..., u_k)]''$$

will not have singularities.

It is important to notice that the Hamiltonian $\tilde{H}$ has well defined ladder operators of order $2k + 1$ [1,15]:

$$L^+ = A^+a^+A, \qquad L^- = A^+a^-A.$$

If $k = 1$ the ladder operators $L^\pm$ are of third order and $\{\tilde{H}, L^+, L^-\}$ directly generates a second order PHA. On the other hand, since A^+ and A are of second order if $k = 2$, then $L^\pm$ will be of fifth order in such a case. It is important to know under which circumstances $L^\pm$ can be 'reduced' to third order ladder operators. The answer is contained in the following result [11]: if the seed solutions $u_1(x)$ and $u_2(x)$ are such that $u_2 = a^-u_1$ and $\epsilon_2 = \epsilon_1 - 1$, then $L^\pm$ can be factorized as

$$L^+ = \left(\tilde{H} - \epsilon_1\right) l^+, \qquad L^- = l^- \left(\tilde{H} - \epsilon_1\right),$$

where l^+ and l^- are third order differential ladder operators of $\tilde{H}$, such that $[\tilde{H}, l^\pm] = \pm l^\pm$, which also satisfy

$$l^+l^- = (\tilde{H} - \epsilon_2)(\tilde{H} - \epsilon_1 - 1)(\tilde{H} - 1/2).$$

Now that we have identified the system having third-order differential ladder operators, we can generate solutions to the PIV equation through its extremal states (see (14)).

3. Truncated harmonic oscillator

In what follows we will be interested in the Hamiltonian $H_0 = -\frac{1}{2}\frac{\mathrm{d}^2}{\mathrm{d}x^2} + V_0(x)$ where

$$V_0(x) = \begin{cases} \frac{x^2}{2} & \text{if } x > 0 \\ \infty & \text{if } x \leq 0. \end{cases}$$

The eigenvalues of H_0 take the form $E_n = 2n + \frac{3}{2}$ with corresponding eigenfunctions

$$\psi_n(x) = C_n\, x\, \mathrm{e}^{-x^2/2}\, {}_1\mathrm{F}_1\Big(-n; \frac{3}{2}; x^2\Big),$$

with $n \in \mathbb{N}$ and $C_n = 2\frac{(2n+1)!}{\pi^{1/4} n!}\sqrt{\frac{2^{-2n}}{(2n+1)!}}$ being normalization constants. These correspond to the odd eigenfunctions of the standard harmonic oscillator normalized in the domain $(0, \infty)$, which are the ones that satisfy the boundary conditions at $x = 0$ and in the limit $x \to \infty$.

Further ahead we will also require the even eigenfunctions of the standard harmonic oscillator normalized in the domain $(0, \infty)$,

$$\chi_n(x) = B_n\, \mathrm{e}^{-x^2/2}\, {}_1\mathrm{F}_1\Big(-n; \frac{1}{2}; x^2\Big),$$

which are associated to $\mathcal{E}_n = 2n + \frac{1}{2}$, where $n \in \mathbb{N}$ and $B_n = \frac{(2n)!}{\pi^{1/4} n!}\sqrt{\frac{2^{1-2n}}{(2n)!}}$ are their normalization constants. Although they satisfy $H_0 \chi_n = \mathcal{E}_n \chi_n$, they do not obey the boundary condition at $x = 0$, and thus they are not eigenfunctions of H_0.

3.1. 1-SUSY

Let us suppose now that H_0 is intertwined with another Hamiltonian $H_1 = -\frac{1}{2}\frac{\mathrm{d}^2}{\mathrm{d}x^2} + V_1$ as in (2), where the interwining operators A^+, A are given by (4). Thus, a single transmformation function $u(x)$ is needed, which must satisfy the stationary Schrödinger equation:

$$-\frac{1}{2}u'' + V_0 u = \epsilon u. \tag{16}$$

For $x > 0$, this equation has a general solution given by

$$u(x) = \mathrm{e}^{-x^2/2}\left[b_1\, {}_1\mathrm{F}_1\left(\frac{1-2\epsilon}{4}; \frac{1}{2}; x^2\right) + b_2\, x\, {}_1\mathrm{F}_1\left(\frac{3-2\epsilon}{4}; \frac{3}{2}; x^2\right)\right], \tag{17}$$

b_1, b_2 being real constants [3]. The boundary conditions required for the eigenfunctions of the new Hamiltonian H_1 will be the same as for H_0, i.e., to vanish at $x = 0$ and for $x \to \infty$. Thus the transformation function in (17) must have a well defined parity and we can very well identify two different cases.

3.2. Odd transformation function

For an odd transformation function let us choose $b_1 = 0$ and $b_2 = 1$ in (17), which together with the expression in (5) yield the potential

$$V_1 = V_0 + \frac{1}{x^2} + 1 - \left\{ \ln\left[{}_1\mathrm{F}_1\Big(\frac{3-2\epsilon}{4}; \frac{3}{2}; x^2\Big)\right]\right\}'', \qquad x > 0. \tag{18}$$

In this expression, the term $\frac{1}{x^2}$ induces in a natural way the vanishing boundary condition at $x = 0$, while the term with the double derivative shows that transformations with $\epsilon > \frac{3}{2}$ are not allowed since they generate additional singularities for $x > 0$.

As is shown in (6), an eigenfunction $\psi_n(x)$ of H_0 asociated to the eigenvalue E_n typically transforms into an eigenfunction $\phi_n(x)$ of H_1 asociated to E_n. In our case this remains true

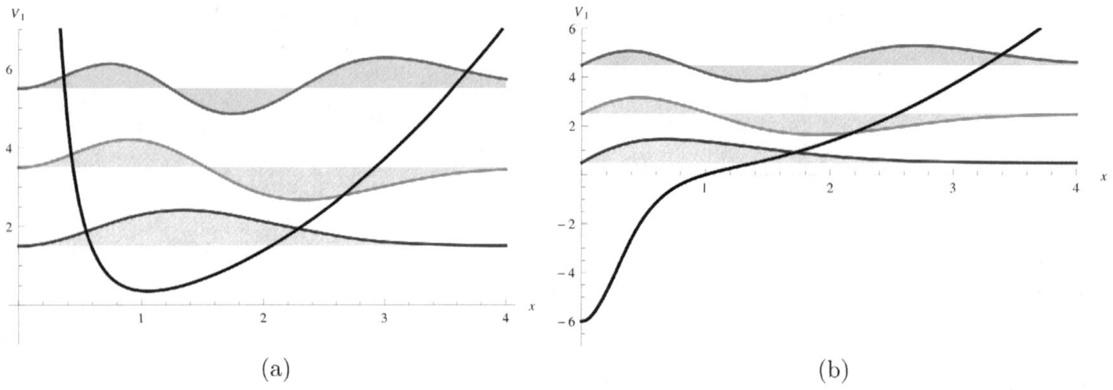

Figure 1. The potential V_1 and its first three eigenfunctions obtained for an odd seed solution (a) and an even one (b), both with factorization energy $\epsilon = -3$.

(see however the next subsection), and when substituting the expressions for $u(x)$ and $\psi_n(x)$ we obtain explicitly the eigenfunctions $\phi_n(x)$:

$$\phi_n(x) = D_n x^2 e^{-x^2/2}\left\{\frac{4n}{3}\,{}_1F_1\left(1-n;\frac{5}{2};x^2\right) + \left(1-\frac{2}{3}\epsilon\right)\left[\frac{{}_1F_1(\frac{7-2\epsilon}{4};\frac{5}{2};x^2)}{{}_1F_1(\frac{3-2\epsilon}{4};\frac{3}{2};x^2)}\right]{}_1F_1\left(-n;\frac{3}{2};x^2\right)\right\},$$

with $D_n = \frac{C_n}{\sqrt{2(E_n-\epsilon)}}$ being normalization constants. The corresponding energies $E_n = 2n+\frac{3}{2}$, $n = 0,1,2,\ldots$ thus belong to the spectrum of H_1. Some eigenfunctions $\phi_n(x)$ along with their corresponding potential have been drawn in figure 1(a). In addition, since $\phi_\epsilon(x) \propto 1/u(x)$ diverges also for $x = 0$, then ϵ does not belong to the spectrum of H_1.

The limit case $\epsilon \to \frac{3}{2}$ is worth of attention, since for this factorization energy the ground state level of H_0 is erased from the spectrum of the new hamiltonian H_1. Although the new spectrum is equivalent to the old one through a finite displacement in the energy, the form of the new potential is different from the initial one due to the singular term $1/x^2$ (see (18)).

3.3. Even transformation function

Now let us choose the even solution of (16) as transformation function; using once again (5) the potential $V_1(x)$ turns out to be

$$V_1(x) = V_0(x) + 1 - \left\{\ln\left[{}_1F_1\left(\frac{1-2\epsilon}{4};\frac{1}{2};x^2\right)\right]\right\}''.$$

This potential also has a singularity at $x = 0$, since $V_0(x)$ includes the infinite potential barrier. Transformations with $\epsilon > \frac{1}{2}$ are not allowed, due to they generate additional singularities for $x > 0$.

This time, the even solutions $\chi_n(x)$ of H_0 are mapped into the eigenfunctions of H_1:

$$\phi_n(x) = D_n x\, e^{\frac{-x^2}{2}}\left\{4n\,{}_1F_1\left(1-n;\frac{3}{2};x^2\right) + \left(1-2\epsilon\right)\frac{{}_1F_1(\frac{5-2\epsilon}{4};\frac{3}{2};x^2)}{{}_1F_1(\frac{1-2\epsilon}{4};\frac{1}{2};x^2)}\,{}_1F_1\left(-n;\frac{1}{2};x^2\right)\right\},$$

with $D_n = \frac{B_n}{\sqrt{2(\mathcal{E}_n-\epsilon)}}$ being normalization constants. Some of them are plotted in figure 1(b) along with the corresponding potential. As in the previous case, the function $\phi_\epsilon(x) \propto 1/u(x)$

Journal of Physics: Conference Series **512** (2014) 012024 doi:10.1088/1742-6596/512/1/012024

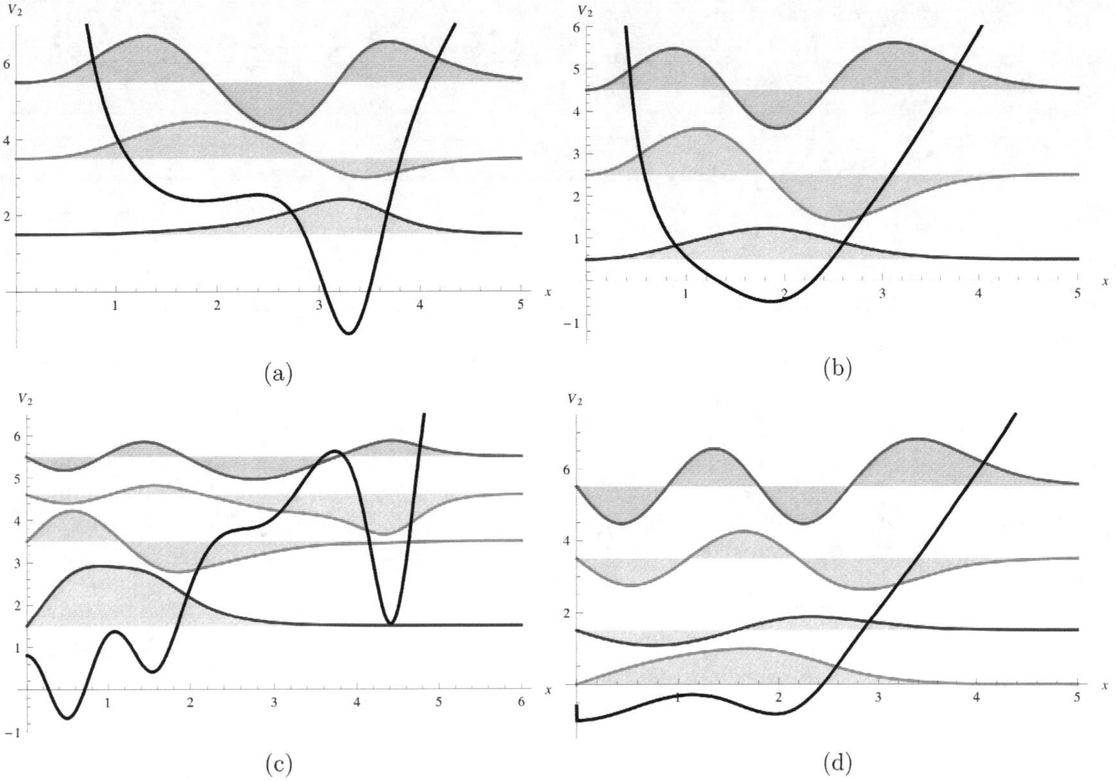

Figure 2. The potential V_2 and its first four eigenfunctions obtained from: (a) two odd seed solutions with factorization energies $\epsilon_1 = \frac{11}{8}$ and $\epsilon_2 = \frac{5}{4}$, (b) two even seed solutions with factorization energies $\epsilon_1 = -\frac{2}{8}$ and $\epsilon_2 = -\frac{3}{8}$, (c) odd and even seed solutions with factorization energies $\epsilon_1 = 5$ and $\epsilon_2 = 4.6$, (d) even and odd seed solutions with factorization energies $\epsilon_1 = 0$ and $\epsilon_2 = -\frac{1}{2}$.

does not obey the boundary condition at $x = 0$ and thus the associated factorization energy ϵ does not belong to the spectrum of H_1, which is composed by the levels $\mathcal{E}_n = 2n + \frac{1}{2}$, $n = 0, 1, \ldots$

Once again, there is a notorious limit $\epsilon \to \frac{1}{2}$, since in this case the otherwise ground state energy level $\mathcal{E}_0 = \frac{1}{2}$ is erased from the spectrum of the new hamiltonian H_1. Notice that in this case the new potential and its spectrum become the same as the initial ones (up to a finite displacement in the energy).

3.4. 2-SUSY

Let us suppose now that H_0 is intertwined with a different Hamiltonian H_2 as in (2), A^+ and A being identified with the second order ones of (7). From (8) we can see that the new potential can be written as

$$V_2 = V_0 - \eta' = V_0 - [\ln W(u_1, u_2)]''.$$

In addition, the eigenfunctions $\psi_n(x)$ of H_0 typically transform into eigenfunctions $\phi_n(x)$ of H_2 through the action of the intertwining operator A^+ as in (9).

According to subsection 2.2, the transformation functions $u_1(x)$ and $u_2(x)$ must satisfy the stationary Schrödinger equation. As it was done previously, we are going to choose $u_1(x)$ and $u_2(x)$ as parity definite solutions for the ordering $\epsilon_1 > \epsilon_2$. Note that there exist four different parity combinations leading to four different kinds of second-order transformations.

3.5. Odd-odd transformation functions

Let us choose $u_1 = x\,e^{-\frac{x^2}{2}}\,{}_1F_1(\frac{3-2\epsilon_1}{4};\frac{3}{2};x^2)$ and $u_2 = x\,e^{-\frac{x^2}{2}}\,{}_1F_1(\frac{3-2\epsilon_2}{4};\frac{3}{2};x^2)$ to attain a new potential with the singularity separated (e.g. [8]), i.e.,

$$V_2(x) = \frac{x^2}{2} + \frac{3}{x^2} + 2 - [\ln w(x)]'' \quad \text{for} \quad x \geq 0,$$

where $w(x)$ turns out to be a continuous function without zeros in $x \geq 0$ as long as the factorization energies satisfy $\epsilon_2 < \epsilon_1 \leq \frac{3}{2} = E_0$ or $E_j = \frac{3+4j}{2} \leq \epsilon_2 < \epsilon_1 \leq \frac{3+4(j+1)}{2} = E_{j+1}$. These are now precisely the conditions to produce a non-singular transformation in said domain.

The eigenfunctions $\phi_n(x)$ of H_2 can be found from those of H_0 in the standard way, $\phi_n(x) \propto A^+\psi_n(x)$, and they satisfy the appropriate boundary conditions so that the eigenvalues E_n in general belong to the spectrum of H_2. In addition, ϕ_{ϵ_1} and ϕ_{ϵ_2} diverge at $x = 0$ and, hence, neither ϵ_1 nor ϵ_2 belong to the spectrum of H_2.

There are several limit cases through which we can delete either one or two levels of H_0 for arriving to H_2. For instance, the initial ground state energy E_0 can be deleted by making $\epsilon_1 = E_0$, $\epsilon_2 < E_0$. On the other hand, in the domain $E_j \leq \epsilon_2 < \epsilon_1 \leq E_{j+1}$ we can delete either E_j or E_{j+1}, by taking $\epsilon_2 = E_j$ with $E_j < \epsilon_1 < E_{j+1}$ in the first case or $\epsilon_1 = E_{j+1}$ and $E_j < \epsilon_2 < E_{j+1}$ in the second. Moreover, the two consecutive levels E_j, E_{j+1} can be deleted by choosing $\epsilon_2 = E_j$ and $\epsilon_1 = E_{j+1}$.

In figure 2(a) we can see an example of the new potential V_2 and several of its eigenfunctions $\phi_n(x)$ for $\epsilon_2 < \epsilon_1 < 3/2$.

3.6. Even-even transformation functions

Let us take now $u_1 = e^{-\frac{x^2}{2}}\,{}_1F_1(\frac{1-2\epsilon_1}{4};\frac{1}{2};x^2)$ and $u_2 = e^{-\frac{x^2}{2}}\,{}_1F_1(\frac{1-2\epsilon_2}{4};\frac{1}{2};x^2)$. We obtain that

$$V_2(x) = \frac{x^2}{2} + \frac{1}{x^2} + 2 - [\ln w(x)]'' \quad \text{for} \quad x \geq 0,$$

where $w(x)$ is a continous function without zeros for $x \geq 0$ as long as the factorization energies satisfy $\epsilon_2 < \epsilon_1 \leq \frac{1}{2} = \mathcal{E}_0$ or $\mathcal{E}_j = \frac{1+4j}{2} \leq \epsilon_2 < \epsilon_1 \leq \frac{1+4(j+1)}{2} = \mathcal{E}_{j+1}$, which are the conditions for the transformation to be non-singular in its domain.

Note that the even solutions $\chi_n(x)$, that do not satisfy the boundary conditions, transform now into the eigenfunctions $\phi_n(x) \propto A^+\chi_n$ of H_2, which do satisfy the boundary conditions and thus, the corresponding eigenvalues $\mathcal{E}_n$ belong to the spectrum of H_2. As in the previous case, solutions $\phi_{\epsilon_{1,2}}$ associated to $\epsilon_{1,2}$ diverge at $x = 0$ and thus $\epsilon_{1,2} \notin \mathrm{Sp}(H_2)$.

The limit cases for which one or two neighbour levels $\mathcal{E}_j$ disappear from $\mathrm{Sp}(H_2)$ work similarly as in the previous case. Thus, by taking $\epsilon_1 = \mathcal{E}_0$, $\epsilon_2 < \mathcal{E}_0$ it turns out that $\mathcal{E}_0 \notin \mathrm{Sp}(H_2)$. On the other hand, if we make either $\epsilon_2 = \mathcal{E}_j$ with $\mathcal{E}_j < \epsilon_1 < \mathcal{E}_{j+1}$ or $\epsilon_1 = \mathcal{E}_{j+1}$ with $\mathcal{E}_j < \epsilon_2 < \mathcal{E}_{j+1}$, it turns out that either $\mathcal{E}_j \notin \mathrm{Sp}(H_2)$ or $\mathcal{E}_{j+1} \notin \mathrm{Sp}(H_2)$ respectively. In addition, if $\epsilon_2 = \mathcal{E}_j$ and $\epsilon_1 = \mathcal{E}_{j+1}$ then both $\mathcal{E}_j, \mathcal{E}_{j+1} \notin \mathrm{Sp}(H_2)$.

In figure 2(b) one can find some examples of the eigenfunctions $\phi_n(x)$ along with the corresponding potential V_2 for $\epsilon_2 < \epsilon_1 < \frac{1}{2}$.

3.7. Odd-even transformation functions

Let $u_1 = x\,e^{-\frac{x^2}{2}}\,{}_1F_1(\frac{3-2\epsilon_1}{4};\frac{3}{2};x^2)$ and $u_2 = e^{-\frac{x^2}{2}}\,{}_1F_1(\frac{1-2\epsilon_2}{4};\frac{1}{2};x^2)$ with $\epsilon_2 < \epsilon_1$. It turns out that now the potential becomes

$$V_2(x) = \frac{x^2}{2} + 2 - [\ln w(x)]'' \quad \text{for} \quad x \geq 0, \tag{19}$$

where $w(x)$ is a continuous function without zeros for $x \geq 0$ as long as the factorization energies satisfy $\mathcal{E}_j = \frac{1+4j}{2} \leq \epsilon_2 < \epsilon_1 \leq \frac{3+4j}{2} = E_j$, i.e., for these conditions the transformation is found to be non-singular for $x > 0$.

Let us note that the eigenfunctions of H_2 are found here by acting the intertwining operator A^+ onto the eigenfunctions ψ_n of H_0, $\phi_n(x) \propto A^+\psi_n(x)$, since they satisfy the boundary conditions so that their corresponding eigenvalues E_n belong to the spectrum of H_2.

Now we need to know if either ϕ_{ϵ_1}, ϕ_{ϵ_2} or both in (15) satisfy the boundary conditions to become also eigenfunctions of H_2. For $\mathcal{E}_j < \epsilon_2 < \epsilon_1 < E_j$ it turns out that ϕ_{ϵ_2} satisfies the boundary conditions while ϕ_{ϵ_1} does not. This implies that $\epsilon_2 \in \text{Sp}(H_2)$ and $\epsilon_1 \notin \text{Sp}(H_2)$, i.e., through the second-order SUSY transformation it can be created a new level at the position ϵ_2. In addition, for $\epsilon_1 = E_j$ with $\mathcal{E}_j < \epsilon_2 < E_j$ the same result is obtained, but now it implies that $\epsilon_1 = E_j \notin \text{Sp}(H_2)$ and $\epsilon_2 \in \text{Sp}(H_2)$. Thus, by employing the second-order SUSY transformation we have deleted the level E_j and at the same time we have created a new one at ϵ_2, so we have effectively 'moved down' E_j to its new position ϵ_2. For $\epsilon_2 = \mathcal{E}_j$ and $\mathcal{E}_j < \epsilon_1 < E_j$ neither ϕ_{ϵ_1} nor ϕ_{ϵ_2} satisfy the boundary conditions so that $\epsilon_{1,2} \notin \text{Sp}(H_2)$. Finally, for $\epsilon_1 = E_j$ and $\epsilon_2 = \mathcal{E}_j$ the same happens, i.e., we have deleted the level E_j in order to produce H_2.

In figure 2(c) one can find an example of the potential V_2 along with some of its eigenfunctions $\phi_n(x)$ for $\mathcal{E}_1 < \epsilon_2 < \epsilon_1 < E_1$.

3.8. Even-odd transformation functions

Finally, let us choose $u_1 = \mathrm{e}^{-\frac{x^2}{2}}\,{}_1\mathrm{F}_1(\frac{1-2\epsilon_1}{4};\frac{1}{2};x^2)$ and $u_2 = x\,\mathrm{e}^{-\frac{x^2}{2}}\,{}_1\mathrm{F}_1(\frac{3-2\epsilon_2}{4};\frac{3}{2};x^2)$ with $\epsilon_2 < \epsilon_1$. As in the previous section, $V_2(x)$ takes the same form of (19), and the eigenfunctions ϕ_n of H_2 are obtained from those of H_0 through $\phi_n(x) \propto A^+\psi_n$, which satisfy the boundary conditions so that the eigenvalues E_n belong to the spectrum of H_2. For this choice of $u_1(x)$ and $u_2(x)$ the transformation is non-singular as long as the factorization energies satisfy $\epsilon_2 < \epsilon_1 \leq \frac{1}{2} = \mathcal{E}_0$ or $E_j = \frac{3+4j}{2} \leq \epsilon_2 < \epsilon_1 \leq \frac{5+4j}{2} = \mathcal{E}_{j+1}$.

By studying once again if the $\phi_{\epsilon_{1,2}}$ of (15) satisfy the boundary conditions we arrive now to the following results: for $\epsilon_2 < \epsilon_1 < \mathcal{E}_0$ or $E_j < \epsilon_2 < \epsilon_1 < \mathcal{E}_{j+1}$ it turns out that ϕ_{ϵ_1} satisfies the boundary conditions while ϕ_{ϵ_2} does not, meaning that $\epsilon_1 \in \text{Sp}(H_2)$ and $\epsilon_2 \notin \text{Sp}(H_2)$, i.e., a new level is created at ϵ_1. For $\epsilon_1 = \mathcal{E}_0$ and $\epsilon_2 < \mathcal{E}_0$ it is obtained that $\epsilon_{1,2} \notin \text{Sp}(H_2)$, namely, there is no additional level in $\text{Sp}(H_2)$. On the other hand, for $\epsilon_2 = E_j$ and $E_j < \epsilon_1 < \mathcal{E}_{j+1}$ once again $\epsilon_1 \in \text{Sp}(H_2)$ and $\epsilon_2 = E_j \notin \text{Sp}(H_2)$, i.e., through the second-order SUSY transformation the level E_j has been 'moved up' to the position ϵ_1. For $\epsilon_1 = \mathcal{E}_{j+1}$ and $E_j < \epsilon_2 < \mathcal{E}_{j+1}$ neither ϕ_{ϵ_1} nor ϕ_{ϵ_2} satisfy the boundary conditions so that $\epsilon_{1,2} \notin \text{Sp}(H_2)$. Finally, for $\epsilon_1 = \mathcal{E}_{j+1}$ and $\epsilon_2 = E_j$ the same happens, which implies that the level E_j is deleted.

Figure 2(d) shows a potential V_2 and some of its eigenfunctions $\phi_n(x)$ for $\epsilon_2 < \epsilon_1 < \frac{1}{2}$.

4. Solutions to the Painlevé IV equation

In section 2 we saw that it is possible to find solutions $g(x)$ to the PIV equation through

$$g(x) = -x - [\ln \phi_{\varepsilon_1}]',$$

where ϕ_{ε_1} is an extremal state for a system having third order ladder operators $l^{\pm}$. Since Hamiltonians generated from the truncated harmonic oscillator through SUSY techniques can have third order ladder operators, hence solutions to the PIV equation can be straightforwardly obtained, as detailed ahead.

4.1. 1-SUSY

Recall that for a first order SUSY transformation there are three extremal states ϕ_1, ϕ_2 and ϕ_3 with eigenvalues ε_1, ε_2 and ε_3 respectively which satisfy $L^+L^-\phi_i = 0$, $i = 1, 2, 3$. Explicit expressions for such extremal states are well known, and we can label them firstly in the way:

$$\phi_1 \propto \frac{1}{u(x)}, \qquad \phi_2 \propto A^+a^+u(x), \qquad \phi_3 \propto A^+\chi_0,$$

where $\{\varepsilon_1 = \epsilon, \varepsilon_2 = \epsilon+1, \varepsilon_3 = \frac{1}{2}\}$. Moreover, the cyclic permutations of the indices of $\{\varepsilon_1, \varepsilon_2, \varepsilon_3\}$ lead immediately to additional solutions of the PIV equation.

It is worth noticing that the solutions to PIV equation depend on our selection of the transformation function $u(x)$, for which there are two different choices (for a fixed ϵ).

For odd $u(x) = x\,e^{-x^2/2}\,{}_1F_1(\frac{3-2\epsilon}{4}; \frac{3}{2}; x^2)$, the above extremal states and the cyclic permutations of $\{\varepsilon_1, \varepsilon_2, \varepsilon_3\}$ lead to the following three solutions $g_i(x) = -x - [\ln \phi_i]'$ of the PIV equation [24]

$$\begin{aligned}
g_1 &= \frac{1}{x} - 2x + \left(1 - \frac{2}{3}\epsilon\right) x\, \frac{{}_1F_1(\frac{7-2\epsilon}{4}; \frac{5}{2}; x^2)}{{}_1F_1(\frac{3-2\epsilon}{4}; \frac{3}{2}; x^2)}, \\
g_2 &= -g_1 - 2x - 2\left[\frac{x + (2\epsilon - x^2)(g_1 + x) + (g_1 + x)^3}{x^2 - 2\epsilon - 1 - (g_1 + x)^2}\right], \\
g_3 &= -\frac{g_1' + 2}{g_1 + 2x} = \frac{g_1^2 + 2xg_1 + 2\epsilon - 1}{g_1 + 2x}.
\end{aligned}$$

Note that g_1 solves the PIV equation with parameters $a_1 = -\epsilon + \frac{1}{2}$, $b_1 = -2\left(\epsilon + \frac{1}{2}\right)^2$, while g_2 and g_3 do it for $a_2 = -\epsilon - \frac{5}{2}$, $b_2 = -2\left(\epsilon - \frac{1}{2}\right)^2$, and $a_3 = 2\epsilon - 1$, $b_3 = -2$, respectively.

For even $u(x) = e^{-x^2/2}\,{}_1F_1(\frac{1-2\epsilon}{4}; \frac{1}{2}; x^2)$ we obtain [24]

$$g_1 = -2x + (1 - 2\epsilon)\, x\, \frac{{}_1F_1(\frac{5-2\epsilon}{4}; \frac{3}{2}; x^2)}{{}_1F_1(\frac{1-2\epsilon}{4}; \frac{1}{2}; x^2)},$$

while the expressions for g_2 and g_3 in terms of g_1 remain the same as in the previous case.

4.2. 2-SUSY

Recall now that, for the second order SUSY partner Hamiltonians generated from the truncated harmonic oscillator by using as transformation function u_1 and $u_2 = a^-u_1$ with $\epsilon_2 = \epsilon_1 - 1$, there are three extremal states ϕ_1, ϕ_2 and ϕ_3 with eigenvalues chosen as $\varepsilon_1 = \epsilon_1 - 1$, $\varepsilon_2 = \epsilon_1 + 1$ and $\varepsilon_3 = \frac{1}{2}$, respectively, which satisfy $l^+l^-\phi_i = 0$, $i = 1, 2, 3$. Their explicit expressions are given by:

$$\phi_1 \propto \frac{u_1}{W[u_1, u_2]}, \qquad \phi_2 \propto A^+a^+u_1, \qquad \phi_3 \propto A^+\chi_0. \tag{20}$$

Once again, we can choose any permutation of the indices of $\{\varepsilon_1, \varepsilon_2, \varepsilon_3\}$ in order to identify ϕ_1 with any of the three extremal states of the system departing from the choice in (20). Hence we will obtain the following three different solutions of the PIV equation [24]:

$$\begin{aligned}
g_1 &= -x - \alpha + 2\left[\frac{x + \alpha}{x^2 + 1 - 2\epsilon - \alpha^2}\right], \\
g_2 &= g_1 + \frac{2\alpha^2 - 2x^2 + 2(2\epsilon + 1)}{\alpha - g_1 - x}, \\
g_3 &= \frac{(x + \alpha)g_1^2 + \left[2\epsilon - 1 + (x + \alpha)^2\right] g_1 + (2\epsilon - 3)(x + \alpha)}{(x + \alpha)^2 + (x + \alpha)g_1 + 2\epsilon - 1}.
\end{aligned}$$

Here we should remember that $\alpha = \frac{u'}{u}$.

Note that g_1 solves the PIV equation with parameters $a_1 = -\epsilon + \frac{5}{2}$, $b_1 = -2\left(\epsilon + \frac{1}{2}\right)^2$, while g_2 and g_3 do it for $a_2 = -\epsilon - \frac{7}{2}$, $b_2 = -2\left(\epsilon - \frac{3}{2}\right)^2$, and $a_3 = 2(\epsilon - 1)$, $b_3 = -8$, respectively.

Journal of Physics: Conference Series **512** (2014) 012024 doi:10.1088/1742-6596/512/1/012024

5. Conclusions

Supersymmetric partners of the truncated harmonic oscillator were obtained from the 1-SUSY technique, both potentials being isospectral, while those obtained through 2-SUSY offered richer possibilities for the spectral design. For the 2-SUSY case it is possible to erase one or two consecutive levels in the energy spectrum. It is also possible to add a new level to the original spectrum in almost any energy position.

We also have found the values for the factorization energies that produce non-singular first and second order SUSY transformations in $(0, \infty)$.

The 1-SUSY transformations with $u(x)$ even and 2-SUSY transformations with both $u_1(x)$ and $u_2(x)$ even behave in a peculiar way, since they transform the eigenfunctions of the harmonic oscillator which correspond to non-physical solutions of the original system into those which are eigenfunctions of the new Hamiltonian while the eigenfunctions of the initial system are transformed into solutions which do not meet the boundary condition at the origin.

A simple and direct procedure to obtain several explicit solutions to the Painlevé IV equation was implemented using the extremal states for the SUSY partners of the truncated harmonic oscillator.

ACKNOWLEDGMENTS

The authors wish to acknowledge the support of CINVESTAV as well as CONACYT through the project 152574 and the PhD scholarship 243374.

References

[1] Mielnik B 1984 *J. Math. Phys.* **25** 3387
[2] Cooper F, Khare A and Sukhatme U 1995 *Phys. Rep.* **251** 267
[3] Junker G and Roy P 1998 *Ann. Phys.* **270** 155
[4] Márquez I F, Negro J and Nieto L M 1998 *J. Phys. A: Math. Gen.* **31** 4115
[5] Quesne C and Vansteenkiste N 1999 *Helv. Phys. Acta* **72** 71
[6] Mielnik B and Rosas-Ortiz O 2004 *J. Phys. A: Math. Gen.* **37** 10007
[7] Carballo J M, Fernández D J, Negro J and Nieto L M 2004 *J. Phys. A: Math. Gen.* **37** 10349
[8] Contreras-Astorga A and Fernández D J 2008 *J. Phys. A: Math. Theor.* **41** 475303
[9] Marquette I 2009 *J. Math. Phys.* **50** 095202
[10] Fernández D J 2010 *AIP Conf. Proc.* **1287** 3
[11] Bermúdez D and Fernández D J 2011 *SIGMA* **7** 025
[12] Andrianov A A and Ioffe M V 2012 *J. Phys. A: Math. Theor.* **45** 503001
[13] Gómez-Ullate D, Grandati Y and Milson R 2013 arXiv:1306.5143 [math-ph]
[14] Mateo J and Negro J 2008 *J. Phys. A: Math. Theor.* **41** 045204
[15] Fernández D J and Hussin V 1999 *J. Phys. A, Math. Gen.* **32** 3603
[16] Fernández A J, Negro J and Nieto L M 2004 *Phys. Lett. A* **324** 139
[17] Veselov A P and Shabat A B 1993 *Funct. Anal. Appl.* **27** 81
[18] Adler V E 1994 *Physica D* **73** 335
[19] Dubov S Y, Eleonskii V M and Kulagin N E 1994 *Chaos* **4** 47
[20] Bassom A P, Clarkson P A and Hicks A C 1995 *Stud. Appl. Math.* **95** 1
[21] Andrianov A A, Cannata F, Ioffe M and Nishnianidze D 2000 *Phys. Lett. A* **266** 341
[22] Bermúdez D, Fernández D J 2011 *Phys. Lett. A* **375** 2974
[23] Bermúdez D 2012 *SIGMA* **8** 069
[24] Fernández D J and Morales-Salgado V S 2013 arXiv:1310.5628 [math-ph]
[25] Aoyama H, Sato M and Tanaka T 2001 *Nucl. Phys. B* **619** 105
[26] Fernández D J and Fernández-García N 2005 *AIP Conference Proceedings* **744** 236
[27] Andrianov A A, Ioffe M V and Spiridonov V P 1993 *Phys. Lett. A* **174** 273
[28] Andrianov A A, Ioffe M V, Cannata F and Dedonder J P 1995 *Int. J. Mod. Phys. A* **10** 2683
[29] Bagrov V G and Samsonov B F 1997 *Phys. Part. Nucl.* **28** 374
[30] Fernández D J, Hussin V and Mielnik B 1998 *Phys. Lett. A* **244** 309
[31] Cariñena J F, Ramos A and Fernández D J 2001 Ann. Phys. **292** 42

Journal of Physics: Conference Series **512** (2014) 012025 doi:10.1088/1742-6596/512/1/012025

Symmetry group of a particle in an impenetrable cubic well potential

A. O. Hernández-Castillo and R. Lemus

Instituto de Ciencias Nucleares,Universidad Nacional Autónoma de México. A. P. 70-543, Circuito Exterior, C. U. , 04510 México, D. F. , Mexico

E-mail: `renato@nucleares.unam.mx`

Abstract. A quantum particle in a impenetrable cubic well potential presents accidental degeneracy when the $\mathcal{O}_h$ group is considered to be the symmetry group of the system. This degeneracy becomes natural when a new symmetry group, embedding the $\mathcal{O}_h$ group, is proposed. This new group turns out to be the semidirect product $\mathcal{G} = \mathcal{T} \wedge \mathcal{O}_h$, where $\mathcal{T}$ is a two-dimensional compact continuous group whose generators correspond to linear combinations of the one-dimensional Hamiltonians. The systematic degeneracy is studied in detail, the new group is identified and its irreducible representations (irreps) are constructed by means of induction, an approach that allows the irreducibility and completeness to be assured. Pythagorean degeneracy as well as the one due to commensurable sides are not considered.

1. Introduction

The degeneracy degree is expected to correspond with the dimension of one of the symmetry group irreducible representations (irrep) [1]. When this is not the case the degeneracy is identified as accidental. Instances of accidental degeneracy are abundant in quantum mechanics [1], but in any case, according to the experience, the existence of this kind of degeneracy suggests an overlooked higher symmetry. The particle in a square or cubic box with impenetrable walls is the most simple quantum mechanical system where accidental degeneracy appears, however this systems receive a very brief mention in the literature[6, 7, 8], this may be explained by the fact that the natural language of symmetry is group theory, a specialized field not included in most textbooks of quantum mechanics. Though, in 1996 *leyvraz et al* [9] derived a new symmetry group to explain the accidental degeneracy of a square box from the group theory point of view.

A more interesting system presenting systematic degeneracy is a quantum particle in an impenetrable cubic well potential, because it provides a deeper insight into the mathematics and physics of the problem. The description of a particle in a cubic box allows the system of square and parallelepiped well potentials to be studied as a symmetry breaking process, in such a way that these cases can be analyzed as a mathematic subduction problem. The simultaneous analysis of these systems provides a clear example of the fact that the greater the symmetry the higher the degeneracy.

A free particle enclosed in an impenetrable three dimensional box of sides a, b and c, as displayed in Figure 1, is described by the eigenstates $\langle \mathbf{r} | \Psi_{n_1 n_2 n_3} \rangle = \psi_{n_1}(x)\psi_{n_2}(y)\psi_{n_3}(z)$, where $\psi_{n_1}(x) = \sqrt{\frac{2}{a}} \sin\left(\frac{n_1 \pi x}{a}\right)$; $\psi_{n_2}(y) = \sqrt{\frac{2}{b}} \sin\left(\frac{n_2 \pi y}{b}\right)$; $\psi_{n_3}(z) = \sqrt{\frac{2}{c}} \sin\left(\frac{n_3 \pi z}{c}\right)$; with n_i positive integers. These states satisfy the condition $\Psi_{n_1 n_2 n_3}(x, y, z) = 0$ at the boundaries of the box.

The corresponding eigenvalues are given by

$$E_{n_1 n_2 n_2} = \frac{\hbar^2 \pi^2}{2\mu}\left(\frac{n_1^2}{a^2} + \frac{n_2^2}{b^2} + \frac{n_3^2}{c^2}\right), \tag{1}$$

where μ is the reduced mass. The energies and eigenstates by itself does not provide information about the kind of degeneracy displayed, this should be established through the proposition of a symmetry group.

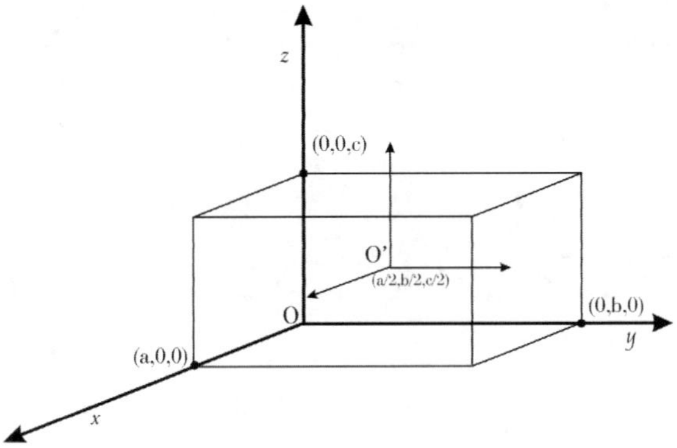

Figure 1. Impenetrable three-dimensional box. The origin $\mathcal{O}$ taken to obtain the wave functions is displaced from the origin $\mathcal{O}'$ where the symmetry elements converge.

When $a = b = c$ we have a cubic box that presents three subspaces of degenerate funtions. For $n_1 = n_2 = n_3$ we have one-dimensional subspaces $\mathcal{L}_1 = \{|\Psi_{nnn}\rangle\}; n_1 = n_2 = n_3 = n$, when only two quantum numbers are equal we have three-dimensional subspaces given by $\mathcal{L}_3 = \{|\Psi_{nnm}\rangle, |\Psi_{nmn}\rangle, |\Psi_{mnn}\rangle\}; n_1 = n_2 = n$, and when all the quantum numbers are different, six dimensional subspaces are described by $\mathcal{L}_6 = \{|\Psi_{n_1 n_2 n_3}\rangle; n_1 \neq n_2 \neq n_3\}$.

2. New symmetry group

To identify the symmetry group we shall first consider the apparent geometrical symmetry of this system, such symmetry corresponds to the point group $\mathcal{O}_h$, hence the eigenstates are expected to carry irreps of this group. The identification of the irreps, however, cannot be achieved straightforwardly since the eigenstates are given with respect to the origin O, while the center of the parallelepiped is located at O'. Both origins are related with the translation $\hat{\mathcal{O}}_T$, with T defined by $\hat{T}(\mathbf{r}_O) = \mathbf{r}_{O'} = \mathbf{r}_O - (\frac{a}{2}, \frac{b}{2}, \frac{c}{2})$. Now, if $\hat{\mathcal{O}}_R$, with $R \in \mathcal{O}_h$, are operators respect to the origin O, the corresponding operators $\hat{\mathcal{O}}'_R$ are obtained through $\hat{\mathcal{O}}_T \hat{\mathcal{O}}_R \hat{\mathcal{O}}_T^{-1} = \hat{\mathcal{O}}'_R$ [14].

The character table for the geometrical symmetry group $\mathcal{O}_h$ is displayed in Table 1 and the reduction assuming this group as the symmetry group is given in Table 2. For $\mathcal{L}_3$ systematic accidental degeneracy appears when n and m are both even or odd. Furthermore the six fold degeneracy appearing in the $\mathcal{L}_6$ subspaces cannot be explained in the context of the $\mathcal{O}_h$ symmetry group [10], because, according to the character table, the expected larger degeneracy degree is three. An interesting peculiarity for this subspace is that a multiplicity of 2 appears for both representations E_g and E_u, Table 2.

Journal of Physics: Conference Series **512** (2014) 012025 doi:10.1088/1742-6596/512/1/012025

Table 1. Character table of the point group O_h. The assignment of the classes is indicated in the second row.

$\mathcal{O}_h$	E	$8C_3$	$3C_2$	$6C_4$	$6C_2'$	I	$8S_6$	$3\sigma_h$	$6S_4$	$6\sigma_d$
	K_1	K_2	K_3	K_4	K_5	K_6	K_7	K_8	K_9	K_{10}
A_{1g}	1	1	1	1	1	1	1	1	1	1
A_{2g}	1	1	1	−1	−1	1	1	1	−1	−1
E_g	2	−1	2	0	0	2	−1	2	0	0
T_{1g}	3	0	−1	1	−1	3	0	−1	1	−1
T_{2g}	3	0	−1	−1	1	3	0	−1	−1	1
A_{1u}	1	1	1	1	1	−1	−1	−1	−1	−1
A_{2u}	1	1	1	−1	−1	−1	−1	−1	−1	−1
E_u	2	−1	2	0	0	−2	1	−2	0	0
T_{1u}	3	0	−1	1	−1	−3	0	1	−1	1
T_{2u}	3	0	−1	−1	1	−3	0	1	1	−1

Table 2. Decomposition of the degenerate states associated with a particle in a cubic box with impenetrable walls. The O_h irreducible representations contained in the degenerate subspaces are display explicitly.

Dimension	Parity	Condition	Representation	Direct product
$1(nnn)$	$n = 2p$	$n_x = n_y = n_z$	A_{2u}	
	$n = 2p + 1$		A_{1g}	
$3(nnm)$	$n = 2p$ $m = 2p$	$n_x = n_y \neq n_z$ $n_x = n_z \neq n_y$ $n_y = n_z \neq n_x$	$A_{2u} \oplus E_u$	$A_{2u} \otimes E_g = E_u$
	$n = 2p + 1$ $m = 2p + 1$		$A_{1g} \oplus E_g$	$A_{1g} \otimes E_g = E_g$
	$n = 2p$ $m = 2p + 1$		T_{2g}	
	$n = 2p + 1$ $m = 2p$		T_{1u}	
$6(nml)$	$n = m = l = 2p$	$n_x \neq n_y \neq n_z$	$A_{1u} \oplus A_{2u} \oplus 2E_u$	$E_g \otimes E_u = A_{1u} \oplus A_{2u} \oplus E_u$
	$n = m = l = 2p + 1$		$A_{1g} \oplus A_{2g} \oplus 2E_g$	$E_g \otimes E_g = A_{1g} \oplus A_{2g} \oplus E_g$
	$n = m = 2p$ $l = 2p + 1$		$T_{1g} \oplus T_{2g}$	$E_g \otimes T_1g = T_{1g} \oplus T_{2g}$
	$n = m = 2p + 1$ $l = 2p$		$T_{1u} \oplus T_{2u}$	$E_g \otimes T_1u = T_{1u} \oplus T_{2u}$

The first thing to establish a new symmetry group is identifying the operator or operators that connect the subspaces spanning $\mathcal{O}_h$ irreps. To achieve this goal we look for an operator $\hat{F}_r^{(\rho)}$, spanning the irrep ρ and satisfies that $\langle \Psi_{\gamma'}^{\Gamma'} | \hat{F}_r^{(\rho)} | \Phi_\gamma^\Gamma \rangle$ vanish unless $\Gamma' \in \rho \otimes \Gamma = \sum_\mu \oplus \mu$ where Γ' and Γ, with components γ' and γ respectively, are the irreps spanned by the kets. It is not difficult to check that such operator must hold the irrep E_g and consequently $\rho = E_g$, [19]. From the $\mathcal{O}_h$ character table [19] we notice that the following Cartesian Harmonics span the irrep E_g:

$$
\begin{aligned}
Y^2_{\cos 2\phi}(x', y', z') &= \frac{1}{4}\sqrt{\frac{15}{\pi}} \frac{(x'^2 - y'^2)}{r'^2}, \\
Y^2_0(x', y', z') &= \frac{1}{4}\sqrt{\frac{5}{\pi}} \frac{(2z'^2 - x'^2 - y'^2)}{r'^2},
\end{aligned}
\tag{2}
$$

The same linear combination in terms of the square momenta does transform according to E_g in both reference systems [9, 10]. We thus have the following operators

$$\begin{aligned}\hat{F}^{(E_g)}_{A_{1g},A_g} &= \frac{1}{4}\sqrt{\frac{5}{\pi}}\,(2\frac{\partial^2}{\partial z^2}-\frac{\partial^2}{\partial x^2}-\frac{\partial^2}{\partial y^2}),\\ \hat{F}^{(E_g)}_{B_{1g},A_g} &= \frac{1}{4}\sqrt{\frac{15}{\pi}}\,(\frac{\partial^2}{\partial x^2}-\frac{\partial^2}{\partial y^2}),\end{aligned} \tag{3}$$

where we have added the components according to the chain $\mathcal{O}_h \supset \mathcal{D}_{4h} \supset \mathcal{D}_{2h}$. The operators (3) explain the accidental degeneracy, because they connect degenerate states with different $\mathcal{O}_h$ irreps.

Once that the accidental degeneracy as well as the operators that connect the degenerate states have been identified, we are ready to propose a new symmetry group for this system. The operators generate the continuous group $\mathcal{T}$, whose elements are obtained by exponentiation and form a two dimensional group with elements

$$U(\alpha,\beta) = U_1(\alpha)\otimes U_2(\beta) = e^{i\alpha\hat{F}^{(E_g)}_{A_{1g}}+i\beta\hat{F}^{(E_g)}_{B_{1g}}}. \tag{4}$$

The corresponding Casimir operator of this group is the Hamiltonian itself.

Considering the transformation of the elements (4) under the action of $R\in\mathcal{O}_h$, we realize that the subgroup $\mathcal{T}$ is invariant in the context of the new group $\mathcal{G}$. This fact allows the new group to be expressed as

$$\mathcal{G} = \mathcal{T}\wedge\mathcal{O}_h, \tag{5}$$

and in terms of left cosets

$$\mathcal{G} = \sum_{\lambda=1}^{|\lambda|} S_\lambda\,\mathcal{T}; \quad |\lambda| = \frac{|\mathcal{G}|}{|\mathcal{T}|} = |\mathcal{O}_h|; \quad S_\lambda\in\mathcal{O}_h. \tag{6}$$

Therefore all the elements $g\in\mathcal{G}$ can be written in the form

$$g = \hat{O}_R\,U(\alpha,\beta); \qquad R\in\mathcal{O}_h, \quad g\in\mathcal{G}, \tag{7}$$

with the product of elements $gg' = g''$.

3. Representations of the group $\mathcal{G}$

Once we have identified the new symmetry group, we proceed to construct the irreps following the induction approach, this method will assure the irreducibility and completeness of the representations [11]. According to this, we have first to construct the irreps of the invariant subgroup $\mathcal{T}$. The representation of the $\mathcal{T}$ elements is diagonal in the basis of the cubic box eigenfunctions.

For the six dimensional space $\mathcal{L}_6 = \{|\Psi_{n_1n_2n_3}\rangle\}$. The action of the subgroup $\mathcal{T}$ elements over the state $|\Psi_{n_1n_2n_3}\rangle$ is depicted by

$$\hat{U}(\alpha,\beta)|\Psi_{n_1n_2n_3}\rangle = D^{(\mathbf{k_n})}(U(\alpha,\beta))\,|\Psi_{n_1n_2n_3}\rangle, \tag{8}$$

where

$$D^{(\mathbf{k_n})}(U(\alpha,\beta)) = e^{i\alpha k^{(1)}_{n_1n_2n_3}+i\beta k^{(2)}_{n_1n_2n_3}} = e^{i\mathbf{k_n}\cdot\mathbf{t}}, \tag{9}$$

where we define the vectors,

$$\begin{aligned}\mathbf{k_n} &\equiv (k^{(1)}_{n_1n_2n_3}, k^{(2)}_{n_1n_2n_3}),\\ \mathbf{t} &\equiv (\alpha,\beta),\end{aligned} \tag{10}$$

Journal of Physics: Conference Series **512** (2014) 012025 doi:10.1088/1742-6596/512/1/012025

The vectors (10) will be considered to be written in an orthonormal basis.

It is appropriate to add the label $\mathbf{k_n}$ to the state in the form $|\Psi^{\mathbf{k_n}}_{n_1 n_2 n_3}\rangle$. If we apply the elements of $\mathcal{T}$ to the other five states of $\mathcal{L}_6$ we obtain five additional representations associated with the functions of permuted indices in accordance to the permutations of S_3. In this context, each function in the space $\mathcal{L}_6$ span a different representation. These irreps are the starting point to obtain the irreps of the complete group $\mathcal{G}$ by induction.

The $\mathcal{T}$ irrep $\mathbf{k}_1$ is invariant under the action of the elements of $\mathcal{T}$ as well as the elements of the subgroup $\mathcal{D}_{2h}$. This set of transformations form a group, the so called *little group* of $\mathbf{k}_1$, denoted by $\mathcal{K}(\mathbf{k}_1)$, in this case given by

$$\mathcal{K}(\mathbf{k}_1) = \mathcal{T} \wedge \mathcal{D}_{2h}. \tag{11}$$

The *little group* of $\mathbf{k}_1$, however, is still infinite. To have a group of finite order we note that

$$\frac{\mathcal{K}(\mathbf{k}_1)}{\mathcal{T}} \approx \mathcal{D}_{2h}. \tag{12}$$

On the other hand

$$\frac{\mathcal{G}}{\mathcal{T}} \approx \mathcal{O}_h, \tag{13}$$

showing that the left coset expansion

$$\mathcal{O}_h = \sum_{\lambda}^{|\lambda|} p_\lambda \, \mathcal{D}_{2h}; \qquad |\lambda| = \frac{|\mathcal{O}_h|}{|\mathcal{D}_{2h}|}, \tag{14}$$

is the relevant expansion in constructing the irreps, since the action of the elements of the group $\mathcal{T}$ is diagonal over the basis $\mathcal{L}_6$. The expression (14) indicates that every element $g \in \mathcal{O}_h$, may be expressed in terms of a product of the form

$$g = p_\lambda \; h; \quad h \in \mathcal{D}_{2h}, \tag{15}$$

in accordance with the explicit coset expansion

$$\mathcal{O}_h = \; \mathcal{D}_{2h} + C_4(x)\, \mathcal{D}_{2h} + C_4(y)\, \mathcal{D}_{2h} + C_4(z)\, \mathcal{D}_{2h} + C_3^I\, \mathcal{D}_{2h} + (C_3^I)^2\, \mathcal{D}_{2h}. \tag{16}$$

To obtain the irreducible representations by induction, we start projecting the state $|\phi_{\mathbf{k}_1}\rangle$ to irreps of the *little cogroup* $\mathcal{D}_{2h}$, giving rise to the states $|\phi_{\mathbf{k}_1}; \Gamma, \gamma\rangle$. Finally, the induction is carried out and the representations so obtained are irreducible and complete with the form

$${}^{\Gamma}_{\mathbf{k}_1}\mathcal{G} = \; {}^{\Gamma}K({}^{\mathbf{k}_1}\mathcal{T}) \uparrow \mathcal{G} \tag{17}$$

There are two labels, the prong corresponding to the irrep $\mathbf{k}_1$ of $\mathcal{T}$, and the irrep Γ of the *little cogroup* ${}_{k_1}\mathcal{K} = \mathcal{D}_{2h}$.

We now proceed to generate the irreps of the group $\mathcal{G}$. For the elements of $\mathcal{T}$ the matrix representation in the new basis is diagonal as we have already noted,

$$D(U(\mathbf{t})) \;\; = \;\; \begin{pmatrix} e^{i\mathbf{t}\cdot\mathbf{k}_1} & 0 & 0 & 0 & 0 & 0 \\ 0 & e^{i\mathbf{t}\cdot\mathbf{k}_6} & 0 & 0 & 0 & 0 \\ 0 & 0 & e^{i\mathbf{t}\cdot\mathbf{k}_5} & 0 & 0 & 0 \\ 0 & 0 & 0 & e^{i\mathbf{t}\cdot\mathbf{k}_4} & 0 & 0 \\ 0 & 0 & 0 & 0 & e^{i\mathbf{t}\cdot\mathbf{k}_3} & 0 \\ & 0 & 0 & 0 & 0 & e^{i\mathbf{t}\cdot\mathbf{k}_2} \end{pmatrix}. \tag{18}$$

Let us now consider the point operations. In order to obtain the representation of the generator $C_4(x)$, for instance, the basal representation should be obtained to generate:

$$D(C_4(x)) = \begin{pmatrix} 0 & \chi^\mu(C_2(x)) & 0 & 0 & 0 & 0 \\ \chi^\mu(E) & 0 & 0 & 0 & 0 & 0 \\ 0 & 0 & 0 & 0 & \chi^\mu(C_2(z)) & 0 \\ 0 & 0 & 0 & 0 & 0 & \chi^\mu(C_2(z)) \\ 0 & 0 & \chi^\mu(E) & 0 & 0 & 0 \\ 0 & 0 & 0 & \chi^\mu(C_2(x)) & 0 & 0 \end{pmatrix}. \tag{19}$$

where $\chi^\mu(h)$ is the irrep, in this case the character, of the element $h \in \mathcal{D}_{2h}$. It should be clear that $\chi^\mu(E) = 1, \forall\mu$. In the same way we obtain for the generator $C_4(y)$:

$$D(C_4(y)) = \begin{pmatrix} 0 & 0 & \chi^\mu(C_2(y)) & 0 & 0 & 0 \\ 0 & 0 & 0 & 0 & 0 & \chi^\mu(C_2(x)) \\ \chi^\mu(E) & 0 & 0 & 0 & 0 & 0 \\ 0 & 0 & 0 & 0 & \chi^\mu(C_2(x)) & 0 \\ 0 & 0 & 0 & \chi^\mu(E) & 0 & 0 \\ 0 & \chi^\mu(C_2(y)) & 0 & 0 & 0 & 0 \end{pmatrix}. \tag{20}$$

Finally, the representation $D(I)$ is diagonal with elements $\chi^\mu(I) = (-1)^{n_1+n_2+n_3+1}$. We have thus constructed the representation of the group $\mathcal{G}$ associated with the $\mathcal{L}_6$ subspaces. Formally we have induced the representations of $\mathcal{T}$ through the little cogroup [11].

Let us now consider the subspace $\mathcal{L}_3$. We proceed to identify the *little group* of $\mathbf{k}_1$. Besides the elements of the subgroups $\mathcal{T}$ and $\mathcal{D}_{2h}$, the transformation σ_d also keeps the irrep $\mathbf{k}_1$ invariant. We note that $\mathcal{D}_{4h} = \mathcal{D}_{2h} + \sigma_d\mathcal{D}_{2h}$, and consequently

$$\mathcal{K}(\mathbf{k}_1) = \mathcal{T} \wedge \mathcal{D}_{4h}. \tag{21}$$

In explicit form we select the expansion

$$\mathcal{O}_h = \mathcal{D}_{4h} + C_4(x)\mathcal{D}_{4h} + C_4(y)\mathcal{D}_{4h}. \tag{22}$$

Again, the basal representation is needed to obtain,

$$D(C_4(x)) = \begin{pmatrix} 0 & \mathbf{D}^\mu(C_2(x)) & 0 \\ \mathbf{D}^\mu(E) & 0 & 0 \\ 0 & 0 & \mathbf{D}^\mu(C_4(z)) \end{pmatrix}, \tag{23}$$

$$D(C_4(y)) = \begin{pmatrix} 0 & 0 & \mathbf{D}^\mu(C_2(y)) \\ 0 & \mathbf{D}^\mu(C_4^3(z)) & 0 \\ \mathbf{D}^\mu(E) & 0 & 0 \end{pmatrix}, \tag{24}$$

with a diagonal matrix for the representation $D(I)$ with elements $\chi^\mu(I) = (-1)^{m+1}$. Finally, for the representation of the elements of $\mathcal{T}$ we have the trivial result

$$D(\mathbf{t}) = \begin{pmatrix} e^{i\mathbf{t}\cdot\mathbf{k}_1} & 0 & 0 \\ 0 & e^{i\mathbf{t}\cdot\mathbf{k}_2} & 0 \\ 0 & 0 & e^{i\mathbf{t}\cdot\mathbf{k}_3} \end{pmatrix}. \tag{25}$$

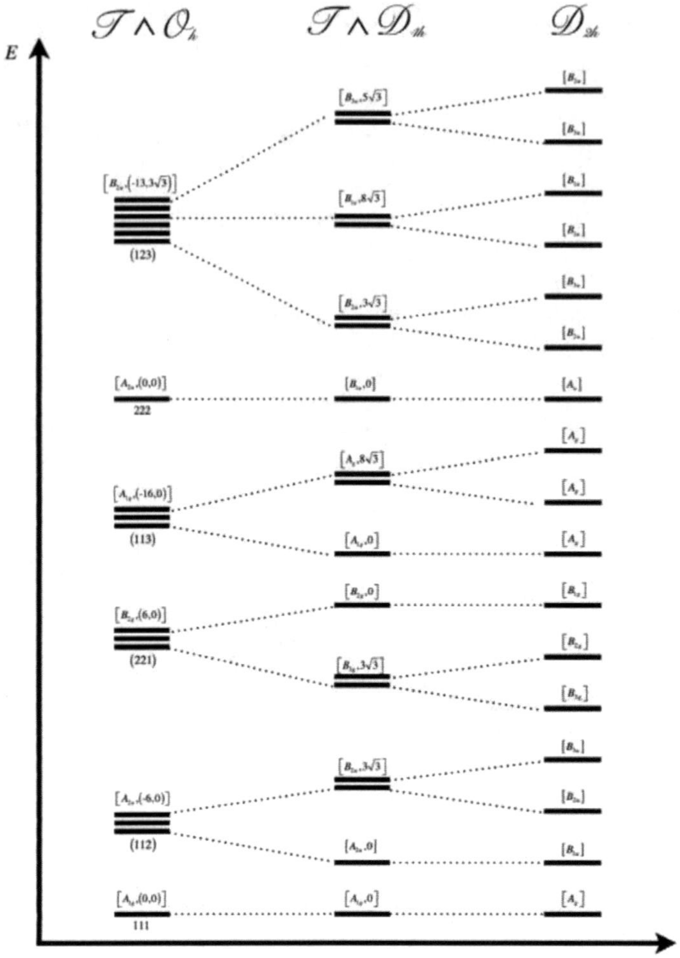

Figure 2. Schematic energy diagram depicting the symmetry breaking from the cubic box to the parallelepiped box. Degenerate levels are shown together, the irreps associated with the symmetry group are indicated above each set. According to these groups the degeneracy of all these levels are natural. Notice that the greater the symmetry the higher the degeneracy. Here we have taken $c > b > a$.

We have also the one dimensional space, for this case the associated representation of $U(\mathbf{t})$ carries the irrep $\mathbf{k} = \mathbf{0}$, which means that $\hat{U}(\mathbf{t})|\Psi^{\mathbf{0}}_{nnn}\rangle = |\Psi^{\mathbf{0}}_{nnn}\rangle$, and consequently the *little group* of $\mathbf{k}$ coincides with the group $\mathcal{G}$ itself. Then the *little cogroup* is given by $\mathcal{O}_h$ with the basal representation given by the one dimensional unit matrix. Hence the states $|\Psi^{\mathbf{0}}_{nnn}\rangle$ span the irreps of the octahedral group.

The analysis of the system when the symmetry is broken is presented in Figure 2. If in our previous analysis of the cubic box we introduce the condition $a = b \neq c$, then some of the transformations of the group $\mathcal{T} \wedge \mathcal{O}_h$ stop keeping invariant the Hamiltonian. We thus have to identify the new symmetry group by subduction. For a rectangular box the z- axis is not equivalent anymore and the operator $\hat{F}^{E_g}_{A_{1g},A_g}$ is not relevant. On the other hand the symmetry is diminished from $\mathcal{O}_h$ to $\mathcal{D}_{4h}$. These considerations lead to the reduction

$$\mathcal{T} \wedge \mathcal{O}_h \supset \mathcal{T}_1 \wedge \mathcal{D}_{4h}, \tag{26}$$

In the case of $a \neq b \neq c$ $k = 0$ the states are labeled by the group $\mathcal{D}_{2h}$.

4. Conclusions

The symmetry group of a particle inside an impenetrable cubic well potential has been identified to be $\mathcal{G} = \mathcal{T} \wedge \mathcal{O}_h$, where $\mathcal{T}$ is a compact continuous group. Reducing the representations generated by the eigenstates allowed us to identify the systematic degeneracy. The continuous group was obtained identifying the operators that connect degenerate states spanning different $\mathcal{O}_h$ irreps, by means of the coupling coefficients. The elements of the group $\mathcal{T}$ are constructed by exponentiation of the operators.

We have thus obtained the irreps of the symmetry group by the process of induction, an approach that assures the irreducibility and completeness of the representations as well as their completeness. From this perspective the systematic accidental degeneracy become natural, then by introducing the continuous group whose Casimir operator is the Hamiltonian all levels are distinguished, as can be observed in Figure 2.

The states of this system have been classified according with the chain of subgroups consistent with the reduction of symmetry to obtain the square and rectangular boxes, allowing us to analyze the symmetry breaking in a natural form [20].

Acknowledgments

This work is partially supported by DGAPA-UNAM under project No. IN109113.

References

[1] McIntosh H V 1971 *Symmetry and Degeneracy* in Group Theory and Its Applications. Edited by E.M.Loebl. Academic Press, New York. Vol I. 75-144.
[2] Wybourne B G 1974 *Classical groups for physisits.*Wiley, , New York.
[3] Fock V 1935 *Z.Physik* **98** 145-154.
[4] Bargmann V 1936 *Z.Physik* **99** 576-582.
[5] McIntosh H V 1959 *Am. J. Phys..* **27** 620-625.
[6] Atkins P W and Friedman R S 2000, *Molecular Quantum Mechanics.* Oxford University Press. Oxford. Page 56.
[7] Cohen-Tannoudji C, Dui B and Laloe F 1977 *Quantum Mechanics.* Wiley. New York. Volume I. Page 199.
[8] Harris D C and Bertolucci M D 1978 *Symmetry and spectroscopy. An introduction to vibrational and electronic spectroscopy.* Dover Publications, INC., New York.
[9] Leyvraz F, Frank A, Lemus R and Andrés M V 1997, *Am.J.Phys.* **65** 1087-1094.
[10] Shaw G B 1974 *J.Phys.A.* **7** 1537-1546.
[11] Altmann S L 1977 *Induced Representations in Crystals and Molecules. Point, space and nonrigid molecule groups.* Academic Press.
[12] Wai-Kee Li 1982 *Am. J. Phys..* **50** 666.
[13] Lemus R, Frank A, Andrés M V and Leyvraz F 1998 *Am.J.Phys.* **66** 629-631.
[14] Hamermesh M1962 *Group theory and its application to physical problems.* Dover Publications, Inc. New York. 1962.
[15] Chen J Q 1989 *Group Representation Theory for Physicists.* World Scientific, Singapore.
[16] Lemus R 2003 *Mol.Phys.* **101** 2511-2528.
[17] Álvarez-Bajo O, Lemus R, Carvajal M and érez-Bernal F 2011 *Mol.Phys.* **109** 797-812.
[18] Lemus R 2012 *Symmetry* **4** 1-19.
[19] Altmann S L and Herzig P 1994 *Point-Group theory Tables.* Clarendon Press, Oxford.
[20] Hernández-Castillo A O and Lemus R 2013 *J.Phys.A:Math. Theor.* **46** 465201 (20pp).

Journal of Physics: Conference Series **512** (2014) 012026 doi:10.1088/1742-6596/512/1/012026

Euclidean and super Euclidean algebras and Localizations of $U_q(\mathfrak{sl}(2))$ and $U_q(\mathfrak{osp}(1|2))$[1]

Patrick Moylan

Dept. of Physics, Pennsylvania State University, The Abington College, Abington, PA 19001 USA

E-mail: pjm11@psu.edu

Abstract. In Ref. 1 we described homomorphisms of $U_q(\mathfrak{sl}(2))$ and $U(\mathfrak{iso}(2))$ into localizations of $U(\mathfrak{iso}(2))$ and $U_q(\mathfrak{sl}(2))$, respectively, with $U(\mathfrak{iso}(2))$ being the universal enveloping algebra of the Lie algebra of $SO(2) \times_s T^2$, the Euclidean group in the plane, with T^2 denoting translations of the plane and $\times_s$ semidirect product. In particular, we obtained explicit expressions for the generators of $U_q(\mathfrak{sl}(2))$ and $U(\mathfrak{iso}(2))$, respectively as irrational functions of $U(\mathfrak{iso}(2))$ and $U_q(\mathfrak{sl}(2))$. Herewith we further develop our results extending them to $U_q(\mathfrak{osp}(1|2))$ and $U(\widetilde{\mathfrak{iso}}\,(2))$ with $\widetilde{\mathfrak{iso}}\,(2)$ being the super Euclidean algebra.

1. Introduction

Let G be a connected Lie group with Lie algebra $\mathfrak{g}$ and H a closed subgroup with subalgebra $\mathfrak{h}$. Let $\mathfrak{h}$ have an $Ad_G(H)$ invariant complement V in $\mathfrak{g}$ i.e. a subspace $V \subset \mathfrak{g}$ such that

$$\mathfrak{g} = \mathfrak{h} \oplus V \quad \text{and} \qquad Ad_G(H)V \subset V.$$

In this case we can form the semidirect product $G_0 = H \times_s V$. An interesting problem is to try to relate representations of G to representations of G_0, a problem which was studied by Mackey [2]. More recent developments along these lines are given in [3] and [4]. Inspired by the Gell-Mann formula, which is defined below, we have attacked the problem at the level of enveloping algebras. For certain physically important cases we have obtained a rather complete description, and it has led us to useful information on the relationship between representations of the Poincaré group and representations of $SO(1,4)$ and $SO(2,3)$ [5], [6], [7]. We have also extended this method to q deformations, in particular, to $U_q(\mathfrak{sl}(2))$ in [1] and also to $U_q(\mathfrak{so}(4))$ and other low dimensional cases in [8] and [9], and in this present article to the q deformed superalgebra $U_q(\mathfrak{osp}(1|2))$. We have also established some general results for $SO(p,q)$ groups and their associated motion groups [10]. For related results on the $SL(n)$ and $SU(n)$ cases see [11].

Since we work at the enveloping algebra level, our results, although being specific to a particular algebra, are quite general. We try to express the generators of G as irrational functions of the generators of G_0. Specifically, for $P_0 \in V$, $\Sigma^2 \in Z(U(\mathfrak{g}_0))$ and $\Delta \in Z(U(\mathfrak{h}))$, we define

$$B_0 = \frac{1}{\sqrt{\Sigma^2}}[\Delta, P_0] + P_0,$$

[1] Dedicated to Professor Miloslav Havlíček on the occasion of his 75th birthday.

which we call the Gell-Mann Formula [12]. ($[\cdot,\cdot]$ denotes commutator and Δ is usually a Casimir operator of $\mathfrak{h}$ or some function of it.) For certain H and V and for a particular choice of P_0, Δ and Σ^2, B_0 together with the generators of H may close under the Lie bracket in G_0 to generate a new Lie algebra isomorphic to the Lie algebra $\mathfrak{g}$ of G.

On the other hand, we may think of P_0 in the Gell-Mann formula as the variable and view the Gell-Mann formula as an algebraic equation for it with coefficients from a noncommutative quotient ring consisting of fractions made out of elements from the enveloping algebra of $\mathfrak{g}$ or from an algebraic extension thereof. It is possible, for certain cases, to solve the Gell-Mann formula for P_0 exactly and to obtain a homomorphic image of G_0 inside this noncommutative quotient ring. The precise mathematical description of the Gell-Mann formula and its inversion, i.e. the solution P_0 to the Gell-Mann formula viewed as an algebraic equation in P_0, involves the notion of noncommutative localization to which we now turn.

Note on notation: except for elements of the Cartan subalgebras for which we always use plain faced letters, quantities made out of elements of $U(\mathfrak{iso}(2))$ and $U\left(\widetilde{\mathfrak{iso}(2)}\right)$ and of their localizations are usually denoted with bold faced letters and we use plain faced letters to denote elements of $\mathfrak{sl}(2)_q$ and $\mathfrak{osp}(1|2)_q$ and their localizations. Elements of superalgebras are always denoted with tildes placed over the letters.

2. Localization

R is a ring with unity and $S \neq \emptyset$ a multiplicatively closed subset of R such that $0 \notin S$, $I_R \in S$ where $I_R(=I)$ the identity in R. A nonzero element a in a ring R is said to be a left [resp. right] zero divisor if there exists a nonzero $b \in R$ such that $ab = 0$ [resp. $ba = 0$]. A zero divisor is an element of R which is both a left and a right zero divisor.

Definition 2.1 *A ring Q is said to be a left quotient ring of R with respect to S if there exists a ring homomorphism $\varphi : R \longrightarrow Q$ such that the following conditions are satisfied:*

1) $\varphi(s)$ is a unit in Q for all $s \in S$ (This means that $a = \varphi(s)$ is both left and right invertible i.e. $\exists c \in Q$(resp. $b \in Q$) such that $ca = I_Q$(resp. $ab = I_Q$).);

2) every element of Q is in the form $(\varphi(s))^{-1}\varphi(r)$, for some $r \in R,\ \ s \in S$;

3) $\ker\varphi = \{r \in R :\ sr = 0,$ for some $s \in S\}$.

The left (resp. right) quotient ring of R w.r.t. S, if it exists, is called the left (resp. right) localization of R at S and it is denoted by $S^{-1}R$ (resp. RS^{-1}). If $S = R$, the localization $S^{-1}R$ is the left skew field of fractions of R i.e.the left quotient field of R.

Note that if $rs = 0$, for some $r \in R,\ \ s \in S$, then $s'r = 0$, for some $s' \in S$. This is because $0 = \varphi(rs) = \varphi(r)\varphi(s)$ and thus $\varphi(r) = 0$, since, by condition 1), $\varphi(s)$ is a unit of Q. We need to multiply fractions like $(s^{-1}a)(s'^{-1}b)$ so we must be able to move s'^{-1} to the other side of a. This leads to the Ore condition: $Ra \cap Rs \neq \emptyset$ for $a \in R$ and $s \in S$. It is a necessary and sufficient condition for the existence of localizations [13].

Integral (no zero divisors) Noetherian rings satisfy the Ore condition, so that we can construct localizations [14]. Examples of such rings include enveloping algebras of finite dimensional Lie algebras and, for semisimple ones, their q-deformations [15].

The following important Lemma tells us when a given representation of a ring R lifts to a representation of its localization.

Lemma 1.1 Suppose $f : R \longrightarrow R_1$ is a ring homomorphism and $Q = S^{-1}R\ \left(RS^{-1}\right)$ is a left (right) quotient ring of R with respect to S. If $f(s)$ is a unit in R_1 for every $s \in S$, then there exists a (unique) ring homomorphism $g : Q \longrightarrow R_1$ which extends f.

Proof: If Q is a left quotient ring of R and φ is the map in Definition 2.1, then for all $r \in R$, $s \in S$ we define $g((\varphi(s))^{-1}\varphi(r)) := (f(s))^{-1}f(r)$ and similarly for right quotient rings. A proof that such defined map g is well defined, unique, and is a ring homomorphism is given in [16].

3. The Algebras $U_q(\mathfrak{sl}(2))$ and $U(\mathfrak{iso}(2))$

The q-deformation $U_q(\mathfrak{sl}(2)) \simeq U_q(\mathfrak{so}(3,\mathbb{C}))$ of the simple Lie algebra $\mathfrak{sl}(2)$ is defined as the unital associative $\mathbb{Q}(q)$ algebra with generators E, F, K, K^{-1} and relations [17]

$$KK^{-1} = K^{-1}K = I, KEK^{-1} = q^2E, KFK^{-1} = q^{-2}F$$

$$EF - FE = \frac{K - K^{-1}}{q - q^{-1}} \quad .$$

Equivalently, with $K = q^H$ we have generators H, $X^{\pm}$ with relations:

$$[H, X^{\pm}] \; = \; \pm 2X^{\pm} \; , \tag{1}$$

$$[X^+, X^-] \; = \; [H]_{q^2} \tag{2}$$

where $[x]_q = \frac{(q^{x/2} - q^{-x/2})}{(q^{1/2} - q^{-1/2})}$ and $E = X^+$, $F = X^-$. The Casimir operator is

$$\Delta_q = X^+X^- + \left(\left[\frac{H-1}{2}\right]_{q^2}\right)^2 - \frac{1}{4}\cdot I = X^-X^+ + \left(\left[\frac{H+1}{2}\right]_{q^2}\right)^2 - \frac{1}{4}\cdot I \quad . \tag{3}$$

Define a real form $U_q(\mathfrak{so}(2,1))$ with generators L_{ij} specified by $X^{\pm} = L_{13} \pm iL_{32}$, $L_{21} = \frac{i}{2}\, H$. The L_{ij} are preserved under the following antilinear, anti-involution (star structure) [17]: $\omega(H) = H$, $\omega(X^{\pm}) = \; - \; X^{\mp}$.

A basis for the Euclidean Lie algebra $\mathfrak{iso}(2)$ is $\mathbf{L}_{12}$ and $\mathbf{P}_i$ (i = 1, 2). They satisfy the following commutation relations:

$$[\mathbf{L}_{12} \; , \; \mathbf{P}_2] = - \; \mathbf{P}_1 \quad , \quad [\mathbf{L}_{12} \; , \; \mathbf{P}_1] = \; \mathbf{P}_2 \quad ,$$

$$[\mathbf{P}_1 \; , \; \mathbf{P}_2] = 0 \quad .$$

Complexified translations generators are $\mathbf{P}^{\pm} = -\mathbf{P}_1 \pm i\mathbf{P}_2$. We also define as above $H = -2i\mathbf{L}_{21}$. We have:

$$[H, \mathbf{P}^{\pm}] \; = \; \pm 2\mathbf{P}^{\pm} \; , \; [\mathbf{P}^+, \mathbf{P}^-] \; = \; 0 \; . \tag{4}$$

The enveloping algebra of $\mathfrak{iso}(2)$ is $U(\mathfrak{iso}(2))$. The center $Z(U(\mathfrak{iso}(2)))$ of $U(\mathfrak{iso}(2))$ is generated by

$$Y^2 = \mathbf{P}^+\mathbf{P}^- = \mathbf{P}^-\mathbf{P}^+. \tag{5}$$

4. Localizations of $U(\mathfrak{iso}(2))$ and $U_q(\mathfrak{sl}(2))$

In what follows we make use of identities for q numbers such as

$$[x \; + \; 1]_q \; - \; [x \; - \; 1]_q = \frac{[2(x)]_q}{[x]_q} \; , \tag{6}$$

Proofs of this and all other q identities which we require, such as $\frac{[x]_q}{[2]_q} = [\frac{x}{2}]_{q^2}$, can be found in Ref. [18]. We shall need quite often the following result and it is readily established by using Eqs. (4) and the Maclaurin series formula: let f be any analytic function, then

$$\mathbf{P}^{\pm} f\left(\left[\frac{H}{2}\right]_q\right) \; = \; f\left(\left[\frac{H \mp 2}{2}\right]_q\right)\mathbf{P}^{\pm} \; . \tag{7}$$

We shall also make use of a similar equation for $U_q(\mathfrak{sl}(2))$; it is the same as Eq. (7) but with $\mathbf{P}^{\pm}$ replaced by $X^{\pm}$.

Define

$$\mathbf{X}^{\pm} = \mathbf{L}_{13} \pm i\mathbf{L}_{32} = \left\{ \pm\frac{1}{Y}\left[\frac{H \mp 1}{2}\right]_{q^2} + I \right\} \mathbf{P}^{\pm} = \mathbf{P}^{\pm} \left\{ \pm\frac{1}{Y}\left[\frac{H \pm 1}{2}\right]_{q^2} + I \right\} \tag{8}$$

where Y is a solution of the algebraic equation $Y^2 - \mathbf{P}^+\mathbf{P}^- = 0$ in $U(\mathfrak{iso}(2))$. (Note that we used Eq. (7) to obtain the last equality.) We refer the reader to Ref. [1] to see how Eq. (8) can be related to the Gell-Mann formula for the real form of $U_q(\mathfrak{sl}(2))$ introduced above.

Proposition 4.1 *The $\mathbf{X}^{\pm}$ defined by Eqs. (8) together with H satisfy the relations, Eqs. (1) and (2), of the generators of $U_q(\mathfrak{sl}(2))$. Furthermore, let Δ_q be defined by Eq. (3) but with $\mathbf{X}^{\pm}$ replacing $X^{\pm}$. Then $\Delta_q = Y^2 - \frac{1}{4}\cdot I$.*

For the proof of this proposition we refer the reader to Ref. [1]. We now solve Eqs. (8) for the $\mathbf{P}^{\pm}$ and we obtain the following:

Proposition 4.2 *Let $P^{\pm} = (D_L^{\pm})^{-1}X^{\pm} = X^{\pm}(D_R^{\pm})^{-1}$ with $D_L^{\pm} = \left(\pm\frac{1}{Y}\left[\frac{H\mp1}{2}\right]_{q^2} + I\right)$ and $D_R^{\pm} = \left(\pm\frac{1}{Y}\left[\frac{H\pm1}{2}\right]_{q^2} + I\right)$. If Y is such that it commutes with all elements of $U_q(\mathfrak{sl}(2))$ and satisfies the equation*

$$Y^2 = \Delta_q + \frac{1}{4}\cdot I , \tag{9}$$

then $P^{\pm}$ and H satisfy the commutation relations, Eqs. (4), of $\mathfrak{iso}(2)$ and $P^+P^- = Y^2$.

Proof: It suffices to show $[P^+, P^-] = 0$ and that $P^+P^- = Y^2$, the rest of the Proposition being trivial. We let $\{\cdot,\cdot\}$ denote anticommutator and we have:

$$\begin{aligned}\{P^+, P^-\} &= \left(\frac{1}{Y}\left[\frac{H-1}{2}\right]_{q^2} + I\right)^{-1} X^+X^- \left(-\frac{1}{Y}\left[\frac{H-1}{2}\right]_{q^2} + I\right)^{-1} + \\ &+ \left(-\frac{1}{Y}\left[\frac{H+1}{2}\right]_{q^2} + I\right)^{-1} X^-X^+ \left(\frac{1}{Y}\left[\frac{H+1}{2}\right]_{q^2} + I\right)^{-1} = \\ &= \frac{Y^2}{\left(Y^2 - \left(\left[\frac{H-1}{2}\right]_{q^2}\right)^2\right)} X^+X^- + \frac{Y^2}{\left(Y^2 - \left(\left[\frac{H+1}{2}\right]_{q^2}\right)^2\right)} X^-X^+ = \\ &= \frac{Y^2\left(\Delta_q + \frac{1}{4}\cdot I - \left(\left[\frac{H-1}{2}\right]_{q^2}\right)^2\right)}{\left(Y^2 - \left(\left[\frac{H-1}{2}\right]_{q^2}\right)^2\right)} + \frac{Y^2\left(\Delta_q + \frac{1}{4}\cdot I - \left(\left[\frac{H+1}{2}\right]_{q^2}\right)^2\right)}{\left(Y^2 - \left(\left[\frac{H+1}{2}\right]_{q^2}\right)^2\right)}.\end{aligned}$$

In obtaining the last line we have used the expressions given in Eq. (3) for the q-Casimir operator, Δ_q. Using $Y^2 = \Delta_q + \frac{1}{4}\cdot I$ gives $\{P^+, P^-\} = 2Y^2$. Similar calculations show $[P^+, P^-] = 0$. From these two results the desired conclusions follow.

Journal of Physics: Conference Series **512** (2014) 012026 doi:10.1088/1742-6596/512/1/012026

5. The Algebras $U_{\tilde{q}}(\mathfrak{osp}(1|2))$ and $U(\widetilde{\mathfrak{iso}}(2))$

The $\tilde{q}$-deformation $U_{\tilde{q}}(\mathfrak{osp}(1|2)) = \mathfrak{osp}(1|2)_{\tilde{q}}$ of the orthosymplectic Lie superalgebra $\mathfrak{osp}(1|2)$ is defined as the unital associative $\mathbb{Q}(\tilde{q})$ algebra with generators e, f, k , k^{-1} and relations [17]:

$$kk^{-1} = k^{-1}k = I, kek^{-1} = \tilde{q}e, kfk^{-1} = \tilde{q}^{-1}f, ef + fe = \frac{K - K^{-1}}{\tilde{q} - \tilde{q}^{-1}} \quad .$$

The $\mathbb{Z}_2$ grading on $U_{\tilde{q}}(\mathfrak{osp}(1|2))$ is $d(e) = d(f) = 1, d(k) = 0$ where $d(x)$ is the parity of x. Let $k = \tilde{q}^{\tilde{H}}$, $e = \tilde{X}^+$, $f = \tilde{X}^-$ and we obtain generators $\tilde{H}$, $\tilde{X}^\pm$ with relations:

$$[\tilde{H}, \tilde{X}^\pm] \;=\; \pm\tilde{X}^\pm \;, \qquad (1^{bis})$$

$$\{\tilde{X}^+, \tilde{X}^-\} \;=\; [\tilde{H}]_{\tilde{q}^2} \;. \qquad (2^{bis})$$

The Casimir operator of $U_{\tilde{q}}(\mathfrak{osp}(1|2))$ is $\tilde{\Delta}_{\tilde{q}} \;=\; \tilde{S}_{\tilde{q}}^{\,2} + 2\cdot I$ with [19]

$$\tilde{S}_{\tilde{q}} = \frac{\tilde{q}^{1/2}k - \tilde{q}^{-1/2}k^{-1}}{\tilde{q} - \tilde{q}^{-1}} - (\tilde{q}^{1/2} + \tilde{q}^{-1/2})fe = [\tilde{H} + \frac{1}{2}]_{\tilde{q}^2} - [2]_{\tilde{q}}\tilde{X}^-\tilde{X}^+ =$$

$$-\frac{\tilde{q}^{-1/2}k - \tilde{q}^{1/2}k^{-1}}{(\tilde{q} - \tilde{q}^{-1})} + (\tilde{q}^{1/2} + \tilde{q}^{-1/2})ef = -[\tilde{H} - \frac{1}{2}]_{\tilde{q}^2} + [2]_{\tilde{q}}\tilde{X}^+\tilde{X}^-. \qquad (3^{bis})$$

It is straightforward to show that $\tilde{S}_{\tilde{q}}$ anticommutes with $\tilde{X}^+$ and $\tilde{X}^-$ and commutes with $\tilde{H}$. A star structure (or real form) is specified as follows. Let $\tilde{\omega}$ be such that $\tilde{\omega}(\tilde{H}) = \tilde{H}$, $\tilde{\omega}(\tilde{X}^\pm) = -\tilde{X}^\mp$. $\tilde{\omega}$ is, as in the $\mathfrak{sl}(2)$ case, an antilinear, anti-involution.

A basis for the three dimensional "super Euclidean Lie algebra" $\widetilde{\mathfrak{iso}}(2)$ is given by $\tilde{\mathbf{L}}_{12}$ and $\tilde{\mathbf{P}}_i$ $(i \;=\; 1,\ 2)$ with commutation relations

$$[\tilde{H}, \tilde{\mathbf{P}}^\pm] \;=\; \pm\tilde{\mathbf{P}}^\pm \;,\; \{\tilde{\mathbf{P}}^+, \tilde{\mathbf{P}}^-\} \;=\; 0 \;. \qquad (4^{bis})$$

The universal enveloping algebra of $\widetilde{\mathfrak{iso}}(2)$ is $U(\widetilde{\mathfrak{iso}}(2))$. Let $\tilde{Y}^2 \in Z(U(\widetilde{\mathrm{iso}}(2)))$ be given by

$$\tilde{Y}^2 \;=\; -\,i(-1)^{\tilde{H}}\;\tilde{\mathbf{P}}^+\tilde{\mathbf{P}}^- \;=\; i(-1)^{\tilde{H}}\tilde{\mathbf{P}}^-\tilde{\mathbf{P}}^+. \qquad (5^{bis})$$

6. Localizations of $U(\widetilde{\mathfrak{iso}}(2))$ and $U_{\tilde{q}}(\mathfrak{osp}(1|2))$

We start by establishing a direct relationship between $U(\widetilde{\mathfrak{iso}}(2))$ and $U(\mathfrak{iso}(2))$.

Proposition 6.1 *Let* $H \;=\; -\,2\tilde{H}$, $\mathbf{P}^+ \;=\; -\,\tilde{\mathbf{P}}^-$, $\mathbf{P}^- \;=\; e^{-i\pi\tilde{H}}\tilde{\mathbf{P}}^+$, *then*

H, $\mathbf{P}^\pm$ *satisfy Eqs. (4)* $\iff$ $\tilde{H}$, $\tilde{\mathbf{P}}^\pm$ *satisfy Eqs.* (4bis).

We now describe homomorphisms of $U_{\tilde{q}}(\mathfrak{osp}(1|2))$ and $U(\widetilde{\mathfrak{iso}}(2))$ into localizations of $U(\widetilde{\mathfrak{iso}}(2))$ and $U_{\tilde{q}}(\mathfrak{osp}(1|2))$, respectively, i.e. the analogs of Propositions 4.1 and 4.2.

Proposition 6.2 *Let*

$$\tilde{\mathbf{X}}^\pm = \left(\frac{1}{\tilde{Y}}\sqrt{(-1)^{\tilde{H}}\left[\tilde{H} \mp \frac{1}{2}\right]_{\tilde{q}^2}} + \sqrt{\pm 1}\; I\right)\frac{\tilde{\mathbf{P}}^\pm}{\sqrt{[2]_{\tilde{q}}}} = \frac{\tilde{\mathbf{P}}^\pm}{\sqrt{[2]_{\tilde{q}}}}\left(\frac{\pm i}{\tilde{Y}}\sqrt{(-1)^{\tilde{H}}\left[\tilde{H} \pm \frac{1}{2}\right]_{\tilde{q}^2}} + \sqrt{\pm 1}\; I\right) \qquad (8^{bis})$$

where I is the identity in $U(\widetilde{\mathfrak{iso}}(2))$ and by $(-1)^{\tilde{H}}$ we mean $e^{-i\pi\tilde{H}}$. $\tilde{Y}$ is such that it commutes with all elements of $U(\widetilde{\mathfrak{iso}}(2))$ and satisfies Eq. (5bis). Then the $\tilde{\mathbf{X}}^\pm$ defined by Eqs. (8bis) together with $\tilde{H}$ satisfy the relations, Eqs. (1bis) and (2bis), of the generators of $U_{\tilde{q}}(\mathfrak{osp}(1|2))$. Furthermore, let $\tilde{\mathbf{S}}_{\tilde{q}}$ be defined by Eq. (3bis) but with $\tilde{\mathbf{X}}^\pm$ replacing $\tilde{X}^\pm$, then $\tilde{\mathbf{S}}_{\tilde{q}} = -(-1)^{-\tilde{H}}\tilde{Y}^2$.

The proof of this proposition is straightforward and is very similar to the proof of Proposition 4.1. In obtaining the last term of Eqs. (8bis) we used the equations $\sqrt{[\tilde{H}\mp\frac{1}{2}]_{\tilde{q}^2}}\tilde{\mathbf{P}}^{\pm}=\tilde{\mathbf{P}}^{\pm}\sqrt{[\tilde{H}\pm\frac{1}{2}]_{\tilde{q}^2}}$ and $(-1)^{\frac{\tilde{H}}{2}}\tilde{\mathbf{P}}^{\pm}=\pm i\tilde{\mathbf{P}}^{\pm}(-1)^{\frac{\tilde{H}}{2}}$ which equations follow easily from Eqs. (4bis). (Note that quantities like $e^{-i\pi\tilde{H}}$ and $(-1)^{\frac{\tilde{H}}{2}}:=e^{-i\pi\frac{\tilde{H}}{2}}$ require clarification in precisely defining them at the abstract algebra level in terms of power series in $\tilde{H}$ e.g. we should work in an extension of $U\big(\widetilde{\mathfrak{iso}}(2)\big)$ [20]. However, in representations where $\tilde{H}$ acts diagonally the actions of $e^{-i\pi\tilde{H}}$ and $e^{-i\pi\frac{\tilde{H}}{2}}$ have precise meanings.)

Proposition 6.3 *Let* $\tilde{P}^{\pm}=(\tilde{D}_L^{\pm})^{-1}\tilde{X}^{\pm}=\tilde{X}^{\pm}(\tilde{D}_R^{\pm})^{-1}$ *with* $\sqrt{[2]_{\tilde{q}}}\tilde{D}_L^{\pm}=\frac{1}{\tilde{Y}}\sqrt{(-1)^{\tilde{H}}\left[\tilde{H}\mp\frac{1}{2}\right]_{\tilde{q}^2}}+\sqrt{\pm 1}\ I$ *and* $\sqrt{[2]_{\tilde{q}}}\tilde{D}_R^{\pm}=\frac{\pm i}{\tilde{Y}}\sqrt{(-1)^{\tilde{H}}\left[\tilde{H}\pm\frac{1}{2}\right]_{\tilde{q}^2}}+\sqrt{\pm 1}\ I$ *where now* I *is the identity in* $U_{\tilde{q}}(\mathfrak{osp}(1|2))$. *If* $\tilde{Y}$ *is such that it commutes with all elements of* $U_{\tilde{q}}(osp(1|2))$ *and satisfies the equation*

$$\tilde{Y}^2+(-1)^{\tilde{H}}\tilde{S}_{\tilde{q}}=0, \qquad (9^{bis})$$

then $\tilde{P}^{\pm}$ *and* $\tilde{H}$ *satisfy the commutation relations* (4bis) *of* $U\big(\widetilde{\text{iso}}(2)\big)$ *and* $\tilde{P}^+\tilde{P}^-=i(-1)^{-\tilde{H}}\tilde{Y}^2$.

Proof. We shall establish that $\{\tilde{P}^+,\tilde{P}^-\}=0$, since the rest of the proof is straightforward. From the proposition we have: $[2]_{\tilde{q}}(\tilde{P}^+\tilde{P}^-+\tilde{P}^-\tilde{P}^+)=$

$$=\frac{1}{\left(\frac{1}{\tilde{Y}}\sqrt{(-1)^{\tilde{H}}\left[\tilde{H}-\frac{1}{2}\right]_{\tilde{q}^2}}+I\right)}\tilde{X}^+\tilde{X}^-\frac{1}{\left(\frac{-i}{\tilde{Y}}\sqrt{(-1)^{\tilde{H}}\left[\tilde{H}-\frac{1}{2}\right]_{\tilde{q}^2}}+i\ I\right)}$$
$$+\frac{1}{\left(\frac{1}{\tilde{Y}}\sqrt{(-1)^{\tilde{H}}\left[\tilde{H}+\frac{1}{2}\right]_{\tilde{q}^2}}+i\ I\right)}\tilde{X}^-\tilde{X}^+\frac{1}{\left(\frac{i}{\tilde{Y}}\sqrt{(-1)^{\tilde{H}}\left[\tilde{H}+\frac{1}{2}\right]_{\tilde{q}^2}}+I\right)}=$$
$$=i\ \tilde{Y}^2\tilde{X}^+\tilde{X}^-\frac{1}{\left(\sqrt{(-1)^{\tilde{H}}\left[\tilde{H}-\frac{1}{2}\right]_{\tilde{q}^2}}+\tilde{Y}\right)\left(\sqrt{(-1)^{\tilde{H}}\left[\tilde{H}-\frac{1}{2}\right]_{\tilde{q}^2}}-\tilde{Y}\right)}$$
$$-i\ \tilde{Y}^2\tilde{X}^-\tilde{X}^+\frac{1}{\left(\sqrt{(-1)^{\tilde{H}}\left[\tilde{H}+\frac{1}{2}\right]_{\tilde{q}^2}}+i\ \tilde{Y}\right)\left(\sqrt{(-1)^{\tilde{H}}\left[\tilde{H}+\frac{1}{2}\right]_{\tilde{q}^2}}-i\ \tilde{Y}\right)}=$$
$$=i\ \tilde{Y}^2\left\{\frac{\tilde{X}^+\tilde{X}^-}{\left((-1)^{\tilde{H}}\left[\tilde{H}-\frac{1}{2}\right]_{\tilde{q}^2}-\tilde{Y}^2\right)}-\frac{\tilde{X}^-\tilde{X}^+}{\left((-1)^{\tilde{H}}\left[\tilde{H}+\frac{1}{2}\right]_{\tilde{q}^2}+\tilde{Y}^2\right)}\right\}=$$
$$=i\ \tilde{Y}^2\left\{\frac{\tilde{X}^+\tilde{X}^-}{\left((-1)^{\tilde{H}}\left[\tilde{H}-\frac{1}{2}\right]_{\tilde{q}^2}+(-1)^{\tilde{H}}\tilde{S}_q\right)}-\frac{\tilde{X}^-\tilde{X}^+}{\left((-1)^{\tilde{H}}\left[\tilde{H}+\frac{1}{2}\right]_{\tilde{q}^2}-(-1)^{\tilde{H}}\tilde{S}_q\right)}\right\}=$$
$$=\frac{i\ \tilde{Y}^2}{(-1)^{\tilde{H}}}\left\{\frac{\tilde{X}^+\tilde{X}^-}{\left(\left[\tilde{H}-\frac{1}{2}\right]_{\tilde{q}^2}+\tilde{S}_q\right)}-\frac{\tilde{X}^-\tilde{X}^+}{\left(\left[\tilde{H}+\frac{1}{2}\right]_{\tilde{q}^2}-\tilde{S}_q\right)}\right\}$$

where in obtaining the second to last line we used Eq. (9bis). Now use Eq. (3bis) twice to obtain the desired result.

7. Representations

Using Propositions 6.1 and 6.2 it is possible to construct representations of $U_{\tilde{q}}(\mathfrak{osp}(1|2))$ out of representations of $U(\mathfrak{iso}(2))$ and $U(\widetilde{\mathfrak{iso}}(2))$. For example, let us start with the positive mass representations of the Euclidean Lie algebra. They are characterized by a real number ρ and an integer ϵ which is either 0 or $\frac{1}{2}$. They are described as follows [21]. The representation space is $\mathcal{D}^{i\rho,\epsilon} = \sum_m {}^{\oplus}\mathcal{H}_{(m,\epsilon)}$ where $\mathcal{H}_{(m,\epsilon)}$ is the one dimensional vector space $\mathbb{C}e_m$ with $m = n+\epsilon, n = 0, \pm 1, \pm 2, \ldots$ for fixed $\epsilon = 0$ or $\frac{1}{2}$. The actions of the generators of $U(\mathfrak{iso}(2))$ on $\mathcal{D}^{i\rho,\epsilon}$ are given by

$$d\pi^{\rho,\epsilon}(\mathbf{P}^{\pm})\, e_m = \; - \; (i\;\rho)\; e_{m\pm 1} \tag{10}$$

and

$$d\pi^{\rho,\epsilon}(H)\, e_m \;=\; 2m\; e_m. \tag{11}$$

Using Proposition 6.1 we readily obtain the following representation of $U(\widetilde{\mathfrak{iso}}(2))$ on $\mathcal{D}^{i\rho,\epsilon}$:

$$d\tilde{\pi}^{\rho,\epsilon}(\tilde{\mathbf{P}}^{\pm})\, e_m = \;\mp\; (i\;\rho)\; e^{-i\frac{\pi}{2}\{(m-1)\pm(m-1)\}}\; e_{m\mp 1} \tag{12}$$

and

$$d\tilde{\pi}^{\rho,\epsilon}(\tilde{H})\, e_m = -m\; e_m \quad . \tag{13}$$

Now use Proposition 6.2 together with Lemma 1.1 to obtain the representation of $U_{\tilde{q}}(\mathfrak{osp}(1|2))$ on $\mathcal{D}^{i\rho,\epsilon}$. We claim that the conditions of the Lemma are satisfied if zero is in the resolvent set of $d\tilde{\pi}^{\rho,\epsilon}(\tilde{Y})$. It is easy to see that this is always the case for any nonzero ρ and any ϵ, since from Eq. (5bis) we have $d\tilde{\pi}^{\rho,\epsilon}(\tilde{Y}^2)e_m = -i(-1)^{d\tilde{\pi}^{\rho,\epsilon}(\tilde{H})}d\tilde{\pi}^{\rho,\epsilon}(\tilde{\mathbf{P}}^+)d\tilde{\pi}^{\rho,\epsilon}(\tilde{\mathbf{P}}^-)e_m$, and using Eqs. (12) and (13) we easily obtain

$$d\tilde{\pi}^{\rho,\epsilon}(\tilde{Y}^2) = -i(-1)^{2\epsilon}\rho^2 I \tag{14}$$

where I is the identity operator on $\mathcal{D}^{i\rho,\epsilon}$. Hence, for $\rho \neq 0$, $d\tilde{\pi}^{\rho,\epsilon}(\tilde{Y}^2)$ is invertible and so also its square root

$$d\tilde{\pi}^{\rho,\epsilon}(\tilde{Y}) = i\sqrt{i}(-1)^{\epsilon}\rho\; I. \tag{15}$$

Clearly the image of $\tilde{Y}^2$ and its square root are units in the localization of the algebraic extension of $U(\widetilde{\mathfrak{iso}}(2))$ obtained by adjoining the square root of $\tilde{Y}^2$ and from Eqs. (14) and (15) we see that the conditions of Lemma 1.1 are satisfied. Using (8bis) together with Eqs. (12), (13) and (15) we can explicitly construct the representation of $U_{\tilde{q}}(\mathfrak{osp}(1|2))$ on the above representation space. We obtain:

$$\tilde{X}^+ e_m = \frac{1}{\sqrt{[2]_{\tilde{q}}}}(-i\rho)\left\{\frac{(-1)^{m/2}\sqrt{[m-\frac{1}{2}]_{q^2}}}{i\sqrt{i}(-1)^{\epsilon}\rho} - 1\right\}(-1)^m e_{m-1} \tag{16}$$

$$\tilde{X}^- e_m = \frac{1}{\sqrt{[2]_{\tilde{q}}}}(i\rho)\left\{\frac{(-1)^{-m/2}\sqrt{[m+\frac{1}{2}]_{q^2}}}{i\sqrt{i}(-1)^{\epsilon}\rho} + i\right\} e_{m+1} \tag{17}$$

with the action of $\tilde{H}$ on the representation space being given by Eq. (13). To our knowledge these representations of $U_{\tilde{q}}(\mathfrak{osp}(1|2))$ are new.

On the other hand, we may use Propositions 4.2 and 6.3 to construct representations of $U(\mathfrak{iso}(2))$ and $U(\widetilde{\mathfrak{iso}}(2))$ out of representations of $U_q(\mathfrak{sl}(2))$ and $U_{\tilde{q}}(\mathfrak{osp}(1|2))$, respectively. In [9] we constructed representations of $U(\mathfrak{iso}(2))$ out of irreducible star representations of $U_q(\mathfrak{sl}(2))$ with the star structure given by the involution ω defined in Sec. 3, i.e. out of representations of $U_q(\mathfrak{so}(2,1))$. We considered cases for q generic and also at roots of unity. Using a slightly different version of Proposition 4.2, we showed, in that paper, that only for the principal series representations of $U_q(\mathfrak{so}(2,1))$ did we obtain infinitesimally unitarizable representations of $U(\mathfrak{iso}(2))$. In particular no finite dimensional irreducible representations of $U_q(\mathfrak{so}(2,1))$ led to representations of $U(\mathfrak{iso}(2))$, and we expect similar results for the supersymmetric case. Let us now show that this indeed so.

Finite dimensional irreducible representations of $U_{\tilde{q}}(\mathfrak{osp}(1|2))$ are given in [22]. They are characterized by a real number ℓ which takes positive integer or half integer values and a parameter λ which takes the values 0 and 1. We shall consider here only the case of integer ℓ, and such representations are described as follows [22]: Let $V^{(\ell)} = l.s.\{e^{\ell}_m(\lambda)|m = \ell, \ell-1, ..., -\ell\}$. For ℓ integer the action of the generators of $U_{\tilde{q}}(\mathfrak{osp}(1|2))$ on the basic vectors $e^{\ell}_m(\lambda)$ are:

$$\tilde{H}e^{\ell}_m(\lambda) = -me^{\ell}_m(\lambda) \tag{14}$$

$$\tilde{X}^{+}e^{\ell}_m(\lambda) = (-1)^{\ell-m-1}\left(\frac{1}{\{2\}_{\tilde{q}}}\{\ell+m\}_{\tilde{q}}\{\ell-m+1\}_{\tilde{q}}\right)^{\frac{1}{2}} e^{\ell}_{m-1}(\lambda) \tag{15}$$

$$\tilde{X}^{-}e^{\ell}_m(\lambda) = (-1)^{\ell-m-1}\left(\frac{1}{\{2\}_{\tilde{q}}}\{\ell-m\}_{\tilde{q}}\{\ell+m+1\}_{\tilde{q}}\right)^{\frac{1}{2}} e^{\ell}_{m+1}(\lambda) \tag{16}$$

where $\{m\}_{\tilde{q}} = \frac{\tilde{q}^{-m/2}-(-1)^m\tilde{q}^{m/2}}{\tilde{q}^{-1/2}+\tilde{q}^{1/2}}$. These representations are grade star for $\tilde{q}\in\mathbb{R}$, where the grade adjoint operation is given by

$$\tilde{H}^{*} = H, \qquad \tilde{X}^{\pm *} = \pm(-1)^{\tilde{\epsilon}}\tilde{X}^{\mp}\ , \tag{17}$$

with $\tilde{\epsilon} = \lambda + 1(\text{mod } 2)$. To see that these representations do not lead to representations of $U(\widetilde{\mathfrak{iso}}(2))$ on $V^{(\ell)}$, we take for the action of $\tilde{Y}$ in the representation the positive square root of $\tilde{Y}^2$ as determined by Eq. (9bis), and it is sufficient to show that the action of $\tilde{D}^{-}_{R}$ vanishes on the basic vector $e^{\ell}_{\ell}(\lambda)$, since then it is not an invertible operator and thus the conditions of Lemma 1.1 are not satisfied. Using Eqs. (14), (15) and (16) together with the definitions of $\tilde{Y}$ and $\tilde{D}^{-}_{R}$ given in Proposition 6.3 we obtain:

$$\tilde{D}^{-}_{R}e^{\ell}_{\ell}(\lambda) = \frac{1}{\sqrt{[2]_{\tilde{q}}}\,\tilde{Y}}\left(-i\sqrt{(-1)^{\tilde{H}}\left[\tilde{H}-\frac{1}{2}\right]_{\tilde{q}^2}} + i\tilde{Y}\right)e^{\ell}_{\ell}(\lambda) =$$

$$= \frac{1}{\sqrt{[2]_{\tilde{q}}}\,\tilde{Y}}\left(\sqrt{(-1)^{-\ell}\left[\ell+\frac{1}{2}\right]_{\tilde{q}^2}} - (-1)^{\ell/2}\sqrt{\left[\ell+\frac{1}{2}\right]_{\tilde{q}^2}}\right)e^{\ell}_{\ell}(\lambda) = 0. \tag{18}$$

Acknowledgments
The author would like to acknowledge useful discussions with Profs. V.K. Dobrev, M. Havlíček, J. van der Jeugt, K. W. Johnson and V. N. Tolstoy.

References
[1] Moylan P 1998 *Czech J. Phys.* **47** 1251-1258
[2] Mackey G 1975 On the analogy between semisiple Lie groups and certain related semi-direct product groups *Lie Groups and their Representations* ed I M Gelfand (New York: Halsted Press/J Wiley & Sons) 339-363
[3] Šťovíček P 1988 *J. Math. Phys.* **29** 1300
[4] Higson N 2008 The Mackey analogy and K-theory *Contemp. Math.* **449** (Providence: Amer. Math. Soc.) 149-172
[5] Božek P Havlíček and Navrátil O 1985 A new relationship between the Lie algeras of Poincaré and de Sitter groups *Preprint Universitas Carolina Pragensis NCITF/1*

[6] Havlíček M and Moylan P 1993 *J. Math. Phys.* **34** 11 5320-5332
[7] Moylan P 2004 Representations of the Poincaré group from positive energy representations of SO(2,3) *Proc. 3rd Int. Sympos. Quantum Theory and Symmetries* ed P C Argyres et al (New Jersey: World Scientific) 105-110
[8] Moylan P 1998 *Czech J. Phys.* **48** 1457-1464
[9] Moylan P 2003 *Group 24: Physical and Mathematical Aspects of Symmetries* (Inst. of Phys. Conf. Series vol 123) ed J-P Gazeau (London: Inst. of Phys. Publishing) 230-235.
[10] Moylan P 2004 *Studies in Mathematical Physics Research* ed V Benton (New York: Nova Science) Chapter 2
[11] Salom I and Šijački D 2013 *Int. J. Geom. Methods Mod. Phys.* **10** 1350017
[12] Hermann R 1966 *Lie groups for Physicists* (New York: Benjamin) 182; Löhmus Ja H 1969 Contractions of Lie Groups *Proceeding of the Second Summer School on Problems of the Theory of Elementary Particles* vol 4 (Tartu: Inst. Fiz. i Astronom. Akad. Nauk Eston. SSR) 67
[13] Cohn P M 1971 *Amer. Math. Monthly* **78** 596-615
[14] Dixmier J 1996 *Enveloping Algebras* (Providence: Amer. Math. Soc.) 122.
[15] Goodearl K R 2013 *Glasgow Math. Jour.* **55A** 75-87
[16] Dixmier J 1996 *Enveloping Algebras* (Providence: Amer. Math. Soc.) 119.
[17] Clark S and Wang W 2013 *Lett. Math. Phys.* **103** 207–231
[18] Kachurik I I 1998 *Ukrainian Math. Jour.* **50** 8 1201-1211
[19] Arnaudon D and Bauer M 1997 *Lett. Math. Phys.* **40** 307-320
[20] N. Jacobson, *Lie Algebras* (Dover, New York, 1979) 171; D. P. Zhelobenko, *Representations of Reductive Lie Algebras* (Nauka, Moscow, 1994) Ch. 3.
[21] Talman J D 1968 *Special Functions: A Group Theoretic Approach* (New York: Benjamin) 199.
[22] Aizawa N et al 2007 *J. Phys. A: Math. Theor.* **40** 14985-15000

Resonances in piecewise potentials and Supersymmetric Quantum Mechanics (SUSY-QM) for the construction of optical potentials.

Luis Fernando Orozco Cortés and Nicolás Fernández García.

Sección de Estudios de Posgrado e Investigación, UPIITA-IPN, Av. IPN 2508, C.P. 07340, Mexico D.F.

Abstract. A method to obtain the general solution of any constant piecewise potential is presented, this is achieved by means of the analysis of the transfer matrices in each cutoff. The resonance phenomenon together with the supersymmetric quantum mechanics technique allow us to construct a wide family of complex potentials which can be used as theoretical models for optical systems. The method is applied to the particular case for which the potential function has six cutoff points.

1. Introduction

The study of one-dimensional potentials in quantum mechanics remains an interest topic to address problems in contemporary physics. Indeed, the square well potential is discussed in each elementary book and it can be used as start model in many fields of science: to describe the motion of an electron in the field of a linear molecule such as acetylene [1], as a simple model of the deuteron [2, 3] and recently for studying the energy levels of the electrons in the graphene [4]. Thus, it is still studying a wide range of one-dimensional potentials that can serve as simple models of physical systems (see for example [5, 6, 7]). On the other hand, the first-order Supersymmetric Quantum Mechanics is a simple but very powerfull technique to generate solvable hamiltonians from a given initial one, whose solution is completely known. This technique is based on the first-order intertwining relationship and it is closely related to the Darboux transformation and with the widely used factorization method [8, 9, 10, 11].

The resonance phenomenon is one of the most surprising phenomena in nature, quantum mechanics is not the exception, in fact, the resonance states (or resonances) are associated with quasi-bound states, that is, systems which have the sufficient energy to break up into two or more subsystems. Mathematically the resonances can be understood as solutions of the Schrödinger equation with complex eigenvalue that fulfill a purely outgoing boundary condition [12, 13] (for the latest information check [14] and references therein). In this work we are interested in obtaining the general solution of piecewise potentials of any number of steps, thus, together with the supersymmetric quantum mechanics, by using as transformation function a resonant state, a wide family of complex potential can be constructed [18, 19]. The structure of this paper is the following: Section 2 is dedicated to obtain the general solution for a general piecewise potential for any number of steps. In Section 3, by means of our method, the discrete spectrum for a piecewise potential of seven steps (asymmetric square well) is obtained. In Section 4 the resonance spectrum is obtained for the same particular case. Section 5 is devoted to construct some complex potentials by means of the first order supersymmetric quantum mechanics; it is important to mention that these complex potentials probably can be used for describing optical systems, this last due that there is an analogy between the Helmholtz equation and the Schrödinger equation (see [20] and references therein). Finally, some concluding remarks are exposed.

2. Piecewise potential for any finite number of steps.

Let us consider the one-dimensional stationary Schrödinger equation, which is defined as follows

$$-\frac{d^2}{dx^2}\psi(x) + V(x)\psi(x) = E\psi(x), \tag{1}$$

where the type of potential we are interested in, for an arbitrary finite number of steps, can be written as follows

$$V(x) = V_1 + \sum_{j=2}^{n}(V_j - V_{j-1})\eta(x - a_{j-1}), \quad \eta(y) = \begin{cases} 1 & y \geq 0 \\ 0 & y < 0 \end{cases} \tag{2}$$

each V_j $(j = 1, ..., n)$ is constant but it changes from one step to the next; and a_j $(j = 1, ..., n-1)$ is the cutoff point which connect the regions j and $j+1$ (a schematic graphic for $n = 7$ is presented in Figure 1). The solution to the equation (1) for the potential (2), in each of the regions, is simply

$$\psi_j(x) = A_j e^{ik_j x} + B_j e^{-ik_j x}, \quad \text{with} \quad k_j^2 = (E - V_j)\,, \, (j = 1, ..., n). \tag{3}$$

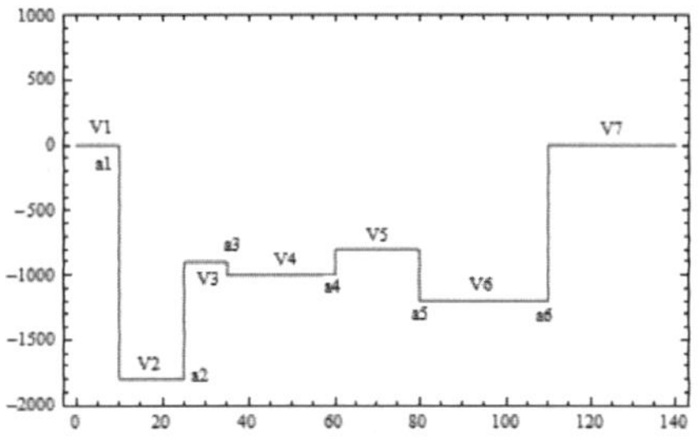

Figure 1. Schematic Figure of the potential (2) for $n = 7$ with all the $a_j > 0$, $V_1 = V_7 = 0$ and $V_j < 0$ $(j = 2, ..., 6)$.

According to the foundations of quantum mechanics, the above solutions must fulfill the following continuity conditions

$$\begin{aligned} \psi_j(x = a_j) &= \psi_{j+1}(x = a_j), \quad j = 1, 2, ..., n-1, \\ \psi_j'(x = a_j) &= \psi_{j+1}'(x = a_j), \quad ' \equiv \frac{d}{dx}. \end{aligned} \tag{4, 5}$$

The last relationships can be written in matrix form as follows:

$$M_j|j> = M_{j+1}|j+1>, \tag{6}$$

where

$$M_j = \begin{pmatrix} e^{ik_j x} & e^{-ik_j x} \\ ik_j e^{ik_j x} & -ik_j e^{-ik_j x} \end{pmatrix}_{x=a_j} \quad \text{and} \quad |j> = \begin{pmatrix} A_j \\ B_j \end{pmatrix}. \tag{7}$$

Hence, the coefficients of each of the solutions may be written in terms of the preceding one through the following relationships

$$|j+1>= T_j|j> \tag{8}$$

where

$$T_j = (M_{j+1})^{-1}M_j = \frac{1}{2}\begin{pmatrix} \left(1+\frac{k_j}{k_{j+1}}\right)e^{-i(k_{j+1}-k_j)x} & \left(1-\frac{k_j}{k_{j+1}}\right)e^{-i(k_{j+1}+k_j)x} \\ \left(1-\frac{k_j}{k_{j+1}}\right)e^{i(k_{j+1}+k_j)x} & \left(1+\frac{k_j}{k_{j+1}}\right)e^{i(k_{j+1}-k_j)x} \end{pmatrix}_{x=a_j} \tag{9}$$

thus, the transfer matrix which connects the last coefficients with the initial ones ($|n>= T_t|1>$) is written as the next matrices product

$$T_t = T_{n-1}T_{n-2}\cdots T_2T_1. \tag{10}$$

In the next sections the discrete and resonance spectrums will be found for an asymmetric square well of seven steps ($n = 7$).

3. Bound states in the asymmetric square well of seven steps.

Let us consider a piecewise constant potential of seven steps with the following constrains on the values of the parameters (see Figure 1):

$$\begin{aligned} a_j &> 0 \text{ for } j = 1, ..., 7. \\ V_1 &= V_7 = 0, \\ V_j &< 0 \text{ for } j = 2, ..., 6. \end{aligned}$$

In this section we restrict our analysis to the case of negative energies. Thus, the kinetic terms in regions 1 and 7 are given by

$$k_1 = k_7 = i\kappa = i\sqrt{|E|}. \tag{11}$$

Therefore, the asymptotic behaviors of the solution can be analyzed as follow

$$\lim_{x\to-\infty}\psi_1(x) = \lim_{x\to-\infty} A_1e^{-\kappa x} + B_1e^{\kappa x} = \lim_{x\to-\infty} A_1e^{-\kappa x} \to \infty, \tag{12}$$

$$\lim_{x\to\infty}\psi_7(x) = \lim_{x\to\infty} A_7e^{-\kappa x} + B_7e^{\kappa x} = \lim_{x\to\infty} B_7e^{\kappa x} \to \infty. \tag{13}$$

Since we are interested in solutions of square integrable, associated with the discrete spectrum of the system, we must pay attention only on those solutions for which it holds that $A_1 = B_7 = 0$. On the other hand, we know that

$$|7>= T_t|1>, \quad \text{with} \quad T_t = T_6\cdots T_1., \tag{14}$$

or equivalently

$$\begin{aligned} A_7 &= (T_t)_{11}A_1 + (T_t)_{12}B_1, \\ B_7 &= (T_t)_{21}A_1 + (T_t)_{22}B_1. \end{aligned} \tag{15}$$

Journal of Physics: Conference Series **512** (2014) 012027 doi:10.1088/1742-6596/512/1/012027

Table 1. Discrete spectrum for the conventional square well ($V_0 = -10$ and width a=4) and for the asymmetric square well with $V_1 = V_7 = 0, V_2 = -12, V_3 = -4, V_4 = -10, V5_= -5, V_6 = -11, a_1 = 1, a_2 = 2, a_3 = 3, a_4 = 3.7, a_5 = 4.3$ and $a_6 = 5$.

Energies	Square well	Asymmetric well
Ground state:	-9.541064761177934	-8.572298254015104
First excited state:	-8.176899605890787	-7.292325563181705
Second excited state:	-5.953924724417183	-5.180401667805235

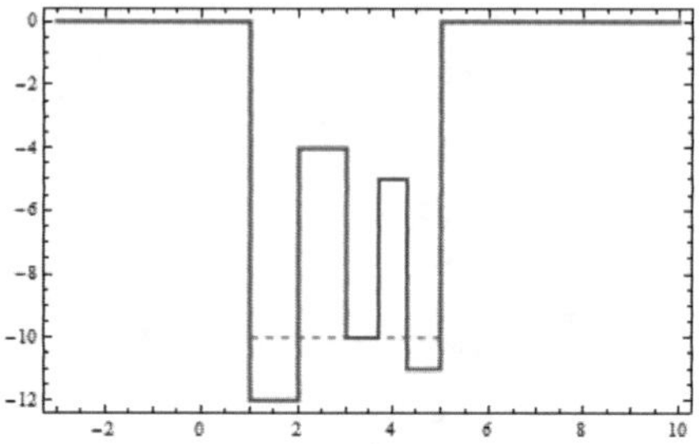

Figure 2. Square well (dashed line) and asymmetric square well (continuous line), for the parameters shown in Table 1.

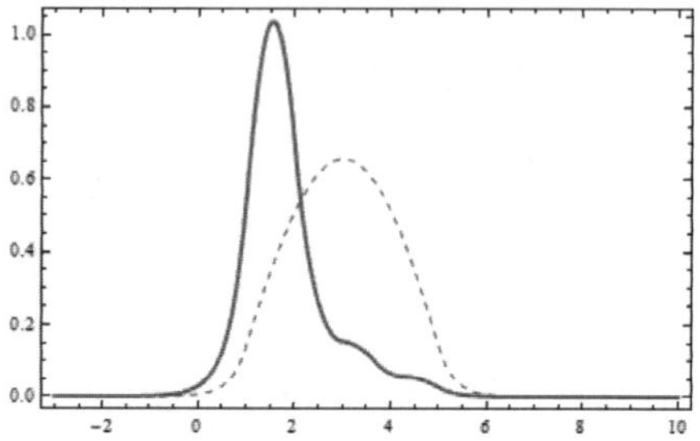

Figure 3. Ground states of the square well (dashed line) and of the asymmetric square well (continuous line).

In accordance with the previous analysis about the square integrable solutions, we take $A_1 = 0$, whence the last equation is reduced to

$$\begin{aligned} A_7 &= (T_t)_{12} B_1, \\ B_7 &= (T_t)_{22} B_1, \end{aligned} \tag{16}$$

but B_7 must be zero, then, the equation that allows us to find the discrete spectrum of the system is simply

$$(T_t)_{22} = 0. \tag{17}$$

The explicit expression for $(T_t)_{22}$ can be calculated from the general expressions for the transfer matrices presented previously. However, this expression is very long and therefore is not presented, instead, the Table 1 shows the negative energies associated with the discrete spectrums of our piecewise constant potential (asymmetric square well) and of the conventional square well for some particular values of the parameters.

Figure 2 shows a comparison between the square well (dashed line) and the asymmetric square well (continuous line) for the parameters showed in Table 1. Furthermore, Figures 3, 4 and 5 make clear that the wave functions corresponding to the discrete spectrum inherited the asymmetry of the piecewise potential. Moreover, the wave function of the ground state shows that any particle in this energy state, has a higher probability to be in the region where the potential is deeper, that is, to the left of the center of the potential. However, as the energy increases the wave function approaches the wave functions of the conventional square well (see Figure 5).

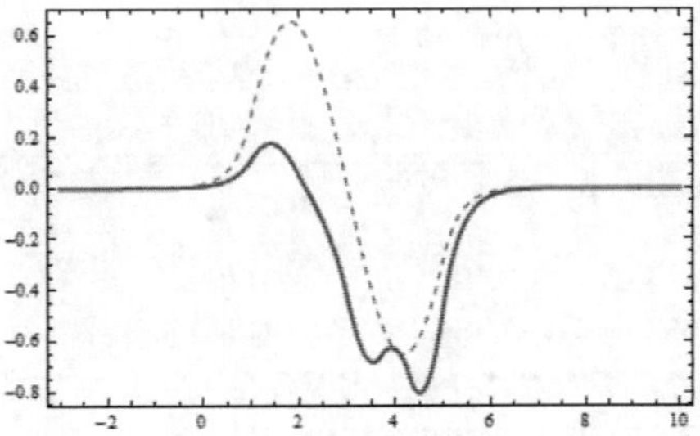

Figure 4. First excited state of the square well (dashed line) and of the asymmetric square well (continuous line).

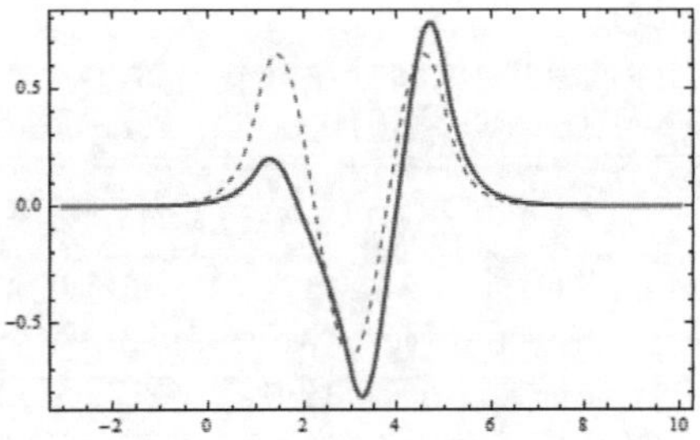

Figure 5. Second excited state of the square well (dashed line) and of the asymmetric square well (continuous line).

4. Resonances in the asymmetric square well of seven steps.

The resonances or resonant states are widely used to describe scattering processes, and they are defined as solutions of the Schrödinger equation with complex eigenvalues, which fulfill a purely outgoing boundary condition. For short range one dimensional potentials, this boundary condition can be written as follows (see [18])

$$\lim_{x\to\pm\infty}[\psi'(x)\mp ik\psi(x)]=0, \quad \text{with} \quad k^2=E. \tag{18}$$

Therefore, as happened to the discrete spectrum, we must analyze the asymptotic behavior of the solution. For this purpose, we start from

$$\begin{aligned} A_7 &= (T_t)_{11}A_1+(T_t)_{12}B_1, \\ B_7 &= (T_t)_{21}A_1+(T_t)_{22}B_1. \end{aligned} \tag{19}$$

If we assume that a single particle source is located far to the left of the interaction region, we have to take $B_7=0$, then

$$B_1=-\frac{(T_t)_{21}}{(T_t)_{22}}A_1 \tag{20}$$

substituting the last expression into (19), a relation of A_7 in terms of A_1 is obtained

$$A_7=\frac{\text{Det}[T_t]}{(T_t)_{22}}A_1. \tag{21}$$

It is not difficult to show that the remaining coefficients A_j and B_j $(j=2,...,6)$, can be written in terms of A_1 and all them have as common factor to $1/(T_t)_{22}$. Thus, the zeros of the function $(T_t)_{22}$ lead to a singular solution. Nevertheless, this singularity can be removed if the solution is renormalized by means of the redefinition of the arbitrary constant A_1 as follows

$$A_1=(T_t)_{22}\,A, \tag{22}$$

where A is the new normalization constant. Thus, with this assumption we have a nonsingular solution and the zeros of the function $(T_t)_{22}$ are justly the complex energy eigenvalues which fulfill the purely outgoing boundary condition (18); moreover, these zeros are simple poles of the transmission coefficient (for more details see for example [18]), which in this case is defined as

$$T=\left|\frac{A_7}{A_1}\right|^2=\left|\frac{Det\,|T_t|}{(T_t)_{22}}\right|^2 \tag{23}$$

Table 2. Resonant states found by the parameters: $V_1 = V_7 = 0, V_2 = -1000, V_3 = -930, V_4 = -980, V_5 = -950, V_6 = -1010, a_1 = 10, a_2 = 15.6, a_3 = 19.3, a_4 = 22, a_5 = 26.7, a_6 = 30$.

Elements	Resonance energies.
First resonance	3.04386926392625 - 0.339066907380203i
Second resonance	12.9822853595348 - 0.704447767786007i
Third resonance	22.8975576232267 - 0.959431583552292i
Fourth resonance	32.6650005179265 - 1.144896448819466i
Fifth resonance	42.8024271749718 - 1.308763165361565i
Sixth resonance	52.7270667854136 - 1.462783086145259i
Seventh resonance	62.8514355221570 - 1.587975668791163i
Eighth resonance	72.9718969834342 - 1.700247690561512i

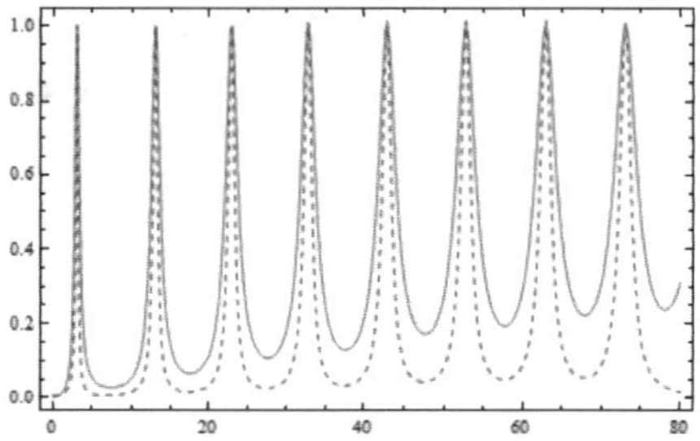

Figure 6. Superposition between the transmission coefficient and the Breit-Wigner distribution.

and can be approximated by a superposition of Breit-Wigner distributions of the form

$$\omega\left(E, \epsilon_R, \Gamma\right) = \frac{\left(\frac{\Gamma}{2}\right)^2}{\left(\epsilon_R - E\right)^2 + \left(\frac{\Gamma}{2}\right)^2}. \tag{24}$$

The complex zeros of $(T_t)_{22}$ for the present case can be compute numerically. The Table 2 shows the resonance spectrum for the asymmetric square well of seven steps, for some specific parameters values. Figure 6 shows a schematic comparison between the transmission coefficient and a superposition of Breit-Wigner distribution, this last defined in terms of the resonances presented in the Table 2. It is important to note that, according to the results presented in this work, both the discrete spectrum as the spectrum of resonances are obtained from the same condition $(T_t)_{22} = 0$, this is why we can understand the resonant states as bound states with finite lifetime and vice versa. In the next section we will use the supersymmetric quantum mechanics technique for constructing a wide family of complex potentials, which can be used as theoretical models of optical systems [20].

5. First-order Supersymmetric Quantum Mechanics.

This technique is based on the first order operatorial intertwining and allows us to generate exact analytical solutions for new potentials from one whose complete solution is known. Then, the starting relationship is

$$H_1 A_1^+ = A_1^+ H_0 \tag{25}$$

with

$$H = -\frac{d^2}{dx^2} + V_i(x) \ , \ i = 0, 1. \tag{26}$$

$$A^+ = \left(-\frac{d}{dx} + \alpha(x)\right). \tag{27}$$

Here, we are assuming that the solution to the Schrödinger equation corresponding to the hamiltonian H_0 is known and $\alpha(x)$ is a function to be determined. Developing both sides of the intertwining relationship and equating coefficients of the same order in the derivatives, the following pair of equations is obtained:

$$V_1 = V_0 - 2\alpha' \tag{28}$$

$$\alpha V_1 - \alpha'' = \alpha V_0 - V_0 \tag{29}$$

after substituting (28) into (29) and integrating, the next Riccati equation for α is obtained

$$\alpha' + \alpha^2 = V_0 - \in . \tag{30}$$

The above equation can be linearized by means of the following change $\alpha = \frac{\psi^{(0)'}}{\psi^{(0)}}$, which leads the Riccati equation to a Schrödinger equation type for the potential V_0

$$\psi^{(0)''} + V_0\psi^{(0)} = \epsilon\psi^{(0)}. \tag{31}$$

Therefore, function α can be found in terms of a general solution of the last equation, in this context, the function $\psi^{(0)}$ is known in the literature as transformation function and ϵ is the factorization constant. From equation (25) can be easily shown that if

$$H_0\psi_\lambda = \lambda\psi_\lambda \tag{32}$$

then

$$H_1\psi_\lambda[1] = \lambda\psi_\lambda[1], \tag{33}$$

where

$$V_1 = V_0 - 2\left(\frac{\psi^{(0)'}}{\psi^{(0)}}\right)' \quad \text{and} \quad \psi_\lambda[1] = A^\dagger\psi_\lambda. \tag{34}$$

The last expressions are also known as Darboux transformation, which is equivalent to the supersymmetric quantum mechanics technique. In this paper we are interested in a special case of this transformation: one in which the transformation function is a resonant state, as this allows us to construct a wide variety of complex potentials that can be used to model optical systems (for details see [19]).

Figures 7 and 8 show respectively the real and imaginary parts of the first-order SUSY partner of the piecewise potential of seven steps, here we have taken the first resonance as transformation function. The real part of the supersymmetric partner shows a sequence of non divergent wells, which are very deep but in turn so narrow. It is not difficul to verify that the number of wells that appears in the supersymmetric partners increases as the resonance we have used for implementing the transformation is more excited (the real part of the complex eigenvalue is bigger). On the other hand, the sign of the imaginary part of the SUSY partner dictates if we have a source (positive values) or sink (negative values) of probability. Furthermore, we can see that there is a symmetry in the system, ie, everything that is absorbed in some region is emmited in other one, so that the net effect of the system is that there is a conservation of probability. This can be demonstrated if the area under the curve of the imaginary part is computed. This last result could be of interest for those involved in the ongoing discussion on PT-symmetry and the possible extension or generalization of Quantum Mechanics, see for example [21].

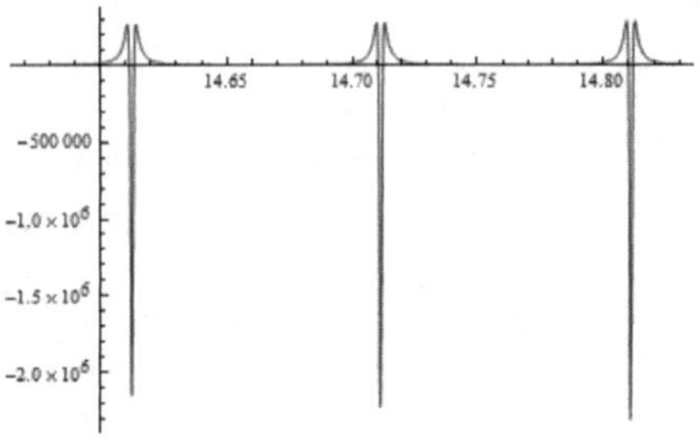

Figure 7. Real part of the first order SUSY partner of the piecewise potential of seven steps, where the transformation function is the first resonant state (see Table 2).

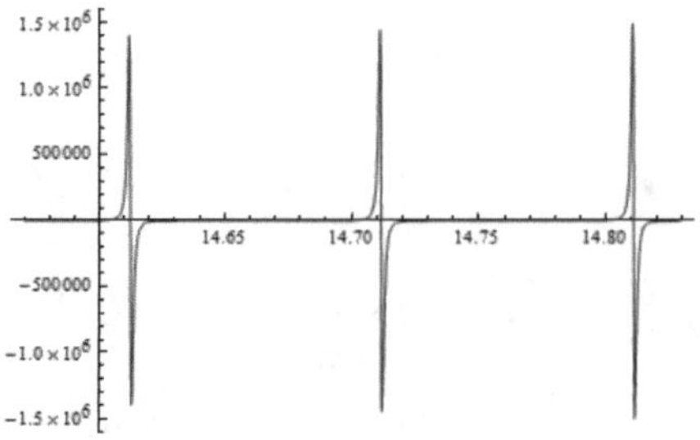

Figure 8. Imaginary part of the first order SUSY partner of the piecewise potential of seven steps, where the transformation function is the first resonant state (see Table 2).

6. Concluding Remarks.

Based on the analysis of the transfer matrices, we have found general expressions for obtaining the complete solution of any piecewise potential. This method is applied to an asymmetric square well of seven steps, where the discrete and resonance spectrums were obtained. Moreover, a general expression for the transmission coefficient was also presented, which gives the dynamics of the scattering process. On the other hand, the supersymmetric quantum mechanics by using a resonance as transformation function, permit us to construct complex potentials, which could be use as theoretical models of optical systems. The method is so general that could be implemented for any number of steps and even keeping as arbitrary all the parameters.

Acknowledgment

The support of IPN projects SIP-20130847 and SIP-SNI-2011/08 is acknowledged.

Bibliography

[1] French A Pand Taylor EF, An introduction to Quantum Physics (New York: W.W. Norton and Company, Inc. 1978).
[2] Park D, Introduction to the Quantum Theory (New York: McGraw-Hill 1992).
[3] Rosas-Ortiz O, Group Theoretical Methods, in Physics Conference Series Number 185, United Kingdom (2005).
[4] Miserev DS and Entin MV, J. of Exp. and Theor. Phys, **115** (2012) 694.
[5] Theocharis G, Kevrekidis PG, Fratzeskakis DJ and Schmelcher P, Phys. Rev. E **74** (2006) 056608.
[6] Gong J, Liang XX and Ban SL, Phys. Stat. Sol. (b) **244** (2007) 2064.
[7] Berezovoj VP and Konchatng MI, J. Phys. A: Math. Theor. **45** (2012) 225302.
[8] Mielnik B and Rosas-Ortiz O, J. Phys. A: Math. Gen. **37**, (2004) 10007.
[9] Fernández DJ and Fernández-García N, AIP Conference Proceedings **744** New York (2005).
[10] Arafeva I, Fernández DJ, et al, J. Phys. A: Math. Gen. Vol. **37**, number 43 (29 october 2004).
[11] Sukumar CV, AIP Conference Proceedings **744**, (New York 2005).
[12] Gamow G, Z Phys. **51** (1928) 204.
[13] Siegert AJF, Phys. Rev. **56** (1939) 750.
[14] Moiseyev N, Non Hermitian Quantum Mechanics, Cambridge University Press 2011.
[15] Ishimaru A, "Electromagnetic Wave Propagation, Radiation, and Scattering", Englewood Cliffs, NJ: Prentice Hall, 1991.
[16] Gadella M and de la Madrid R, Am. J. Phys. **70** (2002) 626-637.
[17] Heitler W and Hu N, Nature **159** (1947) 776-777.
[18] Fernández García N and Rosas-Ortiz O, Ann. Phys. **323** (2008) 1397.
[19] Fernández García N and Rosas-Ortiz O, J. Phys: Conf. Ser. **128** (2008) 012044.

[20] Monika AM Marte and Stenholm S, Phys. Rev. A, **56** (1997) 2940.
[21] Bender CM, Fring A, Günther U, Jons H, J. Phys. A: Math. and Theor. (2013) 45 440301.

Journal of Physics: Conference Series **512** (2014) 012028 doi:10.1088/1742-6596/512/1/012028

A relation between massive scalar field in AdS_{d+1} and diffusion in channels

Juan M. Romero, Alejandro Gaona
Departamento de Matemáticas Aplicadas y Sistemas, Universidad Autónoma Metropolitana-Cuajimalpa, México, D.F 01120, México

E-mail: jromero@correo.cua.uam.mx

Abstract. It is shown that, when the diffusion coefficient is a constant and is taken a particular family of channels, the Fick-Jacobs equation is invariant under conformal symmetry. In addition, using the diffusion coefficient and the geometric parameters of the channels, a representation for the conformal algebra is obtained. Furthermore, it is found that for these systems the Fick-Jacobs equation is equivalent to the Schrödinger equation for the 1-dimensional conformal quantum mechanics. Moreover, using this equivalence, it is found a relation between a massive scalar field equation in AdS_{d+1} background and Fick-Jacobs equation, where the geometric parameter of the channels and the geometric parameters of AdS_{d+1} are identified.

1. Introduction

Anti-de Sitter space AdS_{d+1} is the maximally symmetric solution of Einstein's equations with a negative cosmological constant. This space can be represented as the hyperboloid

$$X_0^2 + X_{d+1}^2 - \sum_{i=1}^{d-1} X^2 = R^2 \tag{1}$$

in the flat $d+2$-dimensional space with metric

$$ds^2 = -dX_0^2 - dX_{d+1}^2 + \sum_{i=1}^{d-1} X_i^2. \tag{2}$$

In Poincaré coordinates, the AdS_{d+1} metric takes the form [1, 2]

$$ds^2 = \frac{R^2}{z^2}\left(\eta_{\mu\nu}dx^\mu dx^\nu + dz^2\right), \qquad \mu = 0, 1, \cdots d \tag{3}$$

where $\eta_{\mu\nu}$ is the Minkowski metric and z is the so call holographic coordinate. In addition the cosmological constant is given by

$$\Lambda = -\frac{d(d-1)}{2R^2}. \tag{4}$$

In this space, the equation of motion for a massive scalar field, $\phi(x, z)$, is

$$\frac{1}{R^2}\frac{\partial^2\phi(x,z)}{\partial z^2} + \frac{1}{R^2}\partial^\mu\partial_\mu\phi(x,z) - \frac{(d-1)}{R^2 z}\frac{\partial\phi(x,z)}{\partial z} - \frac{m^2}{z^2}\phi(x,z) = 0. \tag{5}$$

An amazing result in physics is given by the AdS_{d+1}/CFT_d correspondence, which allows a relation between $(d+1)$-dimensional gravitational theory and certain classes of d-dimensional Yang-Mills theories [1, 2]. Recently this correspondence has been extended, for example there is a correspondence between a gravitational theory and condensed matter AdS/CMT [5] and also there is a duality between a gravitational theory and fluid dynamics *gravity/fluid* [6, 7]. Furthermore, Brownian motion has been studied in this frame-work [8, 9]. In the AdS_{d+1}/CFT_d correspondence, in order to get a conformal field theory the limit $z \to 0$ is taken [1, 2], then the holographic coordinate disappears in the CFT_d. However, the holographic coordinate is important too, for example if we take

$$\phi(x,z) = e^{-ip\cdot x} z^{\frac{(d-1)}{2}} \psi(z), \qquad p^2 = -M_d^2, \tag{6}$$

the equation (5) becomes

$$\left(-\frac{1}{R^2}\frac{\partial^2}{\partial z^2} + \frac{g}{z^2}\right)\psi(z) = \frac{M_d^2}{R^2}\psi(z), \quad g = \frac{1}{R^2}\left(\left(\frac{d-1}{2}\right)\left(\frac{d+1}{2}\right) + m^2R^2\right). \tag{7}$$

Using this last equation, it is possible to get the spectrum of the light hadrons, where z is related with the invariant separation of the quark constituents [3, 4]. Now, the equation (7) is the static Schrödinger equation for the conformal quantum mechanics [34], which appears in different contexts, from black-holes to atomic physics [35, 36, 37]. Notice that, if a system is related with the conformal quantum mechanics, then it is also related with a massive scalar field in AdS_{d+1}.

The conformal group is the symmetry group for both Anti-de Sitter space and conformal field theory, for this reason the conformal symmetry is important in the AdS_{d+1}/CFT_d correspondence [2]. This symmetry appears in other systems, for example the free Schrödinger equation is invariant under a non-relativistic conformal transformations, which is known as the Schrödinger group [10, 11]. In fact this group has been important to study the non-relativistic AdS_{d+1}/CFT_d correspondence [12, 13]. Some work about the Schrödinger group can be seen in [14, 15, 16, 17, 18, 19, 20, 21]. Now, the simplest model of diffusion is described by the Fick equation and Sophus Lie showed that this equation is invariant under the Schrödinger group [22]. Another study about diffusion phenomena and the Schrödinger group can be seen in [23].

When the diffusion is in a channel, which has the shape of surface of revolution with cross sectional area $A(x)$, the Fick equation has to be changed to the Fick-Jacobs equation [24]

$$\frac{\partial C(x,t)}{\partial t} = \frac{\partial}{\partial x}\left[D(x)A(x)\frac{\partial}{\partial x}\left(\frac{C(x,t)}{A(x)}\right)\right], \tag{8}$$

where $C(x,t)$ is the particle concentration and $D(x)$ is the diffusion coefficient. This last equation is important to study diffusion in biological channels, zeolites and nano-channels [25, 26, 27, 28, 29, 30, 31]. In this paper we will show that, when the diffusion coefficient is a constant and the cross sectional area is $A(x) = (b+\lambda x)^{2\nu}$ (where ν, b, and λ are real constants), the Fick-Jacobs equation is equivalent to the conformal quantum mechanics. In addition, a relation between a massive scalar field equation in AdS_{d+1} background and Fick-Jacobs equation is found, in which the diffusion coefficient and cosmological constant for AdS_{d+1} space are associated. Furthermore, it is found that the geometric parameter of the channels and the geometric parameters of AdS_{d+1} space are related too. In this case the axial coordinate of the channel x and the holographic coordinate z of AdS_{d+1} are identified.

This paper is organized in the following way: in section 2 a brief review about the conformal quantum mechanics is given; in section 3 it is shown that the Fick-Jacobs equation is invariant under conformal symmetry for a family of channels; in section 4 it is shown that the Fick-Jacobs equation is equivalent to the conformal quantum mechanics for a set of particular channels and an exact solution for this equation is given. In section 5 a relation between Fick-Jacobs equation and a massive scalar field in AdS_{d+1} is found. Finally, in section 6 a summary is given.

2. Conformal Quantum Mechanics

The Schrödinger equation for the 1-dimensional conformal quantum mechanics is given by

$$i\hbar\frac{\partial\psi(x,t)}{\partial t} = H\psi(x,t) \qquad H = -\frac{\hbar^2}{2m}\frac{\partial^2}{\partial x^2} + \frac{g}{x^2}, \tag{9}$$

which is invariant under the non-relativistic conformal symmetry

$$t' = \frac{\alpha t+\beta}{\gamma t+\delta}, \qquad x' = \frac{ax}{\gamma t+\delta}, \quad a^2 = \alpha\delta - \beta\gamma \neq 0, \tag{10}$$

in order to keep the wave equation invariant under the last transformation the wave function must transforms like

$$\psi'(x',t') = \sqrt{\gamma t+\delta}\, e^{i\frac{m}{2\hbar}\Phi(x,t)}\psi(x,t) \tag{11}$$

where

$$\Phi(x,t) = -\frac{\gamma x^2}{\gamma t+\delta}. \tag{12}$$

In this case the group generators are given by H and

$$K_1 = tH - \frac{1}{2}\left(xp - \frac{i}{2}\right), \tag{13}$$

$$K_2 = t^2 H - t\left(xp - \frac{i}{2}\right) + \frac{mx^2}{2} \tag{14}$$

and the algebra

$$[H,K_1] = iH, \qquad [H,K_2] = 2iK_1, \qquad [K_1,K_2] = iK_2 \tag{15}$$

is satisfied. Using this algebra, it is possible to show that the operator H, K_1, K_2 are conserved.

The classical system with the general potential $V(r) = gr^z$ was first studied by Jacobi [32] and the quantum system with $z = -2$ was originally proposed by Jackiw [33]. The spectrum of the Hamiltonian (9) was found by de Alfaro, Fubini and Furlan [34]. It is worth to mention that the conformal quantum mechanics appears in different contexts, from black-holes to atomic physics [35, 36, 37] and has been proposed as the CFT_1 dual to AdS_2 [38, 39]. We will show that this Hamiltonian appears in diffusion phenomena too.

3. Conformal symmetry and Fick-Jacobs equation

Now, we will study the family of channels with cross sectional area

$$A(x) = (b + \lambda x)^{2\nu}, \qquad (\nu, b, \lambda = \text{constant}). \tag{16}$$

In this case, taking $D(x) = D_0$ as a constant and using the change of variable

$$y = b + \lambda x, \tag{17}$$

the Fick-Jacobs equation (8) becomes

$$\frac{\partial C(y,t)}{\partial t} = \lambda^2 D_0 \left(\frac{\partial^2}{\partial y^2} - \frac{2\nu}{y}\frac{\partial}{\partial y} + \frac{2\nu}{y^2} \right) C(y,t). \tag{18}$$

This equation is invariant under conformal transformation (10), where the concentration transforms as

$$C'(y',t') = (\gamma t + \delta)^{\frac{1-2\nu}{2}} e^{-\frac{1}{4D_0}\Phi(y,t)} C(y,t). \tag{19}$$

and Φ is given by (12). If we take $p = -i\frac{\partial}{\partial y}$ the equation (18) can be written like

$$\frac{\partial C(y,t)}{\partial t} = \mathbf{H} C(y,t), \tag{20}$$

where

$$\mathbf{H} = \lambda^2 D_0 \left(-p^2 - i\frac{2\nu}{y} p + \frac{2\nu}{y^2} \right). \tag{21}$$

Now, using

$$\mathbf{K}_1 = t\mathbf{H} - \frac{1}{2}\left(yp - \frac{i}{2} + i\nu \right), \tag{22}$$

$$\mathbf{K}_2 = t^2\mathbf{H} - t\left(yp - \frac{i}{2} + i\nu \right) - \frac{y^2}{4\lambda^2 D_0}. \tag{23}$$

we obtained the conformal algebra

$$[\mathbf{H}, \mathbf{K}_1] = i\mathbf{H}, \quad [\mathbf{H}, \mathbf{K}_2] = 2i\mathbf{K}_1, \quad [\mathbf{K}_1, \mathbf{K}_2] = i\mathbf{K}_2.$$

However, these operators are not conserved.

4. Conformal quantum mechanics and Fick-Jacobs equation

In the last section we obtained a kind of not hermitian quantum mechanics. However, the equation (18) is related with the usual conformal quantum mechanics. In fact, using the no unitary transformation

$$C(y,t) = (b + \lambda x)^{\nu} \psi(y,t) \tag{24}$$

the equation (18) becomes

$$\frac{\partial \psi(y,t)}{\partial t} = -H\psi(y,t), \tag{25}$$

Journal of Physics: Conference Series **512** (2014) 012028 doi:10.1088/1742-6596/512/1/012028

here

$$H = -\lambda^2 D_0 \frac{\partial^2}{\partial y^2} + \frac{g}{y^2}, \qquad g = \lambda^2 D_0 \nu \left(\nu - 1\right). \tag{26}$$

Notice that H is hermitian and is the Hamiltonian for the conformal quantum mechanics (9). In fact, proposing $\psi(y,t) = e^{-Et}\Psi(y)$, we get the static Schrödinger equation for the conformal quantum mechanics

$$E\Psi(y) = H\Psi(y). \tag{27}$$

Then, the family of channels (16) is associated with the family of conformal Hamiltonians (26). Observe that for each Hamiltonian (26) we have two channels, namely each Hamiltonian is associated with two ν values. For example, $\nu = 0$ represents a cylindrical channel while $\nu = 1$ represents a conical channel, but both gave the Hamiltonian

$$H = -\lambda^2 D_0 \frac{\partial^2}{\partial y^2}. \tag{28}$$

Now, using the change of variable (17), we found that the exact solution for the Fick-Jacobs equation is given by

$$C_\nu(x,t) = Be^{-Et} \left(b + \lambda x\right)^{\frac{2\nu+1}{2}} J_{\pm\left(\frac{2\nu-1}{2}\right)} \left(\pm\sqrt{\frac{E}{\lambda^2 D_0}} \left(b + \lambda x\right) \right), \tag{29}$$

here B is a constant. The parameter E can be obtained from the boundary condition. For example, if $C_\nu(L,t) = 0$, the values

$$E_n = \frac{\rho_n^2 \lambda^2 D_0}{b + \lambda L} \tag{30}$$

are obtained, where ρ_n is the nth root for the Bessel function of the order $\left(\frac{2\nu-1}{2}\right)$. In this case the initial condition $C_\nu(x,0) = 0$ can be written as a Fourier-Bessel series.

5. AdS/Fick-Jacobs

Now, if the holographic coordinate $z \neq 0$, then the massive scalar field in AdS_{d+1} can be expressed as the equation (6), where the field $\psi(z)$ satisfies the static Schrödinger equation for the conformal quantum mechanics. Then, if the cross sectional area is given by $A(x) = (b + \lambda x)^{2\nu}$, with the mapping

$$b + \lambda x \longleftrightarrow z \tag{31}$$

$$\lambda^2 D_0 \longleftrightarrow R^{-2} = -\frac{2\Lambda}{d(d+1)}, \tag{32}$$

$$\nu(\nu - 1) \longleftrightarrow \left(\frac{d-1}{2}\right)\left(\frac{d+1}{2}\right) + m^2 R^2, \tag{33}$$

$$E \longleftrightarrow \frac{M_d^2}{R^2}, \tag{34}$$

a relation between the Fick-Jacobs equation (26) and a massive scalar field in AdS_{d+1} is obtained. Notice that the cosmological constant Λ leads the expansion of AdS_{d+1} universe, while the diffusion coefficient D_0 leads the diffusion of particles in the channel and these both parameters are related. In addition, it can be seen that ν and d, R are related too, it is worth to mention that ν drives the geometric properties of the channel, while d and R drive the geometric properties of the AdS_{d+1} space.

In the usual AdS_{d+1}/CFT_d duality, the limit $z \to 0$ is taken [1]. However there is a correspondence with z no constant, like AdS/QCD correspondence [3]. In this paper a correspondence between the wave equation for a massive scalar field in AdS_{d+1} and Fick-Jacobs equation was gotten. Even more the axial coordinate x of the channel and the holographic coordinate z of AdS_{d+1} are identified.

6. Summary

Recently mathematical techniques developed in quantum physics have been employed to study biological systems [40]. Now, symmetries are important in quantum systems, then symmetries might be important in biological systems too. In this paper was studied the Fick-Jacobs equation which models diffusion in biological channels, zeolites and nano-channels. It was shown that, when the diffusion coefficient is a constant and is taken a particular family of channels, this equation is invariant under conformal symmetry. In addition, using the diffusion coefficient and the geometric parameters of the channels, a representation for the conformal algebra was obtained. Furthermore, it was found that for these systems the Fick-Jacobs equation is equivalent to the Schrödinger equation for the 1-dimensional conformal quantum mechanics. Moreover, using this equivalence, it was found a relation between a massive scalar field equation in AdS_{d+1} background and Fick-Jacobs equation, where the geometric parameter of the channels and the geometric parameters of AdS_{d+1} are identified.

It is well known that symmetry methods are very useful in quantum mechanics. However, in biological systems these methods are few employed. The Fick-Jacobs equation is an example where symmetries are important and useful. In a future work we will study other equations that describe biological systems using these methods.

Acknowledgments

We would like to thank P.A Horvathy for his comments about the history of Schrödinger symmetry.

[1] J. Maldacena, *The Large N limit of Superconformal Field Theories and Supergravity,* Adv. Theor. Math. Phys. **2**, 231 (1998).
[2] O. Aharony, S. S. Gubser, J. Maldacena, H. Ooguri, *Large N field theories, string theory and gravity*, Phys. Rept. **323**, 183-386 (2000).
[3] S. J. Brodsky, G. F. de Teramond, *Light-Front Dynamics and AdS/QCD Correspondence: The Pion Form Factor in the Space- and Time-Like Regions,* Phys. Rev. D **77**, 056007 (2008).
[4] G. F. de Teramond, Stanley J. Brodsky, *Light-Front Holography: A First Approximation to QCD,* Phys.Rev.Lett. **102** (2009) 081601.
[5] S. A. Hartnoll, *Lectures on holographic methods for condensed matter physics,* Class. Quant. Grav. **26**, 224002 (2009) .
[6] S. Bhattacharyya, V. E. Hubeny, S. Minwalla, M. Rangamani, *Nonlinear Fluid Dynamics from Gravity,* JHEP **0802**, 045 (2008).
[7] M. Rangamani, *Gravity & Hydrodynamics: Lectures on the fluid-gravity correspondence,* Class. Quant. Grav. **26**, 224003 (2009).
[8] J. de Boer, V. E. Hubeny, M. Rangamani, M. Shigemori, *Brownian motion in AdS/CFT,* JHEP **0907** 094 (2009).
[9] W. Fischler, J. F. Pedraza, W. T. Garcia, *Holographic Brownian Motion in Magnetic Environments,* JHEP **1212**, 002 (2012).

[10] C. R. Hagen, *Scale and conformal transformations in galilean-covariant field theory,* Phys. Rev. D **5**, 377 (1972) .
[11] U. Niederer, *The maximal kinematical invariance group of the free Schrodinger equation,* Helv. Phys. Acta. **45**, 802 (1972).
[12] D. T. Son, *Toward an AdS/cold atoms correspondence: a geometric realization of the Schroedinger symmetry,* Phys. Rev. D **78**, 046003 (2008).
[13] K. Balasubramanian, J. McGreevy, *Gravity duals for non-relativistic CFTs,* Phys. Rev. Lett. **101**, 061601 (2008).
[14] C. Duval, P. A. Horvathy, *Non-relativistic conformal symmetries and Newton-Cartan structures,* J.Phys. A **42**, 465206 (2009).
[15] C. Duval, G. W. Gibbons. P. A. Horvathy, *Celestial mechanics, conformal structures and gravitational waves,* Phys. Rev. D **43**, 3907-3922 (1991).
[16] C. Duval, M. Hassaine. P. A. Horvathy, *The geometry of Schrödinger symmetry in non-relativistic CFT,* Annals Phys. **324**, 1158 (2009) .
[17] C. Leiva, M. S. Plyushchay, *Conformal symmetry of relativistic and nonrelativistic systems and AdS/CFT correspondence,* Annals Phys. **307**, 372-391 (2003).
[18] C. Leiva, M. S. Plyushchay, *Superconformal mechanics and nonlinear supersymmetry,* JHEP **0310**, 069 (2003).
[19] A. Anabalon, M. S. Plyushchay, *Interaction via reduction and nonlinear superconformal symmetry,* Phys. Lett. B **572**, 202-209 (2003).
[20] A. Galajinsky, *Remark on quantum mechanics with conformal Galilean symmetry,* Phys. Rev. D **78**, 087701 (2008).
[21] A. Galajinsky, I.V. Masterov, *Remark on quantum mechanics with N=2 Schrödinger supersymmetry,* Phys. Lett. B **675**, 116 (2009).
[22] S. Lie, *Arch. Math. Nat.* vid. (Kristiania) **6**, 328 (1882).
[23] S. Stoimenov, M. Henkel, *Dynamical symmetries of semi-linear Schrödinger and diffusion equations,* Nucl. Phys. B **723**, 205-233 (2005).
[24] M. Jacobs, *Diffusion Processes,* Springer, New York, (1967).
[25] D. Reguera, J. M. Rubí, *Kinetic equations for diffusion in the presence of entropic barriers,* Phys. Rev. E **64**, 061106 (2001).
[26] D. Reguera, G. Schmid, P. S. Burada, J. M. Rubí, P. Reimann, P. Hänggi, *Entropic Transport: Kinetics, Scaling, and Control Mechanisms,* Phys. Rev. Lett. **96**, 130603 (2006).
[27] P. S. Burada, P. Hänggi, F. Marchesoni, G. Schmid, P. Talkner, *Diffusion in Confined Geometries,* Chem. Phys. Chem. **10**, 45 (2009) .
[28] A. M. Berezhkovskii, M. A. Pustovoit, S. M. Bezrukov, *Diffusion in a tube of varying cross section: Numerical study of reduction to effective one-dimensional description,* J. Chem. Phys. **126**, 134706 (2007) .
[29] A. Fulinski, I. Kosinska, Z Siwy, *Transport properties of nanopores in electrolyte solutions: the diffusional model and surface currents,* New J. Phys. **7**, 132 (2005).
[30] R. Zwanzig, *Diffusion past an entropy barrier,* J. Chem. Phys. **96**, 3926 (1992).
[31] P. Kalinay, J. K. Percus, *Corrections to the Fick-Jacobs equation,* Phys. Rev. E **74**, 041203 (2006).
[32] C. G. J. Jacobi, *Vorlesungen uber Dynamik* (Univ. Königsberg 1842-43). Her- ausg. A. Clebsch. Vierte Vorlesung: Das Princip der Erhaltung der lebendigen Kraft. Zweite ausg. C. G. J. Jacobis Gesammelte Werke. Supplementband. Herausg. E. Lottner. Berlin Reimer (1884).
[33] R. Jackiw, *Introducing Scale Symmetry,* Physics Today **25**, 23 (1972).
[34] V. de Alfaro, S. Fubini, G. Furlan, *Conformal Invariance in Quantum Mechanics,* Nuovo Cimento A **34**, 569 (1976).
[35] R. Britto-Pacumio, J. Michelson, A. Strominger, A. Volovich, *Lectures on superconformal quantum mechanics and multiblack hole moduli spaces,* (Contributed to NATO Advanced Study Institute on Quantum Geometry) (Published in Progress in string theory and M-theory) 235-264 (1999), [hep-th/9911066].
[36] H. E. Camblong, C. R. Ordonez, *Anomaly in conformal quantum mechanics: From molecular physics to black holes,* Phys. Rev. D **68**, 125013 (2003).
[37] K. Sakamoto, K. Shiraishi, *Conformal Quantum Mechanics in Two Black Hole Moduli Space,* Phys. Rev. D **66** 024004(2002).
[38] C. Chamon, R. Jackiw R, S-Y. Pi, L. Santos, *Conformal quantum mechanics as the CFT_1 dual to AdS_2,* Phys. Lett. B **701**, 503 (2011) .
[39] R. Jackiw, S-Y. Pi S, *Conformal Blocks for the 4-Point Function in Conformal Quantum Mechanics,* Phys. Rev. D **86**, 045017, Erratum-ibid. D **86** 089905 (2012).
[40] L. Abbott, E. Farhi, S. Gutmann, *The path integral for dendritic trees,* Biol. Cybern. **66** 49 (1991).

Journal of Physics: Conference Series **512** (2014) 012029 doi:10.1088/1742-6596/512/1/012029

Conformal symmetry in quantum finance

Juan M. Romero[1], Elio Martínez Miranda[2] and Ulises Lavana[1]

[1]Departamento de Matemáticas Aplicadas y Sistemas, Universidad Autónoma Metropolitana-Cuajimalpa, México, D.F 01120, México

[2] Facultad de Economía, Universidad Nacional Autónoma de México, México, D.F 04510, México

E-mail: jromero@correo.cua.uam.mx

Abstract. The quantum finance symmetries are studied. In order to do this, the one dimensional free non-relativistic particle and its symmetries are revisited and the particle mass is identified as the inverse of square of the volatility. Furthermore, using financial variables, a Schrödinger algebra representation is constructed. In addition, it is shown that the operators of this last representation are not hermitian and not conserved.

1. Introduction

Lately, mathematical techniques developed in physics have been employed to study systems from other areas. For example, the Black-Scholes-Merton equation [1, 2] is important in theoretical finance and it can be mapped to the one dimensional free Schrödinger equation [3]. Then, mathematical techniques that arise in quantum mechanics can be used to study financial phenomena, this fact allowed the birth of a new discipline, the so call Quantum Finance [3]. Now, it is well known that when symmetries are present in a physical system we can get at the properties of a system without completely solving all the equations that describe the system, in fact symmetries imply conserved quantities. The conformal symmetry is important in physics, for example the relativistic conformal group is the largest symmetry group of special relativity [4]. In addition, the free Schrödinger equation is invariant under the Schrödinger group, which is a non-relativistic conformal group [5, 6]. It is worth mentioning that in 1882 Sophus Lie showed that the Fick equation, which describes diffusion, is invariant under the Schrödinger group [7].

In this paper, it will be shown that the Black-Scholes-Merton equation is invariant under the Schrödinger group. In order to do this, the one dimensional free non-relativistic particle and its symmetries will be revisited. To get the Black-Scholes-Merton equation symmetries, the particle mass is identified as the inverse of square of the volatility. Furthermore, using financial variables, a Schrödinger algebra representation is constructed. In addition, it is shown that the operators of this last representation are not hermitian and not conserved.

This paper is organized as follows: in section 2 we provide a brief overview of the one dimensional non-relativistic free particle and its symmetries; in section 3 the one dimensional free Schrödinger equation and its symmetries are studied; in section 4 the Black-Scholes-Merton equation and its symmetries are studied. Finally, in section 5 a summary is given.

2. Free particle

The action for one-dimensional non-relativistic free particle is given by

$$S = \int dt \frac{m}{2}\left(\frac{dx}{dt}\right)^2, \tag{1}$$

this is the simplest mechanical system. Now, using the coordinates transformation

$$t' = \frac{\alpha t + \beta}{\gamma t + \delta}, \qquad x' = \frac{ax + vt + c}{\gamma t + \delta}, \qquad a^2 = \alpha\delta - \beta\gamma \neq 0, \tag{2}$$

where $\alpha, \beta, \gamma, \delta, a, v, c$ are constants, the action (1) transforms as

$$S' = \int dt' \frac{m}{2}\left(\frac{dx'}{dt'}\right)^2 = S + \frac{m}{2}\int dt \left(\frac{d\phi(x,t)}{dt}\right), \tag{3}$$

where

$$\phi(x,t) = \frac{1}{a^2}\left(2avx + v^2 t - \frac{\gamma\,(ax + vt + c)^2}{\gamma t + \delta}\right). \tag{4}$$

Then, the equation of motion for one dimensional free particle is invariant under the conformal coordinate transformations (2). This coordinates transformation includes temporal translations

$$t' = t + \beta, \qquad x' = x, \tag{5}$$

spatial translations

$$t', \qquad x' = x + c, \tag{6}$$

Galileo‘s transformations

$$t', \qquad x' = x + vt, \tag{7}$$

anisotropic scaling

$$t' = a^2 t, \qquad x' = ax \tag{8}$$

and the special conformal transformations

$$t' = \frac{1}{\gamma t + 1}, \qquad x' = \frac{x}{\gamma t + 1}. \tag{9}$$

It is shown below that the conformal coordinate transformations (2) and the quantity (4) are useful to study Black-Scholes-Merton equation symmetries.

2.1. *Conservative quantities*

For the one dimensional non-relativistic particle, the following quantities

$$P = m\dot{x}, \tag{10}$$

$$H = \frac{P^2}{2m}, \tag{11}$$

$$G = tP - mx, \tag{12}$$

$$K_1 = tH - \frac{1}{2}xP, \tag{13}$$

$$K_2 = t^2 H - txP + \frac{m}{2}x^2 \tag{14}$$

are conserved.

The momentum P is associated with spatial translations (6). The Hamiltonian H is associated with temporal translations (5) . The quantity G is associated with Galileo's transformations (7). While K_1 is associated with anisotropic scaling (8) and K_2 is associated with the special conformal transformations (9).

Furthermore, using the Poisson brackets, it can be shown that the following relations

$$\{P, H\} = 0, \tag{15}$$
$$\{P, K_1\} = \frac{1}{2}P, \tag{16}$$
$$\{P, K_2\} = G, \tag{17}$$
$$\{P, G\} = m, \tag{18}$$
$$\{H, K_1\} = H, \tag{19}$$
$$\{H, G\} = P, \tag{20}$$
$$\{H, K_2\} = 2K_1, \tag{21}$$
$$\{K_1, K_2\} = K_2, \tag{22}$$
$$\{K_1, G\} = \frac{1}{2}G, \tag{23}$$
$$\{K_2, G\} = 0 \tag{24}$$

are satisfied.

2.2. *Schrödinger group*

In quantum mechanics, if a particle is in x_0 at time t_0, the amplitude to travel to x in a time $T = t - t_0$ is given by

$$U(x, t; x_0, t_0) = \left(\frac{m}{2\pi i(t - t_0)}\right)^{\frac{1}{2}} e^{\frac{i}{\hbar}\int_{t_0}^{t} d\tilde{t}\frac{m}{2}\left(\frac{dx}{dt}\right)^2}, \tag{25}$$

which satisfies the Schrödinger equation

$$i\hbar\frac{\partial U(x, t; x_0, t_0)}{\partial t} = -\frac{\hbar^2}{2m}\frac{\partial^2 U(x, t; x_0, t_0)}{\partial x^2}. \tag{26}$$

Now, in another system with coordinates x', t', x'_0, t'_0 we have the amplitude

$$U'(x', t'; x'_0, t'_0) = \left(\frac{m}{2\pi i(t' - t'_0)}\right)^{\frac{1}{2}} e^{\frac{i}{\hbar}\int_{t'_0}^{t'} d\tilde{t}'\frac{m}{2}\left(\frac{dx'}{d\tilde{t}'}\right)^2}, \tag{27}$$

which satisfies the Schrödinger equation (26) on primed coordinates (x', t'). In addition, using the conformal transformations (2) we have

$$t' = \frac{\alpha t + \beta}{\gamma t + \delta}, \qquad x' = \frac{ax + vt + c}{\gamma t + \delta}, \tag{28}$$
$$t'_0 = \frac{\alpha t_0 + \beta}{\gamma t_0 + \delta}, \qquad x'_0 = \frac{ax_0 + vt_0 + c}{\gamma t_0 + \delta}. \tag{29}$$

Journal of Physics: Conference Series **512** (2014) 012029 doi:10.1088/1742-6596/512/1/012029

Notice that, without loss of generality, we can take $x_0 = 0$ and $t_0 = 0$, in this case

$$t' - t'_0 = \frac{a^2 t}{\delta(\gamma t + \delta)}. \tag{30}$$

Then, using the equation (3), we have

$$U'(x', t'; x'_0, t'_0) = \sqrt{\frac{\delta}{a^2}} \left(\sqrt{\gamma t + \delta}\right) e^{\frac{im}{2\hbar}\phi(x,t)} U(x, t; 0, 0), \tag{31}$$

where $\phi(x, t)$ is given by (4). Now, due that $U(x, t; x_0, t_0)$ is solution for the Schrödinger equation (26) on coordinates (x, t), and $U'(x', t'; x'_0, t'_0)$ satisfies the same equation on primed coordinates (x', t'), the expression (31) implies that the Schrödinger equation for the one dimensional non-relativistic free particle

$$i\hbar \frac{\partial \psi(x, t)}{\partial t} = -\frac{\hbar^2}{2m} \frac{\partial^2 \psi(x, t)}{\partial x^2} \tag{32}$$

is invariant under the conformal coordinate transformations (2), where the wave function transforms as

$$\psi'(x', t') = \left(\sqrt{\gamma t + \delta}\right) e^{\frac{im}{2\hbar}\phi(x,t)} \psi(x, t), \tag{33}$$

and $\phi(x, t)$ is given by (4).

Using other methods, the conformal symmetry for the free Schrödinger equation was found by Niederer and Hagen in 1972 [6, 5]. However, this symmetry was obtained by S. Lie in 1882 while he was studying the Fick equation [7].

Furthermore, according to quantum mechanics, the quantities (10)-(14) are represented by the operators

$$\begin{aligned}
\hat{P} &= -i\hbar \frac{\partial}{\partial x}, && (34)\\
\hat{H} &= \frac{\hat{P}^2}{2m}, && (35)\\
\hat{G} &= t\hat{P} - mx, && (36)\\
\hat{K}_1 &= t\hat{H} - \frac{1}{4}\left(x\hat{P} + \hat{P}x\right), && (37)\\
\hat{K}_2 &= t^2\hat{H} - \frac{t}{2}\left(x\hat{P} + \hat{P}x\right) + \frac{m}{2}x^2. && (38)
\end{aligned}$$

These operators satisfy the Schrödinger algebra

$$\begin{aligned}
\left[\hat{P}, \hat{H}\right] &= 0, && (39)\\
\left[\hat{P}, \hat{K}_1\right] &= \frac{i\hbar}{2}\hat{P}, && (40)\\
\left[\hat{P}, \hat{K}_2\right] &= i\hbar\hat{G}, && (41)\\
\left[\hat{P}, \hat{G}\right] &= i\hbar m, && (42)
\end{aligned}$$

$$\left[\hat{H}, \hat{K}_1\right] = i\hbar\hat{H}, \tag{43}$$

$$\left[\hat{H}, \hat{G}\right] = i\hbar\hat{P}, \tag{44}$$

$$\left[\hat{H}, \hat{K}_2\right] = 2i\hbar\hat{K}_1, \tag{45}$$

$$\left[\hat{K}_1, \hat{K}_2\right] = i\hbar\hat{K}_2, \tag{46}$$

$$\left[\hat{K}_1, \hat{G}\right] = \frac{i\hbar}{2}\hat{G}, \tag{47}$$

$$\left[\hat{K}_2, \hat{G}\right] = 0, \tag{48}$$

which is similar to the algebra (15)-(24). It is possible to show that the operators (34)-(38) are conserved.

In the next section, it will be shown that Black-Scholes-Merton equation is invariant under Schrödinger symmetry.

3. The Black-Scholes-Merton equation and the Black-Scholes Formula

Today, an esential question in financial world refers to how obtain the price of an option. Financial markets use the Black-Scholes option pricing model and it is the basis of sophisticated methods of options valuation. A remarkable result in this field is given by the Black-Scholes-Merton equation [1, 2]

$$\frac{\partial C(s,t)}{\partial t} = -\frac{\sigma^2}{2}s^2\frac{\partial^2 C(s,t)}{\partial s^2} - rs\frac{\partial C(s,t)}{\partial s} + rC(s,t), \tag{49}$$

where C is the price of an option, s is the price of the stock, σ is the volatility and r is the annualized risk-free interest rate. The Black-Scholes-Merton equation have to be solved with condition

$$C(s,T) = (s-K)\theta(s-K), \tag{50}$$

where K is the strike price of an option, T is the time to expiration and $\theta(x)$ is the Heaviside function.

Amazingly, the Black-Scholes-Merton equation (49) is equivalent to the Schrödinger equation [3]. In fact, using the change of variable

$$s = e^x \tag{51}$$

in the equation (49), the following result

$$\frac{\partial C(x,t)}{\partial t} = -\frac{\sigma^2}{2}\frac{\partial^2 C(x,t)}{\partial x^2} + \left(\frac{\sigma^2}{2} - r\right)\frac{\partial C(x,t)}{\partial x} + rC(x,t). \tag{52}$$

is gotten. Additionally, if

$$C(x,t) = e^{\left[\frac{1}{\sigma^2}\left(\frac{\sigma^2}{2}-r\right)x+\frac{1}{2\sigma^2}\left(\frac{\sigma^2}{2}+r\right)^2 t\right]}\Psi(x,t) \tag{53}$$

the following equation

$$\frac{\partial \Psi(x,t)}{\partial t} = -\frac{\sigma^2}{2}\frac{\partial^2 \Psi(x,t)}{\partial x^2} \tag{54}$$

is obtained, which is Schrödinger-like wave equation (32). Notice that $1/\sigma^2$ has the role of particle mass m. In addition, with the change of variable $\tau = T - t$, we get

$$\frac{\partial \Psi(x,\tau)}{\partial \tau} = \frac{\sigma^2}{2}\frac{\partial^2 \Psi(x,\tau)}{\partial x^2}, \tag{55}$$

which is the heat equation.

Now, using the solution for the free Schrödinger equation, we can get the solution for the Black-Sholes-Merton equation. In fact, using the initial condition (50) we get

$$C(s,t) = sN(d_+) - Ke^{-r(T-t)}N(d_-) \tag{56}$$

where

$$N(z) = \int_{-\infty}^{z} du \frac{e^{-\frac{u^2}{2}}}{\sqrt{2\pi}}, \quad d_\pm = \frac{\ln\left(\frac{s}{K}\right) + (T-t)\left(r \pm \frac{\sigma^2}{2}\right)}{\sigma\sqrt{T-t}}.$$

The equation (56) is the so call Black-Scholes formula and it can be obtained using quantum mechanics techniques.

4. The Schrödinger group and the Black-Scholes-Merton equation

Due that the Schrödinger equation (32) is invariant under conformal transformation (2), the equation (55) is invariant under the same transformation. In this case the function $\psi(x,t)$ transforms as

$$\Psi'(x',t') = \left(\sqrt{\gamma t + \delta}\right) e^{\frac{1}{2\sigma^2}\phi(x,t)} \Psi(x,t), \tag{57}$$

where $\phi(x,t)$ is given by (4). Notice that the particle mass m is changed for $1/\sigma^2$.

Using the change of variable (51), the coordinate transformations (2) can be written as

$$t' = \frac{\alpha t + \beta}{\gamma t + \delta}, \qquad s' = e^{\left(\frac{vt+c}{\gamma t+\delta}\right)} s^{\left(\frac{a}{\gamma t+\delta}\right)}. \tag{58}$$

Through a long but straightforward calculation, it can be shown that the Black-Scholes-Merton equation (49) is invariant under this last transformation, where the price $C(s,t)$ transforms as

$$C'(s',t') = \left(\sqrt{\gamma t + \delta}\right) s^{\Phi_1(s,t)} e^{\Phi_2(s,t)} C(s,t), \tag{59}$$

here

$$\Phi_1(s,t) = \frac{-2a^2\gamma\left(\frac{\sigma^2}{2} - r\right)t + 2a(v\delta - \gamma c) + 2a^2(a-\delta)\left(\frac{\sigma^2}{2} - r\right)}{2a^2\sigma^2(\gamma t + \delta)} - \frac{\gamma a^2 (\ln s)}{2a^2\sigma^2(\gamma t + \delta)}$$

and

$$\Phi_2(s,t) = \left(\frac{a^2\left(\frac{\sigma^2}{2}+r\right)^2(\alpha-\delta)+2a^2v\left(\frac{\sigma^2}{2}-r\right)+v\left(v\delta-2\gamma c\right)}{2\sigma^2a^2(\gamma t+\delta)}\right)t + \frac{a^2\beta\left(\frac{\sigma^2}{2}+r\right)^2+2a^2\left(\frac{\sigma^2}{2}-r\right)c-\gamma c^2-\gamma a^2\left(\frac{\sigma^2}{2}+r\right)^2t^2}{2\sigma^2a^2(\gamma t+\delta)}.$$

Now, the Black-Scholes-Merton equation (49) can be written as

$$\frac{\partial C(s,t)}{\partial t} = \hat{\mathbf{H}}C(s,t), \tag{60}$$

where

$$\hat{\mathbf{H}} = -\frac{\sigma^2}{2}s^2\frac{\partial^2}{\partial s^2} - rs\frac{\partial}{\partial s} + r. \tag{61}$$

Moreover, using the operator

$$\hat{\Pi} = -is\frac{\partial}{\partial s} + \frac{i}{\sigma^2}\left(\frac{\sigma^2}{2} - r\right), \tag{62}$$

the operator $\hat{\mathbf{H}}$ can be rewritten as

$$\hat{\mathbf{H}} = \frac{\sigma^2}{2}\hat{\Pi}^2 + \frac{1}{2\sigma^2}\left(\frac{\sigma^2}{2}+r\right)^2. \tag{63}$$

Notice that the operator $\hat{\mathbf{H}}$ is similar to the Hamiltonian operator $\hat{H}$ (35), where the particle mass m is associated with $1/\sigma^2$. However, the operator $\hat{\mathbf{H}}$ is not hermitian.

Additionally, using the operator (62) it is possible construct quantities related with the non-relativistic free particle conserved quantities (34)-(38). In fact, the operators

$$\hat{\Pi} = -is\frac{\partial}{\partial s} + \frac{i}{\sigma^2}\left(\frac{\sigma^2}{2} - r\right), \tag{64}$$

$$\hat{\mathbf{H}}_0 = \frac{\sigma^2}{2}\hat{\Pi}^2, \tag{65}$$

$$\hat{\mathbf{G}} = t\hat{\Pi} - \frac{1}{\sigma^2}\ln s, \tag{66}$$

$$\hat{\mathbf{K}}_1 = t\hat{\mathbf{H}}_0 - \frac{1}{4}\left(\ln s\hat{\Pi} + \hat{\Pi}\ln s\right), \tag{67}$$

$$\hat{\mathbf{K}}_2 = t^2\hat{\mathbf{H}}_0 - \frac{t}{2}\left(\ln s\hat{\Pi} + \hat{\Pi}\ln s\right) + \frac{1}{2\sigma^2}\left(\ln s\right)^2 \tag{68}$$

can be constructed, which are similar to the quantities (34)-(38). Now, using the relation

$$[\ln s, \hat{\Pi}] = i, \tag{69}$$

Journal of Physics: Conference Series **512** (2014) 012029 doi:10.1088/1742-6596/512/1/012029

the algebra

$$\left[\hat{\Pi}, \hat{\mathbf{H}}_0\right] = 0, \tag{70}$$

$$\left[\hat{\Pi}, \hat{\mathbf{K}}_1\right] = \frac{i}{2}\hat{\Pi}, \tag{71}$$

$$\left[\hat{\Pi}, \hat{\mathbf{K}}_2\right] = i\hat{\mathbf{G}}, \tag{72}$$

$$\left[\hat{\Pi}, \hat{\mathbf{G}}\right] = \frac{i}{\sigma^2}, \tag{73}$$

$$\left[\hat{\mathbf{H}}_0, \hat{\mathbf{K}}_1\right] = i\hat{\mathbf{H}}_0 \tag{74}$$

$$\left[\hat{\mathbf{H}}_0, \hat{\mathbf{G}}\right] = i\hat{\Pi}, \tag{75}$$

$$\left[\hat{\mathbf{H}}_0, \hat{\mathbf{K}}_2\right] = 2i\hat{\mathbf{K}}_1, \tag{76}$$

$$\left[\hat{\mathbf{K}}_1, \hat{\mathbf{K}}_2\right] = i\hat{\mathbf{K}}_2, \tag{77}$$

$$\left[\hat{\mathbf{K}}_1, \hat{\mathbf{G}}\right] = \frac{i}{2}\hat{\mathbf{G}}, \tag{78}$$

$$\left[\hat{\mathbf{K}}_2, \hat{\mathbf{G}}\right] = 0. \tag{79}$$

is satisfied. Then the operators (64)-(68) satisfy the Schrödinger algebra. However, the operators (64)-(68) are not hermitian and are not conserved. Then, the equivalence between the Black-Scholes-Merton and the free Schrödinger equation is not exactly, this last happen because the transformation (53) is not unitary.

Using other methods, the Black-Scholes-Merton symmetries have been studied in [8].

5. Summary

It was shown that the Black-Scholes-Merton equation is invariant under the Schrödinger group. In order to do this, the one dimensional free non-relativistic particle and its symmetries were revisited. To get the Black-Scholes-Merton equation symmetries, the particle mass was identified as the inverse of square of the volatility. Besides, using financial variables, a Schrödinger algebra representation was constructed. However, the operators of this last representation are not hermitian and not conserved.

References

[1] F. Black and M. Scholes, *The pricing options and corporate liabilities,* Journal of Political Economy **81** (1973), 637-659.

[2] R.C. Merton, *Theory of Rational Option Pricing,* Bell J. Econ. and Management Sci. **4** (1973) 141-183.

[3] B. E. Baaquie, *Quantum Finance,* Cambridge University Press (2004).

[4] O. Aharony, S. S. Gubser, J. Maldacena and H. Ooguri, *Large N field theories, string theory and gravity,* Phys. Rept. **323** (2000) 183, [hep-th/9905111].

[5] C. R. Hagen, *Scale and conformal transformations in galilean-covariant field theory,* Phys. Rev. D **5** (1972) 377 .

[6] U. Niederer, *The maximal kinematical invariance group of the free Schrodinger equation,* Helv. Phys. Acta **45** (1972) 802.

[7] S. Lie, Arch. Math. Nat. vid. (Kristiania) 6, 328 (1882).

[8] R. K. Gazizov, N. H. Ibragimov, *Lie Symmetry Analysis of Differential Equations in Finance,* Nonlinear Dynamics, (1998) **17**, 387-407.

8th International Symposium on Quantum Theory and Symmetries (QTS8) IOP Publishing
Journal of Physics: Conference Series **512** (2014) 012030 doi:10.1088/1742-6596/512/1/012030

Chiral symmetry and classical solution of the NLSUSY model

Kazunari Shima[1], **Motomu Tsuda**[2]

[1] Laboratory of Physics, Saitama Institute of Technology Fukaya, Saitama 369-0293, Japan
[2] Department of Mathematics, Saitama Institute of Technology Fukaya, Saitama 369-0293, Japan

E-mail: shima@sit.ac.jp, tsuda@sit.ac.jp

Abstract. The classical solution of the equation of the motion for the Nambu-Goldstone fermion of the nonlinear representation of supersymmetry and its physical significance are discussed, which gives a new insight into the chiral symmetry of the standard model for the low energy particle physics.

Supersymmetry (SUSY) [1], which constitutes the space-time symmetry, may be the most promising notion for unifying two Standard Models(tSM), i.e., the SM for the low energy particle physics and Einstein General Relativity theory(GR) for space-time.The linear representation of SUSY (LSUSY) [2] is realized on the multiplet of fields.

Various models based upon $N = 1$ LSUSY have been studied extensively and remarkable phenomenological and field theoretical results have been obtained [1]. The minimal supersymmetric standard model (MSSM), which may stabilize the mass of light Higgs particle, looks severely constrained by the LHC experiment. The unpleasant point of LSUSY model is that the mechanism and the physical meaning of the spontaneous SUSY breaking, e.g., the origin of D term, is not clear. While the nonlinear representation of SUSY (NLSUSY) [3] is given by the Nambu-Goldstone fermion corresponding to [superPoincaré/Poincaré], which is realized *geometrically* on specific *flat* space-time equipped with the robust spontaneous SUSY breaking encoded in the nature of space-time itself. The action containing higher orders of the stress-energy-momentum tensors of the fermion field is almost unique up to the higher-order derivative terms.

Despite the apparent non-renormalizability it may be worth exploring in the NLSUSY framework a new paradigm for the unified theory of nature in terms of the fermion field, which is the long standing challenge. To incorporate the space-time dynamics we assume that the ultimate shape of nature is the *four* dimensional (Riemann) manifold possessing the NLSUSY degrees of freedom on *tangent* space, i.e., *tangent* space is specified by the Grassmann coordinates ψ_α besides the Minkowski coordinates x_a. NLSUSY can be easily fused with the general relativity (GR) principle, which produces the nonlinear supersymmetric general relativity theory (NLSUSYGR) [4, 5] in the form of Einstein-Hilbert (EH) action of the ordinary GR equipped with the NLSUSY cosmological term indicating the NLSUSY nature of tangent space. NLSUSYGR *would* break down spontaneously (*Big Decay*) to EH

Journal of Physics: Conference Series **512** (2014) 012030 doi:10.1088/1742-6596/512/1/012030

action for ordinary Riemann space-time (graviton) and NLSUSY action for massless Nambu-Goldstone(NG) fermion (*superon*) corresponding to the spontaneous space-time SUSY breaking; [super $GL(4,\mathbf{R})/GL(4,\mathbf{R})$] [4, 5] called *superon graviton model (SGM)*.

Although NLSUSY is the non-renormalizable highly nonlinear theoryfor the fermion,it is shown by the systematic linearizations of NLSUSYGR in flat space-time that the NLSUSY cosmological term is recasted (equivalent) to the familiar broken LSUSY theory, i.e., the familiar broken LSUSY theory emerges in the true vacuum of NLSUSY theory as the low energy (effective) theory, where all fields including D term are composed of NG fermion degrees of freedom [5, 6].

The NLSUSYGR scenario has a potential to give new insights into the SUSY effects which can be tested in cosmology and the low energy particle physics [5, 6] In this talk we study the classical solution of the equation of motion for the fermion of NLSUSY moel regarded as an asymptotic flat-space limit of NLSUSYGR, which may explain the chiral eigenstates for fermions in the SM(MSSM) for the low energy particle physics.

The $N = 1$ NLSUSY model $L_{\rm NLSUSY}$ in flat space-time is given by Volkov and Akulov as follows [3];

$$L_{\rm NLSUSY} = -\frac{1}{2\kappa^2}|w| = -\frac{1}{2\kappa^2}\left[1 + t^a{}_a + \frac{1}{2}(t^a{}_a t^b{}_b - t^a{}_b t^b{}_a) + \cdots\right], \tag{1}$$

where

$$|w| = \det w^a{}_b = \det(\delta^a_b + t^a{}_b), \quad t^a{}_b = -i\kappa^2\bar{\psi}\gamma^a\partial_b\psi \tag{2}$$

and ψ is a four components Majorana spinor.[1] L_{NLSUSY} is invariant under $N = 1$ NLSUSY transformation,

$$\delta_\zeta\psi = \frac{1}{\kappa}\zeta - i\kappa\bar{\zeta}\gamma^a\psi\partial_a\psi, \tag{3}$$

where ζ is a global Majorana spinor parameter, and ψ is the coordinates of the coset space $\frac{Super-Poincare}{Poincare}$ and subsequently recasted as the NG fermion in the vacuum of $L_{\rm NLSUSY}$. The constant κ with the dimension $[length]^{+2}$ gives the strength of the coupling of NG fermion to the vacuum, which is analogous to the pion decay constant f_π.

The variation of (1) with respect to $\bar{\psi}$ gives the following equation of motion for ψ;

$$\begin{aligned} &\partial\!\!\!/\psi - i\kappa^2\left\{T^a{}_a\partial\!\!\!/\psi - T^a{}_b\gamma^b\partial_a\psi + \frac{1}{2}(\partial_a T^b{}_b - \partial_b T^b{}_a)\gamma^a\psi\right\} \\ &-\frac{1}{2}(-i\kappa^2)^2\epsilon_{abcd}\epsilon^{efgd}(T^a{}_e T^b{}_f\gamma^c\partial_g\psi + T^a{}_e\partial_g T^b{}_f\gamma^c\psi) = 0, \end{aligned} \tag{4}$$

where $T^a{}_b = i\kappa^{-2}t^a{}_b = \bar{\psi}\gamma^a\partial_b\psi$. The highest order term with $\mathcal{O}(T^4)$ in the NLSUSY action (1) is known to vanish identically for $N = 1$ SUSY [7].

It is not easy to solve (4) in general. However, considering ψ is the NG fermion it is natural to argue the (free) massless solution, namely, we consider the case,

$$\partial\!\!\!/\psi(x) = 0, \tag{5}$$

which eliminates the lowest order term of (4). In NLSUSYGR scenario, the NG fermion $\psi(x)$ is besides x^a the Grassmann degrees of freedom on tangent space of ultimate space-time $(x^a, \psi\,; x^\mu)$.

[1] Minkowski space-time indices are denoted by $a, b, \cdots = 0, 1, 2, 3$ and the flat metric is $\eta^{ab} = {\rm diag}(+,-,-,-)$. Gamma matrices satisfy $\{\gamma^a, \gamma^b\} = 2\eta^{ab}$ and we define $\sigma^{ab} = \frac{i}{4}[\gamma^a, \gamma^b]$.

Journal of Physics: Conference Series **512** (2014) 012030 doi:10.1088/1742-6596/512/1/012030

This means that the exact solution of $\psi(x)$ is dictated by the nontrivial geometry of space-time, even in asymptotic flat space-time.

In fact, we have shown in supergravity [8, 9] that the nontrivial exact classical solution of the massless (free) gravitino under the Coulomb type condition $\psi_\mu = \delta_{0\mu}\psi$ exists in flat space-time of the Kerr solution of GR, not in that of Schwarzshild solution [10]. Therefore, without ristricting to the specific (plane wave) case, the condition (5) is worth considering seriously from the physical point of view, which allow the non-trivial (localized) solution depending on the geometry of space-time. We will see throughout the discussion that Eq.(4) is satisfied identically at each order in κ, provided ψ is *a single* chiral state which satisfies (5), i.e. $\partial\!\!\!/\psi_L(x) = 0$ *or* $\partial\!\!\!/\psi_R(x) = 0$. We show below some details of the computation.

First let us study the term $T^a{}_b\gamma^b\partial_a\psi$ in the second term at $\mathcal{O}(\kappa^2)$ of Eq.(4), for the first and the third terms at $\mathcal{O}(\kappa^2)$ vanish due to $\partial\!\!\!/\psi$ (5). By using the Fierz transformation and considering Eq.(5) and the (anti)commuting properties characteristic of the Majorana spinor in tensor products, many terms containing the product of γ matrices with (pseudo)scalar and (axial) tensor structures vanish provided ψ is *chiral* and also terms containig $\partial\!\!\!/\psi(x)$ vanish by (5). Consequently, the following (axial)vector-type terms survive appearently in the second term;

$$T^a{}_b\gamma^b\partial_a\psi \sim \frac{i}{2}\epsilon^{abcd}(\partial_a\bar\psi\gamma_b\psi\ \gamma_5\gamma_c\partial_d\psi - \partial_a\bar\psi\gamma_5\gamma_b\psi\ \gamma_c\partial_d\psi). \tag{6}$$

Again by the Fierz transformation and the subsequent similar arguments the terms (6) become,

$$-\frac{i}{4}\epsilon^{abcd}(\partial_a\bar\psi\gamma_b\gamma_e\partial_d\psi\ \gamma_5\gamma_c\gamma^e\psi - \partial_a\bar\psi\gamma_5\gamma_b\gamma_e\partial_d\psi\ \gamma_c\gamma^e\psi), \tag{7}$$

which vanishes for the *chiral* eigenstate of ψ. Next we discuss the terms at $\mathcal{O}(\kappa^4)$ in Eq.(4). From the relation,

$$\begin{aligned}&\epsilon_{abcd}\epsilon^{efgd}T^a{}_eT^b{}_f\gamma^c\partial_g\psi\\ &= -(T^a{}_aT^b{}_b - T^a{}_bT^b{}_a)\partial\!\!\!/\psi - 2(T^a{}_cT^c{}_b - T^c{}_cT^a{}_b)\gamma^b\partial_a\psi,\end{aligned} \tag{8}$$

we give an example of calculations for $T^a{}_cT^c{}_b\gamma^b\partial_a\psi$ in the third term of Eq.(8). By using repeatedly the Fierz transformation and performing the similar arguments the following term survives,

$$T^a{}_cT^c{}_b\gamma^b\partial_a\psi \sim \frac{1}{4}\gamma_a\sigma_{de}\sigma_{fg}\partial^a\psi\ \bar\psi\gamma_b\sigma^{de}\partial^b\psi\ \bar\psi\gamma_c\sigma^{fg}\partial^c\psi. \tag{9}$$

However, it can be shown that the term (9) is expressed by only terms proportional to $\partial\!\!\!/\psi$ by using Clifford algebra of γ matrices and becomes trivial. As for the last term in Eq.(4), the similar argument shows that it vanishes identically for the single chiral state satisfying Eq.(5). This means that the NG fermion with higher order self-interactions of NLSUSY is the massless chiral fermion which satisfies either of the equation $\partial\!\!\!/\psi_L(x) = 0$ or $\partial\!\!\!/\psi_R(x) = 0$. This situation is interesting, for the lowest order of $T_a{}^b$, i.e., the kinetic term does not constrain the chirality of the (left-right symmetric) solution by itself, however higher order self-interaction terms with NLSUSY structure *as a whole* constrain the chirality of the solution to *one* (left *or* right) chiral eigenstate, i.e. ψ should be recasted as the chiral eigenstate. If NLSUSYGR is the underlying principle beyond the SM, this mechanism may explain the chiral eigenstate for all massless fermions in the SM, which is a basic assumption within the SM framework. Considering that the observed fermion in the SM is inconsistent with the NG fermion, the SM may be the composite (effective) theory of the NG fermion as in NLSUSYGR scenario.

Journal of Physics: Conference Series **512** (2014) 012030 doi:10.1088/1742-6596/512/1/012030

In fact, in $N = 2$ minimal SGM scenario ($N = 1$ is unphysical [11] in this scenario), LSUSY QED (and probably the MSSM as well) emerges in the *true* vacuum of NLSUSY, where all particles for the LSUSY multiplet are composites of superons.

For example, the massless fermion λ^I ($I = 1, 2$ for leptons and quarks) in Wess-Zumino gauge is composed of superon ψ^I as follows [6];

$$\lambda^I = \left\{ \psi^I |w| - \frac{i}{2}\kappa^2 \partial_a (\gamma^a \psi^I \bar{\psi}^J \psi^J |w|) \right\}, \tag{10}$$

which shows that quarks and leptons λ^I are *chiral* and *composites* fermions, because the constituent ψ^I (superon) is chiral. The above results do not depend on the representation of γ-matrices. Here it may be useful to see an explicit form of the chiral eigenstate in a specific basis of γ-matrices. For example, for a real four-component $\psi = (\psi_1, \psi_2, \psi_3, \psi_4)^T$ Majorana spinor in the Majorana basis [1] of γ-matrices, ψ_L is expressd as follows,

$$\psi_L = \frac{1}{2} \begin{pmatrix} \psi_1 - i\psi_3 \\ \psi_2 - i\psi_4 \\ \psi_3 + i\psi_1 \\ \psi_4 + i\psi_2 \end{pmatrix} = \frac{1}{2} \begin{pmatrix} u_a \\ iu_a \end{pmatrix}, (a = 1, 2) \tag{11}$$

which is equivalent to a complex two-components Weyl spinor u_a.

Finally before conclusions, we mention that NLSUSYGR (SGM) scenario constrains the dimensions of space-time to *four*, provided we require $SO(D-1, 1) \cong SL(d, C)$, i.e., $\frac{D(D-1)}{2} = 2(d^2 - 1)$ for the Lorentz invariance, which holds for only $D = 4$, $d = 2$ besides the exceptional case $D = 2, d = 1$.

Now we summarize the result as follows. The classical solution of NLSUSY model predicts the chiral eigenstates for the NG fermion, where higher order self-interaction terms of the NG fermion constrain the chirality of itself. The chiral eigenstate for quarks and leptons in the SM (MSSM), i.e. the chiral symmetry, is understood naturally, if quarks and leptons are composites of the chiral fermion(superon), which achieves simultaneously the true vacuum of NLSUSY as demonstrated in the SGM scenario of NLSUSYGR [4, 5].

References

[1] For a review, see J. Wess and J. Bagger, *Supersymmetry and Supergravity* (Princeton University Press, Princeton, New Jersey, 1992).
[2] J. Wess and B. Zumino, *Phys. Lett.* B49, 52 (1974).
[3] D.V. Volkov and V.P. Akulov, *Phys. Lett.* B46, 109 (1973).
[4] K. Shima, *European Phys. J.* **C7**, 341 (1999);
K. Shima, *Phys. Lett.* B501, 237 (2001);
K. Shima and M. Tsuda, *Phys. Lett.* B507, 260 (2001).
[5] K. Shima and M. Tsuda, *J. Phys.* **343**, 012111 (2012);
K. Shima and M. Tsuda, *Nonlinear SUSY General Relativity Theory and Significances*, Proceedings of the 7th International Conference *Quantum Theory and Symmetries*, 07-13, August, 2011, Czech Technical University, Prague, Czech Republic; arXiv:1112.3098 [hep-th].
[6] K. Shima and M. Tsuda, *Phys. Lett.* **B645**, 455 (2007).
[7] S. M. Kuzenko and S. A. McCarthy, *JHEP* **0505** (2005) 012.
[8] D. Z. Freedman, P. van Nieuwenhuisen and S. Ferrara, *Phys. Rev. D* **13**, 3214 (1976).
[9] S. Deser and B. Zumino, *Phys. Lett.* **B62**, 335 (1976).
[10] K. Shima and M. Kasuya, *Phys. Rev.* **D22**, 290 (1980).
[11] K. Shima, Y. Tanii and M. Tsuda, *Phys. Lett.* **B546**, 162 (2002).

8th International Symposium on Quantum Theory and Symmetries (QTS8) IOP Publishing
Journal of Physics: Conference Series **512** (2014) 012031 doi:10.1088/1742-6596/512/1/012031

Kravchuk oscillator revisited

Mesuma K. Atakishiyeva[1], **Natig M. Atakishiyev**[2] **and Kurt Bernardo Wolf**[3]

1 Facultad de Ciencias, Universidad Autónoma del Estado de Morelos, Av. Universidad s/n, Cuernavaca, Morelos 62250, México

2 Instituto de Matemáticas, Unidad Cuernavaca, Universidad Nacional Autónoma de México, Av. Universidad s/n, Cuernavaca, Morelos 62250, México

3 Instituto de Ciencias Físicas, Universidad Nacional Autónoma de México, Av. Universidad s/n, Cuernavaca, Morelos 62250, México

E-mail: natig@matcuer.unam.mx, bwolf@fis.unam.mx

Abstract. The study of irreducible representations of Lie algebras and groups has traditionally considered their action on functions of a continuous manifold (e.g. the 'rotation' Lie algebra $so(3)$ on functions on the sphere). Here we argue that functions of a discrete variable —Kravchuk functions— are on equal footing for that study in the case of $so(3)$. They lead to a discrete quantum model of the harmonic oscillator, and offer a corresponding set of special function relations. The technique is applicable to other special function families of a discrete variable, which stem from low-dimensional Lie algebras and are stationary solutions for the corresponding discrete quantum models.

1. Introduction

The representation theory of Lie algebras and groups has provided a royal road to study families of special functions [1, 2]. Here we follow this road in the reverse direction: starting from a family of discrete special functions —Kravchuk polynomials— that satisfy a three-term recurrence relation and a difference equation, we find the Lie algebra in which they appear as bases for irreducible representations —in this case the rotation algebra $so(3)$. On the way, seen from the perspective of Hamiltonian systems, we identify the discrete physical system for which this family provides the stationary wave functions.

The strategy developed here opens the way to propose other known discrete special function families of (as yet unknown) Lie-algebraic import, and find the discrete (or continuous) Hamiltonian system that will realize them as proper wave functions. In Sect. 2 we condense the definition and relevant properties of the Kravchuk polynomials, and the Kravchuk functions that are orthonormal and complete in finite-dimensional vector spaces. Their recurrence relation and difference properties are made explicit in Sect. 3, and used in Sect. 4 to endow the family with a Hamiltonian and a position operator realized by matrices; when subject to obey the Hamilton equations, their commutator produces a momentum operator and provides the potential of the discrete Hamiltonian system whose stationary solutions are the proposed family of discrete functions. They will inherit a rich set of properties: in this $so(3)$ case they participate with the Wigner little-d functions in the fractional Fourier-Kravchuk transform [3] and define in Sect. 5 a *canonical* dual basis of momentum. In Sect. 6 we offer some applications, conclusions and comments on other discrete special functions that may be subject to this analysis.

Journal of Physics: Conference Series **512** (2014) 012031 doi:10.1088/1742-6596/512/1/012031

2. Kravchuk polynomials and functions

Kravchuk polynomials were originally introduced as a generalization of Hermite polynomials [4], replacing their orthogonality relation under an integral over the real line x with Gaussian weight $\exp(-x^2)$, by a summation over $N+1$ points with a binomial distribution for weight. Here we find it more useful to represent them as terminating hypergeometric series [5, pp. 237–241], i.e., polynomials of degree n in x,

$$K_n(x;\, p, N) \;:=\; {}_2F_1\left(-n, -x;\, -N; p^{-1}\right) \tag{1}$$

$$= \left(\frac{p-1}{p}\right)^n {}_2F_1\left(-n, x{-}N;\, -N;\, \frac{1}{1{-}p}\right) \tag{2}$$

$$= \left(\frac{p-1}{p}\right)^n K_n(N-x;\, 1-p, N), \tag{3}$$

for $n \in \{0, 1, 2, ..., N\}$, $0 < p < 1$, where $N \in \mathsf{Z}_0^+$ is some nonnegative integer number, and where we passed from (1) to (2) by using the Euler transformation formula [5, Eq. (1.7.2)].

Since the hypergeometric function in (1) is symmetric with respect to first two parameters it follows that the Kravchuk polynomials are *self-dual*:

$$K_n(m;\, p, N) = K_m(n;\, p, N), \qquad m, n \in \{0, 1, 2, ..., N\}. \tag{4}$$

For generic values of the parameter p, the Kravchuk polynomials do not exhibit any special property under reflection of the argument m across the midpoint of its interval, $m \leftrightarrow N{-}m$. Only for $p = \frac{1}{2}$ do they possess definite parity:

$$K_n(m;\, \tfrac{1}{2}, N) = (-1)^n\, K_n(N{-}m;\, \tfrac{1}{2}, N). \tag{5}$$

We are also informed [5] of the discrete orthogonality relation satisfied by the Kravchuk polynomials over the range of their argument:

$$\sum_{m=0}^{N} \binom{N}{m} K_n(m;\, \tfrac{1}{2}, N)\, K_{n'}(m;\, \tfrac{1}{2}, N) = 2^N \binom{N}{n}^{-1} \delta_{n,n'}, \tag{6}$$

where $\binom{r}{s} := r!/s!(r-s)!$ is the binomial coefficient. Indeed, there is a corresponding orthogonality relation for $K_n(m;\, p, N)$ when the parameter p has a generic value [5, Eq. (9.11.2)]. Of course, being polynomials in m, the Kravchuk polynomials are defined for any real or complex value of m. To have explicit parity, we will shift the orthogonality interval so that it be symmetric under inversions, denoting $x_m := \frac{1}{2}N - m$ for the $N{+}1$ equidistant points in the symmetric interval $x \in [-\frac{1}{2}N, \frac{1}{2}N]$. When N grows without bound while that interval and the density of points grow as $\sim \sqrt{N}$, one recovers the Hermite polynomials [5, Eq. (9.11.15)],

$$H_n(x) = \lim_{N\to\infty} (2N)^{n/2}\, K_n\Big(\sqrt{\tfrac{1}{2}N}\,\Big(\sqrt{\tfrac{1}{2}N} - x\Big);\, \tfrac{1}{2}, N\Big). \tag{7}$$

Out of the Kravchuk polynomials with definite parity $K_n(m;\, \frac{1}{2}, N)$ and orthogonality (6), it is convenient to define the Kravchuk *functions*:

$$\psi_n^{(M)}(x_m) := c_{n,m}^{(M)}\, K_n(x_m{+}M;\, \tfrac{1}{2}, 2M), \tag{8}$$

of the argument x_m that ranges over the symmetric set of points $x_m = m - \frac{1}{2}N$, and where we denote $M := \frac{1}{2}N$; the coefficients $c_{n,m}^{(M)}$ we choose so as to have the orthonormality relation in its simplest form,

$$\sum_{x_m=-M}^{M} \psi_n^{(M)}(x_m)\, \psi_{n'}^{(M)}(x_m) = \delta_{n,n'}. \tag{9}$$

To find $c_{n,m}^{(M)}$ in (8), we note that $K_0(m;\frac{1}{2},N)=1$, so we can sum (9) using $\sum_{m=0}^{N}\binom{N}{m}=2^N$, and conclude that $\psi_0^{(M)}(x)$ is the square root of the binomial distribution,

$$\psi_0^{(M)}(x)=\frac{1}{2^M}\sqrt{\binom{2M}{M+x}}=\frac{1}{2^M}\sqrt{\frac{(2M)!}{\Gamma(M+x+1)\Gamma(M-x+1)}}. \tag{10}$$

We shall call $\psi_0^{(M)}(x)$ the *ground state* in analogy with the quantum harmonic oscillator case. As we argued above, the argument x need not be an integer, because the ratio of gamma functions analytically extends the binomial to the complex plane, with poles on real line at $\pm x_k=M+k$, $k\in\{1,2,\ldots\}$; to avoid them, let us keep $x\in[-M,M]$.

Now, writing $c_{n,m}^{(M)}=d_n^{(M)}\,\psi_0^{(M)}(m)$ and replacing again in (9) to find $d_n^{(M)}$, we conclude that for $n\in\{0,1,\ldots,2M\}$,

$$\psi_n^{(M)}(x_m)=\frac{(-1)^n}{2^M}\sqrt{\binom{2M}{n}\binom{2M}{M+x_m}}\;K_n(M+x_m;\,\tfrac{1}{2},\,2M) \tag{11}$$

provides an orthonormal basis for the $(2M+1)$-dimensional vector space of functions of the discrete variable $x_m=M-m$ in the symmetric interval $x_m\in\{-M,\,-M+1,\ldots,\,M\}$; the sign in (11) has been added so that $\psi_n^{(M)}(-M)>0$, following the standard sign convention for the quantum harmonic oscillator wavefunctions $\Psi_n(x)$. Indeed, it has been known [6] that as M grows without bound, while the orthogonality interval and density of points grow as $\sim\sqrt{M}$ with $x\sqrt{M}$ integer, their limit is

$$\lim_{M\to\infty}(-1)^{x\sqrt{M}}M^{1/4}\psi_n^{(M)}(x\sqrt{M})=\frac{1}{\sqrt{2^n n!\sqrt{\pi}}}e^{-x^2/2}H_n(x)=\Psi_n(x). \tag{12}$$

A consequence of the self-duality (4) of the symmetric Kravchuk polynomials and of their parity (5) is that, for $0\le n,m\le N$,

$$\psi_n^{(M)}(x_m)=\psi_{M-m}^{(M)}(M-n)=(-1)^n\psi_n^{(M)}(-x_m), \tag{13}$$

and also that the basis of Kravchuk functions $\psi_n^{(M)}(x_m)$ is *complete*:

$$\sum_{n=0}^{2M}\psi_n^{(M)}(x_m)\,\psi_n^{(M)}(x_{m'})=\delta_{m,m'}. \tag{14}$$

3. Three-term recurrence relations and difference equations

Two important properties satisfied by the Kravchuk polynomials is a three-term recurrence relation, which is inherited to the Kravchuk functions $\psi_n^{(M)}(x_m)$ in (11), and a difference equation, which in discrete systems takes the place of the Schrödinger equation. The factorization of this difference equation leads to the raising and lowering operators, which in turn determine the *mother* Lie algebra of the discrete system.

The three-term recurrence relation for Kravchuk polynomials [5, Eq. (9.11.3)] yields the corresponding relation for the Kravchuk functions (11),

$$-2x_m\,\psi_n^{(M)}(x_m)=\sqrt{(n+1)(2M-n)}\,\psi_{n+1}^{(M)}(x_m)+\sqrt{n(2M-n+1)}\,\psi_{n-1}^{(M)}(x_m). \tag{15}$$

Similarly, the difference equation for Kravchuk polynomials [5, Eq. (9.11.5)], with $x_m=M-m$ so that $x_{m\mp1}=x_m\pm1$, defines the ground state

$$\psi_0^{(M)}(x_m\pm1)=\sqrt{\frac{M-x_m}{M+x_m\mp1}}\,\psi_0^{(M)}(x_m), \tag{16}$$

and a relation between three neighboring points x_m of the Kravchuk functions,

$$\begin{aligned} 2(M-n)\,\psi_n^{(M)}(x_m) &= \sqrt{(M+x_m+1)(M-x_m)}\,\psi_n^{(M)}(x_m+1) \\ &+ \sqrt{(M+x_m)(M-x_m+1)}\,\psi_n^{(M)}(x_m-1), \end{aligned} \tag{17}$$

which we identify as the 'Schrödinger' difference equation that *defines* the *discrete* model of the quantum harmonic oscillator, as we shall proceed to show below.

The distinction between the three-term recurrence relation (15) and the difference equation (17), is that the latter can be *slid* along x, i.e., it remains valid if we replace $x_m \mapsto x_m + \varepsilon$, for any $-1 < \varepsilon < 1$; the limits are imposed by the poles that $\psi_n^{(M)}(x)$ inherits from (16) at $\pm x = M+k$, $k \in \{1, 2, \ldots\}$. On the other hand, the recurrence relation (15) is between three successive states of this oscillator model, n and $n \pm 1$, except for the self-duality (4).

4. Geometry and dynamics in the $so(3)$ algebra

We can regard the difference equation (17) as an eigenvector equation for the Kravchuk functions $\psi_n^{(M)}(x_m)$, when these are accommodated into $2M+1$ column vectors $\boldsymbol{\psi}_n^{(M)}$ numbered by n, of rows $x_m = M-m$, with $n, m \in \{0, 1, \ldots, 2M\}$. This is then a matrix equation that reads

$$\mathbf{H}^{(M)}\,\boldsymbol{\psi}_n^{(M)} = (M-n)\,\boldsymbol{\psi}_n^{(M)}, \qquad M-n \in \{-M, -M+1, \ldots, M\}, \tag{18}$$

governed by the $(2M+1) \times (2M+1)$ *shifted Hamiltonian* matrix $\mathbf{H}^{(M)} = \{H_{m,m'}^{(M)}\}_{m,m'=-M}^{M}$,

$$\mathbf{H}^{(M)} = \frac{1}{2}\begin{pmatrix} 0 & \alpha_{1-M}^{(M)} & 0 & \cdots & 0 & 0 \\ \alpha_{1-M}^{(M)} & 0 & \alpha_{2-M}^{(M)} & \cdots & 0 & 0 \\ 0 & \alpha_{2-M}^{(M)} & 0 & \cdots & 0 & 0 \\ \cdots & \cdots & \cdots & \cdots & \cdots & \cdots \\ 0 & 0 & 0 & \cdots & 0 & \alpha_M^{(M)} \\ 0 & 0 & 0 & \cdots & \alpha_M^{(M)} & 0 \end{pmatrix}, \tag{19}$$

where from (17) we identify

$$\alpha_m^{(M)} := \sqrt{(x_m+M)(M-x_m+1)}, \qquad x_m \in \{-M, -M+1, \ldots, M\}. \tag{20}$$

Since the matrix (19) is symmetric and real, $H_{m,m'}^{(M)} = H_{m',m}^{(M)}$, it is self-adjoint. We have called it '*shifted Hamiltonian*' because it is a second-difference operator, thus a finite analogue of the quantum second-differential Hamiltonian operators, and its equally-spaced spectrum is *shifted* from $\{\frac{1}{2}, \frac{3}{2}, \ldots\}$ to $[-M, M]$. For short, we will henceforth call it simply 'Hamiltonian'.

Having written the difference equation for Kravchuk functions as an eigenvalue problem for a Hamiltonian matrix, we can easily define a *position* matrix $\mathbf{X}$ such that its eigenvalues count the rows of the vectors in that basis,

$$\mathbf{X}^{(M)}\,\boldsymbol{\psi}^{(M)}(x_m) = x_m\,\boldsymbol{\psi}^{(M)}(x_m), \qquad x_m = M-m \in \{-M, -M+1, \ldots, M\}. \tag{21}$$

This matrix is of course diagonal,

$$\mathbf{X}^{(M)} = \begin{pmatrix} -M & 0 & 0 & \cdots & 0 & 0 \\ 0 & 1-M & 0 & \cdots & 0 & 0 \\ 0 & 0 & 2-M & \cdots & 0 & 0 \\ \cdots & \cdots & \cdots & \cdots & \cdots & \cdots \\ 0 & 0 & 0 & \cdots & M-1 & 0 \\ 0 & 0 & 0 & & 0 & M \end{pmatrix}, \tag{22}$$

with elements x_m, and is also obviously self-adjoint and real.

Now that we have Hamiltonian and position matrices, $\mathbf{H}$ and $\mathbf{X}$, we can investigate the geometry and dynamics afforded by the Lie commutator form of the Hamilton equations of mechanics. The first Hamilton equation is geometric: it defines the *momentum* matrix $\mathbf{P}^{(M)}$ as the vector tangent to the evolution trajectory,

$$\mathbf{P}^{(M)} := \mathrm{i}\,[\mathbf{H}^{(M)},\, \mathbf{X}^{(M)}], \tag{23}$$

where the factor i is required for the momentum matrix to be self-adjoint. It is

$$\mathbf{P}^{(M)} = \frac{\mathrm{i}}{2}\begin{pmatrix} 0 & \alpha^{(M)}_{1-M} & 0 & \cdots & 0 & 0 \\ -\alpha^{(M)}_{1-M} & 0 & \alpha^{(M)}_{2-M} & \cdots & 0 & 0 \\ 0 & -\alpha^{(M)}_{2-M} & 0 & \cdots & 0 & 0 \\ \cdots & \cdots & \cdots & \cdots & \cdots & \cdots \\ 0 & 0 & 0 & \cdots & 0 & \alpha^{(M)}_{M} \\ 0 & 0 & 0 & \cdots & -\alpha^{(M)}_{M} & 0 \end{pmatrix}, \tag{24}$$

with elements also given in terms of the $\alpha^{(M)}_m$'s in (20). This matrix is an analogue of the momentum operator in quantum mechanics, $-\mathrm{i}\,d/dx$; instead of a derivative, (24) acts as a weighted finite difference on $f(x_m)$ that yields $-\mathrm{i}[\alpha^{(M)}_{m+1}f(x_{m+1}) - \alpha^{(M)}_{m-1}f(x_{m-1})]$.

The second Hamilton equation is dynamic and determines how the tangent momentum of the trajectory evolves under the Hamiltonian of the system. One finds that the Lie-commutator form of the second Hamilton equation returns the diagonal position matrix (22),

$$[\mathbf{H}^{(M)},\, \mathbf{P}^{(M)}] = \mathrm{i}\,\mathbf{X}^{(M)}. \tag{25}$$

The two Hamilton equations, (23) and (25), are those of a harmonic oscillator —classical or quantum— and combine to yield its Newton equation,

$$[\mathbf{H}^{(M)}, [\mathbf{H}^{(M)},\, \mathbf{X}^{(M)}]] = \mathbf{X}^{(M)}. \tag{26}$$

The generated trajectories of position and momentum will thus follow harmonic motion.

Granted that we have a Hamiltonian system that is harmonic, we should lastly investigate the remaining commutator between the above position and momentum matrices. Instead of the quantum-mechanically expected unit matrix, we find

$$[\mathbf{X}^{(M)},\, \mathbf{P}^{(M)}] = -\mathrm{i}\,\mathbf{H}^{(M)}. \tag{27}$$

This is a consequence of the difference equation (17) satisfied by the Kravchuk functions $\psi^{(M)}_n(x)$ in (11), and the construction of the diagonal position matrix in (22). The three commutators (23), (25) and (27) close with the structure constants that characterize the rotation Lie algebra $so(3)$. This algebra is different from the 4-parameter noncompact *oscillator* algebra (which includes $\mathbf{1}$) that we would obtain had we started with the Hermite functions $\Psi_n(x)$ and their differential equation.

The orthogonality and completeness of the Kravchuk functions over a $(2M{+}1)$-dimensional vector space, Eqs. (9) and (14), insure that the matrices are self-adjoint on that finite-dimensional space. We identify the self-adjoint irreducible representations to which the three matrices belong through the value of the Casimir invariant,

$$(\mathbf{H}^{(M)})^2 + (\mathbf{X}^{(M)})^2 + (\mathbf{P}^{(M)})^2 = M(M{+}1)\,\mathbf{1}^{(M)}, \tag{28}$$

i.e., it corresponds to generic 'spin' M.

5. The algebra $so(3)$ and dual canonical bases

In the previous section we presented matrices that we called Hamiltonian, position, and momentum, and which were built to act as difference operators on the vectors (that can be called states, or signals, or wavefunctions) in a complex vector space of integer dimension $2M+1$. We can identify these operators with the common notation for the generators of $so(3)$ as follows:

$$J_1 \leftrightarrow H, \qquad J_2 \leftrightarrow P, \qquad J_3 \leftrightarrow X, \tag{29}$$

(dropping the index $^{(M)}$) with commutation relations $[J_i, J_j] = \mathrm{i}\, J_k$ (where i, j, k is a cyclic permutation of $1, 2, 3$) that characterize $so(3) = su(2)$ and generate the corresponding Lie groups of spin, $SO(3)$ if M is integer and $SU(2)$ if it is half-integer.

The elements of $SO(3)$ are, with Euler angle parametrization,

$$\begin{aligned} \mathcal{R}(\alpha,\beta,\gamma) &= \exp(-\mathrm{i}\alpha J_3)\exp(-\mathrm{i}\beta J_2)\exp(-\mathrm{i}\gamma J_3) \\ &\leftrightarrow \exp(-\mathrm{i}\alpha X)\exp(-\mathrm{i}\beta P)\exp(-\mathrm{i}\gamma X), \end{aligned} \tag{30}$$

and their action $\mathcal{R} : \boldsymbol{\phi} = \boldsymbol{\phi}_R$ on any vector $\boldsymbol{\phi} = \{\phi(x_m)\}_{m=0}^{2M}$ is given by a linear combination with Wigner *big-D* coefficients [7, Sect. 3.6],

$$\phi_R(x_m) := \Big(\mathcal{R}(\alpha,\beta,\gamma) : \boldsymbol{\phi}\Big)(x_m) = \sum_{x_{m'}=-M}^{M} D^M_{x_m,x_{m'}}(\alpha,\beta,\gamma)\,\phi(x_{m'}). \tag{31}$$

In particular, an eigenvector of the position generator $J_3 \leftrightarrow X$, namely $\boldsymbol{\xi}_{m'} = \{\delta_{x_{m'},x_m}\}_{x_m=-M}^{M}$ (with fixed $x_{m'} = M - m' \in \{-M, -M+1, \ldots M\}$) can be transformed into the Kravchuk eigenvector of the Hamiltonian $J_1 \leftrightarrow H$, namely $\boldsymbol{\psi}_{m'} = \{\psi_{m'}(x_m)\}_{x_m=-M}^{M}$ through a rotation of $\frac{1}{2}\pi$ around the axis of momentum, $J_2 \leftrightarrow P$. The Wigner matrix function $D^M_{x_m,x_{m'}}(0, \frac{1}{2}\pi, 0)$ in (31) reduces then to a *little-d* Wigner matrix function $d^M_{x_m,x_{m'}}(\frac{1}{2}\pi)$ while the sum over x_m collapses to the single $x_{m'}$. Thus we find the relation

$$\psi^{(M)}_{m'}(x_m) = d^M_{x_m,x_{m'}}(\tfrac{1}{2}\pi), \tag{32}$$

between the Kravchuk functions and the Wigner d-functions of angle $\frac{1}{2}\pi$.

Now a rotation $\mathcal{R}(0, 0, \frac{1}{2}\pi)$ by $\frac{1}{2}\pi$ around the $J_3 \leftrightarrow X$ axis, will bring the Hamiltonian $J_1 \leftrightarrow H$ to momentum $J_2 \leftrightarrow P$. The momentum eigenvectors are obtained from the Kravchuk eigenvectors of the Hamiltonian through the Wigner *big-D* matrix $D^M_{x_m,x_{m'}}(0, 0, \frac{1}{2}\pi) = \delta_{m,m'} \exp(-\mathrm{i}\frac{1}{2}\pi x_{m'})$. This defines the momentum eigenvectors in the position basis to be

$$\widetilde{\psi}^{(M)}_{m'}(x_m) := (-\mathrm{i})^{x_m}\psi^{(M)}_{m'}(x_m). \tag{33}$$

This basis is the *canonical dual* to the position eigenbasis of Kronecker deltas, $\boldsymbol{\xi}_{m'}$ at $x_m = x'_m$. Performing the rotation (33) twice, the $J_1 \leftrightarrow H$ axis is inverted, which inverts the Hamiltonian eigenvalues; this exchanges the bottom $\boldsymbol{\psi}_0$ and top $\boldsymbol{\psi}_N$ states, and generally the states $n \in \{0, 1, \ldots, N\}$ and $N - n$, with an alternating sign between every two positions,

$$\psi^{(M)}_{N-n}(x_m) := (-1)^{x_m}\psi^{(M)}_n(x_m). \tag{34}$$

Finally, a rotation by θ around the $J_1 \leftrightarrow H$ axis represents Hamiltonian evolution, rotating the momentum axis $J_2 \leftrightarrow P$ towards the position axis $J_3 \leftrightarrow X$, exactly as the harmonic oscillator does in quantum mechanics [8], which is the (inverse) fractional Fourier transform. This rotation is

$$\mathcal{F}(-\theta) := \exp(-\mathrm{i}\theta H) = \mathcal{R}(-\tfrac{1}{2}\pi, \theta, \tfrac{1}{2}\pi), \tag{35}$$

and its action on any vector in the position basis $\phi(x_m)$, is then given by

$$\phi_\theta(x_m) := (\mathcal{F}(\theta) : \boldsymbol{\phi})(x_m) = \sum_{x_{m'}=-M}^{M} \exp(-i\theta(x_m - x_{m'})\, d^M_{x_m, x_{m'}}(\theta)\, \phi(x_{m'}). \tag{36}$$

This is the fractional Fourier-Kravchuk transform [3]. In particular, acting on the Kravchuk functions,

$$(\mathcal{F}(\theta) : \boldsymbol{\psi}_n^{(M)})(x_m) = \exp(-in\theta)\, \psi_n^{(M)}(x_m), \tag{37}$$

which is the same property enjoyed by the quantum harmonic oscillator wavefunctions under the fractional Fourier integral transform [8].

We close this section by reminding the reader that coherent states in the finite oscillator model can be obtained by rotating the ground state $\psi_0^{(M)}(x_m)$ to any (θ, ϕ) on the sphere with (31) [9]. These coherent states will perform harmonic motion under evolution by $H \leftrightarrow J_1$, although they are not shape-invariant as their continuous counterparts are. The rotation of points on the sphere projected on any plane is of course a harmonic motion, which is geometric rather than physical under the oscillator potential.

6. Conclusions

Starting with the family of Kravchuk functions, with their recurrence and finite difference properties, we have built the generators of the group of rotations of three-dimensional space $SO(3)$. The three generators thus define the discrete and finite model of the quantum harmonic oscillator, which shares with the continuous model both the geometry and dynamics of a Hamiltonian system. We should point out that the necessary prerequisites to follow this strategy were presumably available at least since the appearance of the monograph by Gel'fand, Minlos and Shapiro [10], published in Russian in 1953 —and of the Kravchuk polynomials [4], which were known since 1929.

It seems to us that the discrete Kravchuk oscillator did not attract attention earlier because it realizes *discrete* function bases for the finite-dimensional spaces of the irreducible representations of the rotation group $SO(3)$, while the attention of the authors of Ref. [10], and many contemporary and later authors, has been on bases of functions such as spherical harmonics, that are continuous functions over the manifold of the sphere.

It should be also pointed out that the Kravchuk polynomials *do* appear in the three-volume encyclopedic monograph by N.Ja. Vilenkin and A.U. Klimyk via the matrix elements of irreducible representations of the Lie group $SU(2)$, treated as functions of the column index (see [11, page 346]). Observe that in our terminology this corresponds to matrix elements in the dual canonical basis; this provides an algebraic reasoning on why matrix elements, as functions of the column index, are of clear interest and explains how they emerge from a group-theoretic point of view.

The approach followed in this work seems to be quite general. In particular, it can be used for constructing a discrete model for free Hamiltonian systems associated with the irreducible representations of the Euclidean group $ISO(2)$, of the Lie group $SU(1,1)$ in the discrete and continuous series of irreducible representations for models of discrete radial harmonic and repulsive oscillators [12]. This will be dealt with elsewhere.

One final remark must be made, however. During the discussion of this work at the VIII Symposium on Quantum Theory and Symmetries (Mexico City, August 5–9, 2013) we became aware, thanks to W. Miller, Jr., about the recent progress in establishing close links between the generic 3-parameter two-dimensional 2nd-order superintegrable system $S9$ [13, 14] and hypergeometric orthogonal polynomials from the Askey scheme [5]. It turns out that various function space realizations of the quadratic Racah–Wilson algebra, which is the symmetry algebra behind this superintegrable model $S9$, can be put into correspondence with *all* hypergeometric polynomials in the Askey scheme. Of course, these remarkable works [13, 14] thus reveal the physical interpretation and group-theoretic context of such intricate orthogonal families as the Wilson and Racah polynomials, which satisfy 2nd-order difference equations with the quadratic spectra [5]. But the Kravchuk polynomials are known to be solutions of the 2nd-order difference equation with the linear spectrum [5] and this fact is essential for constructing unitary irreducible representations of the rotation Lie algebra $so(3)$ in terms of the Kravchuk polynomials. Therefore it seems to us that in this particular case when one aims at revealing a physical interpretation and fundamental group-theoretic properties of the Kravchuk polynomials it quite suffices to exploit the much simpler Lie algebra $so(3)$ rather than the more elaborate algebraic approach of Kalnins, Miller, Jr. and Post.

Acknowledgments

Discussions with E.I. Jafarov, W. Miller, Jr. and L. Vinet are gratefully acknowledged. The participation of MKA in this work was partially supported by the SEP-CONACYT project 168104 *Operadores Integrales y Pseudodiferenciales en Problemas de Física Matemática*. NMA and KBW have been supported by the DGAPA-UNAM IN105008-3 and SEP-CONACYT 79899 *Óptica Matemática* projects.

References

[1] J. Talman, *Special Functions: A Group Theoretic Approach* (W.A. Benjamin, New York, 1968)

[2] W. Miller Jr., *Symmetry and Separation of Variables*, Encyclopedia of Mathematics and Its Applications, Vol. 4, Ed. by G.-C. Rota (Addison-Wesley, Reading, MA, 1977).

[3] N.M. Atakishiyev and K.B. Wolf, Fractional Fourier–Kravchuk transform, *J. Opt. Soc. Amer. A*, Vol. **14**, No. 7, pp. 1467–1477, 1997.

[4] M. Krawtchouk, Sur une généralisation des polynômes d'Hermite, *Comptes Rendus de l'Académie des Sciences*, Series I - Mathématique, Vol. **189**, pp. 620–622, 1929.

[5] R. Koekoek, P.A. Lesky and R.F. Swarttouw, *Hypergeometric Orthogonal Polynomials and Their q-Analogues*, Springer Monographs in Mathematics (Springer-Verlag, Berlin Heidelberg, 2010).

[6] N.M. Atakishiyev, G.S. Pogosyan and K.B. Wolf, Contraction of the finite one-dimensional oscillator, *Int. J. Mod. Phys. A*, Vol. **18**, pp. 317–327, 2003.

[7] L.C. Biedenharn and J.D. Louck, *Angular Momentum in Quantum Mechanics*, Encyclopedia of Mathematics and Its Applications, Vol. 8, Ed. by G.-C. Rota (Addison-Wesley, Reading MA, 1981).

[8] H.M. Ozaktas, Z. Zalevsky, and M. Alper Kutay, *The Fractional Fourier Transform with Applications in Optics and Signal Processing* (Wiley, Chichester, 2001).

[9] N.M. Atakishiyev, G.S. Pogosyan, and K.B. Wolf, Finite models of the oscillator, *Phys. Part. Nuclei*, Suppl. #3, Vol. **36**, pp. 521–555, 2005.

[10] I.M. Gel'fand, R.A. Minlos and Z.Ya. Shapiro, *Representations of the Rotation and Lorentz Groups and Their Applications* (Macmillan, New York, 1963).

[11] N.Ja. Vilenkin and A.U. Klimyk, *Representations of Lie Groups and Special Functions* Vol. I (Kluwer Acad. Publ., Dordrecht, The Netherlands, 1991).

[12] K.B. Wolf, Discrete systems and signals on phase space, *Appl. Math. & Information Science*, Vol. 4, pp. 141–181, 2010.

[13] E.G. Kalnins, W. Miller, Jr. and S. Post, Contractions of 2D 2nd order quantum superintegrable systems and the Askey scheme for hypergeometric orthogonal polynomials, *Symmetry, Integrability and Geometry: Methods and Applications*, Vol. **9**, Art. # 057, 28 p., 2013.

[14] E.G. Kalnins, W. Miller, Jr. and S. Post, Wilson polynomials and the generic superintegrable system on the 2-sphere, *J. Phys. A: Math. Theor., Vol.* **40**, *No. 38, pp. 11525–11538, 2007.*

Journal of Physics: Conference Series **512** (2014) 012032 doi:10.1088/1742-6596/512/1/012032

Quantizations from (P)OVM's

H. Bergeron[a], **E.M.F. Curado**[b,c], **J.P. Gazeau**[b,d], **Ligia M.C.S. Rodrigues**[b]

[a] Univ Paris-Sud, ISMO, UMR 8214, 91405 Orsay, France
[b] Centro Brasileiro de Pesquisas Fisicas; [c] Instituto Nacional de Ciência e Tecnologia - Sistemas Complexos, Rua Xavier Sigaud 150, 22290-180 - Rio de Janeiro, RJ, Brazil
[d] APC, UMR 7164, Univ Paris Diderot, Sorbonne Paris Cité, 75205 Paris, France

E-mail: herve.bergeron@u-psud.fr; evaldo@cbpf.br; gazeau@apc.univ-paris7.fr; ligia@cbpf.br

Abstract. We explain the powerful rôle that operator-valued measures can play in quantizing any set equipped with a measure, for instance a group (resp. group coset) with its invariant (resp. quasi-invariant) measure. Coherent state quantization is a particular case. Such integral quantizations are illustrated with two examples based on the Weyl-Heisenberg group and on the affine group respectively. An interesting application of the affine quantization in quantum cosmology is mentioned, and we sketch a construction of new coherent states for the hydrogen atom.

1. What is really quantization?

The word "quantization" is widely used in Physics, in Mathematics and in Signal Analysis. While the meaning of quantization is perfectly clear to signal practitioners, the word singularly lacks precision in the two former domains, as was stated in the following quotation:

> *First quantization is a mystery. It is the attempt to get from a classical description of a physical system to a quantum description of the "same" system. Now it doesn't seem to be true that God created a classical universe on the first day and then quantized it on the second day. So it's unnatural to try to get from classical to quantum mechanics. Nonetheless we are inclined to do so since we understand classical mechanics better. So we'd like to find a way to start with a classical mechanics problem - that is, a phase space and a Hamiltonian function on it - and cook up a quantum mechanics problem - that is, a Hilbert space with a Hamiltonian operator on it. It has become clear that there is no utterly general systematic procedure for doing so* [1] (see also the recent Todorov's review [2]).

In a nutshell, we might claim that " *One quantizes things one does not really know to obtain things most of which one is unable to measure*".

Let us recall the basic procedure, which rests upon the canonical commutation rules (ccr). Starting from a n-dimensional phase space or symplectic manifold, it is summarized by the map ($n = 2$):

$$(q,p)\,, \quad \{q,p\} = 1 \mapsto (Q,P)\,, \quad [Q,P] = i\hbar I\,,$$
$$f(q,p) \mapsto f(Q,P) \mapsto (\mathrm{Sym}f)(Q,P)\,.$$

But then what about singular f, e.g. the phase $\arctan(p/q)$? What about barriers or other impassable boundaries? Even the elegant Weyl-Wigner integral quantization [3, 4, 5, 6, 7], which we revisit in this paper, is subject to serious difficulties when the geometry of the phase space is not euclidean.

Journal of Physics: Conference Series **512** (2014) 012032 doi:10.1088/1742-6596/512/1/012032

Let us propose here a definition resting on three minimal requirements, linearity, identity, self-adjointness. More precisely, quantization is

(i) a linear map

$$\mathfrak{Q} : \mathcal{C}(X) \mapsto \mathcal{A}(\mathfrak{H})$$

where $\mathcal{C}(X)$ is a vector space of complex-valued functions $f(x)$ on a set X and $\mathcal{A}(\mathfrak{H})$ is a vector space of linear operators

$$\mathfrak{Q}(f) \equiv A_f$$

in some complex Hilbert space $\mathfrak{H}$ such that

(ii) $f = 1$ is mapped to the identity operator I on $\mathfrak{H}$,

(iii) a real f is mapped to an (essentially) self-adjoint operator A_f in $\mathfrak{H}$.

One is free to add further requirements on X and $\mathcal{C}(X)$, e.g., measure, topology, manifold, closure under algebraic operations ...; to add physical interpretation about measurement of spectra of classical $f \in \mathcal{C}(X)$ or quantum $\mathcal{A}(\mathfrak{H})$ to which are given the status of *observables*; to add dynamical considerations as time evolution; finally to add the requirement of unambiguous classical limit of the quantum physical quantities, the limit operation being associated with a change of scale.

2. Integral quantization[8, 9, 10]

2.1. Integral quantization: general setting

We start from a measure space (X, ν) where X is the set of things we "do not know well". Suppose we were able to build an X-labelled family of bounded operators,

$$X \ni x \mapsto \mathsf{M}(x) \in \mathcal{L}(\mathfrak{H}), \tag{1}$$

on Hilbert space $\mathfrak{H}$, resolving the identity I:

$$\int_X \mathsf{M}(x)\, \mathrm{d}\nu(x) = I\,, \quad \text{in a weak sense.} \tag{2}$$

If the $\mathsf{M}(x)$'s are positive and unit trace, we will use the standard notation for density matrices :

$$\mathsf{M}(x) \equiv \rho(x)\,.$$

In this case, if X is equipped with a suitable topology, then the map

$$\mathcal{B}(X) \ni \Delta \mapsto \int_\Delta \rho(x)\, \mathrm{d}\nu(x)$$

may define a normalized positive operator-valued measure (POVM) $\mathfrak{m}$ on the σ-algebra $\mathcal{B}(X)$ of Borel sets

$$\mathcal{B}(X) \ni \Delta \mapsto \mathfrak{m}(\Delta) = \int_\Delta \rho(x)\, \mathrm{d}\nu(x)\,.$$

Then the integral quantization of complex-valued functions $f(x)$ on X is the linear map:

$$f \mapsto A_f = \int_X f(x)\, \mathsf{M}(x)\, \mathrm{d}\nu(x)\,. \tag{3}$$

If this map makes sense, the operator A_f in $\mathfrak{H}$ has to be understood as the sesquilinear form,

$$B_f(\psi_1, \psi_2) = \int_X f(x)\, \langle \psi_1 | \mathsf{M}(x) | \psi_2 \rangle\, \mathrm{d}\nu(x)\,, \tag{4}$$

Journal of Physics: Conference Series **512** (2014) 012032 doi:10.1088/1742-6596/512/1/012032

defined on a dense subspace of $\mathfrak{H}$. Note that if f is real and at least semi-bounded, the Friedrich's extension [11] of B_f univocally defines a self-adjoint operator. If f is not semi-bounded, there is no natural choice of a self-adjoint operator associated with B_f (see for instance [12, 13]). A first observation about the above construction is its probabilistic content when $\mathsf{M}(x) = \rho(x)$. Pick another (or the same) family of positive unit trace operators $X \ni x \mapsto \widetilde{\rho}(x) \in \mathcal{L}^+(\mathfrak{H})$. Then we return to the classical world through the construction of the so-called lower (Lieb) or covariant (Berezin) symbol:

$$A_f \mapsto \check{f}(x) := \int_X f(x')\,\mathrm{tr}(\widetilde{\rho}(x)\rho(x'))\,\mathrm{d}\nu(x')\,,\ \text{“lower symbol”}\,, \tag{5}$$

provided the integral be defined. Due to the fact that $x' \mapsto \mathrm{tr}(\widetilde{\rho}(x)\rho(x'))$ is also a probability distribution, the map $f \mapsto \check{f}$ should be viewed as a local averaging (a “blurring”) of the original f. It is also a generalization of the so-called Bargmann-Segal transform (see for instance [14, 15]). Quantization issues, e.g. spectral properties of A_f, may be derived from functional properties of $\check{f}$.

Moreover, equipped with such a map, a classical limit condition can be considered in the following sense: given a scale parameter ϵ and a distance $d(f, \check{f})$:

$$d(f, \check{f}) \to 0 \quad \text{as} \quad \epsilon \to 0\,. \tag{6}$$

Another interesting aspect of the integral quantization map is the granted possibility of quantizing constraints: suppose that (X, ν) is a smooth n-dim. manifold on which is defined space $\mathcal{D}'(X)$ of distributions as the topological dual of compactly supported n-forms on X [16]. Some of these distributions, e.g. $\delta(u(x))$, express geometrical constraints. Extending the map $f \mapsto A_f$ to these objects yields the quantum version $A_{\delta(u(x))}$ of these constraints. There exists a different starting point, more in Dirac's spirit [17] and used in (Loop) Quantum Gravity and Quantum Cosmology. It would consist in determining the kernel of the operator A_u issued from integral quantization $u \mapsto A_u$. Both methods are obviously not *mathematically* equivalent, except for a few cases. They are possibly *physically* equivalent, i.e. indistinguishable in terms of measurement.

2.2. *Covariant integral quantization based on Unitary Irreducible Representation*

Let G be a Lie group with left Haar measure $\mathrm{d}\mu(g)$, and let $g \mapsto U(g)$ be a unitary irreducible representation (UIR) of G in a Hilbert space $\mathfrak{H}$. Let M a bounded operator on $\mathfrak{H}$. Suppose that the operator

$$R := \int_G \mathsf{M}(g)\,\mathrm{d}\mu(g)\,, \quad \mathsf{M}(g) := U(g)\,\mathsf{M}\,U^{\dagger}(g)\,, \tag{7}$$

is defined in a weak sense. From the left invariance of $\mathrm{d}\mu(g)$ we have $U(g_0)\,R\,U^{\dagger}(g_0) = \int_G \mathrm{d}\mu(g)\,\mathsf{M}(g_0 g) = R$ and so R commutes with all operators $U(g)$, $g \in G$. Thus, from Schur's Lemma, $R = c_{\mathsf{M}} I$ with

$$c_{\mathsf{M}} = \int_G \mathrm{tr}\left(\rho_0\,\mathsf{M}(g)\right)\,\mathrm{d}\mu(g)\,, \tag{8}$$

where the unit trace positive operator ρ_0 is chosen in order to make the integral convergent. The resolution of the identity follows:

$$\int_G \mathsf{M}(g)\,\mathrm{d}\nu(g) = I\,, \quad \mathrm{d}\nu(g) := \mathrm{d}\mu(g)/c_{\mathsf{M}}\,. \tag{9}$$

If UIR U is square integrable with $|\eta\rangle$ being an admissible (unit) vector for U, i.e. $c(\eta) := \int_G \mathrm{d}\mu(g)\,|\langle\eta|U(g)\eta\rangle|^2 < \infty$, then the resolution of the identity is obeyed by *coherent states* for G: $|\eta_g\rangle = U(g)|\eta\rangle$, i.e. by $\rho(g) = U(g)\,\rho\,U^{\dagger}(g)$, where $\rho = |\eta\rangle\langle\eta|$. (A nice and meaningful example of

Journal of Physics: Conference Series **512** (2014) 012032 doi:10.1088/1742-6596/512/1/012032

group having square integrable UIR's is the affine group, see Section 4). This allows a *covariant* integral quantization of complex-valued functions on the group $f \mapsto A_f = \int_G \rho(g)\, f(g)\, \mathrm{d}\nu(g)$:

$$U(g) A_f U^\dagger(g) = A_{U(g)f}\,, \tag{10}$$

where $(U(g)f)(g') := f(g^{-1}g')$ (regular representation if $f \in L^2(G, \mathrm{d}\mu(g))$). The Bargmann-Berezin-Segal or "heat kernel" transform on G is then defined as $\check{f}(g) := \int_G \mathrm{tr}(\rho(g)\, \rho(g'))\, f(g')\, \mathrm{d}\nu(g')$.

In the absence of square-integrability of U over G, there may exist a definition of square-integrability and related covariant coherent states with respect to a left coset manifold $X = G/H$, with H a closed subgroup of G, equipped with a quasi-invariant measure ν. The most known example is the Weyl-Heisenberg group (see below). For more details and examples see [9].

3. Weyl-Heisenberg covariant integral quantization(s)

W-H quantization(s)

We first recall the basic material and notations for the Weyl-Heisenberg (W-H) algebra and its Fock or number representation Let $\mathcal{H}$ be a separable (complex) Hilbert space with orthonormal basis $e_0, e_1, \dots, e_n \equiv |e_n\rangle, \dots$. Lowering and raising operators a and $a^\dagger$ are defined by their action on the basis:

$$a\,|e_n\rangle = \sqrt{n}|e_{n-1}\rangle\,, \quad a|e_0\rangle = 0\,, \quad a^\dagger\,|e_n\rangle = \sqrt{n+1}|e_{n+1}\rangle\,, \tag{11}$$

and the triplet $\{a, a^\dagger, I\}$ generate the Weyl-Heisenberg algebra characterized by the canonical commutation rule

$$[a, a^\dagger] = \mathbf{I}\,. \tag{12}$$

There results from (11) that the number operator $N := a^\dagger a$ is diagonal with spectrum $\mathbb{N}$, $N|e_n\rangle = n|e_n\rangle$. It is well known that there exists an essentially unique UIR of the W-H algebra or group, at the root of quantum mechanics. Its square integrability holds with respect to the center $C \sim \mathbb{R}$ of the W-H group, and the measure space which has to be considered here is the euclidean or complex plane $X = G_{\mathrm{WH}}/C \sim \mathbb{C}$ with measure d^2z/π. To each $z \in \mathbb{C}$ corresponds the (unitary) displacement operator $D(z)$:

$$\mathbb{C} \ni z \mapsto D(z) = e^{za^\dagger - \bar{z}a}\,, \quad D(-z) = (D(z))^{-1} = D(z)^\dagger\,. \tag{13}$$

The ccr (12) or QM non commutativity is encoded by the addition formula

$$D(z)D(z') = e^{\frac{1}{2}(z\bar{z}' - \bar{z}z')} D(z + z')\,. \tag{14}$$

The standard (i.e., Schrödinger-Klauder-Glauber-Sudarshan) CS [18] are then obtained by

$$|z\rangle = D(z)|e_0\rangle\,. \tag{15}$$

Let $\varpi(z)$ be a function on the complex plane obeying $\varpi(0) = 1$ and chosen in such a way that the operator-valued integral

$$\int_{\mathbb{C}} D(z)\, \varpi(z) \frac{\mathrm{d}^2 z}{\pi} := \mathsf{M}$$

defines (in a weak sense) a bounded operator M on $\mathfrak{H}$. Then the family of displaced $\mathsf{M}(z) := D(z)\mathsf{M}D(z)^\dagger$ under the unitary action $D(z)$ resolves the identity

$$\int_{\mathbb{C}} \mathsf{M}(z) \frac{\mathrm{d}^2 z}{\pi} = I\,. \tag{16}$$

This is a direct consequence of $D(z)D(z')D(z)^\dagger = e^{z\bar{z}' - \bar{z}z'} D(z')$, of $\int_{\mathbb{C}} e^{z\bar{\xi} - \bar{z}\xi} \frac{\mathrm{d}^2\xi}{\pi} = \pi\delta^2(z)$, and of $\varpi(0) = 1$ with $D(0) = I$.

The resulting quantization map is given by

$$f \mapsto A_f = \int_{\mathbb{C}} f(z)\, \mathsf{M}(z)\, \frac{\mathrm{d}^2 z}{\pi} = \int_{\mathbb{C}} \hat{f}(-z)\, D(z)\, \varpi(z) \frac{\mathrm{d}^2 z}{\pi}\,, \tag{17}$$

where is involved the symplectic Fourier transform $\hat{f}(z) = \int_{\mathbb{C}} f(\xi)\, e^{z\bar{\xi}-\bar{z}\xi}\, \frac{\mathrm{d}^2\xi}{\pi}$. The covariance with respect to the translations reads as

$$A_{f(z-z_0)} = D(z_0) A_{f(z)} D(z_0)^{\dagger}\,. \tag{18}$$

Note the properties of A_f resulting from properties of the weight function:

$$A_{f(-z)} = \mathsf{P} A_{f(z)} \mathsf{P}, \forall f \iff \varpi(z) = \varpi(-z),\, \forall z\,,$$
$$A_{\overline{f(z)}} = A^{\dagger}_{f(z)}, \forall f \iff \overline{\varpi(-z)} = \varpi(z),\, \forall z\,,$$

where $\mathsf{P} = \sum_{n=0}^{\infty}(-1)^n |e_n\rangle\langle e_n|$ is the parity operator.

Now, the ccr is almost always the rule when ϖ is chosen *real* and *even*. Indeed we have in this case

$$A_z = a\,, \quad A_{\overline{f(z)}} = A^{\dagger}_{f(z)}\,, \tag{19}$$

or equivalently, with $z = (q+ip)/\sqrt{2}$,

$$A_q = \frac{a + a^{\dagger}}{\sqrt{2}} := Q\,,\; A_p = \frac{a - a^{\dagger}}{i\sqrt{2}} := P\,,\; [Q,P] = iI\,. \tag{20}$$

Morover, iff $|\varpi(z)| = 1$,

$$\mathrm{tr}(A_f^{\dagger} A_f) = \int_{\mathbb{C}} |f(z)|^2 \frac{\mathrm{d}^2 z}{\pi}\,, \tag{21}$$

which means that the map $L^2(\mathbb{C}, d^2z/\pi) \ni f \mapsto A_f \in \mathfrak{H}_{\mathrm{Hilbert-Schmidt}}$ is invertible through a trace formula.

Cahill-Glauber weight

For instance let us choose as a weight function the exponential

$$\varpi_s(z) = e^{s|z|^2/2}\,, \quad \mathrm{Re}\, s < 1\,. \tag{22}$$

This yields a diagonal $\mathsf{M} \equiv \mathsf{M}_s$ with

$$\langle e_n|\mathsf{M}_s|e_n\rangle = \frac{2}{1-s}\left(\frac{s+1}{s-1}\right)^n\,, \tag{23}$$

and so

$$\mathsf{M}_s = \int_{\mathbb{C}} D(z)\, \varpi_s(z)\, \frac{\mathrm{d}^2 z}{\pi} = \frac{2}{1-s} \exp\left[\ln\left(\frac{s+1}{s-1}\right) a^{\dagger} a\right]\,. \tag{24}$$

The value $s = -1$ corresponds to the CS (anti-normal) quantization, since

$$\mathsf{M} = \lim_{s \to -1} \frac{2}{1-s} \exp\left(\ln \frac{s+1}{s-1} a^{\dagger} a\right) = |e_0\rangle\langle e_0|\,, \tag{25}$$

Journal of Physics: Conference Series **512** (2014) 012032 doi:10.1088/1742-6596/512/1/012032

and so

$$A_f = \int_{\mathbb{C}} D(z)\mathsf{M}D(z)^\dagger\, f(z)\, \frac{\mathrm{d}^2 z}{\pi} = \int_{\mathbb{C}} |z\rangle\langle z|\, f(z)\, \frac{\mathrm{d}^2 z}{\pi}\,. \tag{26}$$

The choice $s = 0$ implies $\mathsf{M} = 2\mathsf{P}$ and corresponds to the Wigner-Weyl quantization. Then

$$A_f = \int_{\mathbb{C}} D(z)\, 2\mathsf{P}\, D(z)^\dagger\, f(z)\, \frac{\mathrm{d}^2 z}{\pi}\,. \tag{27}$$

The case $s = 1$ is the normal quantization obtained in an asymptotic sense.

The function (22) was originally introduced by Cahill and Glauber [19, 20] in view of discussing the problem of expanding an arbitrary operator as an ordered power series in a and $a^\dagger$, a typical question encountered in quantum field theory, specially in quantum optics. Actually, they were not interested in the question of quantization itself. Now, a very interesting feature of (22) is that the operator M_s is positive unit trace class for $s \leq -1$ (and only trace class if $\mathrm{Re}\, s < 0$), i.e., is a density operator, $\mathsf{M}_s \equiv \rho_s$. Therefore, in the range $s \leq -1$ the corresponding quantization has a consistent probabilistic content in the sense that the operator-valued measure

$$\mathbb{C} \supset \Delta \mapsto \int_{\Delta \in \mathcal{B}(\mathbb{C})} D(z)\mathsf{M}_s D(z)^\dagger\, \frac{\mathrm{d}^2 z}{\pi}$$

is a POVM. Moreover, a Boltzmann-Planck expression with temperature T can be associated with this lack of knowledge of the classical $X = \mathbb{C}$, say a kind of noise temperature as we have in electronics. Given an elementary quantum energy, say $\hbar\omega$ and with the temperature T-dependent $s = -\coth \dfrac{\hbar\omega}{2k_B T}$ the density operator quantization is Boltzmann-Planck

$$\rho_s = \left(1 - e^{-\frac{\hbar\omega}{k_B T}}\right) \sum_{n=0}^{\infty} e^{-\frac{n\hbar\omega}{k_B T}} |e_n\rangle\langle e_n|\,. \tag{28}$$

Thus, the temperature-dependent operators $\rho_s(z) = D(z)\, \rho_s\, D(z)^\dagger$ defines a Weyl-Heisenberg covariant family of POVM's on the phase space $\mathbb{C}$, the null temperature limit case being the POVM built from standard CS. We notice that the Weyl-Wigner integral quantization ($s = 0$) and its associated Wigner function are out of the scope of this thermodynamic consideration.

Quantum harmonic oscillator according to ϖ

We saw that any real even ϖ defining a bounded operator through (3) yields the ccr for the quantization of the canonical pairs $(z, \bar{z})$ or, equivalently (q, p). Actually it yields also the correct energy spectrum for the harmonic oscillator. Indeed, we have

$$A_{q^2} = Q^2 - \partial_z \partial_{\bar{z}} \varpi\big|_{z=0} + \frac{1}{2}\left(\partial_z^2 \varpi\big|_{z=0} + \partial_{\bar{z}}^2 \varpi\big|_{z=0}\right),$$
$$A_{p^2} = P^2 - \partial_z \partial_{\bar{z}} \varpi\big|_{z=0} - \frac{1}{2}\left(\partial_z^2 \varpi\big|_{z=0} + \partial_{\bar{z}}^2 \varpi\big|_{z=0}\right),$$

and so

$$A_{|z|^2} \equiv A_J = a^\dagger a + \frac{1}{2} - \partial_z \partial_{\bar{z}} \varpi\big|_{z=0}\,. \tag{29}$$

where $|z|^2 (= J)$ is the energy (or action variable) for the H.O. The difference between the ground state energy $E_0 = 1/2 - \partial_z \partial_{\bar{z}} \varpi\big|_{z=0}$, and the minimum of the quantum potential energy $E_m = [\min(A_{q^2}) + \min(A_{p^2})]/2 = -\partial_z \partial_{\bar{z}} \varpi\big|_{z=0}$ is $E_0 - E_m = 1/2$. So it is the (experimentally verified, see for instance [21]) half quantum (in appropriate units), independently of the particular quantization which has been chosen. In the exponential Cahill-Glauber case $\varpi_s(z) = e^{s|z|^2/2}$ the above operators reduce to

$$A_{|z|^2} = a^\dagger a + \frac{1-s}{2}\,, \quad A_{q^2} = Q^2 - \frac{s}{2}\,, \quad A_{p^2} = P^2 - \frac{s}{2}\,.$$

More details and relevant references on this question are found in [22] where it is proven that these constant shifts in energy are inaccessible to measurement.

Weyl-Heisenberg integral quantization with action-angle variables

With $z = \sqrt{J}\, e^{i\gamma}$ in action-angle (J, γ) notations for the harmonic oscillator, the quantization of $f(J, \gamma)$, 2π-periodic in γ, yields formally

$$A_f = \int_0^{+\infty} \mathrm{d}J \int_0^{2\pi} \frac{\mathrm{d}\gamma}{2\pi} f(J,\gamma) \mathsf{M}\left(\sqrt{J} e^{i\gamma}\right)\,. \tag{30}$$

We define the unitary representation $\theta \mapsto U_{\mathbb{T}}(\theta)$ of the unit circle $\mathbb{S}^1$ on the Hilbert space $\mathcal{H}$ as $U_{\mathbb{T}}(\theta)|e_n\rangle = e^{i(n+\nu)\theta}|e_n\rangle$, where ν is arbitrary real. If the operator M is diagonal, then one has the angular covariance property:

$$U_{\mathbb{T}}(\theta) A_f U_{\mathbb{T}}(-\theta) = A_{T(\theta)f}\,, \quad T(\theta) f(J,\gamma) = f(J, \gamma - \theta)\,. \tag{31}$$

In particular, let us quantize with coherent states, $\mathsf{M}(z) = \rho_{-1}(z) = |z\rangle\langle z|$, the discontinuous 2π-periodic angle function $\daleth(\gamma) = \gamma$ for $\gamma \in [0, 2\pi)$. In terms of the action-angle variables these CS read as

$$|z\rangle \equiv |J,\gamma\rangle = \sum_n \sqrt{p_n(J)} e^{in\gamma} |e_n\rangle\,, \tag{32}$$

where $n \mapsto p_n(J) = e^{-J} J^n/n!$ is the Poisson distribution. Since the angle function is real and bounded, its quantum counterpart $A_\daleth$ is a bounded self-adjoint operator, and it is covariant in the above sense. In the basis $|e_n\rangle$, it is given by the infinite matrix:

$$A_\daleth = \pi\, 1_{\mathcal{H}} + i \sum_{n \neq n'} \frac{\Gamma\left(\frac{n+n'}{2} + 1\right)}{\sqrt{n! n'!}} \frac{1}{n' - n} |e_n\rangle\langle e_{n'}|\,. \tag{33}$$

This quantum angle has spectral measure with support $[0, 2\pi]$. Of course, plenty of similar quantum angles are made possible with that freedom we have in choosing the weight function. It is an interesting question to be considered from different viewpoints, particularly from measurement viewpoints based on POVM, as they are described in [23, 24].

4. Affine quantization

General setting

This is the second basic illuminating example of the method. Like the above complex plane and abelian group $X = \mathbb{C}$, is viewed as the phase space for the motion of a particle on the line, the half-plane is a group which can be viewed as the phase space for the motion of a particle on the half-line. Let us be more precise. Our measure space (X, ν) is the upper half-plane $X \equiv \Pi_+ := \{(q,p)\,|\, p \in \mathbb{R}\,,\, q > 0\}$ equipped with the left invariant measure $\mathrm{d}q\,\mathrm{d}p$. Equipped with the multiplication $(q,p)(q_0,p_0) = (qq_0, p_0/q + p)$, $q \in \mathbb{R}_+^*$, $p \in \mathbb{R}$, Π_+ is viewed as the affine group $\mathrm{Aff}_+(\mathbb{R})$ of the real line. $\mathrm{Aff}_+(\mathbb{R})$ has two non-equivalent UIR, $U_\pm$ [25, 26]. Both are square integrable and this is the rationale backing the *continuous wavelet analysis* [27, 28, 29, 9]. The UIR $U_+ \equiv U$ is carried on by Hilbert space $\mathcal{H} = L^2(\mathbb{R}_+^*, dx)$:

$$U(q,p)\psi(x) = (e^{ipx}/\sqrt{q})\psi(x/q)\,. \tag{34}$$

As we did for the Weyl-Heisenberg group, we pick a suitably localized weight function $\mathsf{w}(q,p)$ on the half-plane such that the integral

$$\int_{\Pi_+} U(q,p)\,\mathsf{w}(q,p)\; \mathrm{d}q\; \mathrm{d}p := \mathsf{M} \tag{35}$$

defines a bounded operator in a weak sense . Proceeding with the same construction as in (7), (8) and (9) yields the resolution of the identity on

$$\int_{\Pi_+} \mathsf{M}(q,p)\,\frac{\mathrm{d}q\; \mathrm{d}p}{c_{\mathsf{w},\mathsf{M}}} = I\,, \quad \mathsf{M}(q,p) = U(q,p)\mathsf{M}U^{\dagger}(q,p)\,, \tag{36}$$

and the resulting covariant quantization based on the affine group:

$$f \mapsto A_f = \int_{\Pi_+} f(q,p)\,\mathsf{M}(q,p)\,\frac{\mathrm{d}q\; \mathrm{d}p}{c_{\mathsf{w},\mathsf{M}}}\,. \tag{37}$$

Due to the square-integrability of U, the simplest choice to be made is $\mathsf{M} = |\psi\rangle\langle\psi| = \rho_0$ where the unit-norm state $\psi \in L^2(\mathbb{R}_+^{\dagger},\, \mathrm{d}x) \cap L^2(\mathbb{R}_+^{\dagger},\, \mathrm{d}x/x)$ ("fiducial vector" or "wavelet") produces all affine coherent states, i.e. wavelets, defined as $|q,p\rangle = U(q,p)|\psi\rangle$. Resolution of the identity and resulting covariant quantization now read:

$$\int_{\Pi_+} |q,p\rangle\langle q,p|\,\frac{\mathrm{d}q\,\mathrm{d}p}{2\pi c_{-1}} = I\,,\; c_\gamma := \int_0^\infty |\psi(x)|^2\,\frac{\mathrm{d}x}{x^{2+\gamma}}\,, \tag{38}$$

$$f \;\mapsto\; A_f = \int_{\Pi_+} f(q,p)|q,p\rangle\langle q,p|\,\frac{\mathrm{d}q\,\mathrm{d}p}{2\pi c_{-1}}\,. \tag{39}$$

(Had we chosen U_- we would produce identical results. It is just a matter of taste between negative or positive half-line.) The quantization is canonical for q and p, in the sense that the ccr gives i times a constant:

$$A_p = P = -i\partial/\partial x\,,\; A_{q^\beta} = (c_{\beta-1}/c_{-1})\,Q^\beta\,,\; Qf(x) = xf(x)\,. \tag{40}$$

Note the multiplicative factor of Q^β, absent in the W-H quantization: this is the price to pay for dealing with dilations. Now, the important point concerns the quantization of the kinetic energy:

$$A_{p^2} = P^2 + KQ^{-2}\,, \quad K = K(\psi) = \int_0^\infty (\psi'(u))^2\,u\,\frac{\mathrm{d}u}{c_{-1}}\,. \tag{41}$$

Thus, due to the presence of repulsive ($\sim$ centrifugal) potential this affine or wavelet quantization prevents a quantum free particle moving on the positive line from reaching the origin. We know (see [11]) that the operator $P^2 = -d^2/dx^2$ alone, in $L^2(\mathbb{R}_+^*,\, \mathrm{d}x)$, is not essentially self-adjoint whereas the above regularized operator, defined on the domain $C_0^\infty(0,\infty)$ of smooth compactly supported functions, is essentially self-adjoint for $K \geq 3/4$. Then quantum dynamics of the free motion is possible.

Quantum states and their dynamics have phase space representation through lower symbols. With a state $|\phi\rangle$ is associated the probability distribution on the phase space:

$$\rho_\phi(q,p) = \frac{1}{2\pi c_{-1}}|\langle q,p|\phi\rangle|^2\,.$$

With energy eigenstates at our disposal, we can compute the time evolution for any state and the formula for the associated time-dependent probability distribution follows.

The integral quantization based on the affine group has recently found interesting applications in quantum cosmology, where the singularity at zero volume of the universe is naturally regularized with such a scheme [30, 31]. Note that proceeding in quantum theory with an "affine" quantization instead of the Weyl-Heisenberg quantization was already present in Klauder's work devoted to the question of dealing with singularities in quantum gravity [32, 33]. The procedure rests on the representation of the affine Lie algebra. In this sense, it remains closer to the canonical one and it is not of the integral type.

Journal of Physics: Conference Series **512** (2014) 012032 doi:10.1088/1742-6596/512/1/012032

As a byproduct: affine CS for central potentials, e.g. Coulomb-Kepler
We sketch now the content of a work in progress [34] where we partly use the material presented in this paper. We know that the construction of coherent states for the hydrogen atom which would have as much rich properties as the standard CS have for the harmonic oscillator is still an open (if solvable!) problem. Several systems of CS have been proposed since Schrödinger's original work, derived mainly from the groups SU(1,1) and SO(4,2), although in a regularized sense (the latter group is the full dynamical group of the H-atom; the former is the subgroup describing its radial motion only), but none of them is fully convincing (see [35] for a list of references). The construction that we propose is based on the above affine CS for the radial part and on the Kowalski-Rembielinski-Hall-Mitchell CS for the angular part. Indeed, we expect that both have good localization properties in the phase space for the motion of a charged particle submitted to the Coulomb potential. We know that the quantum hamiltonian $H = -\Delta + k/r^2 - g/r$ with domain $C_0^\infty(\mathbb{R}^3)$ in $L^2(\mathbb{R}^3,\, \mathrm{d}^3\mathbf{r})$ is self-adjoint [11]. Using spherical coordinates $\mathbf{r} = (r, \hat{\mathbf{y}})$, a state $\Psi(\mathbf{r})$ in $L^2(\mathbb{R}^3,\, \mathrm{d}^3\mathbf{r})$ factorizes as

$$\Psi(\mathbf{r}) = r\psi(r)\mathcal{Y}(\hat{\mathbf{y}})\,, \tag{42}$$

where $\psi \in L^2(\mathbb{R}^+,\, \mathrm{d}r)$ and $\mathcal{Y} \in L^2(\mathbb{S}^2,\, \mathrm{d}\hat{\mathbf{y}})$, where $\hat{\mathbf{y}}$ is the SO(3) invariant measure on the 2-sphere. Use the material above with a suitable fiducial vector $\psi(r)$ yields affine CS for the radial part which are labeled by points (q_r, p_r) of the upper half-plane:

$$\mathcal{R}_{q_r,p_r}(r) = (U(q_r, p_r)\psi)\,(r)\,.$$

Now, the phase space for the motion on sphere $\mathbb{S}^2 = \{\mathbf{x} \in \mathbb{R}^3\,,\, x^2 = \sum_k x_k^2 = 1\}$ is realized as the complexified sphere $\mathbb{S}^2_C$

$$T^*(\mathbb{S}^2) \simeq \mathbb{S}^2_c = \{\mathbf{a} = (a_1, a_2, a_3) \in \mathbb{C}^3 : a^2 = \sum_k a_k^2 = 1\}\,, \tag{43}$$

with, in suitable units,

$$\mathbf{a} = (\cosh J)\,\mathbf{x} + i\frac{1}{J}(\sinh J)\,\mathbf{p}\,, \quad J = \|\mathbf{x} \wedge \mathbf{p}\|\,, \quad \mathbf{x}\cdot\mathbf{p} = 0\,. \tag{44}$$

The Kowalski-Rembieliński coherent states $|\zeta_{\mathbf{a}}\rangle$ are realized as elements of the Hilbert space $L^2(\mathbb{S}^2,\, \mathrm{d}\hat{\mathbf{y}})$ as follows

$$\zeta_{\mathbf{a}}(\hat{\mathbf{y}}) == (2\pi)^{-1}\frac{e^{1/8}}{\sqrt{\pi}}\int_{\Omega}^{\pi}\frac{\mathrm{d}\phi}{\sqrt{\cos\Omega - \cos\phi}}\,, \tag{45}$$

which depends only on the complex angle $\Omega = \Omega(\mathbf{a}, \hat{\mathbf{y}})$ defined through analytic continuation by

$$\cos\Omega = \mathbf{a}\cdot\hat{\mathbf{y}} = (\cosh J)\,\mathbf{x}\cdot\hat{\mathbf{y}} + i\frac{1}{J}(\sinh J)\,\mathbf{p}\cdot\hat{\mathbf{y}}\,. \tag{46}$$

They solve the identity in $L^2(\mathbb{S}^2_c,\, \mathrm{d}\mu_h)$ with an explicit measure given in [36, 37]. Finally, the coherent states for the particle read

$$\mathcal{Z}_{q_r,p_r,\mathbf{a}}(\mathbf{r}) = r\mathcal{R}_{q_r,p_r}(r)\,\zeta_{\mathbf{a}}(\hat{\mathbf{y}})\,. \tag{47}$$

5. Conclusion

Beyond the freedom (think to the analogy with Signal Analysis where different techniques are employed in a complementary way) allowed by integral quantization, the advantages of the method with regard to other quantization procedures in use are of four types.

(i) The minimal amount of constraints imposed to the classical objects to be quantized.

(ii) Once a choice of (positive) operator-valued measure has been made, which must be consistent with experiment, there is no ambiguity in the issue, contrarily to other method(s) in use (think in particular of the ordering problem). To one classical object corresponds one and only one quantum object. Of course different choices are requested to be physically equivalent

(iii) The method produces in essence a regularizing effect, at the exception of certain choices, like the Weyl-Wigner integral quantization.

(iv) The method, through POVM choices, offers the possibility to keep a full probabilistic content. As a matter of fact, the Weyl-Wigner integral quantization does not rest on a POVM.

But what is the real meaning of that freedom granted to us in the choice of POVM or others? Such a freedom is governed by our degree of confidence in localizing a pure classical state (q, p) in phase space. The latter is usually viewed as an ideal continuous manifold where all points are physically accessible. As everybody knows, such a view is physically untenable ... However, and this is the paradoxical paradigm of contemporary physics, one needs such a leibnizian mathematical ideality (*natura non saltum facit*) to build a more realistic, though still highly mathematical, representation of the physical world.

[1] J. Baez, Categories, quantization and much more, available at http://math.ucr.edu/home/baez/categories.html (2006)
[2] I. Todorov, *Quantization is a mystery*, IHES/P/12/01 (2012)
[3] H. Weyl, *Gruppentheorie und Quantenmechanik* (Hirzel, Leipzig,1928); H. Weyl, *The Theory of Groups and Quantum Mechanics* (Dover, New York, 1931)
[4] A. Grossmann, Parity operator and quantization of δ-functions, *Commun. Math. Phys.* **48** (1976) 191–194
[5] I. Daubechies, On the distributions corresponding to bounded operators in the Weyl quantization, *Commun. Math. Phys.* **75** 229–238 (1980)
[6] I. Daubechies and A. Grossmann, An integral transform related to quantization. I. *J. Math. Phys.* **21** 2080–2090 (1980)
[7] I. Daubechies, A. Grossmann, and J. Reignier, An integral transform related to quantization. II. *J. Math. Phys.* **24** 239–254 (1983)
[8] H. Bergeron and J.-P. Gazeau, *Integral quantizations with two basic examples*; arXiv:1308.2348 [quant-ph]
[9] S. T. Ali, J.-P. Antoine, and J.-P. Gazeau, *Coherent States, Wavelets and their Generalizations* (Graduate Texts in Mathematics, Springer, New York, 2000), 2nd Edition in *Theoretical and Mathematical Physics*, Springer, New York, to appear, November 2013.
[10] J.-P. Gazeau, *Coherent States in Quantum Physics*, Wiley-VCH Verlag (2009)
[11] Reed M. and Simon B., Methods of Modern Mathematical Physics, II. Fourier Analysis, Self-Adjointness Volume 2, Academic Press, New York, 1975
[12] H. Bergeron, J.P. Gazeau, P. Siegl, and A. Youssef, Semi-classical behavior of Pöschl-Teller coherent states, *Eur. Phys. Lett.* **92** (2010) 60003
[13] H. Bergeron, P. Siegl, and A. Youssef, New SUSYQM coherent states for Pöschl-Teller potentials: a detailed mathematical analysis, *J. Phys. A: Math. Theor.* **45** (2012) 244028
[14] M. B. Stenzel, The Segal-Bargmann transform on a symmetric space of compact type, *J. Funct. Analysis* **165** (1994) 44–58
[15] B. C. Hall and J. J. Mitchell. The SegalBargmann transform for noncompact symmetric spaces of the complex type, *J. Funct. Analysis* **227** (2005) 338-371
[16] M. Grosser, A note on distribution spaces on manifolds, Novi Sad Math. **38** (2008) 121–128
[17] P. A. M. Dirac, *Lectures on Quantum Mechanics* (Dover, New York, 2001)
[18] A. M. Perelomov, *Generalized Coherent States and their Applications* (Springer-Verlag, Berlin, 1986)
[19] K.E. Cahill and R. Glauber, Ordered expansion in Boson Amplitude Operators, *Phys. Rev.* **177** 1857-1881 (1969)
[20] K.E. Cahill and R. Glauber, Density Operators and Quasiprobability Distributions, *Phys. Rev.* **177** 1882-1902 (1969)
[21] G. Herzberg, *Molecular Spectra and Molecular Structure: Spectra of Diatomic Molecules, 2nd. ed.* (Krieger Pub., Malabar, FL, 1989)
[22] H. Bergeron, J.P. Gazeau, A. Youssef, Phys. Lett. A **377** 598605 (2013).
[23] P. Busch, M. Grabowski, and P. J. Lahti, Operational Quantum Physics, Lecture Notes in Physics Monographs (Springer 1995)
[24] A. Royer, Phase states and phase operators for the quantum harmonic oscillator, *Phys. Rev. A* **53** (1996) 70–108
[25] I.M. Gel'fand and M.A. N'aimark, *Dokl. Akad. Nauk SSSR* **55** (1947) 570.
[26] E. W. Aslaksen and J. R. Klauder, Unitary Representations of the Affine Group, *J. Math. Phys.* **15** (1968) 206–211
[27] A. Grossmann and J. Morlet, Decomposition of Hardy functions into square integrable wavelets of constant shape, *SIAM J. Math. Anal.* **15** (1984) 723–736.

[28] A. Grossmann, J. Morlet and T. Paul, Integral transforms associated to square integrable representations. I. General results, *J. Math. Phys.* **26** (1985) 2473–2479.

[29] A. Grossmann, J. Morlet and T. Paul, Integral transforms associated to square integrable representations. II. Examples, *Ann. Inst. H. Poincaré* **45** (1986) 293–309.

[30] H Bergeron, A Dapor, J.-P. Gazeau and P Małkiewicz, *Wavelet Quantum Cosmology* (2013); arXiv:1305.0653 [gr-qc]

[31] H Bergeron, A Dapor, J-P G and P Małkiewicz, *Towards singularity-free cosmology: coherent state quantization* submitted, (2013)

[32] J. R. Klauder, An Affinity for Affine Quantum Gravity, *Proceedings of the Steklov Institute of Mathematics* **272** (2011)169-176; gr-qc/1003.261

[33] M. Fanuel and S. Zonetti, Affine Quantization and the Initial Cosmological Singularity, *Eur. Phys. Lett.* **101** (2013) 10001; gr-qc/1203.4936.

[34] H. Bergeron, E. M. F. Curado, J.P. Gazeau and Ligia M. C. S. Rodrigues *Coherent states for the hydrogen atom: a new approach*, in preparation.

[35] J. R. Klauder, Coherent states for the hydrogen atom, *J. Phys. A: Math. Gen.* **29** (1996) L293–L296

[36] K. Kowalski and J. Rembieliński, Quantum mechanics on a sphere and coherent states, *J. Phys. A: Math. Gen.* **33** (2000) 6035–6048; The Bargmann representation for the quantum mechanics on a sphere, *J. Math. Phys.* **42** (2001) 4138–4147

[37] B. Hall and J. J. Mitchell, Coherent states on spheres, *J. Math. Phys.* **43** (2002) 1211–1236

Comparison of laser-assisted charge transfer of symmetric and asymmetric colliding systems

F. J. Domínguez-Gutiérrez and R. Cabrera-Trujillo
Instituto de Ciencias Físicas, Universidad Nacional Autónoma de México, Ap. Postal 43-8, Cuernavaca, Morelos, 62251, México
E-mail: javier@fis.unam.mx, trujillo@fis.unam.mx

Abstract. We study the effect of an intense, ultra-short, and ultra-fast (1 and 2 fs) pulse with the aim to manipulate the charge transfer process of symmetric and asymmetric colliding systems. We report the total and state-selective cross section for charge transfer for the symmetric H^{+}+ H and for the asymmetric He^{2+}+H system by means of the Crank-Nicolson method in the energy collision range for 1-10 keV/amu and 0.25-10 keV/amu, respectively. In this work, we show that the laser assistance do minimal contribution to charge transfer process in the symmetric system and that the laser pulse increases the charge transfer process at low impact energies for asymmetric system up to an order of magnitude. To asses the validity of our results, we compare our numerical results for the case of collisions with no laser to available experimental data in the literature showing good agreement for both systems.

1. Introduction

The charge transfer process in collisions between bare ions as projectiles and one electron atomic targets is an important process in determining the radiation losses and neutral beam heating efficiencies of plasmas in Tokamak [1]. Nowadays, laser assisted have been suggested to improve the charge transfer process in ion-atom collisions [2, 3, 4, 5, 6] with the aim to manipulate and control the electron capture process. The laser assisted collision for resonant symmetric systems have been studied by Ferrante et al. [7] by using the impact parameter method and the atomic two-state approximation for the resonant systems H^{+}+H and Rb^{+}+Rb finding that the effect of the laser on the total charge transfer process is to enhance it in a small order. Meanwhile a numerically treatment of the symmetric system H^{+}+H with a laser assistance is done by Niederhausen et al. [5] solving the time dependent Schrödinger equation finding a strong laser-phase dependence for charge transfer processes, and weak evidence for a charge-resonant enhanced ionization process for a circular polarized laser pulse. A study of the asymmetric system He^{2+}+H considering a laser assisted collision is done by Kirchner [9] finding that the laser enhances the total charge transfer process at low impact energy collision with a dependence on the initial phase of the laser.

The purpose of this work is to compare the effect of an ultra-fast, ultra-short and intense laser pulse on the state selective charge transfer process considering a symmetric H^{+}+H and asymmetric He^{2+}+H collisions systems at intermediate impact energy collisions.

Our work is organized as follows. In Sec. 2, we define the Hamiltonian that describes the dynamics of the laser-assisted ion-atom collision. In Sec. 2.1 we describe our method to obtain the electronic states for the projectile and target. In Sec. 2.2, the dynamics of the electron during the collision is obtained by solving the time-dependent Schrödinger equation by means the finite difference method. In Sec. 2.3, we describe the numerical implementation of our finite-difference procedure. In Sec. 2.4, we show the procedure to obtain the charge transfer cross section within the finite-differences approach. At the end of this section, we describe the implementation of the initial conditions of the dynamics of the collision with laser assisted collision to obtain the charge transfer probability. In Sec. 3, we report our numerical results for the state selective charge transfer cross-section of the symmetric and asymmetric collision systems with and without laser assistance. There we show a good agreement between our laser-free numerical data and the results reported in the literature by other theoretical methods and experimental data for both collision systems. Finally a brief summary is given in Sec. 4. Atomic units are used throughout unless otherwise noted.

2. Theory

We consider a system with a bare projectile ion (Z_p) moving on a straight-line trajectory colliding with an atomic hydrogen target. The Hamiltonian that describes the dynamics of the electron is

$$\mathbf{H} = -\frac{1}{2}\nabla^2 - \frac{1}{|\mathbf{r}|} - \frac{Z_p}{|\mathbf{r} - \mathbf{R}(t)|} - \mathbf{r} \cdot \mathcal{E}(t), \tag{1}$$

where the first term in Eq. (1) corresponds to the kinetic energy operator $\mathbf{T} = -\frac{1}{2}\nabla^2$, the second term is the interaction potential between the electron and the target, and the third term is the interaction potential between the electron and the projectile. The straight-line trajectory projectile is defined by $\mathbf{R}(t) = (b, 0, v_p t)$, where b is the impact parameter and $v_p = \sqrt{2E_p/m_p}$ is the projectile velocity. Finally, the fourth term is the semi-classical description of the electron-laser electric field coupling in the dipole-length representation. We model the laser pulse by a Gaussian envelope as

$$\mathcal{E}(t) = \mathcal{E}_0 \exp\left(-\frac{t^2}{\tau^2}\right) \cos(\omega t + \phi), \tag{2}$$

where $\mathcal{E}_0$ defines the polarization vector of the laser pulse which in our case is placed in the plane of the collision along the projectile velocity. The intensity at full width half-maximum (FWHM) is given by $\tau\sqrt{4\ln 2}$, such that τ defines the duration of the pulse, ω is the carrier frequency and ϕ is the relative collision laser phase (RCLP). The RCLP combines both the carrier-envelope phase of the laser as well as the synchronization of the laser pulse with the collision.

2.1. Stationary states of the projectile bare ion

We use the finite-differences method to solve the time-independent Schrödinger equation (TISE) to obtain the neutral hydrogen atom and He^+ ion stationary states at the end of the collisions, with the following Hamiltonian

$$\mathbb{H} = -\frac{1}{2}\nabla^2 - \frac{Z}{|\mathbf{r}|}, \tag{3}$$

where $Z = 1$ or 2 for the H or He^+ atom, respectively. To solve the previous equation, we propouse a solution in spherical coordinates as $\Phi(\mathbf{r}) = D(r)Y_{l,m}(\theta, \phi)$. For the radial part, $D(r)$, we use the change of variable $D(r) = \zeta(r)/r$ obtaining the following differential equation

$$\left(-\frac{1}{2}\frac{d^2}{dr^2} - \frac{l(l+1)}{2r^2} - \frac{Z}{r}\right)\zeta(r) = E\zeta(r). \tag{4}$$

Now, we discretize the function $\zeta(r)$ and the second derivative of the function $\zeta(r)$ of Eq. (4) in a numerical grid as $\zeta(r) \to \zeta_k$ when $r \to r_k$ on the k-points in the numerical grid, with an uniform spacing Δr. By considering the ζ_k points as the elements of a N-dimensional vector, we rewrite the previous equations as an eigenvalue problem, that is

$$\mathbb{B}\vec{\zeta} = E\vec{\zeta} \tag{5}$$

where the matrix $\mathbb{B}$ has the following elements

$$\begin{pmatrix} 2B+V_{\text{eff}_1} & -B & 0 & \cdots & 0 \\ -B & 2B+V_{\text{eff}_2} & -B & \cdots & 0 \\ \vdots & \vdots & \vdots & \ddots & \vdots \\ 0 & \cdots & 0 & -B & 2B+V_{\text{eff}_N} \end{pmatrix}. \tag{6}$$

Here V_{eff_i} is the discretization of the effective potential $V_{\text{eff}} = -l(l+1)/2r^2 - Z/r$ and $B = 1/2\Delta r^2$. From Eq. (5) we can obtain the values for the bound-energies and the numerical eigen-functions of the neutral H atom or He^+ ion.

2.2. Dynamics of the electron during the collision

The total wave function $\Psi(\mathbf{r}, t)$ describes the dynamics of the electron cloud in the collision and it is obtained by solving the time-dependent Schrödinger equation (TDSE)

$$\mathbf{H}\Psi(\mathbf{r}, t) = i\frac{\partial}{\partial t}\Psi(\mathbf{r}, t), \tag{7}$$

with the Hamiltonian in Eq. (1). The solution to Eq. (7) is given by the unitary propagator operator as

$$\Psi(\mathbf{r}, t) = \exp(-i\mathbf{H}\Delta t)\Psi(\mathbf{r}, t_0). \tag{8}$$

where t_0 and t are the initial and final time, respectively, and $\Delta t = t - t_0$. The propagation in time is obtained by implementing the Crank-Nicolson and the finite-differences methods. Then, we consider the Hamiltonian as $\mathbf{H} = \mathbf{T} + \mathbf{V}$ with the kinetic energy operator as $\mathbf{T} = \sum_\eta \mathbf{T}_\eta$ where $\eta = x, y, z$, such that the unitary propagator operator becomes

$$\prod_\eta \exp\left(\frac{i}{2}\mathbf{T}_\eta \Delta t\right)\Psi = \prod_\eta \exp\left(-\frac{i}{2}\mathbf{T}_\eta \Delta t\right)\Xi, \tag{9}$$

where $\Xi(\mathbf{r}, t) = \exp(-i\mathbf{V}\Delta t)\Psi(\mathbf{r}, t_0)$. The previous equation is thus equivalent to the separation of variables method such that x, y, and z become independent and allows a natural separation in matrix form.

We approximate the exponential function to first order obtaining the following equation

$$\prod_{\eta=x,y,z}\left(1 - \frac{i\Delta t}{4}\frac{\partial^2}{\partial \eta^2}\right)\Psi = \prod_{\eta=x,y,z}\left(1 + \frac{i\Delta t}{4}\frac{\partial^2}{\partial \eta^2}\right)\Xi, \tag{10}$$

In Eq. (10), we discretize the wave function Ψ as $\Psi(\mathbf{r}, t) \to \Psi(x_i, y_j, z_k; t^n) \to \Psi^n_{i,j,k}$ and the second derivatives of each axis onto a three dimensional Cartesian numerical grid, where the points of the grid η are discretized to η_k with $k = 0, \ldots, N$ and $\eta = x, y, z$. We consider Dirichlet boundary conditions, that is $\Psi(\mathbf{r} \to \infty) = 0$, at the boundary of the numerical grid and by arranging in matrix form, we rewrite the time-dependent Schrödinger equation as a linear algebra problem

$$\mathsf{A}^+_x \mathsf{A}^+_y \mathsf{A}^+_z \Psi^{n+1} = \mathsf{A}^-_x \mathsf{A}^-_y \mathsf{A}^-_z \Xi^n \tag{11}$$

where the matrices $\mathsf{A}_\eta^{\pm}$ are given by

$$\begin{pmatrix} 1 \pm 2\nu_\eta & \mp\nu_\eta & 0 & \dots & 0 \\ \mp\nu_\eta & 1 \pm 2\nu_\eta & \mp\nu_\eta & \dots & 0 \\ 0 & \ddots & \ddots & \ddots & 0 \\ \vdots & \ddots & \ddots & \ddots & \vdots \\ 0 & \dots & \mp\nu_\eta & 1 \pm 2\nu_\eta & \mp\nu_\eta \\ 0 & \dots & 0 & \mp\nu_\eta & 1 \pm 2\nu_\eta \end{pmatrix}, \tag{12}$$

where $\nu_\eta = i\Delta t/(4\Delta\eta^2)$. We require that $|\nu_\eta| < 1$ for the stability of the numerical method. At the end of the dynamics of the collision, we obtain the total evolved wave function.

2.3. Numerical implementation and initial conditions

We consider $R(0) = (b, 0, z_0)$ as the point where the incident bare ion projectile starts its straight-line trajectory. We propagate the ion on the z axis, and we consider the impact parameter range as $b \in [0.2, 10]$ and we increase it in steps of $\Delta b = 0.2$. Initially, the atomic hydrogen target is in the ground state and it is placed at the origin of the Cartesian numerical grid. The position of the incident bare ion projectile starts at a distance such that the electric field strength of the laser pulse is negligible. We have chosen $\mathcal{E}(t)$ to be 1% of the value at the peak of the pulse, i.e., $\mathcal{E}(t_{\min})/\mathcal{E}_0 = 0.01$, such that we reduce Stark effects when we do the projections at the final time of the collision. So, for the symmetric collisions system we consider $\tau = 1$ fs (24.835 a.u.) obtaining $t_{min} = 53.3$ a.u. and since the H^+ ion is placed at $z_0 = -v_p t_{\min}$ a.u. from the target hydrogen atom, the distance where the H^+ ion starts the collision is $z_0 = -34$ a.u. for 10 keV/amu ($v_p = 0.6328$ a.u.) impact energy collision. These assumptions fixed the size of the numerical grid as $[-10, 20]_x \times [-15, 15]_y \times [-45, 45]_z$. For the asymmetric system we consider $\tau = 2$ fs (49.67 a.u.) and we get that $t_{min} = 106.59$ a.u. and the distance $z_0 = -68.0$ a.u. at 10 keV/amu impact energy collision is where the He^{2+} ion starts the straight-line trajectory, so the size of the numerical grid becomes $[-10, 20]_x \times [-15, 15]_y \times [-75, 75]_z$ for this case. Also, we use a spacing of $\Delta\eta = 0.2$ with $\eta = x, y, z$ and $\Delta t = 0.06$ in both numerical grids. The initial ground state wave-function of the hydrogen atom is obtained by relaxation of the TDSE by means of imaginary time technique. We obtain the $1s$ state in the Cartesian numerical grid with a total energy $E(1s) = -0.4980$ a.u. for both numerical grids. The laser pulse parameters that we consider are the following: the wavelength is $\lambda = 800$ nm ($\omega = 0.0569$) because this value is suggested as a common value in an ultra-fast and intense lasers [8]. The RCLP, ϕ, in the laser pulse, is an important parameter during the collision processes [6, 9], so, we consider the effect of the RCLP on the charge transfer process starting with $\phi = 0$, then, we increase the phase in quarters of π up to the value $\phi = 2\pi$. We report our numerical results in the Sec. 3 by performing an average on the RCLP.

For the stationary states, we use a numerical grid $[0, 100]$ and a spacing of $\Delta r = 0.05$. to obtain a good description of the wave function cusp and eigen-energies.

2.4. State selective charge transfer cross-section

Once the solution for the $n = 2$ for the hydrogen atom and $n = 2, 3$ for the He^+ ion are obtained, we calculate the state selective charge transfer probabilities by projecting the final wave function as follows

$$P_{n,l}(E, b) = \sum_{m=-l}^{l} \left| \int_V D_{n,l}(r) Y_{l,m}^* \Psi(x, y, z; t_f) dV \right|^2. \tag{13}$$

Here $r = \sqrt{x^2 + y^2 + z^2}$ and we use a linear interpolation to match the numerical data of the radial wave functions associated to the stationary states of the hydrogen atom or He^+ ion

Table 1. Energy eigenvalues (in eV) and expectation value $\langle r \rangle$ (in a.u.) for hydrogen atom and He^+ ion as obtained with our numerical method for $n = 2$, and 3. The results are compared to the exact analytic result $\langle r \rangle^* = [3n^2 - l(l+1)]/2Z$ and $-13.606Z^2/n^2$ for a hydrogen atom.

	Hydrogen atom				He^+ ion			
State	$\langle r \rangle$	$\langle r \rangle^*$	E_{nl}	E_{exact}	$\langle r \rangle$	$\langle r \rangle^*$	E_{nl}	E_{exact}
$2s$	6.000	6.0	-3.400	-3.402	3.000	3.00	-13.604	-13.606
$2p$	4.999	5.0	-3.400	-3.402	2.499	2.50	-13.606	-13.606
$3s$	13.500	13.5	-1.511	-1.512	6.751	6.75	-6.047	-6.047
$3p$	12.499	12.5	-1.511	-1.512	6.248	6.25	-6.047	-6.047
$3d$	10.499	10.5	-1.511	-1.512	5.249	5.25	-6.046	-6.047

with the total wave function Ψ at the final time of the dynamics in Cartesian coordinates. We calculate the state selective charge transfer cross-section as

$$\sigma_{nl}(E) = 2\pi \int_0^\infty P_{n,l}(E, b) b db, \tag{14}$$

where the upper limit is replaced by a cut-off value determined by the size of the numerical grid.

3. Results

3.1. Stationary states of the hydrogen atom and He^+ ion

To asses the validity of our numerical approach, the electron position expectation value $\langle r \rangle$ and eigen-energies of the stationary states of the neutral hydrogen atom and He^+ are obtained by solving numerically the TISE. In table 1, we show our numerical results for the expectation value $\langle r \rangle$ and compare with the analytic result $\langle r \rangle = [3n^2 - l(l+1)]/2Z$ a.u. for the $n = 2$ and 3 states of atomic hydrogen. Also, the eigen-energy values are compared to the analytical value of $-13.606Z^2/n^2$ eV. We have a good agreement with all analytic quantities. Thus, within our numerical grid, we are confident that we have a proper wave function description of the hydrogen atom and He^+ ion states to calculate the projections to obtain the charge transfer probabilities.

3.2. Charge transfer cross-section into the $n = 2$ states of H^++H

The H^++H collision system is a symmetric system, the target and the projectile have the same energy levels. In Fig. 1, we show the electron transfer cross section results for the $n = 2$ shell. We observe in the Fig. 1(a) that our $2s$ state laser-free results (solid line) have a good agreement with the theoretical results reported by Winter [10], Lüdde et al. [11], Cheshire et al. [12] and Fritsch et al. [13] in the high energy region for up to 3 keV/amu. For energies lower than 3 keV/amu we have a discrepancy when comparing to the experimental data that we attribute to the straight-line trajectory approximation. For the $2p$ state we show, in Fig. 1(b), the good agreement between our laser-free numerical results to those reported by the theoretical methods of Winter [10], Lüdde et al. [11], Cheshire et al. [12] and Fritsch et al. [13], and with experimental data by Barnett et al. [16] and Morgan et al. [14]. Also, in the same figures, we compare our numerical results for the $2s$ and $2p$ states for the laser case for a 1 fs laser pulse width. We notice a minimal effect for all the collision energies on the charge transfer process in agreement to Ferrante et al. [7] within the error bars of the experiment. As from the results reported in figure 1, we observe that the laser pulse affects the $2p_{\pm 1}$ states of the projectile, and the laser pulse increases the probability of charge transfer, but for the $2s$ state the effect is negligible.

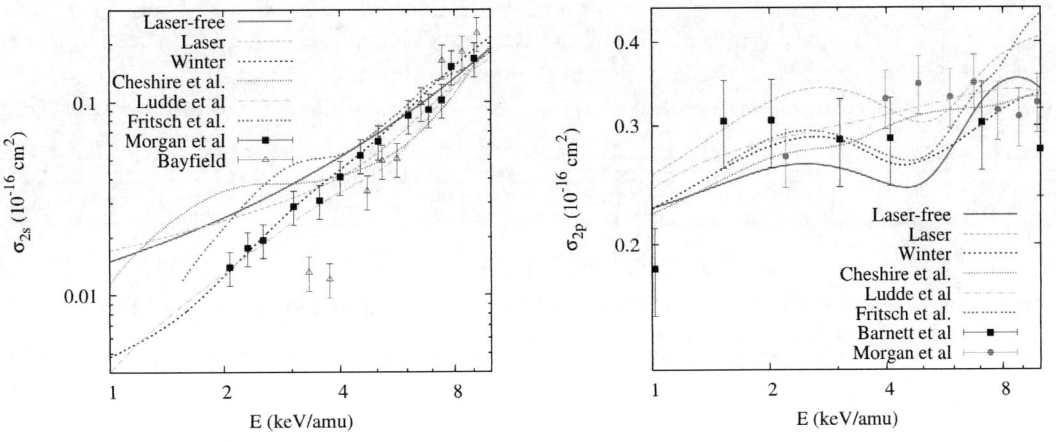

(a) 2s-state (b) 2p-state

Figure 1. We show the effect of the laser pulse on the 2s (Fig. 1(a)) and 2p (Fig. 1(b)) state charge transfer process of the symmetric system H^{+}+ H by comparing our laser-free and laser cross-section. We compare with the laser-free theoretical methods by Winter [10], Lüdde et al. [11], Cheshire et al. [12] and Fritsch et al. [13]. Also, we compare with laser-free experimental data by Morgan et al. [14], Bayfield [15] and Barnett et al. [16].

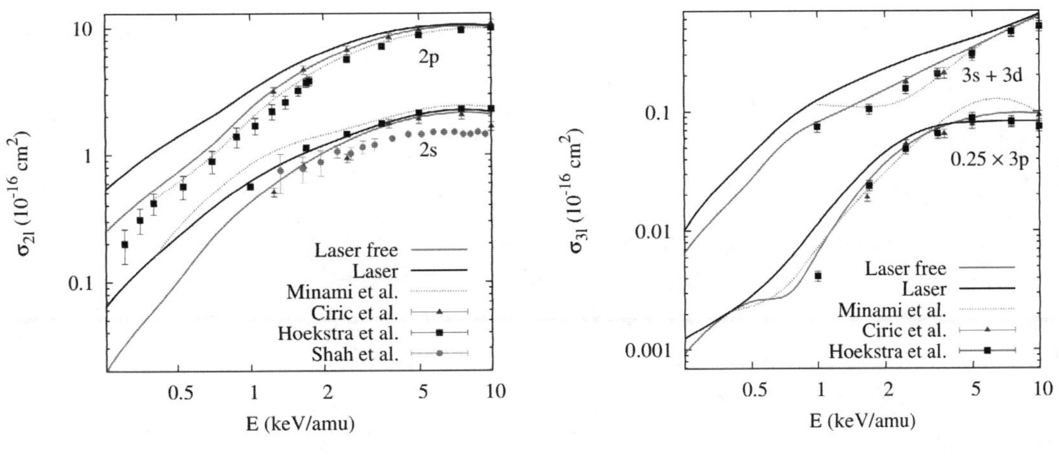

(a) $n = 2$ states (b) $n = 3$ states

Figure 2. In Fig. 2(a) and Fig. 2(b) we report the $n = 2$ and $n = 3$ charge transfer cross section, respectively, for the He^{2+}+H collision system as a function of the projectile kinetic energy for the laser and laser-free cases. We compare our laser-free results with the theoretical results by Minami et al. [17] and the experimental measurements for laser free collisions by Ćirić et al. [18] and Hoekstra et al. [19].

3.3. $n = 2$ and 3 states charge transfer cross-section of He^{2+}+H system

Finally, we obtain the probability for $n = 2, 3$ charge transfer of the asymmetric system He^{2+}+H using Eq. (14), for the laser-free and for the laser assisted collision. In Fig. 2(a) we report our numerical results for the $n = 2, 3$ charge transfer cross-section for both cases. For the laser-free case, we compare our numerical results with the theoretical results by Minami et al. [17] and experimental measurements by Ćirić et al. [18] and Hoekstra et al. [19]. We notice a good agreement between our numerical results and those reported in the literature. In Fig. 2(b), we notice an increase of the $n = 3$ charge transfer cross-section in the range 0.25-1 keV. In particular, the value of the $3s + 3d$ state charge transfer cross-section is increased by a factor of two in the range 0.25-3 keV. The laser pulse has a greater effect on this system because the coupling of the projectile and the target are different and the Stark effect in the collisions is not the same for both particles as in the symmetrical collision system case.

4. Summary

In this paper we have presented our first results of the study of laser-assisted collision of asymmetric and symmetric collision systems. We used the finite-differences method to solve the TISE finding a good agreement between our numerical results for the energy eigenvalues of the Hydrogen atom and He^{+} ion with the hydrogen analytical values. We have analyzed the effect of an ultra-fast, short and intense laser pulse on the state-selective charge transfer process of the both collisions systems finding a minimal effect on the symmetrical system and a larger effect on the asymmetrical system. We have showed that our numerical results for the $n = 2$ charge transfer cross-section of the H^{+}+H for the laser-free case and for the laser assistance, finding a good agreement with the laser-free experimental data. We have found an appreciable effect of the laser assisted charge transfer process of the asymmetric system He^{2+}+H for low impact energy collisions.

In conclusion, for the symmetric collision systems the laser effect is minimal on the total and $n = 2$ charge transfer process, but for the He^{2+}+H collision systems the laser enhances the total [6, 9] and $n = 2$ charge transfer process in the low impact energy collision range. Further work is in progress to find the effect of the laser wave-length and intensity, as well as the polarization to characterize in full the effect of laser assisted collision in the symmetry of the laser-assisted ion-atom collision.

We acknowledge support from grant PAPIIT IN 101-611 and CONACyT (Ph.D. scholarship). We thank the ICF-UNAM computer center for assistance.

References

[1] Isler R C and Crume E C 1978 *Phys. Rev. Lett.* **41**(19) 1296–1300
[2] Hsu Y P, Kimura M and Olson R E 1985 *Phys. Rev. A* **31** 576
[3] Madsen L B, Hansen J P and Kocbach L 2002 *Phys. Rev. Lett.* **89** 093202
[4] Kirchner T 2004 *Phys. Rev. A* **69** 063412
[5] Niederhausen T and Thumm U 2006 *Phys. Rev. A* **73**(4) 041404
[6] Anis F, Roudnev V, Cabrera-Trujillo R and Esry B D 2006 *Phys. Rev. A* **73** 043414
[7] Ferrante G, Casico L L and Spagnolo B 1981 *Journal of Physics B: Atomic and Molecular Physics* **14** 3961
[8] Wall K A and Sanchez A 1990 *The Lincoln laboratory journal* **3**
[9] Kirchner T 2002 *Phys. Rev. Lett.* **89** 093203
[10] Winter T G 2009 *Phys. Rev. A* **80** 032701
[11] Ludde H J and Dreizler R M 1982 *Journal of Physics B: Atomic and Molecular Physics* **15** 2703

[12] Cheshire I M, Gallaher D F and Taylor A J 1970 *Journal of Physics B: Atomic and Molecular Physics* **3** 813

[13] Fritsch W and Lin C D 1982 *Phys. Rev. A* **26** 762–769

[14] Morgan T J, Geddes J and Gilbody H B 1973 *Journal of Physics B: Atomic and Molecular Physics* **6** 2118

[15] Bayfield J E 1969 *Phys. Rev.* **185** 105–112

[16] Barnett C F 1990 *Oak Ridge National Laboratory Report* **6068**

[17] Minami T, Lee T G, Pindzola M S and Schultz D R 2008 *Journal of Physics B: Atomic, Molecular and Optical Physics* **41** 135201

[18] Ćirić D, Dijkkamp D, Vlieg E and de Heer F J 1985 *Journal of Physics B: Atomic and Molecular Physics* **18** 4745

[19] Hoekstra R, de Heer F J and Morgenstern R 1991 *Journal of Physics B: Atomic, Molecular and Optical Physics* **24** 4025

Journal of Physics: Conference Series **512** (2014) 012034 doi:10.1088/1742-6596/512/1/012034

Quantum oscillator models with a discrete position spectrum in the framework of Lie superalgebras

E.I. afarov[1] and . an der eugt

Department of Applied Mathematics,Computer Science and Statistics, Ghent University, Krijgslaan 281-S9, B-9000 Gent, Belgium

E-mail: ejafarov@physics.ab.az, Joris.VanderJeugt@UGent.be

Abstract. We present some algebraic models for the quantum oscillator based upon Lie superalgebras. The Hamiltonian, position and momentum operator are identified as elements of the Lie superalgebra, and then the emphasis is on the spectral analysis of these elements in Lie superalgebra representations. The first example is the Heisenberg-Weyl superalgebra $\mathfrak{sh}(2|2)$, which is considered as a "toy model". The representation considered is the Fock representation. The position operator has a discrete spectrum in this Fock representation, and the corresponding wavefunctions are in terms of Charlier polynomials. The second example is $\mathfrak{sl}(2|1)$, where we construct a class of discrete series representations explicitly. The spectral analysis of the position operator in these representations is an interesting problem, and gives rise to discrete position wavefunctions given in terms of Meixner polynomials. This model is more fundamental, since it contains the paraboson oscillator and the canonical oscillator as special cases.

1. Introduction

In all textbooks on quantum mechanics, it is described how the position wavefunctions of the one-dimensional canonical quantum oscillator are given in terms of Hermite polynomials. This is a first example of the relation between oscillator models and special functions. One of the alternative oscillator models is the so-called paraboson oscillator [1, 2], and here the position wavefunctions are given in terms of Laguerre polynomials [3, 4].

Both of these well-known oscillator models have an algebraic description as well: for the canonical oscillator this is in terms of the oscillator Lie algebra (and its Fock representation); for the paraboson oscillator this is in terms of the Lie superalgebra $\mathfrak{osp}(1|2)$ and its positive discrete series representations.

During the last years there has been increasing interest in other algebraic oscillator models, some of them inspired by applications in quantum optics or signal analysis. For these new oscillator models, the spectrum of the position operator can be continuous or discrete (with finite or infinite support). In the case of a discrete spectrum, one is faced with the interesting phenomenon of "discrete position wavefunctions", and there is a relation with discrete orthogonal polynomials.

The paper that triggered most of the work in this area is [5], where an oscillator model based on the Lie algebra $\mathfrak{su}(2)$ was introduced. In this model, the position wavefunctions are

[1] Permanent address: Institute of Physics, Azerbaijan National Academy of Sciences, Javid av. 33, AZ-1143 Baku, Azerbaijan. E.I. Jafarov kindly acknowledges support from Research Grant EIF-2012-2(6)-39/08/1 of the Science Development Foundation under the President of the Republic of Azerbaijan.

indeed discrete and expressed in terms of symmetric Krawtchouk polynomials. This model was extended by an extra parameter in [6, 7], with an underlying deformed $\mathfrak{su}(2)$ algebra, and position wavefunctions given by means of Hahn polynomials.

In the current contribution, we wish to present some examples of quantum oscillator models where the underlying algebra is a Lie superalgebra. For the first example, the underlying Lie superalgebra is the Heisenberg-Weyl superalgebra $\mathfrak{sh}(2|2)$, an algebra generated by one boson and one fermion pair. We study the spectral analysis of a position operator $\hat{q}$ in the Fock space representation of $\mathfrak{sh}(2|2)$. This operator has an infinite but discrete spectrum, and the corresponding position wavefunctions are given in terms of Charlier polynomials. We observe that also for an underlying orthosymplectic Lie superalgebra $\mathfrak{osp}(3|2)$, one would get the same position wavefunctions. This $\mathfrak{sh}(2|2)$ model is interesting because of its simplicity, but from the physical point of view it is maybe just a "toy model". The second example presented here has $\mathfrak{sl}(2|1)$ as underlying Lie superalgebra. In this model, the spectrum of a position operator can be discrete or continuous, depending on the value of a parameter γ. In the discrete case, the position wavefunctions are related to Meixner polynomials. In the continuous case, the position wavefunction reduce to the mentioned paraboson wavefunctions and thus, for a specific value of a representation parameter, to the canonical oscillator. Because of these special cases, the canonical oscillator is naturally embedded in the $\mathfrak{sl}(2|1)$ model, and it can be considered as more fundamental.

The examples presented here have been discussed in two previous papers. The $\mathfrak{sh}(2|2)$ toy model was analyzed in [8], and the $\mathfrak{sl}(2|1)$ in [9]. The purpose of the current contribution is to briefly present these examples again, compare them, give a relation to an $\mathfrak{osp}(3|2)$ model, and discuss some of the most significant properties.

To finalize this introduction, let us briefly recall the context of "algebraic oscillator models." For these models, one requires the same dynamics as for the classical or quantum oscillator, but the operators corresponding to position, momentum and Hamiltonian can be elements of some algebra different from the traditional Heisenberg (or oscillator) Lie algebra [5, 10, 11, 6, 12]. In the one-dimensional case, there are three (essentially self-adjoint) operators involved: the position operator $\hat{q}$, its corresponding momentum operator $\hat{p}$ and the Hamiltonian $\hat{H}$ which is the generator of time evolution. The main requirement is that these operators should satisfy the Hamilton-Lie equations (or the compatibility of Hamilton's equations with the Heisenberg equations):

$$[\hat{H}, \hat{q}] = -\mathrm{i}\hat{p}, \qquad [\hat{H}, \hat{p}] = \mathrm{i}\hat{q}, \tag{1}$$

in units with mass and frequency both equal to 1, and $\hbar = 1$. Contrary to the canonical case, the commutator $[\hat{q}, \hat{p}] = \mathrm{i}$ is not required. Apart from (1) and the self-adjointness, it is then common to require the following conditions [5]:

- all operators $\hat{q}$, $\hat{p}$, $\hat{H}$ belong to some Lie algebra or Lie superalgebra $\mathcal{A}$;
- the spectrum of $\hat{H}$ in (unitary) representations of $\mathcal{A}$ is equidistant.

2. The $\mathfrak{sh}(2|2)$ oscillator model

Consider a bosonic creation and annihilation operator $b^{\pm}$ with commutation relation

$$[b^-, b^+] = 1, \tag{2}$$

and a fermionic creation and annihilation operator $a^{\pm}$ with anticommutation relations

$$\{a^-, a^-\} = \{a^+, a^+\} = 0, \qquad \{a^-, a^+\} = 1. \tag{3}$$

Let us also assume that the two sets of operators commute with each other:

$$[b^{\xi}, a^{\eta}] = 0, \qquad \xi, \eta \in \{+, -\}. \tag{4}$$

Journal of Physics: Conference Series **512** (2014) 012034 doi:10.1088/1742-6596/512/1/012034

Then the Lie superalgebra generated by the even elements $1, b^+, b^-$ and the odd elements a^+, a^- is known as the Heisenberg-Weyl superalgebra $\mathfrak{sh}(2|2)$ [13, 14]. It will be more convenient to work with a subalgebra of the enveloping algebra $\mathcal{U}(\mathfrak{sh}(2|2))$, namely the Lie superalgebra $\mathcal{S}$ with four odd basis elements

$$F^+ = a^+, \quad F^- = a^-, \quad Q^+ = b^+a^-, \quad Q^- = b^-a^+, \tag{5}$$

and even basis elements

$$E^+ = b^+, \quad E^- = b^-, \quad H = b^+b^- + a^+a^- \text{ and } 1. \tag{6}$$

The complete set of (anti)commutation relations among the elements of $\mathcal{S}$ is easy to determine [8]. The common Fock representation V of $\mathfrak{sh}(2|2)$ is generated by a vacuum vector $|0\rangle$ and by the relations $b^-|0\rangle = a^-|0\rangle = 0$. An orthonormal basis of V is given by the vectors

$$(a^+)^j \frac{(b^+)^m}{\sqrt{m!}}|0\rangle, \qquad m \in \{0,1,2,\ldots\},\ j \in \{0,1\}.$$

It will be convenient to write these vectors as $|n\rangle$, with $n = 0,1,2,\ldots$, so V can be identified with the Hilbert space $\ell^2(\mathbb{Z}_+)$:

$$|2m\rangle = \frac{(b^+)^m}{\sqrt{m!}}|0\rangle, \qquad |2m+1\rangle = a^+\frac{(b^+)^m}{\sqrt{m!}}|0\rangle. \tag{7}$$

Then the action of all basis elements of $\mathcal{S}$ on these basis vectors is easy to determine. In particular, for the odd elements one finds:

$$F^+|n\rangle = \frac{1}{2}(1+(-1)^n)|n+1\rangle, \qquad F^-|n\rangle = \frac{1}{2}(1-(-1)^n)|n-1\rangle,$$
$$Q^+|n\rangle = \frac{1}{2}(1-(-1)^n)\sqrt{\frac{n+1}{2}}|n+1\rangle, \qquad Q^-|n\rangle = \frac{1}{2}(1+(-1)^n)\sqrt{\frac{n}{2}}|n-1\rangle. \tag{8}$$

Note also that the usual conjugacy for the creation and annihilation operators leads to a natural $\star$-structure for $\mathcal{S}$, and the Fock representation is unitary with respect to this $\star$-structure.

In [8], we constructed an algebraic oscillator model with $\mathcal{S}$ as underlying Lie superalgebra. More precisely, using the reflection operator R with action $R|n\rangle = (-1)^n|n\rangle$, the Hamiltonian, position and momentum operators of the model are given by

$$\hat{H} = 2H + \frac{1}{2}R, \tag{9}$$
$$\hat{q} = \gamma F^+ + Q^+ + \gamma F^- + Q^-, \tag{10}$$
$$\hat{p} = \mathrm{i}\gamma F^+ + \mathrm{i}Q^+ - \mathrm{i}\gamma F^- - \mathrm{i}Q^-. \tag{11}$$

The spectrum of $\hat{H}$ in V coincides with that of the canonical oscillator, $\hat{H}|n\rangle = (n+\frac{1}{2})|n\rangle$, and the three self-adjoint operators satisfy (1). Herein, γ is a free parameter. Up to an arbitrary factor, (10) is the most general self-adjoint odd element in the Lie superalgebra $\mathcal{S}$. The main effort then goes to the spectral analysis of the position operator $\hat{q}$, which is represented by an infinite symmetric tridiagonal matrix in the ordered basis $\{|n\rangle,\ n = 0,1,2,\ldots\}$ of the Fock space V:

$$\hat{q} = \begin{pmatrix} 0 & \gamma & & & & & & & \\ \gamma & 0 & \sqrt{1} & & & & & & \\ & \sqrt{1} & 0 & \gamma & & & & & \\ & & \gamma & 0 & \sqrt{2} & & & & \\ & & & \sqrt{2} & 0 & \gamma & & & \\ & & & & \gamma & 0 & \sqrt{3} & & \\ & & & & & \sqrt{3} & 0 & \ddots & \\ & & & & & & \ddots & \ddots & \end{pmatrix}. \tag{12}$$

For $\gamma > 0$, such a matrix is a acobi matrix, and its spectral theory is related to orthogonal polynomials [15, 16, 17]. One should construct polynomials $p_n(x)$ of degree n in x, with $p_{-1}(x) = 0$, $p_0(x) = 1$, and recurrence relation determined by the matrix (12),

$$\begin{aligned} xp_{2n}(x) &= \sqrt{n}p_{2n-1}(x) + \gamma p_{2n+1}(x), \\ xp_{2n+1}(x) &= \gamma p_{2n}(x) + \sqrt{n+1}p_{2n+2}(x), \qquad (n = 0, 1, 2, \ldots). \end{aligned} \tag{13}$$

If certain conditions are satisfied, the support of the weight function $w(x)$ for these polynomials determines the spectrum of the operator $\hat{q}$. Furthermore, for a real value x belonging to this support, the corresponding formal eigenvector of $\hat{q}$ is given by

$$v(x) = \sum_{n=0}^{\infty} p_n(x) \, |n\rangle. \tag{14}$$

In [8] we have shown that the solution of this recurrence relation leads to Charlier polynomials $C_n(x; a)$, defined by [18, 19, 20]:

$$C_n(x; a) = {}_2F_0 \left(\begin{matrix} -n, -x \\ - \end{matrix} ; -\frac{1}{a} \right). \tag{15}$$

More particularly, for $\gamma \neq 0$, the solution of (13) is

$$p_{2n}(x) = \frac{(-\gamma)^n}{\sqrt{n!}} C_n(x^2; \gamma^2), \qquad p_{2n+1}(x) = -\frac{(-\gamma)^{n-1}}{\sqrt{n!}} \, x \, C_n(x^2 - 1; \gamma^2). \tag{16}$$

Then, using the orthogonality of Charlier polynomials [18], one deduces that the polynomials $p_n(x)$ satisfy a discrete orthogonality relation:

$$\sum_{x \in \mathbb{S}} w(x) p_n(x) p_m(x) = \mathrm{e}^{\gamma^2} \delta_{mn}, \tag{17}$$

where

$$\mathbb{S} = \{\pm\sqrt{k} \mid k \in \mathbb{Z}_+\} = \{\ldots, -\sqrt{3}, -\sqrt{2}, -1, 0, 1, \sqrt{2}, \sqrt{3}, \ldots\}, \tag{18}$$

and where the weight function is given by

$$w(x) = \frac{1}{2} \frac{\gamma^{2k}}{k!} \qquad \text{for } x = \pm\sqrt{k} \quad (k = 1, 2, 3, \ldots) \tag{19}$$

and by $w(x) = 1$ for $x = 0$. So (18) gives the position spectrum.

Journal of Physics: Conference Series **512** (2014) 012034 doi:10.1088/1742-6596/512/1/012034

It will be convenient to normalize the polynomials, and consider the corresponding orthonormal functions $\tilde{p}_n(x)$ satisfying $\sum_{x\in\mathbb{S}} \tilde{p}_m(x)\tilde{p}_n(x) = \delta_{mn}$. Then the normalized eigenvectors of $\hat{q}$, for an eigenvalue $x \in \mathbb{S}$, are given by

$$\tilde{v}(x) = \sum_{n=0}^{\infty} \tilde{p}_n(x)|n\rangle. \tag{20}$$

In such an expression, the overlap between the normalized $\hat{q}$-eigenvectors (20) and the $\hat{H}$-eigenvectors $|n\rangle$ can be interpreted as the position wavefunctions and will be denoted by $\varphi_n^{(\gamma)}(x)$. So these are discrete position wavefunctions, with support $\mathbb{S}$, given by

$$\varphi_n^{(\gamma)}(x) = \tilde{p}_n(x), \tag{21}$$

where x belongs to $\mathbb{S}$. The parameter γ originates from the freedom in choosing $\hat{q}$, in (10). In Figure 1 of [8], one can find example plots of $\varphi_n^{(\gamma)}(x)$, for certain values of n and of γ. So these are discrete plots, with the wavefunction consisting of "dots" positioned at $\pm\sqrt{k}$ ($k = \{0, 1, 2, 3, \ldots\}$). The shape of these wavefunctions depends on the value of γ. For $\gamma < 1$, the shape is close to a discrete version of the normalized Hermite wavefunctions of the canonical oscillator. As γ increases ($\gamma > 1$), the shape is closer to a discrete version of the paraboson wavefunctions. Note that these observation are just descriptive. For actual limits of the wavefunctions we refer to [8].

An additional remark relates the current structure also to the Lie superalgebra $\mathfrak{osp}(3|2)$. Indeed, consider the superalgebra generated by the same elements $a^\pm$ and $b^\pm$ satisfying (3) and (2), but with a different grading: $a^\pm$ are even elements, $b^\pm$ are odd elements, and they mutually anticommute:

$$\{b^\xi, a^\eta\} = 0, \qquad \xi, \eta \in \{+, -\}. \tag{22}$$

It is known that the superalgebra generated by these elements and relations is the enveloping algebra of $\mathfrak{osp}(3|2)$ [21]. So for $\mathfrak{osp}(3|2)$ one can construct a similar Fock space with the same basis vectors (7). In this representation space, the operator $\gamma a^+ + \gamma a^- + b^+a^- + a^+b^-$ has the same action as (10), so its spectrum and spectral analysis is the same. Note, however, that we are now working in a completely different superalgebra, namely $\mathfrak{osp}(3|2)$. In this context, the operator $\gamma a^+ + \gamma a^- + b^+a^- + a^+b^-$ does not have a direct meaning as a position operator: in fact, it is a linear combination of both even and odd elements of $\mathfrak{osp}(3|2)$.

To conclude this section, note that we are certainly not the first to relate Charlier polynomials to quantum oscillators. They have been related to the Weyl algebra in [22, 23], and appear as overlap coefficients between eigenvectors of a number operator and a "shifted" number operator [23]. In [24] some finite shift operators were defined closely related to $\hat{q}$ and $\hat{p}$, with a spectral analysis also in terms of Charlier polynomials. Still, it should be noticed that the operators (10)-(11) studied here are different from the ones investigated earlier in the literature, and the solution in terms of Charlier polynomials uses quite different properties of these polynomials.

3. The $\mathfrak{sl}(2|1)$ oscillator model

The example described in this section has been analyzed in detail in [9]. The idea is the same as in the previous section, but now we work with a different Lie superalgebra, which turns out to be more fundamental. This superalgebra is $\mathfrak{sl}(2|1)$, which has a basis consisting of four odd (or fermionic') elements F^+, F^-, G^+, G^- and four even (or bosonic') elements H, E^+, E^-, Z,

Journal of Physics: Conference Series **512** (2014) 012034 doi:10.1088/1742-6596/512/1/012034

given by

$$F^+ = e_{32},\ G^+ = e_{13},\ F^- = e_{31},\ G^- = e_{23}, \tag{23}$$

$$H = \frac{1}{2}(e_{11} - e_{22}),\ E^+ = e_{12},\ E^- = e_{21},\ Z = \frac{1}{2}(e_{11} + e_{22}) + e_{33} \tag{24}$$

in terms (graded) 3×3 Weyl matrices e_{ij}. From this matrix form, the Lie superalgebra brackets (i.e. commutators and anti-commutators) can be deduced (see [25, p. 261], [26] or [12]).

$\mathfrak{sl}(2|1)$ has both finite and infinite-dimensional irreducible representations, and the unitarity of these representations depends on the choice of a $\star$-structure (or an adjoint operation) on the Lie superalgebra. Here this is:

$$Z^\dagger = Z, \quad H^\dagger = H, \quad (E^\pm)^\dagger = -E^\mp, \quad (F^\pm)^\dagger = \mp G^\mp, \quad (G^\pm)^\dagger = \pm F^\mp, \tag{25}$$

and for this $\star$-structure the finite-dimensional representations are not unitary. But one can construct a class of infinite-dimensional positive discrete series representations of $\mathfrak{sl}(2|1)$, Π_β, labeled by $\beta > 0$, which are unitary [9]. The representation space is $\ell^2(\mathbb{Z}_+)$ equipped with an orthonormal basis $|\beta, n\rangle$ $(n = 0, 1, 2, \ldots)$, i.e. $\langle\beta, m|\beta, n\rangle = \delta_{m,n}$. For the actions of the $\mathfrak{sl}(2|1)$ basis elements on these vectors, it is handy to use the following "even" and "odd" functions, defined on integers n:

$$\mathcal{E}(n) = 1 \text{ if } n \text{ is even and } 0 \text{ otherwise}; \qquad \mathcal{O}(n) = 1 \text{ if } n \text{ is odd and } 0 \text{ otherwise}. \tag{26}$$

Let us just give the actions of the odd generators here:

$$F^+|\beta, n\rangle = \mathcal{E}(n)\sqrt{\beta + \frac{n}{2}}\,|\beta, n+1\rangle, \qquad F^-|\beta, n\rangle = \mathcal{E}(n)\sqrt{\frac{n}{2}}\,|\beta, n-1\rangle,$$

$$G^+|\beta, n\rangle = \mathcal{O}(n)\sqrt{\frac{n+1}{2}}\,|\beta, n+1\rangle, \qquad G^-|\beta, n\rangle = -\mathcal{O}(n)\sqrt{\beta + \frac{n-1}{2}}\,|\beta, n-1\rangle. \tag{27}$$

The verification that this leads to a unitary irreducible representation is given in [9].

ust as before, we can construct a model by identifying three self-adjoint operators

$$\hat{H} = 2H + \frac{1}{2} - \beta, \tag{28}$$

$$\hat{q} = F^+ + \gamma G^+ - G^- + \gamma F^-, \tag{29}$$

$$\hat{p} = \mathrm{i}(F^+ + \gamma G^+ + G^- - \gamma F^-). \tag{30}$$

In the representation Π_β, the spectrum of $\hat{H}$ coincides with that of the canonical oscillator, and the three operators satisfy (1). Again, in (29), γ is a free parameter; the choice of the position operator $\hat{q}$ is such that it is (up to a factor) the most general self-adjoint odd element of $\mathfrak{sl}(2|1)$. Then the form of $\hat{p}$ follows from (1). In the (ordered) basis $\{|\beta, n\rangle,\ n = 0, 1, 2, \ldots\}$, the operator $\hat{q}$ is represented by an infinite symmetric tridiagonal matrix M_q:

$$M_q = \begin{pmatrix} 0 & R_0 & & & & \\ R_0 & 0 & S_1 & & & \\ & S_1 & 0 & R_1 & & \\ & & R_1 & 0 & S_2 & \\ & & & S_2 & 0 & \ddots \\ & & & & \ddots & \ddots \end{pmatrix}, \tag{31}$$

where

$$R_n = \sqrt{\beta + n}, \qquad S_n = \gamma\sqrt{n} \qquad (n = 0, 1, 2, \ldots). \tag{32}$$

For $\gamma > 0$, such a matrix is a acobi matrix, and its spectral theory has been discussed in [9]. Like in the previous section, one should first construct polynomials with a recurrence relation governed by the matrix (31), i.e. with $p_{-1}(x) = 0$, $p_0(x) = 1$, and

$$\begin{aligned} xp_{2n}(x) &= S_n p_{2n-1}(x) + R_n p_{2n+1}(x), \\ xp_{2n+1}(x) &= R_n p_{2n}(x) + S_{n+1} p_{2n+2}(x), \qquad (n = 0, 1, 2, \ldots). \end{aligned} \tag{33}$$

Such polynomials are orthogonal for some positive weight function $w(x)$, and the spectrum of M_q (or of $\hat{q}$) is the support of this weight function. Furthermore, for a real value x belonging to this support, the corresponding formal eigenvector of $\hat{q}$ is given by

$$v(x) = \sum_{n=0}^{\infty} p_n(x)\, |\beta, n\rangle. \tag{34}$$

In the current case, the solution of (33) depends on the value of γ. When $\gamma^2 \neq 1$, the solution of the recurrence relations (33) is given by [9]

$$\begin{aligned} p_{2n}(x) &= (-\gamma)^{-n} \sqrt{\frac{(\beta)_n}{n!}}\; {}_2F_1\left({-n, \frac{x^2}{1-\gamma^2} \atop \beta} ; 1 - \gamma^2 \right), \\ p_{2n+1}(x) &= x(-\gamma)^{-n} \sqrt{\frac{(\beta+1)_n}{n!\beta}}\; {}_2F_1\left({-n, \frac{x^2}{1-\gamma^2} + 1 \atop \beta + 1} ; 1 - \gamma^2 \right). \end{aligned} \tag{35}$$

When $\gamma^2 = 1$, the solution is of a different type and given by [9]

$$\begin{aligned} p_{2n}(x) &= (-\gamma)^{n} \sqrt{\frac{(\beta)_n}{n!}}\; {}_1F_1\left({-n \atop \beta} ; x^2 \right), \\ p_{2n+1}(x) &= x(-\gamma)^{n} \sqrt{\frac{(\beta+1)_n}{n!\beta}}\; {}_1F_1\left({-n \atop \beta + 1} ; x^2 \right). \end{aligned} \tag{36}$$

In the first case, these polynomials can be related to the Meixner polynomial $M_n(k; \beta, c)$ of degree n in k, with parameters β and c [18, 19, 20]:

$$M_n(k; \beta, c) = {}_2F_1\left({-n, -k \atop \beta} ; 1 - \frac{1}{c} \right), \tag{37}$$

and these satisfy a discrete orthogonality. In the second case ($\gamma^2 = 1$) these polynomials are related to Laguerre polynomials, and their orthogonality is continuous. For technical reasons, the case $\gamma^2 \neq 1$ should be split in two subcases ($|\gamma| < 1$ and $|\gamma| > 1$), and we shall consider only one of them. For $|\gamma| > 1$, the polynomials $p_n(x)$ satisfy a discrete orthogonality relation:

$$\sum_{x \in \mathbb{S}_1} w(x) p_n(x) p_m(x) = \left(\frac{\gamma^2}{\gamma^2 - 1} \right)^{\beta} \delta_{mn}, \tag{38}$$

where

$$\mathbb{S}_1 = \{ \pm\sqrt{\gamma^2 - 1}\sqrt{k} \mid k \in \mathbb{Z}_+ \}, \tag{39}$$

and where the weight function is given by

$$w(x) = \frac{1}{2}(1+\delta_{k,0})\frac{(\beta)_k}{k!}\gamma^{-2k} \qquad \text{for } x = \pm\sqrt{\gamma^2-1}\sqrt{k} \quad (k=0,1,2,\ldots). \tag{40}$$

So in this case, the spectrum of the position operator $\hat{q}$ is discrete and given by (39). For $|\gamma| = 1$, the orthogonality relation of Laguerre polynomials implies that the polynomials $p_n(x)$ satisfy a continuous orthogonality relation:

$$\int_{-\infty}^{+\infty} w(x)p_n(x)p_m(x)dx = \Gamma(\beta)\delta_{mn}, \tag{41}$$

where

$$w(x) = \mathrm{e}^{-x^2}|x|^{2\beta-1}. \tag{42}$$

For $|\gamma| < 1$ the polynomials satisfy again a discrete orthogonality relation, deduced from the one for Meixner polynomials.

For an interpretation as position wavefunction, it will be useful to normalize the corresponding polynomials $p_n(x)$, in such a way that the normalized ones $\tilde{p}_n(x)$ satisfy $\sum_{x\in\mathbb{S}_1}\tilde{p}_m(x)\tilde{p}_n(x) = \delta_{mn}$ (or $\int \tilde{p}_m(x)\tilde{p}_n(x)dx = \delta_{mn}$ in case $\gamma^2 = 1$), and then

$$\tilde{v}(x) = \sum_{n=0}^{\infty} \tilde{p}_n(x)\,|\beta, n\rangle. \tag{43}$$

It will be convenient to write these coefficients as

$$\tilde{v}(x) = \sum_{n=0}^{\infty} \Phi_n^{(\beta,\gamma)}(x)\,|\beta, n\rangle. \tag{44}$$

Herein, x belongs to the spectrum of $\hat{q}$, and in the notation of the coefficients (or overlaps) $\Phi_n^{(\beta,\gamma)}(x)$ we have emphasized the dependence on the representation parameter β and the model parameter γ.

The overlaps $\Phi_n^{(\beta,\gamma)}(x)$ have again an interpretation as position wavefunctions. In this $\mathfrak{sl}(2|1)$ model, some interesting phenomena occur. For $\gamma^2 > 1$, these wavefunctions satisfy a discrete orthogonality, and their plots consist of discrete "dots" at positions determined by the support (39). Note that these dots get "closer" as γ gets closer to 1. When $\gamma^2 = 1$, the support becomes $\mathbb{R}$, and we get continuous wavefunctions. Plots of these wavefunctions can be found in Figures 1 and 2 of [9]. Note that the model parameter determines the type of the spectrum: discrete or continuous; and when discrete γ also determines how close the dots of the spectrum are to each other. The representation parameter β determines the shape of the wavefunctions: for $\beta = 1/2$ the shape (whether discrete or continuous) is close to that of the wavefunction of the canonical oscillator, for $\beta > 1/2$ the shape is close to that of the paraboson oscillator. More precisely, for $\gamma = 1$ and arbitrary $\beta > 0$, the wavefunctions $\Phi_n^{(\beta,1)}(x)$ *coincide* with the well known wavefunctions of the paraboson oscillator [9]. Since the canonical oscillator (boson oscillator) is a special case of the paraboson oscillator (namely with $\beta = 1/2$), one finds

$$\Phi_n^{(1/2,1)}(x) = \frac{1}{2^{n/2}\pi^{1/4}\sqrt{n!}}\mathrm{e}^{-x^2/2}H_n(x), \tag{45}$$

with $H_n(x)$ the common Hermite polynomial.

It should be mentioned that this relation to the paraboson oscillator (and hence to the canonical oscillator) also follows from the underlying algebra. Indeed, $\mathfrak{sl}(2|1)$ contains the paraboson algebra $\mathfrak{osp}(1|2)$ as a subalgebra, and the representations Π_β correspond to irreducible representations of $\mathfrak{osp}(1|2)$. In this context, the position operator for the paraboson oscillator equals the expression (29) with $\gamma = 1$. This, by the way, explains why we have chosen – in the $\mathfrak{sl}(2|1)$ case – as position operator the most general self-adjoint odd element of the superalgebra.

4. Concluding remarks

We have presented here two algebraic models for the quantum oscillator based on Lie superalgebras. The first model comes from the Lie superalgebra $\mathfrak{sh}(2|2)$, and is rather a toy model. It is easy to analyze, and serves mainly as supporting example to tackle more complicated models. For the $\mathfrak{sh}(2|2)$ case, there is one model parameter γ which arises by allowing the "most general" expression for a possible position operator. There is only one natural unitary representation (the Fock space), so this makes the spectral analysis not too difficult.

The second model comes from the Lie superalgebra $\mathfrak{sl}(2|1)$, and is more fundamental. For the $\mathfrak{sl}(2|1)$ case, there is again one model parameter γ by allowing the "most general" expression for a possible position operator. But now there is a natural class of unitary representations, each representation Π_β characterized by a positive real number β. The spectral analysis of the position operator and its corresponding position wavefunctions is more subtle. The model parameter γ plays an important role in the support of the position wavefunctions (which can be discrete or continuous), and the representation parameter β plays a role in the "shape" of the wavefunctions (closer to the canonical case or to the paraboson case). The $\mathfrak{sl}(2|1)$ model includes, as special cases (i.e. for special values of the parameters γ and β) the paraboson oscillator and the canonical oscillator.

As a final remark, let us mention that the current examples are both sufficiently simple to compute other relevant physical quantities, such as the action of $[\hat{q},\hat{p}]$, in the relevant representations. We refer to [9, 8] for these computations.

References

[1] Wigner E P 1950 *Phys. Rev.* **77** 711
[2] Green H S 1953 *Phys. Rev.* **90** 270
[3] Mukunda N, Sudarshan E C G, Sharma J K and Mehta C L 1980 *J. Math. Phys.* **21** 2386
[4] Jafarov E I, Lievens S and Van der Jeugt J 2008 *J. Phys. A* **41** 235301
[5] Atakishiyev N M, Pogosyan G S, Vicent L E and Wolf K B 2001 *J. Phys. A* **34** 9381
[6] Jafarov E I, Stoilova N I and Van der Jeugt J 2011 *J. Phys. A* **44** 265203
[7] Jafarov E I, Stoilova N I and Van der Jeugt J 2011 *J. Phys. A* **44** 355205
[8] Jafarov E I and Van der Jeugt J 2013 *J. Math. Phys.* **54** 103506
[9] Jafarov E I and Van der Jeugt J 2012 *J. Phys. A* **45** 485201
[10] Atakishiyev N M, Pogosyan G S and Wolf K B 2005, *Phys. Part. Nuclei* **36** 247
[11] Klimyk A U 2006 *Ukr. J. Phys.* **51** 1019
[12] Jafarov E I and Van der Jeugt J 2012 *J. Phys. A* **45** 275301
[13] Alvarez-Moraga N and Hussin V 2004 *Int. J. Theor. Phys.* **43** 179
[14] D'Hoker E and Vinet L 1984 *Phys. Lett. B* **137** 72
[15] Berezanskiĭ Yu M 1968 *Expansions in eigenfunctions of selfadjoint operators* (Providence: American Mathematical Society)
[16] Koelink H T and Van der Jeugt J 1998 *SIAM J. Math. Anal.* **29** 794
[17] Koelink H T 2004 *Laredo Lectures on Orthogonal Polynomials and Special Functions* eds. R Alvarez-Nodarse, F Marcellán and W Van Assche (New York: Nova Science Publishers), p 45
[18] Koekoek R, Lesky P A and Swarttouw R F 2010 *Hypergeometric orthogonal polynomials and their q-analogues* (Berlin: Springer-Verlag)
[19] Ismail M E H 2005, *Classical and quantum orthogonal polynomials in one variable* (Cambridge: Cambridge University Press)
[20] Andrews G E, Askey R and Roy R 1999 *Special functions* (Cambridge: Cambridge University Press)
[21] Palev T D 1993 *J. Math. Phys.* **34** 4872
[22] Floreanini R, LeTourneux J and Vinet L 1993 *Ann. Physics* **226** 331
[23] Vinet L and Zhedanov A 2011 *J. Phys.: Conf. Series* **284** 012060
[24] Szafraniec F H 1995 *Operator Theory: Advances and Applications* vol 80, eds I Gohberg and H Langer (Basel: Birkhäuser) p 266
[25] Frappat L, Sciarrino A and Sorba P 2000 *Dictionary on Lie Algebras and Superalgebras* (London: Academic Press)
[26] Scheunert M, Nahm W and Rittenberg V 1977 *J. Math. Phys.* **18** 155

Quantum Control: discovered, repeated and reformulated ideas mark the progress

Bogdan Mielnik

Depto. de Física, Centro de Investigación y de Estudios Avanzados del IPN,
A.P. 14-740, México, DF 07000, Mexico. e-mail: bogdan@fis.cinvestav.mx

Keywords: quantum control, oscillator kicks, soft operations.

Abstract

The dynamical manipulation of physical systems is necessary to explain the structure of physical theories. However, the manipulation techniques in quantum theories progress slowly, giving still no certainty that all theoretically described states and all unitary transformations can be indeed achieved. Below, we report some elements of idealized control of the evolution processes in time dependent quadratic potentials, with attention focused on the abstract δ-pulses of the elastic forces. Some of them, if empirically approximated, could become essential for a sequence of difficult interpretation problems.

PACS numbers: 03.65.Fd, 03.67.Ta

1 Introduction

One of nontrivial challenges of the present day quantum theories are the dynamical manipulations of particle systems by variable external conditions. While multiple experiences already exist, they are still incomplete. However, if the physical theories were at all created it is only because the physical systems can be actively controlled by flexible external forces making possible to perform experiments. Indeed, without the interaction with variable surrounding, not even the theory of freely propagating particles in an empty space could be formulated.

In spite of the limited scope of the dynamical operations, the modern quantum theories show spectecular achievements (the description of the anomalous magnetic moments, the probable detection of the Higgs boson, etc.). However, almost all methods applied are borrowed from the theory of stationary oscillations or just static fields. The mathematical concepts too are repetitive, with pure states described always by vectors in the linear (Hilbert) spaces and the time evolution (in absence of dissipation) given by linear, unitary

Journal of Physics: Conference Series **512** (2014) 012035 doi:10.1088/1742-6596/512/1/012035

operations. The picture persists in the theories of coupled systems and quantum field theories (QFT) where the states are represented in tensor product spaces - in an obvious intent to conserve the linearity of basic laws at the cost of multiplying the number of variables.

Yet, some doubts remain. Can all theoretically predicted states be indeed created [1]? Can all unitary operations be indeed achieved (or at least approached) by physical evolution? Can all measurements described by self-adjoint operators be truly performed? (The close interrelation between the last two questions can be noticed [2]).

One of the main obstacles in checking the 'obligatory beliefs' of quantum theories, are the perturbative complications and the difficulties of extrapolating to 'the little' and to 'the great' [3]. What could help are the exact algorithms, though they still require exceptional effort and patience [4, 5, 6, 7, 8, 9, 10]. The purpose of this talk is to show that even the simplest cases of the exactly soluble evolution, for particles in time dependent quadratic potentials, have some consequences which are not generally known. To them belong the possibility manipulating or even inverting the free evolution (i.e. the 'quantum time machine'), modeling the repulsive oscillator pulses by the attractive ones, and programming the operational schemes which challenge the time-energy uncertainty [11].

2 Quadratic Hamiltonians: the simplest approach

The quantum theory obeys the *correspondence principle*, turning classical for $\hbar \to 0$. However, in some dynamical theories the evolution laws are shared by classical and quantum systems without the need of any limiting transition. This occurs for the non-relativistic quadratic Hamiltonians - where the motion equations for canonical variables are exactly the same in classical and quantum levels. While the fundamental problems were carefully revised by Barry Simon [12], what is still worth attention are the algebraic aspects of the evolution generated by the quadratic Hamiltonians:

$$H(t) = \frac{1}{2}\Sigma\beta_{k,l}(t)Q_kQ_l \tag{1}$$

where $(Q_1...Q_{2n}) = (q_1q_2...q_n, p_1, p_2, ...p_n)$ are the canonical position and momentum variables and $\beta_{k,l}(t) = \beta_{l,k}(t) \in \mathbb{R}$ the time dependent coefficients defining the external forces. While the most general dynamical operations of these Hamiltonians remain of significant interest, the subject of the present report will be some exact solutions of the evolution problem (1) for $n \leq 3$ (i.e., taking place in $\mathbb{R}^n, n \leq 3$), with the special attention dedicated to the quadratic Hamiltonians in one space dimension:

$$H(t) = \frac{p^2}{2} + \beta(t)\frac{q^2}{2} \tag{2}$$

typically representing the separable parts of the evolution process in the traditional, quadrupole ion traps. The only challenge we shall attend is strictly combinatorial: how should one program the oscillations of the c-number amplitude $\beta(t)$, to generate some useful quantum operations? On purely mathematical level, the problem is elementary, though not always simple.

In the classical theory and for nonsingular $\beta(t)$ the canonical equations lead to the linear evolution of the canonical variables, represented by the 2×2 symplectic *evolution matrices* $u(t, t_0)$:

$$\begin{pmatrix} q(t) \\ p(t) \end{pmatrix} = u(t, t_0) \begin{pmatrix} q(t_0) \\ p(t_0) \end{pmatrix} \tag{3}$$

given by the matrix equations:

$$\frac{d}{dt} u(t, t_0) = \Lambda(t) u(t, t_0) \qquad \frac{d}{dt_0} u(t, t_0) = -u(t, t_0) \Lambda(t_0) \tag{4}$$

with:

$$\Lambda(t) = \begin{pmatrix} 0 & 1 \\ -\beta(t) & 0 \end{pmatrix} \tag{5}$$

and:

$$u(t, \theta) u(\theta, t_0) = u(t, t_0) \qquad u(t_0, t_0) = \mathbb{1} \tag{6}$$

In quantum case, the corresponding evolution operators $U(t, t_0)$, are given by:

$$i \frac{d}{dt} U(t, t_0) = H(t) U(t, t_0), \qquad i \frac{d}{dt_0} U(t, t_0) = -U(t, t_0) H(t_0) \tag{7}$$

with:

$$U(t, \theta) U(\theta, t_0) = U(t, t_0) \qquad U(t_0, t_0) = 1 \tag{8}$$

defining the evolution of the observables q and p in Heisenberg's picture (the *"Heisenberg's trajectory"*), generating the same symplectic matrices (3-5):

$$U(t, t_0)^\dagger \begin{pmatrix} q \\ p \end{pmatrix} U(t, t_0) = u(t, t_0) \begin{pmatrix} q \\ p \end{pmatrix} \tag{9}$$

though now q, p are operators with $[q, p] = i$ (we put for simplicity $\hbar = 1$). The exact formal correspondence of the classical and quantum cases was explored by multiple research groups, quite frequently without knowing of each other. In all these approaches, a useful observation is

Proposition 1. In absence of spin or any additional degrees of freedom, each unitary evolution operator $U(t, t_0)$ in $L^2(\mathbb{R})$ generated by the quadratic, time dependent Hamiltonians (2) is determined up to a phase factor by the classical motion trajectories.

Journal of Physics: Conference Series **512** (2014) 012035 doi:10.1088/1742-6596/512/1/012035

Proof. Instead of any sophisticated arguments, notice that if two unitary operators U_1 and U_2 produce the same transformation of the canonical variables i.e., $U_1^\dagger q U_1 = U_2^\dagger q U_2$ and $U_1^\dagger p U_1 = U_2^\dagger p U_2$, then $U_1 U_2^\dagger$ commutes with both q and p. Hence, it commutes also with any function of q and p. Since in $L^2(\mathbb{R})$ the functions of q and p generate an irreducible algebra, then $U_1 U_2^\dagger$ must be a c-number and since it is unitary, it can be only a phase factor, $U_1 U_2^\dagger = e^{i\varphi} \Rightarrow U_1 = e^{i\varphi} U_2$ where $\varphi \in \mathbb{R}$ [13, 12, 14, 15]. Note that the same statement holds for the systems (1) with any number of canonical variables. □

Any two unitary operators which differ only by a c-number phase can generate different transformations of the *state vectors*, but the same transformation of *quantum states*, so we shall call them *equivalent* and write $U_1 \equiv U_2$. The fact that they might wear different phase factors $e^{i\varphi}$ can be of interest for the linear representation theory studying the *metaplectic groups* [16, 17, 18] but does not affect the operations on physical states, the only subject of our interest. From now on, our report will thus concern the *manipulations* of quantum states, skipping almost all contributions to the group representation as well as the pictures of "passive" unitary transformations (meaning the change of the observation frame). Moreover, when describing the non-relativistic, time dependent Hamiltonians our review (with little exceptions) will consider only the evolution generated by physical fields, which, at least in principle can be created by some controllable external conditions. We thus omit the reports on hypothetical media with the time dependent dielectric or magnetic constants, or (with one exception) the time dependent mass. The so defined subject might seem narrow, but precisely, most interesting for the problem of quantum manipulation.

Below, our attention will be focused on the evolution caused by quadratic, time dependent Hamiltonians (2). The possibility of deducing the quantum state evolution from the transformations of the canonical variables (9) permits one to program ample families of the classical/quantum control operations. To classify them, the algebraic types of matrices (3-9) are quite essential. Since every evolution matrix $u = u(t, t_0)$ is symplectic ($\mathrm{Det}u = 1$) its algebraic type is defined just by one invariant $\mathrm{Tr}u$. The characteristic equation:

$$\mathrm{D}(\lambda) = \mathrm{Det}(\lambda - u) = \lambda^2 - \mathrm{Tr}(u)\lambda + 1 = 0 \tag{10}$$

has two roots $\lambda_\pm = \frac{1}{2}\,\mathrm{Tr}\,u \pm i\sqrt{\Delta}$, where $\Delta = 1 - \frac{1}{4}(\mathrm{Tr}\,u)^2$, permitting to distinguish three types of evolution matrices:

(I) If $|\,\mathrm{Tr}(u)| < 2$, then u has two complex eigenvalues $\lambda_\pm = e^{\pm i\sigma}$ where $0 \neq \sigma \in \mathbb{R}$

(II) If $|\,\mathrm{Tr}(u)| = 2$ then u is on the threshold: $\lambda_+ = \lambda_- = \pm 1$

(III) If $|\,\mathrm{Tr}(u)| > 2$, then u has two real eigenvalues $\lambda_\pm = e^{\pm\sigma}$ where $0 \neq \sigma \in \mathbb{R}$

The classification turns specially relevant if the function $\beta(t)$ in (2) is periodic, $\beta(t + T) = \beta(t)$, defining a *Floquet process*. The (crucial) *Floquet matrices* $u(t_0 + T, t_0)$ describe then the repeated evolution incidents. One easily shows that their types do not depend on t_0. Choosing $t_0 = 0$ and denoting for simplicity $u(t) = u(t, 0)$ one sees that the results of the evolution in the sequence of expanding intervals $[0, nT]$ are given by repetitions of $u(T)$, i.e., $u(nT) = u(T)^n$. Now, if $u(T)$ is in the class (**I**) the evolution is oscillatory. The eigenvectors of $u(T)$ define a pair of variables $A_\pm$ (the global *creation* and *annihilation* operators for the entire periodicity intervals) which for $t = nT$ perform just the phase rotation $U(t)^\dagger A_\pm U(t) = \mathrm{e}^{\pm i\sigma t} A_\pm$. However, if $u(T)$ is in the class (**III**) then the equilibrium is lost: the eigenvectors of $u(T)$ define now two real canonical variables $A_\pm$ which are multiplied by $\mathrm{e}^{\pm\sigma t}$, $(0 \neq \sigma \in \mathbb{R})$ i.e., endlessly squeezed or endlessly amplified as $t = nT \to \infty$. In turn, the threshold cases (**II**) offer some exceptional manipulation techniques which might deserve attention.

3 The early branches of quantum control

We still comment the evolution caused by the Hamiltonians (2). Even if elementary, the problem has a considerable history. The subject was extensively studied already in 1967-1969 by Lewis and Riesenfeld [19, 20, 21] via the concept of *invariants*, then by Malkin, Manko, Trifonov [22, 23] leading later to an important techniques of *quantum tomography* [7, 8, 10]. The subsequent steps are no less interesting; they form a colorful story of achievements and obstacles. The concepts of *invariants* in [19, 20], though suggestive, were not immediate to apply. As the matter of fact, the problem seemed to obey some law of 'difficulty conservation'. In order to solve the initial 2-nd order differential eq. for the canonical coordinate q, i.e.: $\ddot{q} + \beta(t)q(t) = 0$, the techniques of invariants required the solution of an equivalent 2-nd order equation (see eqs. (45) or (101) in [20])

$$\frac{d^2}{dt^2}\rho + \beta(t)\rho + \frac{1}{\rho^3} = 0 \tag{11}$$

not simpler than the original one.[1] A stumble was also, that the model used by Lewis and Riesenfeld [20] while mathematically correct, was unphysical. It described the free motion of a charged particle along the z-axis, while on the orthogonal plane x, y the evolution was obeying the homogeneous time dependent magnetic field $\mathbf{B}(t)$ parallel to z and the time dependent scalar potential $\phi = e\eta(t)(x^2 + y^2)$). It was enough to apply the Laplace operator Δ to see, that the electromagnetic model in [20] involved a homogeneous, but time dependent charge density all over the space (acknowledged by the authors), awaking some uneasy thoughts about the charge conservation.[2].The same *scenario* was adopted later by Baseia et al [24], apparently without perceiving any problem. Of course, one can

[1] By consulting [20] one can even get an impression that the 'invariants' complicate rather than facilitate the solution.

[2] Indeed, it is not an accident that the time dependent quadratic potentials in the ion traps in 3D have the form $\eta(t)(x^2 + y^2 - 2z^2)$ but not $\eta(t)(x^2 + y^2)$

Journal of Physics: Conference Series **512** (2014) 012035 doi:10.1088/1742-6596/512/1/012035

agree that the creative achievement in [19, 20] consists not so much in the realistic control models but rather in understanding better the mathematical structure. Moreover, their works initiated the whole manipulation trend. (One can notice though, that the authors of subsequent papers, while developing the method, abstain from assuming the unphysical scenarios [22, 23]).

A historical anecdote is also that Lewis and Riesenfeld in their eq. (11) arrived (without knowing) at the transformed Schrödinger equation discovered already by Milne in 1930 [25], used to determine the energy spectra of the Schrödinger electron [26, 27] the works widely known (with 269, 462 and 651 citations respectively) in an epoch when the computer techniques were not yet available. In 1950 the same equation was again discovered in a short (1/2 page) note by Pinney [28], without presenting any application, but since that time (perhaps in honor of Pinney) the eq. (11) was known as the Milne-Pinney equation. Yet, the *quid pro quo* does not end up here.

After the work of Maamache [29] it was suddenly discovered that the Ukrainian mathematician Ermakov from Kiev University, has written around 1880 a handbook for students presenting the transformed versions the 2-nd order ordinary differential equation $\frac{d^2}{dt^2}\rho + \beta(t)\rho = 0$. One of them was precisely the Milne-Pinney eq. (11) used later by Lewis and Riesenfeld. Of course, Ermakov could not know that he made a work in the Schrödinger's quantum mechanics; neither Milne, Pinney, Lewis and Riesenfeld could know that in some distant past, behind the curtain of the Cyrillic alphabet, somebody had already written the eq. (11). Soon, a sequence of papers started to "rewrite the past" (*not* in the sense of Orwell [30], since Ermakov indeed existed!). Now, however, the name of Milne completely disappeared, though Pinney curiously survived: in a recent trend of publications the authors tell about "Ermakov-Pinney" equation. Moya and Leach [31] tell delicately that the Ermakov invariant was "revisited" by Lewis in 1967 and hence, they call it the "Ermakov-Lewis invariant". Yet, in all this labyrinth of results, the inspirations of Lewis and Riesenfeld still occupy a key position. Meanwhile, the dynamical control theory climbed already on different branches.

4 The evolution controlled by sharp pulses

In spite of the development of invariants [19, 20, 31] and the abundant works on group representations, the exact solutions of (2) were still absent. In their 1973 paper [23] Malkin, Man'ko and Trifonov modified the method of [20], developing the step-by-step approximations in terms of the *adiabatic invariants.* Yet, they admit: "*Attractive as these results may seem at first sight, there is, however, one difficulty: The point is that the exact invariants are expressed in terms of the solutions of linear differential equations (5) or (38), but neither (5) nor (38) can be solved exactly for every $R(t)$ or $\Omega(t)$*" ([23], p.580, l. 4-9 of Sec.III). It was henceforth obvious that the problem of explicit solutions was still open.

An alternative possibility of driving the quantum states by $\delta(t)$-pulses of the external fields was first considered by Lamb Jr. [32], but the idea was not applied for some years. Yet, an extremely simple class of exact though formal "pulse solutions" of (2) in $L^2(\mathbb{R})$ (though ignoring the paper of Lamb [32]), was obtained in a sequence of our papers [14, 33, 34, 35, 36] by superposing two types of elementary operations: the incidents of free evolution and the effects of the sharp pulses of oscillator potentials. Since it seems that they brought some operational possibilities, let me describe them shortly.

Each free evolution incident in any interval $[t_0, t_1]$ produces the unitary evolution operator $e^{-i\tau\frac{p^2}{2}}$, where $\tau = t_1 - t_0$. In turn, the result of each sudden δ-kick of the quadratic potential $V(q,t) = a\delta(t-t_0)\frac{q^2}{2}$ (where a is the pulse amplitude) is most easily described by adopting the rectangular δ-model defined by $\delta_\epsilon(t) = \frac{1}{\epsilon}$ in a narrow interval $[t_0, t_0+\epsilon]$ and vanishing outside. The evolution in $[t_0, t_0+\epsilon]$ is then generated by the constant Hamiltonian $H_\epsilon = \frac{p^2}{2} + \frac{a}{\epsilon}\frac{q^2}{2}$ and the corresponding unitary operator $U_\epsilon = \mathrm{e}^{-i\epsilon[\frac{p^2}{2}+\frac{a}{\epsilon}\frac{q^2}{2}]} = \mathrm{e}^{-i[\epsilon\frac{p^2}{2}+a\frac{q^2}{2}]} \rightarrow \mathrm{e}^{-ia\frac{q^2}{2}}$ for $\epsilon \rightarrow 0$. In agreement with the Baker formula [37]:

$$e^{\lambda A}B\mathrm{e}^{-\lambda A} = B + \lambda[A,B] + \frac{\lambda^2}{2!}[A,[A,B]] + \frac{\lambda^3}{3!}[A,[A,[A,B]]] + \ldots, \tag{12}$$

which for $[A,[A,B]] = 0$ reduces just to the two first terms, both operations lead to extremely simple transformations of the canonical pair q,p. The free evolution incidents generate:

$$\mathrm{e}^{i\tau\frac{p^2}{2}}\begin{pmatrix}q\\p\end{pmatrix}\mathrm{e}^{-i\tau\frac{p^2}{2}} = \begin{pmatrix}q+\tau p\\p\end{pmatrix} = \begin{pmatrix}1 & \tau\\0 & 1\end{pmatrix}\begin{pmatrix}q\\p\end{pmatrix} \tag{13}$$

while the potential shocks:

$$\mathrm{e}^{ia\frac{q^2}{2}}\begin{pmatrix}q\\p\end{pmatrix}\mathrm{e}^{-ia\frac{q^2}{2}} = \begin{pmatrix}q\\p-aq\end{pmatrix} = \begin{pmatrix}1 & 0\\-a & 1\end{pmatrix}\begin{pmatrix}q\\p\end{pmatrix} \tag{14}$$

Within this scheme, an interesting operation is performed by a pair of free evolution steps separated by an oscillator pulse:

$$\mathrm{e}^{-i\tau\frac{p^2}{2}}\mathrm{e}^{-i\frac{1}{\tau}\frac{q^2}{2}}\mathrm{e}^{-i\tau\frac{p^2}{2}} \equiv F_\tau \tag{15}$$

It generates:

$$\begin{aligned} q &\rightarrow \tau p\\ p &\rightarrow -\frac{1}{\tau}q \end{aligned} \tag{16}$$

which might be called the *squeezed Fourier transformation*. Curiously, an equivalent operation is performed by:

Journal of Physics: Conference Series **512** (2014) 012035 doi:10.1088/1742-6596/512/1/012035

$$e^{-i\frac{1}{\tau}\frac{q^2}{2}} e^{-i\tau\frac{p^2}{2}} e^{-i\frac{1}{\tau}\frac{q^2}{2}} \equiv F_\tau \tag{17}$$

Henceforth the following product of the 6 unitary operations yields the transformation $q \to -q$ and $p \to -p$ (the parity operator)

$$e^{-i\tau\frac{p^2}{2}} e^{-i\frac{1}{\tau}\frac{q^2}{2}} \dots e^{-i\tau\frac{p^2}{2}} e^{-i\frac{1}{\tau}\frac{q^2}{2}} \equiv P, \tag{18}$$

whereas the sequence of 12 unitary terms produces an evolution loop [14]:

$$\underbrace{e^{-i\tau\frac{p^2}{2}} e^{-i\frac{1}{\tau}\frac{q^2}{2}} \dots e^{-i\tau\frac{p^2}{2}} e^{-i\frac{1}{\tau}\frac{q^2}{2}}}_{\text{12 terms}} \equiv 1 \tag{19}$$

An intriguing property of (19) is that all 6 free evolution exponents arise with the same signs and so do the exponents of the kick operations (a kind of non-perturbative Baker-Campbell-Hausdorff effect [37]). More important aspect of (19) is that it contains the free evolution intervals (see Fig.1). It means that the remaining eleven unitary operations must cause the free evolution inversion:

$$\underbrace{e^{-i\frac{1}{\tau}\frac{q^2}{2}} e^{-i\tau\frac{p^2}{2}} \dots e^{-i\tau\frac{p^2}{2}} e^{-i\frac{1}{\tau}\frac{q^2}{2}}}_{\text{11 terms}} \equiv e^{+i\tau\frac{p^2}{2}}, \tag{20}$$

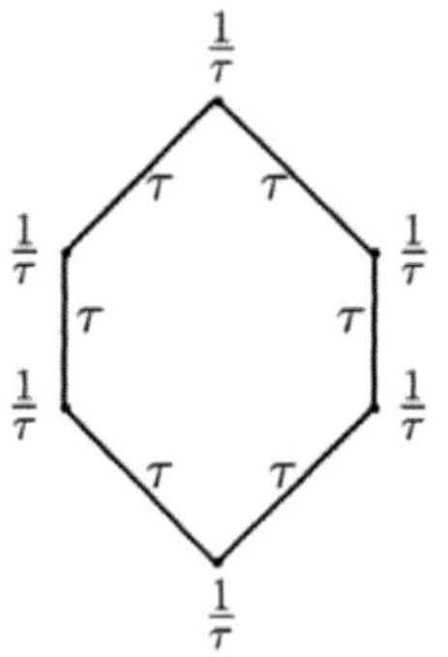

Figure 1: The evolution loop formed by 12 elementary evolution operators. The δ-pulses of the attractive oscillator potential of amplitudes $1/\tau$ are represented by the hexagon vertices's, while the 6 sides symbolize the τ-intervals of the free evolution. Each 3 consecutive operators yield the squeezed Fourier operation (15-17). Each 11 operations (6 consecutive oscillator kicks separated by 5 free evolution intervals) invert the free evolution.

If the idealized pulses could be indeed applied, the effect would be generated for every wave packet independently of its initial shape [14]. (The effect seemed strange and was objected by Marcos Moshinski though accepted by Bernardo Wolf as a case of the Fourier engineering [38]). Notice also that in this way, the loop mechanism (19, 20) predicted a part of the 1990 hypothesis about the quantum *time machine* [39].

The loop effects can be also caused by elastic pulses with alternating signs. Their basic fragment might be the sequence of 4 operators $S = e^{-i\tau\frac{p^2}{2}} e^{-ia\frac{q^2}{2}} e^{-i\tau\frac{p^2}{2}} e^{ia\frac{q^2}{2}}$, represented by the nilpotent matrices $Q = \begin{pmatrix} 0 & 0 \\ 1 & 0 \end{pmatrix}$, $Q^\dagger = \begin{pmatrix} 0 & 1 \\ 0 & 0 \end{pmatrix}$:

$$s = e^{\tau Q^\dagger} e^{-aQ} e^{\tau Q^\dagger} e^{aQ} = \begin{pmatrix} 1 & \tau \\ 0 & 1 \end{pmatrix} \begin{pmatrix} 1 & 0 \\ -a & 1 \end{pmatrix} \begin{pmatrix} 1 & \tau \\ 0 & 1 \end{pmatrix} \begin{pmatrix} 1 & 0 \\ a & 1 \end{pmatrix} \tag{21}$$

The algebraic properties of S depend again on the $\mathrm{Tr}s = 2 - \tau^2 a^2$. If $\tau^2 a^2 > 4 \Rightarrow \mathrm{Tr}s < -2$, then the matrix s is of the type (**III**) and the multiple pulse repetitions generate an unstable motion of squeezing type. However, if $0 \neq \tau^2 a^2 < 4 \Rightarrow -2 < \mathrm{Tr}s < 2$ the matrix s is of type (**I**) and the repeated pulses produce a confined motion, including the possibility of the evolution loops [33, 15]. One of simplest cases occurs for $\tau^2 a^2 = 2 \Rightarrow \mathrm{Tr}s = 0$; the Hamilton Cayley eq. for s then implies $s^2 = -\mathbb{1} \Rightarrow S^2 \equiv -1$ and so, the pulse pattern S, if repeated, creates an octagonal evolution loop [33], in which the sum of the oscillator pulses cancels, but the effects don't (see Fig.2):

$$[e^{-i\frac{\sqrt{2}}{\tau}\frac{q^2}{2}} e^{-i\tau\frac{p^2}{2}} e^{+i\frac{\sqrt{2}}{\tau}\frac{q^2}{2}} e^{-i\tau\frac{p^2}{2}}]^4 \equiv 1 \tag{22}$$

The possibility of *non-adiabatic* loops generated by time dependent oscillator forces was noticed already in 1970 by Malkin, Man'ko and Trifonov [40], though without elaborating the dynamical consequences. Yet, as shown subsequently, the existence of such phenomenon leads to a certain operational hypothesis. It states that one of the simplest ways to control the dynamical evolution consists in generating a closed evolution pattern (*evolution loop*) and then considering their perturbed or deformed versions [14, 33, 34]. If the unperturbed evolution is represented by a certain family of unitary operators $U_0(t)$, then the perturbed evolution operators split into $U(t) = U_0(t)W(t)$, where $U_0(t)$ represent the basic dynamical process and $W(t)$ is the correction (the evolution in the *interaction frame*). If now at some moment T the basic evolution closes to a loop $U_0(T) \equiv 1$, then the full evolution reduces just to the pure deformation, $U(T) \equiv W(T)$, see Fig.3, in general, much easier to manipulate by the external fields [14, 33, 34, 15]. The most elementary evolution loops occur in the time independent oscillator potentials, leading e.g., to the *non-demolishing quantum measurements* [41], but do not exhaust the manipulation techniques.

In fact, the same mechanism predicts an ample family of elementary effects such as the 'time squeezing', i.e., accelerating or slowing down the free evolution [33, 34]. Moreover,

Journal of Physics: Conference Series **512** (2014) 012035 doi:10.1088/1742-6596/512/1/012035

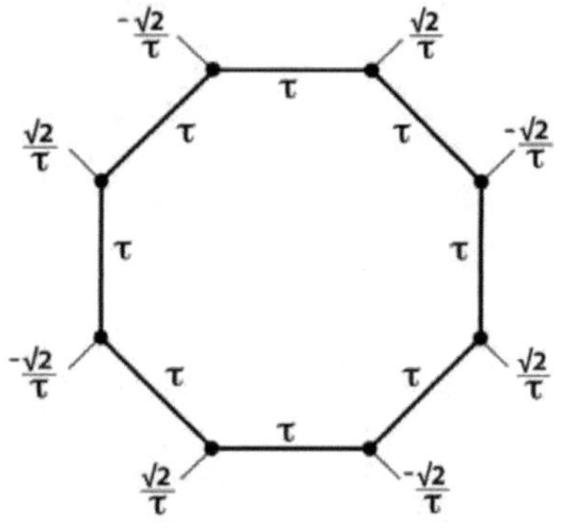

Figure 2: The 8 intervals of free evolution interrupted by 4 attractive and 4 repulsive oscillator pulses close the dynamical process to an octagonal evolution loop. Similarly as for Fig.1, the incomplete fragments of the process generate the free evolution inversions.

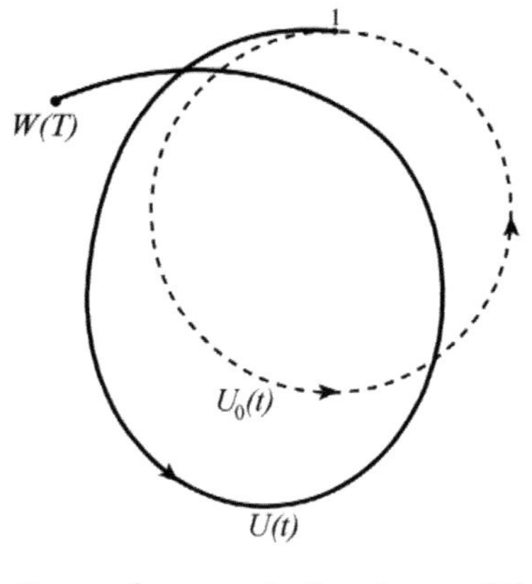

Figure 3: The applications of an evolution loop. The basic and perturbed evolution processes are represented by $U_0(t)$ and $U(t) = U_0(t)W(t)$, where the $W(t)$ is the *evolution in the interaction frame.* If for some T, the $U_0(T)$ closes to the evolution loop, $U_0(T) \equiv 1$, then the whole process reduces to $W(T)$ alone, the *precession operator*, sensitive to manipulation programs.

some simple, asymmetric sequences of the oscillator pulses can produce the squeezing and/or magnification of canonical variables. The most elementary such effects can be produced by two different *squeezed Fourier* operations, F_α and F_β [15]:

$$F_\alpha F_\beta \to \begin{pmatrix} 0 & \alpha \\ -\frac{1}{\alpha} & 0 \end{pmatrix} \begin{pmatrix} 0 & \beta \\ -\frac{1}{\beta} & 0 \end{pmatrix} = \begin{pmatrix} -\frac{\alpha}{\beta} & 0 \\ 0 & -\frac{\beta}{\alpha} \end{pmatrix} = \begin{pmatrix} -\sigma & 0 \\ 0 & -\frac{1}{\sigma} \end{pmatrix} \tag{23}$$

where $\sigma = \frac{\alpha}{\beta}$. Some of these phenomena were independently observed in [42, 43]. An ample collection of most general squeezing effects was described by Dodonov [44]; the manipulations by complex Hamiltonians were also considered [45, 46, 47, 48, 49]. Some "exotic" effects are described in [15]. All this does not yet exhaust the early history.

5 The Optical Prehistory

While the oscillator kicks have some aspects of Platonic ideas, it was discovered by Wolf [17, 50] that their empirical equivalents were known since long time in geometrical optics describing images on the optical bench. In particular, the 2×2 matrix transformations (13, 14, 23) of the canonical variables q, p correspond to the application of some typical optical instruments (microscopes, telescopes etc). Here, q is the distance of a light ray from the bench axis z and $p = c\sin\theta$, where θ is an angle between the light ray and the bench axis, c meaning the light velocity (or simply $p = sin\theta$, in dimensionless variables where $c = 1$). The matrix (13) then describes the propagation of images formed by the light rays on the planes orthogonal to the bench axis. In turn, (14) describes the action of a thin optical lens placed on the bench, the amplitude a meaning the (positive or negative) lens curvature [51]). A slightly different optical representation of symplectic matrices is proposed by Hiley [52], who adopts (the dimensionless) $p = \theta$, perhaps thinking about the spherical screens. The most detailed and complete report on the optical interpretation of the symplectic matrices is given by Collins [53].

Both interpretations have some advantages. Thus, what in the dynamical language is an *evolution loop* (useful for the manipulation algorithms), in optical terms is the simple reproduction of an optical image. The squeezing/amplification mechanisms produced by two or more oscillator shocks in quantum mechanics are easily interpretable as the applications of the microscope (or telescope) in geometrical optics. (So, in a sense, the effects of the oscillator kicks were known already to Galileo). Moreover, what was not so easy to predict in the dynamical language, i.e., the asymmetry of $\beta(t)$ needed to produce the squeezing, is immediately obvious at the optical level. (In fact, the symmetric apparatus could not produce amplified or reduced images).

Both have also some limits. In the area of optics, the applications of too close lenses with too big $|a|$ would mean that a part of one lens (if not the whole) must overlap with the interior of the other. Some more exact equivalents of the elastic pulses (14) appeared

Journal of Physics: Conference Series **512** (2014) 012035 doi:10.1088/1742-6596/512/1/012035

later in works on optical signals in dispersive fibers [54] (more recently, c.f. also [55]). All this does not yet explain how to create the elastic pulses in the ion traps. Of course, an ideal application of the δ kicks is practicably impossible. Forgetting even about the finite resistance of the trap walls, no infinite energy shocks can be engineered. Yet, despite all their imperfections, the idealized dynamical operations might be of interest for some unfinished fundamental discussions.

6 The uncertainties about the Time-Energy uncertainty

Indeed, it is enough to remind the polemics about the "time-energy uncertainty principle". The first objections against the (too narrow) interpretations of Landau and Peierls [56] supporting this principle, appeared already in the study of Aharonov and Bohm in 1961 [57] (see also [58]). Further counter-arguments were collected by Aharonov, Massar and Popescu [11] who argue that an arbitrarily exact measurement of the energy of a quantum system can be performed in an arbitrarily short time, provided that the Hamiltonian is known and "the measurement is brutal". Their example was the spin measurement. As it seems, the pulse patterns of Sec.4 bring the next illustration to the same idea for the continuous canonical variables.

Indeed, suppose a δ-pulse of the oscillator potential surrounded by two short intervals of the free evolution performs the "squeezed Fourier transformation" (15) of a free particle with an energy $\frac{p^2}{2m}$. After the operation, the unknown (classical or quantum) momentum p is converted into the new particle position $\tilde{q} = \tau p$. Could such transformation be produced, it would be no longer necessary to detect directly the particle energy, e.g., by observing its collision with a heavier microobject [56]. It would be enough to measure its new position $\tilde{q}$. Whatever the technical difficulties, there is no fundamental law which would forbid to determine $\tilde{q}$ in an arbitrarily short time.

An objection can still arise, that our prescription (15) offers an inconvenient relation between the final $\tilde{q}$ and the initial momentum p. The $\tilde{q} = \tau p$ implies $\Delta\tilde{q} = \tau\Delta p$, so if the operation time τ is very short, then little errors in $\tilde{q}$ will correspond to much greater errors in p. It looks almost as a revenge of the time-energy uncertainty. Yet, it is not. The parameter $\tau \in \mathbb{R}$ marks the time of an external operation (i.e., an *external time* of Aharonov and Bohm [57]), while $\Delta\tilde{q}$ *is not* limited by any universal constant. Moreover, (still accepting the δ-pulses), the dynamics of (2) can offer also much better measurement methods which we proposed to call the *"Fourier microscopes"* [15]. A simple option can be to apply three elastic shocks divided by two free evolution incidents (i.e., an incomplete case of the triangular evolution loop [33]):

$$u = e^{-aQ} e^{\tau Q^\dagger} e^{-bQ} e^{\tau Q^\dagger} e^{-cQ} = \begin{pmatrix} 1 & 0 \\ -a & 1 \end{pmatrix} \begin{pmatrix} 1 & \tau \\ 0 & 1 \end{pmatrix} \begin{pmatrix} 1 & 0 \\ -b & 1 \end{pmatrix} \begin{pmatrix} 1 & \tau \\ 0 & 1 \end{pmatrix} \begin{pmatrix} 1 & 0 \\ -c & 1 \end{pmatrix}$$

$$= \begin{pmatrix} 1 + b\tau + 2c\tau + (b\tau)(c\tau) & \tau(2 - b\tau) \\ \frac{1}{\tau}\Big(a\tau + b\tau + c\tau + (a\tau)(b\tau) + 2(a\tau)(c\tau) \\ +(b\tau)(c\tau) + (a\tau)(b\tau)(c\tau)\Big) & 1 + 2a\tau + b\tau + (a\tau)(b\tau) \end{pmatrix} \quad (24)$$

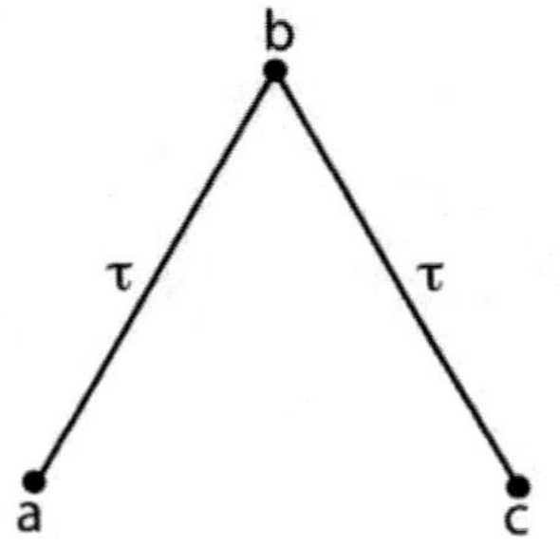

Figure 4: Depending on the values of the elastic pulses $a, b.c$ the incomplete version of the triangular evolution loop can generate the distorted cases of the inverted free evolution or else, the cases of Fourier microscope with an arbitrary magnification coefficient achieved in an arbitrarily short time.

The above simple pattern offers a variety of dynamical operations. By asking $u_{11} = u_{22} = \pm 1$, one can obtain the free evolution with 'squeezed' or even inverted time. In turn, by asking that $u_{11} = u_{22} = 0$, one obtains the 'squeezed Fourier' with $\Gamma = |u_{12}| = |\tau(2 - b\tau)|$ arbitrarily great, i.e., a "Fourier microscope" yielding the final position $\tilde{q} = \Gamma p$ large enough to assure very little errors of $\Delta p = \frac{1}{\Gamma}\Delta\tilde{q}$ even if the measurement of $\tilde{q}$ is far from perfect. While technical difficulties still exists, they do not seem a fundamental obstacle. This appears to confirm the idea of the "brutal measurement" in an arbitrarily short time [58].

Is there the "time operator"? The whole discussion could not avoid an interrelated question, as to, whether the quantum theories should adopt the *time of event* as a legitimate quantum observable, represented by a certain hermitian operator $\mathbf{t}$ in a Hilbert space, a canonical counterpart of the energy $\mathbf{E}$ with $[\mathbf{t}, \mathbf{E}] = i\hbar$. The idea, though, was not so easy

Journal of Physics: Conference Series **512** (2014) 012035 doi:10.1088/1742-6596/512/1/012035

to accept. As many other controversies it produced an enormous sequence of papers, but no final solution. Apart of difficulties with defining consistently the wave function on the time axis, the main obstacle was the theorem of Pauli, stating that then $\mathbf{E}$ must have a continuous, displacement invariant spectrum extending over the entire $\mathbb{R}$.

Trying to avoid the obstacle, Kijowski [59] proposed in 1974 a simplified 1D model replacing the free energy operator $\mathbf{E} = \frac{\mathbf{p}^2}{2}$ by $\tilde{\mathbf{E}} = \frac{\mathbf{p}|\mathbf{p}|}{2}$ with an infinite spectrum covering $\mathbb{R}$, and then he proved a theorem finding the canonically conjugated 'time' $\tilde{\mathfrak{t}}$. The further development has again some aspects of a historical anecdote. The construction of Kijowski remained almost unperceived for more than 2 decades (Not so surprising. The "publish or perish" culture produces such mountains of publications, that neither the XXII century archeologists will have patience to read all of them!). Yet, around 2000, a group of authors noticed that Kijowski results may be useful to determine the *time of arrival* into a waiting detector (or screen) for the wave packets which belong only to the positive (or only to the negative) subspace of the momentum operator. The observation produced a trend of papers describing either the wave packets localized initially to the left of the detector and moving to the right (the 'right movers'), or vice versa [60, 61]. Yet, to restrict the problem to the 'left component' is not the same as to consider the 'right movers' (and vice versa). Moreover, the method did not describe the interference of both, left and right components (if they both exist), converging from opposite sides to the detection point (the difficulty, as far as I know, discovered first by Leavens [62, 63, 64, 65, 66])

In spite of subsequent works on POV measures and some new but still partial results [67, 68, 69], it looks that the idea of time-operator has too many gaps to grant the *time-energy* uncertainty (except if some additional limitations of quantum measurements proved real [70]).

Here again a historical anecdote. In the summer of 2001 we have finished with Gabino Torres Vega an article containing those and other critical remarks about the time operator. Our text was ended up with a little story which (we thought) exclude completely the existence of an orthodox quantum mechanical time operator at least for a class of physical phenomena, such as the radioactive decay of unstable atoms. We argued:

> "Note that this negative result might imply some good news. Indeed, imagine a simple experiment which consists in registering the time moment in which an unstable particle decays. The result is a real number (the decay time); yet, an attempt to describe it in terms of the orthodox scheme (i.e. as an eigenvalue of a self-adjoint *time operator*) would lead to a wrong conclusion: about the existence of initial states for which the moment of the decay can be predicted with certainty! Should this be true, the consequences would be quite dramatic. They would include a suspense story about a suitcase full of radioactive atoms, smuggled safely through the airport security, with all atoms programmed to decay tomorrow! So, perhaps we should not regret that the *time of events* does

not obey the axioms of Dirac and v. Neumann?"

After submitting the work to Phys.Rev.A, we thought that the truth of our little story must conquer the referees. However, it was not so. Our article circulated until November, meeting finally some counter-anecdotes. Since the discussion is fundamental, some of them deserve to be cited. The First Referee has written:

> *"... As a general comment, I also have problems to be sympathetic with the general negative tone of this article. While criticism and finding difficulties are inherent (...) for the progress of science, as much or even more effort should be put in finding the solution to the problems. This positive component (and effort?) is essentially absent in this article (...). The article would much improve if it could balance negative aspects with something positive. As a minor point, the final anecdote in the paragraph on the "good news" is possibly not a very happy idea considering the present situation of international affairs. I admit of course that these are largely matters of taste".*

The Second Referee wrote:

> *... "It is problematic that the authors limit themselves to a sharp criticism of the Kijowski proposal without delivering any ideas on how to actually construct a feasible operator of time. However, this referee liked the fact that the authors bothered to give numerical examples illustrating the action of Kijowski's operator (this down-to-earth approach is certainly required to achieve some progress in the field). But I also strongly oppose the aggressive language used throughout this contribution".*

The first objection could be answered on logical grounds. Our paper was dedicated to show that the "time operator " does not exist. So, the request that we should complete our argument by constructing the time operator (both referees) was indeed demanding too much! Besides, the first referee objection against our anecdote seemed incomprehensible: our argument was correct! But suddenly, we understood the chronological complication. Our paper was send to Phys.Rev.A at the beginning of August 2001, then it started to circulate, and the referee must have obtained it in October or November 2001. He thought that we were joking of September 11 terrorist attack! We understood that the defense might be difficult and we sent the paper to the open access Journal in Lodz University in Poland, see [71].

Another slight surprise in this discussion was, however, the voice of the 3-rd referee, which was Kijowski himself. He defended his result, as a mathematical theorem, but not necessarily as a physical fact. He quoted the following fragment of his 1999 paper [72]:

Journal of Physics: Conference Series **512** (2014) 012035 doi:10.1088/1742-6596/512/1/012035

"Although there is no room to modify the mathematically unique definition of probability (...) within the standard mathematical framework of Quantum Mechanics (spectral measures, self-adjoint operators etc) the problem is still open from the physical point of view.". This may be taken as a hint, that the 'standard framework' is not an absolute truth, but still requires careful studies. But then, how much we really know?

7 Are the operations of δ-pulses real?

The operations considered until now concern the states in 1 space dimension and are singular. Their generalizations in 2D and 3D, employing the harmonic δ-pulses (without an unphysical charge density of [20]), were designed by Fernandez, who discussed the equivalents of all previously predicted effects, including the comments on the *time-energy* uncertainty. The paper was 'softly' rejected by a referee in Phys.Rev.A who did not question the mathematical results but decided that the δ-pulses are not acceptable, as they do not reflect the empirical reality required in Phys.Rev.A (yet, the papers on *kicked rotators* were accepted without any problems, perhaps by referees impressed by the luggage of special functions?). Distasted by the suspicious attitude in Phys.Rev.A, Fernadez made a quick decision to publish the results in *Nuovo Cimento* [73], where they are now waiting for some new followers of the control problem.

But, what's about the reality of δ-pulses? They don't exactly exist in nature. Yet, their effects could be idealized forms of some real phenomena. In fact, the strong, but bounded fields (e.g. in form of rectangular steps) allowed already to design a family of dynamical operations generating the "distorted free evolution", rigid displacement of the wave packet, etc. The possibility of obtaining still more general operations for softly varying, differentiable fields (without any steps!) was considered in [74, 34], then confirmed by computer studies [75, 76, 77, 33, 34, 35, 78]).

Here, let me also mention some manifestly unphysical results, but distinguished by mathematical elegance. I am referring to papers Caldirolla [79], Kanai [80] and specially Baseia et al. [81, 82] dedicated to the time dependent mass. Note an essential difference: while the *effective mass* depending on space points can describe the average particle propagation in some inhomogeneous media [83, 84, 85], the Schrödinger's particles with the *time dependent* mass do not seem to have any empirical equivalent. Yet, the Caldirolla-Kanai Hamiltonian with an exponentially varying mass:

$$H(t) = \frac{1}{2m(t)}p^2 + \frac{m(t)\omega^2}{2}q^2, \quad m(t) = m_0 e^{\lambda t} \tag{25}$$

was persistently used as a description of a quantum system affected by friction: a conviction hardly acceptable in spite of the work of Bateman [86].

Indeed, the evolution operations defined by any Hamiltonian with a time dependent mass (in classical or quantum case) cannot describe the friction, since they yield always

Journal of Physics: Conference Series **512** (2014) 012035 doi:10.1088/1742-6596/512/1/012035

the invertible transformations of pure into pure states. For quadratic Hamiltonians of two canonical variables, this is typically either the trapped motion, or the squeezing of one canonical variable at the cost of another [81, 87], but not a dissipative process, which needs a more general evolution laws for mixed ensembles (see Lindblad [88], Gisin and Percival [89]). This fact found a nice illustration in [82], for the Hamiltonian (25) with $m(t)$ forming a sequence of sharp, rectangular steps. Even if operationally impossible, it offers a clean mathematical image of a *squeezing* which affects just a pair of canonical variables q, p. Moreover, the authors are able to design a programme of steps which invert the operation. The composition of both closes up to a clean *evolution loop*, illustrating nicely the absence of any dissipative process for Hamiltonians of type (25). The story does not end up here.

Some surprise was, that the merits of the evolution loops were finally noticed in Phys.Rev.Lett., in 1999 work of Harel and Akulin on the control problem in finite dimensional spaces [45]. They wrote:

> Our approach is based on the following idea: We should first find an "identity map $T_1, T_2, ..., T_{N^2}$ for the particular case $\hat{H}_{eff} = 0$. If we succeed, the operator of the evolution over the period $T = \Sigma_{n=1}^{N^2} T_n$ is the identity transformation
> $$\hat{U}(T) = e^{-i\hat{B}T_{N^2}} e^{-i\hat{A}T_{N^2-1}} ... e^{-i\hat{B}T_2} . e^{-i\hat{A}T_1} = 1$$
> For small variations (...) the expansion of the exponential factors (...) yields (...)

This repeats almost (though not completely) our early *evolution loops* typically composed of two alternative operations. However, since Harel and Akulin work in finite dimensional Hilbert spaces and their Hamiltonians are not quadratic, it looks that their inspiration was independent. Quite simply, in the *Platonic heaven* some predatory ideas are circulating looking all the time for some new victims! Anyhow, our loop doctrine was recognized by a group of quantum control in Texas [47]. In further development, the quantum control by sharp pulses have found new forms and new leaders (or victims?). The pulses have now the form of general Hamiltonians interrupting the continuous evolution, called *decouplers* [46]; the evolution affected by sequences of their sharp kicks are studied for the spin systems, as one of the most promising quantum control methods [45, 46] (obeying an equally dangerous Platonic idea, known as the *"bam-bam*-control"), though less violent methods are also considered [48, 90]. The question arises, what is the future of their and our approaches? The 'pulse operations' of course must have some natural limits. Hence, the soft alternatives of the potential kicks can be no less interesting. This brings us back to the manipulation problems for the quadratic Hamiltonians (2).

Journal of Physics: Conference Series **512** (2014) 012035 doi:10.1088/1742-6596/512/1/012035

8 The soft operations

As recently found, the control operations are significantly simplified if the field amplitude $\beta(t)$ is symmetric with respect to the center of the operation interval [90, 87, 91, 15]. We shall show now, that in the symmetric generation program the matrices (3-9), together with the corresponding 'driving amplitudes' $\beta(t)$ can be expressed exactly (without perturbations!) in terms of a single real function which, appart of details, may be fixed at will [15]. Without pretending too much, it facilitates enormously the task of programing the dynamical operations. Indeed, one has:

Proposition 2. Consider a nontrivial operation interval $[-\mathrm{T}, \mathrm{T}]$ with the quadratic potential $\beta(t)\frac{q^2}{2}$ and suppose, $\beta(t)$ is bounded, piecewise continuous and symmetric, $\beta(t) = \beta(-t)$. Then the evolution matrix $u(t,-t)$ and the driving amplitude $\beta(t)$ in the expanding family of intervals $[-t, t]$ (where $t \leq \mathrm{T}$) can be written explicitly in terms of $\theta(t) = u_{12}(t,-t)$, which may be choosen at will everywhere except its zero points. Moreover, whenever $u = u(t,-t)$, for $t \in [0, \mathrm{T}]$, reaches the stability threshold with $\mathrm{Tr}u = \pm 2$, then either u or -u adopts one of the forms (13) or (14), imitating the results of a simple or distorted free evolution, or else, of a sharp oscillator kick.

Proof. In what follows, whenever there will be no reasonable doubt, we shall simplify the notation, writing just $u(t)$ instead $u(t,-t)$ and $u_{kl}(t)$ instead of $u_{kl}(t,-t)$ Due to the symmetry of the Hamiltonians $H(t) = H(-t)$, the unitary evolution operators $U(t,-t)$ for the expanding intervals $[-t, t]$ satisfy:

$$i\frac{d}{dt}U(t,-t) = H(t)U(t,-t) + U(t,-t)H(t) \tag{26}$$

Hence, the corresponding evolution matrix $u(t) = u(t,-t)$ fulfills:

$$\frac{du}{dt} = \Lambda(t)u + u\Lambda(t) \tag{27}$$

Since $\Lambda(t)$ is given by (5), this becomes

$$\begin{aligned}\frac{du}{dt} &= \begin{pmatrix} u_{21} - \beta u_{12} & \mathrm{Tr}u \\ -\beta\mathrm{Tr}u & u_{21} - \beta u_{12} \end{pmatrix} \\ &= (u_{21} - \beta u_{12})\mathbb{1} + \mathrm{Tr}u \begin{pmatrix} 0 & 1 \\ -\beta & 0 \end{pmatrix}\end{aligned} \tag{28}$$

Therefore,

$$\frac{d}{dt}(u_{12}u_{21}) = \mathrm{Tr}u(u_{21} - \beta u_{12}) = \mathrm{Tr}u\frac{1}{2}\frac{d}{dt}\mathrm{Tr}u = \frac{1}{4}\frac{d}{dt}(\mathrm{Tr}u)^2 \tag{29}$$

and integrating:

$$\frac{d}{dt}\left[u_{12}u_{21} - \frac{1}{4}(\mathrm{Tr}u)^2\right] = 0 \Rightarrow u_{12}u_{21} - \frac{1}{4}(\mathrm{Tr}u)^2 = C = const. \tag{30}$$

Journal of Physics: Conference Series **512** (2014) 012035 doi:10.1088/1742-6596/512/1/012035

To determine C it is enough to take $t = 0$. The initial condition $u(0,0) = \mathbb{1}$ then tells that $C = -1$, and so:

$$u_{12}u_{21} = \frac{1}{4}(\mathrm{Tr}u)^2 - 1 \quad t \in [-\mathrm{T}, \mathrm{T}] \tag{31}$$

The above eqs. (28-31) provide an elementary solution of the inverse evolution problem, permitting to reconstruct the entire evolution matrices $u(t,-t)$ together with the driving pulse $\beta(t)$ in terms of one function $\theta = u_{1,2}(t)$ in the expanding intervals $[-t, t]$.

Indeed, (28) implies $\frac{du_{11}}{dt} = \frac{du_{22}}{dt} = u_{21} - \beta u_{12}$ and since the initial condition at $t = 0$ is $u_{11}(0) = u_{22}(0) = 1$, then $u_{11} = u_{22}$ in all intervals $[-t,t]$ $(t \leq \mathrm{T})$. In view of (28) this means that $u_{11}(t) = u_{22}(t) = \frac{1}{2}\,\mathrm{Tr}\,u = \frac{1}{2}\theta'(t)$. In turn, since u is symplectic, then $(\frac{1}{2}\theta')^2 - \theta u_{2,1} = 1$ and the remaining matrix element $\alpha = u_{2,1}$ is determined as:

$$\alpha = \frac{(\frac{1}{2}\theta')^2 - 1}{\theta} \tag{32}$$

with the pulse shape $\beta(t)$ defined in terms of θ as well:

$$\beta = -\frac{\theta''}{2\theta} + \frac{(\frac{1}{2}\theta')^2 - 1}{\theta^2} \tag{33}$$

If no singularity of β occurs for $t = 0$ (or in other points where θ might vanish) these expressions grant (28), yielding an exact solution of the symmetric evolution problem for the family of the expanding intervals $[-t,t]$. □

As one can notice, whenever the symmetric evolution matrix $u(t) = u(t,-t)$ reaches the threshold values $\mathrm{Tr}u = \pm 2$ (the case II of our classification), (31) implies that either $u_{1,2}$ or $u_{2,1}$ (or both) must vanish, leading to the canonical transformations (28) which simulate the oscillator kicks, the incidents of the free evolution with distorted, squeezed or negative time, or just one of the evolution loops (c.f. [87, 91]), all of them with or without the simultaneous parity transformation [87, 91]. The operations in the stability areas might provide also the softly achieved "squeezed Fourier" operations, including the "Fourier microscopes" - which can be generated by strong though differentiable amplitudes $\beta(t)$ in arbitrarily short time intervals. An essential consequence of the soft imitations is that they bring several ideas one step closer to the laboratory experience. This includes our symbolic results in [14, 33, 34], then Fernandez 3D proposals [73] and the critiques of the time-energy uncertainty [11].

To complete this report, let me describe a kind of NO-effect: a driving amplitude which does not vanish in $[-\mathrm{T},\mathrm{T}]$, though vanishes outside, and moreover, it represents an "hidden pulse" completely undetectable out of the time interval in which it is acting. This happens always if the action of a (symmetric) $\beta(t)$ in an interval $[-\mathrm{T},\mathrm{T}]$ produces a matrix $u = u(-\mathrm{T},\mathrm{T})$ with $u_{11} = u_{22} = \theta'(\mathrm{T}) = 1$ and $u_{12} = \theta(\mathrm{T}) = 2T$, i.e., the u will mimic exactly the free evolution case, in spite of $\beta \neq 0$ in $[-\mathrm{T},\mathrm{T}]$.

Journal of Physics: Conference Series **512** (2014) 012035 doi:10.1088/1742-6596/512/1/012035

Examples of several invisible pulses obtained for simple polynomial models $\theta(t) = 2t - \theta_3 t^3 - \theta_5 t^5 - \theta_7 t^7 - \theta_9 t^9 - \theta_{11} t^{11}$ are represented in Fig. 5. Our mathematical algorithm used to construct them is not identical, though interrelated with the invisibility studies employing the complex variables [92]. The importance of the subject may extend from the single particle behavior, to the particle tunneling in solid state wells, or even to the apparently unrelated phenomena[3] But what is the precision of our manipulation effects?

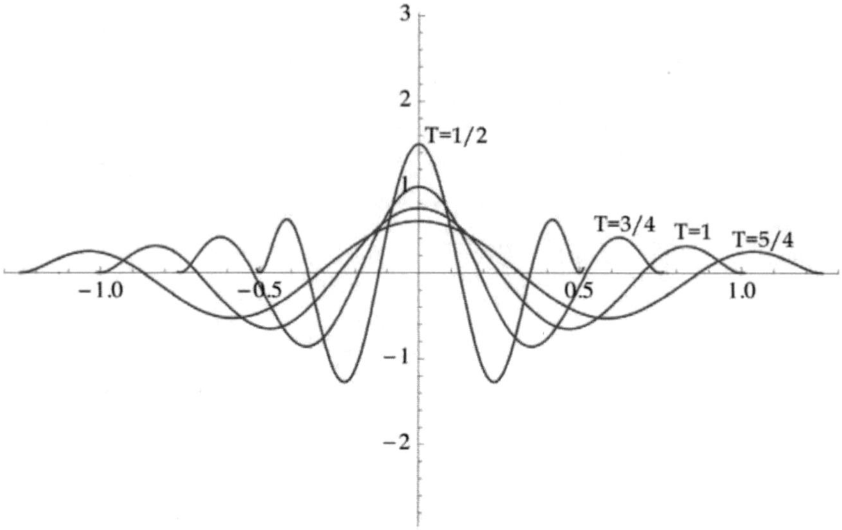

Figure 5: The four "invisible pulses" in the expanding [-T,T] intervals for T=1/2, 3/4, 1 and 5/4. While they can be arbitrarily strong in their action intervals, their consequences are undetectable outside.

9 Are there natural precision limits?

The difficulties are still abundant. If accepting the validity of the present day quantum field theories, then the described unitary operations are merely approximate results which may be easily spoiled by "hostile" external environments. In fact, the use of the quantum mechanical quadratic Hamiltonians (1) depends on the semiclassical picture interpretable as the *low frequency approximation* in which the charged microparticle interacts with a huge ensemble of extremely tiny quanta. It suggests that the driving pulse $\beta(t)$ should be composed only of weak and long waves (of course if the role of the Fourier analysis was not exaggerated in quantum theories.). The semiclassical models actually used include

[3]Indeed, the 2-nd order differential equations related to Hamiltonians (2) appear also in astronomy and cosmology; see [94, 95, ?, 97, 98, 15].

the microwave frequencies, but what about the infrared components? Our description would be obviously broken for microparticles interacting not with some coherent clouds of practically imperceptible quanta, but with one or more "high energy quanta", leading to the scattering picture rather than the smooth unitary evolution. Some other aspects of our solutions may also need a critical revision.

The exact solutions (28-31) describe the particle evolution driven by time dependent $\beta(t)$, basically applicable in the ion traps, but they do not take into account the packet reflections from the trap walls. To neglect them, the trap should be ample enough. Yet, if too huge, the variable potentials on the trap walls will propagate to the trap interior with relativistic delays, causing the timing errors $\delta t = \frac{|\Delta \mathbf{x}|}{c}$ of the field values, (where $|\Delta \mathbf{x}|$ are the distances between the points in the trap interior). In the dimensionless time variable $\frac{t}{\tau}$ used in the expression (2) (where τ is a certain conventional time unit), the delays would be therefore $\delta(\frac{t}{\tau}) = \frac{|\Delta \mathbf{x}|}{\tau c}$. Small indeed, perhaps allowing the traps of the size of few cm (or more?).

A difficult problem is, however, how to assure that the scalar potential on the trap surfaces will evolve exactly according to $\phi(\mathbf{x}, t) = \beta(t)\phi_0(\mathbf{x})$ where $\phi_0(\mathbf{x})$ represents the static case. This would mean that the little potential changes $\delta\phi = \delta\beta(t)\phi_0(\mathbf{x})$ should arrive at the surface points in a perfect synchronization, the demand not so easy to satisfy [15]. In real experiments, the potential changes are typically implemented just at some particular surface points; they need some time, extremely short, but still non-vanishing, to propagate all over the trap walls. To get rid of this effect for wider ion traps, some method of introducing the simultaneous potential changes in a certain (dense?) net of the wall points would be needed, though the technology is still missing.

An interesting remark is, however, that if only such simultaneity could be achieved on the trap surfaces, then the relativistic (scalar) potential in the trap interior, obtained by substituting t in $\beta(t)$ by the retarded time $t_r = \frac{t}{\tau} - \frac{|\mathbf{x}-\mathbf{x}'|}{\tau c}$ [99] would not produce the retarded corrections of the orders of magnitude $\approx \frac{|\Delta \mathbf{x}|}{\tau c}$ (post-Newtonian), but paradoxically, only the corrections of the 2-nd order, $(\delta\frac{t}{\tau}) \approx \frac{|\Delta \mathbf{x}|^2}{(\tau c)^2}$, post-post-Newtonian in the EIH hierarchy (see [15, 93]). The problem, of how to approximate this situation is still open. Yet, it seems that the classical electrodynamics does not preclude the existence of such retarded fields. As an example, it might be interesting to consider a time dependent field with $\beta(t)$ generalising the traditional Paul's formula [100] in the simplest, cylindrical trap.

Assuming the absence of charge and current densities in the trap interior, a simple (though idealized) expressions for the softly changing electromagnetic fields can be constructed by applying the step by step approximations, independent of any particular gauge. In particular, for a cilindric Paul's trap with perfectly hyperbolic surfaces, one might define a sequence of vector-functions $\mathbf{E}_1, \mathbf{E}_3, \mathbf{E}_5, ...$ and $\mathbf{B}_2$, $\mathbf{B}_4$, $\mathbf{B}_6, \ldots$ such that: $\mathbf{E}_1 = \nabla \times \mathbf{B}_2$, $\mathbf{B}_2 = -\nabla \times \mathbf{E}_3$, $\mathbf{E}_3 = \nabla \times \mathbf{B}_4$, $\mathbf{B}_4 = -\nabla \times \mathbf{E}_5$, $\mathbf{E}_5 = \nabla \times \mathbf{B}_6, \ldots$; and henceforth: $\mathbf{E}_1 = \Delta\mathbf{E}_3$, $\mathbf{E}_3 = \Delta\mathbf{E}_5$,.. and $\mathbf{B}_2 = \Delta\mathbf{B}_4$, $\mathbf{B}_4 = \Delta\mathbf{B}_6$,... or explicitly:

$$\mathbf{E}_1 = \begin{pmatrix} x \\ -y \\ 0 \end{pmatrix} \quad \mathbf{E}_3 = \begin{pmatrix} \frac{1}{4}xy^2 + \frac{1}{3}\frac{1}{4}x^3 \\ +\frac{-}{1}4x^2y - \frac{1}{3}\frac{1}{4}y^3 \\ 0 \end{pmatrix} \quad \mathbf{E}_5 = \begin{pmatrix} \frac{1}{8}\frac{1}{3}x^3y^2 \\ -\frac{1}{8}\frac{1}{3}y^3x^2 \\ 0 \end{pmatrix} \ldots$$

$$\mathbf{B}_2 = \begin{pmatrix} 0 \\ 0 \\ xy \end{pmatrix} \quad \mathbf{B}_4 = \begin{pmatrix} 0 \\ 0 \\ \frac{1}{4}\frac{1}{3}(x^3y + xy^3) \end{pmatrix} \quad \mathbf{B}_6 = \begin{pmatrix} 0 \\ 0 \\ \frac{1}{8}\frac{1}{3^2}x^3y^3 \end{pmatrix} \ldots$$

which can be used to construct the electric and magnetic fields

$$\mathbf{E}(x, y, t) = \beta(t)\mathbf{E}_1 + \frac{1}{c^2}\beta''(t)\mathbf{E}_3 + \frac{1}{c^4}\beta''''(t)\mathbf{E}_5 + ... \tag{34}$$

$$\mathbf{B}(x, y, t) = \frac{1}{c}\beta'(t)\mathbf{B}_2 + \frac{1}{c^3}\beta'''(t)\mathbf{B}_4 + \frac{1}{c^5}\beta'''''(t)\mathbf{B}_6, \ldots \tag{35}$$

satisfying the Maxwell-Faraday equations, $\nabla \times \mathbf{B} = \frac{1}{c}\frac{\partial \mathbf{E}}{\partial t}$, $\nabla \times \mathbf{E} = -\frac{1}{c}\frac{\partial \mathbf{B}}{\partial t}$, where the principal part would be $\beta(t)\mathbf{E}_1$, i.e., an electric analogue of Paul's field driven by an arbitrary $\beta(t)$ (instead of the trigonometric $\beta = \beta_0 + 2\beta_1 cos\omega t$ leading to the Mathieu functions [100]). In turn, $\frac{1}{c}\beta'(t)\mathbf{B}_2$ is the main magnetic part which must unavoidably appear in any cylindric trap. It will generate the contributions $\sim \frac{1}{c^2}$ (post-post-Newtonian) to the Lorentz force, acting on the charged particles. The whole rest, are just the magnetic and electric relativistic corrections of increasing powers in $\frac{1}{c^k}(k \geq 2)$. Yet, the expressions (34) and (35) cannot represent the exact time dependent fields of the cylindric ion trap. since they ignore the boundary conditions on the trap surfaces. If taken to the letter, they can be just a hint, that the simultaneity of the fields on the trap surfaces cannot be improved beyond the $\frac{1}{c^2}$ order of EIH [93].

So, what is the sense of our constructions? Are they a kind of a daydream? Or perhaps, an incomplete prediction of the approaching steps of quantum control?

Acknowledgements The author is indebted to David Fernandez, Piotr Kielanowski, Francisco Delgado and Alonso Contreras for helpful comments, to Jesus Fuentes Aguilar for technical assistance. The support of the CONACYT project 152574 is acknowledged.

References

[1] H. Primas *Chemistry, Quantum Mechanics and Reductionism*, Springer (1983)

[2] E. Lubkin, *J. Math. Phys.* **15**, 673 (1974)

[3] R. Penrose *The Large, the Small and the Human Mind*, CUP (1997)

[4] M. Brune, F. Schmidt-Kaler, A. Maali, J. Dreyer, E. Hagley, J.M. Raymond and S. Haroche, *Phys.Rev. Lett.* **76**, 1800 (1996)

[5] S. Haroche, *Phys.Today* **51** (7) 36 (1998)

[6] A.J. Leggett, *J.Phys. Condensed Matter* **14**, R415 (2002)

[7] S. Mancini, V.I. Man'ko, P. Tombesi, *Phys.Lett. A* **213**, 1 (1996); *Found.Phys.* **27**, 801 (1997);

[8] S. Mancini, V.I. Man'ko, P. Tombesi, arXiv:quant-ph/9709012v1 (5 Sep 1997)

[9] G. Marmo, A. Simony, and F. Ventriglia, *J.Phys.Conf.Series* **87**, 012013 (2007)

[10] M. Asorey, P. Facchi, G. Florio, V.I. Man'ko, G. Marmo, S. Pascuazio, E.C.G. Sudarshan, *Phys.Lett. A* **375**, 861 (2011)

[11] Y. Aharonov, S. Massar, S. Popescu, *Phys.Rev. A* **66**, 052107 (2002)

[12] B. Simon, *Quantum Mechanics for Hamiltonians Defined As Quadratic Forms*, Princeton Univ. Press, Princeton, N.J. (1971)

[13] G. Reed, B. Simon, *Methods of Modern mathematical Physics I: Functional Analysis*, New York, Acad. Press. (1975)

[14] B. Mielnik, *Rep. Math. Phys.* **12**, 331 (1977)

[15] B. Mielnik, *J.Phys. A* **46**, 385301 (2013) ; `arXiv:1305.3664 [quant-ph]`

[16] M. Moshinsky, T.H. Seligman, K.B. Wolf *J. Math. Phys.* **13**, 901 (1972)

[17] K.B. Wolf, *SIGMA*, **8** 033 (2012)

[18] D.J. Fernandez C., *Int.J.Theor.Phys.* **33**, 2037 (1994)

[19] H.R. Lewis Jr., *Phys.Rev.Lett.* **18**, 510 (1967)

[20] H.R. Lewis Jr. and W.B. Riesenfeld, *J. Math. Phys.* **10**, 1458 (1969)

[21] P.G.L. Leach, K. Andriopoulos, *Appl.Anal.Discrete.Math.***2**, 146 (2008)

[22] I.A. Malkin, V.I. Man'ko and D.A. Trifonov, *Phys.Lett. A* **30**, 414 (1969)

[23] I.A. Malkin, V.I. Man'ko and D.A. Trifonov, *J.Math.Phys.* **14**, 576 (1973)

[24] B. Baseia, S.S. Mizrahi, M.H.Y. Moussa *Phys.Rev. A* **46**, 5885 (1992)

[25] W.E. Milne, *Phys.Rev.* **35** (1930) 863

[26] W.E. Milne, *Numerical calculus* Princeton Univ. Press (1949)

[27] W.E. Milne, *Numerical solutions of differential equations* J.Wiley and Sons (1953)

[28] E. Pinney, *Proc.Amer.Math.Soc.* **1** (1950) 681

[29] M. Maamache, *J. Math. Phys.* **39**, 161 (1998)

[30] G. Orwell, *NineteenEighty-Four* Reprinted: Penguin Books (1954)

[31] S. Mayo and P.G.L. Leach, *J.Phys. A* **35**, 5333 (2002)

[32] W.E. Lamb Jr, Phys. Today **22**, 23 (1969)

[33] B. Mielnik, *J. Math. Phys.* **27**, 2290 (1986)

[34] D.J. Fernandez C. and B. Mielnik, *J. Math. Phys.* **35**, 2290 (1994)

[35] D.J. Fernandez C., *Geometric Methods in Physics. XXX Workshop 2011* p. 135, Trends in Mathematics, Ed. P. Kielanowski et al., Springer, Basel (2012)

[36] D.J. Fernandez C., *SIGMA* **8** 041 (2012)

[37] B. Mielnik and J. Plebanski, Ann. Inst. H. Poincar, Sec.**A**, Vol **XII** (1970) 215-254

[38] K.B. Wolf *Integral Transforms in Science and Engineering* Plenum, N.Y. (1979)

[39] Y. Aharonov, J. Anandan, S. Popescu and L. Vaidman, *Phys.Rev. Lett.* **64**, 2965 (1990)

[40] I.A. Malkin and V.I. Man'ko and D.A. Trifonov *JETP* **31**, 386 (1970)

[41] K.S. Thorne, R.W.P. Drever, C.M. Caves, M. Zimmermann, and V.D. Sandberg *Phys.Rev.Lett.* **40** 667 (1978)

[42] V.T. Ni *Phys.Rev. A* **33** 2225 (1986)

[43] A. Royer *Phys.Rev. A* **36** 2460 (1987)

[44] V.V. Dodonov, *J. Opt. B* **4** **R**, 1 (2002)

[45] G. Harel and V.M. Akulin, *Phys.Rev.Lett.* **82** 1 (1999)

[46] L. Viola, S. Lloyd, and E. Knill, *Phys.Rev.Lett.* **82** 2417 (1999)

[47] A. Emmanouilidou, X.G. Zhao, P. Ao, and Q. Niu, *Phys.Rev.Lett.* **85** 1626 (2000)

[48] L. Viola, and E. Knill, *Phys.Rev.Lett.* **90** 037901 (2003)

[49] O. Castaños, D. Schuch and O. Rosas-Ortiz, *J. Phys. A* **46**, 075304 (2013)

[50] K.B. Wolf *Phys.Rev.Lett.* **60** 757 (1988)

[51] K.B. Wolf *Geometric optics on phase space*, Springer-Verlag, Berlin (2004)

[52] B.J. Hiley *Non-Commutative Quantum Geometry: A Reaprisal of the Bohm Approach to Quantum Theory* in *Quo Vadis Quantum Mechanics?* Eds. A.Elitzur et al, Springer (2005) 299-324

[53] S.A. Collins *J.Opt.Soc.Am.* **60** 1168 (1970)

[54] D.F.V. James, G.S. Agarwal *J.Soc.Opt.Am. B* **12** 704 (1995)

[55] C.L. Hernandez-Cedillo, S. Bernon, H. Hattermann, J. Fortagh and R. Jauregui, *Phys.Rev. A* **87**, 023404 (2013)

[56] L.D. Landau and R. Peierls, *Z.Phys.* **69** 56 (1931)

[57] Y. Aharonov and D. Bohm, *Phys.Rev.* **66**, 1649 (1961)

[58] Y. Aharonov and B. Reznik, *Phys.Rev. Lett.* **84**, 1368 (2000)

[59] J. Kijowski, *Rep.Math.Phys.* **6** 361 (1974)

[60] N. Grot, C. Rovelli, R.S. Tate, *Phys.Rev. A* **54**, 4676 (1996)

[61] A.D. Baute, R. Sala Mayato, J.P. Palao, J.G. Muga, I.L. Egusquiza, *Phys.Rev. A* **61**, 022118 (2000)

[62] C.R. Leavens, *Phys.Rev. A* **58** 840 (1998)

[63] I.G. Muga, C.R. Leavens, *Phys.Rep.* **338** 353 (2000)

[64] C.R. Leavens, *Phys.Lett. A* **303** 154 (2002)

[65] C.R. Leavens, *Phys.Lett. A* **338** 19 (2005)

[66] C.R. Leavens, *Phys.Lett. A* **345** 251 (2005)

[67] E.A. Galapon, F. Delgado, J.G. Muga, I. Egusquiza *Phys.Rev. A* **72** 042107 (2005)

[68] E.A. Galapon, R.F. Caballar, R.T. Bahague *Phys.Rev. A* **72** 062107 (2005)

[69] G. Torres-Vega, *Phys.Lett. A* **369** 384 (2007)

[70] G. Torres-Vega, *J.Phys. A* **45** 215302 (2012)

[71] B. Mielnik and G. Torres Vega, *Concepts of Phys.* **II** 81 (2005) arXiv:1112.4198v1 [quant.ph]

[72] J. Kijowski, *Phys.Rev. A* **59** 897 (1999)

[73] D.J. Fernandez C., *Nouv.Cim.***107**, 885 (1992)

[74] G. Herrera. C *Manipulacion de una particula cuantica en los potenciales cuadraticos dependientes del tiempo* Master Thesis, Depto de Fisica, Cinvestav, Mexico (1987)

[75] F. Delgado C., *Operaciones selectivas de control cuantico* Ph.D. Thesis, Depto de Fisica, Cinvestav, Mexico (1999)

[76] F. Delgado C., B. Mielnik, *Phys.Lett. A* **249** 369 (1998)

[77] A. Ramirez, *Operaciones magneticas: Una ventana a la geometria no conmutativa?* **PhD** Thesis, Cinvestav, Mexico (2008)

[78] V. Ramakrishna, R.J. Ober, K.L. Flores, H. Rabitz *Phys.Rev. A* **65** 063405 (2002)

[79] P. Caldirola *Nuov.Cimento* **18** 393 (1941)

[80] E. Kanai *Prog.Theor.Phys.* **3** 440 (1948)

[81] A.L. De Brito and B. Baseia, *Phys.Rev. A* **40**, 4097 (1989)

[82] B. Baseia, A.L. De Brito, V.S. Bagnato, *Phys.Rev. A* **45**, 5308 (1992)

[83] S. Cruz y Cruz, and O. Rosas-Ortiz, *J.Phys. A* **42** 185205 (2009);

[84] S. Cruz y Cruz, and O. Rosas-Ortiz, *Int.J.Theor.Phys.* **50** 185205 (2011);

[85] S. Cruz y Cruz, J. Negro and L.M. Nieto, *Phys.Lett. A* **369** 400 (2007);

[86] H. Bateman *Phys.Rev.* **38** 815 (1931)

[87] B. Mielnik and A. Ramirez, *Phys.Scr.* **82** 055002 (2010)

[88] G. Lindblad *Commun.Math.Phys.* **48** 119 (1976)

[89] N. Gisin and I.C. Percival, *J.Phys. A* **25**, 5677 (1992)

[90] S. Pasini, P. Karbach, C. Raas and G.S. Uhrig, *Phys.Rev. A* **80** 022328 (2009)

[91] B. Mielnik and A. Ramirez, *Phys.Scr.* **84** 045008 (2011)

[92] S. Cordero and G. Garcia-Calderon, *Phys.Rev. A* **72** 062107 (2005)

[93] L. Infeld and J. Plebanski, *Motion and Relativity*, Pergamon Press and PWN (1960)

[94] J.E. Avron and B. Simon, *Phys.Rev.Lett.* **2746**, 1166 (1981)

[95] B. Mielnik and M.A. Reyes, *J.Phys. A* **29**, 6009 (1996)

[96] R.M. Hawkins and J.E. Lidsey, *Phys.Rev. D* **66** (2002) 023523

[97] S. Weinberg, *Cosmology*, OUP, Oxford (2008)

[98] M. Barbosa-Cendejas and M.A. Reyes *arXiv:1001.0084v1* [gr-qc] (31 Dec 2009)

[99] D. Griffiths, *Introduction to Electrodynamics*, Addison-Wesley (1999)

[100] W. Paul, *Rev.Mod.Phys.* **62**, 531 (1990)

8th International Symposium on Quantum Theory and Symmetries (QTS8) IOP Publishing
Journal of Physics: Conference Series **512** (2014) 012036 doi:10.1088/1742-6596/512/1/012036

Subsystems of a finite quantum system and Bell-like inequalities

A. Vourdas
Department of Computing,
University of Bradford
Bradford BD7 1DP, UK

E-mail: a.vourdas@bradford.ac.uk

Abstract. The set of subsystems $\Sigma(m)$ of a finite quantum system $\Sigma(n)$ with variables in $\mathbb{Z}(n)$, together with logical connectives, is a Heyting algebra. The probabilities $\tau(m|\rho_n) = \text{Tr}[\mathfrak{P}(m)\rho_n]$ (where $\mathfrak{P}(m)$ is the projector to $\Sigma(m)$) are compatible with associativity of the join in the Heyting algebra, only if the variables belong to the same chain. Consequently, contextuality in the present formalism, has the chains as contexts. Various Bell-like inequalities are discussed. They are violated, and this proves that quantum mechanics is a contextual theory.

1. Introduction

In recent work [1] we have studied the mathematical structure of the set of subsystems of a finite quantum system $\Sigma(n)$ with variables in $\mathbb{Z}(n)$ (the integers modulo n). A subsystem of $\Sigma(n)$, is another finite system $\Sigma(m)$, with $m|n$. In this case the variables of $\Sigma(m)$ take values in $\mathbb{Z}(m)$, which is a subgroup of $\mathbb{Z}(n)$. Also, all states of $\Sigma(m)$ are embedded in $\Sigma(n)$ as described below. We have shown that the set of subsystems of $\Sigma(n)$ with logical connectives, is a distributive lattice $\Lambda(\mathbf{\Sigma}_n)$. All finite distributive lattices are Heyting algebras[2, 3], and therefore $\Lambda(\mathbf{\Sigma}_n)$ is a Heyting algebra.

Probability theory needs the concepts of conjuction, disjunction and negation for its axioms. Kolmogorov probabilities are defined on a Boolean algebra (a powerset 2^{Ω}), where the intersection, union and complement play the role of conjuction, disjunction and negation. Let $q(E_i)$ be the Kolmogorov probability for event E_i. Then

$$q(E_1 \vee E_2) - q(E_1) - q(E_2) + q(E_1 \wedge E_2) = 0. \tag{1}$$

Refs [4, 5, 6] proved this equality, using the associativity property of the lattice. Alternatively, it is introduced as the axiom of additivity of probability (the last term generalizes it to non-exclusive events).

Quantum mechanics uses the orthomodular lattice of closed subspaces of a Hilbert space [7, 8, 9], which has various Boolean algebras as sublattices. Kolmogorov probabilities are defined on them, but the topic of contextuality and Bell inequalities [10, 11, 12, 13, 14, 15, 16, 17, 18, 19, 20], proves that there is inconsistency between the various parts of the full formalism.

Various aspects of contextuality have been studied in the literature. In a bipartite system with a space-like separation, it leads to non-locality which is a special form of contextuality.

Journal of Physics: Conference Series **512** (2014) 012036 doi:10.1088/1742-6596/512/1/012036

In a single spin one particle, contextuality is proved through a logical contradiction. We are interested in another form of contextuality, which appears in the formalism of subsystems of a finite quantum system.

We consider the quantum probabilities, $\tau(m|\rho_n) = \mathrm{Tr}[\mathfrak{P}(m)\rho_n]$, where $\mathfrak{P}(m)$ is the projector to the subsystem $\Sigma(m)$ and ρ_n is a density matrix of the system $\Sigma(n)$. In ref[1] we have shown that these probabilities are incompatible with the associativity of the join in the distributive lattice of subsystems $\Lambda(\mathbf{\Sigma}_n)$, because they do not obey the equality of Eq.(1), but they obey the inequality

$$\tau(m_1 \vee m_2|\rho_n) - \tau(m_1|\rho_n) - \tau(m_2|\rho_n) + \tau(m_1 \wedge m_2|\rho_n) \geq 0. \tag{2}$$

Only if the m_1, m_2 belong to the same chain, Eq(2) is valid as an equality. For this reason, chains play the role of contexts, in the present formalism.

Within a context (chain), the quantum probabilities $\tau(m|\rho_n)$ obey an equality analogous to Eq.(1), and they behave like Kolmogorov probabilities. But when we consider various contexts together, this equality and other important relations which are derived from it (e.g., Boole's inequality $\tau(m_1 \vee m_2|\rho_n) \leq \tau(m_1|\rho_n) + \tau(m_2|\rho_n)$ which is used in the proof of Bell-like inequalities), are no longer valid. In the present formalism, the term non-contextual quantum mechanics, means that the equality of Eq.(1) is assumed to be valid. The term contextual quantum mechanics, means that Eq.(1) is only valid within a context (chain).

Ref[19] has studied 'logical Bell inequalities' for Boolean variables. For Heyting variables, which are relevant in the present formalism, they acquire some extra terms, as discussed in [1]. The validity of these inequalities is based on the assumption that quantum mechanics is non-contextual theory, and their violation in experiments, proves that quantum mechanics is a contextual theory. In this paper we extend the work of [1], and present new Bell-like inequalities, together with examples which violate them. There are many technical details associated with the lattice structure of the set of subsystems, which are important for understanding why the Bell-like inequalities are violated, but which are not presented here (see [1]). In this paper we present a methodology for deriving novel Bell-like inequalities, starting from an equality in Boolean algebra, and using Boole's inequality for probabilities (which is valid in non-contextual quantum mechanics).

In section II we present briefly Heyting algebras in the present formalism, in order to eastablish the notation. In section III we discuss the set of subsystems of $\Sigma(n)$, and probabilities associated with projectors to these subsystems. In section IV we present various Bell-like inequalities and examples of their violation. We conclude in section V, with a discussion of our results.

2. Preliminaries

(1) $r|s$ or $r \prec s$ denotes that r is a divisor of s. GCD(r, s) and LCM(r, s) are the greatest common divisor and least common multiplier correspondingly, of the integers r, s. $\mathbb{D}(n)$ is the set of divisors of n.

(2) *Heyting and Boolean algebras:* A set A viewed as a lattice (i.e., with the operations $\vee$ and $\wedge$) is denoted as $\Lambda(A)$. Throughout the paper we have various finite distributive lattices and for simplicity we use the same symbols $\prec$, $\wedge$, $\vee$, $\neg$, for the 'partial order', 'meet', 'join' and 'negation', correspondingly. We also use the same symbols $\mathcal{O}$ and $\mathcal{I}$ for the smallest and greatest elements.

Every finite distributive lattice is a Heyting algebra. If $a \in \Lambda(A)$, then $\neg a$ is the largest element such that $a \wedge (\neg a) = \mathcal{O}$. The elements of a Heyting algebra obey the relations

$$a \vee (\neg a) \prec \mathcal{I}; \qquad a \prec \neg\neg a \tag{3}$$

Journal of Physics: Conference Series **512** (2014) 012036 doi:10.1088/1742-6596/512/1/012036

A special case of Heyting algebera are the Boolean algebras. In this case the relations of Eq.(3) are valid as equalities:

$$a \vee (\neg a) = \mathcal{I}; \qquad a = \neg\neg a \tag{4}$$

The relation $a \vee (\neg a) = \mathcal{I}$ is the 'law of the excluded middle' and it does not hold in general in Heyting algebras.

(3) *The Heyting algebra of divisors of n:* The set $\mathbb{D}(n)$ with divisibility as partial order, and with

$$k \wedge m = \mathrm{GCD}(k, m); \qquad k \vee m = \mathrm{LCM}(k, m) \tag{5}$$

is a finite distributive lattice and as such it is a Heyting algebra with $\mathcal{O} = 1$ and $\mathcal{I} = n$. We denote it as $\Lambda[\mathbb{D}(n)]$. $\neg k$ is the largest divisor of n, such that $k \wedge (\neg k) = 1$.

(4) *The Heyting algebra of subgroups of* $\mathbb{Z}(n)$: If $m \prec n$ then $\mathbb{Z}(m)$ is a subgroup of $\mathbb{Z}(n)$. The set $\mathfrak{Z}(n) = \{\mathbb{Z}(m) \mid m \in \mathbb{D}(n)\}$ of the subgroups of $\mathbb{Z}(n)$, with subgroup as partial order, and with

$$\mathbb{Z}(k \wedge m) = \mathbb{Z}(k) \wedge \mathbb{Z}(m); \qquad \mathbb{Z}(k \vee m) = \mathbb{Z}(k) \vee \mathbb{Z}(m); \qquad \neg\mathbb{Z}(k) = \mathbb{Z}(\neg k) \tag{6}$$

is the Heyting algebra $\Lambda[\mathfrak{Z}(n)]$, and it is isomorphic to $\Lambda[\mathbb{D}(n)]$. In this case $\mathcal{O} = \mathbb{Z}(1)$ and $\mathcal{I} = \mathbb{Z}(n)$. The elements of $\mathbb{Z}(m)$ can be embedded into a supergroup $\mathbb{Z}(k)$ (where $m \prec k \prec n$), by mapping $a \in \mathbb{Z}(m)$ into $\frac{ka}{m} \in \mathbb{Z}(k)$.

3. Subsystems of $\Sigma(n)$

$\Sigma(n)$ is a quantum system with variables in $\mathbb{Z}(n)$ and Hilbert space $H(n)$. $|X_n; r\rangle$ where $r \in \mathbb{Z}(n)$, is the basis of position states (the X_n in this notation is not a variable, but it simply indicates position states). Through a Fourier transform we get the basis of momentum states:

$$|P_n; r\rangle = n^{-1/2} \sum_{r,s} \omega_n(rs)|X_n; r\rangle.; \qquad \omega_n(r) = \exp\left(\frac{i2\pi r}{n}\right) \tag{7}$$

For $m \prec k \prec n$, the $\Sigma(m)$ is a subsystem of $\Sigma(k)$. In this case the variables of $\Sigma(m)$ take values in $\mathbb{Z}(m)$, which is a subgroup of $\mathbb{Z}(k)$, and all states of $\Sigma(m)$ are embedded in $\Sigma(k)$ as follows:

$$\sum_{r=0}^{m-1} a_r |X_m; r\rangle \;\rightarrow\; \sum_{r=0}^{m-1} a_r |X_k; \frac{kr}{m}\rangle; \qquad m \prec k. \tag{8}$$

The system $\Sigma(1)$ is physically trivial, and it consists of the 'vacuum' state $|X_1; 0\rangle = |P_1; 0\rangle$.

We define the projector to the system $\Sigma(m)$ (as embedded into a supersystem $\Sigma(k)$), as

$$\mathfrak{P}(m) = \sum_{r=0}^{m-1} |X_k; \frac{kr}{m}\rangle\langle X_k; \frac{kr}{m}|; \qquad m \prec k; \qquad m, k \in \mathbb{D}(n). \tag{9}$$

All these projectors commute with each other.

The set $\mathbf{\Sigma}_n$ of all subsystems of $\Sigma(n)$, with partial order 'subsystem', and with the logical connectives

$$\begin{aligned} \Sigma(m) \wedge \Sigma(k) &= \Sigma(m \wedge k) \\ \Sigma(m) \vee \Sigma(k) &= \Sigma(m \vee k) \\ \neg\Sigma(m) &= \Sigma(\neg m); \qquad m, k \in \mathbb{D}(n) \end{aligned} \tag{10}$$

Journal of Physics: Conference Series **512** (2014) 012036 doi:10.1088/1742-6596/512/1/012036

is the Heyting algebra $\Lambda(\mathbf{\Sigma}_n)$, and it is isomorphic to $\Lambda[\mathbb{D}(n)]$. In this case $\mathcal{O} = \Sigma(1)$ and $\mathcal{I} = \Sigma(n)$. The physical meaning of these connectives is discussed in [1]. In similar way we define the logical operations in the set $\mathbf{H}_n$ of the Hilbert spaces of the subsystems of $\Sigma(n)$. This is also a Heyting algebra isomorphic to $\Lambda[\mathbb{D}(n)]$ and $\Lambda(\mathbf{\Sigma}_n)$, which we denote as $\Lambda(\mathbf{H}_n)$.

In ref[1] we have proved that the space $H(m_1 \vee m_2)$ is given by

$$H(m_1 \vee m_2) = \mathrm{span}[H(m_1) \cup H(m_2)] \oplus S(m_1, m_2). \tag{11}$$

The span$[H(m_1) \cup H(m_2)]$ contains all superpositions of states in $H(m_1)$ and $H(m_2)$. The $S(m_1, m_2)$ is orthogonal to it. The

$$\begin{aligned} \mathfrak{T}(m_1, m_2) &= \mathfrak{P}(m_1) + \mathfrak{P}(m_2) - \mathfrak{P}(m_1 \wedge m_2) \\ \mathfrak{S}(m_1, m_2) &= \mathfrak{P}(m_1 \vee m_2) - \mathfrak{P}(m_1) - \mathfrak{P}(m_2) + \mathfrak{P}(m_1 \wedge m_2), \end{aligned} \tag{12}$$

are projectors to the spaces span$[H(m_1) \cup H(m_2)]$ and $S(m_1, m_2)$, correspondingly. In the special case that m_1, m_2 belong to the same chain, $H(m_1 \vee m_2) = \mathrm{span}[H(m_1) \cup H(m_2)]$.

We next define the quantum probabilities

$$\tau(m|\rho_n) = \mathrm{Tr}[\rho_n \mathfrak{P}(m)]; \qquad \sigma(m_1, m_2|\rho_n) = \mathrm{Tr}[\rho_n \mathfrak{S}(m_1, m_2)]; \qquad m, m_1, m_2 \in \mathbb{D}(n), \tag{13}$$

where ρ_n is density matrix describing the system $\Sigma(n)$. From Eq.(12), it follows that

$$\tau(m_1 \vee m_2|\rho_n) - \tau(m_1|\rho_n) - \tau(m_2|\rho_n) + \tau(m_1 \wedge m_2|\rho_n) = \sigma(m_1, m_2|\rho_n). \tag{14}$$

From this follows the inequality of Eq.(2). For variables in a chain, $\sigma(m_1, m_2|\rho_n) = 0$ for all density matrices, and the $\tau(m|\rho_n)$ obey the equality of Eq.(1). For this reason, contexts in the present formalism are the chains in the Heyting algebra.

For later use, we also define the

$$\widetilde{\mathfrak{P}}(m) = \mathfrak{P}(m) - \mathfrak{P}(1); \qquad \widetilde{\tau}(m|\rho_n) = \mathrm{Tr}[\rho_n \widetilde{\mathfrak{P}}(m)]. \tag{15}$$

3.1. Example

We consider the $\Lambda(\mathbf{\Sigma}_{12})$ which comprises of the subsystems of $\Sigma(12)$. The projectors to these subsystems are

$$\begin{aligned} &\mathfrak{P}(1) = |X_{12}; 0\rangle\langle X_{12}; 0| \\ &\mathfrak{P}(2) = |X_{12}; 0\rangle\langle X_{12}; 0| + |X_{12}; 6\rangle\langle X_{12}; 6| \\ &\mathfrak{P}(3) = |X_{12}; 0\rangle\langle X_{12}; 0| + |X_{12}; 4\rangle\langle X_{12}; 4| + |X_{12}; 8\rangle\langle X_{12}; 8| \\ &\mathfrak{P}(4) = \sum_{\nu=0}^{3} |X_{12}; 3\nu\rangle\langle X_{12}; 3\nu| \\ &\mathfrak{P}(6) = \sum_{\nu=0}^{5} |X_{12}; 2\nu\rangle\langle X_{12}; 2\nu| \\ &\mathfrak{P}(12) = \mathbf{1} \end{aligned} \tag{16}$$

In this case we have 3 maximal contexts:

$$\{\Sigma(12), \Sigma(6), \Sigma(3), \Sigma(1)\}; \quad \{\Sigma(12), \Sigma(6), \Sigma(2), \Sigma(1)\}; \quad \{\Sigma(12), \Sigma(4), \Sigma(2), \Sigma(1)\}. \tag{17}$$

Journal of Physics: Conference Series **512** (2014) 012036 doi:10.1088/1742-6596/512/1/012036

In $\Sigma(12)$ we consider the state

$$\rho = |s\rangle\langle s|; \qquad |s\rangle = \sum_{\nu=0}^{11} a_\nu |X_{12};\nu\rangle; \quad \sum_{\nu=0}^{11} |a_\nu|^2 = 1, \tag{18}$$

and we find

$$\begin{array}{l}\tau(12|\rho) = 1; \quad \tau(6|\rho) = \sum_{\nu=0}^{5} |a_{2\nu}|^2; \quad \tau(4|\rho) = \sum_{\nu=0}^{3} |a_{3\nu}|^2 \\ \tau(3|\rho) = |a_0|^2 + |a_4|^2 + |a_8|^2; \quad \tau(2|\rho) = |a_0|^2 + |a_6|^2; \quad \tau(1|\rho) = |a_0|^2. \end{array} \tag{19}$$

We next calculate the $\sigma(m_1, m_2)$ of Eq.(14). We find that

$$\begin{array}{rcl}\sigma(4,6|\rho) & = & |a_1|^2 + |a_5|^2 + |a_7|^2 + |a_{11}|^2; \\ \sigma(4,3|\rho) & = & |a_1|^2 + |a_2|^2 + |a_5|^2 + |a_7|^2 + |a_{10}|^2 + |a_{11}|^2; \\ \sigma(2,3|\rho) & = & |a_2|^2 + |a_{10}|^2, \end{array} \tag{20}$$

and that the rest $\sigma(m_1, m_2|\rho) = 0$.

4. Bell-like inequalities

Ref [19] has studied 'logical Bell inequalities' for Boolean variables. Their violation proves the contextual nature of quantum mechanics. In ref[1] we have generalized them for the case of Heyting variables. In this case, they acquire extra terms which we call 'Heyting factors' and which are zero in the case of Boolean variables.

In the present formalism:

- Quantum mechanics is a non-contextual theory, means that an equality analogous to Eq.(1) holds. In this case we easily prove Boole's inequality for probabilities:

$$\tau(m_1 \vee ... \vee m_N|\rho_n) \le \tau(m_1|\rho_n) + ... + \tau(m_N|\rho_n). \tag{21}$$

- Quantum mechanics is a contextual theory, means that an equality analogous to Eq.(1) is valid only for variables within a context (chain). When we consider various contexts together, then the inequality of Eq.(2) holds. In this case we cannot infer Boole's inequality for probabilities. An example where Boole's inequality is violated, is presented below.

There are many Bell-like inequalities, which are proved using Boole's inequality for probabilities and which would be valid if quantum mechanics were a non-contextual theory.

Proposition 4.1. *We assume that quantum mechanics is a non-contextual theory. If*

$$m_1 \wedge ... \wedge m_N = r; \qquad m_1, ..., m_N \in \mathbb{D}(n) - \{1\}, \tag{22}$$

then

$$\sum_{i=1}^{N} \widetilde{\tau}(m_i|\rho_n) \le N - \tau(r|\rho_n) - \sum_{i=1}^{N} f_i; \qquad f_i = 1 - \tau(m_i \vee \neg m_i|\rho_n). \tag{23}$$

f_i are 'Heyting factors' and are zero for Boolean variables.

Proof. The proof has been presented in [1]. □

The following propositions start from an equality in Boolean algebra, and use Boole's inequality for probabilities, to derive a Bell-like inequality. There are many equalities in Boolean algebra which can be used in a similar way, and they will lead to various Bell-like inequalities. Below we give two of them. Violation of these inequalities, supports the fact that quantum mechanics is a contextual theory.

Proposition 4.2. *We assume that quantum mechanics is a non-contextual theory. Let $q(A), q(B), q(C)$ be probabilities corresponding to quantum events A, B, C, which are assumed to be elements of a Boolean algebra. Then*

$$q(A) \leq q(A \wedge B) + q[A \wedge (\neg B)] \tag{24}$$
$$q(A) \leq q(A \wedge C) + q(B \wedge C) + q[(\neg A) \wedge (\neg B) \wedge C] \tag{25}$$

Proof. Using the basic rules of Boolean algebra we prove that

$$\begin{aligned} A &= (A \wedge B) \vee [A \wedge (\neg B)] \\ C &= [C \wedge (A \vee B)] \vee \{C \wedge [\neg(A \vee B)]\} = [C \wedge A] \vee [C \wedge B] \vee [C \wedge (\neg A) \wedge (\neg B)] \end{aligned} \tag{26}$$

Using Boole's inequality (in a non-contextual quantum mechanics) we get

$$q(A) = q\{(A \wedge B) \vee [A \wedge (\neg B)]\} \leq q(A \wedge B) + q[A \wedge (\neg B)], \tag{27}$$

and also

$$\begin{aligned} q(C) &= q\{[C \wedge A] \vee [C \wedge B] \vee [C \wedge (\neg A) \wedge (\neg B)]\} \\ &\leq q[C \wedge A] + q[C \wedge B] + q[C \wedge (\neg A) \wedge (\neg B)]. \end{aligned} \tag{28}$$

□

In our formalism, $q(A)$ will be the $\tau(m|\rho_n)$. It can be measured with the

$$M = a\mathfrak{P}(m) + b[\mathbf{1}_n - \mathfrak{P}(m)]. \tag{29}$$

We have to perform this von Neumann measurement on many systems in the state ρ_n, and count the number of times the system will collapse into a state which belongs entirely in $H(m)$.

4.1. Example

We consider the example discussed in section 3.1, and we show that Eqs.(21),(24),(25) are violated. An example which shows that Eq.(23) is violated, has been presented in [1]. Therefore quantum mechanics is a contextual theory.

(1) We consider Boole's inequality of Eq.(21), and we take $m_1 = 4$, $m_2 = 3$, $m_3 = 6$. In this case $m_1 \vee m_2 \vee m_3 = 12$ and the inequality becomes

$$\tau(6|\rho) + \tau(4|\rho) + \tau(3|\rho) \geq \tau(12|\rho). \tag{30}$$

We substitute the values from Eq.(19) and we get

$$\sum_{\nu=0}^{5} |a_{2\nu}|^2 + \sum_{\nu=0}^{3} |a_{3\nu}| + |a_0|^2 + |a_4|^2 + |a_8|^2 \geq 1 \tag{31}$$

which reduces to

$$2|a_0|^2 + |a_4|^2 + |a_6|^2 + |a_8|^2 \geq |a_1|^2 + |a_5|^2 + |a_7|^2 + |a_{11}|^2 \tag{32}$$

An example where this is violated is the case $a_0 = a_4 = a_6 = a_8 = 0$.

(2) We consider the inequality of Eq.(24). The events A, B are taken to be projections to the subsystems $\Sigma(12), \Sigma(4)$, correspondingly. Then

$$q(A) = \tau(12|\rho); \qquad q(B) = \tau(4|\rho). \tag{33}$$

The inequality becomes

$$\tau(12|\rho) \leq \tau(12 \wedge 4|\rho) + \tau[12 \wedge (\neg 4)|\rho], \tag{34}$$

which reduces to

$$\tau(12|\rho) \leq \tau(4|\rho) + \tau(3|\rho). \tag{35}$$

We substitute the values from Eq.(19) and we get

$$1 \leq \tau(4|\rho) + |a_0|^2 + |a_4|^2 + |a_8|^2 \tag{36}$$

An example where this is violated is the case $a_0 = a_4 = a_8 = 0$.

(3) We consider the inequality of Eq.(25). The events A, B, C are taken to be projections to the subsystems $\Sigma(12), \Sigma(4), \Sigma(3)$, correspondingly. Then

$$q(A) = \tau(12|\rho); \qquad q(B) = \tau(4|\rho); \qquad q(C) = \tau(3|\rho). \tag{37}$$

The inequality becomes

$$\tau(12|\rho) \leq \tau(12 \wedge 3|\rho) + \tau(4 \wedge 3|\rho) + \tau[(\neg 12) \wedge (\neg 4) \wedge 3|\rho], \tag{38}$$

which reduces to

$$\tau(12|\rho) \leq \tau(3|\rho) + \tau(1|\rho) + \tau(1|\rho). \tag{39}$$

We substitute the values from Eq.(19) and we get

$$1 \leq 3|a_0|^2 + |a_4|^2 + |a_8|^2 \tag{40}$$

An example where this is violated is the case $a_0 = a_4 = a_8 = 0$.

5. Discussion

We have studied the set of subsystems of a finite quantum system $\Sigma(n)$, as a distributive lattice (Heyting algebra). We have shown that the quantum probabilities $\tau(m|\rho_n)$ do not obey the equality of Eq.(1), which is intimately related to the associativity of the join in the lattice, but they obey the inequality of Eq.(2). Only if the variables belong to the same chain, Eq(2) is valid as an equality.

In the present formalism chains play the role of contexts. The $\tau(m|\rho_n)$ can be viewed as Kolmogorov probabilities only within a particular chain. When we consider various contexts together, Eq.(1) and other relations which are derived from it, are not valid. This is related to the fact that the space $H(m_1 \vee m_2)$ is larger than the space of superpositions span$[H(m_1) \cup H(m_2)]$ (Eq.(11)).

In this paper we extended the work of [1], and gave new Bell-like inequalities in proposition 4.2. They combine equalities in Boolean algebra, with Boole's inequality for probabilities. Boole's inequality follows from Eq.(2), but it does not follow from Eq.(1). Consequently, the inequalities in proposition 4.2 are valid for variables within a context (chain), but they are violated when variables take values in different contexts. More Bell-like inequalities can be proved with this methodology. In section 4.1 we gave examples which show that these inequalities are violated. This shows that quantum mechanics is a contextual theory.

The work studies contextuality within the formalism of subsystems of a finite quantum system.

Journal of Physics: Conference Series **512** (2014) 012036 doi:10.1088/1742-6596/512/1/012036

[1] A. Vourdas, J. Math. Phys, 54, 082105 (2013)
[2] G. Birkhoff 'Lattice theory' (Amer. Math. Soc., Rhode Island, 1995)
[3] D.E. Rutherford, 'Introduction to lattice theory' (Oliver and Boyd, Edinburgh, 1965)
[4] R.T. Cox, Am. J. Phys. 14, 1 (1946)
[5] R.T. Cox, 'The algebra of probable inference' (John Hopkis Press, Baltimore, 1961)
[6] E.T. Jaynes, 'Probability theory. The logic of Science' (Cambridge Univ. Press, Cambridge, 2003)
[7] G. Birkhoff, J. von Neumann, Ann. Math. 37, 823 (1936)
[8] C. Piron, 'Foundations of quantum physics', Benjamin, New York, 1976
[9] G.W. Mackey, 'Mathematical foundations of quantum mechanics' Benjamin, New York, 1963
[10] J.S. Bell, Physics 1, 195 (1964)
[11] S. Kochen, E.P.Specker, J. Math. Mech. 17, 59 (1967)
[12] J.F. Clauser, M.A. Horne, A. Shimony, R.A. Holt, Phys. Rev. Lett., 23, 880 (1969)
[13] E.P. Wigner, Amer. J. Phys. 38, 1005 (1970)
[14] N.D. Mermin, Rev. Mod. Phys. 65, 803 (1993)
[15] I. Pitowski, Brit. J. Phil. Sci. 45, 95 (1994)
[16] A.A. Klyachko, M.A. Can, S. Binicioglu, A.S. Shumovsky, Phys. Rev. Lett. 101, 020403 (2008)
[17] Y-C Liang, R.W. Spekkens, H.M. Wiseman, Phys. Rep. 506, 1 (2011)
[18] S. Abramsky, A. Brandenburger, New J. Phys. 13, 113036 (2011)
[19] S. Abramsky, L. Hardy, Phys. Rev. A85, 062114 (2012)
[20] A. Cabello, Phys. Rev. Lett. 110, 060402 (2013)

AUTHOR INDEX